GENOMES 3

GENOMES 3

T. A. Brown

Garland Science
Taylor & Francis Group

NEW YORK AND LONDON

Vice President: Denise Schanck
Senior Publisher: Jackie Harbor
Production Editor: Simon Hill
Assistant Editor: Dominic Holdsworth
Copyeditor: Joanne Clayton
Illustrations: Matthew McClements, Blink Studio Limited
Layout: Georgina Lucas
Cover design: Andy Magee
Proofreader: Sally Livitt
Indexer: Indexing Specialists (UK) Ltd
Printer: Quebecor World Inc.

Front cover image courtesy of Dr Paul Andrews, University of Dundee/Science Photo Library.
Chapter opener image courtesy of Lawrence Lawry/ Science Photo Library.
Part opener image courtesy of Comstock Images/Alamy.

The cover image shows a fluorescence micrograph of a cell during metaphase of mitosis (nuclear division).

ISBN 0 8153 4138 5

Library of Congress Cataloging-in-Publication Data
Brown, T. A. (Terence A.)
    Genomes 3 / T. A. Brown. -- 3rd ed.
            p. ; cm.
    Rev. ed. of: Genomes / edited by T.A. Brown. Oxford ; New York : Wiley-Liss, 2002.
    Includes bibliographical references and index.
    ISBN 0-8153-4138-5 (alk. paper)
  1. Genomes. I. Genomes. II. Title. III. Title: Genomes three.
    [DNLM: 1. Genome. QH 447 B881g 2006]

    QH447.B76 2006
    572.8'6--dc22

                                                    2006009128

Published by Garland Science Publishing, a member of the Taylor & Francis Group, an informa business
270 Madison Avenue, New York, NY 10016, USA and
2 Park Square, Milton Park, Abingdon, Oxon, OX14 4RN, UK

Printed in the United States of America

15 14 13 12 11 10 9 8 7 6 5 4 3 2 1

Taylor & Francis Group, an informa business

Visit our web site at http://www.garlandscience.com

Preface

Four exciting years have elapsed since publication of the Second Edition of *Genomes*.

Finished sequences of human chromosomes have appeared at regular intervals and a draft of the chimpanzee genome has been completed. The number of eukaryotes with partial or complete sequences is increasing dramatically, and new prokaryotic sequences are announced virtually every week. Experimental techniques for studying transcriptomes and proteomes are providing novel insights into genome expression and the new discipline of systems biology is linking genome studies with cellular biochemistry. All of these advances have been incorporated into *Genomes 3*. In particular, what was previously a single chapter on genome anatomies has been expanded into three chapters, and I have greatly increased the material on post-genomics by writing separate chapters on how sequences are analyzed and how transcriptomes and proteomes are studied. I have also taken the opportunity to give greater depth to the descriptions of genome expression, replication and recombination.

The changes have led to an expansion in the length of *Genomes*, and to balance this I have tried to make the book more user friendly. Boxes are now used only for descriptions of techniques, so the text as a whole is less broken up and forms a more continuous narrative. The artwork has been completely redesigned, bringing greater clarity to the figures and giving the book a more attractive appearance. The reading lists and end-of-chapter problem sets have undergone an equally comprehensive re-evaluation.

In making these revisions I have taken account of feedback provided by a number of lecturers and students from various parts of the world. These people are literally "too many to mention" so I would like to say a general thank you. The one person to whom I will extend individual thanks is Daniela Delneri of the University of Manchester, whose comments on the chapters concerning post-genomics and molecular evolution were so comprehensive that I found it almost unnecessary to do any research into these areas myself. I am extremely grateful to Ted Lee from the Department of Biology, SUNY Fredonia, NY, for taking on the daunting (at least to me) task of writing comprehensive sets of questions and problems for each chapter – these learning aids greatly enhance the quality of the book. Also, I thank Dominic Holdsworth and Jackie Harbor of Garland Science for the tremendous support that they provided when I was writing *Genomes 3*, and Matthew McClements for his excellent redesigns of the artwork. Finally, *Genomes 3* would not have appeared without the support of my wife, Keri. In the Acknowledgements to the First Edition I wrote that "if you find this book useful then you should thank Keri, not me, because she is the one who ensured that it was written", and I am pleased that one or two people have actually taken me up on this.

T.A. Brown
Manchester

A Note to the Reader

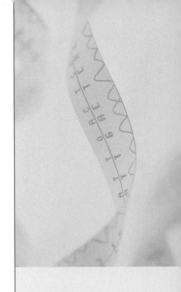

I have tried to make the third edition of *Genomes* as user friendly as possible. The book therefore includes a number of devices intended to help the reader and to make the book an effective teaching aid.

Organization of the Book

Genomes is divided into four parts:

Part 1 – Studying Genomes begins with an orientation chapter that introduces the reader to genomes, transcriptomes, and proteomes, and then moves on to the methods, centered on cloning and PCR, that were used in the pre-genome era to examine individual genes (Chapter 2). The techniques that are more specifically used for studying genomes are then described in the order in which they would be used in a genome project: methods for constructing genetic and physical maps (Chapter 3); DNA sequencing methodology and the strategies used to assemble a contiguous genome sequence (Chapter 4); methods for identifying genes in a genome sequence and determining the functions of those genes in the cell (Chapter 5); and approaches for studying transcriptomes and proteomes (Chapter 6). The Human Genome Project forms a continuous thread throughout Part 1, but this is not to the exclusion of all else and I have tried to give adequate coverage to the strategies that have been used, and are being used, to understand the genomes of other organisms.

Part 2 – Genome Anatomies surveys the anatomies of the various type of genome that are found on our planet. Chapter 7 covers eukaryotic nuclear genomes with primary emphasis on the human genome. Chapter 8 deals with the genomes of prokaryotes and of eukaryotic organelles, the latter included here because of their prokaryotic origins, and Chapter 9 describes virus genomes and mobile genetic elements, these being grouped together because some types of mobile element are related to virus genomes.

Part 3 – How Genomes Function covers the material that in the past has been described inadequately as "DNA goes to RNA goes to protein". Chapter 10 addresses the increasingly important issue of how chromatin structure influences genome expression. Chapter 11 then describes the assembly of the transcription initiation complexes of prokaryotes and eukaryotes, and includes a detailed discussion of DNA-binding proteins, these playing the central roles in the initial stages of genome expression. Chapters 12 and 13 give details of the synthesis of the transcriptome and proteome, and Chapter 14 surveys the regulation of genome activity. Keeping Chapter 14 to a manageable length is difficult, as many different topics are relevant to genome regulation, but I hope that by using specific examples to illustrate general themes I have managed to achieve a satisfactory balance between conciseness and breadth of coverage.

Part 4 – How Genomes Replicate and Evolve links DNA replication, mutation and recombination with the gradual evolution of genomes over time. In Chapters 15–17 the molecular processes responsible for replication, mutation, repair and recombination are described, and in Chapter 18 the ways in which these processes are thought to have shaped the structures and genetic contents of genomes over evolutionary time are considered. Finally, Chapter 19 is devoted to the increasingly informative use of molecular phylogenetics to infer the evolutionary relationships between DNA sequences.

Organization of Chapters

Learning outcomes

Each chapter starts with a set of learning outcomes. These have been phrased very carefully. They are not merely a series of synopses of the factual content of each chapter, but instead indicate the level and type of knowledge that the student should gain from reading the chapter. Therefore, the learning outcomes state what the student should be able to describe, draw, discuss, explain, evaluate, each verb having been selected to convey precisely what it is that the student is expected to be able to do. The intention is that the student is left in no doubt about what they should get out of each chapter, and hence is in no doubt about whether they have dealt satisfactorily with the material.

Figures

A good diagram is certainly worth a thousand words but a bad one can confuse the reader and a superfluous one is merely distracting. I have therefore tried to ensure that every figure is necessary and fulfills a purpose beyond simply breaking up the text and making the book look pretty. I have also tried to make figures reproducible because in my opinion this makes them much more useful as a learning aid for the student. I have never understood the penchant for making textbook diagrams into works of art because if the student cannot redraw a diagram then it is merely an illustration and does not help the student learn the information that it is designed to convey. The figures in *Genomes 3* are as clear, simple, and uncluttered as possible.

Technical Notes

The main text is supported and extended by a series of technical notes, contained in separate boxes. Each technical note is a self-contained description of a technique or a group of techniques important in the study of genomes. The technical notes are designed to be read in conjunction with the main text, each one being located at the place in the book where an application of that technique is described for the first time.

Questions, Problems, and Figure Tests

Four different types of self-study exercise are given at the end of each chapter:

- **Multiple choice questions** cover the key points from the chapter and test the student's basic understanding of the material. Traditionalists sometimes debate the value of multiple choice questions in formal assessments, but there can be no doubt of their value as a review aid: if a student can accurately answer every one of them then they almost certainly have an excellent knowledge of the factual content of the chapter.

- **Short answer questions** require 50- to 300-word answers, or occasionally ask for an annotated diagram or a table. The questions cover the entire content of each chapter in a fairly straightforward manner, and most can be marked simply by checking each answer against the relevant part of the text. A student can use the short answer questions to work systematically through a chapter, or can select individual ones in order to evaluate their ability to answer questions on specific topics. The short answer questions could also be used in closed-book tests.

- **In-depth problems** require a more detailed answer. They vary in nature and in difficulty, the simplest requiring little more than a literature survey, the intention with these problems being that the student advances his or her learning a few stages from where *Genomes 3* leaves off. Other problems require that the student evaluates a statement or a hypothesis, based on their understanding of the material in the book, possibly supplemented by reading around the subject. These problems will, hopefully, engender a certain amount of thought and critical awareness. A few problems are difficult, in some cases to the extent that there is no solid answer to the question posed. These are designed to stimulate debate and speculation, which stretches the knowledge of each student and forces them to think carefully about their statements. The in-depth problems can be tackled by students working individually, or alternatively can form the starting point for a group discussion.

- **Figure tests** are similar to short answer questions, but use selected figures from the preceding chapter as the focal point of the exercise. These tests are valuable as a means of linking factual information gained from reading the text with the structures and processes that are illustrated by the figures. A good diagram is indeed worth a thousand words, but only if the diagram is studied carefully and fully understood. The figure tests help to provide this type of understanding.

For the multiple choice, short answer questions and figure tests, answers to odd-numbered questions are given in the Appendix. Upon request, answers to all questions will be provided to instructors who adopt the book via Garland Science Classwire™. For the in-depth problems, guidance rather than answers are provided.

Further Reading

The reading lists at the end of each chapter include those research papers, reviews and books that I look on as the most useful sources of additional material. My intention throughout *Genomes 3* has been that students should be able to use the reading lists to obtain further information when writing extended essays or dissertations on particular topics. Research papers are therefore included, but only if their content is likely to be understandable to the average reader of the book. Emphasis is placed on accessible reviews such as *Science* Perspectives, *Nature* News and Views and articles in the *Trends* journals, one strength of these general articles being the context and relevance that they provide to a piece of work. Most reading lists are divided into sections reflecting the organization of information in the chapter, and in some cases I have appended a few words summarizing the particular value of each item, to help the reader decide which ones he or she wishes to seek out. The lists are not all-inclusive and I encourage readers to spend some time searching the shelves of their own libraries for other books and articles. Browsing is an excellent way to discover interests that you never realized you had!

Glossary

I am very much in favor of glossaries as learning aids and I have provided an extensive one for this third edition of *Genomes*. Every term that is highlighted in bold in the text is defined in the Glossary, along with a number of additional terms that the reader might come across when referring to books or articles in the reading lists. Each term in the Glossary also appears in the index, so the reader can quickly gain access to the relevant pages where the Glossary term is covered in more detail.

The Art of Genomes 3

The CD-ROM packaged with the text contains the images from the book, available in two convenient formats: PowerPoint® and JPEG. The images have been pre-loaded into PowerPoint® presentations, one presentation for each chapter of the book. The images are also available as individual JPEG files, which are contained in separate folders from the PowerPoint® presentations. The individual JPEG files have been optimized for printing and Web display.

For Instructors

Garland Science Classwire™ located at http://www.classwire.com/garlandscience, offers instructional resources and course management tools for adopters. It contains the images from *Genomes 3* in JPEG and PowerPoint® formats. Multiple choice questions, short answer questions, in-depth problems and figure tests for which no answers or guidance are given in the Appendix are useful for homework assignments and as exam questions. Answers and guidance for these exercises will be provided to instructors upon request via Classwire™. Instructors who adopt *Genomes 3* can additionally access resources from our other textbooks. Classwire™ is also a flexible and easy-to-use course management tool that allows instructors to build web sites for their classes. It offers features such as a syllabus builder, a course calendar, a message center, a course planner, virtual office hours and a resource manager. No programming or technical skills are needed.

List of Reviewers

The Author and Publisher of Genomes 3 gratefully acknowledge the contribution of the following reviewers in the development of this edition.

Dean Danner, Emory University School of Medicine
Daniela Delneri, University of Manchester
Yuri Dubrova, University of Leicester
Bart Eggen, University of Groningen
Robert Fowler, San Jose State University
Adrian Hall, Sheffield Hallam University
Glyn Jenkins, University of Aberystwyth
Torsten Kristensen, University of Aarhus
Mike McPherson, University of Leeds
Andrew Read, University of Manchester
Darcy Russell, Baker College
Amal Shervington, University of Central Lancashire
Robert Slater, University of Hertfordshire
Klaas Swart, Wageningen University
John Taylor, University of Newcastle
Guido van den Ackerveken, Utrecht University
Vassie Ware, Lehigh University
Matthew Upton, University of Manchester

Contents in Brief

Contents

μm	micrometer	Col	colicin
5-bU	5-bromouracil	CPSF	cleavage and polyadenylation specificity factor
A	adenine; alanine		
ABF	ARS binding factor	CRM	chromatin remodeling machine
Ac/Ds	activator/dissociation	CstF	cleavage stimulation factor
ADAR	adenosine deaminase acting on RNA	CTAB	cetyltrimethylammonium bromide
ADP	adenosine 5′-diphosphate	CTD	C-terminal domain
AIDS	acquired immune deficiency syndrome	CTP	cytidine 5′-triphosphate
ala	alanine	cys	cysteine
AMP	adenosine 5′-monophosphate	D	aspartic acid
ANT-C	Antennapedia complex	DAG	1,2-diacylglycerol
AP	apurinic/apyrimidinic	Dam	DNA adenine methylase
arg	arginine	DAPI	4,6-diamino-2-phenylindole dihydrochloride
ARMS	amplification refractory mutation system		
ARS	autonomously replicating sequence	DASH	dynamic allele-specific hybridization
A site	acceptor site	dATP	2′-deoxyadenosine 5′-triphosphate
asn	asparagine	DBS	double-stranded DNA binding site
ASO	allele-specific oligonucleotide	Dcm	DNA cytosine methylase
asp	aspartic acid	dCTP	2′-deoxycytidine 5′-triphosphate
ATP	adenosine 5′-triphosphate	ddATP	2′,3′-dideoxyadenosine 5′-triphosphate
ATPase	adenosine 5′-triphosphatase	ddCTP	2′,3′-dideoxycytidine 5′-triphosphate
BAC	bacterial artificial chromosome	ddGTP	2′,3′-dideoxyguanosine 5′-triphosphate
bis	N,N′-methylenebisacrylamide	ddNTP	2′,3′-dideoxynucleoside 5′-triphosphate
BLAST	Basic Local Alignment Search Tool	ddTTP	2′,3′-dideoxythymidine 5′-triphosphate
bp	base pair	Dfd	Deformed
BSE	bovine spongiform encephalopathy	dGTP	2′-deoxyguanosine 5′-triphosphate
BX-C	Bithorax complex	DMSO	dimethylsulfoxide
C	cysteine; cytosine	DNA	deoxyribonucleic acid
cAMP	cyclic AMP	DNase	deoxyribonuclease
CAP	catabolite activator protein	Dnmt	DNA methyltransferase
CASP	CTD-associated SR-like protein	dNTP	2′-deoxynucleoside 5′-triphosphate
cDNA	complementary DNA	DPE	downstream promoter element
CEPH	Centre d'études du Polymorphisme Humaine	DSB	double-strand break
		DSP1	Dorsal switch protein 1
cGMP	cyclic GMP	dsRAD	double-stranded RNA adenosine deaminase
CHEF	contour-clamped homogeneous electric fields		
		dsRBD	double-stranded RNA binding domain
CJD	Creutzfeldt-Jakob disease	dTTP	2′-deoxythymidine 5′-triphosphate

E	glutamic acid		HNPCC	hereditary nonpolyposis colorectal cancer
EDTA	ethylenediamine tetraacetate		hnRNA	heterogenous nuclear RNA
eEF	eukaryotic elongation factor		HOM-C	homeotic gene complex
EEO	electroendosmosis value		HPLC	high-performance liquid chromatography
EF	elongation factor		HPRT	hypoxanthine phosphoribosyl transferase
eIF	eukaryotic initiation factor			
EMS	ethylmethane sulfonate		HTH	helix-turn-helix
eRF	eukaryotic release factor		I	isoleucine
ERV	endogenous retrovirus		ICAT	isotope coded affinity tag
ES	embryonic stem		ICF	immunodeficiency, centromere instability, and facial anomalies
ESE	exonic splicing enhancer			
E site	exit site		IF	initiation factor
ESS	exonic splicing silencer		Ig	immunoglobulin
EST	expressed sequence tag		IHF	integration host factor
F	fertility; phenylalanine		ile	isoleucine
FEN	flap endonuclease		Inr	initiator
FIGE	field inversion gel electrophoresis		Ins(1,4,5)P$_3$	inositol-1,4,5-trisphosphate
FISH	fluorescent *in situ* hybridization		IPTG	isopropylthiogalactoside
FRAP	fluorescence recovery after photobleaching		IRE-PCR	interspersed repeat element PCR
			IRES	internal ribosome entry site
G	glycine; guanine		IS	insertion sequence
G1	gap phase 1		ITF	integration host factor
G2	gap phase 2		ITR	inverted terminal repeat
GABA	γ-aminobutyric acid		JAK	Janus kinase
GAP	GTPase activating protein		K	lysine
Gb	gigabase		kb	kilobase
GDP	guanosine 5′-diphosphate		kcal	kilocalorie
GFP	green fluorescent protein		kDa	kilodalton
gln	glutamine		L	leucine
glu	glutamic acid		LCR	locus control region
gly	glycine		leu	leucine
GMP	guanosine 5′-monophosphate		LINE	long interspersed nuclear element
GNRP	guanine nucleotide releasing protein		lod	logarithm of the odds
GTF	general transcription factor		LTR	long terminal repeat
GTP	guanosine 5′-triphosphate		lys	lysine
H	histidine		M	methionine; mitosis phase
HAT	hypoxanthine + aminopterin + thymidine		MALDI-TOF	matrix-assisted laser desorption ionization time-of-flight
HBS	heteroduplex binding site		MAP	mitogen activated protein
HDAC	histone deacetylase		MAR	matrix-associated region
his	histidine		Mb	megabase
HIV	human immunodeficiency virus		MeCP	methyl-CpG-binding protein
HLA	human leukocyte antigen		met	methionine
HMG	high mobility group			

MGMT	O^6-methylguanine-DNA methyltransferase
miRNA	microRNA
mol	mole
mRNA	messenger RNA
MudPIT	multi-dimensional protein identification technique
MULE	*Mutator*-like transposable element
Myr	million years
N	2′-deoxynucleoside 5′-triphosphate; asparagine
NAD	nicotinamide adenine dinucleotide
NADH	reduced nicotinamide adenine dinucleotide
ng	nanogram
NHEJ	nonhomologous end-joining
NJ	neighbor-joining
nm	nanometer
NMD	nonsense-mediated RNA decay
NMR	nuclear magnetic resonance
NTP	nucleoside 5′-triphosphate
OFAGE	orthogonal field alternation gel electrophoresis
OLA	oligonucleotide ligation assay
Omp	outer membrane protein
ORC	origin recognition complex
ORF	open reading frame
OTU	operational taxonomic unit
P	proline
PAC	P1-derived artificial chromosome
PADP	polyadenylate-binding protein
PAUP	Phylogenetic Analysis Using Parsimony
PCNA	proliferating cell nuclear antigen
PCR	polymerase chain reaction
pg	picogram
phe	phenylalanine
PHYLIP	Phylogeny Inference Package
PIC	preinitiation complex
PNA	peptide nucleic acid
PNPase	polynucleotide phosphorylase
pro	proline
PSE	proximal sequence element
PSI-BLAST	position-specific iterated Basic Local Alignment Search Tool

P site	peptidyl site
PtdIns(4,5)P$_2$	phosphatidylinositol-4,5-bisphosphate
PTRF	polymerase I and transcript release factor
Pu	purine
Py	pyrimidine
Q	glutamine
R	arginine; purine
RACE	rapid amplification of cDNA ends
RAM	random access memory
RBS	RNA binding site
RC	replication complex
RF	release factor
RFC	replication factor C
RFLP	restriction fragment length polymorphism
RHB	Rel homology domain
RISC	RNA induced silencing complex
RLF	replication licensing factor
RMP	replication mediator protein
RNA	ribonucleic acid
RNAi	RNA interference
RNase	ribonuclease
RNP	ribonucleoprotein
RPA	replication protein A
RRF	ribosome recycling factor
rRNA	ribosomal RNA
RT-PCR	reverse transcriptase PCR
RTVL	retroviral-like element
S	serine; synthesis phase
SAGE	serial analysis of gene expression
SAP	stress activated protein
SAR	scaffold attachment region
SCAF	SR-like CTD-associated factor
scRNA	small cytoplasmic RNA
SCS	specialized chromatin structure
SDS	sodium dodecyl sulfate
SeCys	selenocysteine
ser	serine
SINE	short interspersed nuclear element
siRNA	small interfering RNA
SIV	simian immunodeficiency virus
SL RNA	spliced leader RNA
SMAD	SMA/MAD related

snoRNA	small nucleolar RNA	T_m	melting temperature
SNP	single nucleotide polymorphism	tmRNA	transfer-messenger RNA
snRNA	small nuclear RNA	Tn	transposon
snRNP	small nuclear ribonucleoprotein	TOL	toluene
SRF	serum response factor	TPA	tissue plasminogen activator
SSB	single-strand binding protein	TRAP	*trp* RNA-binding attenuation protein
SSLP	simple sequence length polymorphism	tRNA	transfer RNA
STAT	signal transducer and activator of transcription	trp	tryptophan
		tyr	tyrosine
STR	simple tandem repeat	U	uracil
STS	sequence tagged site	UCE	upstream control element
T	threonine; thymine	UTP	uridine 5′-triphosphate
TAF	TBP-associated factor	UTR	untranslated region
TAP	tandem-affinity purification	UV	ultraviolet
TBP	TATA-binding protein	val	valine
TEMED	*N,N,N′,N′*-tetramethylethylenediamine	VNTR	variable number of tandem repeats
TF	transcription factor	W	adenine or thymine; tryptophan
TGF	transforming growth factor	X-gal	5-bromo-4-chloro-3-indolyl-β-D-galactopyranoside
thr	threonine		
Ti	tumor inducing	Y	pyrimidine; tyrosine
TIC	TAF- and initiator-dependent cofactor	YAC	yeast artificial chromosome
TK	thymidine kinase	YIp	yeast integrative plasmid

Studying Genomes

Part 1 – Studying Genomes describes the techniques and scientific approaches that underlie our knowledge of genomes. We begin with an orientation chapter that introduces genomes, transcriptomes, and proteomes, and then, in Chapter 2, move on to the methods, centered on DNA cloning and the polymerase chain reaction, that are used to study short segments of DNA, such as individual genes. Chapter 3 begins our examination of genomics by describing how genetic and physical maps are constructed, and Chapter 4 makes the link between mapping and sequencing. As you read Chapter 4 you will realize that although a map can be a valuable aid to assembly of a long DNA sequence, mapping is not always an essential prerequisite to genome sequencing. In Chapter 5, we look at the various approaches that are used to understand a genome sequence, and in Chapter 6 we examine the methods used to study how a genome functions by directing synthesis of a transcriptome and proteome and, through these, specifying the biochemical capability of the cell.

Genomes, Transcriptomes, and Proteomes

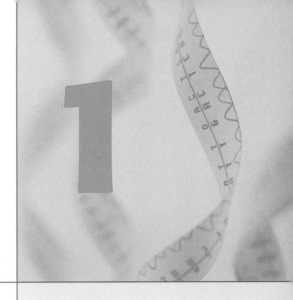

1

When you have read Chapter 1, you should be able to:

Define the terms "genome," "transcriptome," and "proteome," and state how these are linked in the process of genome expression.

Describe the two experiments that led molecular biologists to conclude that genes are made of DNA, and explain the limitations of those experiments.

Give a detailed description of the structure of a polynucleotide, and summarize the chemical differences between DNA and RNA.

Discuss the evidence that Watson and Crick used to deduce the double helix structure of DNA and describe the key features of this structure.

Distinguish between coding and functional RNA and give examples of each type.

Describe in outline how RNA is synthesized and processed in the cell.

Give a detailed description of the various levels of protein structure, and explain why amino acid diversity underlies protein diversity.

Describe the key features of the genetic code.

Explain why the function of a protein is dependent on its amino acid sequence.

List the major roles of proteins in living organisms and relate this diversity to the function of the genome.

Life as we know it is specified by the **genomes** of the myriad organisms with which we share the planet. Every organism possesses a genome that contains the **biological information** needed to construct and maintain a living example of that organism. Most genomes, including the human genome and those of all other cellular life forms, are made of **DNA** (deoxyribonucleic acid) but a few viruses have **RNA** (ribonucleic acid) genomes. DNA and RNA are polymeric molecules made up of chains of monomeric subunits called **nucleotides**.

The human genome, which is typical of the genomes of all multicellular animals, consists of two distinct parts (Figure 1.1):

- The **nuclear genome** comprises approximately 3,200,000,000 nucleotides of DNA, divided into 24 linear molecules, the shortest 50,000,000 nucleotides in length and the longest 260,000,000 nucleotides, each contained in a different **chromosome**. These 24 chromosomes consist of 22 **autosomes** and the two sex chromosomes, X and Y. Altogether, some 35,000 **genes** are present in the human nuclear genome.

- The **mitochondrial genome** is a circular DNA molecule of 16,569 nucleotides, multiple copies of which are located in the energy-generating organelles called mitochondria. The human mitochondrial genome contains just 37 genes.

Figure 1.1 The nuclear and mitochondrial components of the human genome.

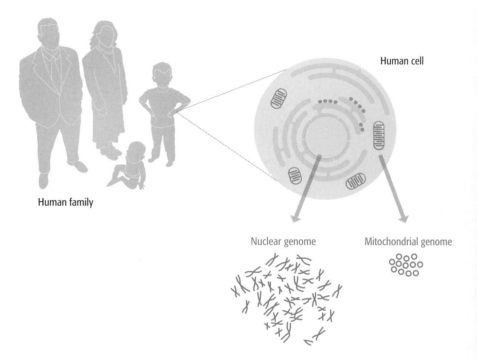

Each of the approximately 10^{13} cells in the adult human body has its own copy or copies of the genome, the only exceptions being those few cell types, such as red blood cells, that lack a nucleus in their fully differentiated state. The vast majority of cells are **diploid** and so have two copies of each autosome, plus two sex chromosomes, XX for females or XY for males—46 chromosomes in all. These are called **somatic cells,** in contrast to **sex cells,** or **gametes**, which are **haploid** and have just 23 chromosomes, comprising one of each autosome and one sex chromosome. Both types of cell have about 8000 copies of the mitochondrial genome, 10 or so in each mitochondrion.

The genome is a store of biological information but on its own it is unable to release that information to the cell. Utilization of the biological information contained in the genome requires the coordinated activity of enzymes and other proteins, which participate in a complex series of biochemical reactions referred to as **genome expression** (Figure 1.2). The initial product of genome expression is the **transcriptome**, a collection of RNA molecules derived from those protein-coding genes whose biological information is required by the cell at a particular time. The transcriptome is maintained by the process called

transcription, in which individual genes are copied into RNA molecules. The second product of genome expression is the **proteome**, the cell's repertoire of **proteins**, which specifies the nature of the biochemical reactions that the cell is able to carry out. The proteins that make up the proteome are synthesized by **translation** of the individual RNA molecules present in the transcriptome.

This book is about genomes and genome expression. It explains how genomes are studied (Part 1), how they are organized (Part 2), how they function (Part 3), and how they replicate and evolve (Part 4). It was not possible to write this book until very recently. Since the 1950s, molecular biologists have studied individual genes or small groups of genes, and from these studies have built up a wealth of knowledge about how genes work. But only during the last 10 years have techniques been available that make it possible to examine entire genomes. Individual genes are still intensively studied, but information about individual genes is now interpreted within the context of the genome as a whole. This new, broader emphasis applies not just to genomes but to all of biochemistry and cell biology. No longer is it sufficient to understand individual biochemical pathways or subcellular processes. The challenge now is provided by **systems biology**, which attempts to link together these pathways and processes into networks that describe the overall functioning of living cells and living organisms.

This book will lead you through our knowledge of genomes and show you how this exciting area of research is underpinning our developing understanding of biological systems. First, however, we must pay attention to the basic principles of molecular biology by reviewing the key features of the three types of biological molecule involved in genomes and genome expression: DNA, RNA, and protein.

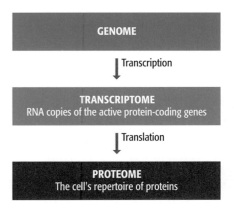

Figure 1.2 **The genome, transcriptome, and proteome.**

1.1 DNA

DNA was discovered in 1869 by Johann Friedrich Miescher, a Swiss biochemist working in Tübingen, Germany. The first extracts that Miescher made from human white blood cells were crude mixtures of DNA and chromosomal proteins, but the following year he moved to Basel, Switzerland (where the research institute named after him is now located), and prepared a pure sample of **nucleic acid** from salmon sperm. Miescher's chemical tests showed that DNA is acidic and rich in phosphorus, and also suggested that the individual molecules are very large, although it was not until the 1930s, when biophysical techniques were applied to DNA, that the huge lengths of the polymeric chains were fully appreciated.

1.1.1 Genes are made of DNA

The fact that genes are made of DNA is so well known today that it can be difficult to appreciate that for the first 75 years after its discovery the true role of DNA was unsuspected. As early as 1903, W.S. Sutton had realized that the inheritance patterns of genes parallel the behavior of chromosomes during cell division, an observation that led to the **chromosome theory**, the proposal that genes are located in chromosomes. Examination of cells by **cytochemistry**, after staining with dyes that bind specifically to just one type of biochemical, showed that chromosomes are made of DNA and protein, in roughly equal amounts. Biologists at that time recognized that billions of different genes must exist and the genetic material must therefore be able to take many different forms. But this requirement appeared not to be satisfied by DNA, because in the early part of the twentieth century it was thought that all DNA molecules were the same. On the other hand, it was known, correctly, that proteins are highly variable, polymeric molecules, each one made up of a different combination of 20 chemically distinct amino-acid monomers (Section 1.3.1). Hence genes simply had to be made of protein, not DNA.

The errors in understanding DNA structure lingered on, but by the late 1930s it had become accepted that DNA, like protein, has immense variability. The notion that protein was the genetic material initially remained strong, but was eventually overturned by the results of two important experiments:

- Oswald Avery, Colin MacLeod, and Maclyn McCarty showed that DNA is the active component of the **transforming principle**, a bacterial cell extract which, when mixed with a harmless strain of *Streptococcus pneumoniae*, converts these bacteria into a virulent form capable of causing pneumonia when injected into mice (Figure 1.3A). In 1944, when the results of this experiment were published, only a few microbiologists appreciated that transformation involves transfer of genes from the cell extract into the living bacteria. However, once this point had been accepted, the true meaning of the "Avery experiment" became clear: bacterial genes must be made of DNA.

- Alfred Hershey and Martha Chase used **radiolabeling** to show that, when a bacterial culture is infected with **bacteriophages** (a type of **virus**), DNA is the major component of the bacteriophages that enters the cells (Figure 1.3B). This was a vital observation because it was known that, during the infection cycle, the genes of the infecting bacteriophages are used to direct synthesis of new bacteriophages, and this synthesis occurs within the bacteria. If it is only the DNA of the infecting bacteriophages that enters the cells, then it follows that the genes of these bacteriophages must be made of DNA.

Although from our perspective these two experiments provide the key results that tell us that genes are made of DNA, biologists at the time were not so easily convinced. Both experiments have limitations that leave room for sceptics to argue that protein could still be the genetic material. For example, there were worries about the specificity of the **deoxyribonuclease** enzyme that Avery and colleagues used to inactivate the transforming principle. This result, a central part of the evidence for the transforming principle being DNA, would be invalid if, as seemed possible, the enzyme contained trace amounts of a contaminating **protease** and hence was also able to degrade protein. Neither is the bacteriophage experiment conclusive, as Hershey and Chase stressed when they published their results: "Our experiments show clearly that a physical separation of phage T2 into genetic and nongenetic parts is possible…The chemical identification of the genetic part must wait, however, until some questions…have been answered." In retrospect, these two experiments are important not because of what they tell us but because they alerted biologists to the fact that DNA *might* be the genetic material and was therefore worth studying. It was this that influenced Watson and Crick to work on DNA and, as we will see below, it was their discovery of the **double helix** structure, which solved the puzzling question of how genes can replicate, which really convinced the scientific world that genes are made of DNA.

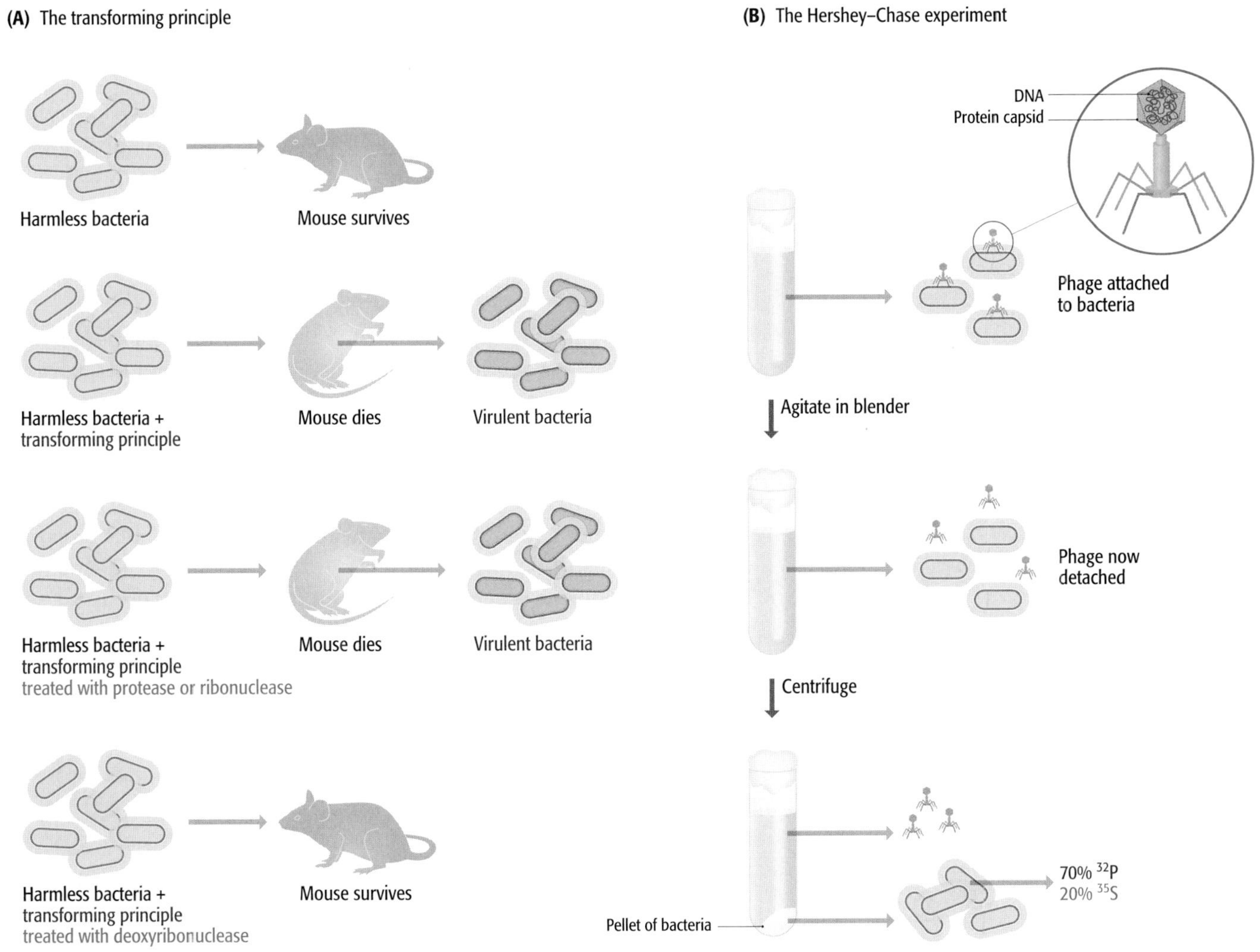

(A) The transforming principle

Harmless bacteria — Mouse survives

Harmless bacteria +
transforming principle — Mouse dies — Virulent bacteria

Harmless bacteria +
transforming principle
treated with protease or ribonuclease — Mouse dies — Virulent bacteria

Harmless bacteria +
transforming principle
treated with deoxyribonuclease — Mouse survives

(B) The Hershey–Chase experiment

DNA
Protein capsid

Phage attached
to bacteria

Agitate in blender

Phage now
detached

Centrifuge

70% ³²P
20% ³⁵S

Pellet of bacteria

Figure 1.3 The two experiments that suggested that genes are made of DNA.

(A) Avery and colleagues showed that the transforming principle is made of DNA. The top two panels show what happens when mice are injected with harmless *Streptococcus pneumoniae* bacteria, with or without addition of the transforming principle, a cell extract obtained from a virulent strain of *S. pneumoniae*. When the transforming principle is present, the mouse dies, because the genes in the transforming principle convert the harmless bacteria into the virulent form, these virulent bacteria subsequently being recovered from the lungs of the dead mouse. The lower two panels show that treatment with protease or ribonuclease has no effect on the transforming principle, but that the transforming principle is inactivated by deoxyribonuclease.

(B) The Hershey–Chase experiment used T2 bacteriophages, each of which comprises a DNA molecule contained in a protein capsid attached to a "body" and "legs" that enable the bacteriophage to attach to the surface of a bacterium and inject its genes into the cell. The DNA of the bacteriophages was labeled with ^{32}P, and the protein with ^{35}S. A few minutes after infection, the culture was agitated to detach the empty phage particles from the cell surface. The culture was then centrifuged, which collects the bacteria plus phage genes as a pellet at the bottom of the tube, but leaves the lighter phage particles in suspension. Hershey and Chase found that the bacterial pellet contained most of the ^{32}P-labeled component of the phages (the DNA) but only 20% of the ^{35}S-labeled material (the phage protein). In a second experiment, Hershey and Chase showed that new phages produced at the end of the infection cycle contained less than 1% of the protein from the parent phages. For more details of the bacteriophage infection cycle, see Figure 2.19.

1.1.2 The structure of DNA

The names of James Watson and Francis Crick are so closely linked with DNA that it is easy to forget that, when they began their collaboration in October 1951, the detailed structure of the DNA polymer was already known. Their contribution was not to determine the structure of DNA *per se*, but to show that in living cells two DNA chains are intertwined to form the double helix. First, therefore, we should examine what Watson and Crick knew before they began their work.

Nucleotides and polynucleotides

DNA is a linear, unbranched polymer in which the monomeric subunits are four chemically distinct nucleotides that can be linked together in any order in chains hundreds, thousands, or even millions of units in length. Each nucleotide in a DNA polymer is made up of three components (Figure 1.4):

- **2′-Deoxyribose**, which is a **pentose**, a type of sugar composed of five carbon atoms. These five carbons are numbered 1′ (spoken as "one-prime"), 2′, and so on. The name "2′-deoxyribose" indicates that this particular sugar is a derivative of ribose, one in which the hydroxyl (–OH) group attached to the 2′-carbon of ribose has been replaced by a hydrogen (–H) group.

- A **nitrogenous base**, one of **cytosine**, **thymine** (single-ring **pyrimidines**), **adenine**, or **guanine** (double-ring **purines**). The base is attached to the 1′-carbon of the sugar by a **β-*N*-glycosidic bond** attached to nitrogen number one of the pyrimidine or number nine of the purine.

- A **phosphate group**, comprising one, two, or three linked phosphate units attached to the 5′-carbon of the sugar. The phosphates are designated α, β, and γ, with the α-phosphate being the one directly attached to the sugar.

A molecule made up of just the sugar and base is called a **nucleoside**; addition of the phosphates converts this to a nucleotide. Although cells contain nucleotides with one, two, or three phosphate groups, only the nucleoside triphosphates act as substrates for DNA synthesis. The full chemical names of the four nucleotides that polymerize to make DNA are:

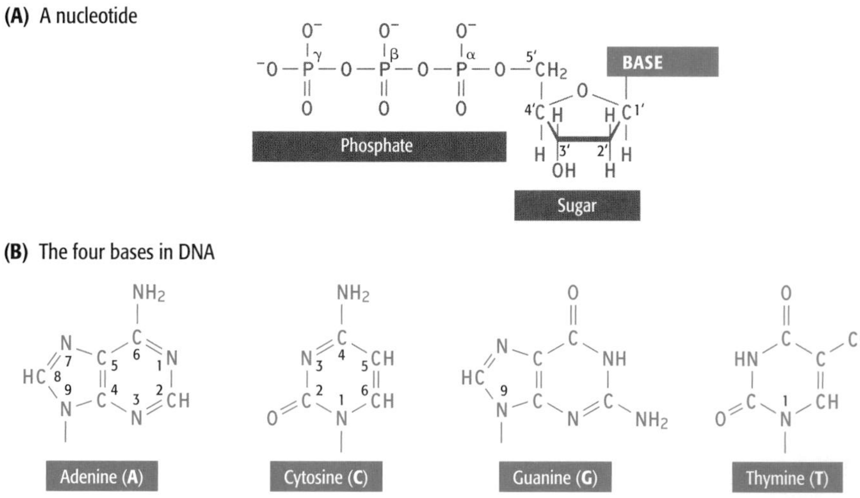

Figure 1.4 **The structure of a nucleotide.** (A) The general structure of a deoxyribonucleotide, the type of nucleotide found in DNA. (B) The four bases that occur in deoxyribonucleotides.

- 2'-Deoxyadenosine 5'-triphosphate.
- 2'-Deoxycytidine 5'-triphosphate.
- 2'-Deoxyguanosine 5'-triphosphate.
- 2'-Deoxythymidine 5'-triphosphate.

The abbreviations of these four nucleotides are dATP, dCTP, dGTP, and dTTP, respectively, or when referring to a DNA sequence, A, C, G, and T, respectively.

In a polynucleotide, individual nucleotides are linked together by **phosphodiester bonds** between their 5'- and 3'-carbons (Figure 1.5). From the structure of this linkage we can see that the polymerization reaction (Figure 1.6) involves removal of the two outer phosphates (the β- and γ-phosphates) from one nucleotide and replacement of the hydroxyl group attached to the 3'-carbon of the second nucleotide. Note that the two ends of the polynucleotide are chemically distinct, one having an unreacted triphosphate group attached to the 5'-carbon (the **5'** or **5'-P terminus**), and the other having an unreacted hydroxyl attached to the 3'-carbon (the **3'** or **3'-OH terminus**). This means that the polynucleotide has a chemical direction, expressed as either 5'→3' (down in Figure 1.5) or 3'→5' (up in Figure 1.5). An important consequence of the polarity of the phosphodiester bond is that the chemical reaction needed to extend a DNA polymer in the 5'→3' direction is different to that needed to make a 3'→5' extension. All natural **DNA polymerase** enzymes are only able to carry out 5'→3' synthesis, which adds significant complications to the process by which double-stranded DNA is replicated (Section 15.2).

The evidence that led to the double helix

In the years before 1950, various lines of evidence had shown that cellular DNA molecules are comprised of two or more polynucleotides assembled together in some way. The possibility that unraveling the nature of this assembly might provide insights into how genes work prompted Watson and Crick, among others, to try to solve the structure. According to Watson in his book *The Double Helix*, their work was a desperate race against the famous American biochemist, Linus Pauling, who initially proposed an incorrect triple helix model, giving Watson and Crick the time they needed to complete

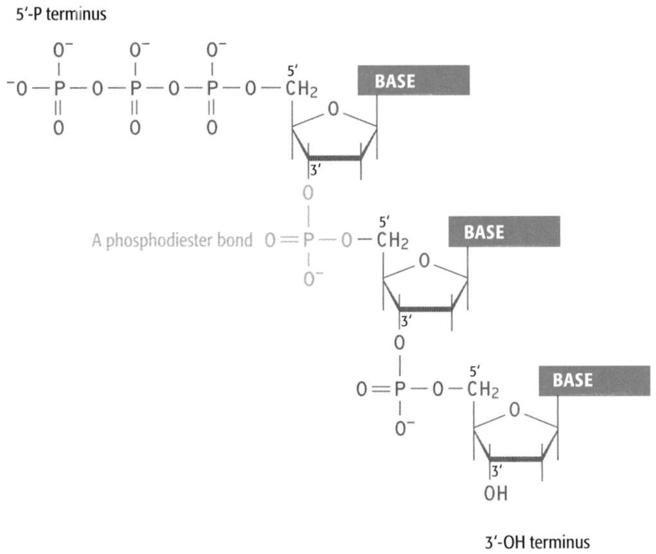

Figure 1.5 A short DNA polynucleotide showing the structure of the phosphodiester bond. Note that the two ends of the polynucleotide are chemically distinct.

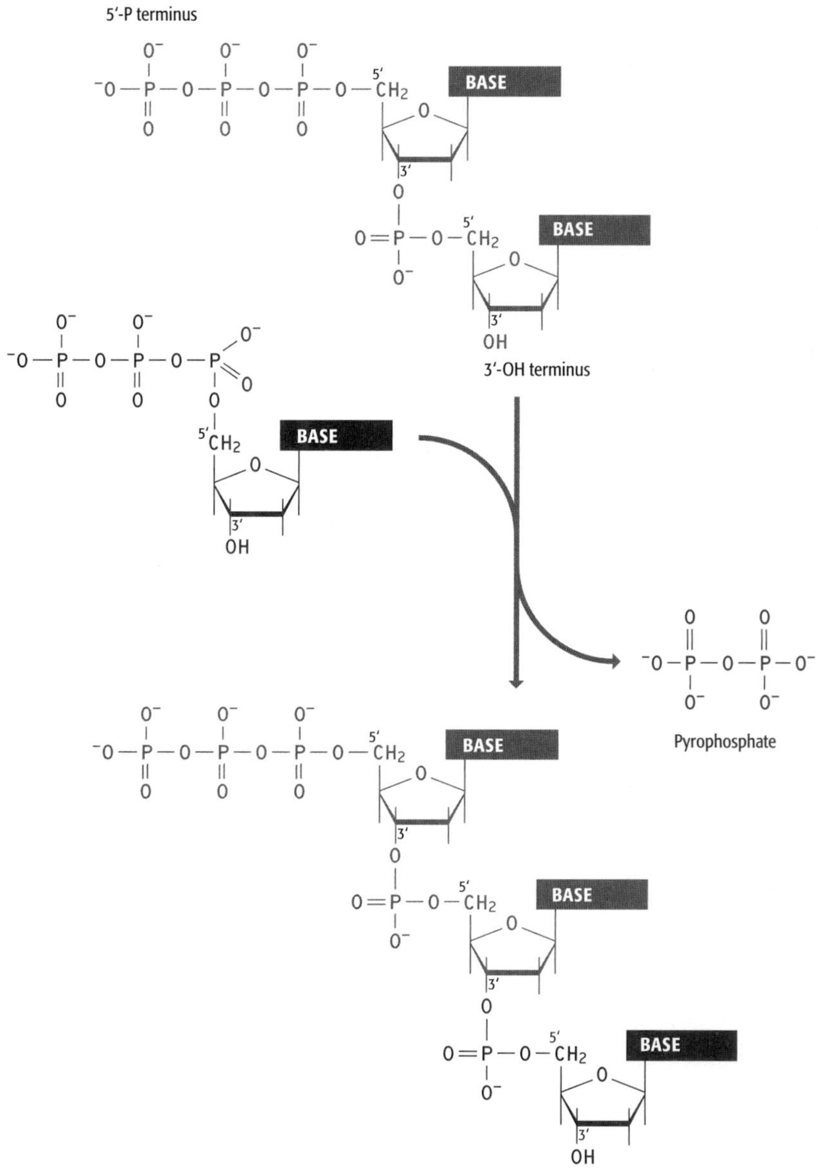

Figure 1.6 The polymerization reaction that results in synthesis of a DNA polynucleotide. Synthesis occurs in the 5′→3′ direction, with the new nucleotide being added to the 3′-carbon at the end of the existing polynucleotide. The β- and γ-phosphates of the nucleotide are removed as a pyrophosphate molecule.

the double helix structure. It is now difficult to separate fact from fiction, especially regarding the part played by Rosalind Franklin, whose **X-ray diffraction** studies provided the bulk of the experimental data in support of the double helix and who was herself very close to solving the structure. The one thing that is clear is that the double helix, discovered by Watson and Crick on Saturday 7 March 1953, was the single most important breakthrough in biology during the twentieth century.

Watson and Crick used four types of information to deduce the double helix structure:

● Biophysical data of various kinds. The water content of DNA fibers was particularly important because it enabled the density of the DNA in a fiber to be estimated. The number of strands in the helix and the spacing between the nucleotides had to be compatible with the fiber density. Pauling's triple helix model was based on an incorrect density measurement that suggested that the DNA molecule was more closely packed than it actually is.

- **X-ray diffraction patterns** (Technical Note 11.1), most of which were produced by Rosalind Franklin and which revealed the helical nature of the structure and indicated some of the key dimensions within the helix.

- The **base ratios**, which had been discovered by Erwin Chargaff of Columbia University, New York. Chargaff carried out a lengthy series of chromatographic studies of DNA samples from various sources and showed that, although the values are different in different organisms, the amount of adenine is always the same as the amount of thymine, and the amount of guanine equals the amount of cytosine (Figure 1.7). These base ratios led to the **base-pairing** rules, which were the key to the discovery of the double helix structure.

- Model building, which was the only major technique that Watson and Crick performed themselves. Scale models of possible DNA structures enabled the relative positioning of the various atoms to be checked, to ensure that pairs that formed bonds were not too far apart, and that other atoms were not so close together as to interfere with one another.

The key features of the double helix

The double helix is right-handed, which means that if it were a spiral staircase and you were climbing upwards then the rail on the outside of the staircase would be on your right-hand side. The two strands run in opposite directions (Figure 1.8A). The helix is stabilized by two types of chemical interaction:

- **Base pairing** between the two strands involves the formation of **hydrogen bonds** between an adenine on one strand and a thymine on the other strand, or between a cytosine and a guanine (Figure 1.8B). Hydrogen bonds are weak electrostatic attractions between an electronegative atom (such as oxygen or nitrogen) and a hydrogen atom attached to a second electronegative atom. Hydrogen bonds are longer than covalent bonds and are much weaker, typical bond energies being 1–10 kcal mol^{-1} at 25°C, compared with up to 90 kcal mol^{-1} for a covalent bond. As well as their role in the DNA double helix, hydrogen bonds stabilize protein secondary structures. The two base-pair combinations—A base-paired with T, and G base-paired with C—explain the base ratios discovered by Chargaff. These are the only pairs that are permissible, partly because of the geometries of the nucleotide bases and the relative positions of the atoms that are able to participate in hydrogen bonds, and partly because the pair must be between a purine and a pyrimidine: a purine–purine pair would be too big to fit within the helix, and a pyrimidine–pyrimidine pair would be too small.

- **Base stacking**, sometimes called **π–π interactions**, involves hydrophobic interactions between adjacent base pairs and adds stability to the double helix once the strands have been brought together by base pairing. These hydrophobic interactions arise because the hydrogen-bonded structure of water forces hydrophobic groups into the internal parts of a molecule.

Both base pairing and base stacking are important in holding the two polynucleotides together, but base pairing has added significance because of its biological implications. The limitation that A can only base-pair with T, and G can only base-pair with C, means that DNA replication can result in perfect copies of a parent molecule through the simple expedient of using the sequences of the preexisting strands to dictate the sequences of the new strands. This is **template-dependent DNA synthesis** and it is the system used by all cellular DNA polymerases (Section 15.2.2). Base pairing therefore enables DNA molecules to

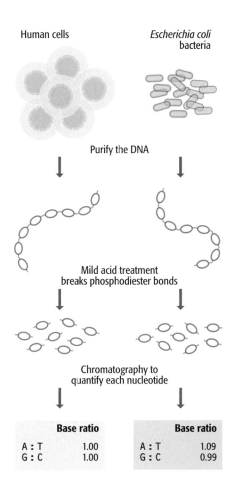

Human cells		Escherichia coli bacteria

Purify the DNA

Mild acid treatment breaks phosphodiester bonds

Chromatography to quantify each nucleotide

	Base ratio			Base ratio
A : T	1.00		A : T	1.09
G : C	1.00		G : C	0.99

Figure 1.7 The base ratio experiments performed by Chargaff. DNA was extracted from various organisms and treated with acid to hydrolyze the phosphodiester bonds and release the individual nucleotides. Each nucleotide was then quantified by chromatography. The data show some of the actual results obtained by Chargaff. These indicate that, within experimental error, the amount of adenine is the same as that of thymine, and the amount of guanine is the same as that of cytosine.

Figure 1.8 The double helix structure of DNA. (A) Two representations of the double helix. On the left the structure is shown with the sugar–phosphate "backbones" of each polynucleotide drawn as a gray ribbon with the base pairs in green. On the right the chemical structure for three base pairs is given. (B) A base-pairs with T, and G base-pairs with C. The bases are drawn in outline, with the hydrogen bonding indicated by dotted lines. Note that a G–C base pair has three hydrogen bonds whereas an A–T base pair has just two.

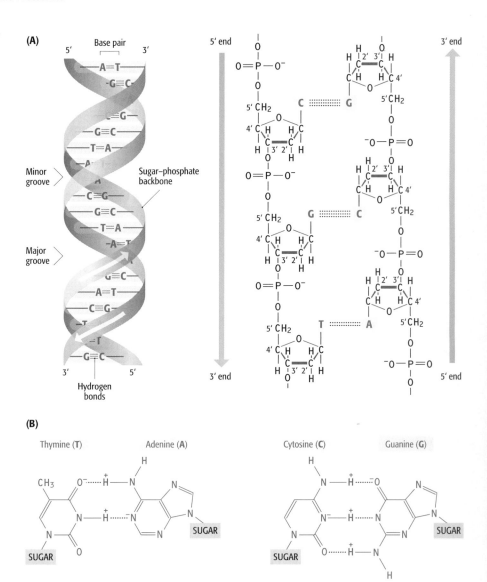

be replicated by a system that is so simple and elegant that as soon as the double helix structure was publicized by Watson and Crick, every biologist became convinced that genes really are made of DNA.

The double helix has structural flexibility

The double helix described by Watson and Crick, and shown in Figure 1.8A, is called the B-form of DNA. Its characteristic features lie in its dimensions: a helical diameter of 2.37 nm, a rise of 0.34 nm per base pair, and a pitch (i.e., distance taken up by a complete turn of the helix) of 3.4 nm, this corresponding to ten base pairs per turn. The DNA in living cells is thought to be predominantly in this B-form, but it is now clear that genomic DNA molecules are not entirely uniform in structure. This is mainly because each nucleotide in the helix has the flexibility to take up slightly different molecular shapes. To adopt these different conformations, the relative positions of the atoms in the nucleotide must change slightly. There are a number of possibilities but the most important conformational changes involve rotation around the β-*N*-glycosidic bond, changing the orientation of the base relative to the sugar, and rotation around the bond between the 3'- and 4'-carbons. Both rotations have a significant effect on the double helix: changing the base orientation influences the

Table 1.1 Features of different conformations of the DNA double helix

Feature	B-DNA	A-DNA	Z-DNA
Type of helix	Right-handed	Right-handed	Left-handed
Helical diameter (nm)	2.37	2.55	1.84
Rise per base pair (nm)	0.34	0.29	0.37
Distance per complete turn (pitch) (nm)	3.4	3.2	4.5
Number of base pairs per complete turn	10	11	12
Topology of major groove	Wide, deep	Narrow, deep	Flat
Topology of minor groove	Narrow, shallow	Broad, shallow	Narrow, deep

relative positioning of the two polynucleotides, and rotation around the 3′–4′ bond affects the conformation of the sugar–phosphate backbone.

Rotations within individual nucleotides therefore lead to major changes in the overall structure of the helix. It has been recognized since the 1950s that changes in the dimensions of the double helix occur when fibers containing DNA molecules are exposed to different relative humidities. For example, the modified version of the double helix called the A-form (Figure 1.9) has a diameter of 2.55 nm, a rise of 0.29 nm per base pair, and a pitch of 3.2 nm, corresponding to 11 base pairs per turn (Table 1.1). Other variations include B′-, C-, C′-, C″-, D-, E- and T-DNAs. All these are right-handed helices like the B-form. A more drastic reorganization is also possible, leading to the left-handed Z-DNA (Figure 1.9), a slimmer version of the double helix with a diameter of only 1.84 nm.

The bare dimensions of the various forms of the double helix do not reveal what are probably the most significant differences between them. These relate not to diameter and pitch, but the extent to which the internal regions of the helix are accessible from the surface of the structure. As shown in Figures 1.8 and 1.9, the B-form of DNA does not have an entirely smooth surface: instead, two grooves spiral along the length of the helix. One of these grooves is relatively wide and deep and is called the **major groove**; the other is narrow and less deep and is called the **minor groove**. A-DNA also has two grooves (Figure 1.9), but with this conformation the major groove is even deeper, and the minor groove shallower and broader compared with B-DNA. Z-DNA is different again, with one groove virtually nonexistent but the other very narrow and deep. In each form of DNA, part of the internal surface of at least one of the grooves is formed by chemical groups attached to the nucleotide bases. In Chapter 11 we will see that expression of the biological information contained within a genome is mediated by DNA-binding proteins that attach to the double helix and regulate the activity of the genes contained within it. To carry out their function, each DNA-binding protein must attach at a specific position, near to the gene whose activity it must influence. This can be achieved, with at least some degree of accuracy, by the protein reaching down into a groove, within which the DNA sequence can be "read" without the helix being opened up by breaking the base pairs. A corollary of this is that a DNA-binding protein whose structure enables it to recognize a specific nucleotide sequence within B-DNA, for example, might not be able to recognize that sequence if the DNA has taken up a different conformation.

Figure 1.9 **The structures of B-DNA (left), A-DNA (center) and Z-DNA (right).** Space-filling models (top) and structural models (bottom) depicting different conformations of DNA molecules. Note the differences in helical diameter, number of base pairs per complete turn, and topology of the major and minor grooves between these molecules. Reprinted with permission from Kendrew, J. (Ed.), *Encyclopaedia of Molecular Biology.* © 1994 Blackwell Publishing.

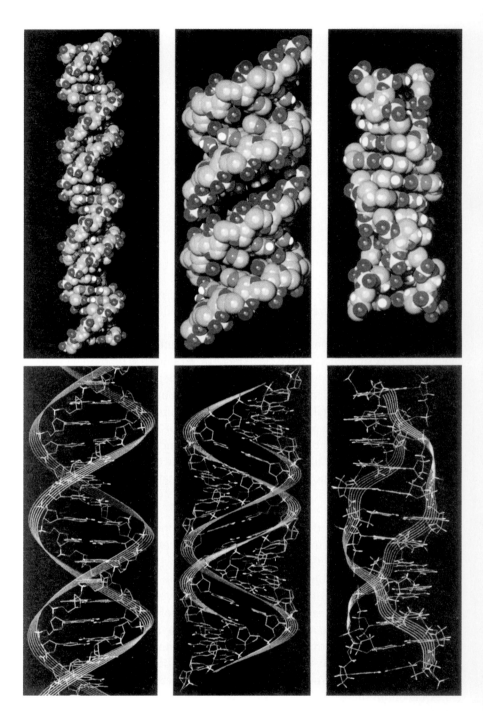

As we will see in Chapter 11, conformational variations along the length of a DNA molecule, together with other structural polymorphisms caused by the nucleotide sequence, could be important in determining the specificity of the interactions between the genome and its DNA-binding proteins.

1.2 RNA and the Transcriptome

The initial product of genome expression is the transcriptome (see Figure 1.2), the collection of RNA molecules derived from those protein-coding genes whose biological information is required by the cell at a particular time. The RNA molecules of the transcriptome, as well as many other RNAs derived from genes that do not code for proteins, are synthesized by the process called transcription. In this section we will examine the structure of RNA and then look more closely at the various types of RNA molecule that are present in living cells.

1.2.1 The structure of RNA

RNA is a polynucleotide similar to DNA but with two important chemical differences (Figure 1.10). First, the sugar in an RNA nucleotide is **ribose** and, second, RNA contains **uracil** instead of thymine. The four nucleotide substrates for synthesis of RNA are therefore:

- Adenosine 5'-triphosphate.
- Cytidine 5'-triphosphate.
- Guanosine 5'-triphosphate.
- Uridine 5'-triphosphate.

These nucleotides are abbreviated to ATP, CTP, GTP, and UTP, or A, C, G, and U, respectively.

As with DNA, RNA polynucleotides contain 3'–5' phosphodiester bonds, but these phosphodiester bonds are less stable than those in a DNA polynucleotide because of the indirect effect of the hydroxyl group at the 2'-position of the sugar. RNA molecules are rarely more than a few thousand nucleotides in length, and although many form *intra*molecular base pairs (for example, see Figure 13.2), most are single- rather than double-stranded.

The enzymes responsible for transcription of DNA into RNA are called **DNA-dependent RNA polymerases**. The name indicates that the enzymatic reaction that they catalyze results in polymerization of RNA from ribonucleotides and occurs in a DNA-dependent manner, meaning that the sequence of nucleotides in a DNA template dictates the sequence of nucleotides in the RNA that is made (Figure 1.11). It is permissible to shorten the enzyme name to **RNA polymerase**, as the context in which the name is used means that there is rarely confusion with the **RNA-dependent RNA polymerases** that are involved in replication and expression of some virus genomes. The chemical basis of the template-dependent synthesis of RNA is equivalent to that shown for the synthesis of DNA in Figure 1.6. Ribonucleotides are added one after another to the growing 3' end of the RNA transcript, the identity of each nucleotide being specified by the base-pairing rules: A base-pairs with T or U; G base-pairs with C. During each nucleotide addition, the β- and γ-phosphates are removed from the incoming nucleotide, and the hydroxyl group is removed from the 3'-carbon of the nucleotide at the end of the chain, precisely the same as for DNA polymerization.

1.2.2 The RNA content of the cell

A typical bacterium contains 0.05–0.10 pg of RNA, making up about 6% of its total weight. A mammalian cell, being much larger, contains more RNA, 20–30 pg in all, but this represents only 1% of the cell as a whole. The best way to understand the RNA content of a cell is to divide it into categories and subcategories depending on function. There are several ways of doing this, the most informative scheme being the one shown in Figure 1.12. The primary division is between **coding RNA** and **noncoding RNA**. The coding RNA comprises the transcriptome and is made up of just one class of molecule, the **messenger RNAs (mRNAs)**, which are transcripts of protein-coding genes

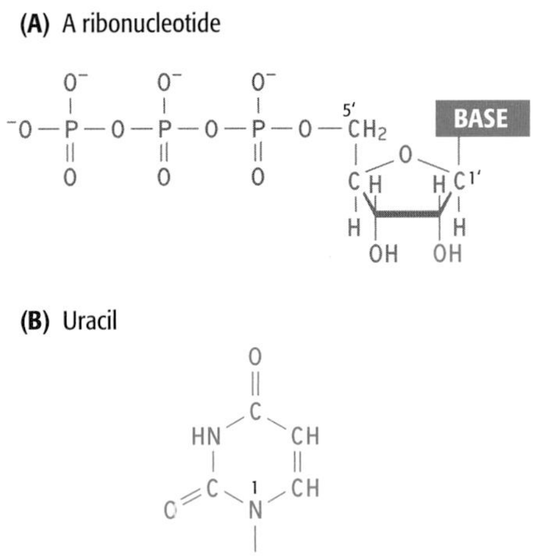

(A) A ribonucleotide

(B) Uracil

Figure 1.10 The chemical differences between DNA and RNA. (A) RNA contains ribonucleotides, in which the sugar is ribose rather than 2'-deoxyribose. The difference is that a hydroxyl group, rather than hydrogen atom, is attached to the 2'-carbon. (B) RNA contains the pyrimidine called uracil instead of thymine.

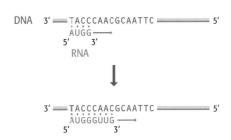

Figure 1.11 Template-dependent RNA synthesis. The RNA transcript is synthesized in the 5'→3' direction, reading the DNA in the 3'→5' direction, with the sequence of the transcript determined by base-pairing to the DNA template.

and hence are translated into protein in the second stage of genome expression. Messenger RNAs rarely make up more than 4% of the total RNA and are short-lived, being degraded soon after synthesis. Bacterial mRNAs have half-lives of no more than a few minutes, and in eukaryotes most mRNAs are degraded a few hours after synthesis. This rapid turnover means that the composition of the transcriptome is not fixed and can quickly be restructured by changing the rate of synthesis of individual mRNAs.

The second type of RNA is referred to as "noncoding" as these molecules are not translated into protein. However, a better name is **functional RNA**, as this emphasizes that, although not part of the transcriptome, the noncoding RNAs still have essential roles within the cell. There are several diverse types of functional RNA, the most important being as follows:

- **Ribosomal RNAs** (**rRNAs**) are present in all organisms and are usually the most abundant RNAs in the cell, making up over 80% of the total RNA in actively dividing bacteria. These molecules are components of ribosomes, the structures on which protein synthesis takes place (Section 13.2).

- **Transfer RNAs** (**tRNAs**) are small molecules that are also involved in protein synthesis and, like rRNA, are found in all organisms. The function of tRNAs is to carry amino acids to the ribosome and ensure that the amino acids are linked together in the order specified by the nucleotide sequence of the mRNA that is being translated (Section 13.1).

- **Small nuclear RNAs** (**snRNAs**; also called **U-RNAs** because these molecules are rich in uridine nucleotides) are found in the nuclei of eukaryotes. These molecules are involved in **splicing**, one of the key steps in the processing events that convert the primary transcripts of protein-coding genes into mRNAs (Section 12.2.2).

- **Small nucleolar RNAs** (**snoRNAs**) are found in the nucleolar regions of eukaryotic nuclei. They play a central role in the chemical modification of rRNA molecules by directing the enzymes that perform the modifications to the specific nucleotides where alterations, such as addition of a methyl group, must be carried out (Section 12.2.5).

- **MicroRNAs** (**miRNAs**) and **short interfering RNAs** (**siRNAs**) are small RNAs that regulate the expression of individual genes (Section 12.2.6).

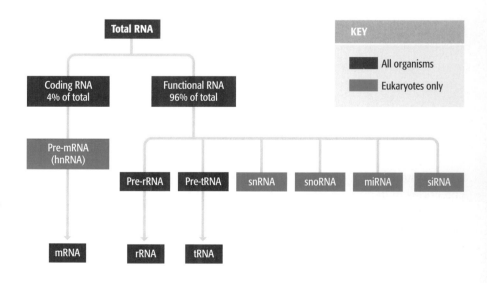

Figure 1.12 The RNA content of a cell.
This scheme shows the types of RNA present in all organisms and those categories found only in eukaryotic cells.

1.2.3 Processing of precursor RNA

As well as the mature RNAs described above, cells also contain precursor molecules. Many RNAs, especially in eukaryotes, are initially synthesized as precursor or **pre-RNA**, which has to be processed before it can carry out its function. The various processing events, all of which are described in Chapter 12, include the following (Figure 1.13):

- **End-modifications** occur during the synthesis of eukaryotic mRNAs, most of which have a single, unusual nucleotide called a **cap** attached at the 5′ end and a **poly(A) tail** attached to the 3′ end.

- **Splicing** is the removal of segments from within a precursor RNA. Many genes, especially in eukaryotes, contain internal segments that contain no biological information. These are called **introns** and they are copied along with the information-containing **exons** when the gene is transcribed. The introns are removed from the **pre-mRNA** by cutting and joining reactions. Unspliced pre-mRNA forms the nuclear RNA fraction called **heterogenous nuclear RNA (hnRNA)**.

- **Cutting events** are particularly important in the processing of rRNAs and tRNAs, many of which are initially synthesized from transcription units that specify more than one molecule. The **pre-rRNAs** and **pre-tRNAs** must therefore be cut into pieces to produce the mature RNAs. This type of processing occurs in both prokaryotes and eukaryotes.

- **Chemical modifications** are made to rRNAs, tRNAs, and mRNAs. The rRNAs and tRNAs of all organisms are modified by addition of new chemical groups, these groups being added to specific nucleotides within each RNA. Chemical modification of mRNA, called **RNA editing**, occurs in many eukaryotes.

1.2.4 The transcriptome

Although the transcriptome makes up less than 4% of the total cell RNA, it is the most significant component because it contains the coding RNAs that are used in the next stage of genome expression. It is important to note that the transcriptome is never synthesized *de novo*. Every cell receives part of its parent's transcriptome when it is first brought into existence by cell division, and maintains a transcriptome throughout its lifetime. Even quiescent cells in bacterial spores or in the seeds of plants have a transcriptome, although translation of that transcriptome into protein may be completely switched off. Transcription of individual protein-coding genes does not, therefore, result in *synthesis* of the transcriptome but instead *maintains* the transcriptome by replacing mRNAs that have been degraded, and brings about

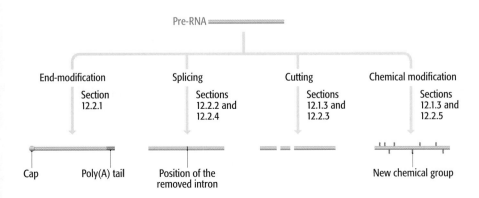

Figure 1.13 Schematic representation of the four types of RNA processing event. Not all events occur in all organisms.

changes to the composition of the transcriptome via the switching on and off of different sets of genes.

Even in the simplest organisms, such as bacteria and yeast, many genes are active at any one time. Transcriptomes are therefore complex, containing copies of hundreds, if not thousands, of different mRNAs. Usually each mRNA makes up only a small fraction of the whole, with the most common type rarely contributing more than 1% of the total mRNA. Exceptions are those cells that have highly specialized biochemistries, reflected by transcriptomes in which one or a few mRNAs predominate. Developing wheat seeds are an example: these synthesize large amounts of the gliadin proteins, which accumulate in the dormant grain and provide a source of amino acids for the germinating seedling. Within the developing seeds, the gliadin mRNAs can make up as much as 30% of the transcriptomes of certain cells.

1.3 Proteins and the Proteome

The second product of genome expression is the proteome (see Figure 1.2), the cell's repertoire of proteins, which specifies the nature of the biochemical reactions that the cell is able to carry out. These proteins are synthesized by **translation** of the mRNA molecules that make up the transcriptome.

1.3.1 Protein structure

A protein**,** like a DNA molecule, is a linear, unbranched polymer. In proteins, the monomeric subunits are called **amino acids** (Figure 1.14) and the resulting polymers, or **polypeptides**, are rarely more than 2000 units in length. As with DNA, the key features of protein structure were determined in the first half of the twentieth century, this phase of protein biochemistry culminating in the 1940s and early 1950s with the elucidation by Pauling and Corey of the major conformations, or **secondary structures**, taken up by polypeptides. In recent years, interest has focused on how these secondary structures combine to produce the complex, three-dimensional shapes of proteins.

The four levels of protein structure

Proteins are traditionally looked upon as having four distinct levels of structure. These levels are hierarchical, the protein being built up stage-by-stage, with each level of structure depending on the one below it:

- The **primary structure** of the protein is formed by joining amino acids into a polypeptide. The amino acids are linked by **peptide bonds** that are formed by a condensation reaction between the carboxyl group of one amino acid and the amino group of a second amino acid (Figure 1.15). Note that, as with a polynucleotide, the two ends of the polypeptide are chemically distinct: one has a free amino group and is called the **amino**, **NH₂–**, or **N terminus**; the other has a free carboxyl group and is called the **carboxyl**, **COOH–**, or **C terminus**. The direction of the polypeptide can therefore be expressed as either N→C (left to right in Figure 1.15) or C→N (right to left in Figure 1.15).

- The **secondary structure** refers to the different conformations that can be taken up by the polypeptide. The two main types of secondary structure are the **α-helix** and **β-sheet** (Figure 1.16). These are stabilized mainly by hydrogen bonds that form between different amino acids in the

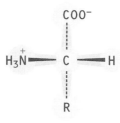

Figure 1.14 The general structure of an amino acid. All amino acids have the same general structure, comprising a central α-carbon attached to a hydrogen atom, a carboxyl group, an amino group, and an R group. The R group is different for each amino acid (see Figure 1.18).

polypeptide. Most polypeptides are long enough to be folded into a series of secondary structures, one after another along the molecule.

- The **tertiary structure** results from folding the secondary structural components of the polypeptide into a three-dimensional configuration (Figure 1.17). The tertiary structure is stabilized by various chemical forces, notably hydrogen bonding between individual amino acids, electrostatic interactions between the R groups of charged amino acids (see Figure 1.18), and hydrophobic forces, which dictate that amino acids with nonpolar ("water-hating") side-groups must be shielded from water by embedding within the internal regions of the protein. There may also be covalent linkages called **disulfide bridges** between cysteine amino acid residues at various places in the polypeptide.

- The **quaternary structure** involves the association of two or more polypeptides, each folded into its tertiary structure, into a multisubunit protein. Not all proteins form quaternary structures, but it is a feature of many proteins with complex functions, including several involved in genome expression. Some quaternary structures are held together by disulfide bridges between the different polypeptides, resulting in stable multisubunit proteins that cannot easily be broken down to the component parts. Other quaternary structures comprise looser associations of subunits stabilized by hydrogen bonding and hydrophobic effects, which means that these proteins can revert to their component polypeptides, or change their subunit composition, according to the functional requirements of the cell.

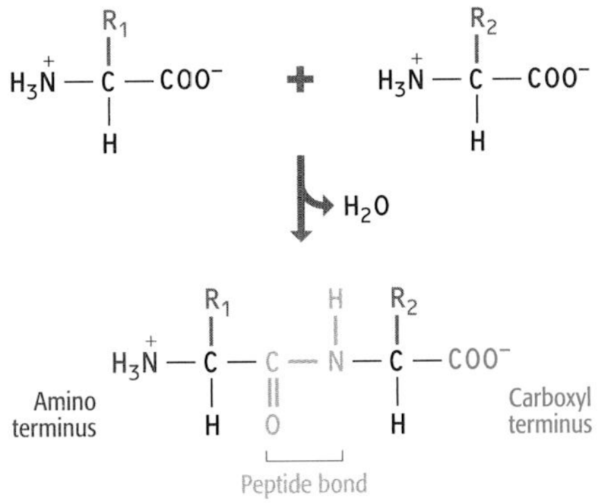

Amino acid diversity underlies protein diversity

Proteins are functionally diverse because the amino acids from which proteins are made are themselves chemically diverse. Different sequences of amino acids therefore result in different combinations of chemical reactivities, these combinations dictating not only the overall structure of the resulting protein but also the positioning on the surface of the structure of reactive groups that determine the chemical properties of the protein.

Amino acid diversity derives from the R group because this part is different in each amino acid and varies greatly in structure. Proteins are made up from a set of 20 amino acids (Figure 1.18; Table 1.2). Some of these have R groups that are small, relatively simple structures, such as a single hydrogen atom (in the amino acid called glycine) or a methyl group (alanine). Other R groups are large, complex aromatic side chains (phenylalanine, tryptophan, and tyrosine). Most amino acids are uncharged, but two are negatively charged (aspartic acid and glutamic acid) and three are positively charged (arginine, histidine, and lysine). Some amino acids are polar (e.g., glycine, serine, and threonine), others are nonpolar (e.g., alanine, leucine, and valine).

The 20 amino acids shown in Figure 1.18 are the ones that are conventionally looked upon as being specified by the genetic code (Section 1.3.2). They are therefore the amino acids that are linked together when mRNA molecules are translated into proteins. However, these 20 amino acids do not, on their own, represent the limit of the chemical diversity of proteins. The diversity is even greater because of two factors:

- At least two additional amino acids—selenocysteine and pyrrolysine (Figure 1.19)—can be inserted into a polypeptide chain during protein synthesis, their insertion directed by a modified reading of the genetic code (Section 13.1.1).

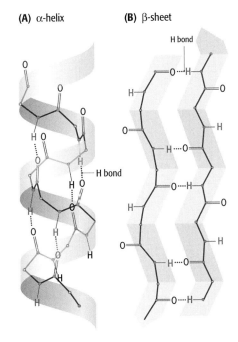

Figure 1.15 In polypeptides, amino acids are linked by peptide bonds. The drawing shows the chemical reaction that results in two amino acids becoming linked together by a peptide bond. The reaction is called a condensation because it results in elimination of water.

Figure 1.16 The two main secondary structural units found in proteins: (A) the α-helix, and (B) the β-sheet. The polypeptide chains are shown in outline. The R groups have been omitted for clarity. Each structure is stabilized by hydrogen (H) bonds between the C=O and N–H groups of different peptide bonds. The β-sheet conformation that is shown is antiparallel, the two chains running in opposite directions. Parallel β-sheets also occur.

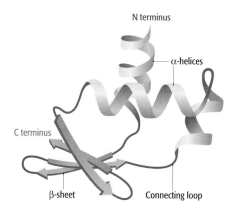

Figure 1.17 The tertiary structure of a protein. This imaginary protein structure comprises three α-helices, shown as coils, and a four-stranded β-sheet, indicated by the arrows.

• During protein processing, some amino acids are modified by the addition of new chemical groups, for example by acetylation or phosphorylation, or by attachment of large side chains made up of sugar units (Section 13.3.3).

Proteins therefore have an immense amount of chemical variability, some of this directly specified by the genome, the remainder arising by protein processing.

1.3.2 The proteome

The proteome comprises all the proteins present in a cell at a particular time. A "typical" mammalian cell, for example a liver hepatocyte, is thought to contain 10,000–20,000 different proteins, about 8×10^9 individual molecules in all, representing approximately 0.5 ng of protein or 18%–20% of the total cell weight. The copy numbers of individual proteins vary enormously, from less than 20,000 molecules per cell for the rarest types to 100 million copies for the commonest ones. Any protein that is present at a copy number of greater than 50,000 per cell is considered to be relatively abundant, and in the average mammalian cell some 2000 proteins fall into this category. When the proteomes of different types of mammalian cell are examined, very few differences

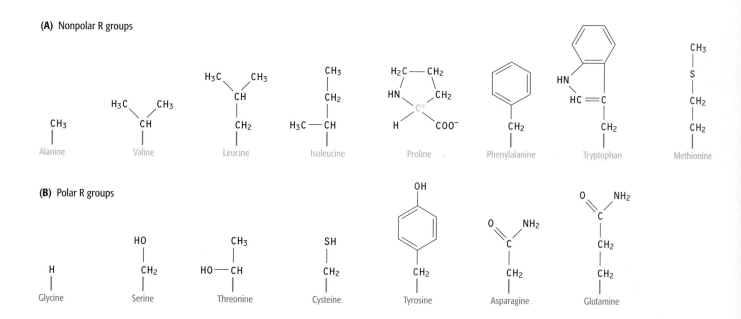

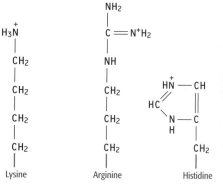

Figure 1.18 Amino acid R groups. These 20 amino acids are the ones that are conventionally looked upon as being specified by the genetic code.

Table 1.2 Amino acid abbreviations

Amino acid	Abbreviation Three-letter	One-letter
Alanine	Ala	A
Arginine	Arg	R
Asparagine	Asn	N
Aspartic acid	Asp	D
Cysteine	Cys	C
Glutamic acid	Glu	E
Glutamine	Gln	Q
Glycine	Gly	G
Histidine	His	H
Isoleucine	Ile	I
Leucine	Leu	L
Lysine	Lys	K
Methionine	Met	M
Phenylalanine	Phe	F
Proline	Pro	P
Serine	Ser	S
Threonine	Thr	T
Tryptophan	Trp	W
Tyrosine	Tyr	Y
Valine	Val	V

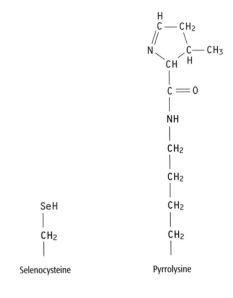

Figure 1.19 **The structures of selenocysteine and pyrrolysine.** The parts shown in brown indicate the differences between these amino acids and cysteine and lysine, respectively.

are seen among these abundant proteins, suggesting that most of them are **housekeeping** proteins that perform general biochemical activities that occur in all cells. The proteins that provide the cell with its specialized function are often quite rare, although there are exceptions, such as the vast amounts of hemoglobin that are present only in red blood cells.

The link between the transcriptome and the proteome

The flow of information from DNA to RNA by transcription does not provide any conceptual difficulty. DNA and RNA polynucleotides have very similar structures and we can easily understand how an RNA copy of a gene can be made by template-dependent synthesis using the base-pairing rules with which we are familiar. The second phase of genome expression, during which the mRNA molecules of the transcriptome direct synthesis of proteins, is less easy to understand simply by considering the structures of the molecules that are involved. In the early 1950s, shortly after the double helix structure of DNA had been discovered, several molecular biologists attempted to devise ways in which amino acids could attach to mRNAs in an ordered fashion, but in all of these schemes at least some of the bonds had to be shorter or longer than was possible according to the laws of physical chemistry, and each idea was quietly dropped. Eventually, in 1957, Francis Crick cut a way through the confusion by predicting the existence of an adaptor molecule that would form a bridge between the mRNA and the polypeptide being synthesized. Soon afterwards it was realized that the tRNAs are these adaptor molecules, and once this fact had been established, a detailed understanding of the mechanism by which proteins are synthesized was quickly built up. We will examine this process in Section 13.1.

The other aspect of protein synthesis that interested molecular biologists in the 1950s was the **informational problem**. This refers to the second important component of the link between the transcriptome and proteome: the **genetic code**, which specifies how the nucleotide sequence of an mRNA is translated into the amino acid sequence of a protein. It was recognized in the 1950s that a triplet genetic code—one in which each codeword, or **codon,** comprises three nucleotides—is required to account for all 20 amino acids found in proteins. A two-letter code would have only $4^2 = 16$ codons, which is not enough to account for all 20 amino acids, whereas a three-letter code would give $4^3 = 64$ codons. The genetic code was worked out in the 1960s, partly by analysis of polypeptides arising from translation of artificial mRNAs of known or predictable sequence in cell-free systems, and partly by determining which amino acids associated with which RNA sequences in an assay based on purified ribosomes. When this work was completed, it was realized that the 64 codons fall into groups, the members of each group coding for the same amino acid (Figure 1.20). Only tryptophan and methionine have just a single codon each: all other amino acids are coded by two, three, four, or six codons. This feature of the code is called **degeneracy**. The code also has four **punctuation codons**, which indicate the points within an mRNA where translation of the nucleotide sequence should start and finish (Figure 1.21). The **initiation codon** is usually 5′–AUG–3′, which also specifies methionine (so most newly synthesized polypeptides start with methionine), although with a few mRNAs other codons such as 5′–GUG–3′ and 5′–UUG–3′ are used. The three **termination codons** are 5′–UAG–3′, 5′–UAA–3′, and 5′–UGA–3′.

The genetic code is not universal

It was originally thought that the genetic code must be the same in all organisms. The argument was that, once established, it would be impossible for the code to change because giving a new meaning to any single codon would result in widespread disruption of the amino acid sequences of proteins. This reasoning seems sound, so it is surprising that, in reality, the genetic code is not universal. The code shown in Figure 1.20 holds for the vast majority of genes in the vast majority of organisms, but deviations are widespread. In particular, mitochondrial genomes often use a nonstandard code (Table 1.3). This was first discovered in 1979 by Frederick Sanger's group in Cambridge, UK, who found that several human mitochondrial mRNAs contain the sequence 5′–UGA–3′, which normally codes for termination, at internal positions where protein synthesis was not expected to stop. Comparisons with the amino acid sequences of the proteins coded by these mRNAs showed that 5′–UGA–3′ is a tryptophan codon in human mitochondria, and that this is just one of four code deviations in this particular genetic system. Mitochondrial genes in other organisms also display code deviations, although at least one of these— the use of 5′–CGG–3′ as a tryptophan codon in plant mitochondria—is probably corrected by RNA editing (Section 12.2.5) before translation occurs.

Nonstandard codes are also known for the nuclear genomes of lower eukaryotes. Often a modification is restricted to just a small group of organisms and

UUU	phe	UCU		UAU	tyr	UGU	cys
UUC		UCC	ser	UAC		UGC	
UUA	leu	UCA		UAA	stop	UGA	stop
UUG		UCG		UAG		UGG	trp
CUU	leu	CCU		CAU	his	CGU	arg
CUC		CCC	pro	CAC		CGC	
CUA		CCA		CAA	gln	CGA	
CUG		CCG		CAG		CGG	
AUU	ile	ACU		AAU	asn	AGU	ser
AUC		ACC	thr	AAC		AGC	
AUA		ACA		AAA	lys	AGA	arg
AUG	met	ACG		AAG		AGG	
GUU	val	GCU		GAU	asp	GGU	gly
GUC		GCC	ala	GAC		GGC	
GUA		GCA		GAA	glu	GGA	
GUG		GCG		GAG		GGG	

Figure 1.20 The genetic code. Amino acids are designated by the standard, three-letter abbreviations (see Table 1.2).

frequently it involves reassignment of the termination codons (Table 1.3). Modifications are less common among prokaryotes, but one example is known in *Mycoplasma* species. A more important type of code variation is **context-dependent codon reassignment**, which occurs when the protein to be synthesized contains either selenocysteine or pyrrolysine. Proteins containing pyrrolysine are rare, and are probably only present in the group of prokaryotes called the **archaea** (Chapter 8), but selenoproteins are widespread in many organisms, one example being the enzyme glutathione peroxidase, which helps protect the cells of humans and other mammals against oxidative damage. Selenocysteine is coded by 5′–UGA–3′ and pyrrolysine by 5′–UAG–3′. These codons therefore have a dual meaning because they are still used as termination codons in the organisms concerned (Table 1.3). A 5′–UGA–3′ codon that specifies selenocysteine is distinguished from true termination codons by the presence of a hairpin loop structure in the mRNA, positioned just downstream of the selenocysteine codon in prokaryotes, and in the 3′ untranslated region (i.e., the part of the mRNA after the termination codon) in eukaryotes. Recognition of the selenocysteine codon requires interaction between the hairpin structure and a special protein that is involved in translation of these mRNAs. A similar system probably operates for specifying pyrrolysine.

The link between the proteome and the biochemistry of the cell

The biological information encoded by the genome finds its final expression in a protein whose biological properties are determined by its folded structure and by the spatial arrangement of chemical groups on its surface. By specifying proteins of different types, the genome is able to construct and maintain a proteome whose overall biological properties form the underlying basis of life. The proteome can play this role because of the huge diversity of protein structures that can be formed, the diversity enabling proteins to carry out a variety of biological functions. These functions include the following:

- Biochemical catalysis is the role of the special type of proteins called enzymes. The central metabolic pathways, which provide the cell with energy, are catalyzed by enzymes, as are the biosynthetic processes that

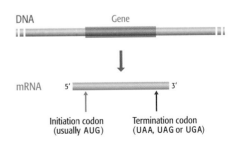

Figure 1.21 The positions of the punctuation codons in an mRNA.

Table 1.3 Examples of deviations from the standard genetic code

Organism	Codon	Should code for	Actually codes for
Mitochondrial genomes			
Mammals	UGA	Stop	Trp
	AGA, AGG	Arg	Stop
	AUA	Ile	Met
Drosophila	UGA	Stop	Trp
	AGA	Arg	Ser
	AUA	Ile	Met
Saccharomyces cerevisiae	UGA	Stop	Trp
	CUN	Leu	Thr
	AUA	Ile	Met
Fungi	UGA	Stop	Trp
Maize	CGG	Arg	Trp
Nuclear and prokaryotic genomes			
Several protozoa	UAA, UAG	Stop	Gln
Candida cylindracea	CUG	Leu	Ser
Micrococcus sp.	AGA	Arg	Stop
	AUA	Ile	Stop
Euplotes sp.	UGA	Stop	Cys
Mycoplasma sp.	UGA	Stop	Trp
	CGG	Arg	Stop
Context-dependent codon reassignments			
Various	UGA	Stop	Selenocysteine
Archaea	UAG	Stop	Pyrrolysine

Abbreviation: N, any nucleotide.

result in construction of nucleic acids, proteins, carbohydrates, and lipids. Biochemical catalysis also drives genome expression through the activities of enzymes such as RNA polymerase.

- Structure, which at the cellular level is determined by the proteins that make up the cytoskeleton, is also the primary function of some extracellular proteins. An example is collagen, which is an important component of bones and tendons.

- Movement is conferred by contractile proteins, of which actin and myosin in cytoskeletal fibers are the best-known examples.

- Transport of materials around the body is an important protein activity: for example, hemoglobin transports oxygen in the bloodstream, and serum albumin transports fatty acids.

- Regulation of cellular processes is mediated by signaling proteins such as STATs (signal transducers and activators of transcription, Section 14.1.2), and by proteins such as **activators** that bind to the genome and influence the expression levels of individual genes and groups of genes (Section 11.3). The activities of groups of cells are regulated and coordinated by extracellular hormones and cytokines, many of which are proteins (e.g., insulin, the hormone that controls blood sugar levels, and the interleukins, a group of cytokines that regulate cell division and differentiation).

- Protection of the body and of individual cells is the function of a range of proteins, including the antibodies and those proteins involved in the blood-clotting response.

- Storage functions are performed by proteins such as ferritin, which acts as an iron store in the liver, and the gliadins, which store amino acids in dormant wheat seeds.

This multiplicity of protein function provides the proteome with its ability to convert the blueprint contained in the genome into the essential features of the life process.

Summary

The genome is the store of biological information possessed by every organism on the planet. The vast majority of genomes are made of DNA, the few exceptions being those viruses that have RNA genomes. Genome expression is the process by which the information contained in the genome is released to the cell. The first product of genome expression is the transcriptome, the collection of RNAs derived from the protein-coding genes that are active in the cell at a particular time. The second product is the proteome, the cell's repertoire of proteins that specify the nature of the biochemical reactions that the cell is able to carry out. Experimental evidence showing that genes are made of DNA was first obtained between 1945 and 1952, but it was the discovery of the double helix structure by Watson and Crick in 1953 that convinced biologists that DNA is, indeed, the genetic material. A DNA polynucleotide is an unbranched polymer made up of multiple copies of four chemically different nucleotides. In the double helix, two polynucleotides are wound around one another, with the nucleotide bases on the inside of the molecule. The polynucleotides are linked by hydrogen bonding between the bases, with A always base-paired to T, and G always base-paired to C. RNA is also a polynucleotide but the individual nucleotides have different structures compared with those found in DNA, and RNA is usually single-stranded. DNA-dependent RNA polymerases are responsible for copying genes into RNA by the process called transcription, which results in synthesis not only of the transcriptome but also of a range of functional RNA molecules, which do not code for proteins but still play vital roles in the cell. Many RNAs are initially synthesized as precursor molecules that are processed by cutting and joining reactions, and by chemical modifications, to give the mature forms. Proteins are also unbranched polymers, but in these the units are amino acids linked by peptide bonds. The amino acid sequence is the primary structure of a protein. The higher levels of structure—secondary, tertiary, and quaternary—are formed by folding the primary structure into three-dimensional conformations and by association of individual polypeptides into multiprotein structures. Proteins are functionally diverse because individual amino acids have different chemical properties that, when combined in different ways, result in proteins with a range of chemical features. Proteins are synthesized by translation of mRNAs, with the rules of the genetic code specifying which triplet of nucleotides codes for which amino acid. The genetic code is not universal, variations occurring in mitochondria and in lower eukaryotes, and some codons can have two different meanings in a single gene.

Multiple Choice Questions

1.1.* Which of the following statements about an organism's genome is FALSE?

    a. The genome contains the genetic information to construct and maintain a living organism.

    b. The genomes of cellular organisms are composed of DNA.

    c. The genome is able to express its own information without the activity of enzymes and proteins.

    d. Eukaryotic genomes are composed of both nuclear and mitochondrial DNA.

1.2. Somatic cells are those that:

    a. Contain a haploid set of chromosomes.

    b. Give rise to the gametes.

    c. Lack mitochondria.

    d. Contain a diploid set of chromosomes and make up the majority of human cells.

1.3.* The flow of genetic information in cells is which of the following?

    a. DNA is transcribed into RNA, which is then translated into protein.

    b. DNA is translated into protein, which is then transcribed into RNA.

    c. RNA is transcribed into DNA, which is then translated into protein.

    d. Proteins are translated into RNA, which is then transcribed into DNA.

1.4. In the early twentieth century it was thought that proteins might carry genetic information. This reasoning was due to which of the following?

    a. Chromosomes are composed of approximately equal amounts of protein and DNA.

    b. Proteins were known to be composed of 20 distinct amino acids whereas DNA is composed of only 4 nucleotides.

    c. Different proteins were known to have unique sequences, whereas it was thought that all DNA molecules have the same sequence.

    d. All of the above.

1.5.* Which type of bonds link the individual nucleotides together in DNA?

    a. Glycosidic.

    b. Peptide.

    c. Phosphodiester.

    d. Electrostatic.

1.6. In solving the structure of DNA, Watson and Crick actively used which of the following techniques?

    a. Model building of DNA molecules to ensure that the atoms were correctly positioned.

    b. X-ray crystallography of DNA.

    c. Chromatographic studies to determine the relative composition of nucleotides from various sources.

    d. Genetic studies that demonstrated that DNA is the genetic material.

1.7.* Erwin Chargaff studied DNA from various organisms and demonstrated that:

    a. DNA is the genetic material.

    b. RNA is transcribed from DNA.

    c. The amount of adenine in a given organism is equal to the amount of thymine (and guanine to cytosine).

    d. The double helix is held together by hydrogen bonding between the bases.

1.8. The transcriptome of a cell is defined as:

    a. All of the RNA molecules present in a cell.

    b. The protein-coding RNA molecules present in a cell.

    c. The ribosomal RNA molecules present in a cell.

    d. The transfer RNA molecules present in a cell.

1.9.* How do DNA-dependent RNA polymerases carry out RNA synthesis?

    a. They use DNA as a template for the polymerization of ribonucleotides.

    b. They use proteins as a template for the polymerization of ribonucleotides.

    c. They use RNA as a template for the polymerization of ribonucleotides.

    d. They require no template for the polymerization of ribonucleotides.

1.10. Which type of functional RNA is a primary component of the structures required for protein synthesis?

    a. Messenger RNA.

    b. Ribosomal RNA.

    c. Small nuclear RNA.

    d. Transfer RNA.

1.11.* The proteome of a cell is defined as:

    a. All of the proteins that a cell is capable of synthesizing.

    b. All of the proteins present in a cell over the cell's lifetime.

    c. All of the proteins present in a cell at a given moment.

    d. All of the proteins that are actively being synthesized in a cell at a given moment.

Multiple Choice Questions (continued) *Answers to odd-numbered questions can be found in the Appendix

1.12. Which level of protein structure describes the folded conformation of a multisubunit protein?

    a. Primary structure.

    b. Secondary structure.

    c. Tertiary structure.

    d. Quaternary structure.

1.13. *Which type of covalent bond is important for linking cysteine residues located at various places in a polypeptide?

    a. Disulfide bridge.

    b. Hydrogen bond.

    c. Peptide bond.

    d. Phosphodiester bond.

1.14. Most of the abundant proteins in a cell are thought to be housekeeping proteins. What is their function?

    a. They are responsible for the specific functions of individual cell types.

    b. They are responsible for regulating genome expression in cells.

    c. They are responsible for removing waste materials from cells.

    d. They are responsible for the general biochemical activities that occur in all cells.

1.15. *The degeneracy of the genetic code refers to which of the following?

    a. Each codon can specify more than one amino acid.

    b. Most amino acids have more than one codon.

    c. There are several initiation codons.

    d. The stop codons can also code for amino acids.

1.16. Which of the following is NOT a biological function of proteins?

    a. Biological catalysis.

    b. Regulation of cellular processes.

    c. Carrying genetic information.

    d. Transport of molecules in multicellular organisms.

Short Answer Questions

*Answers to odd-numbered questions can be found in the Appendix

1.1. * Provide a time line for the discovery of DNA, the discovery that DNA is the genetic material, the discovery of the structure of DNA, and the characterization of the first genome.

1.2. Which two types of chemical interaction stabilize the double helix?

1.3. *Why does the specific base pairing between A with T, and G with C, provide a basis for the fidelity of DNA replication?

1.4. What are the two important chemical differences between RNA and DNA?

1.5. *Why do mRNA molecules have short half-lives compared to other RNA molecules?

1.6. Is the mRNA that is translated in the same form as that synthesized from the DNA template?

1.7. *Do cells ever lack a transcriptome? Explain the significance of your answer.

1.8. How do hydrogen bonds, electrostatic interactions, and hydrophobic forces play important roles in the secondary, tertiary, and quaternary structures of proteins?

1.9. *How can proteins have so many diverse structures and functions when they are all synthesized from just 20 amino acids?

1.10. In addition to the 20 amino acids, proteins have additional chemical diversity because of two factors. What are these two factors, and what is their importance?

1.11. *How can the codon 5'–UCA–3' function as both a stop codon and as a codon for the modified amino acid selenocysteine?

1.12. How does the genome direct the biological activity of a cell?

continued …

In-depth Problems

*Guidance to odd-numbered questions can be found in the Appendix

1.1.* The text (page 10) states that Watson and Crick discovered the double helix structure of DNA on Saturday 7 March 1953. Justify this statement.

1.2. Discuss why the double helix gained immediate universal acceptance as the correct structure for DNA.

1.3.* What experiments led to elucidation of the genetic code in the 1960s?

1.4. The transcriptome and proteome are looked on as, respectively, an intermediate and the end-product of genome expression. Evaluate the strengths and limitations of these terms for our understanding of genome expression.

Figure Tests

*Answers to odd-numbered questions can be found in the Appendix

1.1.* Discuss how each of these experiments helped to demonstrate that DNA, and not proteins, contains genetic information.

(A) The transforming principle

(B) The Hershey–Chase experiment

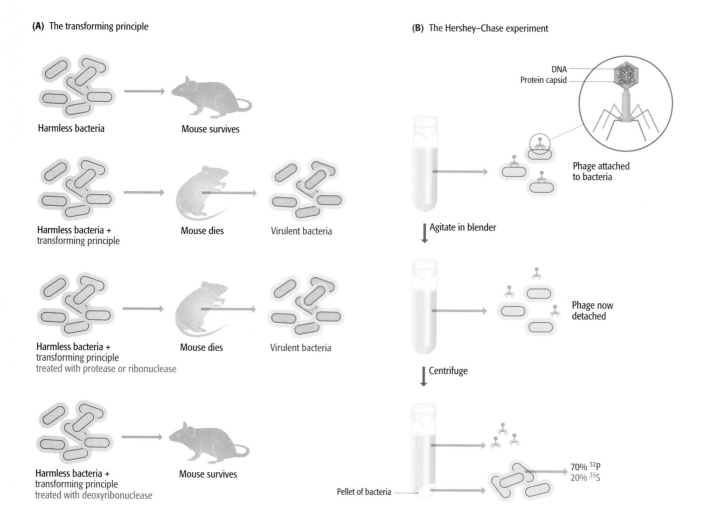

Figure Tests (continued)

*Answers to odd-numbered questions can be found in the Appendix

1.2. Identify the deoxyribose, phosphate groups, and different nitrogenous bases. Can you identify the 1′ through 5′ carbon atoms of the deoxyribose?

1.3.* For this space-filling model of B-DNA, describe the important structural features of the molecule.

1.4. Explain the differences in RNA between prokaryotic and eukaryotic cells.

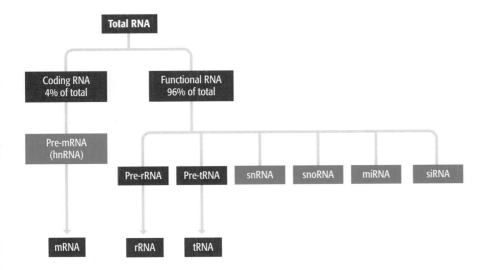

Further Reading

Books and articles on the discovery of the double helix and other important landmarks in the study of DNA

Brock, T.D. (1990) *The Emergence of Bacterial Genetics.* Cold Spring Harbor Laboratory Press, New York. *A detailed history that puts into context the work on the transforming principle and the Hershey–Chase experiment.*

Judson, H.F. (1979) *The Eighth Day of Creation.* Jonathan Cape, London. *A highly readable account of the development of molecular biology up to the 1970s.*

Kay, L.E. (1993) *The Molecular Vision of Life.* Oxford University Press, Oxford. *Contains a particularly informative explanation of why genes were once thought to be made of protein.*

Lander, E.S. and Weinberg, R.A. (2000) Genomics: journey to the center of biology. *Science* **287**: 1777–1782. *A brief description of genetics and molecular biology from Mendel to the human genome sequence.*

Maddox, B. (2002) *Rosalind Franklin: The Dark Lady of DNA.* HarperCollins, London.

McCarty, M. (1985) *The Transforming Principle: Discovering that Genes are Made of DNA.* Norton, London.

Olby, R. (1974) *The Path to the Double Helix.* Macmillan, London *A scholarly account of the research that led to the discovery of the double helix.*

Watson, J.D. (1968) *The Double Helix.* Atheneum, London. *The most important discovery of twentieth century biology, written as a soap opera.*

Studying DNA

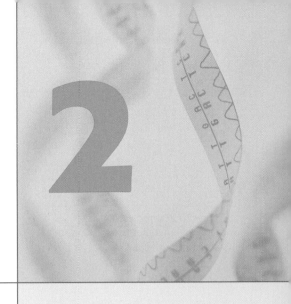

2

When you have read Chapter 2, you should be able to:

Describe the events involved in DNA cloning and the polymerase chain reaction (PCR), and state the applications and limitations of these techniques.

List the activities and main applications of the different types of enzyme used in recombinant DNA research.

Identify the important features of DNA polymerases, and distinguish between the various DNA polymerases used in genomics research.

Describe, with examples, the way that restriction endonucleases cut DNA, and explain how the results of a restriction digest are examined.

Distinguish between blunt- and sticky-end ligation and explain how the efficiency of blunt-end ligation can be increased.

Give details of the key features of plasmid cloning vectors and describe how these vectors are used in cloning experiments.

Describe how bacteriophage λ vectors are used to clone DNA.

Give examples of vectors used to clone long pieces of DNA, and evaluate the advantages and limitations of each type.

Summarize how DNA is cloned in yeast, animals, and plants.

Describe how a PCR is performed, paying particular attention to the importance of the primers and of the temperatures used during the thermal cycling.

Virtually everything we know about genomes and genome expression has been discovered by scientific research: theoretical studies have played very little role in this or any other area of molecular and cell biology. It is possible to learn "facts" about genomes without knowing very much about how those facts were obtained, but in order to gain a real understanding of the subject we must examine in detail the techniques and scientific approaches that have been used to study genomes. The next five chapters cover these research methods. First, we examine the techniques, centered on DNA cloning and the

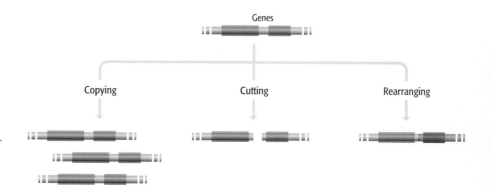

Figure 2.1 Examples of the manipulations that can be carried out with DNA molecules.

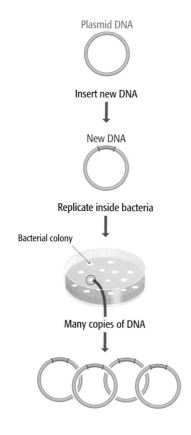

Figure 2.2 DNA cloning. In this example, the fragment of DNA to be cloned is inserted into a plasmid vector, which is subsequently replicated inside a bacterial host.

polymerase chain reaction, that are used to study DNA molecules. These techniques are very effective with short segments of DNA, including individual genes, enabling a wealth of information to be obtained at this level. Chapter 3 moves on to the methods developed to construct maps of genomes, and describes how the techniques of genetic mapping, first developed almost a century ago, have been supplemented with complementary methods for physical mapping of genomes. Chapter 4 makes the link between mapping and sequencing, and shows that although a map can be a valuable aid to assembly of a long DNA sequence, mapping is not always an essential prerequisite to genome sequencing. In Chapter 5, we look at the various approaches that are used to understand a genome sequence, and in Chapter 6 we examine the methods used to study genome expression. As you read Chapter 6 you will begin to appreciate that understanding how a genome specifies the biochemical capability of a living cell is one of the major research challenges of modern biology.

The toolkit of techniques used by molecular biologists to study DNA molecules was assembled during the 1970s and 1980s. Before then, the only way in which individual genes could be studied was by classical genetics, using the procedures that we will meet in Chapter 3. The development of more direct methods for studying DNA was stimulated by breakthroughs in biochemical research that, in the early 1970s, provided molecular biologists with enzymes that could be used to manipulate DNA molecules in the test tube. These enzymes occur naturally in living cells and are involved in processes such as DNA replication, repair, and recombination, which we will meet in Chapters 15, 16 and 17. In order to determine the functions of these enzymes, many of them were purified and the reactions that they catalyze studied. Molecular biologists then adopted the pure enzymes as tools for manipulating DNA molecules in predetermined ways, using them to make copies of DNA molecules, to cut DNA molecules into shorter fragments, and to join them together again in combinations that do not exist in nature (Figure 2.1). These manipulations form the basis of **recombinant DNA technology**, in which new or "recombinant" DNA molecules are constructed from pieces of naturally occurring chromosomes and plasmids.

Recombinant DNA methodology led to the development of **DNA cloning,** or **gene cloning**, in which short DNA fragments, possibly containing a single gene, are inserted into a plasmid or virus chromosome and then replicated in a bacterial or eukaryotic host (Figure 2.2). We will examine exactly how gene cloning is performed, and the reasons why this technique resulted in a revolution in molecular biology, in Section 2.2.

Gene cloning was well established by the end of the 1970s. The next major technical breakthrough came in the mid 1980s when the **polymerase chain reaction (PCR)** was invented. PCR is not a complicated technique—all that it achieves is the repeated copying of a short segment of a DNA molecule (Figure 2.3)—but it has become immensely important in many areas of biological research, not least the study of genomes. PCR is covered in detail in Section 2.3.

2.1 Enzymes for DNA Manipulation

Recombinant DNA technology was one of the main factors that contributed to the rapid advance in knowledge concerning gene expression that occurred during the 1970s and 1980s. The basis of recombinant DNA technology is the ability to manipulate DNA molecules in the test tube. This, in turn, depends on the availability of purified enzymes whose activities are known and can be controlled, and which can therefore be used to make specified changes to the DNA molecules that are being manipulated. The enzymes available to the molecular biologist fall into four broad categories:

- **DNA polymerases** (Section 2.1.1), which are enzymes that synthesize new polynucleotides complementary to an existing DNA or RNA template (Figure 2.4A).

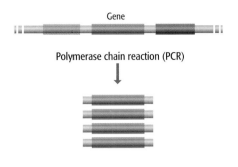

Figure 2.3 The polymerase chain reaction (PCR) is used to make copies of a selected segment of a DNA molecule. In this example, a single gene is copied.

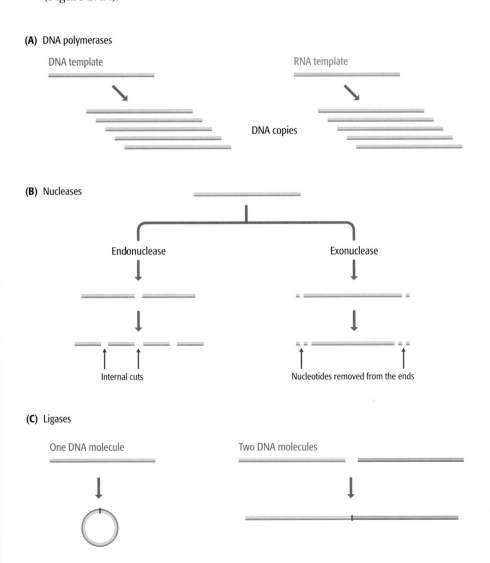

Figure 2.4 The activities of (A) DNA polymerases, (B) nucleases, and (C) ligases. In (A), the activity of a DNA-dependent DNA polymerase is shown on the left, and that of an RNA-dependent DNA polymerase on the right. In (B), the activities of endonucleases and exonucleases are shown. In (C), the gray DNA molecule is ligated to itself (left), and to a second DNA molecule (right).

- **Nucleases** (Section 2.1.2), which degrade DNA molecules by breaking the phosphodiester bonds that link one nucleotide to the next (Figure 2.4B).

- **Ligases** (Section 2.1.3), which join DNA molecules together by synthesizing phosphodiester bonds between nucleotides at the ends of two different molecules, or at the two ends of a single molecule (Figure 2.4C).

- **End-modification enzymes** (Section 2.1.4), which make changes to the ends of DNA molecules, adding an important dimension to the design of ligation experiments, and providing one means of labeling DNA molecules with radioactive and other markers (Technical Note 2.1).

2.1.1 DNA polymerases

Many of the techniques used to study DNA depend on the synthesis of DNA copies of all or part of existing DNA or RNA molecules. This is an essential requirement for PCR (Section 2.3), DNA sequencing (Section 4.1), DNA labeling (Technical Note 2.1), and many other procedures that are central to molecular biology research. An enzyme that synthesizes DNA is called a **DNA polymerase**, and one that copies an existing DNA or RNA molecule is called a **template-dependent DNA polymerase**.

Technical Note 2.1 DNA labeling

Attachment of radioactive, fluorescent, or other types of marker to DNA molecules

DNA labeling is a central part of many molecular biology procedures, including Southern hybridization (Section 2.1.2), fluorescent *in situ* hybridization (FISH; Section 3.3.2), and DNA sequencing (Section 4.1). It enables the location of a particular DNA molecule—on a nitrocellulose or nylon membrane, in a chromosome, or in a gel—to be determined by detecting the signal emitted by the marker. Labeled RNA molecules are also used in some applications (Technical Note 5.1).

Radioactive markers are frequently used for labeling DNA molecules. Nucleotides can be synthesized in which one of the phosphorus atoms is replaced with ^{32}P or ^{33}P, one of the oxygen atoms in the phosphate group is replaced with ^{35}S, or one or more of the hydrogen atoms is replaced with ^{3}H (see Figure 1.4). Radioactive nucleotides still act as substrates for DNA polymerases and so are incorporated into a DNA molecule by any strand-synthesis reaction catalyzed by a DNA polymerase. Labeled nucleotides or individual phosphate groups can also be attached to one or both ends of a DNA molecule by the reactions catalyzed by T4 polynucleotide kinase or terminal deoxynucleotidyl transferase (Section 2.1.4). The radioactive signal can be detected by scintillation counting, but for most molecular biology applications positional information is needed, so detection is by exposure of an X-ray-sensitive film (**autoradiography**: see Figure 2.11 for an example) or a radiation-

sensitive phosphorescent screen (**phosphorimaging**). The choice between the various radioactive labels depends on the requirements of the procedure. High sensitivity is possible with ^{32}P because this isotope has a high emission energy, but sensitivity is accompanied by low resolution because of signal scattering. Low-emission isotopes such as ^{35}S or ^{3}H give less sensitivity but greater resolution.

Health and environmental issues have meant that radioactive markers have become less popular in recent years and for many procedures they are now largely superseded by nonradioactive alternatives. The most useful of these are fluorescent markers, which are central components of techniques such as FISH (Section 3.3.2) and DNA sequencing (Section 4.1.1). Fluorescent labels with various emission wavelengths (i.e., of different colors) are incorporated into nucleotides or attached directly to DNA molecules, and are detected with a suitable film, by fluorescence microscopy, or with a fluorescence detector. Other types of nonradioactive labeling make use of chemiluminescent emissions, but these have the disadvantage that the signal is not generated directly by the label, but instead must be "developed" by treatment of the labeled molecule with chemicals. A popular method involves labeling the DNA with the enzyme alkaline phosphatase, which is detected by applying dioxetane, which the enzyme dephosphorylates to produce the chemiluminescence.

The mode of action of a template-dependent DNA polymerase

A template-dependent DNA polymerase makes a new DNA polynucleotide whose sequence is dictated, via the base-pairing rules, by the sequence of nucleotides in the DNA or RNA molecule that is being copied (Figure 2.5). The new polynucleotide is always synthesized in the 5′→3′ direction: DNA polymerases that make DNA in the other direction are unknown in nature.

An important feature of template-dependent DNA synthesis is that a DNA polymerase is unable to use an entirely single-stranded molecule as the template. In order to initiate DNA synthesis there must be a short, double-stranded region to provide a 3′ end onto which the enzyme will add new nucleotides (Figure 2.6A). The way in which this requirement is met in living cells when the genome is replicated is described in Chapter 15. In the test tube, a DNA-copying reaction is initiated by attaching to the template a short, synthetic **oligonucleotide**, usually about 20 nucleotides in length, which acts as a **primer** for DNA synthesis. At first glance, the need for a primer might appear to be an undesired complication in the use of DNA polymerases in recombinant DNA technology, but nothing could be further from the truth. Because annealing of the primer to the template depends on complementary base pairing, the position within the template molecule at which DNA copying is initiated can be specified by synthesizing a primer with the appropriate nucleotide sequence (Figure 2.6B). A short, specific segment of a much longer template molecule can therefore be copied, which is much more valuable than the random copying that would occur if DNA synthesis did not need to be primed. You will fully appreciate the importance of priming when we deal with PCR in Section 2.3.

A second general feature of template-dependent DNA polymerases is that many of these enzymes are multifunctional, being able to degrade DNA molecules as well as synthesize them. This is a reflection of the way in which DNA polymerases act in the cell during genome replication (Section 15.2). As well as their 5′→3′ DNA synthesis capability, DNA polymerases can also have one or both of the following exonuclease activities (Figure 2.7):

● A **3′→5′ exonuclease** activity enables the enzyme to remove nucleotides from the 3′ end of the strand that it has just synthesized. This is called the **proofreading** activity because it allows the polymerase to correct errors by removing a nucleotide that has been inserted incorrectly.

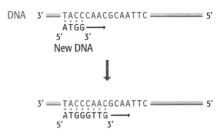

Figure 2.5 The activity of a DNA-dependent DNA polymerase. New nucleotides are added on to the 3′ end of the growing polynucleotide, the sequence of this new polynucleotide being determined by the sequence of the template DNA. Compare with the process of transcription (DNA-dependent RNA synthesis) shown in Figure 1.11.

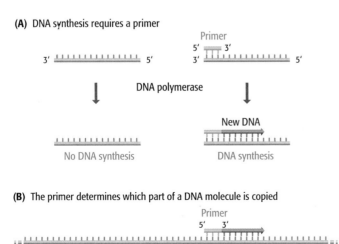

(A) DNA synthesis requires a primer

(B) The primer determines which part of a DNA molecule is copied

Figure 2.6 The role of the primer in template-dependent DNA synthesis. (A) A DNA polymerase requires a primer in order to initiate the synthesis of a new polynucleotide. (B) The sequence of this oligonucleotide determines the position at which it attaches to the template DNA and hence specifies the region of the template that will be copied. When a DNA polymerase is used to make new DNA *in vitro*, the primer is usually a short oligonucleotide made by chemical synthesis. For details of how DNA synthesis is primed *in vivo*, see Section 15.2.2.

(A) 5′→3′ DNA synthesis

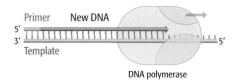

(B) 3′→5′ exonuclease activity

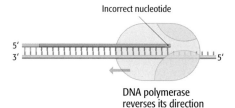

(C) 5′→3′ exonuclease activity

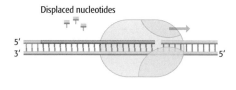

Figure 2.7 The DNA synthesis and exonuclease activities of DNA polymerases. All DNA polymerases can make DNA, and many also have one or both of the exonuclease activities.

● A **5′→3′ exonuclease** activity is less common, but is possessed by some DNA polymerases whose natural function in genome replication requires that they must be able to remove at least part of a polynucleotide that is already attached to the template strand that the polymerase is copying.

The types of DNA polymerases used in research

Several of the template-dependent DNA polymerases that are used in molecular biology research (Table 2.1) are versions of the *Escherichia coli* DNA polymerase I enzyme, which plays a central role in replication of this bacterium's genome (Section 15.2). This enzyme, sometimes called the **Kornberg polymerase**, after its discoverer Arthur Kornberg, has both the 3′→5′ and 5′→3′ exonuclease activities, which limits its usefulness in DNA manipulation. Its main application is in DNA labeling (Technical Note 2.1).

Of the two exonuclease activities, it is the 5′→3′ version that causes most problems when a DNA polymerase is used to manipulate molecules in the test tube. This is because an enzyme that possesses this activity is able to remove nucleotides from the 5′ ends of polynucleotides that have just been synthesized (Figure 2.8). It is unlikely that the polynucleotides will be completely degraded, because the polymerase function is usually much more active than the exonuclease function, but some techniques will not work if the 5′ ends of the new polynucleotides are shortened in any way. In particular, DNA sequencing is based on synthesis of new polynucleotides, all of which share exactly the same 5′ end, marked by the primer used to initiate the sequencing reactions. If any "nibbling" of the 5′ ends occurs, then it is impossible to determine the correct DNA sequence. When DNA sequencing was first introduced in the late 1970s, it made use of a modified version of the Kornberg enzyme called the **Klenow polymerase**. The Klenow polymerase was initially prepared by cutting the natural *E. coli* DNA polymerase I enzyme into two segments using a protease. One of these segments retained the polymerase and 3′→5′ exonuclease activities, but lacked the 5′→3′ exonuclease function of the untreated enzyme. The enzyme is still often called the Klenow *fragment*, in memory of this old method of preparation, but nowadays it is almost always prepared from *E. coli* cells whose polymerase gene has been engineered so that the resulting enzyme has the desired properties. But in fact the Klenow polymerase is now rarely used in sequencing and has its major application in DNA labeling (see Technical Note 2.1). This is because an enzyme called **Sequenase** (see Table 2.1), which has superior properties as far as sequencing is concerned, was developed in the 1980s. We will return to the

Table 2.1 Features of template-dependent DNA polymerases used in molecular biology research

Polymerase	Description	Main use	Cross-reference
DNA polymerase I	Unmodified *E. coli* enzyme	DNA labeling	Technical Note 2.1
Klenow polymerase	Modified version of *E. coli* DNA polymerase I	DNA labeling	Technical Note 2.1
Sequenase	Modified version of phage T7 DNA polymerase I	DNA sequencing	Section 4.1.1
Taq polymerase	*Thermus aquaticus* DNA polymerase I	PCR	Section 2.3
Reverse transcriptase	RNA-dependent DNA polymerase, obtained from various retroviruses	cDNA synthesis	Section 5.1.2

features of Sequenase, and why they make the enzyme ideal for sequencing, in Section 4.1.1.

The *E. coli* DNA polymerase I enzyme has an optimum reaction temperature of 37°C, this being the usual temperature of the natural environment of the bacterium, inside the intestines of mammals such as humans. Test-tube reactions with either the Kornberg or Klenow polymerases, and with Sequenase, are therefore incubated at 37°C, and terminated by raising the temperature to 75°C or above, which causes the protein to unfold, or **denature**, destroying its enzymatic activity. This regimen is perfectly adequate for most molecular biology techniques, but for reasons that will become clear in Section 2.3, PCR requires a **thermostable** DNA polymerase—one that is able to function at temperatures much higher than 37°C. Suitable enzymes can be obtained from bacteria such as *Thermus aquaticus*, which live in hot springs at temperatures up to 95°C, and whose DNA polymerase I enzyme has an optimum working temperature of 72°C. The biochemical basis of protein thermostability is not fully understood, but probably centers on structural features that reduce the amount of protein unfolding that occurs at elevated temperatures.

One additional type of DNA polymerase is important in molecular biology research. This is **reverse transcriptase**, which is an **RNA-dependent DNA polymerase** and so makes DNA copies of RNA rather than DNA templates. Reverse transcriptases are involved in the replication cycles of retroviruses (Section 9.1.2), including the human immunodeficiency viruses that cause acquired immune deficiency syndrome, or AIDS, these viruses having RNA genomes that are copied into DNA after infection of the host. In the test tube, a reverse transcriptase can be used to make DNA copies of mRNA molecules. These copies are called **complementary DNAs** (**cDNAs**). Their synthesis is important in some types of gene cloning and in techniques used to map the regions of a genome that specify particular mRNAs (Section 5.1.2).

2.1.2 Nucleases

A variety of nucleases have found applications in recombinant DNA technology (Table 2.2). Some nucleases have a broad range of activities but most are either **exonucleases**, removing nucleotides from the ends of DNA and/or RNA molecules, or **endonucleases**, making cuts at internal phosphodiester bonds. Some nucleases are specific for DNA and some for RNA, some work only on double-stranded DNA and others only on single-stranded DNA, and

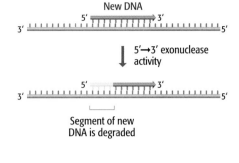

Figure 2.8 The 5′→3′ exonuclease activity of a DNA polymerase can degrade the 5′ end of a polynucleotide that has just been synthesized.

Table 2.2 Features of important nucleases used in molecular biology research

Nuclease	Description	Main use	Cross-reference
Restriction endonucleases	Sequence-specific DNA endonucleases, from many sources	Many applications	Section 2.1.2
Nuclease S1	Endonuclease specific for single-stranded DNA and RNA, from the fungus *Aspergillus oryzae*	Transcript mapping	Section 5.1.2
Deoxyribonuclease	Endonuclease specific for double-stranded DNA and RNA, from *Escherichia coli*	Nuclease footprinting	Section 11.1.2

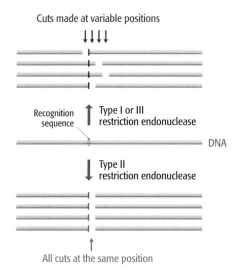

Cuts made at variable positions

Recognition sequence — Type I or III restriction endonuclease

DNA

Type II restriction endonuclease

All cuts at the same position

Figure 2.9 Cuts produced by restriction endonucleases. In the top part of the diagram, the DNA is cut by a Type I or Type III restriction endonuclease. The cuts are made in slightly different positions relative to the recognition sequence, so the resulting fragments have different lengths. In the lower part of the diagram, a Type II enzyme is used. Each molecule is cut at exactly the same position to give exactly the same pair of fragments.

some are not fussy what they work on. We will encounter various examples of nucleases in later chapters when we deal with the techniques in which they are used. Only one type of nuclease will be considered in detail here: the **restriction endonucleases**, which play a central role in all aspects of recombinant DNA technology.

Restriction endonucleases enable DNA molecules to be cut at defined positions

A restriction endonuclease is an enzyme that binds to a DNA molecule at a specific sequence and makes a double-stranded cut at or near that sequence. Because of the sequence specificity, the positions of cuts within a DNA molecule can be predicted, assuming that the DNA sequence is known, enabling defined segments to be excised from a larger molecule. This ability underlies gene cloning and all other aspects of recombinant DNA technology in which DNA fragments of known sequence are required.

There are three types of restriction endonuclease. With Types I and III, there is no strict control over the position of the cut relative to the specific sequence in the DNA molecule that is recognized by the enzyme. These enzymes are therefore less useful because the sequences of the resulting fragments are not precisely known. Type II enzymes do not suffer from this disadvantage because the cut is always at the same place, either within the recognition sequence or very close to it (Figure 2.9). For example, the Type II enzyme called *Eco*RI (isolated from *E. coli*) cuts DNA only at the hexanucleotide 5′–GAATTC–3′. Digestion of DNA with a Type II enzyme therefore gives a reproducible set of fragments whose sequences are predictable if the

Table 2.3 Some examples of restriction endonucleases

Enzyme	Recognition sequence	Type of ends	End sequences
*Alu*I	5′–AGCT–3′ 3′–TCGA–5′	Blunt	5′–AG CT–3′ 3′–TC GA–5′
*Sau*3AI	5′–GATC–3′ 3′–CTAG–5′	Sticky, 5′ overhang	5′– GATC–3′ 3′–CTAG –5′
*Hin*fI	5′–GANTC–3′ 3′–CTNAG–5′	Sticky, 5′ overhang	5′–G ANTC–3′ 3′–CTNA G–5′
*Bam*HI	5′–GGATCC–3′ 3′–CCTAGG–5′	Sticky, 5′ overhang	5′–G GATCC–3′ 3′–CCTAG G–5′
*Bsr*BI	5′–CCGCTC–3′ 3′–GGCGAG–5′	Blunt	5′– NNNCCGCTC–3′ 3′– NNNGGCGAG–5′
*Eco*RI	5′–GAATTC–3′ 3′–CTTAAG–5′	Sticky, 5′ overhang	5′–G AATTC–3′ 3′–CTTAA G–5′
*Pst*I	5′–CTGCAG–3′ 3′–GACGTC–5′	Sticky, 3′ overhang	5′–CTGCA G–3′ 3′–G ACGTC–5′
*Not*I	5′–GCGGCCGC–3′ 3′–CGCCGGCG–5′	Sticky, 5′ overhang	5′–GC GGCCGC–3′ 3′–CGCCGG CG–5′
*Bgl*I	5′–GCCNNNNNGGC–3′ 3′–CGGNNNNNCCG–5′	Sticky, 3′ overhang	5′–GCCNNNN NGGC–3′ 3′–CGGN NNNNCCG–5′

Abbreviation: N, any nucleotide.
Note that most, but not all, recognition sequences have inverted symmetry: when read in the 5′→3′ direction, the sequence is the same in both strands.

sequence of the target DNA molecule is known. Over 2500 Type II enzymes have been isolated, and more than 300 are available for use in the laboratory. Many enzymes have hexanucleotide target sites, but others recognize shorter or longer sequences (Table 2.3). There are also examples of enzymes with degenerate recognition sequences, meaning that they cut DNA at any of a family of related sites. *Hin*fI (from *Haemophilus influenzae*), for example, recognizes 5′–GANTC–3′, where "N" is any nucleotide, and so cuts at 5′–GAATC–3′, 5′–GATTC–3′, 5′–GAGTC–3′, and 5′–GACTC–3′. Most enzymes cut within the recognition sequence, but a few, such as *Bsr*BI, cut at a specified position outside of this sequence.

Restriction enzymes cut DNA in two different ways. Many make a simple double-stranded cut, giving a **blunt** or **flush end**, but others cut the two DNA strands at different positions, usually two or four nucleotides apart, so that the resulting DNA fragments have short, single-stranded overhangs at each end. These are called **sticky** or **cohesive ends** because base-pairing between them can stick the DNA molecule back together again (Figure 2.10A). Some sticky-end cutters give 5′ overhangs (e.g., *Sau*3AI, *Hin*fI) whereas others leave 3′ overhangs (e.g., *Pst*I) (Figure 2.10B). One feature that is particularly important in recombinant DNA technology is that some pairs of restriction enzymes have different recognition sequences but give the same sticky ends, examples being *Sau*3AI and *Bam*HI, which both give a 5′–GATC–3′ sticky end, even though *Sau*3AI has a four-base-pair recognition sequence and *Bam*HI recognizes a six-base-pair sequence (Figure 2.10C).

(A) Blunt and sticky ends

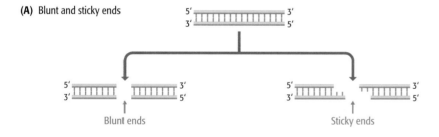

(B) 5′ and 3′ overhangs

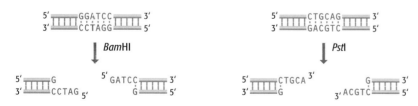

(C) The same sticky end produced by different enzymes

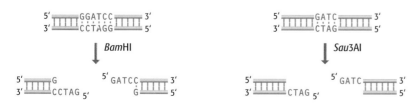

Figure 2.10 **The results of digestion of DNA with different restriction endonucleases.** (A) Blunt ends and sticky ends. (B) Different types of sticky end: the 5′ overhangs produced by *Bam*HI and the 3′ overhangs produced by *Pst*I. (C) The same sticky ends produced by two different restriction endonucleases: a 5′ overhang with the sequence 5′–GATC–3′ is produced by both *Bam*HI (recognizes 5′–GGATCC–3′) and *Sau*3AI (recognizes 5′–GATC–3′).

Technical Note 2.2 Agarose gel electrophoresis

Separation of DNA molecules of different lengths

Gel electrophoresis is the standard method for separating DNA molecules of different lengths. It has many applications in size analysis of DNA fragments and can also be used to separate RNA molecules (see Technical Note 5.1).

Electrophoresis is the movement of charged molecules in an electric field: negatively charged molecules migrate towards the positive electrode, and positively charged molecules migrate towards the negative electrode. The technique was originally carried out in aqueous solution, in which the predominant factors influencing migration rate are the shape of a molecule and its electric charge. This is not particularly useful for DNA separations because most DNA molecules are the same shape (linear) and although the charge of a DNA molecule is dependent on its length, the differences in charge are not sufficient to result in effective separation. The situation is different when electrophoresis is carried out in a gel, because now shape and charge are less important and molecular length is the critical determinant of migration rate. This is because the gel is a network of pores through which the DNA molecules have to travel to reach the positive electrode. Shorter molecules are less impeded by the pores than are longer molecules and so move through the gel more quickly. Molecules of different lengths therefore form bands in the gel.

Two types of gel are used in molecular biology: **agarose** gels, as described here, and **polyacrylamide** gels, which are covered in Technical Note 4.1. Agarose is a polysaccharide that forms gels with pores ranging from 100 nm to 300 nm in diameter, the size depending on the concentration of agarose in the gel. Gel concentration therefore determines the range of DNA fragments that can be separated. The separation range is also affected by the electroendosmosis (EEO) value of the agarose, this being a measure of the amount of bound sulfate and pyruvate anions. The greater the EEO, the slower the migration rate for a negatively charged molecule such as DNA.

An agarose gel is prepared by mixing the appropriate amount of agarose powder in a buffer solution, heating to dissolve the agarose, and then pouring the molten gel onto a Perspex plate with tape around the sides to prevent spillage. A comb is placed in the gel to form wells for the samples. The gel is allowed to set and the electrophoresis then carried out with the gel submerged under buffer. In order to follow the progress of the electrophoresis, one or two dyes of known migration rates are added to the DNA samples before loading. The bands of DNA are visualized by soaking the gel in ethidium bromide solution, this compound intercalating between DNA base pairs and fluorescing when activated with ultraviolet radiation (Figure T2.1). Depending on the concentration of agarose in the gel, fragments between 100 base pairs (bp) and 50 kb in length can be separated into sharp bands after electrophoresis (Figure T2.2). For example, a 0.3% gel can be used for molecules between 5 kb and 50 kb, and a 5% gel for molecules 100–500 bp in length. Fragments less than 150 bp long can be separated in a 4% or 5% agarose gel, making it possible to distinguish bands representing molecules that differ in size by just a single nucleotide. With larger fragments, however, it is not always possible to separate molecules of similar size, even in gels of lower agarose concentration.

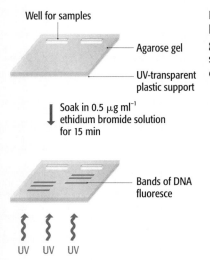

Figure T2.1 DNA bands in an agarose gel are visualized by staining with ethidium bromide.

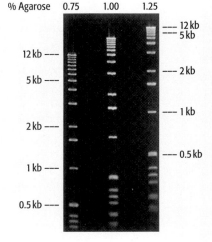

Figure T2.2 The range of fragment sizes that can be resolved depends on the concentration of agarose in the gel. Electrophoresis has been performed with three different concentrations of agarose. The labels indicate the sizes of bands in the left and right lanes. Photograph courtesy of BioWhittaker Molecular Applications.

Examining the results of a restriction digest

After treatment with a restriction endonuclease, the resulting DNA fragments can be examined by agarose gel electrophoresis (Technical Note 2.2) to determine their sizes. If the starting DNA is a relatively short molecule, and twenty or fewer fragments are produced after restriction, then usually it is possible to select an agarose concentration that results in each fragment being visible as a separate band in the gel. If the starting DNA is long, and so gives rise to many fragments after digestion with a restriction enzyme, then whatever agarose concentration is used the gel may simply show a smear of DNA, because there are fragments of every possible length that all merge together. This is the usual result when genomic DNA is cut with a restriction enzyme.

If the sequence of the starting DNA is known, then the sequences, and hence the sizes, of the fragments resulting from treatment with a particular restriction enzyme can be predicted. The band for a desired fragment (for example, one containing a gene) can then be identified, cut out of the gel, and the DNA purified. Even if its size is unknown, a fragment containing a gene or another segment of DNA of interest can be identified by the technique called **Southern hybridization**, providing that some of the sequence within the fragment is known or can be predicted. The first step is to transfer the restriction fragments from the agarose gel to a nitrocellulose or nylon membrane. This is done by placing the membrane on the gel and allowing buffer to soak through, taking the DNA from the gel to the membrane, where it becomes bound (Figure 2.11A). This process results in the DNA bands becoming immobilized in the same relative positions on the surface of the membrane.

The next step is to prepare a **hybridization probe**, which is a labeled DNA molecule whose sequence is complementary to the target DNA that we wish to detect. The probe could, for example, be a synthetic oligonucleotide whose sequence matches part of an interesting gene. Because the probe and target DNAs are complementary, they can base-pair or **hybridize**, the position of the hybridized probe on the membrane being identified by detecting the signal given out by a label attached to the probe. To carry out the hybridization, the membrane is placed in a glass bottle with the labeled probe and some buffer, and the bottle gently rotated for several hours so that the probe has

Figure 2.11 Southern hybridization. (A) Transfer of DNA from the gel to the membrane. (B) The membrane is probed with a radioactively labeled DNA molecule. On the resulting autoradiograph, one hybridizing band is seen in lane 2, and two in lane 3.

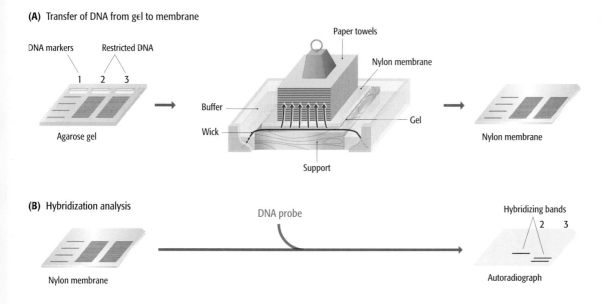

(A) Transfer of DNA from gel to membrane

DNA markers   Restricted DNA

1   2   3

Agarose gel

Paper towels

Nylon membrane

Buffer

Gel

Wick

Support

Nylon membrane

(B) Hybridization analysis

Nylon membrane

DNA probe

Hybridizing bands

2   3

Autoradiograph

(A) The role of DNA ligase *in vivo*

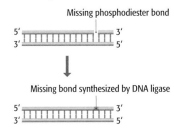

(B) Ligation *in vitro*

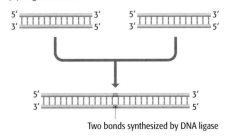

(C) Sticky-end ligation is more efficient

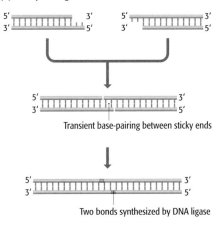

Figure 2.12 Ligation of DNA molecules with DNA ligase. (A) In living cells, DNA ligase synthesizes a missing phosphodiester bond in one strand of a double-stranded DNA molecule. (B) To link two DNA molecules *in vitro*, DNA ligase must make two phosphodiester bonds, one in each strand. (C) Ligation *in vitro* is more efficient when the molecules have compatible sticky ends, because transient base-pairing between these ends holds the molecules together and so increases the opportunity for DNA ligase to attach and synthesize the new phosphodiester bonds. For the role of DNA ligase during DNA replication *in vivo*, see Figure 15.18.

plenty of opportunity to hybridize to its target DNA. The membrane is then washed to remove any probe that has not become hybridized, and the signal from the label is detected (see Technical Note 2.1). In the example shown in Figure 2.11B the probe is radioactively labeled and the signal is detected by **autoradiography**. The band that is seen on the autoradiograph is the one that corresponds to the restriction fragment that hybridizes to the probe and which therefore contains the gene that we are searching for.

2.1.3 DNA ligases

DNA fragments that have been generated by treatment with a restriction endonuclease can be joined back together again, or attached to a new partner, by a DNA ligase. The reaction requires energy, which is provided by adding either ATP or nicotinamide adenine dinucleotide (NAD) to the reaction mixture, depending on the type of ligase that is being used.

The most widely used DNA ligase is obtained from *E. coli* cells infected with T4 bacteriophage. This enzyme is involved in replication of the phage DNA and is encoded by the T4 genome. Its natural role is to synthesize phosphodiester bonds between unlinked nucleotides present in one polynucleotide of a double-stranded molecule (Figure 2.12A). In order to join together two restriction fragments, the ligase has to synthesize two phosphodiester bonds, one in each strand (Figure 2.12B). This is by no means beyond the capabilities of the enzyme, but the reaction can occur only if the ends to be joined come close enough to one another by chance—the ligase is not able to catch hold of them and bring them together. If the two molecules have complementary sticky ends, and the ends come together by random diffusion events in the ligation mixture, then transient base pairs might form between the two overhangs. These base pairs are not particularly stable but they may persist for sufficient time for a ligase enzyme to attach to the junction and synthesize phosphodiester bonds to fuse the ends together (Figure 2.12C). If the molecules are blunt ended, then they cannot base-pair to one another, not even temporarily, and ligation is a much less efficient process, even when the DNA concentration is high and pairs of ends are in relatively close proximity.

The greater efficiency of sticky-end ligation has stimulated the development of methods for converting blunt ends into sticky ends. In one method, short double-stranded molecules called **linkers** or **adaptors** are attached to the blunt ends. Linkers and adaptors work in slightly different ways, but both contain a recognition sequence for a restriction endonuclease and so produce a sticky end after treatment with the appropriate enzyme (Figure 2.13). Another way to create a sticky end is by **homopolymer tailing**, in which nucleotides are added one after the other to the 3′ terminus at a blunt end (Figure 2.14). The enzyme involved is called **terminal deoxynucleotidyl transferase**, which we will meet in the next section. If the reaction mixture contains the DNA, enzyme, and only one of the four nucleotides, then the new stretch of single-stranded DNA that is made consists entirely of just that single nucleotide. It could, for example, be a poly(G) tail, which would enable the molecule to base-pair to other molecules that carry poly(C) tails, created in the same way but with dCTP, rather than dGTP, in the reaction mixture.

2.1.4 End-modification enzymes

Terminal deoxynucleotidyl transferase (see Figure 2.14), obtained from calf thymus tissue, is one example of an end-modification enzyme. It is, in fact, a

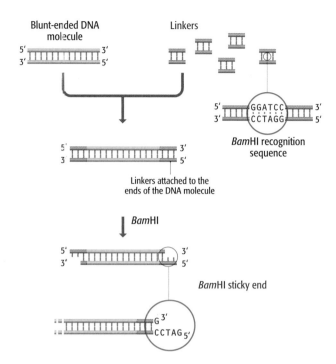

Figure 2.13 Linkers are used to place sticky ends on to a blunt-ended molecule. In this example, each linker contains the recognition sequence for the restriction endonuclease *Bam*HI. DNA ligase attaches the linkers to the ends of the blunt-ended molecule in a reaction that is made relatively efficient because the linkers are present at a high concentration. The restriction enzyme is then added to cleave the linkers and produce the sticky ends. Note that during the ligation the linkers ligate to one another, so a series of linkers (a **concatamer**) is attached to each end of the blunt molecule. When the restriction enzyme is added, these linker concatamers are cut into segments, with half of the innermost linker left attached to the DNA molecule. Adaptors are similar to linkers but each one has one blunt end and one sticky end. The blunt-ended DNA is therefore given sticky ends simply by ligating it to the adaptors: there is no need to carry out the restriction step.

template-independent DNA polymerase, because it is able to synthesize a new DNA polynucleotide without base-pairing of the incoming nucleotides to an existing strand of DNA or RNA. Its main role in recombinant DNA technology is in homopolymer tailing, as described above.

Two other end-modification enzymes are also frequently used. These are **alkaline phosphatase** and **T4 polynucleotide kinase**, which act in complementary ways. Alkaline phosphatase, which is obtained from various sources, including *E. coli* and calf intestinal tissue, removes phosphate groups from the 5′ ends of DNA molecules, which prevents these molecules from being ligated to one another. Two ends carrying 5′ phosphates can be ligated to one another, and a phosphatased end can ligate to a nonphosphatased end, but a link cannot be formed between a pair of ends if neither carries a 5′ phosphate. Judicious use of alkaline phosphatase can therefore direct the action of a DNA ligase in a predetermined way so that only desired ligation products are obtained. T4 polynucleotide kinase, obtained from *E. coli* cells infected with T4 bacteriophage, performs the reverse reaction to alkaline phosphatase, adding phosphates to 5′ ends. Like alkaline phosphatase, the enzyme is used during complicated ligation experiments, but its main application is in the end-labeling of DNA molecules (see Technical Note 2.1).

2.2 DNA Cloning

DNA cloning is a logical extension of the ability to manipulate DNA molecules with restriction endonucleases and ligases. Imagine that an animal gene has been obtained as a single restriction fragment after digestion of a larger molecule with the restriction enzyme *Bam*HI, which leaves 5′–GATC–3′ sticky ends (Figure 2.15). Imagine also that a **plasmid**—a small circle of DNA capable of replicating inside a bacterium—has been purified from *E. coli* and treated with *Bam*HI, which cuts the plasmid in a single position. The circular

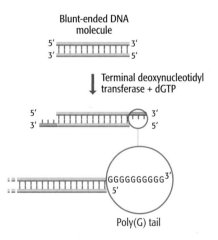

Figure 2.14 Homopolymer tailing. In this example, a poly(G) tail is synthesized at each end of a blunt-ended DNA molecule. Tails comprising other nucleotides are synthesized by including the appropriate dNTP in the reaction mixture.

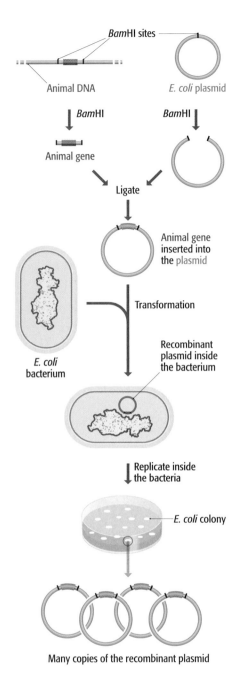

Figure 2.15 An outline of gene cloning.

plasmid has therefore been converted into a linear molecule, again with 5′–GATC–3′ sticky ends. Mix the two DNA molecules together and add DNA ligase. Various recombinant ligation products will be obtained, one of which comprises the circularized plasmid with the animal gene inserted into the position originally taken by the *Bam*HI restriction site. If the recombinant plasmid is now reintroduced into *E. coli*, and the inserted gene has not disrupted its replicative ability, then the plasmid plus inserted gene will be replicated and copies passed to the daughter bacteria after cell division. More rounds of plasmid replication and cell division will result in a colony of recombinant *E. coli* bacteria, each bacterium containing multiple copies of the animal gene. This series of events, as illustrated in Figure 2.15, constitutes the process called DNA or gene cloning.

When DNA cloning was first invented in the early 1970s, it revolutionized molecular biology by making possible experiments that previously had been inconceivable. This is because cloning can provide a pure sample of an individual gene, separated from all the other genes in the cell. Consider the basic procedure drawn in a slightly different way (Figure 2.16). In this example, the DNA fragment to be cloned is one member of a mixture of many different fragments, each carrying a different gene or part of a gene. This mixture could indeed be an entire genome. Each of these fragments becomes inserted into a different plasmid molecule to produce a family of recombinant plasmids, one of which carries the gene of interest. Usually only one recombinant molecule is transported into any single host cell, so that although the final set of clones may contain many different recombinant molecules, each individual clone contains multiple copies of just one. The gene is now separated away from all the other genes in the original mixture. Purification of the recombinant molecule from the bacterial colony, or from a liquid culture grown from the colony, will yield microgram amounts of DNA, sufficient for analysis by DNA sequencing or by one of the myriad other techniques devised for the study of cloned genes, many of which we will encounter in later chapters.

2.2.1 Cloning vectors and the way they are used

In the experiments shown in Figures 2.15 and 2.16, the plasmid acts as a **cloning vector**, providing the replicative ability that enables the cloned gene to be propagated inside the host cell. Plasmids replicate efficiently in bacterial hosts because each plasmid possesses an **origin of replication** that is recognized by the DNA polymerases and other proteins that normally replicate the bacterium's chromosomes (Section 15.2.1). The host cell's replicative machinery therefore propagates the plasmid, plus any new genes that have been inserted into it. Bacteriophage genomes can also be used as cloning vectors because they too possess origins of replication that enable them to be propagated inside bacteria, either by the host enzymes or by DNA polymerases and other proteins specified by phage genes. The next two sections describe how plasmid and phage vectors are used to clone DNA in *E. coli*.

Plasmids are uncommon in eukaryotes, although *Saccharomyces cerevisiae* possesses one that is sometimes used for cloning purposes; most vectors for use in eukaryotic cells are therefore based on virus genomes. Alternatively, with a eukaryotic host, the replication requirement can be bypassed by performing the experiment in such a way that the DNA to be cloned becomes inserted into one of the host chromosomes. These approaches to cloning in eukaryotic cells are described later in the chapter.

Vectors based on E. coli *plasmids*

The easiest way to understand how a cloning vector is used is to start with the simplest *E. coli* plasmid vectors, which illustrate all of the basic principles of DNA cloning. We will then be able to turn our attention to the special features of phage vectors and vectors used with eukaryotes.

One of the most popular plasmid vectors is pUC8, a member of a series of vectors that was first introduced in the early 1980s. The pUC series is derived from an earlier cloning vector, pBR322, which was originally constructed by ligating together restriction fragments from three naturally occurring *E. coli* plasmids: R1, R6.5, and pMB1. pUC8 is a small plasmid, comprising just 2.7 kilobases (kb). As well as its origin of replication, it carries two genes (Figure 2.17):

- A gene for ampicillin resistance. The presence of this gene means that a bacterium containing a pUC8 plasmid is able to synthesize an enzyme, called β-lactamase, that enables the cell to withstand the growth-inhibitory effect of the antibiotic. This means that cells containing pUC8 plasmids can be distinguished from those that do not by plating the bacteria onto agar medium containing ampicillin. Normal *E. coli* cells are sensitive to ampicillin and cannot grow when the antibiotic is present. Ampicillin resistance is therefore a **selectable marker** for pUC8.

- The *lacZ'* gene, which codes for part of the enzyme β-galactosidase. β-galactosidase is one of a series of enzymes involved in the breakdown of lactose to glucose plus galactose. It is normally coded by the gene *lacZ*, which resides on the *E. coli* chromosome. Some strains of *E. coli* have a modified *lacZ* gene, one that lacks the segment referred to as *lacZ'* and coding for the α-peptide portion of β-galactosidase. These mutants can synthesize the enzyme only when they harbor a plasmid, such as pUC8, that carries the missing *lacZ'* segment of the gene.

To carry out a cloning experiment with pUC8, the manipulations shown in Figure 2.15, resulting in construction of a recombinant plasmid, are performed in the test tube with purified DNA. Pure pUC8 DNA can be obtained quite easily from extracts of bacterial cells (Technical Note 2.3), and after manipulation the plasmids can be reintroduced into *E. coli* by **transformation**, the process by which "naked" DNA is taken up by a bacterial cell. This is the system studied by Avery and his colleagues in the experiments that showed that bacterial genes are made of DNA (Section 1.1.1). Transformation is not a particularly efficient process in many bacteria, including *E. coli*, but the rate of uptake can be enhanced significantly by suspending the cells in calcium chloride before adding the DNA, and then briefly incubating the mixture at 42°C. Even after this enhancement, only a very small proportion of the cells take up a plasmid. This is why the ampicillin-resistance marker is so important—it allows the small number of transformants to be selected from the large background of nontransformed cells.

The map of pUC8 shown in Figure 2.17 indicates that the *lacZ'* gene contains a cluster of unique restriction sites. Ligation of new DNA into any one of these sites results in **insertional inactivation** of the gene and hence loss of β-galactosidase activity. This is the key to distinguishing a **recombinant** plasmid—one that contains an inserted piece of DNA—from a nonrecombinant plasmid that has no new DNA. Identifying recombinants is important because the manipulations illustrated in Figures 2.15 and 2.16 result in a variety of ligation products, including plasmids that have recircularized without insertion of

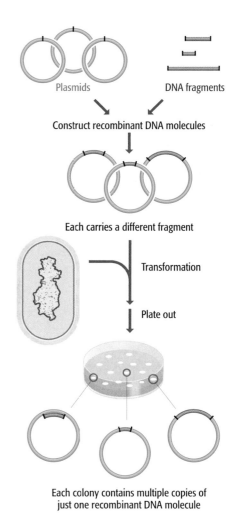

Figure 2.16 Cloning can provide a pure sample of a gene.

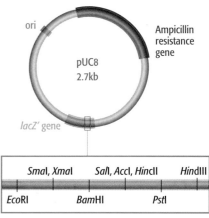

Figure 2.17 pUC8. The map shows the positions of the ampicillin-resistance gene, the lacZ' gene, the origin of replication (ori), and the cluster of restriction sites within the lacZ' gene.

Technical Note 2.3 DNA purification

Techniques for the preparation of pure samples of DNA from living cells play a central role in molecular biology research

The first step in DNA purification is to break open the cells from which the DNA will be obtained. With some types of material this step is easy: cultured animal cells, for example, are broken open simply by adding a detergent such as sodium dodecyl sulfate (SDS), which disrupts the cell membranes, releasing the cell contents. Other types of cell have strong walls and so demand a harsher treatment. Plant cells are usually frozen and then ground using a mortar and pestle, this being the only effective way of breaking their cellulose walls. Bacteria such as *Escherichia coli* can be lysed by a combination of enzymatic and chemical treatment. The enzyme used is **lysozyme**, obtained from egg white, which breaks down the polymeric compounds in the bacterial cell wall, and the chemical is ethylenediaminetetraacetate (EDTA), which chelates magnesium ions, further reducing the integrity of the cell wall. Disruption of the cell membrane, by adding a detergent, then causes the cells to burst.

Once the cells have been broken, two different approaches can be used to purify the DNA from the resulting extract. The first involves degrading or removing all the cellular components other than the DNA, a method that works best if the cells do not contain large amounts of lipid or carbohydrate. The extract is first centrifuged at low speed to remove debris such as pieces of cell wall, which form a pellet at the bottom of the tube (Figure T2.3). The supernatant is transferred to another test tube and mixed with phenol, which causes the protein to precipitate at the interface between the organic and aqueous layers. The aqueous layer, which contains the dissolved nucleic acids, is collected and a ribonuclease enzyme added, which breaks the RNA into a mixture of nucleotides and short oligonucleotides. The DNA polynucleotides, which remain intact, can now be precipitated by adding ethanol, pelleted by centrifugation, and resuspended in an appropriate volume of buffer.

In the second approach to DNA purification, rather than degrading everything other than DNA, the DNA itself is selectively removed from the extract. One way of doing this is by **ion-exchange chromatography**, which separates molecules according to how tightly they bind to electrically charged particles in a chromatography **resin**. DNA and RNA are both negatively charged, as are some proteins, and so bind to a positively charged resin. The simplest way to carry out ion-exchange chromatography is to place the resin in a column and add the cell extract to the top (Figure T2.4). The extract passes through the column, and all the negatively charged molecules bind to the resin. The binding, which involves ionic interactions, can be disrupted by adding a salt solution, with the less tightly bound molecules detaching from the resin at relatively low salt concentrations. This means that if a solution of gradually increasing salt concentration is passed through the column, then different types of molecule **elute** in the sequence protein, RNA and DNA, reflecting their relative binding strengths. In fact, such careful separation is usually not needed so just two salt solutions are used, one containing 1.0 M NaCl at pH 7.0, sufficient to elute the protein and RNA, leaving just the DNA bound, followed by a second solution containing 1.25 M NaCl at pH 8.5, which elutes the DNA, now free from protein and RNA contaminants.

Figure T2.3 Purifying DNA from a cell extract by degrading or removing all the other components.

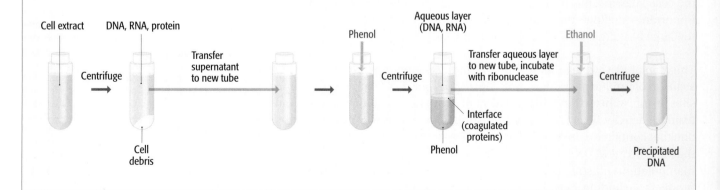

The two approaches described above purify all the DNA in a cell. Special methods are needed if the aim is to obtain just plasmid DNA (for example, recombinant cloning vectors) from bacterial cells. One popular method makes use of the fact that, although both plasmids and the bacterial chromosome are made up of supercoiled DNA, lysis of the bacterial cell inevitably leads to a certain amount of disruption of the bacterial chromosome, leading to loss of supercoiling. A cell extract therefore contains supercoiled plasmid DNA and *nonsupercoiled* chromosomal DNA, and the plasmids can be purified by a method that distinguishes between DNA molecules with these different conformations. One technique involves adding sodium hydroxide until the pH of the cell extract reaches 12.0–12.5, which causes the base pairs in nonsupercoiled DNA to break. The resulting single strands tangle up into an insoluble network that can be removed by centrifugation, leaving the supercoiled plasmids in the supernatant.

Figure T2.4 Purification of DNA by ion-exchange chromatography.

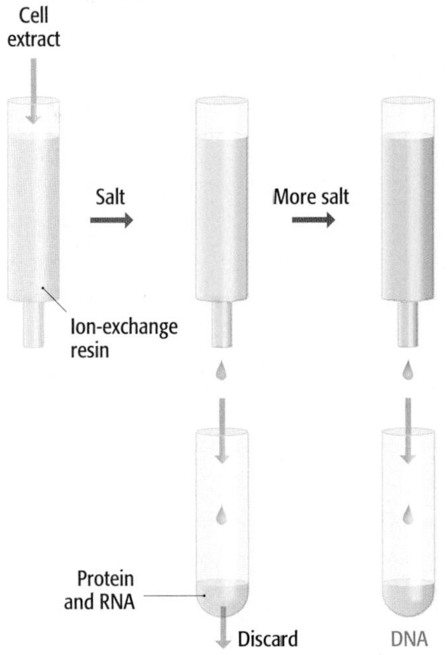

Cell extract

Salt

More salt

Ion-exchange resin

Protein and RNA

Discard

DNA

new DNA. Screening for β-galactosidase presence or absence is, in fact, quite easy. Rather than assay for lactose being split to glucose and galactose, the presence of functional β-galactosidase molecules in the cells is checked by a histochemical test with a compound called X-gal (5-bromo-4-chloro-3-indolyl-β-D-galactopyranoside), which the enzyme converts into a blue product. If X-gal (plus an inducer of the enzyme such as isopropylthiogalactoside, IPTG) is added to the agar, along with ampicillin, then nonrecombinant colonies, the cells of which synthesize β-galactosidase, will be colored blue, whereas recombinants with a disrupted *lacZ'* gene, which are unable to make β-galactosidase, will be white (Figure 2.18). This system is called **Lac selection**.

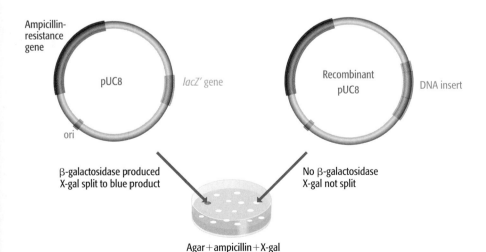

Ampicillin-resistance gene

pUC8

lacZ' gene

Recombinant pUC8

DNA insert

ori

β-galactosidase produced X-gal split to blue product

No β-galactosidase X-gal not split

Agar + ampicillin + X-gal

Figure 2.18 Recombinant selection with pUC8.

Cloning vectors based on E. coli bacteriophage genomes

E. coli bacteriophages were developed as cloning vectors back in the earliest days of the recombinant DNA revolution. The main reason for seeking a different type of vector was the inability of plasmids such as pUC8 to handle DNA fragments greater than about 10 kb in size, larger inserts undergoing rearrangements or interfering with the plasmid replication system in such a way that the recombinant DNA molecules become lost from the host cells. The first attempts to develop vectors able to handle larger fragments of DNA centered on bacteriophage λ.

To replicate, a bacteriophage must enter a bacterial cell and subvert the bacterial enzymes into expressing the information contained in the phage genes, so that the bacterium synthesizes new phages. Once replication is complete,

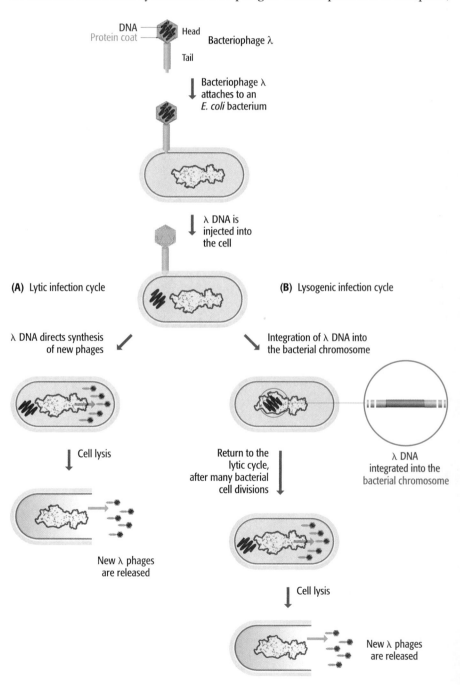

Figure 2.19 The lytic and lysogenic infection cycles of bacteriophage λ. (A) In the lytic cycle, new phages are produced shortly after infection. (B) During the lysogenic cycle, the phage genome becomes inserted into the bacterium's chromosomal DNA, where it can remain quiescent for many generations.

the new phages leave the bacterium, usually causing its death as they do so, and move on to infect new cells (Figure 2.19A). This is called a **lytic infection cycle**, because it results in **lysis** of the bacterium. As well as the lytic cycle, λ (unlike many other types of bacteriophage) can also follow a **lysogenic infection cycle**, during which the λ genome integrates into the bacterial chromosome, where it can remain quiescent for many generations, being replicated along with the host chromosome every time the cell divides (Figure 2.19B).

The size of the λ genome is 48.5 kb, of which some 15 kb or so is "optional" in that it contains genes that are only needed for integration of the phage DNA into the *E. coli* chromosome (Figure 2.20A). These segments can therefore be deleted without impairing the ability of the phage to infect bacteria and direct synthesis of new λ particles by the lytic cycle. Two types of vector have been developed (Figure 2.20B):

- **Insertion vectors**, in which part or all of the optional DNA has been removed, and a unique restriction site introduced at some position within the trimmed-down genome.

- **Replacement vectors**, in which the optional DNA is contained within a **stuffer fragment**, flanked by a pair of restriction sites, that is replaced when the DNA to be cloned is ligated into the vector.

The λ genome is linear, but the two natural ends of the molecule have 12-nucleotide single-stranded overhangs, called ***cos* sites**, which have complementary sequences and so can base-pair to one another. A λ cloning vector can therefore be obtained as a circular molecule that can be manipulated in the test tube in the same way as a plasmid, and reintroduced into *E. coli* by **transfection**, the term used for uptake of naked phage DNA. Alternatively, a more efficient uptake system called ***in vitro* packaging** can be utilized. This procedure starts with the linear version of the cloning vector, the initial restriction cutting the molecule into two segments, the left and right arms, each with a *cos* site at one end. The ligation is carried out with carefully measured quantities of each arm and the DNA to be cloned, the aim being to

(A) The λ genome contains 'optional' DNA

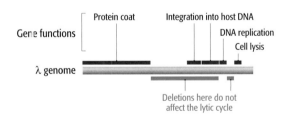

(B) Insertion and replacement vectors

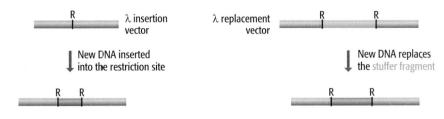

R = restriction site

Figure 2.20 **Cloning vectors based on bacteriophage λ.** (A) In the λ genome, the genes are arranged into functional groups. For example, the region marked as "protein coat" comprises 21 genes coding for proteins that are either components of the phage capsid or are required for capsid assembly, and "cell lysis" comprises four genes involved in lysis of the bacterium at the end of the lytic phase of the infection cycle. The regions of the genome that can be deleted without impairing the ability of the phage to follow the lytic cycle are indicated in green. (B) The differences between a λ insertion vector and a λ replacement vector.

Figure 2.21 Cloning with a λ insertion vector. The linear form of the vector is shown at the top of the diagram. Treatment with the appropriate restriction endonuclease produces the left and right arms, both of which have one blunt end and one end with the 12-nucleotide overhang of the *cos* site. The DNA to be cloned is blunt ended and so is inserted between the two arms during the ligation step. These arms also ligate to one another via their *cos* sites, forming a concatamer. Some parts of the concatamer comprise left arm–insert DNA–right arm and, assuming this combination is 37–52 kb in length, will be enclosed inside the capsid by the *in vitro* packaging mix. Parts of the concatamer made up of left arm ligated directly to right arm, without new DNA, are too short to be packaged.

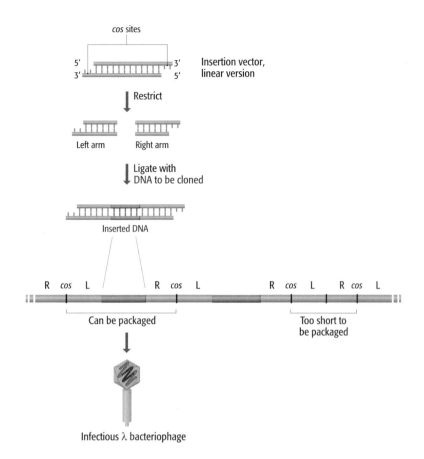

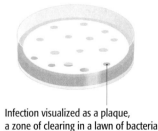

Infection visualized as a plaque, a zone of clearing in a lawn of bacteria

Figure 2.22 Bacteriophage infection is visualized as a plaque on a lawn of bacteria.

produce concatamers in which the different fragments are linked together in the order left arm–new DNA–right arm, as shown in Figure 2.21. The concatamers are then added to an *in vitro* packaging mix, which contains all the proteins needed to make a λ phage particle. These proteins form phage particles spontaneously, and will place inside the particles any DNA fragment that is between 37 kb and 52 kb in length and is flanked by *cos* sites. The packaging mix therefore cuts left arm–new DNA–right arm combinations of 37–52 kb out of the concatamers and constructs λ phages around them. The phages are then mixed with *E. coli* cells, and the natural infection process transports the vector plus new DNA into the bacteria.

After infection, the cells are spread onto an agar plate. The objective is not to obtain individual colonies but to produce an even layer of bacteria across the entire surface of the agar. Bacteria that were infected with the packaged cloning vector die within about 20 minutes because the λ genes contained in the arms of the vector direct replication of the DNA and synthesis of new phages by the lytic cycle, each of these new phages containing its own copy of the vector plus cloned DNA. Death and lysis of the bacterium releases these phages into the surrounding medium, where they infect new cells and begin another round of phage replication and lysis. The end result is a zone of clearing, called a **plaque**, which is visible on the lawn of bacteria that grows on the agar plate (Figure 2.22). With some λ vectors, all plaques are made up of recombinant phages because ligation of the two arms without insertion of new DNA results in a molecule that is too short to be packaged. With other vectors it is necessary to distinguish recombinant plaques from nonrecombinant ones. Various methods are used, including the β-galactosidase system described above for the plasmid vector pUC8 (see Figure 2.18), which is also applicable to those λ vectors that carry a fragment of the *lacZ* gene into which the DNA to be cloned is inserted.

Vectors for longer pieces of DNA

The λ phage particle can accommodate up to 52 kb of DNA, so if the genome has 15 kb removed, then up to 18 kb of new DNA can be cloned. This limit is higher than that for plasmid vectors, but is still very small compared with the sizes of intact genomes. The comparison is important because a **clone library**—a collection of clones whose inserts cover an entire genome—is often the starting point for a project aimed at determining the sequence of that genome (Chapter 4). If a λ vector is used with human DNA, then over half a million clones are needed for there to be a 95% chance of any particular part of the genome being present in the library (Table 2.4). It is possible to prepare a library comprising half a million clones, especially if automated techniques are used, but such a large collection is far from ideal. It would be much better to reduce the number of clones by using a vector that is able to handle fragments of DNA longer than 18 kb. Many of the developments in cloning technology over the last 20 years have been aimed at finding ways of doing this.

One possibility is to use a **cosmid**—a plasmid that carries a λ *cos* site (Figure 2.23). Concatamers of cosmid molecules, linked at their *cos* sites, act as substrates for *in vitro* packaging because the *cos* site is the only sequence that a DNA molecule needs in order to be recognized as a "λ genome" by the proteins that package DNA into λ phage particles. Particles containing cosmid DNA are as infective as real λ phages, but once inside the cell the cosmid cannot direct synthesis of new phage particles and instead replicates as a plasmid. Recombinant DNA is therefore obtained from colonies rather than plaques. As with other types of λ vector, the upper limit for the length of the cloned DNA is set by the space available within the λ phage particle. A cosmid can be 8 kb or less in size, so up to 44 kb of new DNA can be inserted before the packaging limit of the λ phage particle is reached. This reduces the size of

Table 2.4 Sizes of human genomic libraries prepared in different types of cloning vector

Type of vector	Insert size (kb)	Number of clones*	
		P = 95%	*P* = 99%
λ replacement	18	532,500	820,000
Cosmid, fosmid	40	240,000	370,000
P1	125	77,000	118,000
BAC, PAC	300	32,000	50,000
YAC	600	16,000	24,500
Mega-YAC	1400	6850	10,500

*Calculated from the equation:

$$N = \frac{\ln\,(1-P)}{\ln\left(1-\dfrac{a}{b}\right)}$$

where *N* is the number of clones required, *P* is the probability that any given segment of the genome is present in the library, *a* is the average size of the DNA fragments inserted into the vector, and *b* is the size of the genome

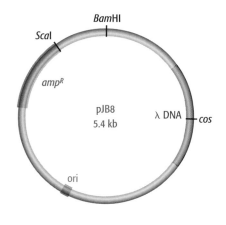

Figure 2.23 A typical cosmid. pJB8 is 5.4 kb in size and carries the ampicillin-resistance gene (*amp^R*), a segment of λ DNA containing the cos site, and an *Escherichia coli* origin of replication (ori).

the human genomic library to about one quarter of a million clones, which is an improvement compared with a λ library, but still a massive number of clones to have to work with.

The first major breakthrough in attempts to clone DNA fragments much longer than 50 kb came with the invention of **yeast artificial chromosomes**, or **YACs**. These vectors are propagated in *S. cerevisiae*, rather than in *E. coli*, and are based on chromosomes, rather than on plasmids or viruses. The first YACs were constructed after studies of natural chromosomes had shown that, in addition to the genes that it carries, each chromosome has three important components (Figure 2.24):

- The **centromere**, which plays a critical role during nuclear division.

- The **telomeres**, the special sequences that mark the ends of chromosomal DNA molecules.

- One or more **origins of replication**, which initiate synthesis of new DNA when the chromosome divides.

In a YAC, the DNA sequences that underlie these chromosomal components are linked together with one or more selectable markers and at least one restriction site into which new DNA can be inserted (Figure 2.25). All of these components can be contained in a DNA molecule of size 10–15 kb. Natural yeast chromosomes range in size from 230 kb to over 1700 kb, so YACs have the potential to clone megabase (Mb)-sized DNA fragments. This potential has been realized, standard YACs being able to clone 600 kb fragments, with special types able to handle DNA up to 1400 kb in length. This is the highest capacity of any type of cloning vector, and several of the early genome projects made extensive use of YACs. Unfortunately, with some types of YAC there have been problems with insert stability, the cloned DNA becoming rearranged into new sequence combinations. For this reason there is also great interest in other types of vectors, ones that cannot clone such large pieces of DNA but which suffer less from instability problems. These vectors include the following:

- **Bacterial artificial chromosomes**, or **BACs**, are based on the naturally occurring F plasmid of *E. coli*. Unlike the plasmids used to construct the early cloning vectors, the F plasmid is relatively large and vectors based on it have a higher capacity for accepting inserted DNA. BACs are designed so that recombinants can be identified by Lac selection (see Figure 2.18) and hence are easy to use. They can clone fragments of 300 kb and longer, and the inserts are very stable. BACs were used extensively in the Human Genome Project (Section 4.3) and they are currently the most popular vectors for cloning large pieces of DNA.

- **Bacteriophage P1 vectors** are very similar to λ vectors, being based on a deleted version of a natural phage genome, the capacity of the cloning vector being determined by the size of the deletion and the space within the phage particle. The P1 genome is larger than the λ genome, and the phage particle is bigger, so a P1 vector can clone larger fragments of DNA than a λ vector, up to 125 kb using current technology.

- **P1-derived artificial chromosomes**, or **PACs**, combine features of P1 vectors and BACs, and have a capacity of up to 300 kb.

- **Fosmids** contain the F plasmid origin of replication and a λ *cos* site. They are similar to cosmids but have a lower copy number in *E. coli*, which means that they are less prone to instability problems.

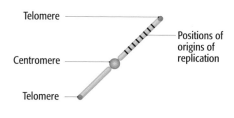

Figure 2.24 The key structural components of a eukaryotic chromosome. For further information on these structures see Sections 7.1.2 (centromeres and telomeres) and 15.2.1 (origins of replication).

The sizes of human genome libraries prepared in these various types of vector are given in Table 2.4.

Cloning in organisms other than E. coli

Cloning is not merely a means of producing DNA for sequencing and other types of analysis. It also provides a means of studying the mode of expression of a gene and the way in which expression is regulated, of carrying out genetic

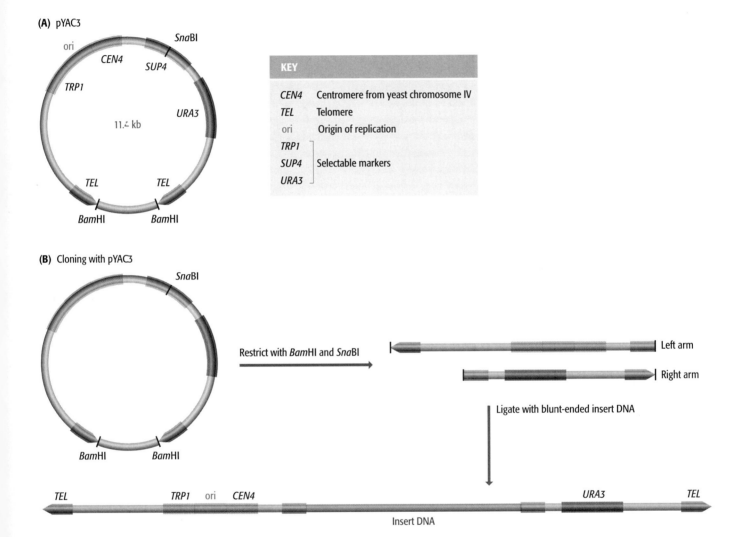

Figure 2.25 Working with a YAC. (A) The cloning vector pYAC3. (B) To clone with pYAC3, the circular vector is digested with *Bam*HI and *Sna*BI. *Bam*HI restriction removes the stuffer fragment held between the two telomeres in the circular molecule. *Sna*BI cuts within the *SUP4* gene and provides the site into which new DNA will be inserted. Ligation of the two vector arms with new DNA produces the structure shown at the bottom. This structure carries functional copies of the *TRP1* and *URA3* selectable markers. The host strain has inactivated copies of these genes, which means that it requires tryptophan and uracil as nutrients. After transformation, cells are plated onto a minimal medium lacking tryptophan and uracil. Only cells that contain the vector, and so can synthesize tryptophan and uracil, are able to survive on this medium and produce colonies. Note that if a vector comprises two right arms, or two left arms, then it will not give rise to colonies because the transformed cells will still require one of the nutrients. The presence of insert DNA in the cloned vector molecules is checked by testing for inactivation of *SUP4*. This is done by a color test: on the appropriate medium, colonies containing recombinant vectors (i.e., with an insert) are white; nonrecombinants (vector but no insert) are red.

engineering experiments aimed at modifying the biological characteristics of the host organism, and of synthesizing important animal proteins, such as pharmaceuticals, in a new host cell from which the proteins can be obtained in larger quantities than is possible by conventional purification from animal tissue. These multifarious applications demand that genes must frequently be cloned in organisms other than *E. coli*.

Cloning vectors based on plasmids or phages have been developed for most of the well-studied species of bacteria, such as *Bacillus*, *Streptomyces*, and *Pseudomonas*, these vectors being used in exactly the same way as the *E. coli* analogs. Plasmid vectors are also available for yeasts and fungi. Some of these carry the origin of replication from the **2 μm circle**, a plasmid present in many strains of *S. cerevisiae*, but other plasmid vectors only have an *E. coli* origin. An example is YIp5, an *S. cerevisiae* vector that is simply an *E.coli* plasmid that contains a copy of the yeast gene called *URA3* (Figure 2.26A). The presence of the *E. coli* origin means that YIp5 is a **shuttle vector** that can be used with either *E. coli* or *S. cerevisiae* as the host. This is a useful feature because cloning in *S. cerevisiae* is a relatively inefficient process, and generating a large number of clones is difficult. If the experiment requires that the desired recombinant be identified from a mixture of clones (as illustrated in Figure 2.16), then it may not be possible to obtain enough recombinants to find the correct one. To avoid this problem, construction of recombinant DNA molecules and selection of the correct recombinant is carried out with *E. coli* as the host. When the correct clones have been identified, the recombinant YIp5 molecules are purified and transferred into *S. cerevisiae*, usually by mixing the DNA with **protoplasts**—yeast cells whose walls have been removed by enzyme treatment. Without an origin of replication, the vector is unable to propagate independently inside yeast cells, but it can survive if it becomes integrated into one of the yeast chromosomes, which can occur by **homologous recombination** (Section 5.2.2) between the *URA3* gene carried by the vector and the chromosomal copy of this gene (Figure 2.26B). "YIp" in fact stands for "yeast integrative plasmid". Once integrated, the YIp, plus any DNA that has been inserted into it, replicates along with the host chromosomes.

Integration into chromosomal DNA is also a feature of many of the cloning systems used with animals and plants, and forms the basis of the construction

Figure 2.26 Cloning with a YIp. (A) YIp5, a typical yeast integrative plasmid. The plasmid contains the ampicillin-resistance gene (*amp^R*), the tetracycline-resistance gene (*tet^R*), the yeast gene *URA3*, and an *Escherichia coli* origin of replication (ori). The presence of the *E. coli* ori means that recombinant YIp5 molecules can be constructed in *E. coli* before their transfer into yeast cells. YIp5 is therefore a **shuttle vector**—it can be shuttled between two species. (B) YIp5 has no origin of replication that can function inside yeast cells, but can survive if it integrates into the yeast chromosomal DNA by homologous recombination between the plasmid and chromosomal copies of the *URA3* gene. The chromosomal gene carries a small mutation that means that it is nonfunctional and the host cells are *ura3⁻*. One of the pair of *URA3* genes that is formed after integration of the plasmid DNA is mutated, but the other is not. Recombinant cells are therefore *ura3⁺* and can be selected by plating onto minimal medium, which does not contain uracil.

(A) YIp5

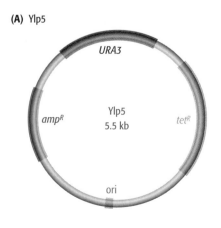

(B) Insertion of YIp5 into yeast chromosomal DNA

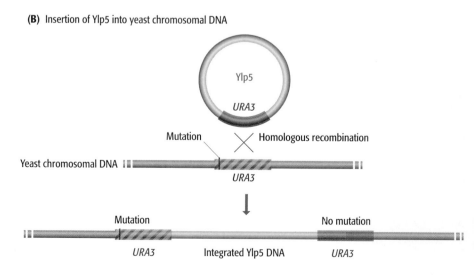

of **knockout mice**, which are used to determine the functions of previously unknown genes that are discovered in the human genome (Section 5.2.2). The vectors are animal equivalents of YIps. Adenoviruses and retroviruses are used to clone genes in animals when the objective is to treat a genetic disease or a cancer by **gene therapy**. A similar range of vectors has been developed for cloning genes in plants. Plasmids can be introduced into plant embryos by bombardment with DNA-coated microprojectiles, a process called **biolistics**. Integration of the plasmid DNA into the plant chromosomes, followed by growth of the embryo, then results in a plant that contains the cloned DNA in most or all of its cells. Some success has also been achieved with plant vectors based on the genomes of caulimoviruses and geminiviruses, but the most interesting types of plant cloning vector are those derived from the **Ti plasmid**, a large bacterial plasmid found in the soil microorganism *Agrobacterium tumefaciens*. Part of the Ti plasmid, the region called the **T-DNA**, becomes integrated into a plant chromosome when the bacterium infects a plant stem and causes crown gall disease. The T-DNA carries a number of genes that are expressed inside the plant cells and induce the various physiological changes that characterize the disease. Vectors such as pBIN19 (Figure 2.27) have been designed to make use of this natural genetic engineering system. The recombinant vector is introduced into *A. tumefaciens* cells, which are allowed to infect a cell suspension or plant callus culture, from which mature, transformed plants can be regenerated.

2.3 The Polymerase Chain Reaction (PCR)

DNA cloning is a powerful technique and its impact on our understanding of genes and genomes has been immeasurable. Cloning does, however, have one major disadvantage: it is a time-consuming and, in parts, difficult procedure. It takes several days to perform the manipulations needed to insert DNA fragments into a cloning vector and then introduce the ligated molecules into the host cells and select recombinants. If the experimental strategy involves generation of a large clone library, followed by screening of the library to identify a clone that contains a gene of interest (see Technical Note 2.4), then several more weeks, or even months, might be needed to complete the project.

PCR complements DNA cloning in that it enables the same result to be achieved—purification of a specified DNA fragment—but in a much shorter time, perhaps just a few hours. PCR is complementary to, not a replacement for, cloning because it has its own limitations, the most important of which is the need to know the sequence of at least part of the fragment that is to be purified. Despite this constraint, PCR has acquired central importance in many areas of molecular biology research. We will examine the technique first, and then survey its applications.

2.3.1 Carrying out a PCR

PCR results in the repeated copying of a selected region of a DNA molecule (see Figure 2.3). Unlike cloning, PCR is a test-tube reaction and does not involve the use of living cells: the copying is carried out not by cellular enzymes but by the purified, thermostable DNA polymerase of *T. aquaticus* (Section 2.1.1). The reason why a thermostable enzyme is needed will become clear when we look in more detail at the events that occur during PCR.

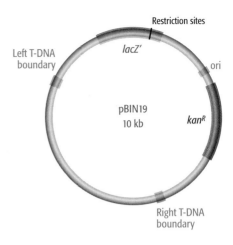

Figure 2.27 The plant cloning vector pBIN19. pBIN19 carries the *lacZ′* gene (see Figure 2.18), the kanamycin-resistance gene (*kan^R*), an *Escherichia coli* origin of replication (ori), and the two boundary sequences from the T-DNA region of the Ti plasmid. These two boundary sequences recombine with plant chromosomal DNA, inserting the segment of DNA between them into the plant DNA. The orientation of the boundary sequences in pBIN19 means that the *lacZ′* and *kan^R* genes, as well as any new DNA ligated into the restriction sites within *lacZ′*, are transferred to the plant DNA. Recombinant plant cells are selected by plating onto kanamycin agar, and then regenerated into whole plants. Note that pBIN19 is another example of a shuttle vector, recombinant molecules being constructed in *E. coli*, using the *lacZ′* selection system, before transfer to *Agrobacterium tumefaciens* and thence to the plant.

Technical Note 2.4 Working with a clone library

Clone collections are used as a source of genes and other DNA segments

Since the 1970s, clone libraries have been prepared from different organisms as a means of obtaining individual genes, and other DNA segments, for further study by sequencing and other recombinant DNA techniques. Libraries can be prepared from either genomic DNA or cDNA, using a plasmid or bacteriophage vector. The clones are usually stored as bacterial colonies or plaques on 23 × 23 cm agar plates, with 100,000–150,000 clones per plate. A complete human library can therefore be contained in just 1–8 plates, depending on the type of cloning vector that has been used (see Table 2.4). Three methods can identify the clone that contains the gene or other piece of DNA that is being sought:

Hybridization analysis can be performed with a labeled oligonucleotide or other DNA molecule that is known to hybridize to the sequence of interest. To do this, a nylon or nitrocellulose membrane is placed on the surface of the agar dish and then carefully removed to "lift off" the colonies or plaques. Treatment with alkali and protease degrades the cellular material, leaving behind the DNA from each clone, which is then bound tightly to the surface of the membrane by heating or ultraviolet irradiation. The labeled probe is now applied to the membrane, in the same way as in Southern hybridization (Figure 2.11), and the position at which the probe attaches determined by the appropriate detection method. The position of the

hybridization signal on the membrane corresponds to the location of the clone of interest on the agar plate.

PCR (Section 2.3) can be used to screen clones for the sequence of interest. This cannot be done *in situ*, so individual clones must be transferred to the wells of microtiter trays. The PCR approach to clone identification is therefore relatively cumbersome because only a few hundred clones can be accommodated in a single tray. PCRs using primers specific for the sequence of interest are performed with each clone in turn, possibly using a combinatorial approach to reduce the number of PCRs that are needed in order to identify the one that gives a positive result (see Figure 4.14).

Immunological techniques can be used if the sequence being sought is a gene that is expressed in the cell in which the clone library has been prepared. If gene expression is occurring then the protein product will be made, and this can be detected by screening the library with a labeled antibody that binds only to that protein. As in hybridization analysis, the clones are first transferred onto a membrane, and then treated to break down the cells and bind the protein to the membrane surface. Exposure of the membrane to the labeled antibody then reveals the position of the clone containing the gene of interest.

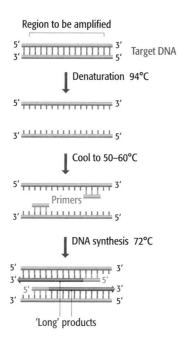

'Long' products

To carry out a PCR experiment, the target DNA is mixed with *Taq* DNA polymerase, a pair of oligonucleotide primers, and a supply of nucleotides. The amount of target DNA can be very small because PCR is extremely sensitive and will work with just a single starting molecule. The primers are needed to initiate the DNA synthesis reactions that will be carried out by the *Taq* polymerase (see Figure 2.6). They must attach to the target DNA at either side of the segment that is to be copied: the sequences of these attachment sites must therefore be known so that primers of the appropriate sequences can be synthesized.

The reaction is started by heating the mixture to 94°C. At this temperature the hydrogen bonds that hold together the two polynucleotides of the double helix are broken, so the target DNA becomes denatured into single-stranded molecules (Figure 2.28). The temperature is then reduced to 50°C–60°C, which results in some rejoining of the single strands of the target DNA, but also allows the primers to attach to their annealing positions. DNA synthesis

Figure 2.28 The first stage of a PCR.

can now begin, so the temperature is raised to 72°C, the optimum for *Taq* polymerase. In this first stage of the PCR, a set of "long" products is synthesized from each strand of the target DNA. These polynucleotides have identical 5′ ends but random 3′ ends, the latter representing positions where DNA synthesis terminates by chance. When the cycle of denaturation–annealing–synthesis is repeated, the long products act as templates for new DNA synthesis, giving rise to "short" products, the 5′ and 3′ ends of which are both set by the primer annealing positions (Figure 2.29). In subsequent cycles, the number of short products accumulates in an exponential fashion (doubling during each cycle) until one of the components of the reaction becomes depleted. This means that after 30 cycles, there will be over 250 million short products derived from each starting molecule. In real terms, this equates to several micrograms of PCR product from a few nanograms or less of target DNA.

The results of a PCR can be determined in various ways. Usually, the products are analyzed by agarose gel electrophoresis, which will reveal a single band if the PCR has worked as expected and has amplified a single segment of the target DNA (Figure 2.30). Alternatively, the sequence of the product can be determined, using techniques described in Section 4.1.1.

2.3.2 The applications of PCR

PCR is such a straightforward procedure that it is sometimes difficult to understand how it can have become so important in modern research. First we will deal with its limitations. In order to synthesize primers that will anneal at the correct positions, the sequences of the boundary regions of the DNA to be amplified must be known. This means that PCR cannot be used to purify fragments of genes, or other parts of a genome, that have never been studied before. A second constraint is the length of DNA that can be copied. Regions of up to 5 kb can be amplified without too much difficulty, and longer amplifications—up to 40 kb—are possible using modifications of the standard technique. However, the fragments greater than 100 kb that are needed for genome sequencing projects, and which can be obtained by cloning in a BAC or other high-capacity vector, are unattainable by PCR.

What are the strengths of PCR? Primary among these is the ease with which products representing a single segment of the genome can be obtained from a number of different DNA samples. We will encounter one important example of this in the next chapter when we look at how DNA markers are typed in genetic mapping projects (Section 3.2.2). PCR is used in a similar way to screen human DNA samples for mutations associated with genetic diseases such as thalassemia and cystic fibrosis. It also forms the basis of genetic profiling, in which variations in microsatellite length are typed (see Figure 7.24).

A second important feature of PCR is its ability to work with minuscule amounts of starting DNA. This means that PCR can be used to obtain sequences from the trace amounts of DNA that are present in hairs, bloodstains, and other forensic specimens, and from bones and other remains preserved at archaeological sites. In clinical diagnosis, PCR is able to detect the presence of viral DNA well before the virus has reached the levels needed to initiate a disease response. This is particularly important in the early identification of viral-induced cancers because it means that treatment programs can be initiated before the cancer becomes established.

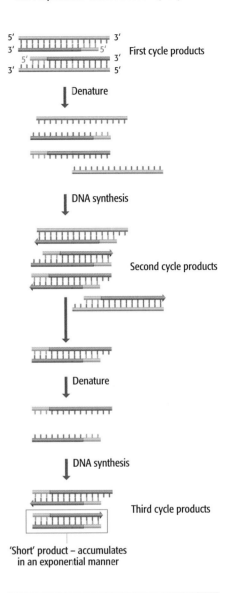

Figure 2.29 The synthesis of "short" products in a PCR. From the first cycle products shown at the top of the diagram, the next cycle of denaturation–annealing–synthesis leads to four products, two of which are identical to the first cycle products and two of which are made entirely of new DNA. During the third cycle, the latter give rise to "short" products that, in subsequent cycles, accumulate in an exponential fashion.

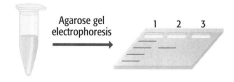

Figure 2.30 Analyzing the results of a PCR by agarose gel electrophoresis. The PCR has been carried out in a microfuge tube. A sample is loaded into lane 2 of an agarose gel. Lane 1 contains DNA size markers, and lane 3 contains a sample of a PCR carried out by a colleague. After electrophoresis, the gel is stained with ethidium bromide (see Technical Note 2.2). Lane 2 contains a single band of the expected size, showing that the PCR has been successful. In lane 3 there is no band—this PCR has not worked.

The above are just a few of the applications of PCR. The technique is now a major component of the molecular biologist's toolkit, and we will discover many more examples of its use as we progress through the remaining chapters of this book.

Summary

Over the last 35 years, molecular biologists have built up a comprehensive toolkit of techniques that can be used to study DNA. These techniques form the basis of recombinant DNA technology and led to the development of DNA cloning and the polymerase chain reaction (PCR). A central feature of recombinant DNA technology is the use of purified enzymes to make specified changes to DNA molecules in the test tube. The four main types of enzyme used in this way are DNA polymerases, nucleases, ligases, and end-modification enzymes. DNA polymerases synthesize new DNA polynucleotides, and are used in procedures such as DNA sequencing, PCR, and DNA labeling. The most important nucleases are the restriction endonucleases, which cut double-stranded DNA molecules at specific nucleotide sequences, and hence cut a molecule into a predicted set of fragments, the sizes of which can be determined by agarose gel electrophoresis. Ligases join molecules together, and end-modification enzymes carry out a variety of reactions including several used to label DNA molecules. DNA cloning is a means of obtaining a pure sample of an individual gene or other segment of a DNA molecule. Many different types of cloning vector have been designed for use with *E. coli* as the host organism, the simplest being based on small plasmids that carry markers such as the *lacZ′* gene. This gene enables recombinant colonies to be identified because they appear white, rather than blue, when X-gal is present in the growth medium. Bacteriophage λ has also been used as the basis for a series of *E. coli* cloning vectors, including the plasmid–phage hybrids called cosmids, which are used to clone fragments of DNA up to 44 kb in length. Other types of vector, such as bacterial artificial chromosomes, can be used to clone even longer pieces of DNA, up to 300 kb in length. These high-capacity vectors are used in construction of clone libraries, collections of clones whose inserts cover an entire genome and which are used to provide material for a genome sequencing project. Organisms other than *E. coli* can also be used as the hosts for DNA cloning. Several types of vector have been designed for *Saccharomyces cerevisiae*, and specialized techniques are available for cloning DNA in animals and plants. PCR provides a complement to DNA cloning by enabling specified segments of DNA to be purified quickly, but at least part of the DNA sequence of this fragment must be known. In PCR, a thermostable DNA polymerase makes repeated copies of the target sequence and of the copies made in earlier rounds of the reaction. Starting with just a single target molecule, over 250 million copies can be made during 30 cycles of a PCR.

Multiple Choice Questions

*Answers to odd-numbered questions can be found in the Appendix

2.1.* Which of the following enzymes are used to degrade DNA molecules?

   a. DNA polymerases.

   b. Nucleases.

   c. Ligases.

   d. Kinases.

2.2. Why does a template-dependent DNA polymerase require a primer to initiate DNA synthesis?

   a. These polymerases require a 5′-phosphate group to add a new nucleotide.

   b. These polymerases require a 3′-hydroxyl group to add a new nucleotide.

   c. The primer is required for the DNA polymerase to bind to the template DNA.

   d. The primer is hydrolyzed to provide the energy required for DNA synthesis.

2.3.* The function of the 3′→5′ exonuclease activity of a DNA polymerase is to:

   a. Remove the 5′ end of the polynucleotide strand that is attached to the template strand that is being copied.

   b. Remove damaged nucleotides from the template strand during DNA synthesis.

   c. Remove nucleotides from the ends of DNA molecules to ensure the generation of blunt ends.

   d. Remove incorrect nucleotides from the newly synthesized strand of DNA.

2.4. The Klenow polymerase version of *E. coli* DNA polymerase I is useful for research as it lacks the 5′→3′ exonuclease activity. This is useful as the 5′→3′ exonuclease activity:

   a. Is more active than the polymerase activity.

   b. Will prevent the incorporation of radioactive or fluorescent labels into the DNA.

   c. May interfere with some research applications by shortening the 5′ ends of the DNA molecules.

   d. Prevents the polymerase from detecting errors in the incorporation of new nucleotides.

2.5.* A temperature of 75°C will terminate DNA synthesis by *E. coli* DNA polymerase I. This is because:

   a. *E. coli* DNA polymerase I is denatured at this temperature.

   b. The DNA is denatured at this temperature.

   c. The primers are denatured at this temperature.

   d. The temperature is too high for enzymatic reactions to occur.

2.6. Which of the following statements accurately describes reverse transcriptases?

   a. They are present in all viruses and are RNA-dependent DNA polymerases.

   b. They are present in all RNA viruses and are DNA-dependent RNA polymerases.

   c. They are present in retroviruses and are RNA-dependent DNA polymerases.

   d. They are present in all viruses and are template-independent DNA polymerases.

2.7.* All three types of restriction enzyme bind to DNA molecules at specific sequences; however, the type II enzymes are favored for research for which of the following reasons?

   a. Type II enzymes cut the DNA at a specific site.

   b. Type II enzymes always cut the DNA to yield blunt-ended molecules.

   c. Type II enzymes always cut the DNA to yield sticky-ended molecules.

   d. Type II enzymes are the only restriction enzymes to cleave double-stranded DNA.

2.8. Which technique is used to resolve the different sizes of DNA fragments following a restriction enzyme digest?

   a. DNA sequencing.

   b. Gel electrophoresis.

   c. Gene cloning.

   d. PCR.

2.9.* DNA ligase synthesizes which type of bond?

   a. The hydrogen bonds between bases.

   b. The phosphodiester bonds between nucleotides.

   c. The bonds between the bases and deoxyribose sugars.

   d. The peptide bonds between amino acids.

2.10. Which of the following polymerases does not require a template?

   a. DNA polymerase I.

   b. Sequenase.

   c. Reverse transcriptase.

   d. Terminal deoxynucleotidyl transferase.

2.11.* *E. coli* cells take up plasmid DNA in laboratory experiments by which of the following methods?

   a. Conjugation.

   b. Electrophoresis.

   c. Transduction.

   d. Transformation.

continued …

Multiple Choice Questions (continued) *Answers to odd-numbered questions can be found in the Appendix

2.12. What is a genomic library?

   a. A collection of recombinant molecules with inserts that contain all of the genes of an organism.

   b. A collection of recombinant molecules with inserts that contain all of an organism's genome.

   c. A collection of recombinant molecules that express all of the genes of an organism.

   d. A collection of recombinant molecules that have been sequenced.

2.13.* Which of the following types of vector would be most suitable for introducing DNA into a human cell?

   a. Plasmid.

   b. Bacteriophage.

   c. Cosmid.

   d. Adenovirus.

2.14. Which of the following is NOT used to introduce recombinant DNA molecules into plants?

   a. Biolistics.

   b. Cosmids.

   c. Ti plasmid.

   d. Viruses.

2.15.* PCR is advantageous to gene cloning for all of the following reasons except:

   a. PCR does not require that the sequence of the gene be known.

   b. PCR is a very rapid technique for the isolation of a gene.

   c. PCR requires very small amounts of starting DNA compared to gene cloning.

   d. PCR is very useful for mapping DNA markers.

Short Answer Questions   *Answers to odd-numbered questions can be found in the Appendix

2.1.* What is meant by the term "gene cloning"?

2.2. How can a researcher identify a single restriction enzyme fragment containing a gene of interest in a digest of genomic DNA that contains thousands of different restriction fragments?

2.3.* Name a useful and quick method for increasing the ligation efficiency of blunt-ended DNA molecules.

2.4. Why are plasmids useful cloning vectors?

2.5.* Why do plasmids contain genes for antibiotic resistance?

2.6. What is X-gal, added to media in cloning experiments?

2.7.* Why is λ bacteriophage useful as a cloning vector?

2.8. Why are vectors that can carry larger DNA inserts beneficial for the creation of clone libraries?

2.9.* Yeast artificial chromosomes must have which three features of normal chromosomes to be maintained in cells?

2.10. Why are the initial PCR products—produced in the first few cycles of the reaction—long and of varying sizes, and the final PCR products all of a shorter and uniform size?

2.11.* How do the primers determine the specificity of a PCR?

2.12. What types of DNA sequence cannot be amplified by PCR?

In-depth Problems   *Guidance to odd-numbered questions can be found in the Appendix

2.1.* Soon after the first gene-cloning experiments were carried out in the early 1970s, a number of scientists argued that there should be a temporary moratorium on this type of research. What was the basis of these scientists' fears and to what extent were these fears justified?

2.2. What would be the features of an ideal cloning vector? To what extent are these requirements met by any of the existing cloning vectors?

2.3.* How might you determine the positions of the restriction sites in a DNA molecule, other than by working out the sequence of the molecule?

In-depth Problems (continued)

*Guidance to odd-numbered questions can be found in the Appendix

2.4. Explore the uses of gene cloning in the production of animal proteins in bacterial cells.

2.5.* The specificity of the primers is a critical feature of a successful PCR. If the primers anneal at more than one position in the target DNA then products additional to the one being sought will be synthesized. Explore the factors that determine primer specificity and evaluate the influence of the annealing temperature on the outcome of a PCR.

Figure Tests

*Answers to odd-numbered questions can be found in the Appendix

2.1.* What is the role of the primer (in blue) in DNA synthesis reactions catalyzed by DNA polymerase?

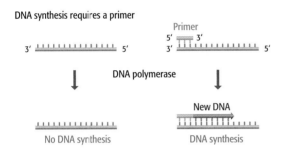

2.2. What if the *Bam*HI-cut plasmid used in this experiment keeps religating to itself at a high frequency and very few recombinant plasmids are isolated? How could you improve the ligation reaction to increase the yield of recombinant plasmids?

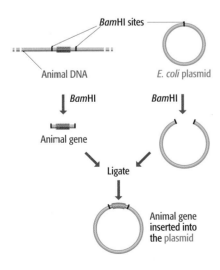

2.3.* This cloning vector possesses a bacterial origin of replication, a selectable marker (antibiotic-resistance gene), and *cos* sites from a bacteriophage. What type of cloning vector is this, and what size insert molecule can it carry?

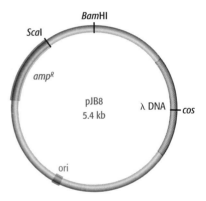

2.4. What is the type of reaction that is shown in this figure? Label the steps in the process and give the temperatures for each step.

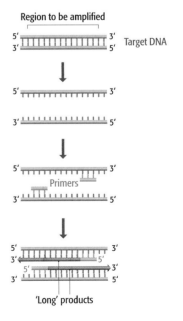

Further Reading

Textbooks and practical guides on the methods used to study DNA

Brown, T.A. (2006) *Gene Cloning and DNA Analysis: An Introduction*, 5th Ed. Blackwell Scientific Publishers, Oxford.

Brown, T.A. (ed.) (2000) *Essential Molecular Biology: A Practical Approach*, Vol. 1 and 2, 2nd Ed. Oxford University Press, Oxford. *Includes detailed protocols for DNA cloning and PCR.*

Dale, J.W. (2004) Molecular Genetics of Bacteria, 4th Ed. Wiley, Chichester. *Provides a detailed description of plasmids and bacteriophages.*

Enzymes for DNA manipulation

Brown, T.A. (1998) *Molecular Biology Labfax. Volume I: Recombinant DNA*, 2nd Ed. Academic Press, London. *Contains details of all types of enzymes used to manipulate DNA and RNA.*

REBASE: http://rebase.neb.com/rebase/ *A comprehensive list of all the known restriction endonucleases and their recognition sequences.*

Smith, H.O. and Wilcox, K.W. (1970) A restriction enzyme from *Haemophilus influenzae. J. Mol. Biol.* **51:** 379–391. *One of the first full descriptions of a restriction endonuclease.*

DNA cloning

Frischauf, A.-M., Lehrach, H., Poustka, A. and Murray, N. (1983) Lambda replacement vectors carrying polylinker sequences. *J. Mol. Biol.* **170:** 827–842.

Hohn, B. and Murray, K. (1977) Packaging recombinant DNA molecules into bacteriophage particles *in vitro. Proc. Natl Acad. Sci. USA* **74:** 3259–3263.

Vieira, J. and Messing, J. (1982) The pUC plasmids, an M13mp7-derived system for insertion mutagenesis and sequencing with synthetic universal primers. *Gene* **19:** 259–268.

High-capacity cloning vectors

Burke, D.T., Carle, G.F. and Olson, M.V. (1987) Cloning of large segments of exogenous DNA into yeast by means of artificial chromosome vectors. *Science* **236:** 806–812. *YACs.*

Ioannou, P.A., Amemiya, C.T., Garnes, J., Kroisel, P.M., Shizuya, H., Chen, C., Batzer, M.A. and de Jong, P.J. (1994) P1-derived vector for the propagation of large human DNA fragments. *Nat. Genet.* **6:** 84–89. *PACs.*

Kim, U.-J., Shizuya, H., de Jong, P.J., Birren, B. and Simon, M.I. (1992) Stable propagation of cosmid and human DNA inserts in an F factor based vector. *Nucleic Acids Res.* **20:** 1083–1085. *Fosmids.*

Monaco, A.P. and Larin, Z. (1994) YACs, BACs, PACs and MACs – artificial chromosomes as research tools. *Trends Biotechnol.* **12:** 280–286. *A good review of high-capacity cloning vectors.*

Shizuya, H., Birren, B., Kim, U.J., Mancino, V., Slepak, T., Tachiiri, Y. and Simon, M. (1992) Cloning and stable maintenance of 300-kilobase-pair fragments of human DNA in *Escherichia coli* using an F-factor-based vector. *Proc. Natl Acad. Sci. USA* **89:** 8794–8797. *The first description of a BAC.*

Sternberg, N. (1990) Bacteriophage P1 cloning system for the isolation, amplification, and recovery of DNA fragments as large as 100 kilobase pairs. *Proc. Natl Acad. Sci. USA* **87:** 103–107. *Bacteriophage P1 vectors.*

Cloning in plants and animals

Bevan, M. (1984) Binary *Agrobacterium* vectors for plant transformation. *Nucleic Acids Res.* **12:** 8711–8721.

Colosimo, A., Goncz, K.K., Holmes, A.R., Kunzelmann, K., Novelli, G., Malone, R.W., Bennett, M.J. and Gruenert, D.C. (2000) Transfer and expression of foreign genes in mammalian cells. *Biotechniques* **29:** 314–321.

Hansen, G. and Wright, M.S. (1999) Recent advances in the transformation of plants. *Trends Plant Sci.* **4:** 226–231.

Kost, T.A. and Condreay, J.P. (2002) Recombinant baculoviruses as mammalian cell gene-delivery vectors. *Trends Biotechnol.* **20:** 173–180.

PCR

Mullis, K.B. (1990) The unusual origins of the polymerase chain reaction. *Sci. Am.* **262 (4):** 56–65.

Saiki, R.K., Gelfand, D.H., Stoffel, S., Scharf, S.J., Higuchi, R., Horn, G.T., Mullis, K.B. and Erlich, H.A. (1988) Primer-directed enzymatic amplification of DNA with a thermostable DNA polymerase. *Science* **239:** 487–491.

Mapping Genomes

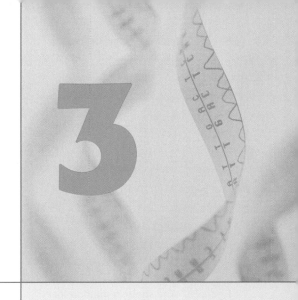

When you have read Chapter 3, you should be able to:

Explain why a map is an important aid to genome sequencing.

Distinguish between the terms "genetic map" and "physical map."

Describe the different types of marker used to construct genetic maps, and state how each type of marker is scored.

Summarize the principles of inheritance as discovered by Mendel, and show how subsequent genetic research led to the development of linkage analysis.

Explain how linkage analysis is used to construct genetic maps, giving details of how the analysis is carried out in various types of organism, including humans and bacteria.

State the limitations of genetic mapping.

Evaluate the strengths and weaknesses of the various methods used to construct physical maps of genomes.

Describe how restriction mapping is carried out.

Describe how fluorescent *in situ* hybridization (FISH) is used to construct a physical map, including the modifications used to increase the sensitivity of this technique.

Explain the basis of sequence tagged site (STS) mapping, and list the various DNA sequences that can be used as STSs.

Describe how radiation hybrids and clone libraries are used in STS mapping.

The next two chapters describe the techniques and strategies used to obtain genome sequences. DNA sequencing is obviously paramount among these techniques, but sequencing has one major limitation: even with the most sophisticated technology it is rarely possible to obtain a sequence of more than about 750 base pairs (bp) in a single experiment. This means that the sequence of a long DNA molecule has to be constructed from a series of shorter sequences. This is done by breaking the molecule into fragments, determining

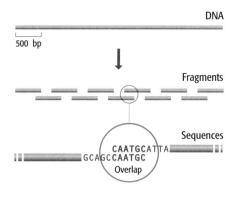

DNA

500 bp

Fragments

Sequences

CAATGCATTA
GCAGCCAATGC
Overlap

Figure 3.1 The shotgun method for sequence assembly. The DNA molecule is broken into small fragments, each of which is sequenced. The master sequence is assembled by searching for overlaps between the sequences of individual fragments. In practice, an overlap of several tens of base pairs would be needed to establish that two sequences should be linked together.

the sequence of each one, and using a computer to search for overlaps and build up the master sequence (Figure 3.1). This **shotgun method** is the standard approach for sequencing small prokaryotic genomes (Section 4.2.1), but is much more difficult with larger genomes because the required data analysis becomes disproportionately more complex as the number of fragments increases (for n fragments, the number of possible overlaps is given by $2n^2 - 2n$). A second problem with the shotgun method is that it can lead to errors when repetitive regions of a genome are analyzed. When a repetitive sequence is broken into fragments, many of the resulting pieces contain the same, or very similar, sequence motifs. It would be very easy to reassemble these sequences so that a portion of a repetitive region is left out, or even to connect together two quite separate pieces of the same or different chromosomes (Figure 3.2).

The difficulties in applying the shotgun method to a large molecule that has a significant repetitive DNA content means that this approach cannot be used on its own to sequence a eukaryotic genome. Instead, a genome **map** must first be generated. A genome map provides a guide for the sequencing experiments by showing the positions of genes and other distinctive features. Once a genome map is available, the sequencing phase of the project can proceed in either of two ways (Figure 3.3):

- By the **whole-genome shotgun method** (Section 4.2.3), which takes the same approach as the standard shotgun procedure but uses the distinctive features on the genome map as landmarks to aid assembly of the master sequence from the huge numbers of short sequences that are obtained. Reference to the map also ensures that regions containing repetitive DNA are assembled correctly. The whole-genome shotgun method is a rapid way of obtaining a draft of a eukaryotic genome sequence.

- By the **clone contig method** (Section 4.2.2). In this method the genome is broken into manageable segments, each a few hundred kilobases or a few megabases in length, which are short enough to be sequenced accurately by the shotgun method. Once the sequence of a segment has been

Figure 3.2 Problems with the shotgun method. (A) The DNA molecule contains a tandemly repeated element made up of many copies of the sequence GATTA. When the sequences are examined, an overlap is identified between two fragments, but these are from either end of the tandem repeat. If the error is not recognized then the internal region of the tandem repeat will be omitted from the master sequence. (B) In the second example, the DNA molecule contains two copies of a repeat element. When the sequences are examined, two fragments appear to overlap, but one fragment contains the left-hand part of one repeat and the other fragment has the right-hand part of the second repeat. In this case, failure to recognize the error would lead to the segment of DNA between the two repeats being left out of the master sequence. If the two repeats were on different chromosomes, then the sequences of these chromosomes might mistakenly be linked together.

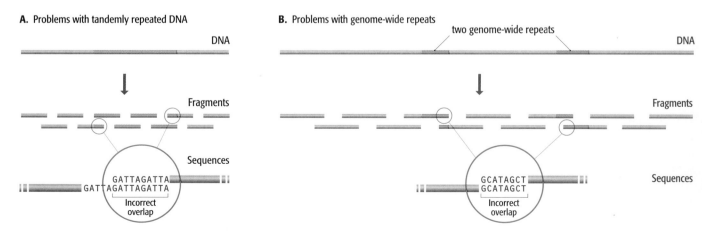

A. Problems with tandemly repeated DNA

DNA

Fragments

Sequences

GATTAGATTA
GATTAGATTAGATTA
Incorrect overlap

B. Problems with genome-wide repeats

two genome-wide repeats

DNA

Fragments

GCATAGCT
GCATAGCT
Incorrect overlap

Sequences

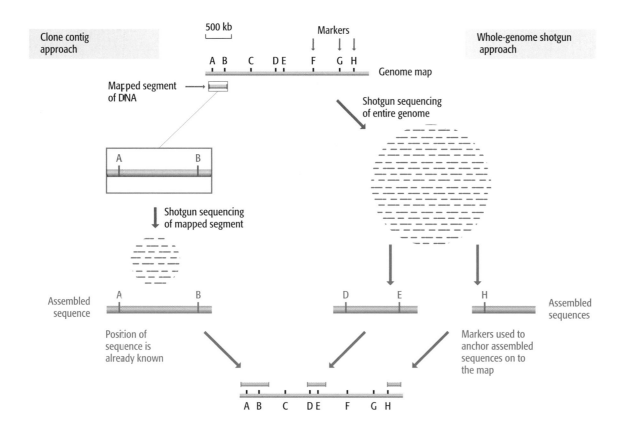

500 kb

Markers

A B   C   D E   F   G H

Genome map

Mapped segment
of DNA

Shotgun sequencing
of entire genome

A   B

Shotgun sequencing
of mapped segment

Assembled
sequence

A   B   D   E   H   Assembled
sequences

Position of
sequence is
already known

Markers used to
anchor assembled
sequences on to
the map

A B   C   D E   F   G H

completed, it is positioned at its correct location on the map. This step-by-step approach takes longer than whole-genome shotgun sequencing, but produces a more accurate and error-free sequence.

With both methods, the map provides the framework for carrying out the sequencing phase of the project. If the map indicates the positions of genes, then it can also be used to direct the initial part of a clone contig project to the interesting regions of a genome, so that the sequences of important genes are obtained as quickly as possible.

3.1 Genetic and Physical Maps

The convention is to divide genome mapping methods into two categories.

- **Genetic mapping** is based on the use of genetic techniques to construct maps showing the positions of genes and other sequence features on a genome. Genetic techniques include cross-breeding experiments or, in the case of humans, the examination of family histories (pedigrees). Genetic mapping is described in Section 3.2.

- **Physical mapping** uses molecular biology techniques to examine DNA molecules directly in order to construct maps showing the positions of sequence features, including genes. Physical mapping is described in Section 3.3.

Figure 3.3 Alternative approaches to genome sequencing. A genome consisting of a linear DNA molecule of size 2.5 Mb has been mapped, and the positions of eight markers (A–H) are known. On the left, the clone contig method starts with a segment of DNA whose position on the genome map has been identified because it contains markers A and B. The segment is sequenced by the shotgun method and the master sequence placed at its known position on the map. On the right, the whole-genome shotgun method involves random sequencing of the entire genome. This results in pieces of contiguous sequence, possibly hundreds of kilobases in length. If a contiguous sequence contains a marker, then it can be positioned on the genome map. Note that with either method, the more markers there are on the genome map the better. For more details of these sequencing strategies, see Section 4.2.

3.2 Genetic Mapping

As with any type of map, a genetic map must show the positions of distinctive features. In a geographic map these **markers** are recognizable components of the landscape, such as rivers, roads and buildings. What markers can we use in a genetic landscape?

3.2.1 Genes were the first markers to be used

The first genetic maps, constructed in the early decades of the twentieth century for organisms such as the fruit fly, used genes as markers. To be useful in genetic analysis, a gene must exist in at least two forms, or **alleles**, each specifying a different phenotype, an example being tall or short stems in the pea plants originally studied by Gregor Mendel. To begin with, the only genes that could be studied were those specifying phenotypes that were distinguishable by visual examination. So, for example, the first fruit-fly maps showed the positions of genes for body color, eye color, wing shape, and suchlike, all of these phenotypes being visible simply by looking at the flies with a low-power microscope or the naked eye. This approach was fine in the early days but geneticists soon realized that there were only a limited number of visual phenotypes whose inheritance could be studied, and in many cases their analysis was complicated because a single phenotype could be affected by more than one gene. For example, by 1922, over 50 genes had been mapped onto the four fruit-fly chromosomes, but nine of these genes were for eye color. In later research, geneticists studying fruit flies had to learn to distinguish between fly eyes that were colored red, light red, vermilion, garnet, carnation, cinnabar, ruby, sepia, scarlet, pink, cardinal, claret, purple, or brown. To make gene maps more comprehensive, it was necessary to find characteristics that were more distinctive and less complex than visual ones.

The answer was to use biochemistry to distinguish phenotypes. This has been particularly important with two types of organisms—microbes and humans. Microbes, such as bacteria and yeast, have very few visual characteristics, so gene mapping with these organisms has to rely on biochemical phenotypes such as those listed in Table 3.1. For humans it is possible to use visual characteristics but, since the 1920s, studies of human genetic variation have been based largely on biochemical phenotypes that can be scored by blood typing. These phenotypes include not only the standard blood groups, such as the ABO series, but also variants of blood serum proteins and of immunological proteins such as the human leukocyte antigens (the HLA system). A big advantage of these markers is that many of the relevant genes have **multiple alleles**. For example, the gene called *HLA-DRB1* has at least 290 alleles and *HLA-B* has

Table 3.1 Typical biochemical markers used for genetic analysis of *Saccharomyces cerevisiae*

Marker	Phenotype	Method by which cells carrying the marker are identified
ADE2	Requires adenine	Grows only when adenine is present in the medium
CAN1	Resistant to canavanine	Grows in the presence of canavanine
CUP1	Resistant to copper	Grows in the presence of copper
CYH1	Resistant to cycloheximide	Grows in the presence of cycloheximide
LEU2	Requires leucine	Grows only when leucine is present in the medium
SUC2	Able to ferment sucrose	Grows if sucrose is the only carbohydrate in the medium
URA3	Requires uracil	Grows only when uracil is present in the medium

over 400. This is relevant because of the way in which gene mapping is carried out with humans (Section 3.2.4). Rather than setting up planned breeding experiments, which is the procedure with experimental organisms such as fruit flies or mice, data on inheritance of human genes have to be gleaned by examining the phenotypes displayed by members of families in which marriages have been arranged for personal reasons rather than for the convenience of a geneticist. If all the members of a family have the same allele for the gene being studied, then no useful information can be obtained. For gene mapping purposes it is therefore necessary to find families in which marriages have occurred, by chance, between individuals with different alleles. This is much more likely if the gene being studied has 290 rather than 2 alleles.

3.2.2 DNA markers for genetic mapping

Genes are very useful markers but they are by no means ideal. One problem, especially with larger genomes such as those of vertebrates and flowering plants, is that a map based entirely on genes is not very detailed. This would be true even if every gene could be mapped because in most eukaryotic genomes the genes are widely spaced out with large gaps between them (see Figure 7.12. The problem is made worse by the fact that only a fraction of the total number of genes exist in allelic forms that can be distinguished conveniently. Gene maps are therefore not very comprehensive. We need other types of marker.

Mapped features that are not genes are called **DNA markers**. As with gene markers, a DNA marker must have at least two alleles to be useful. There are three types of DNA sequence feature that satisfy this requirement: **restriction fragment length polymorphisms (RFLPs)**, **simple sequence length polymorphisms (SSLPs)**, and **single nucleotide polymorphisms (SNPs)**.

Restriction fragment length polymorphisms

RFLPs were the first type of DNA marker to be studied. Recall that restriction enzymes cut DNA molecules at specific recognition sequences (Section 2.1.2). This sequence specificity means that treatment of a DNA molecule with a restriction enzyme should always produce the same set of fragments. This is not always the case with genomic DNA molecules because some restriction sites are polymorphic, existing as two alleles, with one allele displaying the correct sequence for the restriction site and therefore being cut when the DNA is treated with the enzyme, and the second allele having a sequence alteration so the restriction site is no longer recognized. The result of the sequence alteration is that the two adjacent restriction fragments remain linked together after treatment with the enzyme, leading to a length polymorphism (Figure 3.4). This is an RFLP, and its position on a genome map can be worked out by following the inheritance of its alleles, just as is done when genes are used as markers. There are thought to be about 10^5 RFLPs in a mammalian genome.

In order to score an RFLP, it is necessary to determine the size of just one or two individual restriction fragments against a background of many irrelevant fragments. This is not a trivial problem: an enzyme such as *Eco*RI, with a six-nucleotide recognition sequence, should cut approximately once every $4^6 = 4096$ bp and so would give almost 800,000 fragments when used with human DNA. After separation by agarose gel electrophoresis, these 800,000 fragments produce a smear of DNA and the RFLP cannot be distinguished.

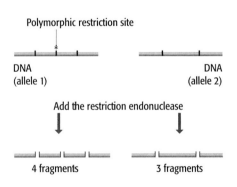

Figure 3.4 A restriction fragment length polymorphism (RFLP). The DNA molecule on the left has a polymorphic restriction site (marked with the asterisk) that is not present in the molecule on the right. The RFLP is revealed after treatment with the restriction enzyme because one of the molecules is cut into four fragments whereas the other is cut into three fragments.

Figure 3.5 Two methods for scoring an RFLP. (A) RFLPs can be scored by Southern hybridization. The DNA is digested with the appropriate restriction enzyme and separated in an agarose gel. The smear of restriction fragments is transferred to a nylon membrane and probed with a piece of DNA that spans the polymorphic restriction site. If the site is absent then a single restriction fragment is detected (lane 2); if the site is present then two fragments are detected (lane 3). (B) The RFLP can also be typed by PCR, using primers that anneal either side of the polymorphic restriction site. After the PCR, the products are treated with the appropriate restriction enzyme and then analyzed by agarose gel electrophoresis. If the site is absent then one band is seen on the agarose gel (lane 2); if the site is present then two bands are seen (lane 3).

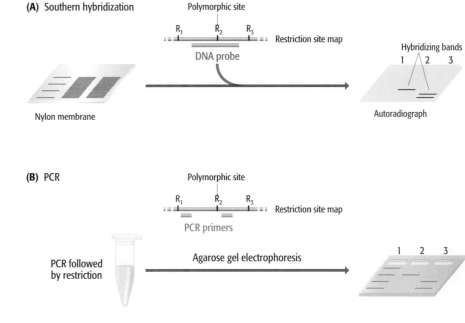

Southern hybridization, using a probe that spans the polymorphic restriction site, provides one way of visualizing the RFLP (Figure 3.5A), but nowadays PCR is more frequently used. The primers for the PCR are designed so that they anneal either side of the polymorphic site, and the RFLP is typed by treating the amplified fragment with the restriction enzyme and then running a sample in an agarose gel (Figure 3.5B).

Simple sequence length polymorphisms

SSLPs are arrays of repeat sequences that display length variations, different alleles containing different numbers of repeat units (Figure 3.6A). Unlike RFLPs, SSLPs can be multiallelic as each SSLP can have a number of different length variants. There are two types of SSLP:

- **Minisatellites**, also known as **variable number of tandem repeats** (**VNTRs**), in which the repeat unit is up to 25 bp in length.

- **Microsatellites,** or **simple tandem repeats** (**STRs**), whose repeats are shorter, usually 13 bp or less.

Microsatellites are more popular than minisatellites as DNA markers, for two reasons. First, minisatellites are not spread evenly around the genome but tend to be found more frequently in the telomeric regions at the ends of chromosomes. In geographic terms, this is equivalent to trying to use a map of lighthouses to find one's way around the middle of an island. Microsatellites

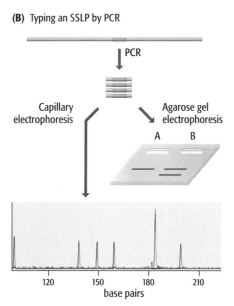

Figure 3.6 STRs and how they are typed. (A) Two alleles of a short tandem repeat, also called a microsatellite. In allele 1 the motif "GA" is repeated three times, and in allele 2 it is repeated five times. (B) Typing an STR by PCR. The STR and part of the surrounding sequence is amplified and the size of the product determined by agarose gel electrophoresis or capillary electrophoresis. In the agarose gel, lane A contains the PCR product and lane B contains DNA markers that show the sizes of the bands given after PCR of the two alleles. The band in lane A is the same size as the larger of the two DNA markers, showing that the DNA that was tested contained allele 2. The results of capillary electrophoresis are displayed as an electrophoretogram, the position of the peak indicating the size of the PCR product. The electrophoretogram is automatically calibrated against size markers so the precise length of the PCR product can be calculated. Image courtesy of Susan Thaw.

are more conveniently spaced throughout the genome. Second, the quickest way to type a length polymorphism is by PCR, but PCR typing is much quicker and more accurate with sequences less than 300 bp in length. Most minisatellite alleles are longer than this because the repeat units are relatively large and there tend to be many of them in a single array, so PCR products several kilobases in length are needed to type them. Microsatellites used as DNA markers typically consist of 10–30 copies of a repeat that is no longer than 6 bp in length, and so are much more amenable to analysis by PCR. There are 5×10^5 microsatellites with repeat units of 6 bp of less in the human genome.

When examined by PCR, the allele present at an STR is revealed by the precise length of the PCR product (Figure 3.6B). The length variations can be visualized by agarose gel electrophoresis, but standard gel electrophoresis is a cumbersome procedure that is difficult to automate, and is hence unsuitable for the high-throughput analyses that are demanded by modern genome research. Instead, STRs are usually typed by **capillary electrophoresis** in a polyacrylamide gel (see Technical Note 4.1). Most capillary electrophoresis systems use fluorescence detection, so a fluorescent label is attached to one or both of the primers before the PCR is carried out (Technical Note 2.1). After PCR, the product is loaded into the capillary system and run past a fluorescence detector. A computer attached to the detector correlates the time of passage of the PCR product with equivalent data for a set of size markers, and hence identifies the precise length of the product.

Single nucleotide polymorphisms

These are positions in a genome where some individuals have one nucleotide (e.g., a G) and others have a different nucleotide (e.g., a C) (Figure 3.7). There are vast numbers of SNPs in every genome (over four million in the human genome), some of which also give rise to RFLPs, but many of which do not because the sequence in which they lie is not recognized by any restriction enzyme.

Any one of the four nucleotides could be present at any one position in the genome, so it might be imagined that each SNP should have four alleles. Theoretically this is possible but in practice most SNPs exist as just two variants. This is because each SNP originates when a **point mutation** (Section 16.1) occurs in a genome, converting one nucleotide into another. If the mutation is in the reproductive cells of an individual, then one or more of that individual's offspring might inherit the mutation and, after many generations, the SNP may eventually become established in the population. But there are just two alleles—the original sequence and the mutated version. For a third allele to arise, a new mutation must occur at the same position in the genome in another individual, and this individual and his or her offspring must reproduce in such a way that the new allele becomes established. This scenario is not impossible but it is unlikely: consequently the vast majority of SNPs are biallelic. This disadvantage is more than outweighed by the huge number of SNPs present in each genome—in most eukaryotes, at least one for every 10 kb of DNA. SNPs therefore enable very detailed genome maps to be constructed.

The importance that SNPs have acquired in genome research has stimulated the development of rapid methods for their typing. Several of these

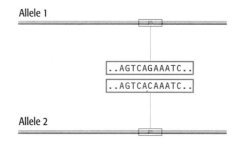

Figure 3.7 A single nucleotide polymorphism (SNP).

methods are based on **oligonucleotide hybridization analysis**. An oligonucleotide is a short, single-stranded DNA molecule, usually less than 50 nucleotides in length, that is synthesized in the test tube. If the conditions are just right, then an oligonucleotide will hybridize with another DNA molecule only if the oligonucleotide forms a completely base-paired structure with the second molecule. If there is a single mismatch—a single position within the oligonucleotide that does not form a base pair—then hybridization does not occur (Figure 3.8A). Oligonucleotide hybridization can therefore discriminate between the two alleles of an SNP. Various screening strategies based on oligonucleotide hybridization have been devised, including the following:

- **DNA chip** technology (Technical Note 3.1) makes use of a wafer of glass or silicon, 2.0 cm² or less in area, carrying many different oligonucleotides in a high-density array. The DNA to be tested is labeled with a fluorescent marker and pipetted onto the surface of the chip. Hybridization is detected by examining the chip with a fluorescence microscope, the positions at which the fluorescent signal is emitted indicating which oligonucleotides have hybridized with the test DNA. Many SNPs can therefore be scored in a single experiment.

- **Solution hybridization techniques** are carried out in the wells of a microtiter tray, each well containing a different oligonucleotide, using a detection system that can discriminate between unhybridized, single-stranded DNA and the double-stranded product that results when an oligonucleotide hybridizes to the test DNA. Several systems have been developed, one of which makes use of a pair of labels comprising a fluorescent dye and a compound that quenches the fluorescent signal when brought into close proximity with the dye. The dye is attached to one end of an oligonucleotide and the quenching compound to the other end. Normally there is no fluorescence because the oligonucleotide is designed in such a way that the two ends base-pair to one another, placing the quencher next to the dye (Figure 3.8B). Hybridization between oligonucleotide and test DNA disrupts this base pairing, moving the quencher away from the dye and enabling the fluorescent signal to be generated.

Other typing methods make use of an oligonucleotide whose mismatch with the SNP occurs at its extreme 5′ or 3′ end. Under the appropriate conditions, an oligonucleotide of this type will hybridize to the mismatched template

(A) Oligonucleotide hybridization is very specific

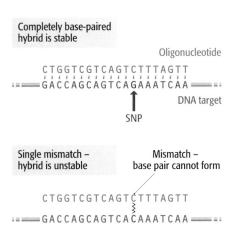

(B) Detecting hybridization by dye-quenching

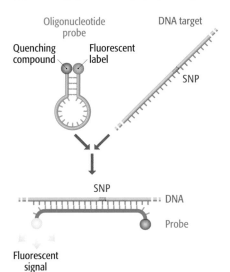

Figure 3.8 SNP typing by oligonucleotide hybridization analysis. (A) Under highly stringent hybridization conditions, a stable hybrid occurs only if the oligonucleotide is able to form a completely base-paired structure with the target DNA. If there is a single mismatch then the hybrid does not form. To achieve this level of stringency, the incubation temperature must be just below the **melting temperature**, or **T_m**, of the oligonucleotide. At temperatures above the T_m, even the fully base-paired hybrid is unstable. At more than 5°C below the T_m, mismatched hybrids might be stable. The T_m for the oligonucleotide shown in the figure would be about 58°C. The T_m in °C is calculated from the formula $T_m = (4 \times$ number of G and C nucleotides) + (2 × number of A and T nucleotides). This formula gives a rough indication of the T_m for oligonucleotides of 15–30 nucleotides in length. (B) One way of typing an SNP by solution hybridization. The oligonucleotide probe has two end-labels. One of these is a fluorescent dye and the other is a quenching compound. The two ends of the oligonucleotide base-pair to one another, so the fluorescent signal is quenched. When the probe hybridizes to its target DNA, the ends of the molecule become separated, enabling the fluorescent dye to emit its signal. The two labels are called "molecular beacons."

Technical Note 3.1 DNA microarrays and chips

High-density arrays of DNA molecules for parallel hybridization analyses

DNA microarrays and chips are designed to allow many hybridization experiments to be performed in parallel. Their main applications have been in the screening of polymorphisms such as SNPs (Section 3.2.2) and comparing the RNA populations of different cells (Section 6.1.2).

Although the terminology is inexact, microarrays and chips are, strictly speaking, two distinct types of matrix. In both architectures, a large number of DNA probes, each one with a different sequence, are immobilized at defined positions on a solid surface. The probes can be synthetic oligonucleotides or other short DNA molecules such as cDNAs or PCR products. These can be spotted onto a glass microscope slide or a piece of nylon membrane to form a **microarray**. With this approach, only a relatively low density can be achieved—typically 6400 spots as an 80 × 80 array in an area of 18 × 18 mm, which is sufficient for examining RNA populations but less applicable to the high-throughput analyses needed to type SNPs.

To prepare really high-density arrays, oligonucleotides are synthesized *in situ* on the surface of a wafer of glass or silicon, resulting in a **DNA chip**. The normal method for oligonucleotide synthesis involves adding nucleotides one-by-one to the growing end of an oligonucleotide, the sequence determined by the order in which the nucleotide substrates are added to the reaction mixture. If used for synthesis on a chip, this method would result in every oligonucleotide having the same sequence. Instead, modified nucleotide substrates are used—ones that have to be light-activated before they will attach to the end of a growing oligonucleotide. The nucleotides are added one after

another to the chip surface, **photolithography** being used to direct pulses of light onto individual positions in the array and hence to determine which of the growing oligonucleotides will be extended by addition of the particular nucleotide added at each step (Figure T3.1). A density of up to 300,000 oligonucleotides per cm^2 is possible, so if used for SNP screening, 150,000 polymorphisms can be typed in a single experiment, presuming there are oligonucleotides for both alleles of each SNP.

Chips and microarrays are not complicated to use. The chip or array is incubated with labeled target DNA to allow hybridization to take place. The positions at which hybridization to the target DNA occurs are determined by scanning the surface and recording the points at which the signal emitted by the label is detected. Radioactive labels can be used, signals being detected electronically by **phosphorimaging**, but this provides only low resolution and is not suitable for high-density chips. Higher resolution can be achieved with fluorescent labeling and detection by laser scanning or fluorescence confocal microscopy (Figure T3.2).

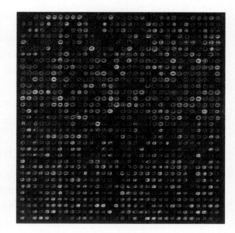

Figure T3.2 Visualization of the hybridization of a fluorescently labeled probe to a microarray. The label has been detected by confocal laser scanning and the signal intensity converted into a pseudocolor spectrum, with red indicating the greatest hybridization, followed by orange, yellow, green, blue, indigo, and violet, the last representing the background level of hybridization. Each spot on the microarray is a different cDNA clone prepared from human blood cell mRNA, and the probe was a cDNA from human bone marrow mRNA. For more information on the use of DNA chips and microarrays to study mRNA populations, see Section 6.1.2. Image courtesy of Tom Strachan, reprinted with permission from *Nature*.

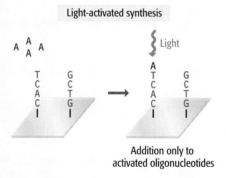

Figure T3.1 Oligonucleotide synthesis on the surface of a DNA chip.

(A) Hybridization with an oligonucleotide with a terminal mismatch

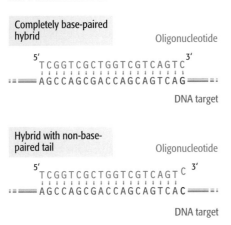

(B) Oligonucleotide ligation assay

(C) The ARMS test

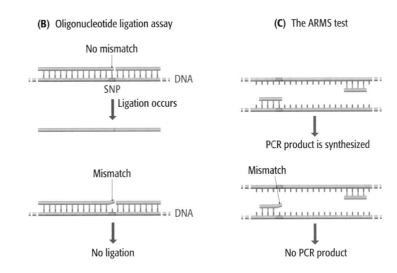

Figure 3.9 Methods for typing SNPs. (A) Under the appropriate conditions, an oligonucleotide whose mismatch with the SNP occurs at its extreme 5′ or 3′ end will hybridize to the mismatched template DNA with a short, non-base-paired "tail." (B) SNP typing by the oligonucleotide ligation assay. (C) The ARMS test.

DNA with a short, non-base-paired "tail" (Figure 3.9A). This feature is utilized in two different ways:

- The **oligonucleotide ligation assay** (**OLA**) makes use of two oligonucleotides that anneal adjacent to one another, with the 3′ end of one of these oligonucleotides positioned exactly at the SNP. This oligonucleotide will form a completely base-paired structure if one version of the SNP is present in the template DNA, and when this occurs the oligonucleotide can be ligated to its partner (Figure 3.9B). If the DNA being examined contains the other allele of the SNP, then the 3′ nucleotide of the test oligonucleotide will not anneal to the template and no ligation occurs. The allele is therefore typed by determining if the ligation product is synthesized, usually by running the postreaction mixture in a capillary electrophoresis system, as described above for STR typing.

- The **amplification refractory mutation system**, or **ARMS test**, is based on the same principle as OLA, but in this method the test oligonucleotide is one of a pair of PCR primers. If the test primer anneals to the SNP then it can be extended by *Taq* polymerase and the PCR can take place, but if it does not anneal, because the alternative version of the SNP is present, then no PCR product is generated (Figure 3.9C).

3.2.3 Linkage analysis is the basis of genetic mapping

Now that we have assembled a set of markers with which to construct a genetic map, we can move on to look at the mapping techniques themselves. These techniques are all based on **genetic linkage**, which in turn derives from the seminal discoveries in genetics made in the mid-nineteenth century by Gregor Mendel.

The principles of inheritance and the discovery of linkage

Genetic mapping is based on the principles of inheritance as first described by Gregor Mendel in 1865. From the results of his breeding experiments with

(A) Self-fertilization of pure-breeding pea plants

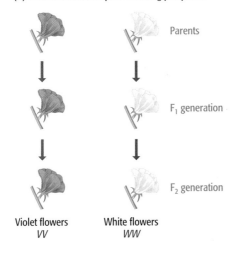

(B) Cross-fertilization of two pure-breeding types

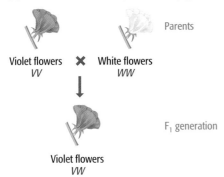

Figure 3.10 Homozygosity and heterozygosity. Mendel studied seven pairs of contrasting characteristics in his pea plants, one of which was violet and white flower color, as shown here. (A) Pure-breeding plants always give rise to flowers with the parental color. These plants are homozygotes, each possessing a pair of identical alleles, denoted here by *VV* for violet flowers and *WW* for white flowers. (B) When two pure-breeding plants are crossed, only one of the phenotypes is seen in the F₁ generation. Mendel deduced that the genotype of the F₁ plants was *VW*, so *V* is the dominant allele and *W* is the recessive allele.

peas, Mendel concluded that each pea plant possesses two alleles for each gene, but displays only one phenotype. This is easy to understand if the plant is pure-breeding, or **homozygous**, for a particular characteristic, as it then possesses two identical alleles and displays the appropriate phenotype (Figure 3.10A). However, Mendel showed that if two pure-breeding plants with different phenotypes are crossed then all the progeny (the F_1 generation) display the same phenotype. These F_1 plants must be **heterozygous**, meaning that they possess two different alleles, one for each phenotype—one allele inherited from the mother and one from the father. Mendel postulated that in this heterozygous condition one allele overrides the effects of the other allele: he therefore described the phenotype expressed in the F_1 plants as being **dominant** over the second, **recessive** phenotype (Figure 3.10B).

Mendel's interpretation of the heterozygous condition is perfectly correct for the pairs of alleles that he studied, but we now appreciate that this simple dominant–recessive rule can be complicated by situations that he did not encounter. These include:

- **Incomplete dominance**, where the heterozygous form displays a phenotype intermediate between the two homozygous forms. Flower color in plants such as carnations (but not peas) is an example: when red carnations are crossed with white ones, the F_1 heterozygotes are neither red nor white, but pink (Figure 3.11A).

- **Codominance**, where the heterozygous form displays both of the homozygous phenotypes. Human blood groups provide several examples of codominance. For example, the two homozygous forms of the MN series are M and N, with these individuals synthesizing only the M or N blood glycoproteins, respectively. Heterozygotes, however, synthesize both glycoproteins and hence are designated MN (Figure 3.11B).

As well as discovering dominance and recessiveness, Mendel carried out additional experiments that enabled him to establish his two Laws of Genetics. The First Law states that **alleles segregate randomly**. In other words, if the parent's alleles are A and a, then a member of the F_1 generation has the same chance of inheriting A as it has of inheriting a. The Second Law is that **pairs of alleles segregate independently**, so that inheritance of the alleles of gene A is independent of inheritance of the alleles of gene B. Because of these laws, the outcomes of genetic crosses are predictable (Figure 3.12).

When Mendel's work was rediscovered in 1900, his Second Law worried the early geneticists because it was soon established that genes reside on chromosomes, and it was realized that all organisms have many more genes than chromosomes. Chromosomes are inherited as intact units, so it was reasoned

(A) Incomplete dominance

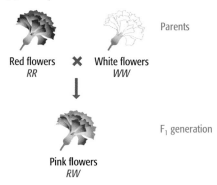

Red flowers   ✖   White flowers
RR   *WW*

Pink flowers
RW

Parents

F_1 generation

(B) Codominance

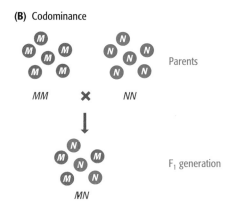

MM   ✖   NN

MN

Parents

F_1 generation

Figure 3.11 Two types of allele interaction not encountered by Mendel. (A) Incomplete dominance and (B) codominance.

Figure 3.12 Mendel's Laws enable the outcome of genetic crosses to be predicted. Two crosses are shown with their predicted outcomes. In a monohybrid cross, the alleles of a single gene are followed, in this case allele *T* for tall pea plants and allele *t* for short pea plants. *T* is dominant and *t* is recessive. The grid shows the predicted genotypes and phenotypes of the F_1 generation based on Mendel's First Law, which states that alleles segregate randomly. When Mendel carried out this cross he obtained 787 tall pea plants and 277 short plants, a ratio of 2.84:1. In the dihybrid cross, two genes are followed. The second gene determines the shape of the peas, the alleles being *R* (round, the dominant allele) and *r* (wrinkled, which is recessive). The genotypes and phenotypes shown are those predicted by Mendel's First and Second Laws, the latter stating that pairs of alleles segregate independently.

MONOHYBRID CROSS

Parents	Tall *Tt* ✕ *Tt* Tall	
F_1 genotypes	*T*	*t*
T	*TT*	*Tt*
t	*Tt*	*tt*
F_1 phenotypes	**3** tall : **1** short	

DIHYBRID CROSS

Parents	Tall round *TtRr* ✕ *TtRr* Tall round			
F_1 genotypes	*TR*	*Tr*	*tR*	*tr*
TR	*TTRR*	*TTRr*	*TtRR*	*TtRr*
Tr	*TTRr*	*TTrr*	*TtRr*	*Ttrr*
tR	*TtRR*	*TtRr*	*ttRR*	*ttRr*
tr	*TtRr*	*Ttrr*	*ttRr*	*ttrr*
F_1 phenotypes	**9** tall round : **3** tall wrinkled : **3** short round : **1** short wrinkled			

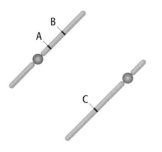

Figure 3.13 Genes on the same chromosome should display linkage. Genes A and B are on the same chromosome and so should be inherited together. Mendel's Second Law should therefore not apply to the inheritance of A and B. Gene C is on a different chromosome, so the Second Law will hold for the inheritance of A and C, or B and C. Mendel did not discover linkage because the seven genes that he studied were each on a different pea chromosome.

that the alleles of some pairs of genes will be inherited together because they are on the same chromosome (Figure 3.13). This is the principle of genetic linkage, and it was quickly shown to be correct, although the results did not turn out exactly as expected. The complete linkage that had been anticipated between many pairs of genes failed to materialize. Pairs of genes were either inherited independently, as expected for genes in different chromosomes, or if they showed linkage, then it was only **partial linkage**: sometimes they were inherited together and sometimes they were not (Figure 3.14). The resolution of this contradiction between theory and observation was the critical step in the development of genetic mapping techniques.

Partial linkage is explained by the behavior of chromosomes during meiosis

The critical breakthrough was achieved by Thomas Hunt Morgan, who made the conceptual leap between partial linkage and the behavior of chromosomes when the nucleus of a cell divides. Cytologists in the late-nineteenth century had distinguished two types of nuclear division: **mitosis** and **meiosis**. Mitosis is more common, being the process by which the diploid nucleus of a somatic cell divides to produce two daughter nuclei, both of which are diploid (Figure 3.15). Approximately 10^{17} mitoses are needed to produce all the cells required during a human lifetime. Before mitosis begins, each chromosome in the nucleus is replicated, but the resulting daughter chromosomes do not immediately break away from one another. To begin with they remain attached at their centromeres. The daughters do not separate until

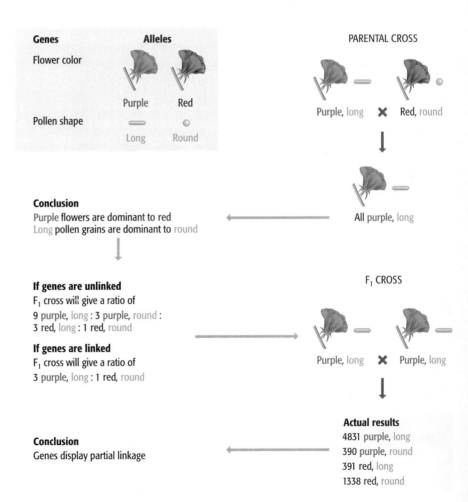

Figure 3.14 Partial linkage. Partial linkage was discovered in the early twentieth century. The cross shown here was carried out by Bateson, Saunders, and Punnett in 1905 with sweet peas. The parental cross gives the typical dihybrid result (see Figure 3.12), with all the F_1 plants displaying the same phenotype, indicating that the dominant alleles are purple flowers and long pollen grains. The F_1 cross gives unexpected results, as the progeny show neither a 9:3:3:1 ratio (expected for genes on different chromosomes) nor a 3:1 ratio (expected if the genes are completely linked). An unusual ratio is typical of partial linkage.

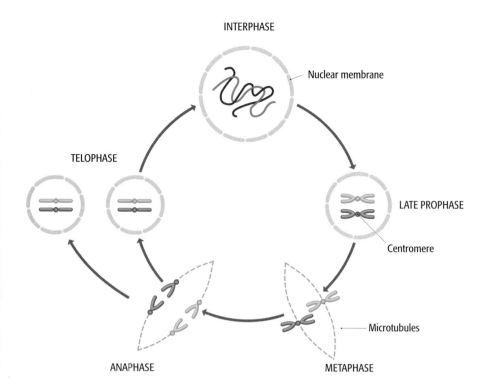

INTERPHASE

Nuclear membrane

TELOPHASE

LATE PROPHASE

Centromere

Microtubules

ANAPHASE

METAPHASE

Figure 3.15 Mitosis. During interphase (the period between nuclear divisions) the chromosomes are in their extended form (Section 7.1.1). At the start of mitosis the chromosomes condense, and by late prophase have formed structures that are visible with the light microscope. Each chromosome has already undergone DNA replication but the two daughter chromosomes are held together by the centromere. During metaphase the nuclear membrane breaks down (in most eukaryotes) and the chromosomes line up in the center of the cell. Microtubules now draw the daughter chromosomes towards either end of the cell. In telophase, nuclear membranes re-form around each collection of daughter chromosomes. The result is that the parent nucleus has given rise to two identical daughter nuclei. For simplicity, just one pair of homologous chromosomes is shown; one member of the pair is red, the other is blue.

later in mitosis when the chromosomes are distributed between the two new nuclei. Obviously it is important that each of the new nuclei receives a complete set of chromosomes, and most of the intricacies of mitosis appear to be devoted to achieving this end.

Mitosis illustrates the basic events occurring during nuclear division, but it is the distinctive features of meiosis that interest us. Meiosis occurs only in reproductive cells, and results in a diploid cell giving rise to four haploid **gametes**, each of which can subsequently fuse with a gamete of the opposite sex during sexual reproduction. The fact that meiosis results in four haploid cells whereas mitosis gives rise to two diploid cells is easy to explain: meiosis involves two nuclear divisions, one after the other, whereas mitosis is just a single nuclear division. This is an important distinction, but the critical difference between mitosis and meiosis is more subtle. Recall that in a diploid cell there are two, separate copies of each chromosome (Chapter 1). We refer to these as pairs of **homologous chromosomes**. During mitosis, homologous chromosomes remain separate from one another, each member of the pair replicating and being passed to a daughter nucleus independently of its homolog. In meiosis, however, the pairs of homologous chromosomes are by no means independent. During prophase I, each chromosome lines up with its homolog to form a **bivalent** (Figure 3.16). This occurs after each chromosome has replicated, but before the replicated structures split, so the bivalent in fact contains four chromosome copies, each of which is destined to find its way into one of the four gametes that will be produced at the end of the meiosis. Within the bivalent, the chromosome arms (the **chromatids**) can undergo physical breakage and exchange of segments of DNA. The process is called **crossing-over**, or **recombination**, and was discovered by the Belgian cytologist Janssens in 1909. This was just two years before Morgan started to think about partial linkage.

How did the discovery of crossing-over help Morgan explain partial linkage? To understand this we need to think about the effect that crossing-over can have

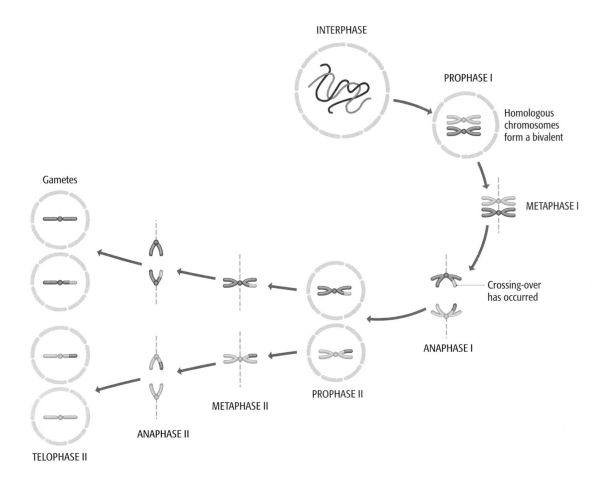

INTERPHASE

PROPHASE I

Homologous chromosomes form a bivalent

METAPHASE I

Crossing-over has occurred

ANAPHASE I

PROPHASE II

METAPHASE II

ANAPHASE II

TELOPHASE II

Gametes

Figure 3.16 Meiosis. The events involving one pair of homologous chromosomes are shown; one member of the pair is red, the other is blue. At the start of meiosis the chromosomes condense and each homologous pair lines up to form a bivalent. Within the bivalent, crossing-over might occur, involving breakage of chromosome arms and exchange of DNA. Meiosis then proceeds by a pair of mitotic nuclear divisions that result initially in two nuclei, each with two copies of each chromosome still attached at their centromeres, and finally in four nuclei, each with a single copy of each chromosome. These final products of meiosis, the gametes, are therefore haploid.

on the inheritance of genes. Let us consider two genes, each of which has two alleles. We will call the first gene A and its alleles *A* and *a*, and the second gene B with alleles *B* and *b*. Imagine that the two genes are located on chromosome number 2 of *Drosophila melanogaster*, the species of fruit fly studied by Morgan. We are going to follow the meiosis of a diploid nucleus in which one copy of chromosome 2 has alleles *A* and *B*, and the second has *a* and *b*. This situation is illustrated in Figure 3.17. Consider the two alternative scenarios:

- A crossover does not occur between genes A and B. If this is what happens, then two of the resulting gametes will contain chromosome copies with alleles *A* and *B*, and the other two will contain *a* and *b*. In other words, two of the gametes have the **genotype** *AB* and two have the genotype *ab*.

- A crossover does occur between genes A and B. This leads to segments of DNA containing gene B being exchanged between homologous chromosomes. The eventual result is that each gamete has a different genotype: one *AB*, one *aB*, one *Ab*, and one *ab*.

Now think about what would happen if we looked at the results of meiosis in one hundred identical cells. If crossovers never occur then the resulting gametes will have the following genotypes:

200 *AB*

200 *ab*

This is complete linkage: genes A and B behave as a single unit during meiosis. But if (as is more likely) crossovers occur between A and B in some of the nuclei, then the allele pairs will not be inherited as single units. Let us say that crossovers occur during 40 of the 100 meioses. The following gametes will result:

160 *AB*

160 *ab*

40 *Ab*

40 *aB*

The linkage is not complete, it is only partial. As well as the two **parental** genotypes (*AB*, *ab*) we see gametes with **recombinant** genotypes (*Ab*, *aB*).

From partial linkage to genetic mapping

Once Morgan had understood how partial linkage could be explained by crossing-over during meiosis, he was able to devise a way of mapping the relative positions of genes on a chromosome. In fact the most important work was done not by Morgan himself but by an undergraduate in his laboratory, Arthur Sturtevant. Sturtevant assumed that crossing-over was a random event, there being an equal chance of it occurring at any position along a pair of lined-up chromatids. If this assumption is correct, then two genes that are close together will be separated by crossovers less frequently than two genes that are more distant from one another. Furthermore, the frequency with which the genes are unlinked by crossovers will be directly proportional to how far apart they are on their chromosome. The **recombination frequency** is therefore a measure of the distance between two genes. If you work out the recombination frequencies for different pairs of genes, you can construct a map of their relative positions on the chromosome (Figure 3.18).

It turns out that Sturtevant's assumption about the randomness of crossovers was not entirely justified. Comparisons between genetic maps and the actual positions of genes on DNA molecules, as revealed by physical mapping and DNA sequencing, have shown that some regions of chromosomes, called **recombination hotspots**, are more likely to be involved in crossovers than others. This means that a genetic map distance does not necessarily indicate the physical distance between two markers (see Figure 3.25). Also, we now realize that a single chromatid can participate in more than one crossover at the same time, but that there are limitations on how close together these crossovers can be, leading to more inaccuracies in the mapping procedure. Despite these qualifications, **linkage analysis** usually makes correct deductions about gene order, and distance estimates are sufficiently accurate to generate genetic maps that are of value as frameworks for genome sequencing projects. We will therefore move on to consider how linkage analysis is carried out with different types of organism.

3.2.4 Linkage analysis with different types of organism

To see how linkage analysis is actually carried out, we need to consider three quite different situations:

- Linkage analysis with species such as fruit flies and mice, with which we can carry out planned breeding experiments.

- Linkage analysis with humans, with whom we cannot carry out planned experiments but instead can make use of family pedigrees.

- Linkage analysis with bacteria, which do not undergo meiosis.

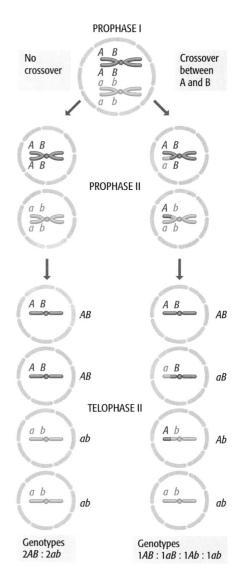

Figure 3.17 The effect of a crossover on linked genes. The drawing shows a pair of homologous chromosomes, one red and the other blue. A and B are linked genes with alleles *A*, *a*, *B*, and *b*. On the left is meiosis with no crossover between A and B: two of the resulting gametes have the genotype *AB* and the other two are *ab*. On the right, a crossover occurs between A and B: the four gametes display all of the possible genotypes—*AB*, *aB*, *Ab*, and *ab*.

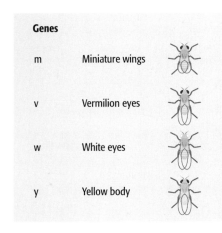

Genes

m	Miniature wings	
v	Vermilion eyes	
w	White eyes	
y	Yellow body	

Recombination frequencies

Between m and v	=	3.0%
Between m and y	=	33.7%
Between v and w	=	29.4%
Between w and y	=	1.3%

Deduced map positions

y	w		v	m
0	1.3		30.7	33.7

Figure 3.18 Working out a genetic map from recombination frequencies. The example is taken from the original experiments carried out with fruit flies by Arthur Sturtevant. All four genes are on the X chromosome of the fruit fly. Recombination frequencies between the genes are shown, along with their deduced map positions.

Linkage analysis when planned breeding experiments are possible

The first type of linkage analysis is the modern counterpart of the method developed by Morgan and his colleagues. The method is based on analysis of the progeny of experimental crosses set up between parents of known genotypes and is, at least in theory, applicable to all eukaryotes. Ethical considerations preclude this approach with humans, and practical problems such as the length of the gestation period and the time taken for the newborn to reach maturity (and hence to participate in subsequent crosses) limit the effectiveness of the method with some animals and plants.

If we return to Figure 3.17 we see that the key to gene mapping is being able to determine the genotypes of the gametes resulting from meiosis. In a few situations this is possible by directly examining the gametes. For example, the gametes produced by some microbial eukaryotes, including the yeast *Saccharomyces cerevisiae*, can be grown into colonies of haploid cells, whose genotypes can be determined by biochemical tests. Direct genotyping of gametes is also possible with higher eukaryotes if DNA markers are used, as PCR can be carried out with the DNA from individual spermatozoa, enabling RFLPs, SSLPs, and SNPs to be typed. Unfortunately, sperm typing is laborious. Routine linkage analysis with higher eukaryotes is therefore carried out not by examining the gametes directly but by determining the genotypes of the diploid progeny that result from fusion of two gametes, one from each of a pair of parents. In other words, a genetic cross is performed.

The complication with a genetic cross is that the resulting diploid progeny are the product not of one meiosis but of two (one in each parent), and in most organisms crossover events are equally likely to occur during production of the male and female gametes. Somehow we have to be able to disentangle from the genotypes of the diploid progeny the crossover events that occurred in each of these two meioses. This means that the cross has to be set up with care. The standard procedure is to use a **test cross**. This is illustrated in Figure 3.19, where we have set up a test cross to map the two genes we met earlier: gene A (alleles *A* and *a*) and gene B (alleles *B* and *b*), both on chromosome 2 of the fruit fly. The critical feature of a test cross is the genotypes of the two parents:

- One parent is a **double heterozygote**. This means that all four alleles are present in this parent: its genotype is *AB/ab*. This notation indicates that one pair of the homologous chromosomes has alleles *A* and *B*, and the other has *a* and *b*. Double heterozygotes can be obtained by crossing two pure-breeding strains, for example *AB/AB × ab/ab*.

- The second parent is a pure-breeding **double homozygote**. In this parent, both homologous copies of chromosome 2 are the same: in the example shown in Figure 3.19, both have alleles *a* and *b* and the genotype of the parent is *ab/ab*.

The double heterozygote has the same genotype as the cell whose meiosis we followed in Figure 3.17. Our objective is therefore to infer the genotypes of the gametes produced by this parent and to calculate the fraction that are recombinants. Note that all the gametes produced by the second parent (the double homozygote) will have the genotype *ab* regardless of whether they are parental or recombinant gametes. Alleles *a* and *b* are both recessive, so meiosis in this parent is, in effect, invisible when the phenotypes of the progeny are examined. This means that, as shown in Figure 3.19, the phenotypes of the diploid progeny can be unambiguously converted into the genotypes of the gametes from the double heterozygous parent. The test cross therefore enables us to make a direct examination of a single meiosis and hence to calculate a recombination frequency and map distance for the two genes being studied.

The power of this type of linkage analysis is enhanced if more than two markers are followed in a single cross. Not only does this generate recombination frequencies more quickly, it also enables the relative order of markers on a chromosome to be determined by simple inspection of the data. This is because two recombination events are required to unlink the central marker from the two outer markers in a series of three, whereas either of the two outer markers can be unlinked by just a single recombination (Figure 3.20). A double recombination is less likely than a single one, so unlinking of the central marker will occur relatively infrequently. A set of typical data from a three-point cross is shown in Table 3.2. A test cross has been set up between a triple heterozygote (*ABC/abc*) and a triple homozygote (*abc/abc*). The most frequent progeny are those with one of the two parental genotypes, resulting from an absence of recombination events in the region containing the markers A, B, and C. Two other classes of progeny are relatively frequent (51 and 63 progeny in the example shown). Both of these are presumed to arise from a single recombination. Inspection of their genotypes shows that in the first of these two classes, marker A has become unlinked from B and C, and in the second class, marker B has become unlinked from A and C. The implication is that A and B are the outer markers. This is confirmed by the number of progeny in which marker C has become unlinked from A and B. There are only two of these, showing that a double recombination is needed to produce this genotype. Marker C is therefore between A and B.

Just one additional point needs to be considered. If, as in Figure 3.19 and Table 3.2, genes whose alleles display dominance and recessiveness are examined in a test cross, then the double or triple homozygous parent must have alleles for the recessive phenotypes. If, on the other hand, codominant markers are used, then the double homozygous parent can have any combination of homozygous alleles (e.g., *AB/AB*, *Ab/Ab*, *aB/aB*, or *ab/ab*). Figure 3.21, which gives an example of this type of test cross, shows the reason for this. Note that DNA markers typed by PCR display what is, in effect, codominance: Figure 3.21 therefore shows a typical scenario encountered when linkage analysis is being carried out with DNA markers.

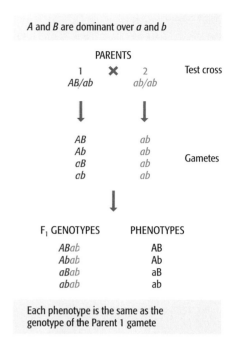

A and B are dominant over a and b

Figure 3.19 A test cross between alleles displaying dominance and recessiveness. A and B are markers with alleles *A*, *a*, *B*, and *b*. The resulting progeny are scored by examining their phenotypes. Because the double homozygous parent (Parent 2) has both recessive alleles—*a* and *b*—it effectively makes no contribution to the phenotypes of the progeny. The phenotype of each individual in the F_1 generation is therefore the same as the genotype of the gamete from Parent 1 that gave rise to that individual.

Table 3.2 Set of typical data from a three-point test cross

Genotypes of progeny	Number of progeny	Inferred recombination events
ABC/abc abc/abc	987	None (parental genotypes)
aBC/abc Abc/abc	51	One, between A and B/C
AbC/abc aBc/abc	63	One, between B and A/C
ABc/abc abC/abc	2	Two, one between C and A, and one between C and B

Single crossover

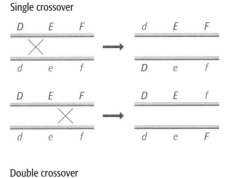

Double crossover

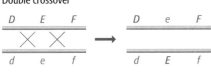

Figure 3.20 The effects of crossovers during a dihybrid cross. Either of the two outer markers can be unlinked by just a single recombination event, but two recombinations are required to unlink the central marker from the two outer markers.

A and B are codominant with a and b

PARENTS

1   ✕   2   Test cross
AB/ab   Ab/Ab

↓   ↓

AB   Ab
Ab   Ab
aB   Ab   Gametes
ab   Ab

↓

F₁ GENOTYPES   ALLELES DETECTED

ABAb   A+B+b
AbAb   A+b
aBAb   A+a+B+b
abAb   A+a+b

Genotypes of Parent 1 gametes are identified from the detected alleles. If only A is detected, then parent 1 gamete was A. If A + a is detected then parent 1 gamete was a, etc.

Figure 3.21 A test cross between alleles displaying codominance. A and B are markers whose allele pairs are codominant. In this particular example, the double homozygous parent has the genotype Ab/Ab. The alleles present in each F₁ individual are directly detected, for example by PCR. These allele combinations enable the genotype of the Parent 1 gamete that gave rise to each individual to be deduced.

Gene mapping by human pedigree analysis

With humans it is, of course, impossible to preselect the genotypes of parents and set up crosses designed specifically for mapping purposes. Instead, data for the calculation of recombination frequencies have to be obtained by examining the genotypes of the members of successive generations of existing families. This means that only limited data are available, and their interpretation is often difficult because a human marriage rarely results in a convenient test cross, and often the genotypes of one or more family members are unobtainable because those individuals are dead or unwilling to cooperate.

The problems are illustrated by Figure 3.22. In this example, we are studying a genetic disease present in a family of two parents and six children. Genetic diseases are frequently used as gene markers in humans, the disease state being one allele and the healthy state being a second allele. The pedigree in Figure 3.22A shows us that the mother is affected by the disease, as are four of her children. We know from family accounts that the maternal grandmother also suffered from this disease, but both she and her husband—the maternal grandfather—are now dead. We can include them in the pedigree, with slashes indicating that they are dead, but we cannot obtain any further information on their genotypes. We know that the disease gene is present on the same chromosome as a microsatellite, which we call M, four alleles of which—M_1, M_2, M_3, and M_4—are present in the living family members. Our aim is to map the position of the disease gene relative to the microsatellite.

To establish a recombination frequency between the disease gene and microsatellite M, we must determine how many of the children are recombinants. If we look at the genotypes of the six children, we see that numbers 1, 3, and 4 have the disease allele and the microsatellite allele M_1. Numbers 2 and 5 have the healthy allele and M_2. We can therefore construct two alternative hypotheses. The first is that the two copies of the relevant homologous chromosomes in the mother have the genotypes Disease–M_1 and Healthy–M_2; therefore, children 1, 2, 3, 4, and 5 have parental genotypes, and child 6 is the one and only recombinant (Figure 3.22B). This would suggest that the disease gene and the microsatellite are relatively closely linked and that crossovers between them occur infrequently. The alternative hypothesis is that the mother's chromosomes have the genotypes Healthy–M_1 and Disease–M_2; this would mean that children 1–5 are recombinants, and child 6 has the parental genotype. This would mean that the gene and microsatellite are relatively far apart on the chromosome. We cannot determine which of these hypotheses is correct: the data are frustratingly ambiguous.

The most satisfying solution to the problem posed by the pedigree in Figure 3.22 would be to know the genotype of the grandmother. Let us pretend that this is a soap-opera family and that the grandmother is not really dead. To everyone's surprise she reappears just in time to save the declining audience ratings. Her genotype for microsatellite M turns out to be M_1M_5 (Figure 3.22C). This tells us that the chromosome inherited by the mother has the genotype Disease–M_1. We can therefore conclude with certainty that Hypothesis 1 is correct and that only child 6 is a recombinant.

Resurrection of key individuals is not usually an option open to real-life geneticists, although DNA can be obtained from old pathology specimens such as slides and Guthrie cards. Imperfect pedigrees are analyzed statistically, using a

(A) The pedigree

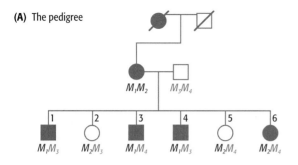

KEY

○ Unaffected female

● Affected female

□ Unaffected male

■ Affected male

╱ Dead

(B) Possible interpretations of the pedigree

MOTHER'S CHROMOSOMES

		Hypothesis 1	Hypothesis 2
		Disease M_1	*Healthy M_1*
		Healthy M_2	*Disease M_2*
Child 1	*Disease M_1*	Parental	Recombinant
Child 2	*Healthy M_2*	Parental	Recombinant
Child 3	*Disease M_1*	Parental	Recombinant
Child 4	*Disease M_1*	Parental	Recombinant
Child 5	*Healthy M_2*	Parental	Recombinant
Child 6	*Disease M_2*	Recombinant	Parental
	Recombination frequency	$1/6 = 16.7\%$	$5/6 = 83.3\%$

(C) Resurrection of the maternal grandmother

Disease allele must be linked to *M1*
HYPOTHESIS 1 IS CORRECT

Figure 3.22 An example of human pedigree analysis. (A) The pedigree shows inheritance of a genetic disease in a family of two living parents and six children, with information about the maternal grandparents available from family records. The disease allele (closed symbols) is dominant over the healthy allele (open symbols). The objective is to determine the degree of linkage between the disease gene and the microsatellite M by typing the alleles for this microsatellite (M_1, M_2, etc.) in living members of the family. (B) The pedigree can be interpreted in two different ways: Hypothesis 1 gives a low recombination frequency and indicates that the disease gene is tightly linked to microsatellite M. Hypothesis 2 suggests that the disease gene and microsatellite are much less closely linked. In (C), the issue is resolved by the reappearance of the maternal grandmother, whose microsatellite genotype is consistent only with Hypothesis 1.

measure called the **lod score**. This stands for logarithm of the odds that the genes are linked, and is used primarily to determine if the two markers being studied lie on the same chromosome—in other words if the genes are linked or not. If the lod analysis establishes linkage then it can also provide a measure of the most likely recombination frequency. Ideally the available data will derive from more than one pedigree, increasing the confidence in the result. The analysis is less ambiguous for families with larger numbers of children, and as we saw in Figure 3.22, it is important that the members of at least three generations can be genotyped. For this reason, family collections have been established, such as the one maintained by the Centre d'Études du Polymorphisme Humaine (CEPH) in Paris. The CEPH collection contains cultured cell lines from families in which all four grandparents as well as at least eight second-generation children could be sampled. This collection is available for DNA marker mapping by any researcher who agrees to submit the resulting data to the central CEPH database.

Genetic mapping in bacteria

The final type of genetic mapping that we must consider is the strategy used with bacteria. The main difficulty that geneticists faced when trying to develop

DONOR	RECIPIENT

(A) Conjugation

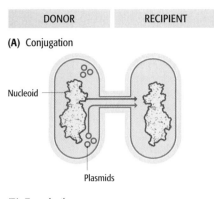

Nucleoid

Plasmids

(B) Transduction

Bacterial DNA

Bacteriophage

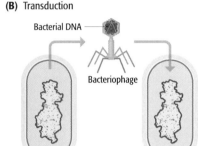

(C) Transformation

Naked DNA

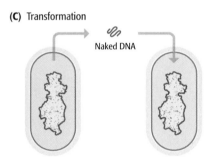

Figure 3.23 **Three ways of achieving DNA transfer between bacteria.** (A) Conjugation can result in transfer of chromosomal or plasmid DNA from the donor bacterium to the recipient. Conjugation involves physical contact between the two bacteria, with transfer thought to occur through a narrow tube called the **pilus**. (B) Transduction is the transfer of a small segment of the donor cell's DNA via a bacteriophage. (C) Transformation is similar to transduction but "naked" DNA is transferred. The events illustrated in (B) and (C) are often accompanied by death of the donor cell. In (B), death occurs when the bacteriophages emerge from the donor cell. In (C), release of DNA from the donor cell is usually a consequence of the cell's death through natural causes.

genetic mapping techniques for bacteria is that these organisms are normally haploid, and so do not undergo meiosis. Some other way therefore had to be devised to induce crossovers between homologous segments of bacterial DNA. The answer was to make use of three natural methods that exist for transferring pieces of DNA from one bacterium to another (Figure 3.23):

- In **conjugation**, two bacteria come into physical contact and one bacterium (the donor) transfers DNA to the second bacterium (the recipient). The transferred DNA can be a copy of some or possibly all of the donor cell's chromosome, or it could be a segment of chromosomal DNA—up to 1 Mb in length—integrated in a plasmid. The latter is called **episome transfer**.

- **Transduction** involves transfer of a small segment of DNA—up to 50 kb or so—from donor to recipient via a bacteriophage.

- In **transformation**, the recipient cell takes up from its environment a fragment of DNA, rarely longer than 50 kb, released from a donor cell.

Biochemical markers are invariably used, the dominant or **wild-type** phenotype being possession of a biochemical characteristic (e.g., ability to synthesize tryptophan) and the recessive phenotype being the complementary characteristic (e.g., inability to synthesize tryptophan). The gene transfer is usually set up between a donor strain that possesses the wild-type allele and a recipient with the recessive allele, transfer into the recipient strain being monitored by looking for acquisition of the biochemical function specified by the gene being studied. This is illustrated in Figure 3.24A, where we see a functional gene for tryptophan biosynthesis being transferred from a wild-type bacterium (genotype described as trp^+) to a recipient that lacks a functional copy of this gene (trp^-). The recipient is called a tryptophan **auxotroph** (the word used to describe a mutant bacterium that can survive only if provided with a nutrient—in this case, tryptophan—not required by the wild type; see Section 16.1.2). After transfer, two crossovers are needed to integrate the transferred gene into the recipient cell's chromosome, converting the recipient from trp^- to trp^+.

The precise details of the mapping procedure depend on the type of gene transfer that is being used. During conjugation, DNA is transferred from donor to recipient in the same way that a string is pulled through a tube. The relative positions of markers on the DNA molecule can therefore be mapped by determining the times at which the markers appear in the recipient cell. In the example shown in Figure 3.24B, markers A, B, and C are transferred 8, 20, and 30 minutes after the beginning of conjugation, respectively. The entire *Escherichia coli* chromosome takes approximately 100 minutes to transfer. In contrast, transduction and transformation mapping enable genes that are relatively close together to be mapped, because the transferred DNA segment is short (<50 kb), so the probability of two genes being transferred together depends on how close together they are on the bacterial chromosome (Figure 3.24C).

3.3 Physical Mapping

A map generated by genetic techniques is rarely sufficient for directing the sequencing phase of a genome project. This is for two reasons:

(A) Transfer of DNA between donor and recipient bacteria

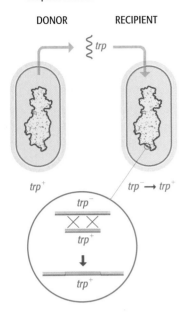

(B) Sequential transfer of markers during conjugation

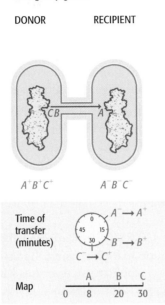

(C) Cotransfer of closely linked markers during transduction or transformation

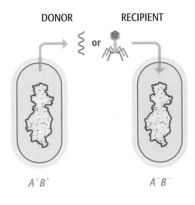

Frequency with which $A^- B^- \rightarrow A^+ B^+$ depends on how close together A and B are on the chromosome

- The resolution of a genetic map depends on the number of crossovers that have been scored. This is not a major problem for microorganisms because these can be obtained in huge numbers, enabling many crossovers to be studied, resulting in a highly detailed genetic map in which the markers are just a few kilobases apart. For example, when the *Escherichia coli* genome sequencing project began in 1990, the latest genetic map for this organism comprised over 1400 markers, an average of 1 per 3.3 kb. This was sufficiently detailed to direct the sequencing program without the need for extensive physical mapping. Similarly, the *Saccharomyces cerevisiae* project was supported by a fine-scale genetic map (approximately 1150 genetic markers, on average 1 per 10 kb). The problem with humans and most other eukaryotes is that it is simply not possible to obtain large numbers of progeny, so relatively few meioses can be studied and the resolving power of linkage analysis is restricted. This means that genes that are several tens of kilobases apart may appear at the same position on the genetic map.

- Genetic maps have limited accuracy. We touched on this point in Section 3.2.3 when we assessed Sturtevant's assumption that crossovers occur at random along chromosomes. This assumption is only partly correct because the presence of recombination hotspots means that crossovers are more likely to occur at some points rather than at others. The effect that this can have on the accuracy of a genetic map was illustrated in 1992 when the complete sequence for *Saccharomyces cerevisiae* chromosome III was published, enabling the first direct comparison to be made between a genetic map and the actual positions of markers as shown by DNA sequencing (Figure 3.25). There were considerable discrepancies, even to the extent that one pair of genes had been ordered incorrectly by genetic analysis. Bear in mind that *S. cerevisiae* is one of the two eukaryotes (fruit fly is the second) whose genomes have been subjected to intensive genetic mapping. If the yeast genetic map is inaccurate, then how precise are the genetic maps of organisms subjected to less-detailed analysis?

Figure 3.24 The basis of gene mapping in bacteria. (A) Transfer of a functional gene for tryptophan biosynthesis from a wild-type bacterium (genotype described as *trp⁺*) to a recipient that lacks a functional copy of this gene (*trp⁻*). (B) Mapping by conjugation. (C) Mapping by transduction and transformation.

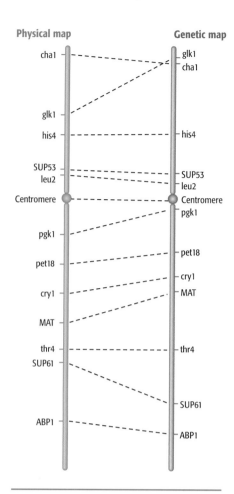

Figure 3.25 **Comparison between the genetic and physical maps of** *Saccharomyces cerevisiae* **chromosome III.** The comparison shows the discrepancies between the genetic and physical maps, the latter determined by DNA sequencing. Note that the order of the upper two markers (glk1 and cha1) is incorrect on the genetic map, and that there are also differences in the relative positioning of other pairs of markers.

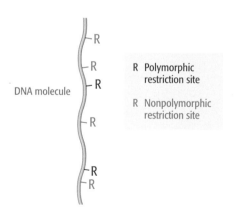

Figure 3.26 **Not all restriction sites are polymorphic.**

These two limitations of genetic mapping mean that for most eukaryotes a genetic map must be checked and supplemented by alternative mapping procedures before large-scale DNA sequencing begins. A plethora of physical mapping techniques has been developed to address this problem, the most important techniques being:

- **Restriction mapping**, which locates the relative positions on a DNA molecule of the recognition sequences for restriction endonucleases.

- **Fluorescent *in situ* hybridization (FISH)**, in which marker locations are mapped by hybridizing a probe containing the marker to intact chromosomes.

- **Sequence tagged site (STS) mapping**, in which the positions of short sequences are mapped by examining collections of genomic DNA fragments by PCR and/or hybridization analysis.

3.3.1 Restriction mapping

Genetic mapping using RFLPs as DNA markers can locate the positions of polymorphic restriction sites within a genome (Section 3.2.2), but very few of the restriction sites in a genome are polymorphic, so many sites are not mapped by this technique (Figure 3.26). Could we increase the marker density on a genome map by using an alternative method to locate the positions of some of the nonpolymorphic restriction sites? This is what restriction mapping achieves, although in practice the technique has limitations that mean it is applicable only to relatively small DNA molecules. We will look first at the technique, and then consider its relevance to genome mapping.

The basic methodology for restriction mapping

The simplest way to construct a restriction map is to compare the fragment sizes produced when a DNA molecule is digested with two different restriction enzymes that recognize different target sequences. An example using the restriction enzymes *Eco*RI and *Bam*HI is shown in Figure 3.27. First, the DNA molecule is digested with just one of the enzymes and the sizes of the resulting fragments measured by agarose gel electrophoresis. Next, the molecule is digested with the second enzyme and the resulting fragments again sized in an agarose gel. The results so far enable the number of restriction sites for each enzyme to be worked out, but do not allow their relative positions to be determined. Additional information is therefore obtained by cutting the DNA molecule with both enzymes together. In the example shown in Figure 3.27, this **double restriction** enables three of the sites to be mapped. However, a problem arises with the larger *Eco*RI fragment because this contains two *Bam*HI sites and there are two alternative possibilities for the map location of the outer one of these. The problem is solved by going back to the original DNA molecule and treating it again with *Bam*HI on its own, but this time preventing the digestion from going to completion by, for example, incubating the reaction for only a short time or using a suboptimal incubation temperature. This is called a **partial restriction** and leads to a more complex set of products, the complete restriction products now being supplemented with partially restricted fragments that still contain one or more uncut *Bam*HI sites. In the example shown in Figure 3.27, the size of one of the partial restriction fragments is diagnostic and the correct map can be identified.

A partial restriction usually gives the information needed to complete a map, but if there are many restriction sites then this type of analysis

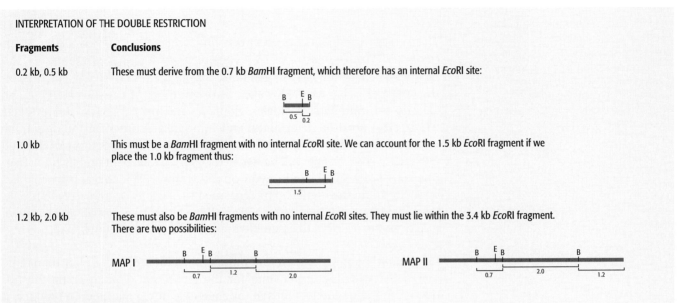

INTERPRETATION OF THE DOUBLE RESTRICTION

Fragments	Conclusions
0.2 kb, 0.5 kb	These must derive from the 0.7 kb *Bam*HI fragment, which therefore has an internal *Eco*RI site:
1.0 kb	This must be a *Bam*HI fragment with no internal *Eco*RI site. We can account for the 1.5 kb *Eco*RI fragment if we place the 1.0 kb fragment thus:
1.2 kb, 2.0 kb	These must also be *Bam*HI fragments with no internal *Eco*RI sites. They must lie within the 3.4 kb *Eco*RI fragment. There are two possibilities:

PREDICTED RESULTS OF A PARTIAL *Bam*HI RESTRICTION

If Map I is correct, then the partial restriction products will include a fragment of 1.2+0.7=1.9 kb
If Map II is correct, then the partial restriction products will include a fragment of 2.0+0.7=2.7 kb

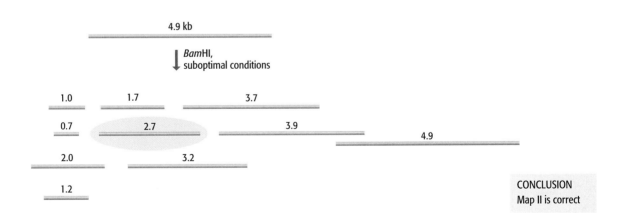

CONCLUSION
Map II is correct

Figure 3.27 Restriction mapping. The objective is to map the *Eco*RI (E) and *Bam*HI (B) sites in a linear DNA molecule of size 4.9 kb. The results of single and double restrictions are shown at the top. The sizes of the fragments given after double restriction enable two alternative maps to be constructed, as explained in the central panel, the unresolved issue being the position of one of the three *Bam*HI sites. The two maps are tested by a partial *Bam*HI restriction (bottom), which shows that Map II is the correct one.

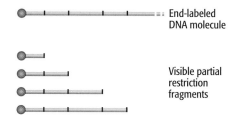

End-labeled
DNA molecule

Visible partial
restriction
fragments

Figure 3.28 Simplifying the analysis of a partial restriction by attaching markers to the ends of the DNA molecule before digestion. One end of an end-labeled DNA molecule is shown. After partial restriction, only those products that include an end-fragment are detected. This greatly simplifies the analysis, enabling the positions of restriction sites to be determined directly from the lengths of the labeled products.

becomes unwieldy, simply because there are so many different fragments to consider. An alternative strategy is simpler because it enables the majority of the fragments to be ignored. This is achieved by attaching a radioactive or other type of marker to each end of the starting DNA molecule before carrying out the partial digestion. The result is that many of the partial restriction products become "invisible" because they do not contain an end-fragment and so do not show up when the agarose gel is screened for labeled products (Figure 3.28). The sizes of the partial restriction products that are visible enable unmapped sites to be positioned relative to the ends of the starting molecule.

The scale of restriction mapping is limited by the sizes of the restriction fragments

Restriction maps are easy to generate if there are relatively few cut sites for the enzymes being used. However, as the number of cut sites increases, so also do the numbers of single-, double-, and partial-restriction products whose sizes must be determined and compared in order for the map to be constructed. Computer analysis can be brought into play but problems still eventually arise. A stage will be reached when a digest contains so many fragments that individual bands merge on the agarose gel, increasing the chances of one or more fragments being measured incorrectly or missed out entirely. If several fragments have similar sizes then even if they can all be identified, it may not be possible to assemble them into an unambiguous map.

Restriction mapping is therefore more applicable to small rather than large molecules, with the upper limit for the technique depending on the frequency of the restriction sites in the molecule being mapped. In practice, if a DNA molecule is less than 50 kb in length it is usually possible to construct an unambiguous restriction map for a selection of enzymes with six-nucleotide recognition sequences. Fifty kilobases is way below the minimum size for bacterial or eukaryotic chromosomes, although it does cover a few viral and organelle genomes, and whole-genome restriction maps have indeed been important in directing sequencing projects with these small molecules. Restriction maps are equally useful after bacterial or eukaryotic genomic DNA has been cloned, if the cloned fragments are less than 50 kb in length, because a detailed restriction map can then be built up as a preliminary to sequencing the cloned region. This is an important application of restriction mapping in projects sequencing large genomes, but is there any possibility of using restriction analysis for the more general mapping of entire genomes larger than 50 kb?

The answer is a qualified "yes," because the limitations of restriction mapping can be eased slightly by choosing enzymes expected to have infrequent cut sites in the target DNA molecule. These "rare cutters" fall into two categories:

- A few restriction enzymes cut at seven- or eight-nucleotide recognition sequences. Examples are *Sap*I (5′–GCTCTTC–3′) and *Sgf*I (5′–GCGATCGC–3′). The enzymes with seven-nucleotide recognition sequences would be expected, on average, to cut a DNA molecule with a GC content of 50% once every $4^7 = 16{,}384$ bp. The enzymes with eight-nucleotide recognition sequences should cut once every $4^8 = 65{,}536$ bp. These figures compare with $4^6 = 4096$ bp for enzymes with six-nucleotide recognition sequences, such as *Bam*HI and *Eco*RI. Cutters with seven- or eight-nucleotide recognition sequences are often used in restriction mapping of large molecules, but the approach is not as useful as it might be simply because not many of these enzymes are known.

- Enzymes can be used whose recognition sequences contain motifs that are rare in the target DNA. Genomic DNA molecules do not have random sequences and some molecules are significantly deficient in certain motifs. For example, the sequence 5′–CG–3′ is rare in the genomes of vertebrates because vertebrate cells possess an enzyme that adds a methyl group to carbon 5 of the C nucleotide in this sequence. Deamination of the resulting 5-methylcytosine gives thymine (Figure 3.29). The consequence is that during vertebrate evolution, many of the 5′–CG–3′ sequences that were originally in these genomes have become converted to 5′–TG–3′. Restriction enzymes that recognize a site containing 5′–CG–3′ therefore cut vertebrate DNA relatively infrequently. Examples are *Sma*I (5′–CC<u>CG</u>GG–3′), which cuts human DNA once every 78 kb on average, and *Bss*HII (5′–G<u>CG</u>CGC–3′), which cuts once every 390 kb. Note that *Not*I, a cutter with an eight-nucleotide recognition sequence, also targets 5′–CG–3′ sequences (recognition sequence 5′–G<u>CG</u>GC<u>CG</u>C–3′) and cuts human DNA very rarely—approximately once every 10 Mb.

The potential of restriction mapping is therefore increased by using rare cutters. It is still not possible to construct restriction maps of the genomes of animals and plants, but it is feasible to use the technique with large cloned fragments, and with the smaller DNA molecules of prokaryotes and lower eukaryotes such as yeast and fungi.

If a rare cutter is used then it may be necessary to employ a special type of agarose gel electrophoresis to study the resulting restriction fragments. This is because the relationship between the length of a DNA molecule and its migration rate in an electrophoresis gel is not linear, the resolution decreasing as the molecules get longer (Figure 3.30A). This means that it is not possible to separate molecules more than about 50 kb in length, because all of these longer molecules run as a single, slowly migrating band in a standard agarose gel. To separate them it is necessary to replace the linear electric field used in conventional gel electrophoresis with a more complex field. An example is provided by **orthogonal field alternation gel electrophoresis** (**OFAGE**), in which the electric field alternates between two pairs of electrodes, each positioned at an angle of 45° to the length of the gel (Figure 3.30B). The DNA molecules still move down through the gel, but each change in the field forces the molecules to realign. Shorter molecules realign more quickly than longer ones and so migrate more rapidly through the gel. The overall result is that molecules much longer than those separated by conventional gel electrophoresis can be resolved. Related techniques include **CHEF** (**contour clamped homogeneous electric fields**) and **FIGE** (**field inversion gel electrophoresis**).

Direct examination of DNA molecules for restriction sites

It is also possible to use methods other than electrophoresis to map restriction sites in DNA molecules. With the technique called **optical mapping**, restriction sites are directly located by looking at the cut DNA molecules with a microscope. The DNA must first be attached to a glass slide in such a way that the individual molecules become stretched out, rather than clumped together in a mass. There are various ways of doing this, including **gel stretching** and **molecular combing**. To prepare gel-stretched DNA

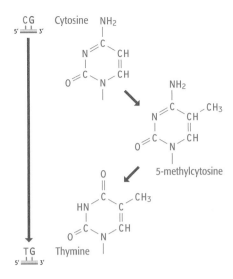

Figure 3.29 The sequence 5′–CG–3′ is rare in vertebrate DNA because of methylation of the C, followed by deamination to give T. Unmethylated cytosine can also be deaminated, but the product—uracil—is detected by the DNA repair system of vertebrate cells (Section 16.2.2) and converted back to cytosine. In contrast, thymine is not recognized efficiently by the repair system, so these nucleotides persist in the genome.

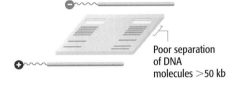

(A) Standard agarose gel electrophoresis

Poor separation of DNA molecules >50 kb

(B) Orthogonal field alternation gel electrophoresis (OFAGE)

Figure 3.30 Conventional and nonconventional agarose gel electrophoresis. (A) In standard agarose gel electrophoresis, the electrodes are placed at either end of the gel and the DNA molecules migrate directly towards the positive electrode. Molecules longer than about 50 kb cannot be separated from one another in this way. (B) In OFAGE, the electrodes are placed at the corners of the gel, with the field pulsing between the A pair and the B pair. OFAGE enables molecules up to 2 Mb in length to be separated.

fibers, chromosomal DNA is suspended in molten agarose and placed on a microscope slide. As the gel cools and solidifies, the DNA molecules become extended (Figure 3.31A). In molecular combing, the DNA fibers are prepared by dipping a silicone-coated cover slip into a solution of DNA, leaving it for 5 minutes (during which time the DNA molecules attach to the cover slip by their ends), and then removing the slip at a constant speed of 0.3 mm s^{-1} (Figure 3.31B). The force required to pull the DNA molecules through the meniscus causes them to line up. Once in the air, the surface of the cover slip dries, retaining the DNA molecules as an array of parallel fibers. After stretching or combing, the immobilized DNA molecules are treated with a restriction enzyme and then visualized by adding a fluorescent dye, such as DAPI (4,6-diamino-2-phenylindole dihydrochloride), which stains the DNA so that the fibers can be seen when the slide is examined with a high-power fluorescence microscope. The restriction sites in the extended molecules gradually become gaps as the degree of fiber extension is reduced by the natural springiness of the DNA, enabling the relative positions of the cuts to be recorded.

Optical mapping was first applied to large DNA fragments cloned in YAC and BAC vectors (Section 2.2.1). More recently, the feasibility of using this

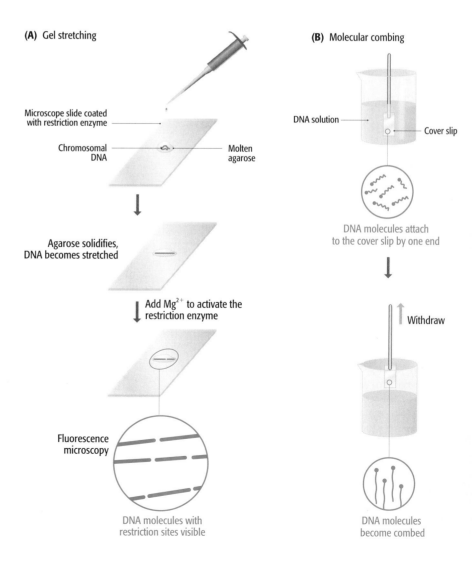

Figure 3.31 Gel stretching and molecular combing. (A) To carry out gel stretching, molten agarose containing chromosomal DNA molecules is pipetted onto a microscope slide coated with a restriction enzyme. As the gel solidifies, the DNA molecules become stretched. It is not understood why this happens, but it is thought that fluid movement on the glass surface during gelation might be responsible. Addition of magnesium chloride activates the restriction enzyme, which cuts the DNA molecules. As the molecules gradually coil up, the gaps representing the cut sites become visible. (B) In molecular combing, a cover slip is dipped into a solution of DNA. The DNA molecules attach to the cover slip by their ends, and the slip is withdrawn from the solution at a rate of 0.3 mm s^{-1}, which produces a "comb" of parallel molecules.

technique with genomic DNA has been proven with studies of a 1 Mb chromosome of the malaria parasite *Plasmodium falciparum*, and the two chromosomes and single megaplasmid of the bacterium *Deinococcus radiodurans* (see Table 8.2).

3.3.2 Fluorescent *in situ* hybridization

The optical mapping method described above provides a link to the second type of physical mapping procedure that we will consider—FISH. As in optical mapping, FISH enables the position of a marker on a chromosome or extended DNA molecule to be directly visualized. In optical mapping, the marker is a restriction site and it is visualized as a gap in an extended DNA fiber. In FISH, the marker is a DNA sequence that is visualized by hybridization with a fluorescent probe.

In situ *hybridization with radioactive or fluorescent probes*

In situ hybridization is a version of hybridization analysis (Section 2.1.2) in which an intact chromosome is examined by probing it with a labeled DNA molecule. The position on the chromosome at which hybridization occurs provides information about the map location of the DNA sequence used as the probe (Figure 3.32). For the method to work, the DNA in the chromosome must be made single-stranded ("denatured") by breaking the base pairs that hold the double helix together. Only then will the chromosomal DNA be able to hybridize with the probe. The standard method for denaturing chromosomal DNA without destroying the morphology of the chromosome is to dry the preparation onto a glass microscope slide and then treat with formamide.

In the early versions of *in situ* hybridization the probe was radioactively labeled, but this procedure was unsatisfactory because, with a radioactive label, it is difficult to achieve both sensitivity and resolution, two critical requirements for successful *in situ* hybridization. Sensitivity requires that the radioactive label has a high emission energy (an example of such a radiolabel is ^{32}P), but if the radiolabel has a high emission energy then it scatters its signal and so gives poor resolution. High resolution is possible if a radiolabel with low emission energy such as ^{3}H is used, but these radiolabels have such low sensitivity that lengthy exposures are needed, leading to a high background and difficulties in discerning the genuine signal.

These problems were solved in the late 1980s by the development of nonradioactive, fluorescent DNA labels. These labels combine high sensitivity with high resolution and are ideal for *in situ* hybridization. Fluorolabels with different colored emissions have been designed, making it possible to hybridize a number of different probes to a single chromosome and distinguish their individual hybridization signals, thus enabling the relative positions of the probe sequences to be mapped (Figure 3.33). To maximize sensitivity, the probes must be labeled as heavily as possible, which in the past has meant that they must be quite lengthy DNA molecules—usually cloned DNA fragments of at least 40 kb. This requirement is less important now that techniques for achieving heavy labeling with shorter molecules have been developed. As far as the construction of a physical map is concerned, a cloned DNA fragment can be looked upon as simply another type of marker, although in practice the use of clones as markers adds a second dimension because the cloned DNA is the material from which the DNA sequence is determined.

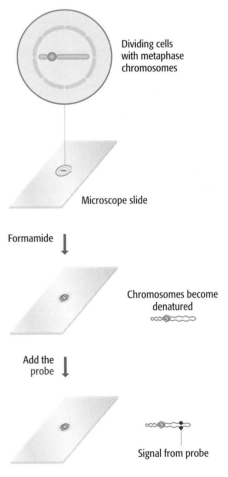

Figure 3.32 Fluorescent *in situ* hybridization. A sample of dividing cells is dried onto a microscope slide and treated with formamide so that the chromosomes become denatured but do not lose their characteristic metaphase morphologies (see Section 7.1.2). The position at which the probe hybridizes to the chromosomal DNA is visualized by detecting the fluorescent signal emitted by the labeled DNA.

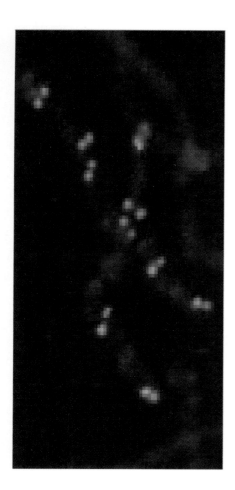

Figure 3.33 **Using FISH in physical mapping.** A series of 18 different cosmid clones have been labeled with different fluorescent markers and hybridized to a single pair of homologous chromosomes. The chromosomes come from a metaphase nucleus and hence are joined at their centromeres. The relative positions of the fluorescent signals enable the map locations of the DNA fragments carried by each cosmid to be determined. Image courtesy of Octavian Henegariu.

Mapping the positions of clones therefore provides a direct link between a genome map and its DNA sequence.

If the probe is a long fragment of DNA then one potential problem, at least with higher eukaryotes, is that it is likely to contain examples of repetitive DNA sequences (Chapter 9) and so may hybridize to many chromosomal positions, not just the specific point to which it is perfectly matched. To reduce this nonspecific hybridization, before use the probe is mixed with unlabeled DNA from the organism being studied. This DNA can simply be total nuclear DNA (i.e., representing the entire genome) but it is better if a fraction enriched for repeat sequences is used. The idea is that the unlabeled DNA hybridizes to the repetitive DNA sequences in the probe, blocking these so that the subsequent *in situ* hybridization is driven wholly by the unique sequences. Nonspecific hybridization is therefore reduced or eliminated entirely (Figure 3.34).

FISH in action

FISH was originally used with metaphase chromosomes (Section 7.1). These chromosomes, prepared from nuclei that are undergoing division, are highly condensed and each chromosome in a set takes up a recognizable appearance, characterized by the position of its centromere and the banding pattern that emerges after the chromosome preparation is stained (see Figure 7.5). With metaphase chromosomes, a fluorescent signal obtained by FISH is mapped by measuring its position relative to the end of the short arm of the chromosome (the **FLpter value**). A disadvantage of this method is that the highly condensed nature of metaphase chromosomes means that only low-resolution mapping is possible, two markers having to be at least 1 Mb apart to be resolved as separate hybridization signals. This degree of resolution is insufficient for the construction of useful chromosome maps, and the main application of metaphase FISH has been in determining the chromosome on which a new marker is located, and providing a rough idea of its map position, as a preliminary to finer-scale mapping by other methods.

For several years these "other methods" did not involve any form of FISH, but since 1995 a range of higher-resolution FISH techniques has been developed. With these techniques, higher resolution is achieved by changing the nature of the chromosomal preparation being studied. If metaphase chromosomes are too condensed for fine-scale mapping then we must use chromosomes that are more extended. There are two ways of doing this:

- **Mechanically stretched chromosomes** can be obtained by modifying the preparative method used to isolate chromosomes from metaphase nuclei. The inclusion of a centrifugation step generates shear forces which can result in the chromosomes becoming stretched to up to 20 times their normal length. Individual chromosomes are still recognizable and FISH signals can be mapped in the same way as with normal metaphase chromosomes. The resolution is significantly improved and markers that are 200–300 kb apart can be distinguished.

- **Nonmetaphase chromosomes** can be used because it is only during metaphase that chromosomes are highly condensed: at other stages of the cell cycle the chromosomes are naturally unpacked. Attempts have been made to use prophase nuclei (see Figure 3.15) because in these the chromosomes are still sufficiently condensed for individual chromosomes to be identified. In practice, however, these preparations provide no advantage over mechanically stretched chromosomes. **Interphase** chromosomes are more useful because this stage of the cell cycle (between nuclear divisions) is when the chromosomes are most unpacked. Resolution down to 25 kb is possible, but chromosome morphology is lost so there are no external reference points against which to map the position of the probe. This technique is therefore used after preliminary map information has been obtained, usually as a means of determining the order of a series of markers in a small region of a chromosome.

Interphase chromosomes contain the most unpacked of all cellular DNA molecules. To improve the resolution of FISH to better than 25 kb it is therefore necessary to abandon intact chromosomes and instead use purified DNA. This approach, called **fiber-FISH**, makes use of DNA prepared by gel stretching or molecular combing (see Figure 3.31), and can distinguish markers that are less than 10 kb apart.

3.3.3 Sequence tagged site mapping

To generate a detailed physical map of a large genome we need, ideally, a high-resolution mapping procedure that is rapid and not technically demanding. Neither of the two techniques that we have considered so far—restriction mapping and FISH—meets these requirements. Restriction mapping is rapid, easy, and provides detailed information, but it cannot be applied to large genomes. FISH can be applied to large genomes, and modified versions such as fiber-FISH can give high-resolution data, but FISH is difficult to carry out and data accumulation is slow, with map positions for no more than three or four markers being obtained in a single experiment. If detailed physical maps are to become a reality then we need a more powerful technique.

At present the most powerful physical mapping technique, and the one that has been responsible for generation of the most detailed maps of large genomes, is STS mapping. A **sequence tagged site**, or **STS**, is simply a short DNA sequence, generally between 100 bp and 500 bp in length, that is easily recognizable and occurs only once in the chromosome or genome being studied. To map a set of STSs, a collection of overlapping DNA fragments from a single chromosome or from the entire genome is needed. In the example shown in Figure 3.35, a fragment collection has been prepared from a single chromosome, with each point along the chromosome represented, on average, five times in the collection. The data from which the map will be derived are obtained by determining which fragments contain which STSs. This can be done by hybridization analysis but PCR is generally used because it is quicker and has proven to be more amenable to automation. The chances of two STSs being present on the same fragment will, of course, depend on how close together they are in the genome. If they are very close then there is a good chance that they will always be on the same fragment; if they are further apart then sometimes they will be on the same fragment and sometimes they will not (Figure 3.35). The data can therefore be used to calculate the distance between two markers, in a manner analogous to the way in which map distances are determined by linkage analysis (Section 3.2.3). Remember that in linkage analysis a map distance is calculated from the frequency at which crossovers occur between two markers. STS mapping is essentially the same, except that the map distance is based on the frequency at which *breaks* occur between two markers.

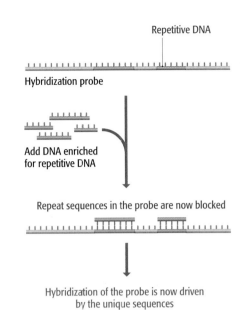

Figure 3.34 A method for blocking repetitive DNA sequences in a hybridization probe. In this example the probe molecule contains two repeat sequences (shown in red). If these sequences are not blocked then the probe will hybridize nonspecifically to any copies of these repeats in the target DNA. To block the repeat sequences, the probe is prehybridized with a DNA fraction enriched for repetitive DNA.

Figure 3.35 A fragment collection suitable for STS mapping. The fragments span the entire length of a chromosome, with each point on the chromosome present in an average of five fragments. The two blue markers are close together on the chromosome map and there is a high probability that they will be found on the same fragment. The two green markers are more distant from one another and so are less likely to be found on the same fragment.

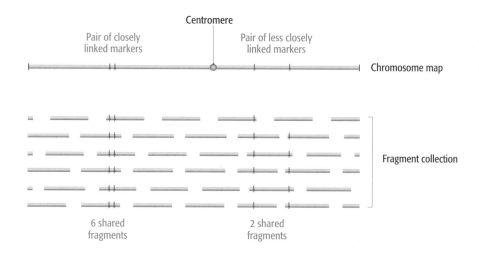

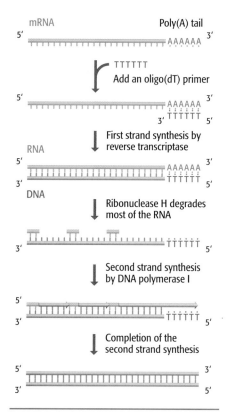

Figure 3.36 One method for preparing cDNA. Most eukaryotic mRNAs have a poly(A) tail at their 3′ end (Section 12.2.1). This series of A nucleotides is used as the priming site for the first stage of cDNA synthesis, carried out by reverse transcriptase—a DNA polymerase that copies an RNA template (Section 2.1.1). The primer is a short, synthetic DNA oligonucleotide, typically 20 nucleotides in length, made up entirely of Ts (an "oligo(dT)" primer). When the first strand synthesis has been completed, the preparation is treated with ribonuclease H, which specifically degrades the RNA component of an RNA–DNA hybrid. Under the conditions used, the enzyme does not degrade all of the RNA, instead leaving short segments that prime the second DNA strand synthesis reaction, this one catalyzed by DNA polymerase I.

The description of STS mapping given above leaves out some critical questions: What exactly is an STS? How is the DNA fragment collection obtained?

Any unique DNA sequence can be used as an STS

To qualify as an STS, a DNA sequence must satisfy two criteria. The first is that its sequence must be known, so that a PCR assay can be set up to test for the presence or absence of the STS on different DNA fragments. The second requirement is that the STS must have a unique location in the chromosome being studied, or in the genome as a whole if the DNA fragment set covers the entire genome. If the STS sequence occurs at more than one position then the mapping data will be ambiguous. Care must therefore be taken to ensure that STSs do not include sequences found in repetitive DNA.

These are easy criteria to satisfy and STSs can be obtained in many ways, the most common sources being **expressed sequence tags (ESTs)**, **SSLPs**, and **random genomic sequences**.

- **Expressed sequence tags.** These are short sequences obtained by analysis of cDNA clones. Complementary DNA is prepared by converting an mRNA preparation into double-stranded DNA (Figure 3.36). Because the mRNA in a cell is derived from protein-coding genes, cDNAs and the ESTs obtained from them represent the genes that were being expressed in the cell from which the mRNA was prepared. ESTs are looked upon as a rapid means of gaining access to the sequences of important genes, and they are valuable even if their sequences are incomplete. An EST can also be used as an STS, assuming that it comes from a unique gene and not from a member of a gene family in which all the genes have the same or very similar sequences.

- **SSLPs.** In Section 3.2.2 we examined the use of microsatellites and other SSLPs in genetic mapping. SSLPs can also be used as STSs in physical mapping. SSLPs that are polymorphic and have already been mapped by linkage analysis are particularly valuable as they provide a direct connection between the genetic and physical maps.

- **Random genomic sequences.** These are obtained by sequencing random pieces of cloned genomic DNA, or simply by downloading sequences that have been deposited in the databases.

Fragments of DNA for STS mapping

The second component of an STS mapping procedure is the collection of DNA fragments spanning the chromosome or genome being studied. This collection is sometimes called the **mapping reagent** and at present there are two ways in which it can be assembled: as a clone library and as a panel of **radiation hybrids**. We will consider radiation hybrids first.

A radiation hybrid is a rodent cell that contains fragments of chromosomes from a second organism. The technology was initially developed with human chromosomes, starting in the 1970s when it was discovered that exposure of human cells to X-ray doses of 3000–8000 rad causes the chromosomes to break up randomly into fragments, larger X-ray doses producing smaller fragments (Figure 3.37A). This treatment is of course lethal for the human cells, but the chromosome fragments can be propagated if the irradiated cells are subsequently fused with nonirradiated cells of a hamster or other rodent. Fusion is stimulated either chemically, with polyethylene glycol, or by exposure to Sendai virus (Figure 3.37B). Not all of the hamster cells take up chromosome fragments, so a means of identifying the hybrids is needed. The routine selection process is to use a hamster cell line that is unable to make either thymidine kinase (TK) or hypoxanthine phosphoribosyl transferase (HPRT), deficiencies in either of these two enzymes being lethal when the cells are grown in a medium containing a mixture of hypoxanthine, aminopterin, and thymidine (HAT medium). After fusion, the cells are placed in HAT medium. Those that grow are hybrid hamster cells that have acquired human DNA fragments that include genes for the human TK and HPRT enzymes, which are synthesized inside the hybrids, enabling these cells to grow in the selective medium. The treatment results in hybrid cells that contain a random selection of human DNA fragments inserted into the hamster chromosomes. Typically the fragments are 5–10 Mb in size, with each cell containing fragments equivalent to 15%–35% of the human genome. The collection of cells is called a radiation hybrid panel and can be used as a mapping reagent in STS mapping, provided that the PCR assay used to identify the STS does not amplify the equivalent region of DNA from the hamster genome.

A second type of radiation hybrid panel, containing DNA from just one human chromosome, can be constructed if the cell line that is irradiated is not a human one but a second type of rodent hybrid. Cytogeneticists have developed a number of rodent cell lines in which a single human chromosome is stably propagated in the rodent nucleus. If a cell line of this type (e.g., a mouse cell line) is irradiated and fused with hamster cells, then the hybrid hamster cells obtained after selection will contain either human or mouse chromosome fragments, or a mixture of both. The ones containing human DNA can be identified by probing with a human-specific repeat sequence, such as the short interspersed nuclear element (SINE) called Alu (Section 9.2.1), which has a copy number of just over 1 million (see Table 9.3) and so occurs on average once every 3 kb in the human genome. Only cells containing human DNA will hybridize to Alu probes, enabling the uninteresting mouse hybrids to be discarded and STS mapping to be directed at the cells containing human chromosome fragments.

Radiation hybrid mapping of the human genome was initially carried out with chromosome-specific rather than whole-genome panels because it was thought that fewer hybrids would be needed to map a single chromosome

(A) Irradiation of chromosomes

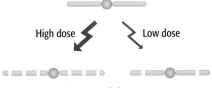

Fragmented chromosomes

(B) Fusion of cells to produce a radiation hybrid

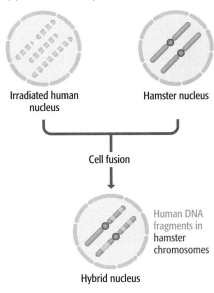

Irradiated human nucleus   Hamster nucleus

Cell fusion

Human DNA fragments in hamster chromosomes

Hybrid nucleus

Figure 3.37 Radiation hybrids. (A) The result of irradiation of human cells: the chromosomes break into fragments, smaller fragments being generated by higher X-ray doses. In (B), a radiation hybrid is produced by fusing an irradiated human cell with an untreated hamster cell. For clarity, only the nuclei are shown.

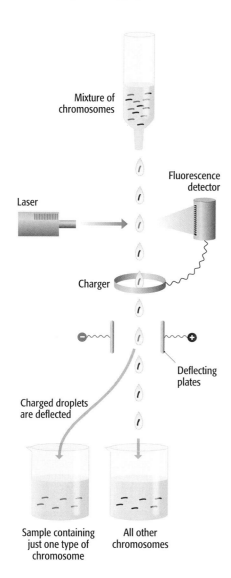

Figure 3.38 Separating chromosomes by flow cytometry. A mixture of fluorescently stained chromosomes is passed through a small aperture so that each drop that emerges contains just one chromosome. The fluorescence detector identifies the signal from drops containing the correct chromosome and applies an electric charge to these drops. When the drops reach the electric plates, the charged ones are deflected into a separate beaker. All other drops fall straight through the deflecting plates and are collected in the waste beaker.

than would be needed to map the entire genome. It turns out that a high-resolution map of a single human chromosome requires a panel of 100–200 hybrids, which is about the most that can be handled conveniently in a PCR screening program. But whole-genome and single-chromosome panels are constructed differently, the former involving irradiation of just human DNA, and the latter requiring irradiation of a mouse cell containing much mouse DNA and relatively little human DNA. This means that the human DNA content per hybrid is much lower in a single-chromosome panel than in a whole-genome panel. It transpires that detailed mapping of the entire human genome is possible with fewer than 100 whole-genome radiation hybrids, so whole-genome mapping is no more difficult than single-chromosome mapping. Once this was realized, whole-genome radiation hybrids became a central component of the mapping phase of the Human Genome Project (Section 4.3). Whole-genome libraries have also been used for STS mapping of other mammalian genomes and for those of the zebra fish and the chicken.

A clone library can also be used as the mapping reagent for STS analysis

A preliminary to the sequencing phase of a genome project is to break the genome or isolated chromosomes into fragments and to clone each one in a high-capacity vector that is able to handle large fragments of DNA (Section 2.2.1). This results in a clone library, a collection of DNA fragments, which in the case of a genome project have an average size of several hundred kilobases. As well as supporting the sequencing work, this type of clone library can also be used as a mapping reagent in STS analysis.

As with radiation hybrid panels, a clone library can be prepared from genomic DNA, and so represents the entire genome, or a chromosome-specific library can be made if the starting DNA comes from just one type of chromosome. The latter is possible because individual chromosomes can be separated by **flow cytometry**. To carry out this technique, dividing cells (ones with condensed chromosomes) are carefully broken open so that a mixture of intact chromosomes is obtained. The chromosomes are then stained with a fluorescent dye. The amount of dye that a chromosome binds depends on its size, so larger chromosomes bind more dye and fluoresce more brightly than smaller ones. The chromosome preparation is diluted and passed through a fine aperture, producing a stream of droplets, each one containing a single chromosome. The droplets pass through a detector that measures the amount of fluorescence and hence identifies which droplets contain the particular chromosome being sought. An electric charge is applied to these drops, and no others (Figure 3.38), enabling the droplets containing the desired chromosome to be deflected and separated from the rest. What if two different chromosomes have similar sizes, as is the case with human chromosomes 21 and 22? These can usually be separated if the dye that is used is not one that binds nonspecifically to DNA, but instead has a preference for AT- or GC-rich regions. Examples of such dyes are Hoechst 33258 and chromomycin A_3, respectively. Two chromosomes that are the same size rarely have identical GC contents, and so can be distinguished by the amounts of AT- or GC-specific dye that they bind.

Compared with radiation hybrid panels, clone libraries have one important advantage for STS mapping. This is the fact that the individual clones can

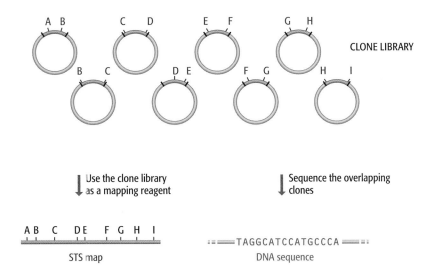

Figure 3.39 The value of clone libraries in genome projects. The small clone library shown in this example contains sufficient information for an STS map to be constructed, and can also be used as the source of the DNA that will be sequenced.

subsequently provide the DNA that is actually sequenced. The data resulting from STS analysis, from which the physical map is generated, can equally well be used to determine which clones contain overlapping DNA fragments, enabling a **clone contig** to be built up (Figure 3.39; for other methods for assembling clone contigs see Section 4.2.2). This assembly of overlapping clones can be used as the base material for a lengthy, continuous DNA sequence, and the STS data can later be used to anchor this sequence precisely onto the physical map. If the STSs also include SSLPs that have been mapped by genetic linkage analysis then the DNA sequence, physical map, and genetic map can all be integrated.

Summary

Genome maps provide the framework for sequencing projects because they indicate the positions of genes and other recognizable features, and hence enable the accuracy of an assembled DNA sequence to be checked. Genetic maps are constructed by cross-breeding experiments and pedigree analysis, and physical maps by direct examination of DNA molecules. In the first genetic maps, the markers were genes whose alleles could be distinguished because they gave rise to easily recognized phenotypes such as different eye colors, or whose alleles could be distinguished by biochemical tests. Today, DNA markers are also extensively used, these including restriction fragment length polymorphisms (RFLPs), simple sequence length polymorphisms (SSLPs), and single nucleotide polymorphisms (SNPs), all of which can be typed quickly and easily by PCR. The relative positions of genes and DNA markers on chromosomes are determined by linkage analysis. This technique is based on the original genetic discoveries made by Mendel and was first developed for use with fruit flies in the early part of the twentieth century. Linkage analysis enables the recombination frequency between a pair of markers to be determined, providing the data needed to deduce the relative positions of the markers on the genetic map. For many organisms, linkage analysis is carried out by following the inheritance of markers in planned breeding experiments, but this is not possible with humans. Instead, genetic mapping of the human genome depends on examination of marker inheritance in large families, a procedure called pedigree analysis. Genetic maps have relatively poor resolution and tend to be inaccurate, and must be refined by physical mapping if the map is to be used in a genome sequencing project. The positions of restriction sites in a small DNA molecule can be determined by restriction mapping, but this is only of limited value with eukaryotic chromosomes. Of greater utility is fluorescent *in situ* hybridization (FISH), in which a preparation of intact chromosomes, possibly ones that have been elongated by mechanical stretching, is probed with a fluorescently labeled marker. The position at which hybridization occurs is determined by examining the preparation by confocal microscopy. The most detailed physical maps are obtained by sequence tagged site (STS) content mapping, which makes use of a mapping reagent, a collection of overlapping DNA fragments that span an entire chromosome or genome. The map position of a marker is determined by identifying which fragments in the collection contain copies of the marker. The mapping reagent can be a library of clones or a radiation hybrid panel.

Multiple Choice Questions

3.1.* A major problem with the computational assembly of DNA sequences of complex eukaryotic genomes is the presence of:

  a. Multiple chromosomes.

  b. Mitochondrial DNA.

  c. Introns within the genome.

  d. Repetitive sequences.

3.2. The first genetic maps used genes as markers because:

  a. The locations of genes on chromosomes could be observed by staining the DNA with dyes.

  b. Phenotypes specified by genes could be identified visually and their inheritance patterns studied.

  c. Individual genes specifying easily identifiable phenotypic traits were easily cloned.

  d. Single nucleotide polymorphisms were used to identify point mutations that resulted in clearly observable phenotypic differences.

3.3.* Which of the following is NOT a reason why biochemical phenotypes were commonly used to create human genetic maps?

  a. Humans have no visual characteristics that are useful for genetic mapping.

  b. There are biochemical phenotypes that are easily screened by blood typing.

  c. Some easily characterized biochemical phenotypes are specified by genes with very large numbers of alleles.

  d. It is unethical to perform controlled breeding experiments with humans.

3.4. Eukaryotic genomes are mapped using DNA markers in addition to genes because:

  a. DNA markers do not require the presence of two or more alleles for mapping.

  b. Gene maps may not cover large regions of the genome.

  c. Most genes contain multiple alleles that can be easily mapped.

  d. DNA markers are less variable than genetic markers.

3.5.* Microsatellites are used more commonly than minisatellites as DNA markers because:

  a. Minisatellites are present in too many locations within the genome.

  b. Restriction enzymes can be used to type microsatellites but not minisatellites.

  c. There are very few microsatellites in eukaryotic genomes so they are easily identified and analyzed.

  d. Microsatellites are present throughout eukaryotic genomes and are easily amplified using PCR.

3.6. Which of the following genetic markers are present in the highest numbers within the human genome?

  a. RFLPs.

  b. Minisatellites.

  c. Microsatellites.

  d. Single nucleotide polymorphisms.

3.7.* The principle of genetic linkage is:

  a. The fact that the different alleles for a given gene will be located at the same position in a chromosome.

  b. The discovery that multiple genes are responsible for some traits (such as eye color in flies).

  c. The observation that some genes will be inherited together if they are located on the same chromosome.

  d. The observation that darkly staining regions of chromosomes do not contain genes.

3.8. The difference between mitosis and meiosis is that mitosis is characterized by:

  a. The production of two diploid cells that are genetically identical to the parental cell.

  b. The exchange of DNA (crossing-over) between homologous chromosomes.

  c. The production of two diploid cells that are genetically distinct from the parental cell.

  d. The production of four haploid cells that are genetically distinct from the parental cell.

3.9.* Which of the following statements correctly describes the recombination frequency between two genes?

  a. The closer two genes are to each other on a chromosome, the higher the frequency of recombination will be between them.

  b. The more distant two genes are to each other on a chromosome, the higher the frequency of recombination will be between them.

  c. If two genes are located on the same chromosome then no recombination events can occur between them.

  d. If two genes are located on different chromosomes then there will be a high frequency of recombination between them.

3.10. In analyzing a human pedigree to determine how closely two genes are linked, it is best to:

  a. Conclude that the most common genotypes in the offspring are the parental genotypes.

  b. Conclude that the most common genotypes in the offspring are the recombinants.

continued …

  c. Perform a test cross to determine the linkage between the genes.

  d. Determine the genotypes of the grandparents.

3.11.* Which of the following is NOT a factor that limits the accuracy of genetic maps for humans and other complex eukaryotic organisms?

  a. It is not possible to obtain enough progeny for many eukaryotic organisms.

  b. Recombination hotspots may interfere with genetic mapping.

  c. Genetic mapping only uses genes and there are not enough genes to map entire genomes.

  d. Genes or markers that are tens of thousands of base pairs apart may appear at the same position on a genetic map.

3.12. Metaphase chromosomes were initially used for fluorescent *in situ* hybridization, but the results were somewhat limiting because:

  a. Many regions of a chromosome are condensed and cannot hybridize to probes.

  b. The probes will hybridize preferentially to repeated sequences present on multiple chromosomes.

  c. The chromosomes are not stable in the condensed state, and the signal diffuses when the chromosomes are relaxed.

  d. Only low-resolution mapping is possible, as the chromosomes are condensed.

3.13.* Interphase chromosomes are useful for fine-scale mapping by fluorescent *in situ* hybridization because they:

  a. Are the least condensed type of chromosome.

  b. Are easily distinguished from each other by their structures.

  c. Have transcriptionally active regions of chromatin that are required for this technique.

  d. Allow for physical mapping of genomes to 1 kb resolution.

3.14. Sequence tagged sites have which of the following properties?

  a. They are present only once within a genome and possess an RFLP site.

  b. They are present only once within a genome and their sequence is known.

  c. Their sequence is known and they must contain repetitive DNA sequences.

  d. They must contain the sequence of a gene, and no repetitive DNA sequences can be present.

3.15.* Which of the following sequences can NOT be used as a sequence tagged site?

  a. Expressed sequence tags.

  b. Random genomic sequences.

  c. Simple sequence length polymorphisms.

  d. Restriction fragment length polymorphisms.

3.16. Radiation hybrid panels provide a useful mechanism for physical mapping because:

  a. Only a portion of the human genome is present in any given hybrid cell.

  b. The host hamster cells lack homologous sequences to human genetic markers.

  c. The host hamster cells are resistant to irradiation.

  d. A complete physical map of the hamster genome is known.

Short Answer Questions

*Answers to odd-numbered questions can be found in the Appendix

3.1.* Why are maps required for the sequencing of genomes? If a map of a genome was unavailable, what would be the major difficulties in obtaining a genome sequence?

3.2. Clearly explain the differences between a genetic and a physical map of a genome.

3.3.* How has PCR made the analysis of RFLPs much faster and easier? What was required to map RFLPs prior to the utilization of PCR?

3.4. How are restriction enzymes used to generate both genetic and physical maps of a genome?

3.5.* How does the linkage between genes provide a critical component to genetic mapping? Discuss how genetic markers can be linked to provide maps of individual chromosomes.

3.6. What are Mendel's two Laws of Genetics? What component of genetic mapping is not covered by Mendel's Laws?

Short Answer Questions (continued)

3.7.* Why is a double homozygote used for test crosses in linkage analysis experiments? Why is it preferable that the homozygote alleles be recessive for the traits being tested?

3.8. Restriction mapping of DNA molecules is often limited to molecules less than 50 kb in size. Why is this the limit for this technique, and how can this limitation be relaxed to study larger DNA molecules?

3.9.* Genetic mapping techniques require at least two alleles for a given marker, while physical mapping techniques do not rely on the presence of alleles to map genomes. Discuss how the technique of fluorescent *in situ* hybridization can be used to map genome locations even if there is no genetic variation present at a given position.

3.10. Why are whole-genome radiation hybrids favored over single-chromosome hybrids for the mapping of genomes?

3.11.* How does a scientist prepare a clone library of DNA from just a single chromosome?

In-depth Problems

3.1.* What are the ideal features of a DNA marker that will be used to construct a genetic map? To what extent can RFLPs, SSLPs, or SNPs be considered "ideal" DNA markers?

3.2. Explore and assess the applications of DNA chip technology in biological research.

3.3.* What features would be desirable for an organism that is to be used for extensive studies of heredity?

3.4. What problems might arise if an attempt were made to sequence a genome before a genetic or physical map had been obtained?

3.5.* Which is more useful—a genetic or a physical map?

continued ...

Figure Tests

3.1.* How does a dye-quenching experiment determine if an oligonucleotide has hybridized to a DNA molecule containing a single nucleotide polymorphism?

(B) Detecting hybridization by dye-quenching

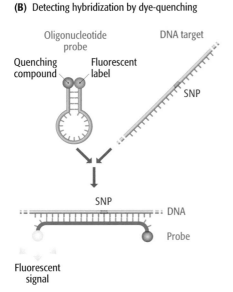
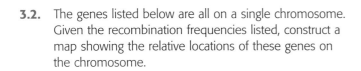

3.2. The genes listed below are all on a single chromosome. Given the recombination frequencies listed, construct a map showing the relative locations of these genes on the chromosome.

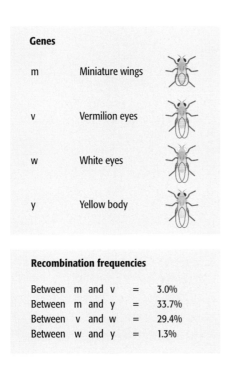

Genes	
m	Miniature wings
v	Vermilion eyes
w	White eyes
y	Yellow body

Recombination frequencies

Between	m and v	=	3.0%
Between	m and y	=	33.7%
Between	v and w	=	29.4%
Between	w and y	=	1.3%

3.3.* The type of electrophoresis shown in this figure is used to separate relatively large DNA molecules (greater than 50 kb). What is the basis to this technique?

3.4. This pair of chromosomes has been hybridized to cloned molecules containing different fluorescent labels. What type of technique is this, and is it an example of genetic or physical mapping?

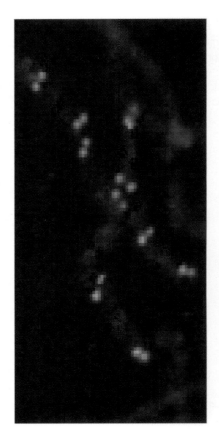

Further Reading

Books on the history of genetics

Orel, V. (1995) *Gregor Mendel: The First Geneticist.* Oxford University Press, Oxford.

Shine, I. and Wrobel, S. (1976) *Thomas Hunt Morgan: Pioneer of Genetics.* University Press of Kentucky, Lexington, Kentucky.

Sturtevant, A.H. (1965) *A History of Genetics.* Harper and Row, New York. *Describes the early gene mapping work carried out by Morgan and his colleagues.*

Genetic and DNA markers

Wang, D.G., Fan, J.-B., Siao, C.-J., et al. (1998) Large-scale identification, mapping, and genotyping of single-nucleotide polymorphisms in the human genome. *Science* **280:** 1077–1082.

Yamamoto, F., Clausen, H., White, T., Marken, J. and Hakamori, S. (1990) Molecular genetic basis of the histo-blood group ABO system. *Nature* **345:** 229–233.

Linkage analysis

Morton, N.E. (1955) Sequential tests for the detection of linkage. *Am. J. Hum. Genet.* **7:** 277–318. *The use of lod scores in human pedigree analysis.*

Strachan, T. and Read, A.P. (2004) *Human Molecular Genetics,* 3rd Ed. Garland, London. *Chapter 13 covers human genetic mapping.*

Sturtevant, A.H. (1913) The linear arrangement of six sex-linked factors in *Drosophila* as shown by mode of association. *J. Exp. Zool.* **14:** 39–45. *Construction of the first linkage map for the fruit fly.*

Restriction mapping

Hosoda, F., Arai, Y., Kitamura, E., et al. (1997) A complete *Not*I restriction map covering the entire long arm of human chromosome 11. *Genes Cells* **2:** 345–357.

Ichikawa, H., Hosoda, F., Arai, Y., Shimizu, K., Ohira, M. and Ohki, M. (1993) *Not*I restriction map of the entire long arm of human chromosome 21. *Nat. Genet.* **4:** 361–366.

Jing, J.P., Lai, Z.W., Aston, C., et al. (1999) Optical mapping of *Plasmodium falciparum* chromosome 2. *Genome Res.* **9:** 175–181.

Lin, J., Qi, R., Aston, C., et al. (1999) Whole-genome shotgun optical mapping of *Deinococcus radiodurans. Science* **285:** 1558–1562.

Michalet, X., Ekong, R., Fougerousse, F., et al. (1997) Dynamic molecular combing: stretching the whole human genome for high-resolution studies. *Science* **277:** 1518–1523.

Zhou, S.G., Kvikstad, E., Kile, A., et al. (2003) Whole-genome shotgun optical mapping of *Rhodobacter sphaeroides* strain 2.4.1 and its use for whole-genome shotgun sequence assembly. *Genome Res.* **13:** 2142–2151.

FISH

Heiskanen, M., Peltonen, L. and Palotie, A. (1996) Visual mapping by high resolution FISH. *Trends Genet.* **12:** 379–382.

Lichter, P. (1997) Multicolor FISHing: what's the catch? *Trends Genet.* **13:** 475–479.

Romanov, M.N., Daniels, L.M., Dodgson, J.B. and Delany, M.E. (2005) Integration of the cytogenetic and physical maps of chicken chromosome 17. *Chromosome Res.* **13:** 215–222. *Describes an application of FISH carried out with BAC probes.*

Tsuchiya, D. and Taga, M. (2001) Application of fibre-FISH (fluorescence *in situ* hybridization) to filamentous fungi: visualization of the rRNA gene cluster of the ascomycete *Cochliobolus heterostrophus. Microbiology* **147:** 1183–1187.

Zelenin, A.V. (2004) Fluorescence *in situ* hybridization in studying the human genome. *Mol. Biol.* **38:** 14–23.

Radiation hybrids

Hudson, T.J., Church, D.M., Greenaway, S., et al. (2001) A radiation hybrid map of the mouse genome. *Nat. Genet.* **29:** 201–205.

Itoh, T., Watanabe, T., Ihara, N., Mariani, P., Beattie, C.W., Sugimoto, Y. and Takasuga, A. (2005) A comprehensive radiation hybrid map of the bovine genome comprising 5593 loci. *Genomics* **85:** 413–424.

McCarthy, L. (1996) Whole genome radiation hybrid mapping. *Trends Genet.* **12:** 491–493.

Walter, M.A., Spillett, D.J., Thomas, P., Weissenbach, J. and Goodfellow, P.N. (1994) A method for constructing radiation hybrid maps of whole genomes. *Nat. Genet.* **7:** 22–28.

Sequencing
Genomes

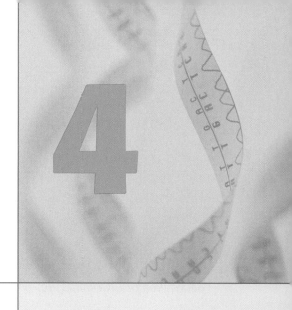

4

When you have read Chapter 4, you should be able to:

Give detailed descriptions of the chain termination and thermal cycle methods for DNA sequencing.

Describe in outline the chemical degradation and pyrosequencing methods, and state their applications.

State the strengths and limitations of the shotgun, whole-genome shotgun, and clone contig methods of genome sequencing.

Describe how a small bacterial genome can be sequenced by the shotgun method, using the *Haemophilus influenzae* project as an example.

Outline the various ways in which a clone contig can be built up.

Explain the basis to the whole-genome shotgun approach to genome sequencing, with emphasis on the steps taken to ensure that the resulting sequence is accurate.

Give an account of the development of the human genome projects up to the publication of the finished chromosome sequences in 2004–2005.

Debate the ethical, legal, and social issues raised by the human genome projects.

The ultimate objective of a genome project is the complete DNA sequence for the organism being studied, ideally integrated with the genetic and/or physical maps of the genome so that genes and other interesting features can be located within the DNA sequence. This chapter describes the techniques and research strategies that are used during the sequencing phase of a genome project, when this ultimate objective is being directly addressed. Techniques for sequencing DNA are clearly of central importance in this context and we will begin the chapter with a detailed examination of sequencing methodology. This methodology is of little value, however, unless the short sequences that result from individual sequencing experiments can be linked together in the correct order to give the master sequences of the chromosomes that make up the genome. The second part of this chapter therefore describes the strategies used to ensure that the master sequences are assembled correctly.

4.1 The Methodology for DNA Sequencing

There are several procedures for DNA sequencing, but by far the most popular is the **chain termination method** first devised by Fred Sanger and colleagues in the mid-1970s. Chain termination sequencing has gained preeminence for several reasons, not least being the relative ease with which the technique can be automated. As we will see later in this chapter, a genome project involves a huge number of individual sequencing experiments and it would take many

---

Technical Note 4.1   Polyacrylamide gel electrophoresis

Separation of DNA molecules differing in length by just one nucleotide

Polyacrylamide gel electrophoresis is used to examine the families of chain-terminated DNA molecules resulting from a sequencing experiment. Agarose gel electrophoresis (Technical Note 2.2) cannot be used for this purpose because it does not have the resolving power needed to separate single-stranded DNA molecules that differ in length by just one nucleotide. Polyacrylamide gels have smaller pore sizes than agarose gels and allow precise separations of molecules from 10 to 1500 bp in length. As well as DNA sequencing, polyacrylamide gels are also used for other applications where fine-scale DNA separations are required, for instance in the examination of amplification products from PCRs directed at microsatellite loci, where the products of different alleles might differ in size by just two or three base pairs (see Figure 3.6). Polyacrylamide gels can be prepared as slabs between two glass plates held apart by spacers, or in long, thin columns suitable for capillary electrophoresis (Figure T4.1).

A polyacrylamide gel consists of chains of acrylamide monomers (CH_2=CH–CO–NH_2) cross-linked with *N, N′*-methylenebisacrylamide units (CH_2=CH–CO–NH–CH_2– NH–CO–CH=CH_2), the latter commonly called "bis." The pore size of the gel is determined by both the total concentration of monomers (acrylamide + bis) and the ratio of acrylamide to bis. In a 1 mm thick slab gel used for DNA sequencing, a 6% gel with an acrylamide:bis ratio of 19:1 is normally used because this allows resolution of single-stranded DNA molecules between 100 and 750 nucleotides in length. About 650 nucleotides of sequence can therefore be read from a single gel. The gel concentration can be increased to 8% in order to read the sequence closer to the primer (resolving molecules 50–400 nucleotides in length) or decreased to 4% to read a more distant sequence (500–1500 nucleotides from the primer). Polymerization of the acrylamide:bis solution is initiated by ammonium persulfate and catalyzed by TEMED (*N, N, N′, N′*-tetramethylethylenediamine). Sequencing gels also contain urea, which is a denaturant that prevents intrastrand base pairs from forming in the chain-terminated molecules. This is important because the change in conformation resulting from base pairing alters the migration rate of a single-stranded molecule, so the strict equivalence between the length of a molecule and its band position, critical for reading the DNA sequence, is lost.

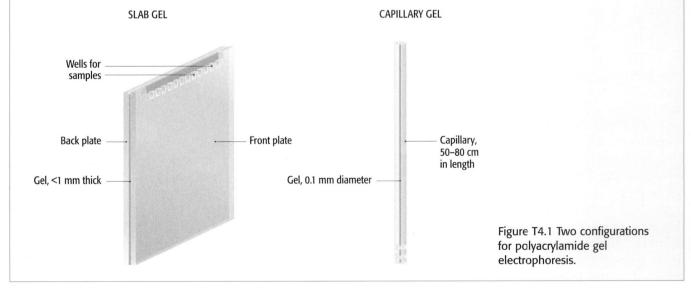

Figure T4.1 Two configurations for polyacrylamide gel electrophoresis.

years to perform all these by hand. Automated sequencing techniques are therefore essential if a project is to be completed in a reasonable timespan.

4.1.1 Chain termination DNA sequencing

Chain termination DNA sequencing is based on the principle that single-stranded DNA molecules that differ in length by just a single nucleotide can be separated from one another by **polyacrylamide gel electrophoresis** (Technical Note 4.1). This means that it is possible to resolve a family of molecules, representing all lengths from 10 to 1500 nucleotides, into a series of bands in a slab or capillary gel (Figure 4.1).

Chain termination sequencing in outline

The starting material for a chain termination sequencing experiment is a preparation of identical, single-stranded DNA molecules. The first step is to anneal a short oligonucleotide to the same position on each molecule, this oligonucleotide subsequently acting as the primer for synthesis of a new DNA strand that is complementary to the template (Figure 4.2A). The strand synthesis reaction, which is catalyzed by a DNA polymerase enzyme (see below) and requires the four deoxyribonucleotide triphosphates (dNTPs—dATP, dCTP, dGTP, and dTTP) as substrates, would normally continue until several thousand nucleotides had been polymerized. This does not occur in a chain termination sequencing experiment because, as well as the four deoxynucleotides, a small amount of each of four **dideoxynucleotide triphosphates** (ddNTPs—ddATP, ddCTP, ddGTP, and ddTTP) is added to the reaction. Each of these dideoxynucleotides is labeled with a different fluorescent marker.

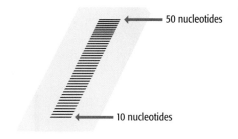

Figure 4.1 Polyacrylamide gel electrophoresis can resolve single-stranded DNA molecules that differ in length by just one nucleotide. The illustration shows the banding pattern obtained after separation of single-stranded DNA molecules by electrophoresis in a denaturing polyacrylamide slab gel. The molecules were labeled with a radioactive marker and the bands have been visualized by autoradiography. In a polyacrylamide gel, the separation between individual DNA molecules increases with their migration toward the positive electrode. Therefore the bands seen on the autoradiograph are further apart toward the bottom of the ladder. In practice, with either a slab gel or capillary system, molecules up to about 1500 nucleotides in length can be separated if the electrophoresis is continued for long enough.

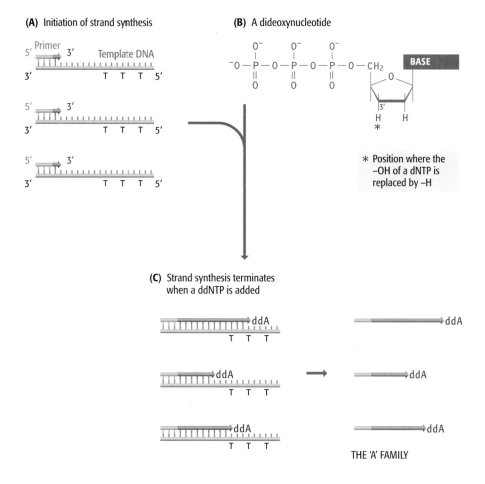

Figure 4.2 Chain termination DNA sequencing. (A) Chain termination sequencing involves the synthesis of new strands of DNA that are complementary to a single-stranded template. (B) Strand synthesis does not proceed indefinitely because the reaction mixture contains small amounts of each of the four dideoxynucleotides, which block further elongation because they have a hydrogen atom rather than a hydroxyl group attached to the 3′-carbon. (C) Incorporation of ddATP results in chains that are terminated opposite Ts in the template. This generates the "A" family of terminated molecules. Incorporation of the other dideoxynucleotides generates the "C," "G," and "T" families.

The polymerase enzyme does not discriminate between deoxy- and dideoxynucleotides, but once incorporated, a dideoxynucleotide blocks further strand elongation because it lacks the 3′–hydroxyl group needed to form a connection with the next nucleotide (Figure 4.2B). Because the normal deoxynucleotides are also present, in larger amounts than the dideoxynucleotides, the strand synthesis does not always terminate close to the primer: in fact several hundred nucleotides may be polymerized before a dideoxynucleotide is eventually incorporated. The result is a set of new molecules, all of different lengths, and each ending in a dideoxynucleotide whose identity indicates the nucleotide—A, C, G, or T—that is present at the equivalent position in the template DNA (Figure 4.2C).

To determine the DNA sequence, all that we have to do is identify the dideoxynucleotide at the end of each chain-terminated molecule. This is where the polyacrylamide gel comes into play. The DNA mixture is loaded into a well of a polyacrylamide slab gel, or into a tube of a capillary gel system, and electrophoresis carried out to separate the molecules according to their lengths. After separation, the molecules are run past a fluorescence detector capable of discriminating the labels attached to the dideoxynucleotides (Figure 4.3A). The detector therefore determines if each molecule ends in an A, C, G, or T. The sequence can be printed out for examination by the operator (Figure 4.3B), or entered directly into a storage device for future analysis. Automated sequencers with multiple capillaries working in parallel can read up to 96 different sequences in a two-hour period, which means that with an average of 750 bp per individual experiment, 864 kb of information

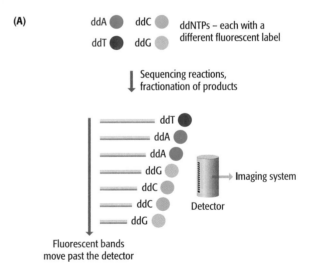

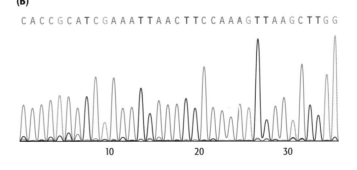

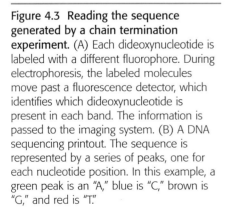

Figure 4.3 Reading the sequence generated by a chain termination experiment. (A) Each dideoxynucleotide is labeled with a different fluorophore. During electrophoresis, the labeled molecules move past a fluorescence detector, which identifies which dideoxynucleotide is present in each band. The information is passed to the imaging system. (B) A DNA sequencing printout. The sequence is represented by a series of peaks, one for each nucleotide position. In this example, a green peak is an "A," blue is "C," brown is "G," and red is "T."

can be generated per machine per day. This, of course, requires round-the-clock technical support, ideally with robotic devices used to prepare the sequencing reactions and to load the reaction products into the sequencers. If such a factory approach can be established and maintained then the data needed to sequence an entire genome can be generated in a period of weeks.

Chain termination sequencing requires a single-stranded DNA template

The template for a chain termination experiment is a single-stranded version of the DNA molecule to be sequenced. There are several ways in which this can be obtained:

- The DNA can be cloned in a plasmid vector (Section 2.2.1). The resulting DNA will be double-stranded so cannot be used directly in sequencing. Instead, it must be converted into single-stranded DNA by denaturation with alkali or by boiling. This is a common method for obtaining template DNA for DNA sequencing, largely because cloning in a plasmid vector is such a routine technique. A shortcoming is that it can be difficult to prepare plasmid DNA that is not contaminated with small quantities of bacterial DNA and RNA, which can act as spurious templates or primers in the DNA sequencing experiment.

- The DNA can be cloned in a bacteriophage M13 vector. Vectors based on M13 bacteriophage are designed specifically for the production of single-stranded templates for DNA sequencing. M13 bacteriophage has a single-stranded DNA genome which, after infection of *Escherichia coli* bacteria, is converted into a double-stranded **replicative form**. The replicative form is copied until over 100 molecules are present in the cell, and when the cell divides the copy number in the new cells is maintained by further replication. At the same time, the infected cells continually secrete new M13 phage particles—approximately 1000 per generation—these phages containing the single-stranded version of the genome (Figure 4.4). Cloning vectors based on M13 vectors are double-stranded DNA molecules equivalent to the replicative form of the M13 genome. They can be manipulated in exactly the same way as a plasmid cloning vector. The difference is that cells that have been transfected with a recombinant M13 vector secrete phage particles containing single-stranded DNA, this DNA comprising the vector molecule plus any additional DNA that has been ligated into it. The phages therefore provide the template DNA for chain termination sequencing. The one disadvantage is that DNA fragments longer than about 3 kb suffer deletions and rearrangements when cloned in an M13 vector, so the system can only be used with short pieces of DNA.

- The DNA can be cloned in a phagemid. This is a plasmid cloning vector that contains, in addition to its plasmid origin of replication, the origin from M13 or another bacteriophage with a single-stranded DNA genome. If an *E. coli* cell contains both a phagemid and the replicative form of a **helper phage**, the latter carrying genes for the phage replication enzymes and coat proteins, then the phage origin of the phagemid becomes activated, resulting in synthesis of phage particles containing the single-stranded version of the phagemid. The double-stranded plasmid DNA is therefore converted into single-stranded template DNA for DNA sequencing. This system avoids the instabilities of M13 cloning and can be used with fragments of lengths up to 10 kb or more.

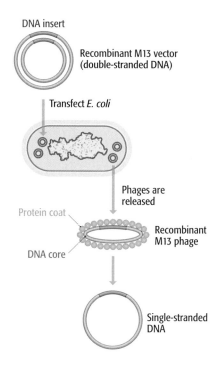

Figure 4.4 Obtaining single-stranded DNA by cloning in a bacteriophage M13 vector. M13 vectors can be obtained in two forms: the double-stranded replicative molecule and the single-stranded version found in bacteriophage particles. The replicative form can be manipulated in the same way as a plasmid cloning vector (Section 2.2.1), with new DNA inserted by restriction followed by ligation. The recombinant vector is introduced into *Escherichia coli* cells by transfection. Once inside an *E. coli* cell, the double-stranded vector replicates and directs synthesis of single-stranded copies, which are packaged into phage particles and secreted from the cell. The phage particles can be collected from the culture medium after centrifuging to pellet the bacteria. The protein coats of the phages are removed by treating with phenol, and the single-stranded version of the recombinant vector is purified for use in DNA sequencing.

DNA polymerases for chain termination sequencing

Any template-dependent DNA polymerase is capable of extending a primer that has been annealed to a single-stranded DNA molecule, but not all polymerases do this in a way that is useful for DNA sequencing. Three criteria in particular must be fulfilled by a sequencing enzyme:

- High **processivity**. This refers to the length of polynucleotide that is synthesized before the polymerase terminates through natural causes. A sequencing polymerase must have high processivity so that it does not dissociate from the template before incorporating a dideoxynucleotide.

- Negligible or zero 5'→3' exonuclease activity. Most DNA polymerases also have exonuclease activities, meaning that they can degrade DNA polynucleotides as well as synthesize them (Section 2.1.1; see Figure 2.7). This is a disadvantage in DNA sequencing because removal of nucleotides from the 5' ends of the newly synthesized strands alters the lengths of these strands, making it impossible to determine the correct sequence.

- Negligible or zero 3'→5' exonuclease activity. This is also desirable so that the polymerase does not remove the dideoxynucleotide at the end of a completed strand. If this happens then the strand might be further extended. The net result will be that there are few short strands in the reaction mixture, and the sequence close to the primer will be unreadable.

These are stringent requirements and are not entirely met by any naturally occurring DNA polymerase. Instead, artificially modified enzymes are generally used. The first of these to be developed was the Klenow polymerase, which is a version of *Escherichia coli* DNA polymerase I from which the 5'→3' exonuclease activity of the standard enzyme has been removed, either by cleaving away the relevant part of the protein or by genetic engineering (Section 2.1.1). The Klenow polymerase has relatively low processivity, limiting the length of sequence that can be obtained from a single experiment to about 250 bp, and giving nonspecific products—strands that have terminated naturally rather than by incorporation of a dideoxynucleotide—in the sequencing reaction. The Klenow enzyme has therefore been superseded by a modified version of the DNA polymerase encoded by bacteriophage T7, this enzyme going under the tradename "Sequenase". Sequenase has high processivity and no exonuclease activity, and also possesses other desirable features such as a rapid reaction rate.

The primer determines the region of the template DNA that will be sequenced

To begin a chain termination sequencing experiment, an oligonucleotide primer is annealed onto the template DNA. The primer is needed because template-dependent DNA polymerases cannot initiate DNA synthesis on a molecule that is entirely single-stranded: there must be a short, double-stranded region to provide a 3' end onto which the enzyme can add new nucleotides (Section 2.1.1).

The primer also plays the critical role of determining the region of the template molecule that will be sequenced. For most sequencing experiments a "universal" primer is used, this being one that is complementary to the part of the vector DNA immediately adjacent to the point into which new DNA is ligated (Figure 4.5A). The same universal primer can therefore give the sequence of any piece of DNA that has been ligated into the vector. Of course

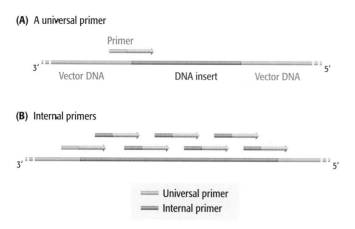

(A) A universal primer

Primer

3′ Vector DNA   DNA insert   Vector DNA   5′

(B) Internal primers

3′   5′

Universal primer
Internal primer

Figure 4.5 **Different types of primer for chain termination sequencing.** (A) A universal primer anneals to the vector DNA adjacent to the position at which new DNA is inserted. A single universal primer can therefore be used to sequence any DNA insert, but only provides the sequence of one end of the insert. (B) One way of obtaining a longer sequence is to carry out a series of chain termination experiments, each with a different internal primer that anneals within the DNA insert.

if this inserted DNA is longer than 750 bp or so then only a part of its sequence will be obtained, but usually this is not a problem because the project as a whole simply requires that a large number of short sequences are generated and subsequently assembled into the contiguous master sequence. It is immaterial whether or not the short sequences are the complete or only partial sequences of the DNA fragments used as templates. If double-stranded plasmid DNA is being used to provide the template then, if desired, more sequence can be obtained from the other end of the insert. Alternatively, it is possible to extend the sequence in one direction by synthesizing a nonuniversal internal primer, designed to anneal at a position within the insert DNA (Figure 4.5B). An experiment with this primer will provide a second short sequence that overlaps the previous one.

Thermal cycle sequencing offers an alternative to the traditional methodology

The discovery of thermostable DNA polymerases, which led to the development of PCR (Sections 2.1.1 and 2.3), has also resulted in new methodologies for chain termination sequencing. In particular, the innovation called **thermal cycle sequencing** has two advantages over traditional chain termination sequencing. First, it uses double-stranded rather than single-stranded DNA as the starting material. Second, very little template DNA is needed, so the DNA does not have to be cloned before being sequenced.

Thermal cycle sequencing is carried out in a similar way to PCR, but just one primer is used and the reaction mixture includes the four dideoxynucleotides (Figure 4.6). Because there is only one primer, only one of the strands of the starting molecule is copied, and the product accumulates in a linear fashion, not exponentially as is the case in a real PCR. The presence of the dideoxynucleotides in the reaction mixture causes chain termination, as in the standard methodology, and the family of resulting strands can be analyzed and the sequence read in the usual way.

4.1.2 Alternative methods for DNA sequencing

Although most sequencing is carried out by the chain termination method, other techniques remain important for specific applications. We will examine two of these alternative techniques: the **chemical degradation method** which, like chain termination sequencing, was devised in the 1970s, and **pyrosequencing**, which is a more recent invention.

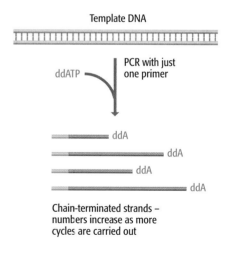

Template DNA

ddATP — PCR with just one primer

ddA
ddA
ddA
ddA

Chain-terminated strands – numbers increase as more cycles are carried out

Figure 4.6 **Thermal cycle sequencing.** PCR is carried out with just one primer and with the four dideoxynucleotides present in the reaction mixture. The result is a set of chain-terminated strands—the "A" family in the part of the reaction shown here. These strands, along with the products of the C, G, and T reactions, are electrophoresed using standard methodology (see Figure 4.3).

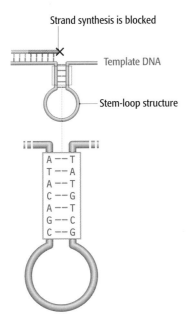

Strand synthesis is blocked

Template DNA

Stem-loop structure

A — — T
T — — A
A — — T
C — — G
A — — T
G — — C
C — — G

Figure 4.7 Intrastrand base pairing can interfere with chain termination sequencing. In this example, the template DNA can form a stem-loop structure because its sequence enables a series of intrastrand base pairs to form. This stem-loop blocks progress of the DNA polymerase, resulting in nonspecific chain termination.

Chemical degradation sequencing

One limitation of chain termination sequencing is that it may not be able to provide an accurate sequence if the template DNA is able to form intrastrand base pairs (Figure 4.7). Intrastrand base pairs can block the progress of the DNA polymerase, reducing the amount of strand synthesis that occurs, and can also alter the mobility of the chain-terminated molecules during electrophoresis, meaning that the order in which the molecules pass the detector is no longer determined solely by their length. Intrastrand base pairs do not hinder chemical degradation sequencing, so this method can be used as an alternative when such problems arise.

The chemical degradation method is similar to chain termination sequencing in that the sequence is determined by examining the lengths of molecules whose terminal nucleotide is known. However, these molecules are generated in a totally different manner, by treatment with chemicals that cut specifically at a particular nucleotide. This means that at least four separate sequencing reactions must be carried out, one for each nucleotide.

The starting material is double-stranded DNA, which is first labeled by attaching a radioactive phosphorus group to the 5′ end of each strand (Figure 4.8A). Dimethylsulfoxide (DMSO) is then added and the DNA heated to 90°C. This breaks the base pairing between the strands, enabling them to be separated from one another by gel electrophoresis, the basis to this being that one of the strands probably contains more purine nucleotides than the other and is therefore slightly heavier and runs more slowly during the electrophoresis. One strand is purified from the gel and divided into four samples, each of which is treated with one of the cleavage reagents. To illustrate the procedure, we will follow the "G" reaction (Figure 4.8B). First, the molecules are treated with dimethyl sulfate, which attaches a methyl group to the purine ring of G nucleotides. Only a limited amount of dimethyl sulfate is added, the objective being to modify, on average, just one G residue per polynucleotide. At this stage the DNA strands are still intact, cleavage not occurring until a second chemical—piperidine—is added. Piperidine removes the modified purine ring and cuts the DNA molecule at the phosphodiester bond immediately upstream of the baseless site that is created. The result is a set of cleaved DNA molecules, some of which are labeled and some of which are not. The labeled molecules all have one end in common and one end determined by the cut sites, the latter indicating the positions of the G nucleotides in the DNA molecules that were cleaved. Similar approaches are used to generate additional families of cleaved molecules, though these are usually not simply "A," "T," and "C" families as problems have been encountered in developing chemical treatments to cut specifically at A or T. The four reactions that are carried out are therefore usually "G," "A + G," "C," and "C + T". This complicates things but does not affect the accuracy of the sequence that is determined.

The family of molecules generated in each reaction is loaded into a lane of a polyacrylamide slab gel and, after electrophoresis, the positions of the bands in the gel are visualized by autoradiography (see Technical Note 2.1). The band that has moved the furthest represents the smallest piece of DNA. In the example shown in Figure 4.8C, this band lies in the "A + G" lane. There is no equivalent-sized band in the "G" lane, so the first nucleotide in the sequence is "A". The next size position is occupied by two bands, one in the "C" lane and one in the "C + T" lane: the second nucleotide is therefore "C" and the

(A) DNA labeling and strand dissociation

(B) The G reaction

(C) Reading the sequence from the autoradiograph

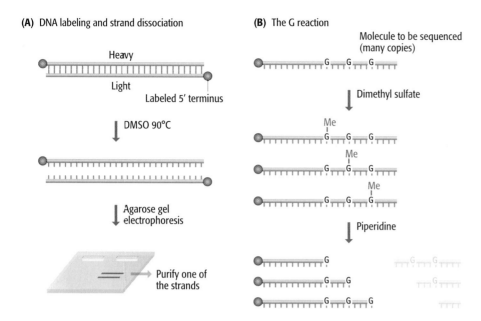

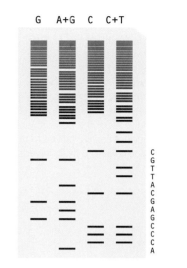

sequence so far is "AC". The sequence reading can be continued up to the region of the gel where individual bands were not separated.

Figure 4.8 Chemical degradation sequencing.

Pyrosequencing is used for rapid determination of very short sequences

Pyrosequencing does not require electrophoresis or any other fragment separation procedure and so is more rapid than either chain termination or chemical degradation sequencing. It can only generate a few tens of base pairs per experiment, but it is becoming an important technique in situations where many short sequences must be generated as quickly as possible, for example in SNP typing (Section 3.2.2).

In pyrosequencing, the template is copied in a straightforward manner without added dideoxynucleotides. As the new strand is being made, the order in which the deoxynucleotides are incorporated is detected, so the sequence can be "read" as the reaction proceeds. The addition of a deoxynucleotide to the end of the growing strand is detectable because it is accompanied by release of a molecule of pyrophosphate, which can be converted by the enzyme sulfurylase into a flash of chemiluminescence. Of course, if all four deoxynucleotides were added at once then flashes of light would be seen all the time and no useful sequence information would be obtained. Each deoxynucleotide is therefore added separately, one after the other, with a nucleotidase enzyme also present in the reaction mixture so that if a deoxynucleotide is not incorporated into the polynucleotide then it is rapidly degraded before the next one is added (Figure 4.9). This procedure makes it possible to follow the order in which the deoxynucleotides are incorporated into the growing strand. The technique sounds complicated, but it simply requires that a repetitive series of additions be made to the reaction mixture, precisely the type of procedure that is easily automated. Detection of the chemiluminescence is very sensitive, so each reaction can be in a very small volume, perhaps just one picoliter. This means that up to 1.6 million reactions can be carried out in parallel on a 6.4 cm^2 slide, enabling 25 million nucleotides of sequence to be obtained in four hours, a rate of sequence generation some 100 times faster than is possible by the chain termination method.

Figure 4.9 Pyrosequencing. The strand synthesis reaction is carried out in the absence of dideoxynucleotides. Each deoxynucleotide is added individually, along with a nucleotidase enzyme that degrades the deoxynucleotide if it is not incorporated into the strand being synthesized. Incorporation is detected by a flash of chemiluminescence induced by the pyrophosphate released from the deoxynucleotide. The order in which deoxynucleotides are added to the growing strand can therefore be followed.

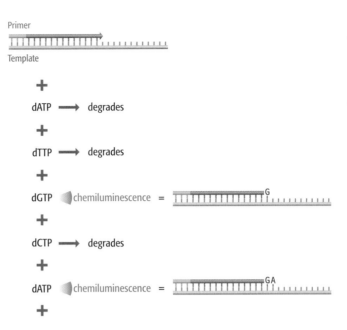

4.2 Assembly of a Contiguous DNA Sequence

The next question to address is how the master sequence of a chromosome, possibly several tens of megabases in length, can be assembled from the multitude of short sequences generated by chain termination sequencing. We addressed this issue at the start of Chapter 3 and established that the relatively short genomes of prokaryotes can be assembled by the *shotgun method*, which involves breaking the DNA molecule into fragments, determining the sequence of each one, and using a computer to search for overlaps from which the master sequence is built up (see Figure 3.1 and Section 4.2.1). This approach has now been used with over 200 prokaryotic genomes, but might lead to errors if applied to larger eukaryotic genomes, mainly because the presence of repetitive sequences in eukaryotic genomes complicates the search for sequence overlaps and could result in segments of the genome being assembled incorrectly (see Figure 3.2). To avoid these errors, the approach referred to as the *whole-genome shotgun method* uses a map to aid assembly of the master sequence (see Figure 3.3 and Section 4.2.3). Whole-genome shotgun sequencing has been used successfully with several eukaryotic genomes, including the fruit-fly and human genomes, but it is generally accepted that the greatest degree of accuracy is achieved with the *clone contig method*. In this approach, the genome is broken down into segments, each with a known position on the genome map, before sequencing is carried out (see Figure 3.3 and Section 4.2.2). We will start by examining how the shotgun method has been applied to prokaryotic genomes.

4.2.1 Sequence assembly by the shotgun method

The straightforward approach to sequence assembly is to build up the master sequence directly from the short sequences obtained from individual sequencing experiments, simply by examining the sequences for overlaps (see Figure 3.1). This is called the shotgun method. It does not require any prior knowledge of the genome and so can be carried out in the absence of a genetic or physical map.

The potential of the shotgun method was proven by the Haemophilus influenzae sequence

During the early 1990s there was extensive debate about whether the shotgun method would work in practice, many molecular biologists being of the opinion that the amount of data handling needed to compare all the mini-sequences and identify overlaps, even with the smallest genomes, would be beyond the capabilities of existing computer systems. These doubts were laid to rest in 1995 when the sequence of the 1830 kb genome of the bacterium *Haemophilus influenzae* was published.

The *H. influenzae* genome was sequenced entirely by the shotgun method and without recourse to any genetic or physical map information. The strategy used to obtain the sequence is shown in Figure 4.10. The first step was to break the genomic DNA into fragments by **sonication**, a technique that uses high-frequency sound waves to make random cuts in DNA molecules. The fragments were then electrophoresed, and those in the size range 1.6–2.0 kb were purified from the agarose gel and ligated into a plasmid vector. From the resulting library, 19,687 clones were taken at random and 28,643 sequencing experiments were carried out, the number of sequencing experiments being greater than the number of plasmids because both ends of some inserts were sequenced. Of these sequencing experiments, 16% were considered to be failures because they resulted in less than 400 bp of sequence. The remaining 24,304 sequences gave a total of 11,631,485 bp, corresponding to six times the length of the *H. influenzae* genome, this amount of redundancy being deemed necessary to ensure complete coverage. Sequence assembly required 30 hours on a computer with 512 Megabytes of random access memory (RAM), and resulted in 140 lengthy, contiguous sequences, each of these **sequence contigs** representing a different, nonoverlapping portion of the genome.

The next step was to join up pairs of contigs by obtaining sequences from the gaps between them (Figure 4.11). First, the library was checked to see if there were any clones whose two end-sequences were located in different contigs. If such a clone could be identified, then additional sequencing of its insert would close the "sequence gap" between the two contigs (Figure 4.11A). In fact, there were 99 clones in this category, so 99 of the gaps could be closed without too much difficulty.

This left 42 gaps, which probably consisted of DNA sequences that were unstable in the cloning vector and therefore not present in the library. To close these "physical gaps" a second clone library was prepared, this one with a different type of vector. Rather than using another plasmid, in which the uncloned sequences would probably still be unstable, the second library was prepared in a bacteriophage λ vector (Section 2.2.1). This new library was

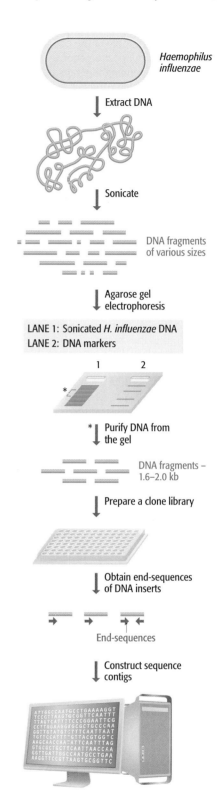

Figure 4.10 The way in which the shotgun method was used to obtain the DNA sequence of the *Haemophilus influenzae* genome. *H. influenzae* DNA was sonicated, and fragments with sizes between 1.6 kb and 2.0 kb were purified from an agarose gel and ligated into a plasmid vector to produce a clone library. End-sequences were obtained from clones taken from this library, and a computer was used to identify overlaps between sequences. This resulted in 140 sequence contigs, which were assembled into the complete genome sequence, as shown in Figure 4.11.

Figure 4.11 Assembly of the complete *Haemophilus influenzae* **genome sequence by spanning the gaps between individual sequence contigs.** (A) "Sequence gaps" are ones that can be closed by further sequencing of clones already present in the library. In this example, the end-sequences of contigs 1 and 2 lie within the same plasmid clone, so further sequencing of this DNA insert with internal primers (see Figure 4.5B) will provide the sequence to close the gap. (B) "Physical gaps" are stretches of sequence that are not present in the clone library, probably because these regions are unstable in the cloning vector that was used. Two strategies for closing these gaps are shown. On the left, a second clone library, prepared with a bacteriophage λ vector rather than a plasmid vector, is probed with oligonucleotides corresponding to the ends of the contigs. Oligonucleotides 1 and 7 both hybridize to the same clone, whose insert must therefore contain DNA spanning the gap between contigs 1 and 4. On the right, PCRs are carried out with pairs of oligonucleotides. Only numbers 1 and 7 give a PCR product, confirming that the contig ends represented by these two oligonucleotides are close together in the genome. The PCR product or the insert from the λ clone could be sequenced to close the gap between contigs 1 and 4.

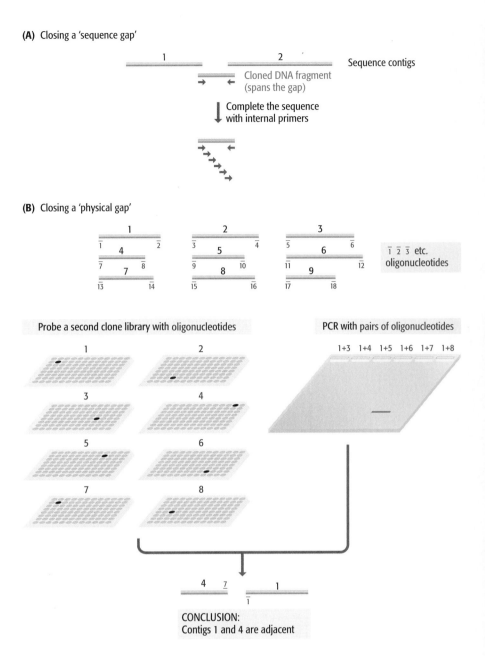

probed with 84 oligonucleotides, one at a time, these 84 oligonucleotides having sequences identical to the sequences at the ends of the unlinked contigs (Figure 4.11B). The rationale was that if two oligonucleotides hybridized to the same λ clone then the ends of the contigs from which they were derived must lie within that clone, and sequencing the DNA in the λ clone would therefore close the gap. Twenty-three of the 42 physical gaps were dealt with in this way.

A second strategy for gap closure was to use pairs of oligonucleotides, from the set of 84 described above, as primers for PCRs of *H. influenzae* genomic DNA. Some oligonucleotide pairs were selected at random, and those spanning a gap were identified simply from whether or not they gave a PCR product (see Figure 4.11B). Sequencing these PCR products closed the relevant gaps. Other primer pairs were chosen on a more rational basis. For example, oligonucleotides were used as Southern hybridization probes (see Figure 2.11) with *H. influenzae* DNA cut with a variety of restriction endonucleases, and pairs that hybridized to similar sets of restriction fragments were identified. The two members of an oligonucleotide pair identified in this way must be contained within the same restriction fragments and so are likely to lie close together on the genome. This means that the pair of contigs that the oligonucleotides are derived from are adjacent, and the gap between the contigs can be spanned by a PCR of genomic DNA using the two oligonucleotides as primers, which will provide the template DNA for gap closure.

The demonstration that a small genome can be sequenced relatively rapidly by the shotgun method led to a sudden plethora of completed microbial genomes. These projects demonstrated that shotgun sequencing can be set up on a production-line basis, with each team member having his or her individual task in DNA preparation, carrying out the sequencing reactions, or analyzing the data. This strategy resulted in the 580 kb genome of *Mycoplasma genitalium* being sequenced by five people in just eight weeks, and it is now accepted that a few months should be ample time to generate the complete sequence of any genome less than about 5 Mb in size, even if nothing is known about the genome before the project begins. The strengths of the shotgun method are therefore its speed and its ability to work in the absence of a genetic or physical map.

4.2.2 Sequence assembly by the clone contig method

The clone contig method is looked on as the conventional method for obtaining the sequence of a eukaryotic genome, and it has also been used with those microbial genomes that have previously been mapped by genetic and/or physical means. In the clone contig method, the genome is broken into fragments of up to 1.5 Mb in length, usually by partial restriction (Section 3.3.1), and these fragments are cloned in a high-capacity vector such as a BAC (Section 2.2.1). A clone contig is built up by identifying clones containing overlapping fragments, which are then individually sequenced by the shotgun method. Ideally the cloned fragments are anchored onto a genetic and/or physical map of the genome, so that the sequence data from the contig can be checked and interpreted by looking for features (e.g., STSs, SSLPs, genes) known to be present in a particular region.

Clone contigs can be built up by chromosome walking, but the method is laborious

The simplest way to build up an overlapping series of cloned DNA fragments is to begin with one clone from a library, identify a second clone whose insert overlaps with the insert in the first clone, then identify a third clone whose insert overlaps with the second clone, and so on. This is the basis of **chromosome walking**, which was the first method devised for assembly of clone contigs.

Chromosome walking was originally used to move relatively short distances along DNA molecules, using clone libraries prepared with λ or cosmid vectors. The most straightforward approach is to use the insert DNA from the starting clone as a hybridization probe to screen all the other clones in the library. Clones whose inserts overlap with the probe give positive hybridization signals, and their inserts can be used as new probes to continue the walk (Figure 4.12).

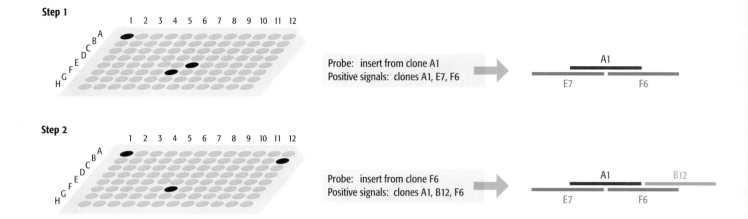

Step 1

Step 2

Probe: insert from clone A1
Positive signals: clones A1, E7, F6

Probe: insert from clone F6
Positive signals: clones A1, B12, F6

Figure 4.12 Chromosome walking. The library comprises 96 clones, each containing a different insert. To begin the walk, the insert from one of the clones is used as a hybridization probe against all the other clones in the library. In the example shown, clone A1 is the probe; it hybridizes to itself and to clones E7 and F6. The inserts from the last two clones must therefore overlap with the insert from clone A1. To continue the walk, the probing is repeated but this time with the insert from clone F6. The hybridizing clones are A1, F6, and B12, showing that the insert from B12 overlaps with the insert from F6.

Figure 4.13 Chromosome walking by PCR. The two oligonucleotides anneal within the end region of insert number 1. They are used in PCRs with all the other clones in the library. Only clone 15 gives a PCR product, showing that the inserts in clones 1 and 15 overlap. The walk would be continued by sequencing the fragment from the other end of clone 15, designing a second pair of oligonucleotides, and using these in a new set of PCRs with all the other clones.

The main problem that arises is that if the probe contains a repeat sequence then it will hybridize not only to overlapping clones but also to nonoverlapping clones whose inserts contain copies of the repeat. The extent of this nonspecific hybridization can be reduced by blocking the repeat sequences by prehybridization with unlabeled genomic DNA (see Figure 3.34), but this does not completely solve the problem, especially if the walk is being carried out with long inserts from a high-capacity vector such as a BAC. For this reason, intact inserts are rarely used for chromosome walks with human DNA and similar DNAs, which have a high frequency of repeat sequences. Instead, a fragment from the end of an insert is used as the probe, there being less chance of a repeat occurring in a short end-fragment compared with the insert as a whole. If complete confidence is required then the end-fragment can be sequenced before use to ensure that no repetitive DNA is present.

If the end-fragment has been sequenced then the walk can be speeded up by using PCR rather than hybridization to identify clones with overlapping inserts. Primers are designed from the sequence of the end-fragment and used in attempted PCRs with all the other clones in the library. A clone that gives a PCR product of the correct size must contain an overlapping insert (Figure 4.13). To speed up the process even more, rather than performing a PCR with each individual clone, groups of clones are mixed together in such a way that unambiguous identification of overlapping inserts can still be made. The method is illustrated in Figure 4.14, in which a library of 960 clones has been prepared in ten microtiter trays, each tray comprising 96 wells in an 8 × 12 array, with one clone per well. PCRs are carried out as follows:

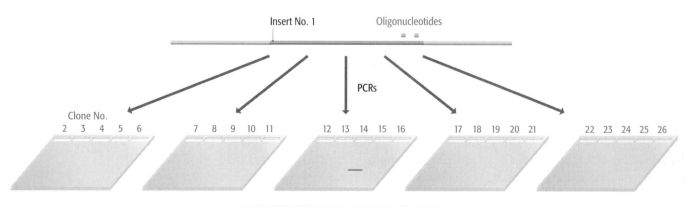

CONCLUSION: Inserts 1 and 15 overlap

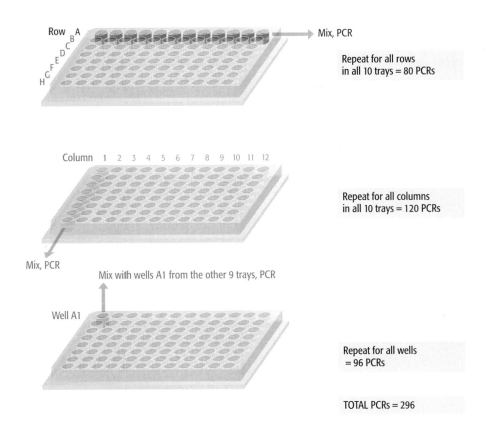

Row A
B
C
D
E
F
G
H

Mix, PCR

Repeat for all rows
in all 10 trays = 80 PCRs

Column 1 2 3 4 5 6 7 8 9 10 11 12

Repeat for all columns
in all 10 trays = 120 PCRs

Mix, PCR

Mix with wells A1 from the other 9 trays, PCR

Well A1

Repeat for all wells
= 96 PCRs

TOTAL PCRs = 296

Figure 4.14 Combinatorial screening of clones in microtiter trays. In this example, a library of 960 clones has to be screened by PCR. Rather than carrying out 960 individual PCRs, the clones are grouped as shown and just 296 PCRs are performed. In most cases, the results enable positive clones to be identified unambiguously. In fact, if there are few positive clones, then sometimes they can be identified by just the "row" and "column" PCRs. For example, if positive PCRs are obtained with tray 2 row A, tray 6 row D, tray 2 column 7, and tray 6 column 9, then it can be concluded that there are two positive clones, one in tray 2 well A7 and one in tray 6 well D9. The "well" PCRs are needed if there are two or more positive clones in the same tray.

- Samples of each clone in row A of the first microtiter tray are mixed together and a single PCR carried out. This is repeated for every row of every tray—80 PCRs in all.

- Samples of each clone in column 1 of the first microtiter tray are mixed together and a single PCR carried out. This is repeated for every column of every tray—120 PCRs in all.

- Clones from well A1 of each of the ten microtiter trays are mixed together and a single PCR carried out. This is repeated for every well—96 PCRs in all.

As explained in the legend to Figure 4.14, these 296 PCRs provide enough information to identify which of the 960 clones give products and which do not. Ambiguities arise only if a substantial number of clones turn out to be positive.

More rapid methods for clone contig assembly

Even when the screening step is carried out by the combinatorial PCR approach shown in Figure 4.14, chromosome walking is a slow process and it is rarely possible to assemble contigs of more than 15 to 20 clones by this method. The procedure has been extremely valuable in **positional cloning**, where the objective is to walk from a mapped site to an interesting gene that is known to be no more than a few megabases distant. It has been less valuable for assembling clone contigs across entire genomes, especially with the complex genomes of higher eukaryotes. So what alternative methods are there?

The main alternative is to use a **clone fingerprinting** technique. Clone fingerprinting provides information on the physical structure of a cloned DNA fragment, this physical information or "fingerprint" being compared with equivalent data from other clones, enabling those with similarities—possibly

Figure 4.15 Four clone fingerprinting techniques.

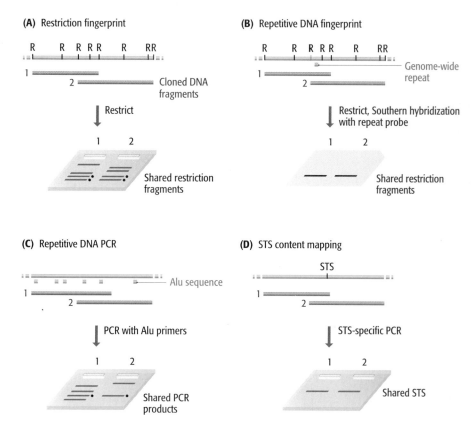

indicating overlaps—to be identified. One or a combination of the following techniques is used (Figure 4.15):

- **Restriction patterns** can be generated by digesting clones with a variety of restriction enzymes and separating the products in an agarose gel. If two clones contain overlapping inserts then their restriction fingerprints will have bands in common, as both will contain fragments derived from the overlap region.

- **Repetitive DNA fingerprints** can be prepared by analyzing a set of restriction fragments by Southern hybridization (Section 2.1.2) with probes specific for one or more types of repeat sequence. As with restriction fingerprints, overlaps are identified by looking for two clones that have some hybridizing bands in common.

- **Repetitive DNA PCR**, or **interspersed repeat element PCR** (**IRE-PCR**), uses primers that anneal within repeat sequences and so amplify the single-copy DNA between two neighboring repeats. Because repeat sequences are not evenly spaced in a genome, the sizes of the products obtained after repetitive DNA PCR can be used as a fingerprint in comparisons with other clones in order to identify potential overlaps. With human DNA, the repeats called Alu elements (Section 9.2.1) are often used because these occur, on average, once every 3 kb. An **Alu-PCR** of a human BAC insert of 150 kb would therefore be expected to give approximately 50 PCR products of various sizes, resulting in a detailed fingerprint.

- **STS content mapping** is particularly useful because it can result in a clone contig that is anchored onto a physical map of STS locations. PCRs directed at individual STSs (Section 3.3.3) are carried out with each member of a clone library. Presuming the STS is single copy in the genome, then all clones that give PCR products must contain overlapping inserts.

As with chromosome walking, efficient application of these fingerprinting techniques requires combinatorial screening of gridded clones, ideally with computerized methodology for analyzing the resulting data.

4.2.3 Whole-genome shotgun sequencing

The whole-genome shotgun method was first proposed by Craig Venter and colleagues as a means of speeding up the acquisition of contiguous sequence data for large genomes such as the human genome and those of other eukaryotes. Experience with conventional shotgun sequencing (Section 4.2.1) had shown that if the total length of sequence that is generated is between 6.5 and 8 times the length of the genome being studied, then the resulting sequence contigs will span over 99.8% of the genome, with a few gaps that can be closed by methods such as those developed during the *Haemophilus influenzae* project (see Figure 4.11). This implies that 70 million individual sequences each 500 bp or so in length, corresponding to a total of 35,000 Mb, would be sufficient if the random approach were taken with a mammalian genome of size between 3000 Mb and 3500 Mb. Seventy million sequences is not an impossibility: in fact, with 60 automatic sequencers, each determining 96 sequences every two-hour period of every day, the task could be achieved in three years.

Could 70 million sequences be assembled correctly? If the conventional shotgun method is used with such a large number of fragments, and no reference is made to a genome map, then the answer is certainly no. The huge amount of computer time needed to identify overlaps between the sequences, and the errors, or at best uncertainties, caused by the extensive repetitive DNA content of most eukaryotic genomes (see Figure 3.2), would make the task impossible. But with reference to a map, it might be possible to assemble the minisequences in the correct way.

Key features of whole-genome shotgun sequencing

The most time-consuming part of a shotgun sequencing project is the phase when individual sequence contigs are joined by closure of sequence gaps and physical gaps (see Figure 4.11). To minimize the amount of gap closure that is needed, the whole-genome shotgun method makes use of at least two clone libraries, prepared with different types of vector. At least two libraries are used because, with any cloning vector, it is anticipated that some fragments will not be cloned due to incompatibility problems that prevent vectors containing these fragments from being propagated. Different types of vector suffer from different problems, so fragments that cannot be cloned in one vector can often be cloned if a second vector is used. Generating sequence from fragments cloned in two different vectors should therefore improve the overall coverage of the genome.

What about the problems that repeat elements pose for sequence assembly? We highlighted this issue in Chapter 3 as the main argument against the use of shotgun sequencing with eukaryotic genomes, because of the possibility that jumps between repeat units will lead to parts of a repetitive region being left out, or an incorrect connection being made between two separate pieces of the same or different chromosomes (see Figure 3.2). Several possible solutions to this problem have been proposed, but the most successful strategy is to ensure that one of the clone libraries contains fragments that are longer than the longest repeat sequences in the genome being studied. For example, one of the plasmid libraries used when this method was applied to the

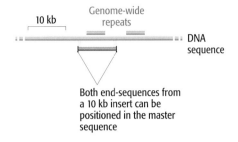

(A) Correct sequence assembly

Both end-sequences from a 10 kb insert can be positioned in the master sequence

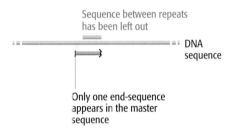

(B) Incorrect sequence assembly

Only one end-sequence appears in the master sequence

Figure 4.16 Avoiding errors when the whole-genome shotgun approach is used. In Figure 3.2B, we saw how easy it would be to "jump" between repeat sequences when assembling the master sequence by the standard shotgun method. The result of such an error would be to lose all the sequence between the two repeats that had mistakenly been linked together. This type of error is avoided in the whole-genome shotgun method by ensuring that the two end-sequences of a cloned DNA fragment (of size 10 kb or so) both appear on the master sequence at their expected positions. If one of the end-sequences is missing, then an error has been made when assembling the master sequence.

(A) Scaffolds

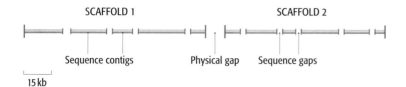

SCAFFOLD 1   SCAFFOLD 2

Sequence contigs   Physical gap   Sequence gaps

15 kb

(B) Closing a sequence gap

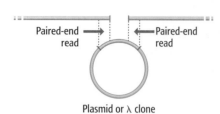

Paired-end →   ← Paired-end
read   read

Plasmid or λ clone

Figure 4.17 The initial result of sequence assembly using the whole-genome shotgun approach. (A) Scaffolds are intermediates in sequence assembly by the whole-genome shotgun approach. Two scaffolds are shown. Each comprises a series of sequence contigs separated by sequence gaps, with the scaffolds themselves separated by physical gaps. (B) Sequence gaps lie between paired-end reads—a pair of minisequences from the two ends of a single cloned fragment—and therefore can be closed by further sequencing of the cloned DNA.

Drosophila genome contained inserts with an average size of 10 kb, because most *Drosophila* repeat sequences are 8 kb or less. Sequence jumps, from one repeat sequence to another, are avoided by ensuring that the two end-sequences of each 10 kb insert are at their appropriate positions in the master sequence (Figure 4.16).

The initial result of sequence assembly is a series of **scaffolds** (Figure 4.17A), each scaffold comprising a set of sequence contigs separated by sequence gaps that lie between **paired-end reads**—the minisequences from the two ends of a single cloned fragment—and so are gaps that can be closed by further sequencing of that fragment (Figure 4.17B). The scaffolds themselves are separated by physical gaps, which are more difficult to close because they represent sequences that are not in the clone libraries. The marker content of each scaffold is used to determine its position on the genome map. For example, if the locations of STSs in the genome map are known then a scaffold can be positioned by determining which STSs it contains. If a scaffold contains STSs from two noncontiguous parts of the genome then an error has occurred during sequence assembly. The accuracy of sequence assembly can be further checked by obtaining end-sequences from fragments of 100 kb or more that have been cloned in a high-capacity vector. If a pair of end-sequences do not fall within a single scaffold at their anticipated positions relative to each other, then again an error in assembly has occurred.

The feasibility of the whole-genome shotgun method has been demonstrated by its application to the fruit-fly and human genomes. But questions still remain about the veracity of genome sequences produced by this method. Comparisons between the two versions of the human genome (Section 4.3) have shown that the sequence generated by the whole-genome shotgun method contains a substantial number of missing segments, totaling 160 Mb, these segments having been lost from the sequence because of the problems caused by repetitive DNA. These errors have resulted in 36 genes being completely deleted, and a further 67 partially deleted. It has also been suggested that a sequence obtained by the whole-genome shotgun method might not have the desired degree of accuracy, even in the regions that have been assembled correctly. Part of the problem is that the random nature of sequence generation means that some parts of the genome are covered by

Figure 4.18 The random nature of sequence generation by the whole-genome shotgun approach means that some parts of the genome are covered by more minisequences than other parts.

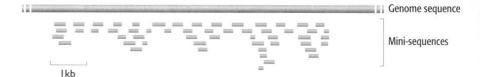

Genome sequence

Mini-sequences

1 kb

several of the minisequences that are obtained, whereas other parts are represented just once or twice (Figure 4.18). It is generally accepted that every part of a genome should be sequenced at least four times to ensure an acceptable level of accuracy, and that this coverage should be increased to 8–10 times before the sequence can be looked upon as being complete. A sequence obtained by the whole-genome shotgun method is likely to exceed this requirement in many regions, but may fall short in other areas. If those areas include genes, then the lack of accuracy could cause major problems when attempts are made to locate the genes and understand their functions (see Chapter 5). These problems were highlighted by analysis of the draft *Drosophila* sequence produced by the whole-genome shotgun method, which suggested that as many as 6500 of the 13,600 genes might contain significant sequence errors.

4.3 The Human Genome Projects

To conclude our examination of mapping and sequencing we will look at how these techniques were applied to the human genome. Although every genome project is different, with its own challenges and its own solutions to those challenges, the human projects illustrate the general issues that have had to be addressed in order to sequence a large eukaryotic genome, and in many ways illustrate the procedures that are currently regarded as state-of-the-art in this area of molecular biology.

4.3.1 The mapping phase of the Human Genome Project

Until the beginning of the 1980s, a detailed map of the human genome was considered to be an unattainable objective. Although comprehensive genetic maps had been constructed for fruit flies and a few other organisms, the problems inherent in analysis of human pedigrees (Section 3.2.4) and the relative paucity of polymorphic genetic markers meant that most geneticists doubted that a human genetic map could ever be achieved. The initial impetus for human genetic mapping came from the discovery of RFLPs, which were the first highly polymorphic DNA markers to be recognized in animal genomes. In 1987 the first human RFLP map was published, comprising 393 RFLPs and 10 additional polymorphic markers. This map, developed from analysis of 21 families, had an average marker density of one per 10 Mb.

In the late 1980s, the Human Genome Project became established as a loose but organized collaboration between geneticists in all parts of the world. One of the goals that the Project set itself was a genetic map with a density of one marker per 1 Mb, although it was thought that a density of one per 2–5 Mb might be the realistic limit. In fact by 1994 an international consortium had met and indeed exceeded the objective, thanks to their use of SSLPs and the large CEPH collection of reference families (Section 3.2.4). The 1994 map contained 5800 markers, of which over 4000 were SSLPs, and had a density of one marker per 0.7 Mb. A subsequent version took the 1994 map slightly further by inclusion of an additional 1250 SSLPs.

Physical mapping did not lag far behind. In the early 1990s, considerable effort was put into the generation of clone contig maps, using STS screening (Section 3.3.3) as well as other clone fingerprinting methods (Section 4.2.2). The major achievement of this phase of the physical mapping project was publication of a clone contig map of the entire genome, consisting of 33,000

DNA

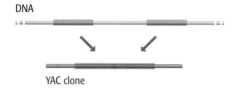

YAC clone

Figure 4.19 Some YAC clones contain segments of DNA from different parts of the human genome.

YACs containing fragments with an average size of 0.9 Mb. However, doubts were raised about the value of YAC contig maps when it was realized that YAC clones can contain two or more pieces of noncontiguous DNA (Figure 4.19). The use of these chimeric clones in the construction of contig maps could result in DNA segments that are widely separated in the genome being mistakenly mapped to adjacent positions. These problems led to the adoption of radiation hybrid mapping of STS markers (Section 3.3.3), largely by the Whitehead Institute/MIT Genome Center in Massachusetts, culminating in 1995 with publication of a human STS map containing 15,086 markers, with an average density of one per 199 kb. This map was later supplemented with an additional 20,104 STSs, most of these being ESTs and hence positioning protein-coding genes on the physical map. The resulting map density approached the target of one marker per 100 kb set as the objective for physical mapping at the outset of the Human Genome Project.

The combined STS maps included positions for almost 7000 polymorphic SSLPs that had also been mapped onto the genome by genetic means. As a result, the physical and genetic maps could be directly compared, and clone contig maps that included STS data could be anchored onto both maps. The net result was a comprehensive, integrated map that could be used as the framework for the DNA sequencing phase of the Human Genome Project.

4.3.2 Sequencing the human genome

The original plan was that the sequencing phase of the Human Genome Project would be based on YAC libraries, because this type of vector can be used with DNA fragments longer than can be handled by any other type of cloning system. This strategy had to be abandoned when it was discovered that some YAC clones contained noncontiguous fragments of DNA. The Project therefore turned its attention to BACs (Section 2.2.1). A library of 300,000 BAC clones was generated and these clones mapped onto the genome, forming a "sequence-ready" map that could be used as the primary foundation for the sequencing phase of the Project, during which the insert from each BAC would be completely sequenced by the shotgun method.

At about the time when the Human Genome Project was gearing itself up to move into the sequence-acquisition phase, the whole-genome shotgun method was first proposed as an alternative to the more laborious clone contig method that had so far been adopted. The possibility that the Human Genome Project would not in fact provide the first human genome sequence stimulated the organizers of the Project to bring forward their planned dates for completion of a working draft. The first draft sequence of an entire human chromosome (number 22) was published in December 1999, and the draft sequence of chromosome 21 appeared a few months later. Finally, on June 26, 2000, accompanied by the President of the United States, Francis Collins and Craig Venter, the leaders of the two projects, jointly announced completion of their working drafts, which appeared in print eight months later.

It is important to understand that the two genome sequences published in 2001 were drafts, not complete final sequences. For example, the version obtained by the clone contig method covered just 90% of the genome, the missing 320 Mb lying predominantly in **constitutive heterochromatin** (Section 10.1.2)—regions of chromosomes in which the DNA is very tightly packaged and which are thought to contain few, if any genes. Within the 90%

of the genome that was covered, each part had been sequenced at least four times, providing an "acceptable" level of accuracy, but only 25% had been sequenced the 8 to 10 times that is necessary before the work is considered to be "finished". Furthermore, this draft sequence had approximately 150,000 gaps, and it was recognized that some segments had probably not been ordered correctly. The International Human Genome Sequencing Consortium, which managed the final phase of the project, set as its goal a **finished sequence** of at least 95% of the **euchromatin**—the part of the genome in which most of the genes are located—with an error rate of less than one in 10^4 nucleotides, and all except the most refractory gaps filled. Achieving this goal required further sequencing of 46,000 BAC, PAC, YAC, fosmid, and cosmid clones (Section 2.2.1). The first finished chromosome sequences began to appear in 2004, with the entire genome sequence being considered complete a year later. This sequence has a total length of 2850 Mb and lacks just 28 Mb of euchromatin, the latter present in 308 gaps that have so far resisted all attempts at closure.

4.3.3 The future of the human genome projects

Completion of a finished sequence is not the only goal of the consortia working on the human genome. Understanding the genome sequence is a massive task that is engaging many groups around the world, making use of various techniques and approaches which will be described in the next two chapters. Important among these is the use of **comparative genomics**, in which two complete genome sequences are compared in order to identify common features that, being conserved, are likely to be important (see Section 5.1.1). With the human genome, comparative genomics has the added value that it may allow the animal versions of human disease genes to be located, paving the way for studies of the genetic basis of these diseases using the animal genes as models for the human condition. Drafts of the mouse and rat genomes were published in 2002, and the chimpanzee draft was completed in 2005. There will also be additional human genome projects aimed at building up a catalog of sequence variability in different populations, the results possibly enabling the ancient origins of these populations to be inferred (Section 19.3.2).

These human diversity projects lead us to the controversial aspects of genome sequencing. Most scientists anticipate that sequence data from different populations will emphasize the unity of the human race, by showing that patterns of genetic variability do not reflect the geographic and political groupings that humans have adopted during the last few centuries. But the outcomes of these projects are still certain to stimulate debate in nonscientific circles. Additional controversies center on the question of who, if anyone, will own human DNA sequences. To many, the idea of ownership of a DNA sequence is a peculiar concept, but large sums of money can be made from the information contained in the human genome, for example by using gene sequences to direct development of new drugs and therapies against cancer and other diseases. Pharmaceutical companies involved in genome sequencing naturally want to protect their investments, as they would for any other research enterprise, and currently the only way of doing this is by patenting the DNA sequences that they discover. Unfortunately, in the past, errors have been made in dealing with the financial issues relating to research with human biological material, the individual from whom the material is obtained not always being a party in the profit sharing. These issues have still to be resolved.

The problems relating to the public usage of human genome sequences are even more contentious. A major concern is the possibility that, once the sequence is understood, individuals whose sequences are considered "substandard," for whatever reason, might be discriminated against. The dangers range from increased insurance premiums for individuals whose sequences include mutations predisposing them to a genetic disease, to the possibility that racists might attempt to define "good" and "bad" sequence features, with depressingly predictable implications for the individuals unlucky enough to fall into the "bad" category.

The two human genome projects, especially in the United States, continue to support research and debate into the ethical, legal, and social issues raised by genome sequencing. In particular, great care is being taken to ensure that the genome sequences that result from the projects cannot be identified with any single individual. The DNA that is being cloned and sequenced is taken only from individuals who have given consent for their material to be used in this way and for whom anonymity can be guaranteed. When this policy was first adopted it required a certain amount of realignment of the research effort because older clone libraries had to be destroyed and the existing physical maps checked with the new material. It was accepted, however, that the extra work was necessary to maintain and enhance public confidence in the projects.

Summary

Procedures for rapid DNA sequencing were first invented in the 1970s. The version that is most frequently used today is the chain termination method, which has become popular because it is easy to automate, enabling a large number of individual experiments to be carried out in a short period of time. This is important because a single experiment gives only 750 bp or less of sequence, so thousands if not millions of individual experiments must be carried out in order to obtain the sequence of an entire genome. Other methods for DNA sequencing, such as the chemical degradation technique and pyrosequencing, have more specialized roles. When a genome is being sequenced the major challenge is assembling all the minisequences, obtained from the multiple sequencing experiments, in the correct order. With a small bacterial genome, sequence assembly is possible by the shotgun method, which simply involves examining the minisequences for overlaps. This approach, which does not need any prior knowledge of the genome, was first used in 1995 for the 1830 kb genome of *Haemophilus influenzae*, and has subsequently become the standard method for sequencing bacterial genomes. Attempts to apply this approach to larger eukaryotic genomes are complicated by the presence of repetitive DNA sequences, which can lead to segments of the genome being assembled incorrectly. The clone contig approach avoids these problems by identifying a series of clones, in a high-capacity vector such as a BAC, that contain overlapping fragments that have been anchored onto a physical and/or genetic map of the genome under study. Short clone contigs can be built up by chromosome walking, but the longer contigs used in sequencing projects are usually assembled by various clone fingerprinting techniques. The fragments present in individual clones are then sequenced by the shotgun method. This is the approach taken by the official Human Genome Project, but as that project neared the end of the sequencing phase, Craig Venter and colleagues showed that human and other larger genomes could be sequenced much more rapidly by the whole-genome shotgun method, which takes the same approach as the standard shotgun method but includes several safeguards, such as close attention to a physical map, to ensure that sequences adjacent to repetitive DNA regions are assembled correctly. Comparison of the two draft human sequences has suggested that the clone contig method provides a more accurate sequence, but that the whole-genome shotgun method, because it is rapid, is the best means of obtaining an initial draft of a genome. The human genome projects have now progressed to the stage where finished chromosome sequences have been published, these covering at least 95% of the euchromatin of each chromosome, with an error rate of less than one in 10^4 nucleotides. The ethical, legal, and social issues raised by completion of the human sequence include questions regarding ownership and patent rights, and the possibility of genetic discrimination.

Multiple Choice Questions

4.1.* What would happen if the concentration of dideoxynucleotides was too high in a chain termination sequencing reaction?

a. The reactions would yield very long molecules and there would be little sequence data close to the primer.

b. The reactions would yield very short molecules.

c. The reactions would not proceed as the high concentrations of the dideoxynucleotides would inhibit the DNA polymerase.

d. The fluorescence of the sequencing products would be too high and difficult to read.

4.2. How are the different nucleotides (A, C, G, or T) labeled in a chain termination sequencing reaction?

a. The primers for the reactions are labeled with fluorescent dyes.

b. The different deoxynucleotides are each labeled with a different fluorescent dye.

c. The different dideoxynucleotides are each labeled with a different fluorescent dye.

d. The different sequencing products are stained with antibodies that detect the different dideoxynucleotides.

4.3.* Why is it advantageous to clone a DNA fragment prior to chain termination sequencing?

a. The chain termination sequencing process requires single-stranded DNA molecules as templates.

b. The chain termination sequencing process requires double-stranded DNA molecules as templates.

c. The chain termination sequencing process requires a vector to stabilize the template DNA.

d. The dideoxynucleotides are only incorporated into cloned DNA fragments.

4.4. Why is the Klenow enzyme a poor choice for chain termination sequencing reactions?

a. The enzyme has high 5′→3′ exonuclease activity and will alter the length of the products.

b. The enzyme has high 3′→5′ exonuclease activity and will remove the 3′ dideoxynucleotides from the products.

c. The enzyme does not incorporate the dideoxynucleotides into the template chain.

d. The enzyme has low processivity which limits the length of the sequence obtained.

4.5.* Which of the following is a problem with chain termination sequencing?

a. The sequence reads are less than 100 bp.

b. The sequences often contain errors.

c. Intrastrand base pairing can block the progress of the DNA polymerase and may also affect the migration of the molecules during electrophoresis.

d. It is not possible to sequence both strands of a DNA molecule.

4.6. What is the purpose of the nucleotidase in the pyrosequencing reaction?

a. It converts the pyrophosphate into a luminescent product.

b. It degrades the DNA molecule, releasing the nucleotides that are detected via chemiluminescence.

c. It stabilizes the short DNA products produced by this technique.

d. It degrades unincorporated nucleotides in the reaction mixture.

4.7.* Many scientists doubted that shotgun sequencing would work, even with the smallest genomes, because:

a. There would not be overlaps between the different mini-sequences.

b. Computers would be unable to handle the huge amount of data generated by a shotgun sequencing project.

c. Small prokaryotic genomes contain large amounts of repetitive DNA.

d. No method existed for breaking genomic DNA into random fragments.

4.8. Why is the clone contig method useful for the sequencing of eukaryotic genomes?

a. The genomes are simply too large to be sequenced by the shotgun method.

b. The repetitive sequences of eukaryotic genomes would make the assembly of contigs generated solely by the shotgun method a difficult and error-prone task.

c. There would simply be too many recombinant plasmids to isolate using the shotgun method.

d. The clone contig method makes it easier for researchers to identify genes.

4.9.* Chromosome walking is best described as:

a. Aligning DNA sequences by computer, to generate contigs.

b. Generating a map along a chromosome in a step-by-step manner.

c. Identifying clones whose inserts overlap to generate a library of clones that cover a given segment of DNA.

Multiple Choice Questions (continued) *Answers to odd-numbered questions can be found in the Appendix

    d. Sequencing a genome one clone at a time to ensure that no gaps are present at the end of the project.

4.10. Positional cloning involves which of the following?

    a. Walking along a chromosome from a marker to a nearby gene.

    b. Assembling clone contigs for an entire genome.

    c. Identifying genes present in genomic sequences.

    d. Fingerprinting a chromosome or DNA fragment to provide a map for sequencing.

4.11.* Which methods do researchers typically use to ensure that only a minimal number of sequence and physical gaps are present in a draft genome sequence?

    a. The total number of nucleotides sequenced is the same as the size of the genome.

    b. The entire genome is cloned by chromosome walking to ensure complete coverage.

    c. It is not necessary to minimize the number of gaps as these are easily filled after the initial sequencing phase.

    d. At least two clone libraries are prepared and sequenced in different vectors.

4.12. What is the most difficult and time-consuming aspect of a shotgun sequencing project?

    a. The generation of the clone libraries.

    b. The sequencing of the clone libraries.

    c. The generation of contigs from the DNA sequences.

    d. The closure of sequence and physical gaps between the sequenced contigs.

4.13.* Which of the following methods yielded the best map of the human genome by the mid-1990s?

    a. Genetic mapping of RFLPs.

    b. Genetic mapping of SSLPs.

    c. Physical mapping of STSs.

    d. Physical mapping by FISH.

4.14. Why were yeast artificial chromosomes (YACs) not used in the sequencing phase of the Human Genome Project?

    a. It was discovered that some YACs contained cloned DNA fragments from different parts of the genome.

    b. The YAC inserts were too large to yield a manageable clone library for sequencing.

    c. Yeast genomic DNA was found to be recombining with human DNA in the YACs.

    d. YACs were found to lose large segments of insert DNA over time.

4.15.* Which of the following statements about the completion of the human genome sequence in 2000 is FALSE?

    a. Only 90% of the genome was sequenced at this time.

    b. The genome sequences released in 2000 were draft versions of the sequence.

    c. All of the euchromatin sequence was finished.

    d. A substantial amount of constitutive heterochromatin had not been sequenced.

Short Answer Questions

*Answers to odd-numbered questions can be found in the Appendix

4.1.* What is the function of the dideoxynucleotides in a chain termination sequencing reaction?

4.2. How are the different products of chain termination sequencing detected during electrophoresis?

4.3.* Would it be possible to sequence a PCR product directly (without cloning the PCR product)? How would this be accomplished?

4.4. DNA polymerases used for DNA sequencing should have which three properties?

4.5.* What are the benefits of the automated versions of the chain termination sequencing procedure?

4.6. What are the applications and limitations of pyrosequencing?

4.7.* Why does the shotgun sequencing method require that the number of nucleotides sequenced is several times larger than the size of the genome?

4.8. Explain the clone contig method of genome sequencing.

continued …

Short Answer Questions (continued)

*Answers to odd-numbered questions can be found in the Appendix

4.9.* What methods can be used to perform DNA fingerprinting with clones from a large DNA fragment that is to be sequenced?

4.10. How are scaffolds used in the assembly of genome sequences?

4.11.* What types of errors are associated with shotgun sequencing of complex eukaryotic genomes?

4.12. What benefits will be gained by comparing the genomes of mice, rats, and chimpanzees to the human genome?

In-depth Problems

*Guidance to odd-numbered questions can be found in the Appendix

4.1.* In the late 1970s, the chain termination and chemical degradation methods for DNA sequencing appeared to be equally efficacious. But today virtually all sequencing is done by the chain termination method. Why did chain termination sequencing become predominant?

4.2. You have isolated a new species of bacterium whose genome is a single DNA molecule of approximately 2.6 Mb. Write a detailed project plan to show how you would obtain the genome sequence for this bacterium.

4.3.* Critically evaluate the clone contig approach as a means of sequencing a large eukaryotic genome.

4.4. Discuss how the Human Genome Project proceeded from the time that researchers first thought about creating a map of the genome to the completion of the genome sequence. What were the critical advances that allowed researchers to create a detailed map of the genome and then to complete the sequence?

4.5.* A pharmaceutical company has invested a great deal of time and money to isolate the gene for a genetic disease. The company is studying the gene and its protein product and is working to develop drugs to treat the disease. Does the company have the right (in your opinion) to patent the gene? Justify your answer.

Figure Tests

*Answers to odd-numbered questions can be found in the Appendix

4.1.* What are the advantages of using universal or internal primers in sequencing projects?

4.2. The thermal cycle method is commonly used for DNA sequencing. What are the benefits of this process?

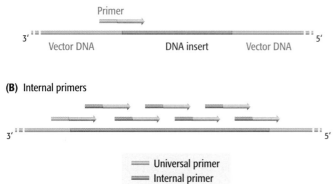

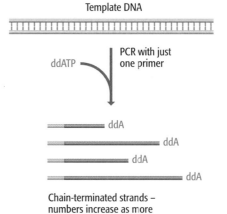

Figure Tests (continued)

4.3.* What type of sequencing reaction is this? What are the advantages of this type of sequencing?

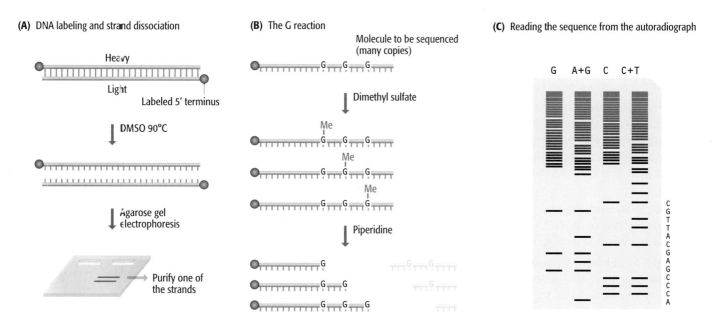

(A) DNA labeling and strand dissociation

(B) The G reaction

(C) Reading the sequence from the autoradiograph

4.4. Discuss how the different methods in the figure can be used to provide clone fingerprints.

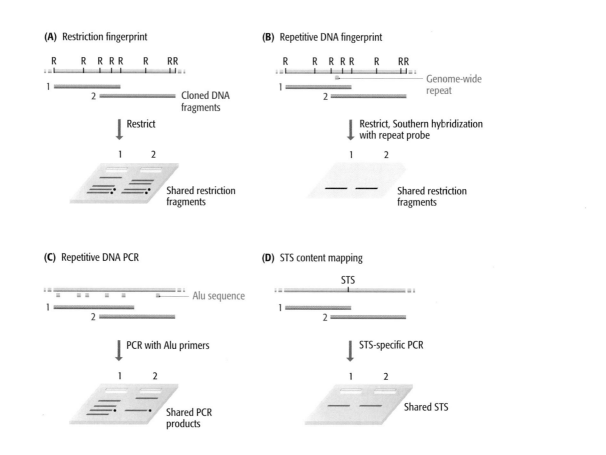

(A) Restriction fingerprint

(B) Repetitive DNA fingerprint

(C) Repetitive DNA PCR

(D) STS content mapping

Further Reading

Methodology for DNA sequencing

Brown, T.A. (ed.) (2000) *Essential Molecular Biology: A Practical Approach,*Vol. 1 and 2, 2nd Ed. Oxford University Press, Oxford. *Includes detailed protocols for DNA sequencing.*

Maxam, A.M. and Gilbert, W. (1977) A new method for sequencing DNA. *Proc. Natl Acad. Sci. USA* **74:** 560–564. *The chemical degradation method.*

Prober, J.M., Trainor, G.L., Dam, R.J., Hobbs, F.W., Robertson, C.W., Zagursky, R.J., Cocuzza, A.J., Jensen, M.A. and Baumeister, K. (1987) A system for rapid DNA sequencing with fluorescent chain-terminating dideoxynucleotides. *Science* **238:** 336–341. *The chain termination method as used today.*

Rogers, Y.-H. and Venter, J.C. (2005) Massively parallel sequencing. *Nature* **437:** 326–327. *Describes how over one million sequencing reactions can be carried out in parallel.*

Ronaghi, M., Ehleen, M. and Nyrn, P. (1998) A sequencing method based on real-time pyrophosphate. *Science* **281:** 363–365. *Pyrosequencing.*

Sanger, F., Nicklen, S. and Coulson, A.R. (1977) DNA sequencing with chain terminating inhibitors. *Proc. Natl Acad. Sci. USA* **74:** 5463–5467. *The first description of chain termination sequencing.*

Sears, L.E., Moran, L.S., Kisinger, C., Creasey, T., Perry-O'Keefe, H., Roskey, M., Sutherland, E. and Slatko, B.E. (1992) CircumVent thermal cycle sequencing and alternative manual and automated DNA sequencing protocols using the highly thermostable Vent (exo⁻) DNA polymerase. *Biotechniques* **13**: 626–633.

Examples of the shotgun method for sequence assembly

Fleischmann, R.D., Adams, M.D., White, O., et al. (1995) Whole-genome random sequencing and assembly of *Haemophilus influenzae* Rd. *Science* **269:** 496–512.

Fraser, C.M., Gocayne, J.D., White, O., et al. (1995) The minimal gene complement of *Mycoplasma genitalium*. *Science* **270:** 397–403.

The clone contig approach

IHGSC (International Human Genome Sequencing Consortium) (2001) Initial sequencing and analysis of the human genome. *Nature* **409:** 860–921.

The whole-genome shotgun approach

Adams, M.A., Celniker, S.E., Holt, R.A., et al. (2000) The genome sequence of *Drosophila melanogaster*. *Science* **287:** 2185–2195.

She, X., Jiang, Z., Clark, R.A., et al. (2004) Shotgun sequence assembly and recent segmental duplications within the human genome. *Nature* **431:** 927–930. *Determines the accuracy of the whole-genome shotgun approach in assembly of sequences containing repetitive DNA.*

Venter, J.C., Adams, M.D., Sutton, G.G., Kerlavage, A.R., Smith, H.O. and Hunkapiller, M. (1998) Shotgun sequencing of the human genome. *Science* **280:** 1540–1542.

Venter, J.C., Adams, M.D., Myers, E.W., et al. (2001) The sequence of the human genome. *Science* **291:** 1304–1351.

Weber, J.L. and Myers, E.W. (1997) Human whole-genome shotgun sequencing. *Genome Res.* **7:** 401–409.

Landmarks in the human genome projects

Donis-Keller, H., Green, P., Helms, C., et al. (1987) A genetic map of the human genome. *Cell* **51:** 319–337. *The first genetic map with a marker density of one per 10 Mb.*

Cohen, D., Chumakov, I. and Weissenbach, J. (1993) A first-generation map of the human genome. *Nature* **366:** 698–701. *The first YAC contig map.*

Murray, J.C., Buetow, K.H., Weber, J.L., et al. (1994) A comprehensive human linkage map with centimorgan density. *Science* **265:** 2049–2054. *The genetic map with a density of one marker per 0.7 Mb.*

Hudson, T.J., Stein, L.D., Gerety, S.S., et al. (1995) An STS-based map of the human genome. *Science* **270:** 1945–1954. *The physical map with a density of one marker per 199 kb.*

Dib, C., Fauré, S., Fizames, C., et al. (1996) A comprehensive genetic map of the human genome based on 5,264 microsatellites. *Nature* **380:** 152–154. *The most comprehensive genetic map.*

Schuler, G.D., Boguski, M.S., Stewart, E.A., et al. (1996) A gene map of the human genome. *Science* **274:** 540–546. *Refinement of the Hudson map, marker density close to one per 100 kb.*

Deloukas, P., Schuler, G.D., Gyapay, G., et al. (1998) A physical map of 30,000 genes. *Science* **282:** 744–746. *The integrated map used as the framework for DNA sequencing.*

IHGSC (International Human Genome Sequencing Consortium) (2001) Initial sequencing and analysis of the human genome. *Nature* **409:** 860–921. *The draft sequence obtained by the "official" Human Genome Project.*

Venter, J.C., Adams, M.D., Myers, E.W., et al. (2001) The sequence of the human genome. *Science* **291:** 1304–1351. *The draft sequence obtained by the whole-genome shotgun approach.*

IHGSC (International Human Genome Sequencing Consortium) (2004) Finishing the euchromatic sequence of the human genome. *Nature* **431:** 931–945. *Review of the finishing process and its outcomes.*

Ross, M.T., Grafham, D.V., Coffey, A.J., *et al.* (2005) The DNA sequence of the human X chromosome. *Nature* **434:** 325–337. *Description of the finished sequence of a human chromosome.*

Issues raised by the human genome sequences

Davies, K. (2001) *Cracking the Genome: Inside the Race to Unlock Human DNA.* Free Press, New York. (Published in the UK as *The Sequence: Inside the Race for the Human Genome.* Weidenfeld and Nicholson, London.) *A history of the human genome projects.*

Garver, K.L. and Garver, B. (1994) The Human Genome Project and eugenic concerns. *Am. J. Hum. Genet.* **54:** 148–158.

Wilkie, T. (1993) *Perilous Knowledge: The Human Genome Project and its Implications.* Faber and Faber, New York. *A view of the social impact of the Human Genome Project.*

Understanding a Genome Sequence

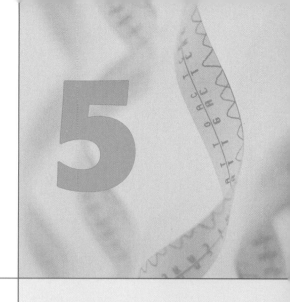

5

When you have read Chapter 5, you should be able to:

Describe the strengths and weaknesses of the computational and experimental methods used to analyze genome sequences.

Describe the basis of open reading frame (ORF) scanning, and explain why this approach is not always successful in locating genes in eukaryotic genomes.

Explain how genes for functional RNA are located in a genome sequence.

Define the term "homology" and explain how homology and comparative genomics are used to locate genes in a genome sequence.

Outline the various experimental methods used to identify parts of a genome sequence that specify RNA molecules.

Evaluate the strengths and limitations of homology analysis as a means of assigning functions to genes.

Describe the methods used to inactivate or overexpress individual genes in yeast and mammals, and explain how these methods can lead to identification of the function of a gene.

Give outline descriptions of techniques that can be used to obtain more detailed information on the activity of a protein coded by an unknown gene.

Summarize the methods used, and progress made, in annotation of the *Saccharomyces cerevisiae* genome sequence.

A genome sequence is not an end in itself. A major challenge still has to be met in understanding what the genome contains and how the genome is expressed. Attempts at understanding what a genome contains make use of a combination of computer analysis and experimentation, with the primary aims of locating the genes and determining their functions. This chapter is devoted to the methods used to address these questions. The second problem—understanding how the genome is expressed—is, to a certain extent, merely a different way of stating the objectives of molecular biology over the last 30 years. The difference is that, in the past, attention has been directed at the expression pathways for individual genes, with groups of genes being considered only when the expression of one gene is clearly linked to that of another. Now the question has become more general and relates to the expression of the genome as a whole. The techniques used to address this topic will be covered in Chapter 6.

5.1 Locating the Genes in a Genome Sequence

Once a DNA sequence has been obtained, whether it is the sequence of a single cloned fragment or of an entire chromosome, then various methods can be employed to locate the genes that are present. These methods can be divided into those that involve simply inspecting the sequence, by eye or more frequently by computer, to look for the special sequence features associated with genes, and those methods that locate genes by experimental analysis of the DNA sequence. The computer methods form part of the methodology called **bioinformatics**, and it is with these that we begin.

5.1.1 Gene location by sequence inspection

Sequence inspection can be used to locate genes because genes are not random series of nucleotides but instead have distinctive features. At present we do not fully understand the nature of all of these specific features, and sequence inspection is therefore not a foolproof way of locating genes, but it is still a powerful tool and is usually the first method that is applied to analysis of a new genome sequence.

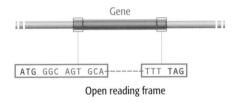

Gene

Open reading frame

Figure 5.1 A protein-coding gene is an open reading frame of triplet codons. The first four and last two codons of the gene are shown. The first four codons specify methionine/initiation–glycine–serine–alanine, and the last two specify phenylalanine–termination.

The coding regions of genes are open reading frames

Genes that code for proteins comprise open reading frames (ORFs) consisting of a series of codons that specify the amino acid sequence of the protein that the gene codes for (Figure 5.1). The ORF begins with an initiation codon—usually (but not always) ATG—and ends with a termination codon: TAA, TAG, or TGA (Section 1.3.2). Searching a DNA sequence for ORFs that begin with an ATG and end with a termination triplet is therefore one way of looking for genes. The analysis is complicated by the fact that each DNA sequence has six **reading frames**, three in one direction and three in the reverse direction on the complementary strand (Figure 5.2), but computers are quite capable of scanning all six reading frames for ORFs. How effective is this as a means of gene location?

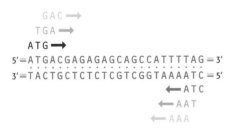

Figure 5.2 A double-stranded DNA molecule has six reading frames. Both strands are read in the 5'→3' direction. Each strand has three reading frames, depending on which nucleotide is chosen as the starting position.

The key to the success of **ORF scanning** is the frequency with which termination codons appear in the DNA sequence. If the DNA has a random sequence and a GC content of 50% then each of the three termination codons —TAA, TAG, and TGA—will appear, on average, once every $4^3 = 64$ bp. If the GC content is greater than 50% then the termination codons, being AT–rich, will occur less frequently, but one will still be expected every 100–200 bp. This means that random DNA should not show many ORFs longer than 50 codons in length, especially if the presence of a starting ATG triplet is used as part of the definition of an ORF. Most genes, on the other hand, are longer than 50 codons: the average lengths are 317 codons for *Escherichia coli*, 483 codons for *Saccharomyces cerevisiae*, and approximately 450 codons for humans. ORF scanning, in its simplest form, therefore takes a figure of, say, 100 codons as the shortest length of a putative gene and records positive hits for all ORFs longer than this.

How well does this strategy work in practice? With bacterial genomes, simple ORF scanning is an effective way of locating most of the genes in a DNA sequence. This is illustrated by Figure 5.3, which shows a segment of the *E. coli* genome with all ORFs longer than 50 codons highlighted. The real genes in the sequence cannot be mistaken because they are much longer than 50 codons in length. With bacteria the analysis is further simplified by the fact that the genes are very closely spaced and hence there is relatively little **intergenic DNA** in the genome (only 11% for *E. coli*; see Section 8.2.1). If we assume that the real genes do not overlap, which is true for most bacterial genes, then it is only in the intergenic regions that there is a possibility of mistaking a short, spurious ORF for a real gene. So if the intergenic component of a genome is small, then there is a reduced chance of making mistakes in interpreting the results of a simple ORF scan.

Simple ORF scans are less effective with DNA of higher eukaryotes

Although ORF scans work well for bacterial genomes, they are less effective for locating genes in DNA sequences from higher eukaryotes. This is partly because there is substantially more space between the real genes in a eukaryotic genome (for example, approximately 62% of the human genome is intergenic), increasing the chances of finding spurious ORFs. But the main problem with the human genome and the genomes of higher eukaryotes in general is that their genes are often split by introns (Section 1.2.3), and so do not appear as continuous ORFs in the DNA sequence. Many exons are shorter than 100 codons, some consisting of fewer than 50 codons, and continuing the reading frame into an intron usually leads to a termination sequence that

Figure 5.3 ORF scanning is an effective way of locating genes in a bacterial genome. The diagram shows 4522 bp of the lactose operon of *Escherichia coli* with all ORFs longer than 50 codons marked. The sequence contains two real genes—*lacZ* and *lacY*—indicated by the red lines. These real genes cannot be mistaken because they are much longer than the spurious ORFs, shown in yellow. See Figure 8.8A for the detailed structure of the lactose operon.

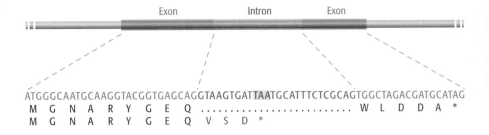

Figure 5.4 ORF scans are complicated by introns. The nucleotide sequence of a short gene containing a single intron is shown. The correct amino acid sequence of the protein translated from the gene is given immediately below the nucleotide sequence: in this sequence, the intron has been left out because it is removed from the transcript before the mRNA is translated into protein. In the lower line, the sequence has been translated without realizing that an intron is present. As a result of this error, the amino acid sequence appears to terminate within the intron. The amino acid sequences have been written using the one-letter abbreviations (see Table 1.2). Asterisks indicate the positions of termination codons. Introns are covered in detail in Section 12.2.2.

appears to close the ORF (Figure 5.4). In other words, the genes of a higher eukaryote do not appear in the genome sequence as long ORFs, and simple ORF scanning cannot locate them.

Solving the problem posed by introns is the main challenge for bioinformaticists writing new software programs for ORF location. Three modifications to the basic procedure for ORF scanning have been adopted:

- **Codon bias** is taken into account. "Codon bias" refers to the fact that not all codons are used equally frequently in the genes of a particular organism. For example, leucine is specified by six codons in the genetic code (TTA, TTG, CTT, CTC, CTA, and CTG; see Figure 1.20), but in human genes leucine is most frequently coded by CTG and is only rarely specified by TTA or CTA. Similarly, of the four valine codons, human genes use GTG four times more frequently than GTA. The biological reason for codon bias is not understood, but all organisms have a bias, which is different in different species. Real exons are expected to display the codon bias whereas chance series of triplets do not. The codon bias of the organism being studied is therefore written into the ORF-scanning software.

- **Exon–intron boundaries** can be searched for as these have distinctive sequence features, although unfortunately the distinctiveness of these sequences is not so great as to make their location a trivial task. The sequence of the upstream exon–intron boundary is usually described as:

  5′–AG↓GTAAGT–3′

  the arrow indicating the precise boundary point. However, only the "GT" immediately after the arrow is invariable: elsewhere in the sequence, nucleotides other than the ones shown are quite often found. In other words, the sequence is a **consensus**, by which we mean that the sequence shows the most frequent nucleotide at each position in all of the upstream exon–intron boundaries that are known, but that in any particular boundary sequence one or more of these positions might have a different nucleotide (Figure 5.5). The downstream intron–exon boundary is even less well defined:

  5′–PyPyPyPyPyPyNCAG↓–3′

  where "Py" means one of the pyrimidine nucleotides (T or C) and "N" is any nucleotide. Simply searching for these consensus sequences will not locate more than a few exon–intron boundaries because most have sequences other than the ones shown. Writing software that takes account of the known variables has proven difficult, and at present locating exon–intron boundaries by sequence analysis is a hit-and-miss affair.

- **Upstream regulatory sequences** can be used to locate the regions where genes begin. This is because these regulatory sequences, like exon–intron boundaries, have distinctive sequence features that they possess in order to carry out their role as recognition signals for the DNA-binding proteins involved in gene expression (Chapter 11). Unfortunately, as with exon–intron boundaries, the regulatory sequences are variable, more so in eukaryotes than in prokaryotes, and in eukaryotes not all genes have the same collection of regulatory sequences. Using these to locate genes is therefore problematic.

These three extensions of simple ORF scanning, despite their limitations, are generally applicable to the genomes of all higher eukaryotes. Additional strategies are also possible with individual organisms, based on the special features of their genomes. For example, vertebrate genomes contain **CpG islands** upstream of many genes, these being sequences of approximately 1 kb in which the GC content is greater than the average for the genome as a whole. Some 40%–50% of human genes are associated with an upstream CpG island. These sequences are distinctive and when one is located in vertebrate DNA, a strong assumption can be made that a gene begins in the region immediately downstream.

Locating genes for functional RNA

ORF scanning is appropriate for protein-coding genes, but what about those genes for functional RNAs such as rRNA and tRNA (Section 1.2.2)? These genes do not comprise open reading frames and hence will not be located by the methods described above. Functional RNA molecules do, however, have their own distinctive features, which can be used to aid their discovery in a genome sequence. The most important of these features is the ability to fold

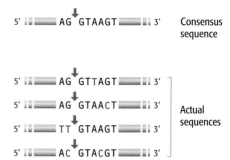

Figure 5.5 The relationship between a consensus sequence for an upstream exon–intron boundary and the actual sequences found in real genes. Differences from the consensus sequence are shown in red. At upstream exon–intron boundaries, only the "GT" immediately after the splice site (shown by the arrow) is invariant.

(A) The tRNA cloverleaf structure

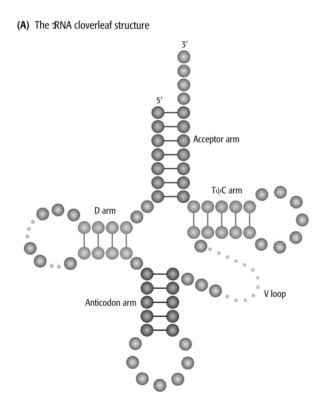

(B) Sequence of one of the *Escherichia coli* tRNA^{leu} genes

5′ GCCGAAGTGGCGAAATCGGTAGTCGCAGTTGATTCAAAATCAACCGTAGAAATACGTGCCGGTTCGAGTCCGGCCTTCGGCACCA 3′

Figure 5.6 The distinctive features of tRNAs aid location of the genes for these functional RNAs. (A) All tRNAs fold into the cloverleaf structure, which is held together by intramolecular base pairing in the four highlighted regions. (B) The DNA sequence of the gene for one of the *Escherichia coli* tRNAs specific for the amino acid leucine. The highlighted segments correspond to the regions of intramolecular base pairing shown in part A. The sequence constraints imposed by the need for these segments to be able to base-pair to one another provide features that can be searched for by computer programs designed to locate tRNA genes. For more information on tRNA structure, see Section 13.1.1.

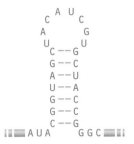

Figure 5.7 A typical RNA stem-loop structure.

into a secondary structure, such as the **cloverleaf** adopted by tRNA molecules (Figure 5.6A). These secondary structures are held together by base pairing not between two separate polynucleotides, as in the DNA double helix, but between different parts of the same polynucleotide—what we call **intramolecular base pairing**. In order for intramolecular base pairs to form, the nucleotide sequences in the two parts of the molecule must be complementary, and to produce a complex structure such as the cloverleaf, the components of these pairs of complementary sequences must be arranged in a characteristic order within the RNA sequence (Figure 5.6B). These features provide a wealth of information that can be used to locate tRNA genes in a genome sequence, and programs designed for this specific purpose are usually very successful.

As well as tRNAs, rRNAs and some of the small functional RNAs (Section 1.2.2) also adopt secondary structures that have sufficient complexity to enable their genes to be identified without too much difficulty. Other functional RNA genes are less easy to locate because the RNAs take up structures that involve relatively little base pairing or the base pairing is not in a regular pattern. Three approaches are being used for location of the genes for these RNAs:

- Although some functional RNAs do not adopt complex secondary structures, most contain one or more **stem-loops** (or **hairpins**), which result from the simplest type of intramolecular base pairing (Figure 5.7). Programs that scan DNA sequences for such structures therefore identify regions where functional RNA genes might be present. These programs incorporate thermodynamic rules that enable the stability of a stem-loop to be estimated, taking into account features such as the size of the loop, the number of base pairs in the stem, and the proportion of G–C base pairs (these being more stable than A–T pairs as they are held together by three rather than two hydrogen bonds; see Figure 1.8). A putative stem-loop structure with an estimated stability above a chosen limit is considered a possible indicator of the presence of a functional RNA gene.

- As with protein-coding genes, a search can be made for regulatory sequences associated with genes for functional RNAs. These regulatory sequences are different to those for protein-coding genes, and may be present *within* a functional RNA gene as well as upstream of it.

- In compact genomes, attention is directed toward regions that remain after a comprehensive search for protein-coding genes. Often these "empty spaces" are not empty at all and a careful examination will reveal the presence of one or more functional RNA genes.

Homology searches and comparative genomics give an extra dimension to sequence inspection

Most of the various software programs available for gene location by ORF scanning can identify up to 95% of the coding regions in a eukaryotic genome, but even the best ones tend to make frequent mistakes in their positioning of the exon–intron boundaries, and identification of spurious ORFs as real genes is still a major problem. These limitations can be offset to a certain extent by the use of a **homology search** to test whether a series of triplets is a real exon or a chance sequence. In this analysis the DNA databases are searched to determine if the test sequence is identical or similar to any genes that have already been sequenced. Obviously if the test sequence is part of a gene that has already been sequenced by someone else then an identical match will be found, but this is not the point of a homology search. Instead the intention is to determine if an entirely new sequence is *similar* to any known genes because, if it is, then there is a chance that the test and match sequences are **homologous**, meaning that they represent genes that are evolutionarily related. The main use of homology searching is to assign functions to newly discovered genes, and we will therefore return to it when we deal with this aspect of genome analysis later in the chapter (Section 5.2.1). The technique is also central to *gene location* because it enables tentative exon sequences located by ORF scanning to be tested for functionality. If the tentative exon sequence gives one or more positive matches after a homology search then it is probably a real exon, but if it gives no match then its authenticity must remain in doubt until it is assessed by one or other of the experiment-based gene location techniques.

A more precise version of homology searching is possible when genome sequences are available for two or more related species. Related species have genomes that share similarities inherited from their common ancestor, overlaid with species-specific differences that have arisen since the species began to evolve independently (Figure 5.8). Because of natural selection (Section 19.3.2), the sequence similarities between related genomes are greatest within the genes and lowest in the intergenic regions. Therefore, when related genomes are compared, homologous genes are easily identified because they have high sequence similarity, and any ORF that does not have a clear homolog in the second genome can be discounted as almost certainly being a chance sequence and not a genuine gene. This type of analysis—called **comparative genomics**—is proving very valuable for locating genes in the *Saccharomyces cerevisiae* genome (Section 5.3), as complete or partial sequences are now available not only for this yeast but also for 16 other members of the Hemiascomycetes, including *Saccharomyces paradoxus*, *Saccharomyces mikatae*, and *Saccharomyces bayanus*, the species most closely related to *S. cerevisiae*. Comparisons between these genomes have confirmed the authenticity of a number of *S. cerevisiae* ORFs, and also enabled almost 500 putative ORFs to be removed from the *S. cerevisiae* catalog on the grounds that they have no equivalents in the related genomes. The

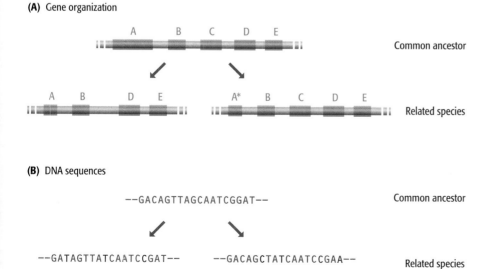

(A) Gene organization

(B) DNA sequences

Figure 5.8 Related species have similar genomes. (A) Illustration of how gene organization might change as two species diverge from their common ancestor. The common ancestor has five genes labeled A to E. In one of the derived species, gene C is no longer present, and in the other species, gene A has become truncated. (B) Related species display DNA sequence similarities. The diagram shows a short section of a gene sequence from an ancestral organism, along with the homologous sequences for this gene segment in the derived species. See Chapter 18 for a more comprehensive description of how genomes evolve.

Figure 5.9 Using comparisons with syntenic genomes to test the authenticity of a short ORF. In this example, the ORF is present in three of the four related genomes, and hence is likely to be a genuine gene.

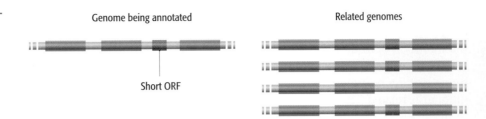

Genome being annotated   Related genomes

Short ORF

analysis is made even more powerful by the **synteny**—conservation of gene order—displayed by the genomes of these related yeasts. Although each genome has undergone its own species-specific rearrangements there are still substantial regions where the gene order in the *S. cerevisiae* genome is the same as in one or more of the related genomes. This makes it very easy to identify homologous genes but, more importantly, enables a spurious ORF, especially a short one, to be discarded with great confidence, because its expected location in a related genome can be searched in detail to ensure that no equivalent is present (Figure 5.9).

Automatic annotation of genome sequences

One great advantage of computer-based approaches to gene location is the possibility of combining a set of analytical programs into a single, integrated system. In this way, different approaches to gene location can be carried out in parallel, and the results automatically compared so that the genome under study is quickly and comprehensively annotated. Genome annotation using these systems begins with sequence analysis using programs that scan for ORFs, exon–intron boundaries, and upstream regulatory sequences, test the ORFs for homology to genes in the databases, and search for repeat sequences and for the characteristic features of the genes for functional RNAs. Databases of cDNA sequences (see below) might also be scanned for any that match segments of the genome sequence, any such matches indicating a

Figure 5.10 The display from a typical genome annotation system. The example shown is the Genotator browser annotation for a 15 kb segment of the human genome containing a tissue factor gene. From the top, the analyses are: locations of possible promoters (upstream regulatory elements); sequences corresponding to proteins in the GenPept database; sequences corresponding to known ESTs (see Section 3.3.3); location of known human repeat sequences; exon predictions by the Genscan, Genefinder, GRAIL, and Genie programs; sequences corresponding to known genes in the GenBank database; and ORFs in each of the three reading frames. Below the black bar the analyses are repeated for the reverse complement of the DNA sequence. The annotation reveals the positions of five possible exons, as indicated by the black arrows at the top of the figure. Image courtesy of Nomi Harris.

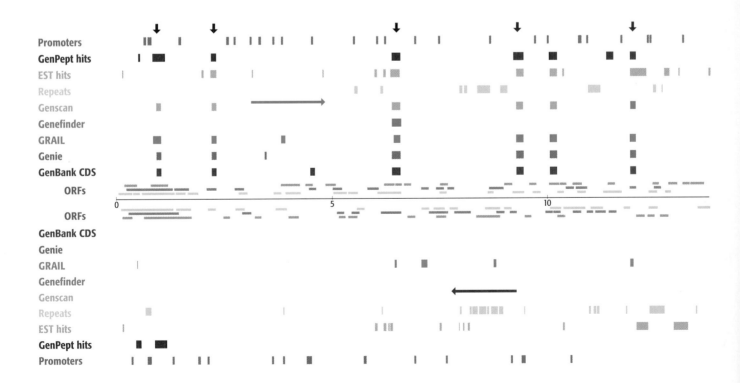

region that is transcribed into mRNA. The information is then integrated into a computer display that shows the locations of the various sequence features as determined by the different programs (Figure 5.10). The resulting information can often be accessed over the internet and hence used by other researchers to plan their own, more detailed computational or experimental studies of specific regions of a genome.

5.1.2 Experimental techniques for gene location

Most experimental methods for gene location are not based on direct examination of DNA molecules but instead rely on detection of the RNA molecules that are transcribed from genes. All genes are transcribed into RNA, and if the gene is discontinuous then the primary transcript is subsequently processed to remove the introns and link up the exons (Sections 1.2.3 and 12.2.2). Techniques that map the positions of transcribed sequences in a DNA fragment can therefore be used to locate exons and entire genes. The only problem to be kept in mind is that the transcript is usually longer than the coding part of the gene because it begins several tens of nucleotides upstream of the initiation codon and continues several tens or hundreds of nucleotides downstream of the termination codon. Transcript analysis does not, therefore, give a precise definition of the start and end of the coding region of a gene, but it does tell you that a gene is present in a particular region and it can locate the exon–intron boundaries. Often this is sufficient information to enable the coding region to be delineated.

Hybridization tests can determine if a fragment contains transcribed sequences

The simplest procedures for studying transcribed sequences are based on hybridization analysis. RNA molecules can be separated by specialized forms of agarose gel electrophoresis, transferred to a nitrocellulose or nylon membrane, and examined by the process called **northern hybridization** (see Technical Note 5.1). This differs from Southern hybridization (Section 2.1.2) only in the precise conditions under which the transfer is carried out, and the fact that it was not invented by a Dr Northern and so does not have a capital "N." If a northern blot of cellular RNA is probed with a labeled fragment of the genome, then RNAs transcribed from genes within that fragment will be detected (Figure 5.11). Northern hybridization is therefore, theoretically, a means of determining the number of genes present in a DNA fragment and the size of each coding region. There are two weaknesses with this approach:

- Some individual genes give rise to two or more transcripts of different lengths because some of their exons are optional and may or may not be retained in the mature RNA (Section 12.2.2). If this is the case, then a fragment that contains just one gene could detect two or more hybridizing bands in the northern blot. A similar problem can occur if the gene is a member of a multigene family (Section 7.2.3).

- With many species, it is not practical to make an mRNA preparation from an entire organism so the extract is obtained from a single organ or tissue. Consequently any genes not expressed in that organ or tissue will not be represented in the RNA population, and so will not be detected when the RNA is probed with the DNA fragment being studied. Even if the whole organism is used, not all genes will give hybridization signals because many are expressed only at a particular developmental stage, and others are weakly expressed, meaning that their RNA products are present in amounts too low to be detected by hybridization analysis.

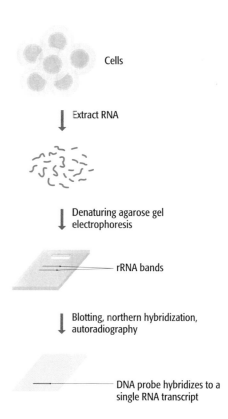

Figure 5.11 Northern hybridization. An RNA extract is electrophoresed under denaturing conditions in an agarose gel (see Technical Note 5.1). After ethidium bromide staining, two bands are seen. These are the two largest rRNA molecules (Section 1.2.2), which are abundant in most cells. The smaller rRNAs, which are also abundant, are not seen because they are so short that they run out the bottom of the gel and, in most cells, none of the mRNAs are abundant enough to form a band visible after ethidium bromide staining. The gel is blotted onto a nylon membrane and, in this example, probed with a radioactively labeled DNA fragment. A single band is visible on the autoradiograph, showing that the DNA fragment used as the probe contains part or all of one transcribed sequence.

Technical Note 5.1 Techniques for studying RNA

Many of the techniques devised for studying DNA molecules can be adapted for use with RNA

Agarose gel electrophoresis of RNA is carried out after denaturation of the RNA so that the migration rate of each molecule is dependent entirely on its length, and is not influenced by the intramolecular base pairs that can form in many RNAs (e.g., Figure 13.2). The denaturant, usually formaldehyde or glyoxal, is added to the sample before it is loaded onto the gel.

Northern hybridization refers to the procedure whereby an RNA gel is blotted onto a nylon membrane and hybridized to a labeled probe (see Figure 5.11). This is equivalent to Southern hybridization (see Figure 2.11) and is done in a similar way.

Labeled RNA molecules are usually prepared by copying a DNA template into RNA in the presence of a labeled ribonucleotide. The RNA polymerase enzymes of SP6, T3, or T7 bacteriophages are used because they can produce up to 30 µg of labeled RNA from 1 µg of DNA in 30 minutes. RNA can also be end-labeled by treatment with purified poly(A) polymerase (Section 12.2.1).

PCR of RNA molecules requires a modification to the first step of the normal reaction. *Taq* polymerase cannot copy an RNA molecule, so the first step is catalyzed by a reverse transcriptase, which makes a DNA copy of the RNA template. This DNA copy is then amplified by *Taq* polymerase. The technique is called **reverse transcriptase-PCR**, or **RT-PCR**. The discovery of thermostable enzymes that make DNA copies of both RNA and DNA templates (e.g., the *Tth* DNA polymerase from the bacterium *Thermus ther-*

mophilus) raises the possibility of carrying out RT-PCR in a single reaction with just one enzyme.

RNA sequencing methods exist but are difficult to perform and are applicable only to small molecules. The methods are similar to chemical degradation sequencing of DNA (Section 4.1.2) but employ sequence-specific endonucleases rather than chemicals to generate the cleaved molecules. In practice, the sequence of an RNA molecule is usually obtained by converting it into cDNA (see Figure 3.36) and sequencing by the chain termination method (Section 4.1.1).

Specialist methods have been developed for mapping the positions of RNA molecules on to DNA sequences, for example to determine the start and end points of transcription and to locate the positions of introns in a DNA sequence. These methods are described in Section 5.1.2.

The only major deficiency in the RNA toolkit is the absence of enzymes with the degree of sequence specificity displayed by the restriction endonucleases that are so important in DNA manipulations. Other than this, the only drawback with RNA work is the ease with which RNAs are degraded by ribonucleases that are released when cells are disrupted (as occurs during RNA extraction), and which are also present on the hands of laboratory workers and which tend to contaminate glassware and solutions. This means that rigorous laboratory procedures (e.g., cleaning of glassware with chemicals that destroy ribonucleases) have to be adopted in order to keep RNA molecules intact.

A second type of hybridization analysis avoids the problems with poorly expressed and tissue-specific genes by searching not for RNAs but for related sequences in the DNAs of other organisms. This approach, like homology searching, is based on the fact that homologous genes in related organisms have similar sequences, whereas the intergenic DNA is usually quite different. If a DNA fragment from one species is used to probe a Southern transfer of DNAs from related species, and one or more hybridization signals are obtained, then it is likely that the probe contains one or more genes (Figure 5.12). This is called **zoo-blotting**.

cDNA sequencing enables genes to be mapped within DNA fragments

Northern hybridization and zoo-blotting enable the presence or absence of genes in a DNA fragment to be determined, but give no positional information relating to the location of those genes in the DNA sequence. The easiest way to obtain this information is to sequence the relevant cDNAs. A cDNA is a copy of an mRNA (see Figure 3.36) and so corresponds to the coding region of a gene, plus any leader or trailer sequences that are also transcribed. Comparing a cDNA sequence with a genomic DNA sequence therefore delineates the position of the relevant gene and reveals the exon–intron boundaries.

In order to obtain an individual cDNA, a cDNA library must first be prepared from all of the mRNA in the tissue being studied. Once the library has been prepared, the success of cDNA sequencing as a means of gene location depends on two factors. The first concerns the frequency of the desired cDNAs in the library. As with northern hybridization, the problem relates to the different expression levels of different genes. If the DNA fragment being studied contains one or more poorly expressed genes, then the relevant cDNAs will be rare in the library and it might be necessary to screen many clones before the desired one is identified. To get around this problem, various methods of **cDNA capture** or **cDNA selection** have been devised, in which the DNA fragment being studied is repeatedly hybridized to the pool of cDNAs in order to enrich the pool for the desired clones. Because the cDNA pool contains so many different sequences, it is generally not possible to discard all the irrelevant clones by these repeated hybridizations, but it is possible to increase significantly the frequency of those clones that specifically hybridize to the DNA fragment. This reduces the size of the library that must subsequently be screened under stringent conditions to identify the desired clones.

A second factor that determines success or failure is the completeness of the individual cDNA molecules. Usually, cDNAs are made by copying RNA molecules into single-stranded DNA with reverse transcriptase and then converting the single-stranded DNA into double-stranded DNA with a DNA polymerase (see Figure 3.36). There is always a chance that one or other of the strand synthesis reactions will not proceed to completion, resulting in a truncated cDNA. The presence of intramolecular base pairs in the RNA can also lead to incomplete copying. Truncated cDNAs may lack some of the information needed to locate the start and end points of a gene and all its exon–intron boundaries.

Methods are available for precise mapping of the ends of transcripts
The problems with incomplete cDNAs mean that more robust methods are needed for locating the precise start and end points of gene transcripts. One possibility is a special type of PCR that uses RNA rather than DNA as the starting material. The first step in this type of PCR is to convert the RNA into cDNA with reverse transcriptase, after which the cDNA is amplified with *Taq* polymerase in the same way as in a normal PCR. These methods go under the collective name of **reverse transcriptase PCR** (**RT-PCR**) but the particular version that interests us at present is **rapid amplification of cDNA ends** (**RACE**). In the simplest form of this method, one of the primers is specific for an internal region close to the beginning of the gene being studied. This primer attaches to the mRNA for the gene and directs the first reverse transcriptase–catalyzed stage of the process, during which a cDNA corresponding to the start of the mRNA is made (Figure 5.13). Because only a small segment of the mRNA is being copied, the expectation is that the cDNA synthesis will not terminate prematurely, so one end of the cDNA will correspond exactly with the start of the mRNA. Once the cDNA has been made, a short

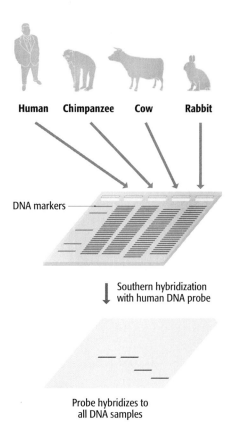

Figure 5.12 Zoo-blotting. The objective is to determine if a fragment of human DNA hybridizes to DNAs from related species. Samples of human, chimpanzee, cow, and rabbit DNAs are therefore prepared, restricted, and electrophoresed in an agarose gel. Southern hybridization is then carried out with a human DNA fragment as the probe. A positive hybridization signal is seen with each of the animal DNAs, suggesting that the human DNA fragment contains an expressed gene. Note that the hybridizing restriction fragments from the cow and rabbit DNAs are smaller than the hybridizing fragments in the human and chimpanzee samples. This indicates that the restriction map around the transcribed sequence is different in cows and rabbits, but does not affect the conclusion that a homologous gene is present in all four species.

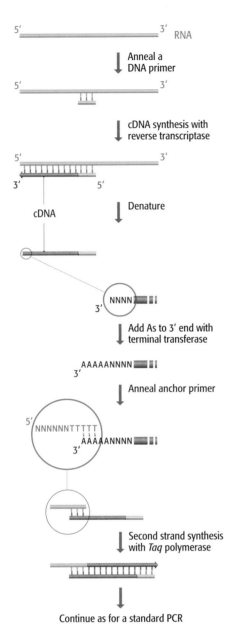

Figure 5.13 RACE—rapid amplification of cDNA ends. The RNA being studied is converted into a partial cDNA by extension of a DNA primer that anneals at an internal position not too distant from the 5′ end of the molecule. The 3′ end of the cDNA is further extended by treatment with terminal deoxynucleotidyl transferase (Section 2.1.4) in the presence of dATP, which results in a series of As being added to the cDNA. This series of As acts as the annealing site for the anchor primer. Extension of the anchor primer leads to a double-stranded DNA molecule, which can now be amplified by a standard PCR. This is 5′-RACE, so-called because it results in amplification of the 5′ end of the starting RNA. A similar method—3′-RACE—can be used if the 3′ end-sequence is desired.

poly(A) tail is attached to its 3′ end. The second primer anneals to this poly(A) sequence and, during the first round of the normal PCR, converts the single-stranded cDNA into a double-stranded molecule, which is subsequently amplified as the PCR proceeds. The sequence of this amplified molecule will reveal the precise position of the start of the transcript.

Other methods for precise transcript mapping involve **heteroduplex analysis**. If the DNA region being studied is cloned as a restriction fragment in an M13 vector (Section 4.1.1) then it can be obtained as single-stranded DNA. When mixed with an appropriate RNA preparation, the transcribed sequence in the cloned DNA hybridizes with the equivalent mRNA, forming a double-stranded heteroduplex. In the example shown in Figure 5.14, the start of this mRNA lies within the cloned restriction fragment, so some of the cloned fragment participates in the heteroduplex, but the rest does not. The single-stranded regions can be digested by treatment with a single-strand-specific nuclease such as S1. The size of the heteroduplex is determined by degrading the RNA component with alkali and electrophoresing the resulting single-stranded DNA in an agarose gel. This size measurement is then used to position the start of the transcript relative to the restriction site at the end of the cloned fragment.

Exon–intron boundaries can also be located with precision
Heteroduplex analysis can also be used to locate exon–intron boundaries. The method is almost the same as that shown in Figure 5.14 with the exception that the cloned restriction fragment spans the exon–intron boundary being mapped rather than the start of the transcript.

A second method for finding exons in a genome sequence is called **exon trapping**. This requires a special type of vector that contains a **minigene** consisting of two exons flanking an intron sequence, the first exon being preceded by the sequence signals needed to initiate transcription in a eukaryotic cell (Figure 5.15). To use the vector, the piece of DNA to be studied is inserted into a restriction site located within the vector's intron region. The vector is then introduced into a suitable eukaryotic cell line, where it is transcribed and the RNA produced from it is spliced. The result is that any exon contained in the genomic fragment becomes attached between the upstream and downstream exons from the minigene. RT-PCR with primers annealing within the two minigene exons is now used to amplify a DNA fragment, which is sequenced. As the minigene sequence is already known, the nucleotide positions at which the inserted exon starts and ends can be determined, precisely delineating this exon.

5.2 Determining the Functions of Individual Genes

Once a new gene has been located in a genome sequence, the question of its function has to be addressed. This is turning out to be an important area of genomics research, because completed sequencing projects have revealed that we know rather less than we thought about the content of individual genomes. *Escherichia coli* and *Saccharomyces cerevisiae*, for example, were studied intensively by conventional genetic analysis before the advent of sequencing projects, and geneticists were at one time fairly confident that most of their genes had been identified. The genome sequences revealed that in fact there are large gaps in our knowledge. Of the 4288 protein-coding genes in the *E. coli* genome sequence, only 1853 (43% of the total) had been previously identified. For *S. cerevisiae* the figure was only 30%.

As with gene location, attempts to determine the functions of unknown genes are made by computer analysis and by experimental studies.

5.2.1 Computer analysis of gene function

We have already seen that computer analysis plays an important role in locating genes in DNA sequences, and that one of the most powerful tools available for this purpose is homology searching, which locates genes by comparing the DNA sequence under study with all the other DNA sequences in the databases. The basis of homology searching is that related genes have similar sequences and so a new gene can be discovered by virtue of its similarity to an equivalent, already-sequenced gene from a different organism. Now we will look more closely at homology analysis and see how it can be used to assign a function to a new gene.

Homology reflects evolutionary relationships

Homologous genes are ones that share a common evolutionary ancestor, revealed by sequence similarities between the genes. These similarities form the data on which molecular phylogenies are based, as we will see in Chapter 19. Homologous genes fall into two categories (Figure 5.16):

- **Orthologous** genes are those homologs that are present in different organisms and whose common ancestor predates the split between the species. Orthologous genes usually have the same, or very similar, functions. For example, the myoglobin genes of humans and chimpanzees are orthologs.

- **Paralogous** genes are present in the same organism, often as members of a recognized multigene family (Section 7.2.3), their common ancestor possibly or possibly not predating the species in which the genes are now found. For example, the myoglobin and β-globin genes of humans are paralogs: they originated by duplication of an ancestral gene some 550 million years ago (Section 18.2.1).

A pair of homologous genes do not usually have identical nucleotide sequences, because the two genes undergo different random changes by mutation, but they have similar sequences because these random changes have operated on the same starting sequence, the common ancestral gene. Homology searching makes use of these sequence similarities. The basis of the analysis is that if a newly sequenced gene turns out to be similar to a previously sequenced gene, then an evolutionary relationship can be inferred and the function of the new gene is likely to be the same, or at least similar, to the function of the known gene.

It is important not to confuse the words *homology* and *similarity*. It is incorrect to describe a pair of related genes as "80% homologous" if their sequences have 80% nucleotide identity (Figure 5.17). A pair of genes are either evolutionarily related or they are not; there are no in-between situations and it is therefore meaningless to ascribe a percentage value to homology.

Homology analysis can provide information on the function of an entire gene or of segments within it

A homology search can be conducted with a DNA sequence but usually a tentative gene sequence is converted into an amino acid sequence before the search is carried out. One reason for this is that there are 20 different amino acids in proteins but only four nucleotides in DNA, so genes that are

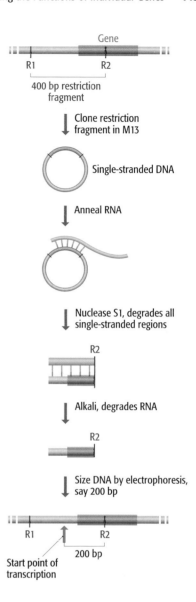

Figure 5.14 Nuclease S1 mapping. This method of transcript mapping makes use of nuclease S1, an enzyme that degrades single-stranded DNA or RNA polynucleotides, including single-stranded regions in predominantly double-stranded molecules, but has no effect on double-stranded DNA or on DNA–RNA hybrids. In the example shown, a restriction fragment that spans the start of a transcription unit is ligated into an M13 vector and the resulting single-stranded DNA hybridized with an RNA preparation. After nuclease S1 treatment, the resulting heteroduplex has one end marked by the start of the transcript and the other by the downstream restriction site (R2). The size of the undigested DNA fragment is therefore measured by gel electrophoresis in order to determine the position of the start of the transcription unit relative to the downstream restriction site.

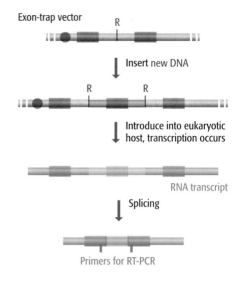

Exon-trap vector

Insert new DNA

Introduce into eukaryotic host, transcription occurs

RNA transcript

Splicing

Primers for RT-PCR

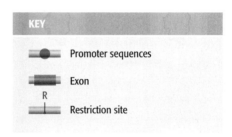

KEY

● Promoter sequences

Exon

R Restriction site

Figure 5.15 Exon trapping. The exon-trap vector consists of two exon sequences preceded by promoter sequences—the signals required for gene expression in a eukaryotic host (Section 11.2.2). New DNA containing an unmapped exon is ligated into the vector and the recombinant molecule introduced into the host cell. The resulting RNA transcript is then examined by RT-PCR to identify the boundaries of the unmapped exon.

unrelated usually appear to be more different from one another when their amino acid sequences are compared (Figure 5.18). A homology search is therefore less likely to give spurious results if the amino acid sequence is used.

A homology search program begins by making alignments between the query sequence and sequences from the databases. For each alignment, a score is calculated from which the operator can gauge the likelihood that the query and test sequences are homologs. There are two ways of generating the score:

- The simplest programs count the number of positions at which the same amino acid is present in both sequences. This number, when converted into a percentage, gives the degree of *identity* between two sequences.

- More sophisticated programs use the chemical relatedness between non-identical amino acids to assign a score to each position in the alignment, a higher score for identical or closely related amino acids (e.g., leucine and isoleucine, or aspartic acid and asparagine) and a lower score for less-related amino acids (e.g., cysteine and tyrosine, or phenylalanine and serine). This analysis determines the degree of *similarity* between a pair of sequences.

To achieve the highest possible score, the algorithm introduces gaps at various positions in one or both sequences, up to limits set by the operator, paralleling processes thought to occur during the evolution of genes, when blocks of nucleotides coding for individual or adjacent amino acids may be inserted into or deleted from a gene.

The practicalities of homology searching are not at all daunting. Several software programs exist for this type of analysis, the most popular being **BLAST** (Basic Local Alignment Search Tool). The analysis can be carried out simply by logging on to the web site for one of the DNA databases and entering the sequence into the online search tool. The standard BLAST program is efficient at identifying homologous genes that have more than 30%–40% sequence similarity, but is less effective at recognizing evolutionary relationships if the similarity is lower than this amount. The modified version called **PSI-BLAST** (position-specific iterated BLAST) identifies more distantly related sequences, by combining the homologous sequences from a standard BLAST search into a profile, the features of which are used to identify additional homologous sequences that were not detected in the initial search.

Homology searching with BLAST and similar programs has gained immense importance in genomics research, but its limitations must be recognized. A growing problem is the presence in the databases of genes whose stated functions are incorrect. If one of these genes is identified as a homolog of the

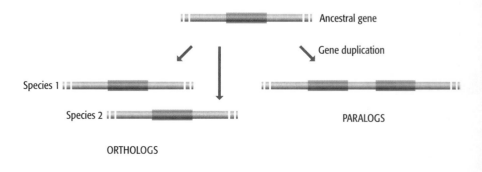

Ancestral gene

Gene duplication

Species 1

Species 2

PARALOGS

ORTHOLOGS

Figure 5.16 Orthologous and paralogous genes.

Sequence 1   GGTGAGGGTATCATCCCATCTGACTACACCTCATCGGGAGACGGAGCAGT
Sequence 2   GGTCAGGATATGATTCCATCACTACACCTTATCCCGAGTCGGAGCAGT
Identities   *** *** *** ** ***** ******** *** *** ********

Figure 5.17 Two DNA sequences with 80% sequence identity.

```
             G   A   P   G   M   W   L   R   L   A   A   G   S   F   E   H   A   G
Sequence 1   GGTGCACCCGGTATGTGACTGCGATTAGCAGCGGGATCATTTCAGCATGCAGGG
             * * ***** **** **** ** *** **** ***** *** ** ** ***** ** *
Sequence 2   GATACACCCCGTATTTGACAGCAATTTGCAGGGGGATGATTGCACCATGGAGCG
             D   T   P   R   I   W   E   E   F   A   G   G   W   L   H   H   G   A
```

Figure 5.18 Lack of homology between two sequences is often more apparent when comparisons are made at the amino acid level. Two nucleotide sequences are shown, with nucleotides that are identical in the two sequences given in green and nonidentities given in red. The two nucleotide sequences are 76% identical, as indicated by the asterisks. This might be taken as evidence that the sequences are homologous. However, when the sequences are translated into amino acids, the identity decreases to 28%. Identical amino acids are shown in yellow, and nonidentities in brown. The comparison between the amino acid sequences suggests that the genes are not homologous, and that the similarity at the nucleotide level was fortuitous. The amino acid sequences have been written using the one-letter abbreviations (see Table 1.2).

query sequence, then the incorrect function will be passed on to this new sequence, adding to the problem. There are also several cases where homologous genes have quite different biological functions, an example being the crystallins of the eye lens, some of which are homologous to metabolic enzymes. Homology between a query sequence and a crystallin therefore does not mean that the query sequence is a crystallin and, similarly, an apparently clear homology between a query sequence and a metabolic enzyme might not mean that the query sequence is a metabolic enzyme.

There are also examples of genes that have similar sequences but no obvious evolutionary relatedness. Sometimes the explanation is that, although the genes are unrelated, their proteins have similar functions and the shared sequence encodes a domain within each protein that is central to that shared function. Although the genes themselves have no common ancestor, the domains do, but with their common ancestor occurring at a very ancient time and the homologous domains having subsequently evolved not only by single nucleotide changes but also by more complex rearrangements that have created new genes within which the domains are found (Section 18.2.1). When recognized, this type of homology can be extremely informative. A typical example is provided by the tudor domain, an approximately 120–amino acid motif which was first identified in the sequence of the *Drosophila melanogaster* gene called *tudor*. The protein coded by the *tudor* gene, whose function is unknown, is made up of ten copies of the tudor domain, one after the other (Figure 5.19). A homology search using the tudor domain as the test revealed that several known proteins contain this domain. The sequences of these proteins are not highly similar to one another and there is no indication that they are true homologs, but they all possess the tudor domain. These proteins include one involved in RNA transport during *Drosophila* oogenesis, a human protein with a role in RNA metabolism, and others whose activities appear to involve RNA in one way or another. The homology analysis therefore suggests that the tudor sequence plays some part in the interaction between the protein and its RNA substrate. The information from the computer analysis is incomplete by itself, but it points the way to the types of experiment that should be done to obtain more clear-cut data on the function of the tudor domain.

Using homology searching to assign functions to human disease genes

To illustrate the importance of homology searching in genomics we will examine how this technique has aided research into human genetic disease. One of the main reasons for sequencing the human genome has been to gain access to genes involved in human disease. The hope is that the sequence of

Drosophila tudor

Drosophila homeless

Human AKAP149

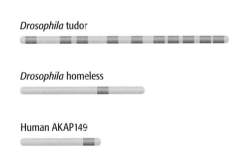

Figure 5.19 **The tudor domain.** The top drawing shows the structure of the *Drosophila* tudor protein, which contains ten copies of the tudor domain. The domain is also found in a second *Drosophila* protein, homeless, and in the human A-kinase anchor protein (AKAP149), which plays a role in RNA metabolism. The proteins have dissimilar structures other than the presence of the tudor domains. The activity of each protein involves RNA in one way or another.

a disease gene will provide an insight into the biochemical basis of the disease and hence indicate a way of preventing or treating the disease. Homology searching has an important role to play in the study of disease genes because the discovery of a homolog of a human disease gene in a second organism is often the key to understanding the biochemical function of the human gene. If the homolog has already been characterized, then the information needed to understand the biochemical role of the human gene may already be in place; if it has not been characterized, then the necessary research can be directed at the homolog.

It is not necessary for the homolog to be present in a closely related species in order to be useful in the study of human disease. *Drosophila* holds great promise in this respect, as the phenotypic effects of many *Drosophila* genes are well known, so the data already exist for inferring the mode of action of human disease genes that have homologs in the *Drosophila* genome. But the greatest success has been with yeast. Several human disease genes have homologs in the *S. cerevisiae* genome (Table 5.1). These disease genes include ones involved in cancer, cystic fibrosis, and neurological syndromes, and in several cases the yeast homolog has a known function that provides a clear indication of the biochemical activity of the human gene. In some cases it has even been possible to demonstrate a physiological similarity between the gene activity in humans and yeast. For example, the yeast gene *SGS1* is a homolog of a human gene involved in the diseases called Bloom's and Werner's syndromes, which are characterized by growth disorders. Yeasts with a mutant *SGS1* gene live for shorter periods than normal yeasts and display accelerated onset-of-aging indicators such as sterility. The yeast gene has been shown to code for one of a pair of related DNA helicases that are required for transcription of rRNA genes and for DNA replication. The link between *SGS1* and the genes for Bloom's and Werner's syndromes, provided by homology searching, has therefore indicated the possible biochemical basis of the human diseases.

5.2.2 Assigning gene function by experimental analysis

It is clear that homology analysis is not a panacea that can identify the functions of all new genes. Experimental methods are therefore needed to complement and extend the results of homology studies. This is proving to be one

Table 5.1 Examples of human disease genes that have homologs in *Saccharomyces cerevisiae*

Human disease gene	Yeast homolog	Function of the yeast gene
Amyotrophic lateral sclerosis	*SOD1*	Protection against superoxide (O_2^-)
Ataxia telangiectasia	*TEL1*	Codes for a protein kinase
Colon cancer	*MSH2, MLH1*	DNA repair
Cystic fibrosis	*YCF1*	Metal resistance
Myotonic dystrophy	*YPK1*	Codes for a protein kinase
Type 1 neurofibromatosis	*IRA2*	Codes for a regulatory protein
Bloom's syndrome, Werner's syndrome	*SGS1*	DNA helicase
Wilson's disease	*CCC2*	Copper transport?

of the biggest challenges in genomics research, and most molecular biologists agree that the methodologies and strategies currently in use are not entirely adequate for assigning functions to the vast numbers of unknown genes being discovered by sequencing projects. The problem is that the objective— to plot a course from gene to function—is the reverse of the route normally taken by genetic analysis, in which the starting point is a phenotype and the objective is to identify the underlying gene or genes. The problem we are currently addressing takes us in the opposite direction: starting with a new gene and hopefully leading to identification of the associated phenotype.

Functional analysis by gene inactivation

In conventional genetic analysis, the genetic basis of a phenotype is usually studied by searching for mutant organisms in which the phenotype has become altered. The mutants might be obtained experimentally, for example by treating a population of organisms (e.g., a culture of bacteria) with ultraviolet radiation or a mutagenic chemical (see Section 16.1.1), or the mutants might be present in a natural population. The gene or genes that have been altered in the mutant organism are then studied by genetic crosses (Section 3.2.4), which can locate the position of a gene in a genome and also determine if the gene is the same as one that has already been characterized. The gene can then be studied further by molecular biology techniques such as cloning and sequencing.

The general principle of this conventional analysis is that the genes responsible for a phenotype can be identified by determining which genes are inactivated in organisms that display a mutant version of the phenotype. If the starting point is the gene, rather than the phenotype, then the equivalent strategy would be to mutate the gene and identify the phenotypic change that results. This is the basis of most of the techniques used to assign functions to unknown genes.

Individual genes can be inactivated by homologous recombination

The easiest way to inactivate a specific gene is to disrupt it with an unrelated segment of DNA (Figure 5.20). This can be achieved by **homologous recombination** between the chromosomal copy of the gene and a second piece of DNA that shares some sequence identity with the target gene. Homologous and other types of recombination are complex events, which we will deal with in detail in Chapter 17. For present purposes it is enough to know that if two DNA molecules have similar sequences, then recombination can result in segments of the molecules being exchanged.

How is gene inactivation carried out in practice? We will consider two examples, the first with *S. cerevisiae*. Since completing the genome sequence in 1996, yeast molecular biologists have embarked on a coordinated, international effort to determine the functions of as many unknown genes as possible (Section 5.3). One technique that is being used is shown in Figure 5.21. The central component is the "deletion cassette," which carries a gene for antibiotic resistance. This gene is not a normal component of the yeast genome but it will work if transferred into a yeast chromosome, giving rise to a transformed yeast cell that is resistant to the antibiotic geneticin. Before using the deletion cassette, new segments of DNA are attached as tails to either end. These segments have sequences identical to parts of the yeast gene that is going to be inactivated. After the modified cassette is introduced into a yeast cell, homologous recombination occurs between the

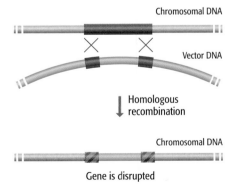

Figure 5.20 Gene inactivation by homologous recombination. The chromosomal copy of the target gene recombines with a disrupted version of the gene carried by a cloning vector. As a result, the target gene becomes inactivated.

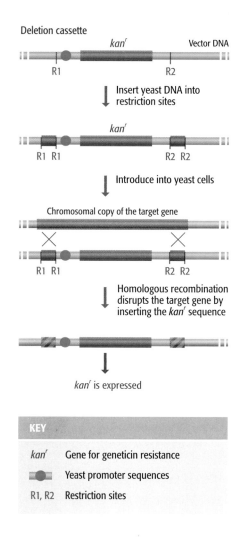

Figure 5.21 The use of a yeast deletion cassette. The deletion cassette consists of an antibiotic-resistance gene preceded by the promoter sequences needed for expression in yeast, and flanked by two restriction sites. The start and end segments of the target gene are inserted into the restriction sites, and the vector is introduced into yeast cells. Recombination between the gene segments in the vector and the chromosomal copy of the target gene results in disruption of the latter. Cells in which the disruption has occurred are identifiable because they now express the antibiotic-resistance gene and so will grow on an agar medium containing geneticin. The gene designation "*kan'*" is an abbreviation for "kanamycin resistance," kanamycin being the family name of the group of antibiotics that include geneticin.

DNA tails and the chromosomal copy of the yeast gene, replacing the latter with the antibiotic-resistance gene. Cells which have undergone the replacement are therefore selected by plating the culture onto agar medium containing geneticin. The resulting colonies lack the target gene activity and their phenotypes can be examined to gain some insight into the function of the gene.

The second example of gene inactivation uses an analogous process but with mice rather than yeast. The mouse is frequently used as a **model organism** for humans because the mouse genome is similar to the human genome, containing many of the same genes. Functional analysis of unknown human genes is therefore being carried out largely by inactivating the equivalent genes in the mouse, these experiments being ethically unthinkable with humans. The homologous recombination part of the procedure is identical to that described for yeast, and once again results in a cell in which the target gene has been inactivated. The problem is that we do not want just one mutated cell, we want a whole mutant mouse, as only with the complete organism can we make a full assessment of the effect of the gene inactivation on the phenotype. To achieve this it is necessary to use a special type of mouse cell: an **embryonic stem** or **ES cell**. Unlike most mouse cells, ES cells are **totipotent**, meaning that they are not committed to a single developmental pathway and can therefore give rise to all types of differentiated cell. The engineered ES cell is therefore injected into a mouse embryo, which continues to develop and eventually gives rise to a **chimera**, a mouse whose cells are a mixture of mutant ones, derived from the engineered ES cells, and nonmutant ones, derived from all the other cells in the embryo. This is still not quite what we want, so the chimeric mice are allowed to mate with one another. Some of the offspring result from fusion of two mutant gametes, and will therefore be nonchimeric, as every one of their cells will carry the inactivated gene. These are **knockout mice**, and with luck their phenotypes will provide the desired information on the function of the gene being studied. This works well for many gene inactivations but some are lethal and so cannot be studied in a homozygous knockout mouse. Instead, a heterozygous mouse is obtained, the product of fusion between one normal and one mutant gamete, in the hope that the phenotypic effect of the gene inactivation will be apparent even though the mouse still has one correct copy of the gene being studied.

Gene inactivation without homologous recombination

Homologous recombination is not the only way to disrupt a gene in order to study its function. One alternative is to use **transposon tagging**, in which inactivation is achieved by the insertion of a transposable element, or transposon, into the gene. Most genomes contain transposable elements (Section 9.2) and although the bulk of these are inactive, there are usually a few that retain their ability to move to new positions in the genome. Under normal circumstances, transposition is a relatively rare event, but it is sometimes possible to use recombinant DNA techniques to make modified transposons that change their position in response to an external stimulus. One way of doing this, involving the yeast transposon *Ty1*, is shown in Figure 5.22. Transposon tagging is also important in analysis of the fruit-fly genome, using the endogenous *Drosophila* transposon called the **P element**. The weakness with transposon tagging is that it is difficult to target individual genes, because transposition is more or less a random event and it is impossible to predict where a transposon will end up after it has jumped. If the intention is to inactivate a particular gene then it is necessary to induce a substantial number of transpositions and then to screen all the resulting organisms to find one with the correct insertion. Transposon tagging is therefore more applicable to global studies of genome function, in which genes are inactivated at random and groups of genes with similar functions identified by examining the progeny for interesting phenotype changes.

A completely different approach to gene inactivation is provided by **RNA interference**, or **RNAi**, one of a series of natural processes by which short RNA molecules influence gene expression in living cells (Section 12.2.6). When used in genomics research, RNAi provides a means of inactivating a target gene, not by disrupting the gene itself, but by destroying its mRNA. This is accomplished by introducing into the cell short, double-stranded RNA molecules whose sequences match that of the mRNA being targeted. The double-stranded RNAs are broken down into shorter molecules, which induce degradation of the mRNA (Figure 5.23). The process was initially shown to work effectively in the worm *Caenorhabditis elegans*, the genome of which has been completely sequenced and which is looked on as an important model organism for higher eukaryotes (Section 14.3.3). Virtually all of the 19,000 predicted genes in the *C. elegans* genome have been individually inactivated by RNA interference. The key step in any RNAi experiment is introducing into the test organism the double-stranded RNA molecule that will give rise to the single-stranded interfering RNAs. With *C. elegans* this can be achieved by feeding the RNA to the worms. *C. elegans* eats bacteria, including *Escherichia coli*, and is often grown on a lawn of bacteria on an agar plate. If the bacteria contain a cloned gene that directs expression of a double-stranded RNA with the same sequence as a *C. elegans* gene then, after ingestion, the RNAi pathway begins to operate. Alternatively, the double-stranded RNA can be directly microinjected into the worm, but this is more time-consuming.

RNA interference is known to occur naturally in a range of eukaryotes, but applying it to mammalian cells was expected to be difficult because these organisms display a parallel response to double-stranded RNA, in which protein synthesis is generally inhibited, resulting in cell death. This problem has been circumvented by using double-stranded RNAs that are just 21–22 bp in length, long enough to be specific for an individual target gene and to activate the RNAi process, but too short to set off the protein synthesis inhibitory response. This approach, using retroviral vectors to deliver cloned genes expressing the RNAs into cultured cells, has enabled over 8000 of the 35,000 human genes to be individually inactivated. The current challenge for mammalian RNAi is to move from cell-based systems, in which only certain phenotypes can be assessed, to knockout mice, in which the full range of phenotypes can be examined. This requires the development of cloning systems for the long-term, stable synthesis of the interfering RNAs, which must be present at a consistently high level in order to keep the target gene inactivated.

Gene overexpression can also be used to assess function

So far we have concentrated on techniques that result in inactivation of the gene being studied ("loss of function"). The complementary approach is to engineer an organism in which the test gene is much more active than normal ("gain of function") and to determine what changes, if any, this has on the phenotype. The results of these experiments must be treated with caution

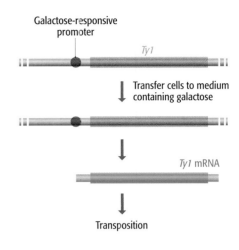

Figure 5.22 Artificial induction of transposition. Recombinant DNA techniques have been used to place a promoter sequence that is responsive to galactose upstream of a *Ty1* element in the yeast genome. When galactose is absent, the *Ty1* element is not transcribed and so remains quiescent. When the cells are transferred to a culture medium containing galactose, the promoter is activated and the *Ty1* element is transcribed, initiating the transposition process. For more information on activation of eukaryotic promoters, see Section 11.3, and for details of the *Ty1* transposition process see Section 17.3.2.

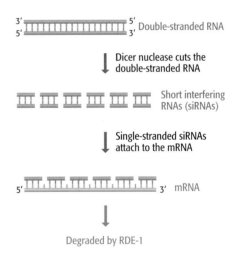

Figure 5.23 RNA interference. The double-stranded RNA molecule is broken down by the Dicer ribonuclease into "short interfering RNAs" (siRNAs) of length 21–25 bp. One strand of each siRNA base-pairs to the target mRNA, which is then degraded by the nuclease RDE-1.

Figure 5.24 Functional analysis by gene overexpression. The objective is to determine if overexpression of the gene being studied has an effect on the phenotype of a transgenic mouse. A cDNA of the gene is therefore inserted into a cloning vector carrying a highly active promoter sequence that directs expression of the cloned gene in mouse liver cells. The cDNA is used rather than the genomic copy of the gene because the former does not contain introns and so is shorter and easier to manipulate in the test tube.

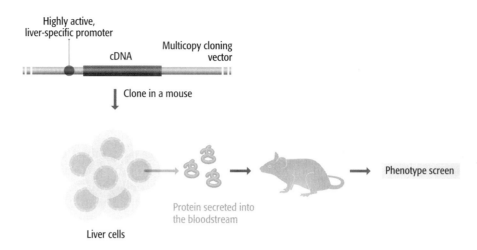

because of the need to distinguish between a phenotype change that is due to the specific function of an overexpressed gene, and a less specific phenotype change that reflects the abnormality of the situation where a single gene product is being synthesized in excessive amounts, possibly in tissues in which the gene is normally inactive. Despite this qualification, overexpression has provided some important information on gene function.

To overexpress a gene, a special type of cloning vector must be used, one designed to ensure that the cloned gene directs the synthesis of as much protein as possible. The vector is therefore **multicopy**, meaning that it multiplies inside the host organism to 40–200 copies per cell, so there are many copies of the test gene. The vector must also contain a highly active promoter (Section 11.2.2) so that each copy of the test gene is converted into large quantities of mRNA, again ensuring that as much protein as possible is made. In the example shown in Figure 5.24, the cloning vector contains a highly active promoter that is expressed only in the liver, so each **transgenic mouse** overexpresses the test gene in its liver. This approach has been used with genes whose sequences suggested that they code for proteins that are secreted into the bloodstream. After synthesis in the liver of the transgenic mouse, the test protein is secreted and the phenotype of the transgenic mouse examined in the search for clues regarding the function of the cloned gene. An interesting discovery was made when it was realized that one transgenic mouse had bones that were significantly more dense than those of normal mice. This was important for two reasons: first, it enabled the relevant gene to be identified as one involved in bone synthesis; second, the discovery of a protein that increases bone density has implications for the development of treatments for human osteoporosis, a fragile-bone disease.

The phenotypic effect of gene inactivation or overexpression may be difficult to discern

The critical aspect of a gene inactivation or overexpression experiment is the need to identify a phenotypic change, the nature of which gives a clue to the function of the manipulated gene. This can be much more difficult than it sounds. With any organism the range of phenotypes that must be examined is immense. Even with a unicellular organism such as yeast, the list is quite lengthy (Table 5.2A) and with multicellular eukaryotes it is much more so (Table 5.2B). In higher organisms some phenotypes (e.g., behavioral ones) are difficult, if not impossible, to assess in a comprehensive fashion.

Table 5.2 Typical phenotypes assessed in screens of *Saccharomyces cerevisiae* or *Caenorhabditis elegans* genes

Phenotype
(A) All *Saccharomyces cerevisiae* genes
DNA synthesis and the cell cycle
RNA synthesis and processing
Protein synthesis
Stress responses
Cell wall synthesis and morphogenesis
Transport of biochemicals within the cell
Energy and carbohydrate metabolism
Lipid metabolism
DNA repair and recombination
Development
Meiosis
Chromosome structure
Cell architecture
Secretion and protein trafficking
(B) Genes involved in the early embryogenesis of *Caenorhabditis elegans*
Sterility/impaired fertility in parent
Osmotic integrity
Polar body extrusion
Passage through meiosis
Entry into interphase
Cortical dynamics
Pronuclear/nuclear appearance
Centrosome attachment
Pronuclear migration
Spindle assembly
Spindle elongation/integrity
Sister chromatid separation
Nuclear appearance
Chromosome segregation
Cytokinesis
Asymmetry of division
Pace of cell division
General pace of development
Severe pleiotropic defects
Integrity of membrane-bound organelles
Egg size
Aberrant cytoplasmic structures
Complex combination of defects

Furthermore, the effect of gene inactivation can be very subtle and may not be recognized when the phenotype is examined. A good example of the problems that occur was provided by the longest gene on yeast chromosome III which, at 2167 codons and with typical yeast codon bias, simply had to be a functional gene rather than a spurious ORF. Inactivation of this gene had no apparent effect, the mutant yeast cells appearing to have an identical phenotype to normal yeast. For some time it was thought that perhaps this gene is dispensable, its protein product either involved in some completely nonessential function, or having a function that is duplicated by a second gene. Eventually it was shown that the mutants die when they are grown at low pH in the presence of glucose and acetic acid, which normal yeasts can tolerate, and it was concluded that the gene codes for a protein that pumps acetate out of the cell. This is definitely an essential function, as the gene plays a vital role in protecting yeast from acetic acid-induced damage, but this essentiality was difficult to track down from the phenotype tests.

Even when the most careful screens are carried out, many gene inactivations appear to give no discernible phenotypic change. Almost 5000 of the 6000 yeast genes can be individually inactivated without causing the cells to die, and inactivation of many of these 5000 genes has no detectable effect on the metabolic properties of the cell under normal growth conditions. With *C. elegans*, large-scale gene inactivation projects have so far assigned phenotypes to less than 10% of the 19,000 predicted genes. The implication is that in both organisms the majority of genes play specialist roles and identifying what this role is for every gene is going to be a lengthy and difficult process.

5.2.3 More detailed studies of the activity of a protein coded by an unknown gene

Gene inactivation and overexpression are the primary techniques used by genome researchers to determine the function of a new gene, but these are not the only procedures that can provide information on gene activity. Other methods can extend and elaborate the results of inactivation and overexpression. These can be used to provide additional information that will aid identification of a gene function, or might form the basis of a more comprehensive examination of the activity of a protein whose gene has already been characterized.

Directed mutagenesis can be used to probe gene function in detail

Inactivation and overexpression can determine the general function of a gene, but they cannot provide detailed information on the activity of a protein coded by a gene. For example, it might be suspected that part of a gene codes for an amino acid sequence that directs its protein product to a particular compartment in the cell, or is responsible for the ability of the protein to respond to a chemical or physical signal. To test these hypotheses it would be necessary to delete or alter the relevant part of the gene sequence, but to leave the bulk unmodified so that the protein is still synthesized and retains the major part of its activity. The various procedures of **site-directed** or *in vitro* **mutagenesis** (Technical Note 5.2) can be used to make these subtle changes. These are important techniques whose applications lie not only with the study of gene activity but also in the area of **protein engineering**, where the objective is to create novel proteins with properties that are better suited for use in industrial or clinical settings.

After mutagenesis, the gene sequence must be introduced into the host cell so that homologous recombination can replace the existing copy of the gene with the modified version. This presents a problem because we must have a way of knowing which cells have undergone homologous recombination. Even with yeast this will only be a fraction of the total, and with mice the fraction will be very small. Normally we would solve this problem by placing a marker gene (e.g., one coding for antibiotic resistance) next to the mutated gene and looking for cells that take on the phenotype conferred by this marker. In most cases, cells that insert the marker gene into their genome also insert the closely attached mutated gene and so are the ones we want. The problem is that in a site-directed mutagenesis experiment we must be sure that any change in the activity of the gene being studied is the result of the specific mutation that was introduced into the gene, rather than the indirect result of changing its environment in the genome by inserting a marker gene next to it. The answer is to use a more complex, two-step gene replacement (Figure 5.25). In this procedure the target gene is first replaced with the marker gene on its own, the cells in which this recombination takes place being identified by selecting for the marker gene phenotype. These cells are then used in the second stage of the gene replacement, when the marker gene is replaced by the mutated gene, success now being monitored by looking for cells that have lost the marker gene phenotype. These cells contain the mutated gene and their phenotypes can be examined to determine the effect of the directed mutation on the activity of the protein product.

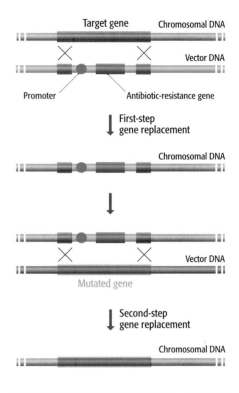

Figure 5.25 Two-step gene replacement.

Reporter genes and immunocytochemistry can be used to locate where and when genes are expressed

Clues to the function of a gene can often be obtained by determining where and when the gene is active. If gene expression is restricted to a particular organ or tissue of a multicellular organism, or to a single set of cells within an organ or tissue, then this positional information can be used to infer the general role of the gene product. The same is true of information relating to the developmental stage at which a gene is expressed. This type of analysis has proved particularly useful in understanding the activities of genes involved in the earliest stages of development in *Drosophila* (Section 14.3.4) and is increasingly being used to unravel the genetics of mammalian development. It is also applicable to those unicellular organisms, such as yeast, that have distinctive developmental stages in their life cycle.

Determining the pattern of gene expression within an organism is possible with a **reporter gene**. This is a gene whose expression can be monitored in a convenient way, ideally by visual examination (Table 5.3), cells that express

Table 5.3 Examples of reporter genes

Gene	Gene product	Assay
lacZ	β-galactosidase	Histochemical test
uidA	β-glucuronidase	Histochemical test
lux	Luciferase	Bioluminescence
GFP	Green fluorescent protein	Fluorescence

Technical Note 5.2 Site-directed mutagenesis

Methods for making a precise alteration in a gene sequence in order to change the structure and possibly the activity of a protein

Changes in protein structure can be engineered by site-directed mutagenesis techniques, which result in defined alterations being made to the nucleotide sequence of the gene coding for a protein of interest. These techniques enable the functions of different parts of a protein to be examined, and also have widespread importance in the development of new enzymes for biotechnological purposes.

Conventional mutagenesis is a random process that introduces changes at unspecified positions in a DNA molecule.

Screening of large numbers of mutated organisms is necessary to find a mutation of interest. Even with microbes, which can be screened in huge numbers, the best that can be hoped for is a range of mutations in the correct gene, one of which might affect the part of the protein being studied. Site-directed mutagenesis offers a means of making much more specific mutations. The most important of these methods are:

- **Oligonucleotide-directed mutagenesis**, in which an oligonucleotide containing the desired mutation is

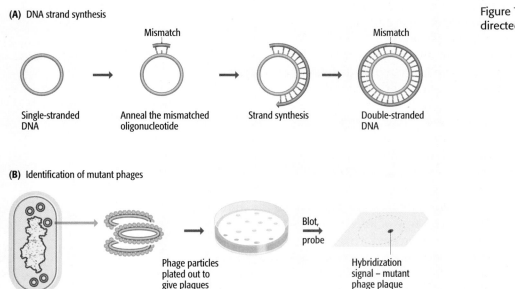

(A) DNA strand synthesis

Mismatch Mismatch

Single-stranded DNA → Anneal the mismatched oligonucleotide → Strand synthesis → Double-stranded DNA

(B) Identification of mutant phages

Infected *E. coli* → Phage particles plated out to give plaques → Blot, probe → Hybridization signal – mutant phage plaque

Figure T5.1 Oligonucleotide-directed mutagenesis.

the reporter gene becoming blue, fluorescing, or giving off some other visible signal. For the reporter gene to give a reliable indication of where and when a test gene is expressed, the reporter must be subject to the same regulatory signals as the test gene. This is achieved by replacing the ORF of the test gene with the ORF of the reporter gene (Figure 5.26). Most of the regulatory signals that control gene expression are contained in the region of DNA upstream of the ORF, so the reporter gene should now display the same expression pattern as the test gene. The expression pattern can therefore be determined by examining the organism for the reporter signal.

As well as knowing in which cells a gene is expressed, it is often useful to locate the position within the cell where the protein coded by the gene is found. For example, key data regarding gene function can be obtained by showing that the annealed to a single-stranded version of the relevant gene, the latter obtained by cloning in an M13 vector (Section 4.1.1). The oligonucleotide primes a strand synthesis reaction that is allowed to continue all the way around the circular template molecule (Figure T5.1A). After introduction into *Escherichia coli*, DNA replication produces numerous copies of this recombinant DNA molecule, half of these being copies of the original strand of DNA, and half copies of the strand that contains the mutated sequence. All of these double-stranded molecules direct synthesis of phage particles, so about half the phages released from the infected bacteria carry a single-stranded version of the mutated molecule. The phages are plated onto solid agar so that plaques are produced, and the mutant ones identified by hybridization probing with the original oligonucleotide (Figure T5.1B). The mutated gene can then be placed back in its original host by homologous recombination, as described in Section 5.2.3, or transferred to an *E. coli* vector designed for synthesis of protein from cloned DNA, so that a sample of the mutated protein can be obtained.

- **Artificial gene synthesis** involves constructing the gene in the test tube, placing mutations at all the desired positions. The gene is constructed by synthesizing a series of partially overlapping oligonucleotides, each up to 150 nucleotides in length. The gene is assembled by filling in the gaps between the overlaps with DNA polymerase, and ligated into a cloning vector prior to introduction into the host organism or into *E. coli*.

- **PCR** can also be used to create mutations in cloned genes, though like oligonucleotide-directed mutagenesis, only one mutation can be created per experiment.

The method shown in Figure T5.2 involves two PCRs, each with one normal primer (which forms a fully base-paired hybrid with the template DNA), and one mutagenic primer (which contains a single base-pair mismatch corresponding to the mutation). This mutation is therefore initially present in two PCR products, each corresponding to one half of the starting DNA molecule. The two PCR products are then mixed together and a final PCR cycle carried out to construct the full-length, mutated DNA molecule.

Figure T5.2 One method for site-directed mutagenesis by PCR.

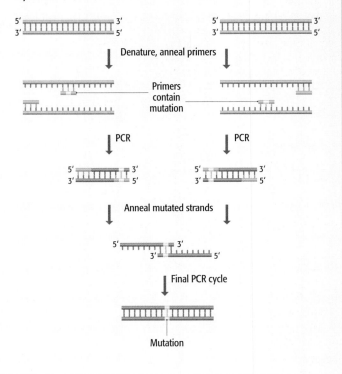

protein product is located in mitochondria, in the nucleus, or on the cell surface. Reporter genes cannot help here because the DNA sequence upstream of the gene—the sequence to which the reporter gene is attached—is not involved in targeting the protein product to its correct intracellular location. Instead it is the amino acid sequence of the protein itself that is important. Therefore the only way to determine where the protein is located is to search for it directly. This can be done by **immunocytochemistry**, which makes use of an antibody that is specific for the protein of interest and so binds to this protein and no other. The antibody is labeled so that its position in the cell, and hence the position of the target protein, can be visualized (Figure 5.27). Fluorescent labeling and confocal microscopy are used for low-resolution studies; alternatively, high-resolution immunocytochemistry can be carried out by electron microscopy using an electron-dense label such as colloidal gold.

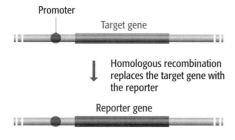

Promoter

Target gene

Homologous recombination replaces the target gene with the reporter

Reporter gene

Figure 5.26 A reporter gene. The open reading frame of the reporter gene replaces the open reading frame of the gene being studied. The result is that the reporter gene is placed under control of the regulatory sequences that usually dictate the expression pattern of the test gene. For more information on these regulatory sequences, see Sections 11.2 and 11.3. Note that the reporter gene strategy assumes that the important regulatory sequences do indeed lie upstream of the gene. This is not always the case for eukaryotic genes.

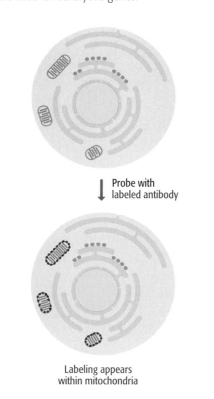

Probe with labeled antibody

Labeling appears within mitochondria

Figure 5.27 Immunocytochemistry. The cell is treated with an antibody that is labeled with a red fluorescent marker. Examination of the cell shows that the fluorescent signal is associated with the inner mitochondrial membrane. A working hypothesis would therefore be that the target protein is involved in electron transport and oxidative phosphorylation, as these are the main biochemical functions of the inner mitochondrial membrane.

5.3 Case Study: Annotation of the *Saccharomyces cerevisiae* Genome Sequence

In this chapter we have looked at a range of techniques, both computer-based and experimental, for locating the genes in a genome sequence and for determining their functions. Not every technique has been used with every organism, the choice often being determined by technical considerations—for example, RNAi has been used extensively with *C. elegans* partly because of the ease with which double-stranded RNAs can be introduced into worms by feeding experiments (Section 5.2.2). To illustrate how these various techniques are combined, we will conclude this chapter by surveying the progress made in annotating the genome of the yeast *S. cerevisiae*. In eukaryotic terms, the yeast genome is relatively noncomplex, in that it contains relatively little intergenic DNA and very few introns. This simplifies the identification of ORFs, but provides no advantage when it comes to determining the function of individual genes.

5.3.1 Annotation of the yeast genome sequence

The *S. cerevisiae* sequencing project was completed in 1996. The initial analysis, which set 100 codons as the minimum cut-off for a potential gene, identified 6274 ORFs, approximately 30% of which were known to be genuine genes because they had previously been identified by conventional genetic analysis before the sequencing project got underway. The remaining 70% were studied by homology analysis when the genome sequence was completed, giving the following results (Figure 5.28):

- Almost 30% of the genes in the genome could be assigned functions after homology searching of the sequence databases. About half of these were clear homologs of genes whose functions had been established previously, and about half had less-striking similarities, including many where the similarities were restricted to discrete domains. For all these genes the homology analysis could be described as successful, but with various degrees of usefulness. For some genes the identification of a homolog enabled the function of the yeast gene to be comprehensively determined: examples included identification of yeast genes for DNA polymerase subunits. For other genes the functional assignment could only be to a broad category, such as "gene for a protein kinase": in other words, the biochemical properties of the gene product could be inferred, but not the exact role of the protein in the cell. Some identifications were initially puzzling, the best example being the discovery of a yeast homolog of a bacterial gene involved in nitrogen fixation. Yeasts do not fix nitrogen so this could not be the function of the yeast gene. In this case, the discovery of the yeast homolog refocused attention on the previously characterized bacterial gene, with the subsequent realization that, although being involved in nitrogen fixation, the primary role of the bacterial gene product was in the synthesis of metal-containing proteins, which have broad roles in all organisms, not just nitrogen-fixing ones.

- About 10% of all the yeast genes had homologs in the databases, but the functions of these homologs were unknown. The homology analysis was therefore unable to help in assigning functions to these yeast genes. These yeast genes and their homologs were called **orphan families**.

- The remaining yeast genes, about 30% of the total, had no homologs in the databases. A proportion of these (about 7% of the total) were questionable ORFs which might not be real genes, being rather short or having an unusual codon bias. The remainder looked like genes but were unique. These were called **single orphans**.

Following the initial annotation of the yeast genome sequence, two questions had to be considered: First, how many of the single orphans are genuine genes? Second, are there any genuine genes less than 100 codons in length and hence not identified by this initial analysis? The latter is an important point: although there are just 6274 ORFs of 100 codons and longer in the yeast genome, there are over 100,000 ORFs of 15 codons or more, most of these displaying a pattern of codon usage that is indistinguishable from genuine yeast genes. The potential for finding new, short genes was therefore huge.

Three approaches were used in refining the yeast gene set, all involving techniques that we have already met in this chapter:

- **Comparative genomics**, making use of the suite of genome sequences for related yeast species, was used to assess the authenticity of many short ORFs.

- Evidence of transcription was sought by sequencing cDNAs, including libraries of expressed sequence tags, which are short and usually incomplete cDNAs derived from the ends of transcribed sequences (Section 3.3.3), and by serial analysis of gene expression (SAGE; Section 6.1.1) and microarray studies (Section 6.1.2).

- **Transposon tagging**, as well as being used to inactivate genes as part of the functional analysis, was used to identify ORFs that are genuine genes. This work made use of a transposon containing a copy of the *lacZ* gene that lacked its initiation codon (Figure 5.29). In normal cells the *lacZ* gene is therefore inactive, so when the X-gal test is applied (Section 2.2.1) the colonies appear white. The *lacZ* gene becomes activated if transposition moves it to a position within a genuine yeast gene and the insertion results in the codons of the yeast gene becoming fused, in frame, with the *lacZ* gene. Now the colonies appear blue. Questionable ORFs can therefore be assessed from the colour of colonies after the ORF has become tagged with the transposon.

These experiments are continuing but their result so far has been to reduce the yeast gene catalog to approximately 6120 ORFs, some 150 fewer than the initial estimate. This reduction comes about by discounting many of the original single orphans, which are no longer looked on as possible genes, but also adding into the list a few ORFs that are shorter than 100 codons but which turn out to be genuine genes.

5.3.2 Assigning functions to yeast genes

S. cerevisiae has two features that aid attempts to assign functions to the unknown genes in its genome. The first of these is a high natural propensity for homologous recombination, which makes it relatively easy to use this approach to inactivate individual genes (see Figure 5.20). The second is the presence in the genome of the *Ty* family of transposons, which enable transposon tagging to be used as an alternative means of gene disruption. Both approaches have been used extensively by yeast researchers. The challenge now lies not in developing methods for inactivation of individual yeast genes, but in devising ways of screening the vast numbers of resulting mutants for novel phenotypic features that might indicate the function of an inactivated gene. Carrying out many parallel experiments, each with a

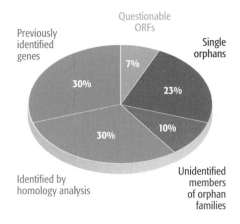

Figure 5.28 A summary of the outcome of the initial annotation of the yeast genome.

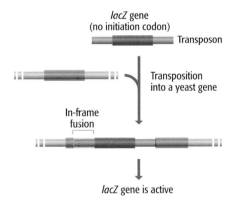

Figure 5.29 Using transposon tagging to identify yeast genes. If the transposon moves into a functional yeast gene, such that an open reading frame is created between the start of the yeast gene and the *lacZ* gene present in the transposon, then the *lacZ* gene can be expressed. The resulting β-galactoside protein has a segment of the protein coded by the yeast gene attached to its N–terminus, but often this will not affect the enzyme activity. Hence transposition into a functional yeast gene is detectable by the X-gal test.

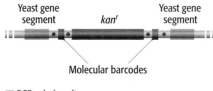

PCR priming sites

Figure 5.30 A deletion cassette used in the barcode strategy. Compare with cassette shown in Figure 5.2.1. The two molecular barcodes are 20-nucleotide-long sequences, different for each cassette, which can be amplified by PCR. During homologous recombination, the barcodes are inserted into the yeast genome along with the kanamycin-resistance gene. The barcodes therefore provide specific tags for each individual gene deletion.

different mutant, takes a long time, especially when there is such a large number of phenotypes to assess. Large-scale screening strategies are therefore needed.

The most successful of these screening methods is the **barcode deletion strategy**. This is a modification of the basic deletion cassette system shown in Figure 5.21, the difference being that the cassette also includes two 20-nucleotide "barcode" sequences, different for each deletion, which act as tags for that particular mutant (Figure 5.30). Each barcode is flanked by the same pair of sequences and so can be amplified by a single PCR. This means that groups of mutated yeast strains, each with a different inactivated gene, can be mixed together and their phenotypes screened in a single experiment. For example, to identify genes required for growth in a glucose-rich medium, a collection of mutants would be mixed together and cultured under these conditions. After incubation, DNA is prepared from the culture and the barcoding PCR carried out. The result is a mixture of PCR products, each representing a different barcode, the relative abundance of each barcode indicating the abundance of each mutant after growth in glucose-rich medium. Those barcodes that are absent or present only at low abundance indicate the mutants whose inactivated genes were needed for growth under these conditions.

As with gene identification, functional characterization of yeast genes is ongoing and it will be several years before this project begins to near completion. But progress is gradually being made. Approximately 55% of all yeast genes now have a well-characterized function that has been assigned by one or more experimental techniques. This is some 1500 genes more than was the case when the genome was first sequenced. Another 2000 genes—33% of the total—have functions that have been assigned on the basis of homology analysis. This leaves just 500 ORFs, which are thought to be genuine genes but have no assigned function, and another 300 questionable ORFs that may not be real genes.

Summary

A variety of methods have been devised for identifying the genes in a genome and determining the functions of these genes. Some of the methods are computer-based, and others involve experimentation. When a genome sequence is first obtained, the initial objective is to locate the positions of all the genes. For protein-coding genes this can be attempted by searching for open reading frames (ORFs), though this is complicated in eukaryotes by the presence of introns, whose boundary sequences are variable and cannot be identified accurately. Genes for functional RNAs can also be located by searching for their characteristic features, primarily the ability of the RNAs to fold into secondary structures based on the formation of base-paired stem-loops. Genes can also be located by homology analysis, which uses the presence of an equivalent gene in a second genome as evidence that a putative gene in the test genome is genuine. Homology analysis is made more powerful if the complete sequence of a related genome is available. Experimental methods for gene location are based on detection of RNA molecules transcribed from the genome. These methods include cDNA sequencing and transcript mapping by reverse-transcriptase PCR (RT-PCR) or heteroduplex analysis. Gene functions can tentatively be assigned by homology analysis, because homologs are evolutionarily related and often, but not always, have similar functions. Most experimental techniques for determining the functions of genes involve examining the effect that gene inactivation has on the phenotype of the organism. Inactivation can be achieved in various ways: by homologous recombination with a defective copy of the gene, by insertion of a transposon into the gene, or by RNA interference, the last of these approaches having been particularly successful with *Caenorhabditis elegans*. Gene overexpression can also be used to assess function, but with both inactivation and overexpression it may be difficult to discern a phenotypic change, and the precise function of the gene may remain elusive. More detailed studies of gene function can be carried out by site-directed mutagenesis, and the cellular location of a protein can be determined by expression of a reporter gene or by immunocytochemistry. When the *Saccharomyces cerevisiae* genome sequence was completed in 1996, 6274 ORFs that might be genes were identified, but this figure has now been revised to 6120 following experimental analysis and comparisons with other yeast genomes. Initially only 30% of these genes had well-characterized functions, but this number has gradually been increased by homology analysis and by the application of high-throughput functional screens such as the barcode deletion strategy.

Multiple Choice Questions

5.1.* What is an open reading frame (ORF)?

  a. All of the nucleotides of a gene that are transcribed into mRNA.

  b. The nucleotides of a gene that make up the codons specifying amino acids.

  c. The nucleotides of an mRNA molecule before the introns have been removed.

  d. The amino acid sequence of a polypeptide.

5.2. Codon bias refers to which of the following?

  a. Some codons for an amino acid are more frequently used in all species.

  b. Some amino acids are rarely used in the proteins of some organisms.

  c. Some codons for an amino acid are more frequently used and the bias varies in different species.

  d. Some codons code for rare amino acids such as selenocysteine in some species.

5.3.* A consensus sequence for an exon–intron boundary or gene promoter refers to:

  a. The exact nucleotide sequence required for the sequence to function.

  b. The sequence of nucleotides found most commonly at these sites.

  c. The shortest sequence needed for the sequence to function.

  d. The sequence of nucleotides surrounding the sites of intron splicing or transcription initiation.

5.4. Why is ORF scanning not applicable when searching for genes coding for functional RNA molecules?

  a. The codons for functional RNA molecules vary from species to species.

  b. Genes for functional RNAs contain introns that make ORF scanning impossible.

  c. The codons in functional RNA genes are each only two nucleotides in length.

  d. Functional RNA genes do not contain codons.

5.5.* What is the purpose of performing a homology search with a DNA sequence?

  a. To determine if any genes with similar sequences are present in the DNA databases.

  b. To determine if the sequence is already in the database.

  c. To search for consensus exon–intron boundaries.

  d. To determine the codon bias for a specific gene.

5.6. Which of the following is the correct definition of synteny?

  a. The percentage of nucleotide sequence identity between two genomes.

  b. The percentage of amino acid sequence identity between two genomes.

  c. The conservation of gene order within two genomes.

  d. The conservation of gene function within two genomes.

5.7.* The amplification of mRNA by PCR is called:

  a. Real time PCR.

  b. Reverse transcriptase PCR.

  c. Transcriptional PCR.

  d. Translational PCR.

5.8. By definition, homologous genes are genes that:

  a. Share a common function.

  b. Share a common evolutionary ancestor.

  c. Are expressed under similar conditions.

  d. Have at least 50% nucleotide sequence identity.

5.9.* The amino acid sequence of the α polypeptide of hemoglobin is more similar to the amino acid sequence of the β polypeptide of hemoglobin than it is to the amino acid sequence of myoglobin. All these genes share a common evolutionary ancestor. Which of the following statements correctly describes the relationship between the genes encoding these polypeptides?

  a. The gene encoding the α polypeptide has higher homology to the β gene than the myoglobin gene.

  b. The genes encoding these three polypeptides are all homologs.

  c. The myoglobin gene is not a homolog of the other two genes.

  d. These genes are homologs only if they are present in the same species.

5.10. Why is inactivation a useful technique to determine the function of a gene?

  a. Gene inactivation provides information about the expression of a gene.

  b. Gene inactivation provides information about the cellular location of a gene product.

  c. Gene inactivation provides an opportunity to identify phenotypic changes associated with the loss of the functional gene.

  d. Gene inactivation provides information on the structure of the gene product.

Multiple Choice Questions (continued) *Answers to odd-numbered questions can be found in the Appendix

5.11.* Mouse embryonic stem cells are used in gene inactivation experiments because they:

    a. Can be cloned to give rise to a stable cell line.

    b. Are chimeric and will produce cells heterozygous for the gene.

    c. Are the only mouse cells that can be genetically engineered to inactivate genes.

    d. Are totipotent and can give rise to all types of differentiated cells.

5.12. RNA interference works by which of the following methods?

    a. Using antisense RNA molecules to block the translation of mRNA molecules.

    b. Using RNA polymerase inhibitors to block the transcription of specific genes.

    c. Using short, double-stranded RNA molecules that will cause the degradation of an mRNA molecule.

    d. Using modified tRNA molecules to block the translation of mRNA molecules.

5.13.* What is the best method to identify the cellular location of a protein?

    a. Place a reporter gene next to the promoter of the gene encoding the protein, and identify the cellular location of the reporter protein.

    b. Use a labeled antibody to identify the cellular location of the protein.

    c. Separate the cellular compartments by centrifugation and screen the different compartments with an antibody.

    d. Tag the protein with fluorescent amino acids and identify the cellular location by fluorescence microscopy.

5.14. Orphan families are families of genes that:

    a. Lack homologs in other species.

    b. Have no known function.

    c. Have no phenotypic change after gene inactivation.

    d. Have not yet been studied.

Short Answer Questions

*Answers to odd-numbered questions can be found in the Appendix

5.1*. Why is it relatively easy to identify ORFs in prokaryotic genomes by computer analysis?

5.2. What are the two major problems that occur when attempting to use computer analysis to identify ORFs in the genomes of higher eukaryotes?

5.3.* What are the three primary modifications that can be used to improve the location of ORFs by computer analysis?

5.4. What structural features of functional RNA molecules, such as tRNA and rRNA, can be searched for in a genome sequence to identify the genes encoding these RNA molecules?

5.5.* What are the two limitations that arise when northern analysis is used to determine the number of genes present in a DNA fragment?

5.6. Describe how the technique of rapid amplification of cDNA ends (RACE) is used to map the transcription initiation site of a gene.

5.7.* What is the difference between orthologous and paralogous genes?

5.8. Why are errors sometimes made when assigning gene function based on the results of a BLAST search?

5.9*. How can the study of homologous genes provide information on human diseases?

5.10. What is the likely explanation when knockout mice are not obtained after a cross between heterozygous parents?

5.11.* When a genome sequence is analyzed, what approaches can be used to identify authentic genes that are 100 codons or less in length?

continued …

In-depth Problems

*Guidance to odd-numbered questions can be found in the Appendix

5.1.* To what extent do you believe it will be possible in future years to use bioinformatics to obtain a complete description of the locations and functions of the protein-coding genes in a eukaryotic genome sequence?

5.2. Gene inactivation studies have suggested that at least some genes in a genome are redundant, meaning that they have the same function as a second gene and so can be inactivated without affecting the phenotype of the organism. What evolutionary questions are raised by genetic redundancy? What are the possible answers to these questions?

5.3.* Perform a BLAST search with the following amino acid sequence: IRLFKGHPETLEKFDKFKHL. What is the protein that contains this amino acid sequence? Are the homologous sequences identified for this search mostly orthologs or paralogs? (A BLAST search can be carried out at www.ncbi.nlm.nih.gov/BLAST/).

5.4. Gene overexpression has so far provided limited but important information on the function of unknown genes. Assess the overall potential of this approach in functional analysis.

Figure Tests

*Answers to odd-numbered questions can be found in the Appendix

5.1.* How can a computer program determine the difference between the stop codon in the intron and the actual stop codon at the end of the exon?

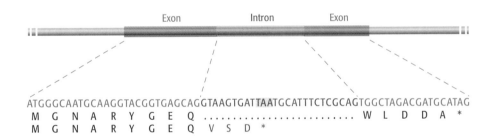

Figure Tests (continued)

5.2. What is the purpose of performing Southern hybridization with genomic DNA from different organisms?

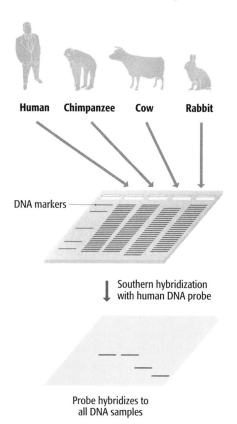

5.3.* What is the purpose of placing the gene for GFP (green fluorescent protein) downstream of a promoter for a gene of interest?

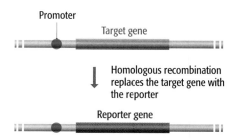

5.4. Explain how the barcode deletion strategy is used to identify the phenotypic properties of yeast deletion mutants.

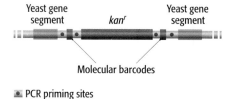

Further Reading

Gene location by computer analysis

Fickett, J.W. (1996) Finding genes by computer: the state of the art. *Trends Genet.* **12:** 316–320.

Kellis, M., Patterson, N., Birren, B. and Lander, E.S. (2003) Sequencing and comparison of yeast species to identify genes and regulatory elements. *Nature* **423:** 241–254. *Using comparative genomics to annotate the yeast genome sequence.*

Ohler, U. and Niemann, H. (2001) Identification and analysis of eukaryotic promoters: recent computational approaches. *Trends Genet.* **17:** 56–60.

Pavesi, G., Mauri1, G., Stefani1, M. and Pesole, G. (2004) RNAProfile: an algorithm for finding conserved secondary structure motifs in unaligned RNA sequences. *Nucleic Acids Res.* **32:** 3258–3269. *Locating functional RNA genes.*

Experimental methods for gene location

Church, D.M., Stotler, C.J., Rutter, J.L., Murrell, J.R., Trofatter, J.A. and Buckler, A.J. (1994) Isolation of genes from complex sources of mammalian genomic DNA using exon amplification. *Nat. Genet.* **6:** 98–105. *Exon trapping.*

Frohman, M.A., Dush, M.K. and Martin, G.R. (1988) Rapid production of full-length cDNAs from rare transcripts: amplification using a single gene-specific oligonucleotide primer. *Proc. Natl Acad. Sci. USA* **85:** 8998–9002. *RT-PCR.*

Lovett, M. (1994) Fishing for complements: finding genes by direct selection. *Trends Genet.* **10:** 352–357. *cDNA capture.*

Assigning function by homology analysis

Altschul, S.F., Gish, W., Miller, W., Myers, E.W. and Lipman, D.J. (1990) Basic local alignment search tool. *J. Mol. Biol.* **215:** 403–410. *The BLAST program.*

Bassett, D.E., Boguski, M.S. and Hieter, P. (1996) Yeast genes and human disease. *Nature* **379:** 589–590. *Studying human disease genes by comparisons with the yeast genome.*

Henikoff, S. and Henikoff, J.G. (1992) Amino acid substitution matrices from protein blocks. *Proc. Natl Acad. Sci. USA* **89:** 10915–10919. *Describes the chemical relationships between amino acids, from which sequence similarity scores are calculated.*

RNA interference studies

Fraser, A.G., Kamath, R.S., Zipperlen, P., Martinez-Campos, M., Sohrmann, M. and Ahringer, J. (2000) Functional genomic analysis of *C. elegans* chromosome I by systematic RNA interference. *Nature* **408:** 325–330.

Kittler, R., Putz, G., Pelletier, L., *et al.* (2004) An endoribonuclease-prepared siRNA screen in human cells identifies genes essential for cell division. *Nature* **432:** 1036–1040.

Novina, C.D. and Sharp, P.A. (2004) The RNAi revolution. *Nature* **430:** 161–164.

Sönnichsen, B., Koski, L.B., Walsh, A., *et al.* (2005) Full-genome RNAi profiling of early embryogenesis in *Caenorhabditis elegans. Nature* **434:** 462–469.

Other methods for gene inactivation

Evans, M.J., Carlton, M.B.L. and Russ, A.P. (1997) Gene trapping and functional genomics. *Trends Genet.* **13:** 370–374. *The use of ES cells.*

Ross-Macdonald, P., Coelho, P.S.R., Roemer, T., *et al.* (1999) Large-scale analysis of the yeast genome by transposon tagging and gene disruption. *Nature* **402:** 413–418.

Wach, A., Brachat, A., Pohlmann, R. and Philippsen, P. (1994) New heterologous modules for classical or PCR-based gene disruptions in *Saccharomyces cerevisiae. Yeast* **10:** 1793–1808. *Gene inactivation by homologous recombination.*

Annotation of the yeast genome sequence

Dujon, B. (1996) The yeast genome project: what did we learn? *Trends Genet.* **12:** 263–270. *A summary of the initial annotation.*

Giaever, G., Chu, A.M., Connelly, C., *et al.* (2002) Functional profiling of the *Saccharomyces cerevisiae* genome. *Nature* **418:** 387–391. *The barcode deletion strategy.*

Snyder, M. and Gerstein, M. (2003) Defining genes in the genomics era. *Science* **300:** 258–260. *Summarizes the methods used to annotate the yeast genome, and the progress made up to 2003.*

Understanding How a Genome Functions

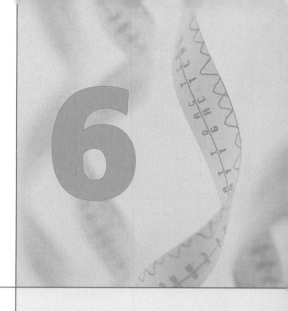

6

When you have read Chapter 6, you should be able to:

Describe how a transcriptome can be studied by sequence analysis of cDNAs.

Evaluate the strengths and weaknesses of microarray and chip technology in the study of transcriptomes, and explain how comparisons are made between gene expression patterns revealed by these studies.

Give examples of the contributions that transcriptome studies have made to our understanding of yeast biology and human cancer.

Distinguish between the different types of information that are obtained by studying transcriptomes and proteomes.

Describe how protein profiling is carried out.

Compare and contrast the methods used to identify pairs and groups of proteins that interact with one another in living cells, in particular distinguishing between methods that identify physical interactions and those that identify functional interactions.

Give examples of protein interaction maps and discuss their important features.

Explain the basis to and importance of biochemical profiling.

Outline the principles and objectives of systems biology.

In the previous chapter we learnt how a variety of computational and experimental techniques can be used to assign functions to genes discovered within a genome sequence, and how application of these techniques to the *Saccharomyces cerevisiae* genome has almost doubled the number of yeast genes for which definite functions are known. This type of genome annotation is a massive undertaking, but even if every gene in a genome can be identified and assigned a function, a challenge still remains. This is to understand how the genome as a whole operates within the cell, specifying and coordinating the various biochemical activities that take place. These global studies of genome activity must address not only the genome itself but also the transcriptome and proteome that are synthesized and maintained by the

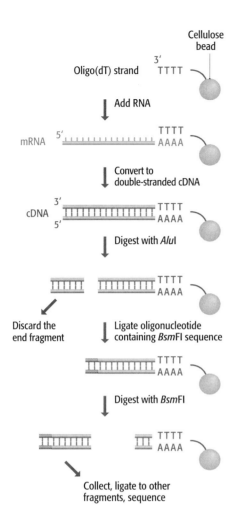

Figure 6.1 SAGE. In this example, the first restriction enzyme to be used is *Alu*I, which recognizes the 4 bp target site 5′–AGCT–3′. The oligonucleotide that is ligated to the cDNA contains the recognition sequence for *Bsm*FI, which cuts 10–14 nucleotides downstream, and so cleaves off a fragment of the cDNA. Fragments of different cDNAs are ligated to produce the concatamer that is sequenced. Using this method, the concatamer that is formed is made up partly of sequences derived from the *Bsm*FI oligonucleotides. To avoid this, and so obtain a concatamer made up entirely of cDNA fragments, the oligonucleotide can be designed so that the end that ligates to the cDNA contains the recognition sequence for a third restriction enzyme. Treatment with this enzyme cleaves the oligonucleotide from the cDNA fragment.

genome. They must also address the way in which the transcriptome and proteome establish and coordinate the ultimate endpoint of genome expression—the network of linked biochemical pathways and processes that constitute a living cell. In this chapter we will look at the methods used in these global studies of genome activity.

6.1 Studying the Transcriptome

The transcriptome comprises the mRNAs that are present in a cell at a particular time. Transcriptomes can have highly complex compositions, with hundreds or thousands of different mRNAs represented, each making up a different fraction of the overall population (Section 1.2.4). To characterize a transcriptome it is therefore necessary to identify the mRNAs that it contains and, ideally, to determine their relative abundances.

6.1.1 Studying a transcriptome by sequence analysis

The most direct way to characterize a transcriptome is to convert its mRNA into cDNA (see Figure 3.36), and then to sequence every clone in the resulting cDNA library. Comparisons between the cDNA sequences and the genome sequence will reveal the identities of the genes whose mRNAs are present in the transcriptome. This approach is feasible but it is laborious, with many different cDNA sequences being needed before a near-complete picture of the composition of the transcriptome begins to emerge. If two or more transcriptomes are being compared then the time needed to complete the project increases. Can any shortcuts be used to obtain the vital sequence information more quickly?

Serial analysis of gene expression (**SAGE**) provides a solution. Rather than studying complete cDNAs, SAGE yields short sequences, as little as 12 bp in length, each of which represents an mRNA present in the transcriptome. The basis of the technique is that these 12 bp sequences, despite their shortness, are sufficient to enable the gene that codes for the mRNA to be identified. The argument is that any particular 12 bp sequence should appear in the genome once every $4^{12} = 16,777,216$ bp. The average size of a eukaryotic mRNA is about 1500 bp, so 4^{12} bp is equivalent to the combined length of over 11,000 transcripts. This number is higher than the number of transcripts expected in all but the most complex transcriptomes, so the 12 bp sequence tags should be able to identify the genes coding for all the mRNAs that are present.

The procedure used to generate the 12 bp tags is shown in Figure 6.1. First, the mRNA is immobilized in a chromatography column by annealing the poly(A) tails present at the 3′ ends of these molecules to oligo(dT) strands that have been attached to cellulose beads. The mRNA is converted into double-stranded cDNA and then treated with a restriction enzyme that recognizes a 4 bp target site and so cuts frequently in each cDNA. The terminal restriction fragment of each cDNA remains attached to the cellulose beads, enabling all the other fragments to be eluted and discarded. A short oligonucleotide is now attached to the free end of each cDNA, this oligonucleotide containing a recognition sequence for *Bsm*FI. This is an unusual restriction enzyme in that rather than cutting within its recognition sequence, it cuts 10–14 nucleotides downstream. Treatment with *Bsm*FI therefore removes a fragment with an average length of 12 bp from the end of each cDNA. The fragments are collected, ligated head-to-tail to produce a concatamer, and sequenced. The individual tag sequences are identified within the concatamer and compared with the sequences of the genes in the genome.

6.1.2 Studying a transcriptome by microarray or chip analysis

DNA chips and microarrays (see Technical Note 3.1) can also be used to study transcriptomes. Recall that the difference between these is that a chip carries an array of immobilized oligonucleotides synthesized *in situ* on the surface of a wafer of glass or silicon, and microarrays comprise DNA molecules—usually PCR products or cDNAs—that have been spotted onto the surface of a glass slide or nylon membrane. Microarrays and chips are both used in the same way (Figure 6.2). The population of mRNAs that make up a transcriptome is converted into a mixture of cDNAs, labeled (usually with a fluorescent marker), and applied to the microarray or chip, and the positions at which hybridization occurs are detected. Compared with SAGE, this approach has the advantage that a rapid evaluation of the differences between two or more transcriptomes can be made by hybridizing different cDNA preparations to identical arrays and comparing the hybridization patterns. A further embellishment can be achieved by probing the array with cDNA that has been prepared from the mRNA fraction that is bound to ribosomes in the cells being studied, rather than from total mRNA. The bound mRNAs correspond to the part of the transcriptome that is actively directing protein synthesis, giving a slightly different picture of genome activity.

First we will consider the technical issues involved in microarray and chip studies, and then we will consider some of the applications of this type of analysis.

Using a microarray or chip to study one or more transcriptomes

When a transcriptome is analyzed, the two key objectives are to identify the genes whose mRNAs are present, and to determine the relative amounts of these various mRNAs. The first of these requirements demands that every relevant gene be represented by at least one probe in the array. With a microarray, this is achieved by using PCR products or cDNAs that are derived from the genes of interest, and with a DNA chip by synthesizing at each position a mixture of oligonucleotides, perhaps as many as 20 different ones in total, the sequences of which match different positions in the relevant gene (Figure 6.3). The second requirement—the ability to determine the relative amounts of individual mRNAs in the transcriptome—is met because each position in

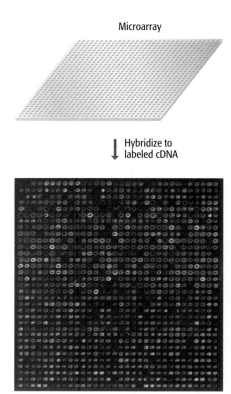

Microarray

↓ Hybridize to labeled cDNA

Figure 6.2 Microarray analysis. A cDNA preparation is labeled with a fluorescent marker and hybridized to the microarray. The label is detected by confocal laser scanning and the signal intensity converted into a pseudocolor spectrum, with red indicating the greatest hybridization, followed by orange, yellow, green, blue, indigo, and violet, the last representing the background level of hybridization. For more information on the preparation and use of a microarray see Technical Note 3.1. Image courtesy of Tom Strachan, reprinted with permission from *Nature*.

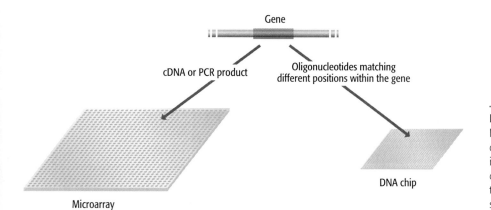

Figure 6.3 Microarrays and DNA chips. Each position in a microarray contains a cDNA or PCR product from a gene of interest, whereas each position in a DNA chip contains a mixture of oligonucleotides, the sequences of these matching different segments of the relevant gene.

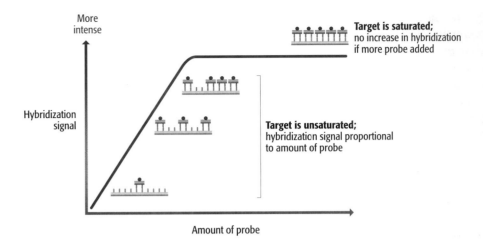

Figure 6.4 The relationship between hybridization intensity and the amount of probe.

the microarray or chip contains up to 10^9 copies of the probe molecules. This is higher than the anticipated copy number for any mRNA in the small amount of the transcriptome that is applied to the array, which means that no position ever becomes saturated (i.e., so much hybridization that every probe molecule is base-paired to a target molecule). The amount of hybridization is therefore variable, the signal intensity at each position depending on the amount of each particular mRNA in the transcriptome (Figure 6.4).

From reading the above, microarray and chip analysis might appear to be straightforward procedures. In practice, a number of complications arise. The first is that, with all but the simplest transcriptomes, hybridization analysis will have insufficient specificity to distinguish between every mRNA that is present. This is because two different mRNAs might have similar sequences and may then cross-hybridize to each other's specific probe on the array. This often happens when two or more paralogous genes (Section 5.2.1) are active in the same tissue. The transcriptome then contains a group of related mRNAs, each of which is able to hybridize to some extent with different members of the gene family. Distinguishing the relative amounts of each mRNA, or even being certain which particular mRNAs are present, can then

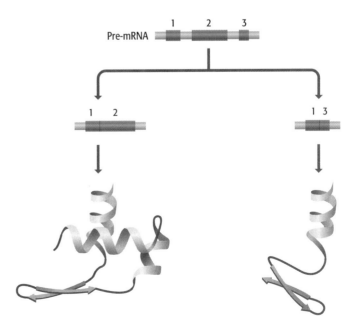

Figure 6.5 **Alternative splicing.** Alternative splicing results in different combinations of exons becoming linked together, resulting in different proteins being synthesized from the same pre-mRNA.

be difficult. A similar problem arises when two or more different mRNAs are derived from the same gene. This is relatively common in vertebrates because of **alternative splicing**, the process by which exons from a pre-mRNA are assembled in different combinations to give a series of related but different mRNAs (Figure 6.5). The array must be designed very carefully if all of these variants are to be detected and accurately quantified.

Further complications arise if the objective is to compare two or more transcriptomes, which, as we will see below, is a frequent scenario. For comparisons to be valid, differences between the hybridization intensities for the same gene with two different microarrays or chips must represent genuine differences in mRNA amount, and not be due to experimental factors such as the amount of target DNA on the array, the efficiency with which the probe has been labeled, or the effectiveness of the hybridization process. Even in a single laboratory, these factors can rarely be controlled with absolute precision, and exact reproducibility between different laboratories is more or less impossible. This means that the data analysis must include normalization procedures that enable results from different array experiments to be accurately compared. The arrays must therefore include negative controls so that the background can be determined in each experiment, as well as positive controls that should always give identical signals. For vertebrate transcriptomes, the actin gene is often used as a positive control as its expression level tends to be fairly constant in a particular tissue, regardless of the developmental stage or disease state. A more satisfactory alternative is to design the experiment so that the two transcriptomes can be directly compared, in a single analysis using a single array. This is done by labeling the cDNA preparations with different fluorescent probes, and then scanning the array at the appropriate wavelengths to determine the relative intensities of the two fluorescent signals at each position, and hence to determine the differences between the mRNA contents of the two transcriptomes (Figure 6.6).

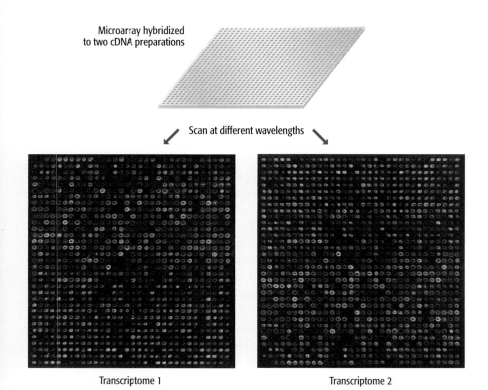

Figure 6.6 Comparing two transcriptomes in a single experiment. Image courtesy of Tom Strachan, reprinted with permission from *Nature*.

Figure 6.7 **Comparing the expression profiles of five genes in seven transcriptomes.** Seven transcriptomes have been prepared from cells at different periods after addition of an energy-rich nutrient to the growth medium. After analysis of the data by hierarchical clustering, a dendrogram is constructed showing the degree of relationship between the expression profiles of the five genes.

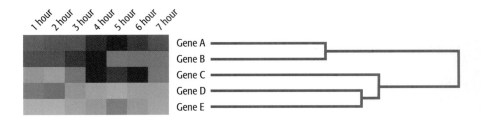

Presuming accurate comparisons can be made between two or more transcriptomes, quite complex differences in gene expression patterns can be distinguished. Genes that display similar expression profiles are likely to be ones with related functions, and a rigorous method is needed for identifying these groups. The standard method, which is called **hierarchical clustering**, involves comparing the expression levels of every pair of genes in every transcriptome that has been analyzed, and assigning a value that indicates the degree of relatedness between those expression levels. These data can be expressed as a **dendrogram**, in which genes with related expression profiles are clustered together (Figure 6.7). The dendrogram gives a clear visual indication of the functional relationships between genes.

Studies of the yeast transcriptome

With just over 6000 genes, the yeast *Saccharomyces cerevisiae* is ideally suited for transcriptome studies, and many of the pioneering projects have been carried out with this organism. One of the first discoveries was that although mRNAs are being degraded and resynthesized all the time, the composition of the yeast transcriptome undergoes very little change if the biochemical features of the environment remain constant. When yeast is grown in a glucose-rich medium, which allows the cells to divide at their maximum rate, the transcriptome is almost completely stable, only 19 mRNAs displaying a greater than twofold change in abundance over a period of two hours. Significant alterations to the transcriptome are seen only when the glucose in the growth medium becomes depleted, forcing the cells to switch from aerobic to anaerobic respiration. During this switch, the levels of over 700 mRNAs increase by a factor of two or more, and another 1000 mRNAs decline to less than half their original amount. The changing environment clearly results in a restructuring of the transcriptome to meet the new biochemical demands of the cell.

The yeast transcriptome also undergoes restructuring during cellular differentiation. This has been established by studying sporulation (spore formation), which is induced by starvation and other stressful environmental conditions. The sporulation pathway can be divided into four stages—early, middle, mid-late, and late—on the basis of the morphological and biochemical events that occur (Figure 6.8). Previous studies have shown, not unexpectedly, that each stage is characterized by expression of a different set of genes. Transcriptome studies have added to our understanding of the sporulation process in several ways. Most significantly, the changes that occur to the composition of the transcriptome indicate that the early stage of sporulation can be subdivided into three distinct phases, called early (I), early (II), and early-middle. The levels of over 250 mRNAs increase significantly during early sporulation, and another 158 mRNAs increase specifically during the middle stage. A further 61 mRNAs increase in abundance during the mid-late period, and 5 more during the late phase. There are also 600 mRNAs that decrease in abundance during sporulation, these presumably coding for proteins that are

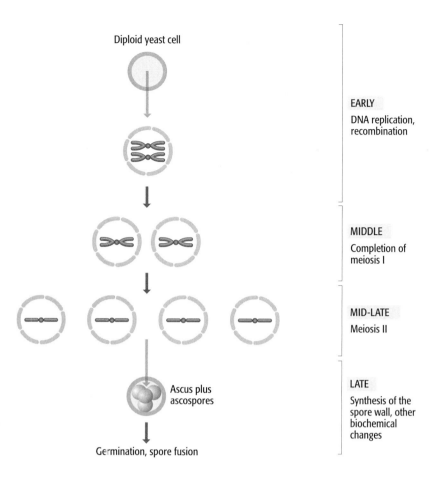

Diploid yeast cell

EARLY
DNA replication, recombination

MIDDLE
Completion of meiosis I

MID-LATE
Meiosis II

LATE
Synthesis of the spore wall, other biochemical changes

Ascus plus ascospores

Germination, spore fusion

Figure 6.8 The sporulation pathway of *Saccharomyces cerevisiae.* The middle three drawings show the nuclear divisions that occur during sporulation. See Figure 3.16 for details of the events involved in meiosis I and meiosis II.

needed during vegetative growth but whose synthesis must be switched off when spores are being formed.

This work on yeast sporulation is important for two reasons. First, by describing the changes in genome expression that occur during sporulation, the transcriptome analyses open the way to studies of the interactions between the genome and the environmental signals that trigger sporulation. Studies of this type, in a relatively simple organism such as yeast, act as an important model for the more complex developmental processes that operate in higher eukaryotes, including humans. Second, several of the mRNAs whose levels change significantly during sporulation are transcripts of genes whose functions were previously unknown. Transcriptome studies therefore help to annotate a genome sequence, aiding identification of genes whose roles in the genome have not been determined by other methods.

The human transcriptome

With five times as many genes, the human transcriptome is substantially more complex than that of yeast, and studies of its composition are still in their infancy. Some interesting results have, nonetheless, been obtained. For example, the transcriptomes of different cell types have been mapped onto the human genome sequence, resulting in global views of the pattern of gene expression along entire chromosomes. This has led to the important discovery that transcripts are made of regions of the genome in which no genes are known to exist. For example, DNA chips have been prepared whose individual oligonucleotide probes target positions occurring, on average, every

Figure 6.9 Transcriptome analysis of human chromosomes 21 and 22. A part of each chromosome is shown with its G-banding pattern (Section 7.1.2) and with the map positions (21p11.2, etc.) indicated. For each chromosome, the upper graph shows where the known exons are located, this information expressed as exon density for every 5.7 Mb "window" of DNA. The lower graph indicates positions where mRNAs were detected in the 11 transcriptomes that were studied, again expressed as density per 5.7 Mb window. Reprinted with permission from Kapranov *et al.*, *Science*, **296**, 916–919. © 2002 AAAS.

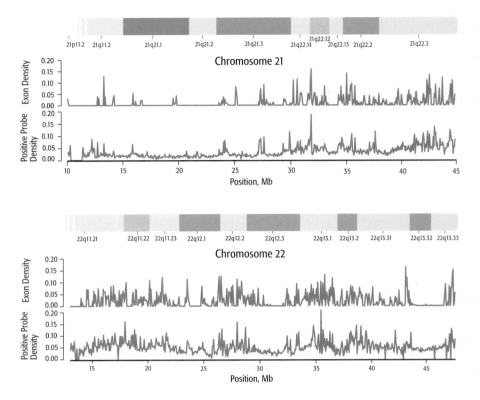

35 nucleotides along chromosomes 21 and 22. These **tiling arrays** comprise over one million probes, but only 26,000 of these lie within the exons present on these chromosomes. However, over 350,000 of the probes detected an mRNA in at least one of 11 human transcriptomes, from different cell lines, that were studied (Figure 6.9). Across the entire genome, some 10,500 transcribed sequences have been identified from regions of the genome not previously thought to contain any genes. This work illustrates the important role that transcriptome analysis can play in genome annotation.

Transcriptome analysis is also having a major impact on studies of human disease, especially cancer. Transcriptome restructuring as a result of cancer was first discovered in 1997, when it was shown that 289 mRNAs are present in significantly different amounts in the transcriptomes of normal colon epithelial cells compared with cancerous colon cells, and that about half of these mRNAs also display an altered abundance in pancreatic cancer cells. These are important observations as understanding the differences between the transcriptomes of normal and cancerous cells points to differences in their biochemistries and hence to new ways of treating the cancers. Transcriptome studies also have applications in cancer diagnosis. The initial breakthrough in this respect came in 1999, when it was shown that the transcriptome of acute lymphoblastic leukemia cells is different from that of acute myeloid leukemia cells. Twenty-seven lymphoblastic and eleven myeloid cancers were studied and, although all the transcriptomes were slightly different, the distinctions between the two types were sufficient for unambiguous identifications to be made. The significance of this work lies with the improved remission rates that are achievable if a cancer is identified accurately at an early stage, before clear morphological indicators are seen. This is not relevant with these two types of leukemia because these can be distinguished by nongenetic means, but it is important with other cancers such as non-Hodgkin lymphoma. The commonest version of this disease is called diffuse large

B-cell lymphoma, and for many years it was thought that all tumors of this type were the same. Transcriptome studies changed this view and showed that B-cell lymphoma can be divided into two distinct subtypes. The distinctions between the transcriptomes of the two subtypes enable each one to be related to a different class of B cells, stimulating and directing the search for specific treatments that are tailored for each lymphoma.

6.2 Studying the Proteome

Proteome studies are important because of the central role that the proteome plays as the link between the genome and the biochemical capability of the cell (Section 1.3.2). Characterization of the proteomes of different cells is therefore the key to understanding how the genome operates and how dysfunctional genome activity can lead to diseases. Transcriptome studies can only partly address these issues. Examination of the transcriptome gives an accurate indication of which genes are active in a particular cell, but gives a less accurate indication of the proteins that are present. This is because the factors that influence protein content include not only the amount of mRNA that is available but also the rate at which the mRNAs are translated into protein and the rate at which the proteins are degraded. Additionally, the protein that is the initial product of translation may not be active, as some proteins must undergo physical and/or chemical modification before becoming functional (Section 13.3). Determining the amount of the *active* form of a protein is therefore critical to understanding the biochemistry of a cell or tissue.

The methodology used to study proteomes is called **proteomics**. Strictly speaking, proteomics is a collection of diverse techniques that are related only in their ability to provide information on a proteome, that information encompassing not only the identities of the constituent proteins that are present but also factors such as the functions of individual proteins and their localization within the cell. The particular technique that is used to study the composition of a proteome is called **protein profiling** or **expression proteomics**.

6.2.1 Protein profiling—methodology for identifying the proteins in a proteome

Protein profiling is based on two techniques—**protein electrophoresis** and **mass spectrometry**—both of which have long pedigrees but which were rarely applied together in the pregenomics era. Today they have been combined into one of the major growth areas of modern research.

Separating the proteins in a proteome

In order to characterize a proteome, it is first necessary to prepare pure samples of its constituent proteins. This is a far from trivial undertaking in view of the complexity of the average proteome: remember that a mammalian cell may contain 10,000–20,000 different proteins (Section 1.3.2).

Polyacrylamide gel electrophoresis (see Technical Note 4.1) is the standard method for separating the proteins in a mixture. Depending on the composition of the gel and the conditions under which the electrophoresis is carried out, different chemical and physical properties of proteins can be used as the basis for their separation. The most frequently used technique makes use of the detergent called sodium dodecyl sulfate, which denatures proteins and

Load the protein sample

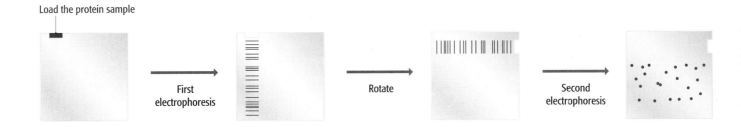

First
electrophoresis

Rotate

Second
electrophoresis

Figure 6.10 Two-dimensional gel electrophoresis.

confers a negative charge that is roughly equivalent to the length of the unfolded polypeptide. Under these conditions, the proteins separate according to their molecular masses, the smallest proteins migrating more quickly towards the positive electrode. Alternatively, proteins can be separated by **isoelectric focusing** in a gel that contains chemicals which establish a pH gradient when the electrical charge is applied. In this type of gel, a protein migrates to its **isoelectric point**, the position in the gradient where its net charge is zero. In protein profiling, these methods are combined in **two-dimensional gel electrophoresis**. In the first dimension, the proteins are separated by isoelectric focusing. The gel is then soaked in sodium dodecyl sulfate, rotated by 90°, and a second electrophoresis, separating the proteins according to their sizes, is carried out at right angles to the first (Figure 6.10). This approach can separate several thousand proteins in a single gel.

After electrophoresis, staining the gel reveals a complex pattern of spots, each one containing a different protein (Figure 6.11). When two gels are compared, differences in the pattern and intensities of the spots indicate differences in the identities and relative amounts of individual proteins in the two proteomes that are being studied. Interesting spots can therefore be targeted for the second stage of profiling in which actual protein identities are determined, as described below. However, before moving on to this stage, we must recognize that two-dimensional gel electrophoresis has limitations that can

Figure 6.11 The result of two-dimensional gel electrophoresis. Mouse liver proteins have been separated by isoelectric focusing in the pH 5–6 range in the first direction and according to molecular mass in the second dimension. The protein spots have been visualized by staining with silver solution. Reprinted with permission from Görg *et al.*, *Electrophoresis*, **21**, 1037–1053. © 2000 Wiley-VCH Verlag.

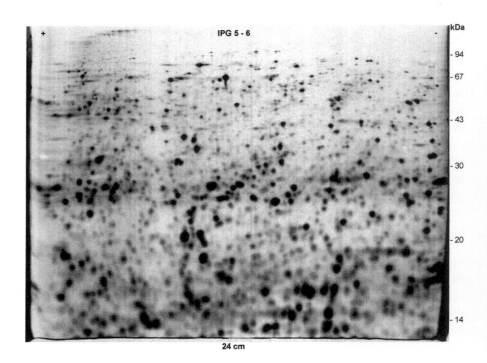

have a significant impact on the overall utility of protein profiling as a means of studying a proteome. The most significant problem is that not all proteins in the proteome will be visible in the gel, and in particular, proteins that are not soluble in an aqueous buffer, such as many of the proteins present in cell membranes, will be absent. In order to study these components of the proteome, special buffers and gel compositions must be used, which means that several parallel experiments must be carried out if the objective is to study a proteome in its entirety. There are also problems with the reproducibility of two-dimensional gel electrophoresis, and the difficulty in devising control procedures that enable the data from such gels to be normalized when two proteomes are compared. For these reasons, alternative separation methods are being sought, with attention currently on high-performance liquid chromatography (HPLC) and free flow isoelectric focusing.

Identifying the proteins in a proteome

Two-dimensional gel electrophoresis results in a complex pattern of spots, each one representing a different protein. How do we identify which protein is present in a spot? This used to be a difficult proposition but advances in mass spectrometry have provided the rapid and accurate identification procedure dictated by the requirements of genome studies. Mass spectrometry was originally designed as a means of identifying a compound from the mass-to-charge ratios of the ionized forms that are produced when molecules of the compound are exposed to a high-energy field. The standard technique could not be used with proteins because they are too large to be ionized effectively, but a new procedure, called **matrix-assisted laser desorption ionization time-of-flight** (**MALDI-TOF**), gets around this problem, at least with peptides of up to 50 amino acids in length. Of course, most proteins are much longer than 50 amino acids, and it is therefore necessary to break them into fragments before examining them by MALDI-TOF. The standard approach is to purify the protein from a spot and then digest it with a sequence-specific protease, such as trypsin, which cleaves proteins immediately after arginine or lysine residues. With most proteins, this results in a series of peptides 5–75 amino acids in length.

Once ionized, the mass-to-charge ratio of a peptide is determined from its "time-of-flight" within the mass spectrometer as it passes from the ionization source to the detector (Figure 6.12). The mass-to-charge ratio enables the molecular mass to be worked out, which in turn allows the amino acid composition of the peptide to be deduced. If a number of peptides from a single protein spot in the two-dimensional gel are analyzed, then the resulting compositional information can be related to the genome sequence in order to identify the gene that specifies that protein. The amino acid compositions of the peptides derived from a single protein can also be used to check that the gene sequence is correct, and in particular to ensure that exon–intron boundaries have been correctly located. This not only helps to delineate the exact position of a gene in a genome (Section 5.1.1), it also allows alternative splicing pathways to be identified in cases where two or more proteins are derived from the same gene.

If two proteomes are being compared then a key requirement is that proteins that are present in different amounts can be identified. If the differences are relatively large, then they will be apparent simply by looking at the stained gels after two-dimensional electrophoresis. However, important changes in the biochemical properties of a proteome can result from relatively minor

Figure 6.12 The use of MALDI-TOF in protein profiling. After two-dimensional gel electrophoresis a protein of interest is excised from the gel and digested with a protease such as trypsin. This cleaves the protein into a series of peptides which can be analyzed by MALDI-TOF (A). In the mass spectrometer the peptides are ionized by a pulse of energy from a laser and then accelerated down the column to the reflector and onto the detector. The time-of-flight of each peptide depends on its mass-to-charge ratio. The data are visualized as a spectrum (B). The computer contains a database of the predicted molecular masses of every trypsin fragment of every protein encoded by the genome of the organism under study. The computer compares the masses of the detected peptides with the database and identifies the most likely source protein.

(A) MALDI-TOF mass spectrometry

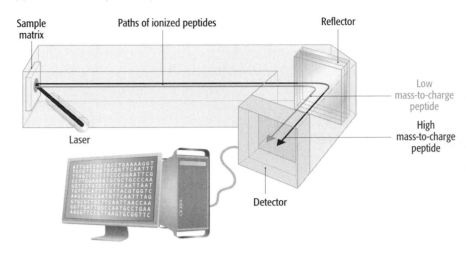

(B) MALDI-TOF spectrum

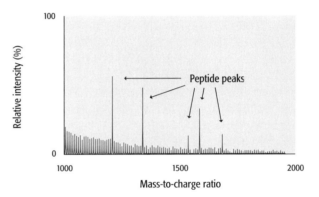

changes in the amounts of individual proteins, and methods for detecting small-scale changes are therefore essential. One possibility is to label the constituents of two proteomes with different fluorescent markers, and then run them together in a single two-dimensional gel. This is the same strategy as is used for comparing pairs of transcriptomes (see Figure 6.6). Visualization of the two-dimensional gel at different wavelengths enables the intensities of equivalent spots to be judged more accurately than is possible when two separate gels are obtained. A more accurate alternative is to label each proteome with an **isotope coded affinity tag** (**ICAT**). These markers can be obtained in two forms, one containing normal hydrogen atoms and the other containing

Figure 6.13 Analyzing two proteomes by ICAT. In the MALDI-TOF spectrum, peaks resulting from peptides containing normal hydrogen atoms are shown in blue, and those from peptides containing deuterium are shown in red. The protein under study is approximately 1.5 times more abundant in the proteome that has been labeled with deuterium.

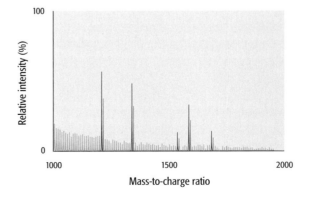

deuterium, the heavy isotope of hydrogen. The normal and heavy versions can be distinguished by mass spectrometry, enabling the relative amounts of a protein in two proteomes that have been mixed together to be determined during the MALDI-TOF stage of the profiling procedure (Figure 6.13).

6.2.2 Identifying proteins that interact with one another

Important data pertaining to genome activity can also be obtained by identifying pairs and groups of proteins that interact with one another. At a detailed level, this information is often valuable when attempts are made to assign a function to a newly discovered gene or protein (Section 5.2) because an interaction with a second, well-characterized protein can often indicate the role of an unknown protein. For example, an interaction with a protein that is located on the cell surface might indicate that an unknown protein is involved in cell–cell signaling (Section 14.1). At a global level, the construction of **protein interaction maps** is looked on as an important step in linking the proteome with the cellular biochemistry.

Identifying pairs of interacting proteins by phage display and two-hybrid studies

There are several methods for studying protein–protein interactions, the two most useful being **phage display** and the **yeast two-hybrid system**. In phage display a special type of cloning vector is used, one based on λ bacteriophage or one of the filamentous bacteriophages such as M13. The vector is designed so that a new gene that is cloned into it is expressed in such a way that its protein product becomes fused with one of the phage coat proteins (Figure 6.14A). The phage protein therefore carries the foreign protein into the phage

(A) Production of a display phage

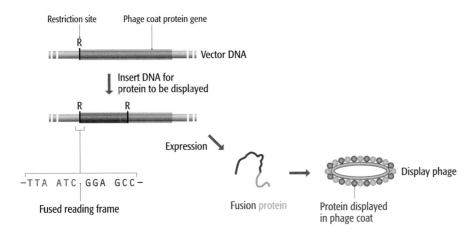

(B) Using a phage display library

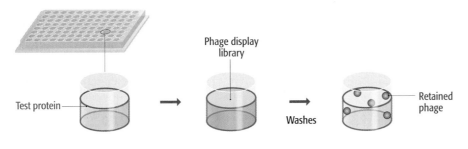

Figure 6.14 Phage display. (A) The cloning vector used for phage display is a bacteriophage genome with a unique restriction site located within a gene for a coat protein. The technique was originally carried out with the gene III coat protein of the filamentous phage called f1, but has now been extended to other phages including λ. To create a display phage, the DNA sequence coding for the test protein is ligated into the restriction site so that a fused reading frame is produced—one in which the series of codons continues unbroken from the test gene into the coat protein gene. After transformation of *Escherichia coli*, this recombinant molecule directs synthesis of a hybrid protein made up of the test protein fused to the coat protein. Phage particles produced by these transformed bacteria therefore display the test protein in their coats. (B) Using a phage display library. The test protein is immobilized within a well of a microtiter tray and the phage display library added. After washing, the phages that are retained in the well are those displaying a protein that interacts with the test protein.

coat, where it is "displayed" in a form that enables it to interact with other proteins that the phage encounters. There are several ways in which phage display can be used to study protein interactions. In one method, the test protein is displayed and interactions sought with a series of purified proteins or protein fragments of known function. This approach is limited because it takes time to carry out each test, so is feasible only if some prior information has been obtained about likely interactions. A more powerful strategy is to prepare a **phage display library**, a collection of clones displaying a range of proteins, and identify which members of the library interact with the test protein (Figure 6.14B).

Figure 6.15 The yeast two-hybrid system. (A) On the left, a gene for a human protein has been ligated to the gene for the DNA-binding domain of a yeast activator. After transformation of yeast, this construct specifies a fusion protein, part human protein and part yeast activator. On the right, various human DNA fragments have been ligated to the gene for the activation domain of the activator: these constructs specify a variety of fusion proteins. (B) The two sets of constructs are mixed and cotransformed into yeast. A colony in which the reporter gene is expressed contains fusion proteins whose human segments interact, thereby bringing the DNA-binding and activation domains into proximity and stimulating the RNA polymerase. See Section 11.3.2 for more information on activators.

The yeast two-hybrid system detects protein interactions in a more complex way. In Section 11.3.2 we will see that proteins called **activators** are responsible for controlling the expression of genes in eukaryotes. To carry out this function an activator must bind to a DNA sequence upstream of a gene and stimulate the RNA polymerase enzyme that copies the gene into RNA. These two abilities—DNA-binding and polymerase activation—are specified by different parts of the activator, and some activators will work even after cleavage into two segments, one segment containing the DNA-binding domain and one containing the activation domain. In the cell, the two segments interact to form the functional activator.

The two-hybrid system makes use of a *Saccharomyces cerevisiae* strain that lacks an activator for a reporter gene. This gene is therefore switched off. An artificial gene that codes for the DNA-binding domain of the activator is ligated to the gene for the protein whose interactions we wish to study. This protein can come from any organism, not just yeast: in the example shown

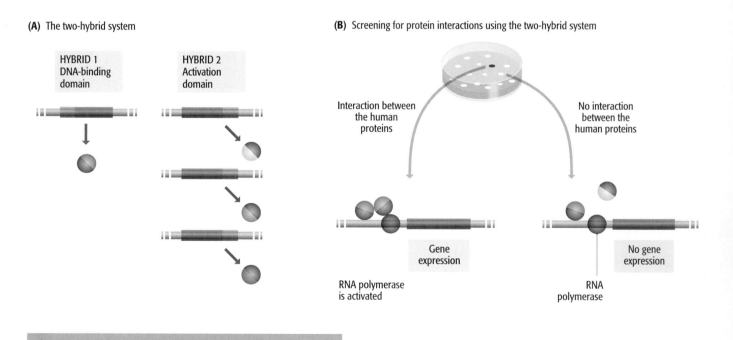

(A) The two-hybrid system

HYBRID 1
DNA-binding domain

HYBRID 2
Activation domain

(B) Screening for protein interactions using the two-hybrid system

Interaction between the human proteins

No interaction between the human proteins

Gene expression

No gene expression

RNA polymerase is activated

RNA polymerase

KEY

Yeast gene
Human gene

Yeast domains

Human domains

in Figure 6.15A it is a human protein. After introduction into yeast, this construct specifies synthesis of a fusion protein made up of the DNA-binding domain of the activator attached to the human protein. The recombinant yeast strain is still unable to express the reporter gene because the modified activator only binds to DNA; it cannot influence the RNA polymerase. Activation only occurs after the yeast strain has been cotransformed with a second construct, one comprising the coding sequence for the activation domain fused to a DNA fragment that specifies a protein able to interact with the human protein that is being tested (Figure 6.15B). As with phage display, if there is some prior knowledge about possible interactions then individual DNA fragments can be tested one by one in the two-hybrid system. Usually, however, the gene for the activation domain is ligated with a mixture of DNA fragments so that many different constructs are made. After transformation, cells are plated out and those that express the reporter gene identified. These are cells that have taken up a copy of the gene for the activation domain fused to a DNA fragment that encodes a protein able to interact with the test protein.

Identifying the components of multiprotein complexes

Phage display and the yeast two-hybrid system are effective methods for identifying pairs of proteins that interact with one another, but identifying such links reveals only the basic level of protein–protein interactions. Many cellular activities are carried out by multiprotein complexes, such as the mediator which plays a central role in regulation of gene transcription (Section 11.3.2) or the spliceosome which is responsible for the removal of introns from pre-mRNA (Section 12.2.2). Complexes such as these typically comprise a set of core proteins, which are present at all times, along with a variety of ancillary proteins that associate with the complex under particular circumstances. Identifying the core and ancillary proteins is a critical step toward understanding how these complexes carry out their functions. These proteins might be identified pair-by-pair by a long series of two-hybrid experiments, but a more direct route to determining the composition of multiprotein complexes is clearly needed.

In principle, a phage display library can be used to identify the members of a multiprotein complex, as in this procedure all proteins that interact with the test protein are identified in a single experiment (see Figure 6.14). The problem is that large proteins are displayed inefficiently as they disrupt the phage replication cycle, and to circumvent this problem it is generally necessary to display a short peptide, representing part of a cellular protein, rather than the entire protein. The displayed peptide may therefore be unable to interact with all members of the complex within which the intact protein is located, because the peptide lacks some of the protein–protein attachment sites present in the intact form (Figure 6.16). A method that avoids this problem, because it works with intact proteins, is **affinity chromatography**. In affinity chromatography, the test protein is attached to a chromatography resin and placed in a column (see Technical Note 2.3). The cell extract is passed through the column in a low-salt buffer, which allows formation of the hydrogen bonds that hold proteins together in a complex (Figure 6.17A). The proteins that interact with the bound test protein are therefore retained in the column, while all the others are washed away. The interacting proteins are then eluted with a high-salt buffer. A disadvantage of this procedure is the need to purify the test protein, which is time-consuming and difficult to use as the basis to a

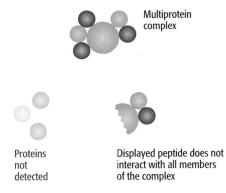

Multiprotein complex

Proteins not detected

Displayed peptide does not interact with all members of the complex

Figure 6.16 Phage display may fail to detect all members of a multiprotein complex. The complex consists of a central protein that interacts with five smaller proteins. In the lower drawing, a peptide from the central protein is used in a phage display experiment. This peptide detects two of the interacting proteins, but the other three proteins are missed because their binding sites lie on a different part of the central protein.

(A) Standard affinity chromatography

Cell extract

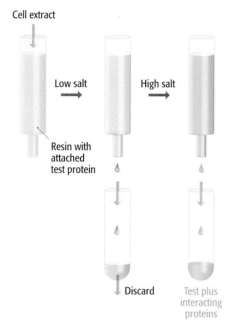

Resin with
attached
test protein

Low salt

High salt

Discard

Test plus
interacting
proteins

(B) Tandem-affinity purification

Cell extract

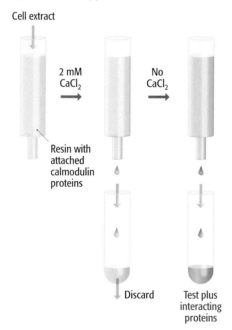

Resin with
attached
calmodulin
proteins

2 mM
CaCl$_2$

No
CaCl$_2$

Discard

Test plus
interacting
proteins

Figure 6.17 Affinity chromatography methods for the purification of multiprotein complexes. (A) In standard affinity chromatography, the test protein is attached to the resin. The cell extract is applied in a low-salt buffer so the other members of the multiprotein complex bind to the test protein. The proteins are then eluted in a high-salt buffer. (B) In TAP, the cell extract is applied in a buffer containing 2 mM CaCl$_2$, conditions which promote attachment of the modified test protein, plus the proteins it interacts with, to the calmodulin molecules attached to the chromatography resin. The proteins are then eluted with a buffer that contains no CaCl$_2$.

large screening programme. In the more sophisticated method called **tandem-affinity purification** (**TAP**), which was developed as a means of studying protein complexes in *S. cerevisiae*, the gene for the test protein is modified so that the protein, when synthesized, has a C–terminal extension that binds to a second protein called calmodulin. The cell extract is prepared under gentle conditions so that multiprotein complexes do not break down, and the extract is then passed through an affinity chromatography column packed with a resin containing attached calmodulin molecules. This results in immobilization both of the test protein and others with which it is associated (Figure 6.17B). In both techniques, the identities of the purified proteins are determined by mass spectrometry. When used in a large-scale screen of 1739 yeast genes, TAP identified 232 multiprotein complexes, providing new insights into the functions of 344 genes, many of which had not previously been characterized by experimental means.

A disadvantage with affinity chromatography methods is that a single member of a multiprotein complex is used as the "bait" for isolation of other proteins from that complex. In practice, if a member of a complex does not interact directly with the bait, then it may not be isolated (Figure 6.18). The methods therefore identify groups of proteins that are present in a complex, but do not necessarily provide the total protein complement of the complex. Developing ways of purifying intact complexes is therefore a major goal of current research. In **coimmunoprecipitation**, a cell extract is prepared under gentle conditions so that complexes remain intact. An antibody specific for the test protein is then added, which results in precipitation of this protein and all other members of the complex within which it is present. More sophisticated is the **multi-dimensional protein identification technique** (**MudPIT**), which combines various chromatography techniques (e.g., reversed-phase liquid chromatography with either cation-exchange or size-exclusion chromatography) in order to isolate intact complexes. The components of a complex can then be identified by mass spectrometry. This method was first used to study the large subunit of the yeast ribosome, and resulted in identification of 11 proteins that had not previously been known to be associated with this complex.

Identifying proteins with functional interactions

Proteins do not need to form physical associations with one another in order to have a functional interaction. For example, in bacteria such as *Escherichia coli*, the enzymes lactose permease and β-galactosidase have a functional interaction in that they are both involved in utilization of lactose as a carbon source. But there is no physical interaction between these two proteins: the permease is located in the cell membrane and transports lactose into the cell, and β-galactosidase, which splits lactose into glucose and galactose, is present in the cell cytoplasm (see Figure 8.8A). Many enzymes that work together in the same biochemical pathway never form physical interactions with one another, and if studies were to be based solely on detection of physical associations between proteins then many functional interactions would be overlooked.

Several methods can be used to identify proteins that have functional interactions. Most of these do not involve direct study of the proteins themselves and hence, strictly speaking, do not come under the general heading of "proteomics." Nonetheless, it is convenient to consider them here because the information they yield is often included, along with the results of proteomics studies, in protein interaction maps. These methods include the following:

- **Comparative genomics** can be used in various ways to identify groups of proteins that have functional relationships. One approach is based on the observation that pairs of proteins that are separate molecules in some organisms are fused into a single polypeptide chain in others. An example is provided by the yeast gene *HIS2*, which codes for an enzyme involved in histidine biosynthesis. In *E. coli*, two genes are homologous to *HIS2*. One of these, itself called *his2*, has sequence similarity with the 5′ region of the yeast gene, and the second, *his10*, is similar to the 3′ region (Figure 6.19). The implication is that the proteins coded by *his2* and *his10* interact within the *E. coli* proteome to provide part of the histidine biosynthesis activity. Analysis of the sequence databases reveals many examples of this type, where two proteins in one organism have become fused into a single protein in another organism. A similar approach is based on examination of bacterial operons. An operon consists of two or more genes that are transcribed together and which usually have a functional relationship (Section 8.2). For example, the genes for the lactose permease and β-galactosidase *of E. coli* are present in the same operon, along with the gene for a third protein involved in lactose utilization. The identities of genes in bacterial operons can therefore be used to infer functional interactions between the proteins coded by homologous genes in a eukaryotic genome.

- Transcriptome studies can identify functional interactions between proteins, as the mRNAs for functionally related proteins often display similar expression profiles under different conditions.

- Gene inactivation studies can be informative. If a change in phenotype is observed only when two or more genes are inactivated together, then it can be inferred that those genes function together in generation of the phenotype.

Protein interaction maps

Protein interaction maps, also called **interactome networks**, display all of the interactions that occur between the components of a proteome. The first of these maps were constructed in 2001 for relatively simple proteomes, almost entirely from two-hybrid experiments. These included maps for the bacterium *Helicobacter pylori*, comprising over 1200 interactions involving almost half of the proteins in the proteome, and for 2240 interactions between 1870 proteins from the *S. cerevisiae* proteome (Figure 6.20A). More recently, the application of additional techniques has led to more detailed versions of the *S. cerevisiae* map, including ones in which the interactions between (not within) multiprotein clusters are illustrated (Figure 6.20B). Maps have also been generated for the first time for more complex species such as *Caenorhabditis elegans*.

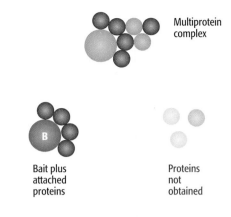

Figure 6.18 A disadvantage with affinity chromatography. If the bait protein (labeled with a "B") does not interact directly with one or more proteins in the complex, then those proteins might not be isolated.

Figure 6.19 Using homology analysis to deduce protein–protein interactions. The 5′ region of the yeast *HIS2* gene is homologous to *Escherichia coli his2*, and the 3′ region is homologous to *E. coli his10*.

(A)

(B)

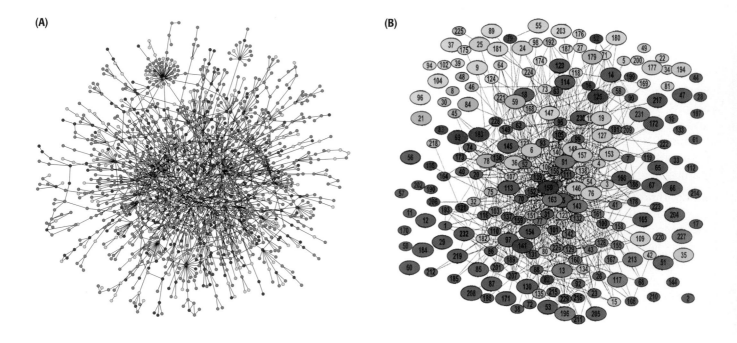

Figure 6.20 Versions of the Saccharomyces cerevisiae protein interaction map. (A) The initial map, first published in 2001. Each dot represents a protein, with connecting lines indicating interactions between pairs of proteins. Red dots are essential proteins: an inactivating mutation in the gene for one of these proteins is lethal. Mutations in the genes for proteins indicated by green dots are nonlethal, and mutations in genes for proteins shown in orange lead to slow growth. The effects of mutations in genes for proteins shown as yellow dots were not known when the map was constructed. (B) A more comprehensive map, published in 2002. In this map each oval is a protein complex, with connections shown between complexes that share at least one protein. Complexes are color coded according to their function as follows: red, cell cycle; dark green, signaling; dark blue, transcription, DNA maintenance, and/or chromatin structure; pink, protein and/or RNA transport; orange, RNA metabolism; light green, protein synthesis and/or turnover; brown, cell polarity and/or structure; violet, intermediate and/or energy metabolism; light blue, membrane biogenesis and/or traffic. Image (A) kindly supplied by Hawoong Jeong. Reprinted with permission from Jeong *et al.*, (2001), *Nature*, **411**, 41–42. Image (B) kindly supplied by Anne Claude Gavin. Reprinted with permission from Gavin *et al.*, (2002), *Nature*, **415**, 141–147.

What interesting features have emerged from these protein interaction maps? The most intriguing discovery is that each network is built up around a small number of proteins that have many interactions, and which form **hubs** in the network, along with a much larger number of proteins with few individual connections (Figure 6.21A). This architecture is thought to minimize the effect on the proteome of the disruptive effects of mutations which might inactivate individual proteins. Only if a mutation affects one of the proteins at a highly interconnected node will the network as a whole be damaged. This hypothesis is consistent with the discovery, from gene inactivation studies (Section 5.2.2), that a substantial number of yeast proteins are apparently redundant, meaning that if the protein activity is destroyed, the proteome as a whole continues to function normally, with no discernible impact on the phenotype of the cell. Examination of the expression profiles of the hub proteins and their direct partners enables these hubs to be divided into two groups. The first group are those hub proteins that interact with all their partners simultaneously. These are called "party" hubs and their removal has little effect on the overall structure of the network (Figure 6.21B). In contrast, removal of the second group, the "date" hubs, which interact with different partners at different times, breaks the network into a series of small subnetworks (Figure 6.21C). The implication is that the party hubs work within individual biological processes, and do not contribute greatly to the overall organization of the proteome. The date hubs, on the other hand, are the key players that provide an organization to the proteome by linking biological processes to one another.

6.3 Beyond the Proteome

The proteome is traditionally looked on as the end-product of genome expression, but this view obscures the true role of the proteome as part of the final link that connects the genome with the biochemistry of the cell (Figure 6.22). Exploring the nature of this link is proving to be one of the most exciting and productive aspects of modern biology.

(A) The complete network

(B) Removal of party hubs

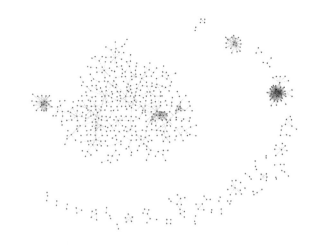

(C) Removal of date hubs

Figure 6.21 Hubs in the *Saccharomyces cerevisiae* **protein interaction map.** This map was published in 2004. The hubs are clearly visible in the complete map (A). After removal of the party hubs the network remains almost intact (B), but after the date hubs are removed the network splits into detached subnetworks (C). Image kindly supplied by Nicolas Bertin. Reprinted with permission from Han *et al.*, (2004) *Nature*, **430**, 38–93.

6.3.1 The metabolome

Often in biology the most important steps forward do not result from some groundbreaking experiment but instead arise because biologists devise a new way of thinking about a problem. The introduction of the concept of the **metabolome** is an example. The metabolome is defined as the complete collection of metabolites present in a cell or tissue under a particular set of conditions. In other words, a metabolome is a biochemical blueprint, and its study, which is called **metabolomics** or **biochemical profiling**, gives a precise description of the biochemistry underlying different physiological states, including disease states, that can be adopted by a cell or tissue. By converting the biochemistry of a cell into an itemized set of metabolites, metabolomics provides a dataset that can be directly linked to the equivalent, itemized information which emerges from proteomics and other studies of genome expression.

A metabolome can be characterized by chemical techniques such as infrared spectroscopy, mass spectrometry, and nuclear magnetic resonance spectroscopy which, individually and in combination, can identify and quantify

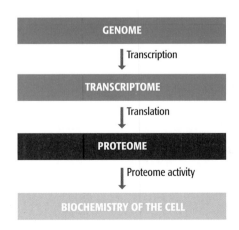

Figure 6.22 The proteome is part of the final link that connects the genome with the biochemistry of the cell.

the various small molecules that make up the metabolites in a cell. When these data are combined with knowledge about the reaction rates for the various steps in well-characterized biochemical pathways such as glycolysis and the tricarboxylic acid (TCA) cycle, it is possible to model the **metabolic flux**, the rate of flow of metabolites through the network of pathways that make up the cellular biochemistry. Changes in the metabolome can then be defined in terms of perturbations in the flux of metabolites through one or more parts of the network, providing a very sophisticated description of the biochemical basis to changes in the physiological state. This leads to the possibility of **metabolic engineering**, in which changes are made to the genome by mutation or recombinant DNA techniques in order to influence the cellular biochemistry in a predetermined way, for example to increase the synthesis of an antibiotic by a microorganism.

At present, metabolomics is most advanced with organisms such as bacteria and yeast whose biochemistries are relatively simple. Considerable research is currently being directed toward the human metabolome, with the objective of describing the metabolic profiles of healthy tissues, of disease states, and of tissues in patients undergoing drug treatments. It is hoped that when these studies reach maturity it will be possible to use the metabolic information to design drugs that reverse or mitigate the particular flux abnormalities that occur in the disease state. Biochemical profiling could also indicate any unwanted side effects of drug treatment, enabling modifications to be made to the chemical structure of the drug, or to its mode of use, so that these side effects are minimized.

6.3.2 Understanding biological systems

The emphasis that is now placed on protein interaction maps and metabolomics leads us to the final aspect of genome function that we must consider. This is the need to describe and understand the expression of a genome not in terms of the molecules—RNAs, proteins, and metabolites—whose synthesis the genome directs, but in terms of the biological systems that result from coordinated activity of those molecules. This is the essence of the leap that has been made in recent years from genes to genomes. One of the underlying principles of the pregenome era of molecular biology was the "one gene, one enzyme" hypothesis first put forward by George Beadle and Edward Tatum in the 1940s. By "one gene, one enzyme", Beadle and Tatum were emphasizing that a single gene codes for a single protein which, if an enzyme, directs a single biochemical reaction. The *trpC* gene of *Escherichia coli*, for example, codes for the enzyme indole-3-glycerol phosphate synthase, which converts 1-(o-carboxyphenylamino)-1′-deoxyribulose-5′-phosphate into indole-3-glycerol phosphate. However, this enzyme does not work in isolation: its activity forms part of the biochemical pathway which results in synthesis of tryptophan, the other enzymes in this pathway being specified by the genes *trpA, B, D*, and *E*, which together with *trpC* form the tryptophan operon of *E. coli* (see Figure 8.8B). The tryptophan biosynthesis pathway is therefore a simple biological system and the tryptophan operon is the set of genes that specify that pathway. But simply transcribing and translating the genes in the operon will not lead to the synthesis of tryptophan. Successful operation of the system requires that the biosynthetic enzymes be present at the appropriate place in the cell, in the appropriate relative amounts, at the appropriate time. The system is therefore dependent on factors such as the rate of synthesis of the proteins coded by the genes, the correct folding of

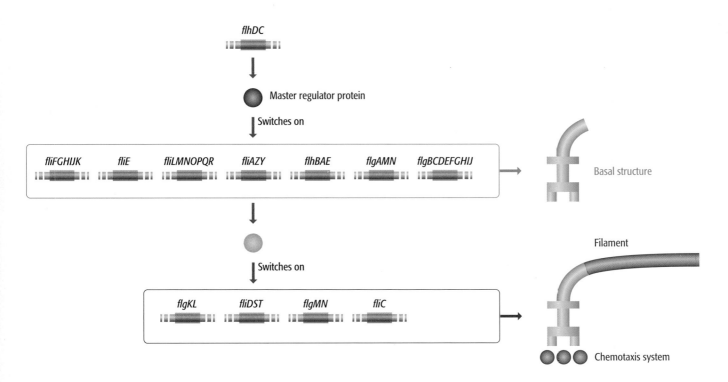

flhDC

Master regulator protein

Switches on

fliFGHIJK   fliE   fliLMNOPQR   fliAZY   flhBAE   flgAMN   flgBCDEFGHIJ

Basal structure

Switches on

Filament

flgKL   fliDST   flgMN   fliC

Chemotaxis system

Figure 6.23 The system responsible for flagellum biosynthesis in *Escherichia coli.*

these proteins into the functional enzymes, the rate of degradation of the enzyme molecules, their localization in the cell, and the presence of the necessary amounts of the metabolites that act as substrates and cofactors for tryptophan synthesis. This simple biological system is starting to assume quite considerable complexity. And yet, with this system we are considering only 5 of the 4405 genes in the *E. coli* genome.

To date, progress in **systems biology** has been made in a number of areas, including understanding the biological system responsible for synthesis of the *E. coli* flagellum. Pregenome studies had shown that flagellar synthesis requires 51 genes organized into 12 operons, which are activated in three groups (Figure 6.23). The first group to be activated comprises a single operon containing two genes which code for a protein that acts as a master regulator, switching on expression of the second group of seven operons, whose genes together specify components of the basal structure of the flagellum. One of these genes codes for a second regulatory protein that switches on the remaining four operons, which direct synthesis of the flagellum filament and the biochemical system that enables the bacterium to respond to chemical stimuli by rotating its flagellum in order to swim towards an attractant. Careful use of reporter genes attached to individual operons has revealed the precise order in which the operons in each group are activated, and has enabled activation coefficients—measures of the relative rates of expression—to be assigned to each operon. The resulting information is sufficient for the system to be modeled on a computer, enabling the detailed roles of the two regulatory proteins to be determined. From the computer models, the effects of subtle changes to the system (such as a change in the properties of one of the regulators) can be predicted and then tested by further experiments with the biological system. This is precisely the type of research that we hope one day to be able to carry out with human cells, in order to understand the exact basis to abnormalities and hopefully to design ways of returning a diseased tissue to its normal state. Scaling-up to these much larger biological systems, with the grand objective of one day understanding how a bacterial or eukaryotic cell works, will test the ingenuity and resourcefulness of biologists for decades to come. The first step has, however, been made with the change in emphasis from the functioning of individual genes to the functioning of entire genomes, and by the establishment of techniques for studying transcriptomes, proteomes, and metabolomes, which together form the components of these biological systems.

Summary

The major challenge of postgenomics is to understand how the genome specifies and coordinates the various biochemical activities that take place within a living cell. Central to this work are studies of the transcriptome and proteome that are synthesized and maintained by the genome. Although transcriptomes can be studied by cDNA sequencing, including use of techniques such as SAGE which provides minisequences from many cDNAs in a single experiment, the most important advances are being made through use of microarray and chip technology. Hybridization of differentially labeled cDNAs prepared from two or more transcriptomes to a microarray or chip gives information on gene expression patterns, which can be analyzed by hierarchical clustering to reveal functional relationships between genes. Transcriptome studies are helping us to understand the genetic basis to developmental pathways and to human diseases including several types of cancer. Proteome studies are equally important because examination of a transcriptome only reveals which genes are expressed in a particular cell, and does not give an accurate picture of the proteins that are present. Protein profiling uses two-dimensional gel electrophoresis followed by MALDI-TOF of isolated peptide fragments to characterize the proteins in a proteome. To understand how a proteome operates within a cell it is useful to know which proteins interact with one another. Phage display and the yeast two-hybrid system are the most frequently used methods for identifying pairs of proteins that form physical associations, and methods such as coimmunoprecipitation can be used to isolate intact multiprotein complexes. Functional interactions, which do not always require that a pair of proteins make physical contact, can be deduced by comparative genomics, analyses of gene expression profiles, and by gene inactivation studies. The resulting information enables protein interaction maps to be constructed, showing all the interactions occurring in a single proteome. These maps are typically structured around a relatively small number of proteins that have many interactions and form hubs in the network, some hubs representing individual biological processes and others linking biological processes to one another. The proteome maintains the metabolome—the complete collection of metabolites present in a cell or tissue—and it is anticipated that metabolome studies will reveal the precise biochemical basis to disease states and to unwanted side effects of drug treatment. Work with transcriptomes, proteomes, and metabolomes is leading biologists towards systems biology, in which attempts are made to understand the expression of a genome not in terms of the molecules whose synthesis the genome directs, but in terms of the biological systems that result from coordinated activity of those molecules.

Multiple Choice Questions

6.1.* What is the advantage of probing a transcriptome with mRNAs bound to ribosomes?

   a. Eukaryotic mRNA molecules are difficult to isolate unless complexed to a ribosome.

   b. mRNA molecules isolated from ribosomes represent those that are actively being translated into proteins.

   c. mRNA molecules isolated from ribosomes are more stable than other mRNA molecules.

   d. mRNA molecules that are not being translated by ribosomes still contain their intron sequences.

6.2. How is it possible for microarrays to be used to measure the expression levels of individual genes?

   a. Each position on the microarray contains more copies of the probe sequence than the anticipated number of identical mRNA molecules in the transcriptome.

   b. Each probe sequence on the microarray is present in multiple positions on the array.

   c. After hybridization, the cDNA molecules are eluted and quantitated from each position on the microarray.

   d. The cDNA molecules are sequenced after hybridization and the fluorescence of the sequencing signals is quantitated.

6.3.* Why is actin used as a control for transcriptome studies in vertebrates?

   a. It is used as a negative control as the gene is not expressed in vertebrates.

   b. It is used as a negative control as the mRNA for actin is rapidly degraded.

   c. It is used as a positive control as actin expression is fairly constant in different cell types.

   d. It is used as a positive control as it is the most highly expressed gene in all cell types.

6.4. How can two different transcriptomes be studied with a single microarray?

   a. One transcriptome is hybridized and studied first and then its sequences are removed and the second transcriptome is studied on the same microarray.

   b. Only one of the transcriptomes is labeled and it competes with the second, unlabeled transcriptome for binding to the probe sequences.

   c. The transcriptomes are hybridized to each other prior to the microarray analysis to remove cDNAs present from both cell types.

   d. The two transcriptomes are labeled with different fluorescent probes and hybridized simultaneously.

6.5.* How are genes grouped together via hierarchical clustering?

   a. By expression patterns.

   b. By homology.

   c. By sequence identity.

   d. By protein domain similarities.

6.6. Studies of *Saccharomyces cerevisiae* revealed that when cells are grown under stable, energy-rich conditions the transcriptome changes in what ways?

   a. There are significant changes in mRNA levels due to varying rates of degradation and synthesis.

   b. Most mRNA levels remain constant, but some fluctuate greatly during the cell cycle.

   c. Nearly all mRNA levels remain constant, only a few change significantly.

   d. All of the mRNA levels remain constant under these conditions.

6.7.* How can transcriptome studies aid in the diagnosis of human cancers?

   a. All cancers exhibit the increased expression of a specific set of genes.

   b. Each cancer possesses its own unique transcriptome.

   c. The genes that cause tumors are not expressed in healthy cells.

   d. Transcriptome studies can indicate the rate of cell division.

6.8. Polyacrylamide gel electrophoresis in the presence of sodium dodecyl sulfate separates proteins on the basis of which of the following?

   a. Charge–mass ratio.

   b. Conformation.

   c. Isoelectric point.

   d. Size.

6.9.* The isoelectric point of a protein is defined as:

   a. The pH at which a protein has no net charge.

   b. The pH at which a protein loses its activity.

   c. The pH at which a protein has maximal activity.

   d. The pH at which a protein's amino acids are all ionized.

6.10. The yeast two-hybrid system is designed to identify which of the following?

   a. All of the components of a multiprotein complex.

   b. Human proteins that are required for binding RNA polymerase.

continued …

Multiple Choice Questions (continued) *Answers to odd-numbered questions can be found in the Appendix

c. Two proteins that directly interact with one another.

d. Two proteins that are involved in the same metabolic pathway.

6.11.* The type of chromatography where a protein is bound to a resin and placed into a column, to determine what proteins bind to it, is called:

a. Gel filtration chromatography.

b. Ion-exchange chromatography.

c. Affinity chromatography.

d. Isoelectric chromatography.

6.12. What are the hubs in a protein interactome network?

a. These are proteins that regulate the activities of the cell.

b. These are proteins that form the scaffolding of the cell.

c. These are proteins that interact with many other proteins in the cell.

d. These are proteins that direct gene expression in the cell.

6.13.* What is the metabolome of a cell?

a. All of the proteins and nucleic acids of the cell.

b. All of the metabolites of the cell under a specific set of conditions.

c. All of the potential metabolites that can be produced by a cell.

d. All of the macromolecules of a cell.

Short Answer Questions

*Answers to odd-numbered questions can be found in the Appendix

6.1.* Why are researchers interested in studying genomes even if all the genes have been assigned a function?

6.2. Explain why a cDNA sequence as short as 12 bp can be used to identify the gene that encoded it.

6.3.* Discuss the problems caused by paralogous genes in microarray studies. What experimental conditions might overcome these problems?

6.4. How can alternative splicing cause difficulties in characterizing the transcriptome of a tissue? What approaches can be used to identify the different splicing products from a single gene?

6.5.* How can transcriptome studies provide information on the functions of genes?

6.6. How are tiling arrays used to screen chromosomes for expressed sequences?

6.7.* Why does the transcriptome not provide a completely accurate indication of the proteome of a cell?

6.8. How can small differences in protein levels be quantitated by two-dimensional gel electrophoresis?

6.9.* How do phage display experiments test for protein–protein interactions?

6.10. What is the difference between proteins that function as "party" hubs compared to proteins that are "date" hubs in an interactome network?

6.11.* In what ways could studies of the metabolome impact on the treatment of human diseases?

6.12. What is the focus of systems biology and how does it compare with molecular studies of gene regulation conducted before genomes were sequenced?

In-depth Problems

*Guidance to odd-numbered questions can be found in the Appendix

6.1.* Researchers are often interested in comparing genome expression in an organism or tissue at different developmental stages or in response to different environmental conditions. What approaches are most useful for this type of comparative study?

6.2. After performing two-dimensional gel electrophoresis on two proteomes of an organism grown under different conditions, you identify a protein that is present in one proteome but absent in the other. What experiments should you carry out to identify the gene that codes for this protein?

6.3.* Under what circumstances might a pair of proteins have a functional relationship but no physical interaction? Are there possible scenarios where the reverse might be true—a pair of proteins displaying a physical interaction but no functional relationship?

6.4. Discuss the role of the hubs in a protein interaction map.

6.5.* Explain why systems biology is the subject of so much attention at the present time.

Figure Tests

*Answers to odd-numbered questions can be found in the Appendix

6.1.* Describe the experimental approach used to obtain the visualization of a transcriptome as shown in this figure.

6.2. The dendrogram in this figure shows genes that are grouped together based on what feature?

Microarray

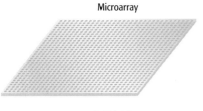

↓ Hybridize to labeled cDNA

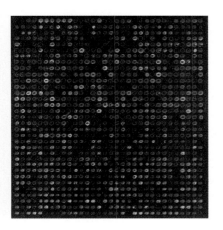

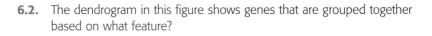

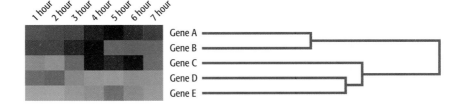

continued ...

Figure Tests (continued)   *Answers to odd-numbered questions can be found in the Appendix

6.3.* Explain how protein molecules are separated in two-dimensional gel electrophoresis.

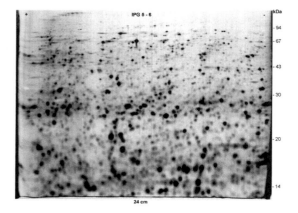

6.4. How does the yeast two-hybrid system test for interactions between two different proteins? Explain the activation of RNA polymerase in this experiment.

Screening for protein interactions using the two-hybrid system

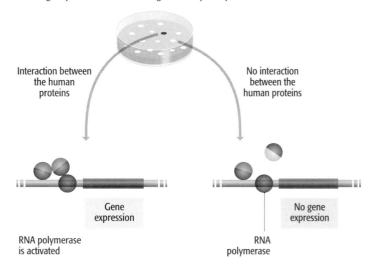

Further Reading

Transcriptome studies—methodology

Leung, Y.F. and Cavalieri, D. (2003) Fundamentals of cDNA microarray data analysis. *Trends Genet.* **19**: 649–659.

Velculescu, V.E., Vogelstein, B. and Kinzler, K.W. (2000) Analysing uncharted transcriptomes with SAGE. *Trends Genet.* **16**: 423–425.

Transcriptome studies—examples

Alizadeh, A.A., Eisen, M.B., Davis, R.E., *et al.* (2000) Distinct types of diffuse large B-cell lymphoma identified by gene expression profiling. *Nature* **403**: 503–511.

Chu, S., DeRisi, J., Eisen, M., Mulholland, J., Botstein, D., Brown, P.O. and Herskowitz, I. (1988) The transcriptional program of sporulation in budding yeast. *Science* **282**: 699–705.

DeRisi, J.L., Iyer, V.R. and Brown, P.O. (1997) Exploring the metabolic and genetic control of gene expression on a genomic scale. *Science* **278**: 680–686. *One of the first studies of the yeast transcriptome.*

Golub, T.R., Slonim, D.K., Tamayo, P., *et al.* (1999) Molecular classification of cancer: class discovery and class prediction by gene expression monitoring. *Science* **286**: 531–537.

Zhang, L., Zhou, W., Velculescu, V.E., Kern, S.E., Hruban, R.H., Hamilton, S.R., Vogelstein, B. and Kinzler, K.W. (1997) Gene expression in normal and cancer cells. *Science* **276**: 1268–1272.

Protein profiling

Fields, S. (2001) Proteomics in genomeland. *Science* **291**: 1221–1224. *Explains the importance of proteomics in understanding the human genome sequence.*

Mann, M., Hendrickson, R.C. and Pandey, A. (2001) Analysis of proteins and proteomes by mass spectrometry. *Annu. Rev. Biochem.* **70**: 437–473.

Phizicky, E., Bastiaens, P.I.H., Zhu, H., Snyder, M. and Fields, S. (2003) Protein analysis on a proteomics scale. *Nature* **422**: 208–215. *Reviews all aspects of proteomics.*

Yates, J.R. (2000) Mass spectrometry: from genomics to proteomics. *Trends Genet.* **16**: 5–8.

Zhu, H., Bilgin, M. and Snyder, M. (2003) Proteomics. *Annu. Rev. Biochem.* **72**: 783–812.

Studying protein interactions

Clackson, T. and Wells, J.A. (1994) *In vitro* selection from protein and peptide libraries. *Trends Biotechnol.* **12**: 173–184. *Phage display.*

Enright, A.J., Iliopoulos, I., Kyrpides, N.C. and Ouzounis, C.A. (1999) Protein interaction maps for complete genomes based on gene fusion events. *Nature* **402**: 86–90. *Using comparative genomics to identify functional interactions.*

Fields, S. and Sternglanz, R. (1994) The two-hybrid system: an assay for protein-protein interactions. *Trends Genet.* **10**: 286–292.

Protein interaction maps

Gavin, A.-C., Bösche, M., Krause, R., *et al.* (2002) Functional organization of the yeast proteome by systematic analysis of protein complexes. *Nature* **415**: 141–147. *A recent yeast protein interaction map.*

Han, J.-D.J., Bertin, N., Hao, T., *et al.* (2004) Evidence for dynamically organized modularity in the yeast protein-protein interaction network. *Nature* **430**: 88–93. *Defines party and date hubs.*

Jeong, H., Mason, S.P., Barabási, A.-L. and Oltvai, Z.N. (2001) Lethality and centrality in protein networks. *Nature* **411**: 41–42. *The first version of the yeast protein interaction map.*

Lee, I., Date, S.V., Adai, A.T. and Marcotte, E.M. (2004) A probabilistic functional network of yeast genes. *Science* **306**: 1555–1558.

Legrain, P., Wojcik, J. and Gauthier, J.-M. (2001) Protein-protein interaction maps: a lead towards cellular functions. *Trends Genet.* **17**: 346–352.

Metabolomics and systems biology

Covert, M.W., Schilling, C.H., Famili, I., Edwards, J.S., Goryanin, I.I., Selkov, E. and Palsson, B.O. (2001) Metabolic modelling of microbial strains *in silico*. *Trends Biochem. Sci.* **26**: 179–186. *Explains the concept of metabolic flux.*

Kalir, S. and Alon, U. (2004) Using a quantitative blueprint to reprogram the dynamics of the flagella gene network. *Cell* **117**: 713–720.

Kirschner, M.W. (2005) The meaning of systems biology. *Cell* **121**: 503–504.

PART 2

Genome Anatomies

Part 2 – Genome Anatomies surveys the information on genome organization that has been revealed, mostly during the last ten years, through use of the techniques described in Part 1. Chapter 7 examines eukaryotic nuclear genomes, with emphasis on the human genome which, as well as being our own genome, is the most complex genome so far sequenced. Chapter 8 investigates the genomes of prokaryotes and eukaryotic organelles, the latter dealt with here because of their prokaryotic origins. Chapter 9 looks at virus genomes and mobile genetic elements, grouped together because some mobile elements are related to virus genomes.

Chapter 7
Eukaryotic Nuclear Genomes

Chapter 8
Genomes of Prokaryotes
and Eukaryotic Organelles

Chapter 9
Virus Genomes
and Mobile Genetic Elements

Eukaryotic Nuclear Genomes

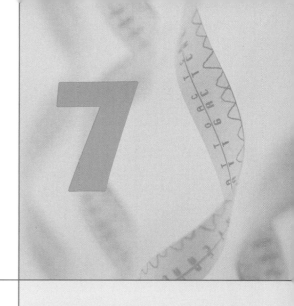

7

When you have read Chapter 7, you should be able to:

Describe the DNA–protein interactions that give rise to nucleosomes, chromatosomes, and the 30 nm chromatin fiber.

State the functions of centromeres and telomeres and describe the specific DNA–protein interactions that occur within these structures.

Explain why chromosome banding patterns and the isochore model suggest that genes are not evenly distributed in eukaryotic chromosomes.

Compare the organization of genes in various eukaryotic nuclear genomes and discuss the relationship between gene organization and genome size.

Summarize the overall content of the human genome.

Describe different ways of categorizing the functions of eukaryotic genes, and outline the important features revealed by comparisons of gene catalogs for different eukaryotes.

Explain, with examples, what is meant by "multigene family."

Distinguish between conventional and processed pseudogenes and other types of evolutionary relic.

Distinguish between tandemly repeated DNA and interspersed repetitive DNA, and describe the important features of satellite, minisatellite, and microsatellite DNA.

In the next three chapters we will survey the anatomies of the various types of genome that are found on our planet. There are three chapters because there are three types of genome to consider:

● **Eukaryotic nuclear genomes** (this chapter), of which the human genome is the one of greatest interest to us.

● **The genomes of prokaryotes and of eukaryotic organelles** (Chapter 8), which we will consider together because eukaryotic organelles are descended from ancient prokaryotes.

● **Virus genomes and mobile genetic elements** (Chapter 9), grouped together because some mobile elements are related to virus genomes.

7.1 Nuclear Genomes are Contained in Chromosomes

The nuclear genome is split into a set of linear DNA molecules, each contained in a chromosome. No exceptions to this pattern are known: all eukaryotes that have been studied have at least two chromosomes and the DNA molecules are always linear. The only variability at this level of eukaryotic genome structure lies with chromosome number, which appears to be unrelated to the biological features of the organism. For example, yeast has 16 chromosomes, four times as many as the fruit fly. Nor is chromosome number linked to genome size: some salamanders have genomes 30 times bigger than the human version but split into half the number of chromosomes. These comparisons are interesting but at present do not tell us anything useful about the genomes themselves; they are more a reflection of the nonuniformity of the evolutionary events that have shaped genome architecture in different organisms.

7.1.1 Packaging of DNA into chromosomes

Chromosomes are much shorter than the DNA molecules that they contain: the average human chromosome has just under 5 cm of DNA. A highly organized packaging system is therefore needed to fit a DNA molecule into its chromosome. We must understand this packaging system before we start to think about how genomes function because the nature of the packaging has an influence on the processes involved in expression of individual genes (Chapter 10).

The important breakthroughs in understanding DNA packaging were made in the early 1970s by a combination of biochemical analysis and electron microscopy. It was already known that nuclear DNA is associated with DNA-binding proteins called **histones** but the exact nature of the association had not been delineated. In 1973–1974, several groups carried out **nuclease**

Figure 7.1 Nuclease protection analysis of chromatin from human nuclei.
Chromatin is gently purified from nuclei and treated with a nuclease enzyme. On the left, the nuclease treatment is carried out under limiting conditions so that the DNA is cut, on average, just once in each of the linker regions between the bound proteins. After removal of the protein, the DNA fragments are analyzed by agarose gel electrophoresis and found to be 200 bp in length, or multiples thereof. On the right, the nuclease treatment proceeds to completion, so all the DNA in the linker regions is digested. The remaining DNA fragments are all 146 bp in length. The results show that in this form of chromatin, protein complexes are spaced along the DNA at regular intervals, one for each 200 bp, with 146 bp of DNA closely attached to each protein complex.

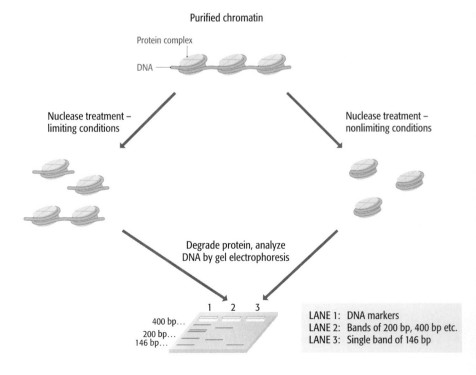

protection experiments on **chromatin** (DNA–histone complexes) that had been gently extracted from nuclei by methods designed to retain as much of the chromatin structure as possible. In a nuclease protection experiment the complex is treated with an enzyme that cuts the DNA at positions that are not "protected" by attachment to a protein. The sizes of the resulting DNA fragments indicate the positioning of the protein complexes on the original DNA molecule (Figure 7.1). After limited nuclease treatment of purified chromatin, the bulk of the DNA fragments have lengths of approximately 200 bp and multiples thereof, suggesting a regular spacing of histone proteins along the DNA.

In 1974 these biochemical results were supplemented by electron micrographs of purified chromatin, which enabled the regular spacing inferred by the protection experiments to be visualized as beads of protein on the string of DNA (Figure 7.2A). Further biochemical analysis indicated that each bead, or **nucleosome**, contains eight histone protein molecules, these being two each of histones H2A, H2B, H3, and H4. Structural studies have shown that these eight proteins form a barrel-shaped **core octamer** with the DNA wound twice around the outside (Figure 7.2B). Between 140 bp and 150 bp of DNA (depending on the species) are associated with the nucleosome particle, and each nucleosome is separated by 50–70 bp of linker DNA, giving the repeat length of 190–220 bp previously shown by the nuclease protection experiments.

As well as the proteins of the core octamer, there is a group of additional histones, all closely related to one another and collectively called **linker histones**. In vertebrates these include histones H1a–e, H1^0, H1t, and H5. A single linker histone is attached to each nucleosome, to form the **chromatosome**, but the precise positioning of this linker histone is not known. Structural studies support the traditional model in which the linker histone acts as a clamp, preventing the coiled DNA from detaching from the nucleosome (Figure 7.2C). However, other results suggest that, at least in some organisms, the linker histone is not located on the extreme surface of the nucleosome–DNA assembly, as would be expected if it really were a clamp, but instead is inserted between the core octamer and the DNA.

The "beads-on-a-string" structure shown in Figure 7.2A is thought to represent an unpacked form of chromatin that occurs only infrequently in living nuclei. Very gentle cell-breakage techniques developed in the mid-1970s resulted in the finding of a more condensed version of the complex, called the **30 nm fiber** (it is approximately 30 nm in width). The exact way in which nucleosomes associate to form the 30 nm fiber is not known, but several models have been proposed, two of which are shown in Figure 7.3. The individual nucleosomes within the 30 nm fiber may be held together by interactions between the linker histones, or the attachments may involve the core histones, whose protein "tails" extend outside the nucleosome (see Figure 10.13). The latter hypothesis is attractive because chemical modification of these tails results in the 30 nm fiber opening up, enabling genes contained within it to be activated (Section 10.2.1).

7.1.2 The special features of metaphase chromosomes

The 30 nm fiber is probably the major type of chromatin in the nucleus during interphase, the period between nuclear divisions. When the nucleus divides, the DNA adopts a more compact form of packaging, resulting in the

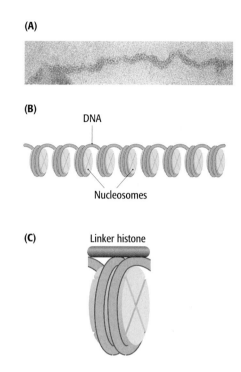

(A)

(B)

DNA

Nucleosomes

(C)

Linker histone

Figure 7.2 Nucleosomes. (A) Electron micrograph of a purified chromatin strand showing the "beads-on-a-string" structure. (B) The model for the "beads-on-a-string" structure, in which each bead is a barrel-shaped nucleosome with the DNA wound twice around the outside. Each nucleosome is made up of eight proteins: a central tetramer of two histone H3 and two histone H4 subunits, plus a pair of H2A–H2B dimers, one above and one below the central tetramer (see Figure 10.13). (C) The precise position of the linker histone relative to the nucleosome is not known but, as shown here, the linker histone may act as a clamp, preventing the DNA from detaching from the outside of the nucleosome. Image (A) courtesy of Dr Barbara Hamkalo.

Figure 7.3 Two models of the 30 nm chromatin fiber. The solenoid model (A) has been in favor for several years but recent experimental evidence supports the helical ribbon (B). Reprinted with permission from Dorigo *et al.*, *Science*, **306**, 1571–1573. Copyright 2004 AAAS.

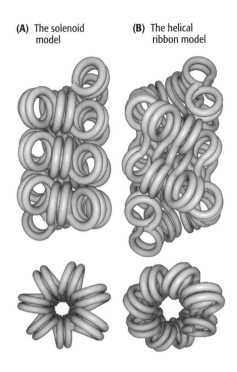

(A) The solenoid model

(B) The helical ribbon model

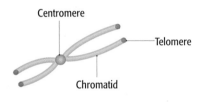

Centromere

Telomere

Chromatid

Figure 7.4 The typical appearance of a metaphase chromosome. Metaphase chromosomes are formed after DNA replication has taken place, so each one is, in effect, two chromosomes linked together at the centromere. The arms are called the chromatids. A telomere is the extreme end of a chromatid.

highly condensed **metaphase chromosomes** that can be seen with the light microscope and which have the appearance generally associated with the word "chromosome" (Figure 7.4). The metaphase chromosomes form at a stage in the **cell cycle** after DNA replication has taken place, and so each one contains two copies of its chromosomal DNA molecule. The two copies are held together at the **centromere**, which has a specific position within each chromosome. The arms of the chromosome, which are called **chromatids** and have terminal structures called **telomeres**, are of different lengths in different chromosomes. Individual chromosomes can therefore be recognized because of the lengths of their chromatids and the location of the centromere relative to the telomeres. Further distinguishing features are revealed when chromosomes are stained. There are a number of different staining techniques (Table 7.1), each resulting in a banding pattern that is characteristic for a particular chromosome. This means that the set of

Table 7.1 Staining techniques used to produce chromosome banding patterns

Technique	Procedure	Banding pattern
G-banding	Mild proteolysis followed by staining with Giemsa	Dark bands are AT-rich Pale bands are GC-rich
R-banding	Heat denaturation followed by staining with Giemsa	Dark bands are GC-rich Pale bands are AT-rich
Q-banding	Stain with quinacrine	Dark bands are AT-rich Pale bands are GC-rich
C-banding	Denature with barium hydroxide and then stain with Giemsa	Dark bands contain constitutive heterochromatin (see Section 10.1.2)

chromosomes possessed by an organism can be represented as a **karyogram**, in which the banded appearance of each one is depicted. The human karyogram is shown in Figure 7.5.

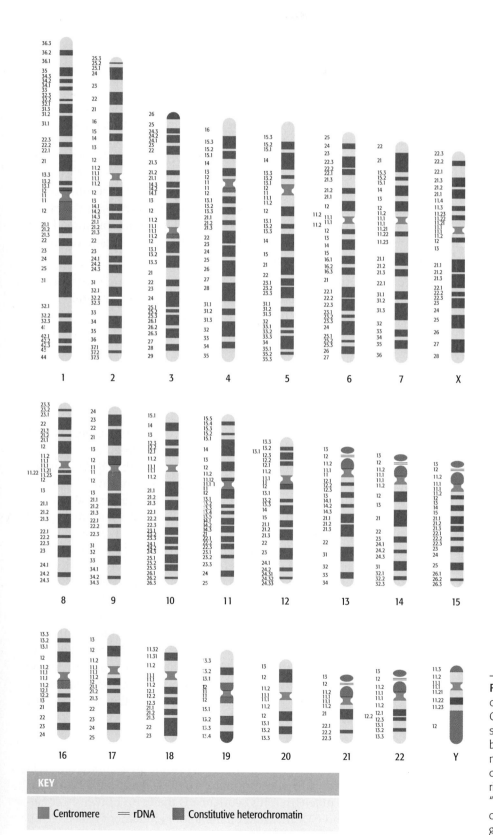

KEY

■ Centromere   ═ rDNA   ■ Constitutive heterochromatin

Figure 7.5 The human karyogram. The chromosomes are shown with the G-banding pattern obtained after Giemsa staining. Chromosome numbers are given below each structure and the band numbers to the left. "rDNA" is a region containing a cluster of repeat units for the ribosomal RNA genes (Section 1.2.2). "Constitutive heterochromatin" is very compact chromatin that has few or no genes (Section 10.1.2).

Figure 7.6 The *Saccharomyces cerevisiae* centromere. CDEI is 9 bp in length, CDEII is 80–90 bp, and CDEIII is 11 bp. Additional sequences flanking the region shown here are looked on as part of the centromeric DNA, whose full length is approximately 125 bp.

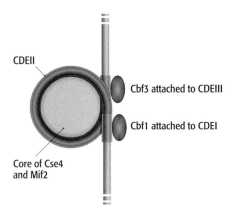

Figure 7.7 DNA–protein interactions in the yeast centromere. The diagram is purely schematic as the precise positioning of the proteins and of the DNA components is unknown.

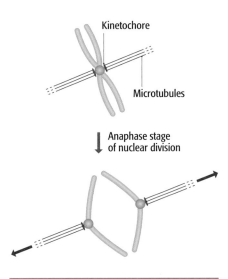

Figure 7.8 The role of the kinetochores during nuclear division. During the anaphase period of nuclear division, individual chromosomes are drawn apart by the contraction of microtubules attached to the kinetochores.

The human karyogram is typical of that of the great majority of eukaryotes, but some organisms display unusual features not displayed by the human version. These include the following:

- **Minichromosomes** are relatively short in length but rich in genes. The chicken genome, for example, is split into 39 chromosomes: six **macrochromosomes** containing 66% of the DNA but only 25% of the genes, and 33 minichromosomes containing the remaining one-third of the genome and 75% of the genes. The gene density in the minichromosomes is therefore some six times greater than that in the macrochromosomes.

- **B chromosomes** are additional chromosomes possessed by some individuals in a population, but not all. They are common in plants and also known in fungi, insects, and animals. B chromosomes appear to be fragmentary versions of normal chromosomes that result from unusual events during nuclear division. Some contain genes, often for rRNAs, but it is not clear if these genes are active. The presence of B chromosomes can affect the biological characteristics of the organism, particularly in plants where they are associated with reduced viability. It is presumed that B chromosomes are gradually lost from cell lineages as a result of irregularities in their inheritance pattern.

- **Holocentric chromosomes** do not have a single centromere but instead have multiple structures spread along their length. The nematode worm *Caenorhabditis elegans* has holocentric chromosomes.

DNA–protein interactions in centromeres and telomeres

The DNA contained within centromeres and telomeres, and the proteins attached to this DNA, have special features related to the particular functions of these structures.

The nucleotide sequence of centromeric DNA in higher eukaryotes is best understood in the plant *Arabidopsis thaliana*, whose amenity to genetic analysis has enabled the positions of the centromeres on the DNA sequence to be located with some precision. Also, a special effort was made to sequence these centromeric regions, which are sometimes excluded from genome sequences because of problems in obtaining an accurate reading through the highly repetitive structures that characterize these regions. *Arabidopsis* centromeres span 0.9–1.2 Mb of DNA and each one is made up largely of 180 bp repeat sequences. In humans the equivalent sequences are 171 bp in length and are called **alphoid DNA**, with 1500–30,000 copies per centromere. Before the *Arabidopsis* sequences were obtained it was thought that these repeat sequences were by far the principal component of centromeric DNA. However, *Arabidopsis* centromeres also contain multiple copies of genome-wide repeats, along with a few genes, the latter at a density of 7–9 per 100 kb compared with 25 genes per 100 kb for the noncentromeric regions of *Arabidopsis* chromosomes. The discovery that centromeric DNA contains genes was a big surprise because it was thought that these regions were genetically inactive.

Arabidopsis and humans display the basic pattern for centromeric DNA as seen in virtually all eukaryotes, but an interesting variation occurs in the yeast *Saccharomyces cerevisiae*, whose centromere is defined by a single sequence, approximately 125 bp in length. This sequence is made up of two short elements, called CDEI and CDEIII, which flank a longer element called CDEII

(Figure 7.6). The sequence of CDEII is variable, though always very rich in A and T nucleotides, whereas both CDEI and CDEIII are highly conserved, meaning that their sequences are very similar in all 16 yeast chromosomes. Mutations in CDEII rarely affect the function of the centromere, but a mutation in CDEI or CDEIII usually prevents the centromere from forming. The short, nonrepetitive nature of the yeast centromeric DNA has enabled progress to be made in understanding how the DNA interacts with proteins to form a functional centromere. A key role is played by a special chromosomal protein called Cse4, which is similar in structure to histone H3 and which, with a second protein called Mif2, forms a core around which the CDEII sequence is wrapped (Figure 7.7). The DNA appears to be held in place by two further proteins: Cbf1, which recognizes and attaches to the CDEI sequence, and Cbf3 (in fact a tetramer of four proteins), which attaches to CDEIII. Cbf1 and Cbf3 also bind to at least some of the 20 or so additional proteins that form the **kinetochore,** the structure that acts as the attachment point for the microtubules which draw the divided chromosomes into the daughter nuclei (Figure 7.8). To what extent this model of the yeast centromere also applies to other eukaryotes is not yet clear. The centromeres of higher eukaryotes are quite different as they contain nucleosomes, similar to those in other regions of the chromosome but some of them containing the protein CENP-A instead of histone H3. CENP-A-containing nucleosomes are more compact and structurally rigid than those containing H3, and it has been suggested that the arrangement of CENP-A and H3 nucleosomes along the DNA is such that the CENP-A versions are located on the surface of the centromere, where they form an outer shell on which the kinetochore is constructed (Figure 7.9).

The second important part of the chromosome is the terminal region or **telomere**. Telomeres are important because they mark the ends of chromosomes and therefore enable the cell to distinguish a real end from an unnatural end caused by chromosome breakage—an essential requirement because the cell must repair the latter but not the former. Telomeric DNA is made up of hundreds of copies of a repeated motif, 5′–TTAGGG–3′ in humans, with a short extension of the 3′ terminus of the double-stranded DNA molecule (Figure 7.10). Two special proteins bind to the repeat sequences in human telomeres. These are called TRF1, which helps to regulate the length of the telomere, and TRF2, which maintains the single-strand extension. If TRF2 is inactivated then this extension is lost and the two polynucleotides fuse together in a covalent linkage. Other telomeric proteins are thought to form a linkage between the telomere and the periphery of the nucleus, the area in which the chromosome ends are localized. Further proteins mediate the enzymatic activity that maintains the length of each telomere during DNA replication. We will return to this last activity in Section 15.2.4: it is critical to the survival of the chromosome and may be a key to understanding cell senescence and death.

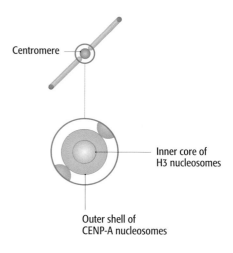

Figure 7.9 Mammalian centromeres contain CENP-A- and H3-nucleosomes. One possibility is that the H3-nucleosomes are located mainly in the central core of the centromere, with the CENP-A versions forming an outer shell onto which the kinetochore is constructed.

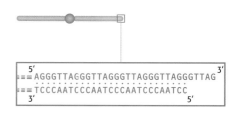

Figure 7.10 Telomeres. The sequence at the end of a human telomere. The length of the 3′ extension is different in each telomere. See Section 15.2.4 for more details about telomeric DNA.

7.2 The Genetic Features of Eukaryotic Nuclear Genomes

In Chapter 5 we examined the range of bioinformatic and experimental methods that can be used to locate the genes in a genome sequence and to determine their functions. Now we turn our attention to what those methods have told us about the genetic features of eukaryotic nuclear genomes.

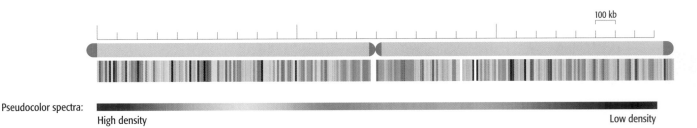

Pseudocolor spectra:

High density   Low density

7.2.1 Where are the genes in a nuclear genome?

Figure 7.11 Gene density along the largest of the five *Arabidopsis thaliana* chromosomes. Chromosome 1, which is 29.1 Mb in length, is illustrated with the sequenced portions shown in light green and the centromere and telomeres in dark green. The gene map below the chromosome gives gene density in pseudocolor, from deep blue (low density) to red (high density). The density varies from 1 to 38 genes per 100 kb. Reprinted with permission from AGI (The Arabidopsis Genome Initiative), *Nature*, **408**, 797–815. © 2000 Macmillan Magazines Limited.

In the previous section we learnt that *Arabidopsis* centromeres contain genes but at a lesser density than that in the rest of the chromosomes. This alerts us to the fact that the genes are not arranged evenly along the length of a chromosome. In most organisms, genes appear to be distributed more at less at random, with substantial variations in gene density at different positions within a chromosome. The average gene density in *Arabidopsis* is 25 genes per 100 kb, but even outside of the centromeres and telomeres the density varies from 1 to 38 genes per 100 kb, as illustrated in Figure 7.11 for the largest of the plant's five chromosomes. The same is true for human chromosomes, where the density ranges from 0 to 64 genes per 100 kb.

The uneven gene distribution within human chromosomes was suspected for several years before the sequence was completed. There were two lines of evidence, one of which related to the banding patterns that are produced when chromosomes are stained. The dyes used in these procedures (see Table 7.1) bind to DNA molecules, but in most cases with preferences for certain base pairs. Giemsa, for example, has a greater affinity for DNA regions that are rich in A and T nucleotides. The dark G-bands in the human karyogram (see Figure 7.5) are therefore thought to be AT-rich regions of the genome. The base composition of the genome as a whole is 59.7% A + T so the dark G-bands must have AT contents substantially greater than 60%. Cytogeneticists therefore predicted that there would be fewer genes in dark G-bands because genes generally have AT contents of 45%–50%. This prediction was confirmed when the genome sequence was compared with the human karyogram.

The second line of evidence pointing to uneven gene distribution derived from the **isochore** model of genome organization. According to this model, the genomes of vertebrates and plants (and possibly of other eukaryotes) are mosaics of segments of DNA, each at least 300 kb in length, with each segment having a uniform base composition that differs from that of the adjacent segments. Support for the isochore model comes from experiments in which genomic DNA is broken into fragments of approximately 100 kb, treated with dyes that bind specifically to AT- or GC-rich regions, and the pieces separated by density gradient centrifugation (Technical Note 7.1). When this experiment is carried out with human DNA, five fractions are seen, each representing a different isochore type with a distinctive base composition: two AT-rich isochores, called L1 and L2, and three GC-rich classes called H1, H2, and H3. The last of these, H3, is the least abundant in the human genome, making up only 3% of the total, but contains over 25% of the genes. This is a clear indication that genes are not distributed evenly through the human genome. In fact, examination of the genome sequence suggests that the isochore theory oversimplifies what is, in reality, a much

Technical Note 7.1   Ultracentrifugation techniques
Methodology for separation of cell components and large molecules

The development of high-speed centrifuges in the 1920s led to techniques for separating organelles and other fractions from disrupted cells. The first technique to be used was **differential centrifugation**, in which pellets of successively lighter cell components are collected by centrifuging cell extracts at different speeds. Intact nuclei, for example, are relatively large and can be collected from a cell extract by centrifugation at 1000 *g* for 10 minutes; mitochondria, being lighter, require centrifugation at 20,000 *g* for 20 minutes. Different cell components can be obtained in fairly pure form by careful manipulation of the centrifugation parameters.

Density gradient centrifugation was used for the first time in 1951. In this procedure, the cell fraction is not centrifuged in a normal aqueous solution. Instead, a sucrose solution is layered into the tube in such a way that a density gradient is formed, the solution being more concentrated and hence denser toward the bottom of the tube. The cell fraction is placed on the top of the gradient and the tube centrifuged at a very high speed: at least 500,000 *g* for several hours. Under these conditions, the rate of migration of a cell component through the gradient depends on

its sedimentation coefficient, which in turn depends on its molecular mass and shape. For example, eukaryotic ribosomes have a sedimentation coefficient of 80S (S stands for Svedberg units, Svedberg being the Swedish scientist who pioneered the biological applications of ultracentrifugation), whereas bacterial ribosomes, being smaller, have a sedimentation coefficient of 70S.

In a second type of density gradient centrifugation, a solution such as 8 M cesium chloride is used, which is substantially denser than the sucrose solution used to measure S values. The starting solution is uniform, the gradient being established during the centrifugation. Cellular components migrate down through the centrifuge tube, but molecules such as DNA and proteins do not reach the bottom; instead each one comes to rest at a position where the density of the matrix equals its own **buoyant density** (see Figure 7.23). This technique has many applications in molecular biology, being able to separate DNA fragments of different base compositions and DNA molecules with different conformations (e.g., supercoiled, circular, and linear DNA). It can also distinguish between normal DNA and DNA labeled with a heavy isotope of nitrogen (Section 15.1.1).

more complex pattern of variations in base composition along the length of each human chromosome. But even if it turns out to be a misconception, the isochore theory will have been a useful misconception as it played an important role in helping molecular biologists of the presequence era to understand genome structure.

7.2.2 How are the genes organized in a nuclear genome?

The variations in gene density that occur along the length of a eukaryotic chromosome mean that it is difficult to identify regions in which the organization of the genes can be looked on as "typical" of the genome as a whole. Despite this difficulty, it is clear that the overall pattern of gene organization varies greatly between different eukaryotes, and we need to understand these differences because they reflect important distinctions between the genetic features and evolutionary histories of these genomes. In order to illustrate the differences we will, in this section, look in detail at a small part of the human genome and compare this segment with equally small parts of the genomes of other organisms. As you work through this material you should focus on the distinctive features that are being drawn out, but remember that the variations that occur along a single chromosome mean that it is impossible to make hard and fast statements about the patterns of gene organization that are present in, or absent from, a particular genome.

The genes make up only a small part of the human genome

How are genes organized in the human nuclear genome? To answer this question we will examine a 50 kb segment of chromosome 12 (Figure 7.12). This segment contains the following genetic features:

- Four genes. These are:
  - PKP2, which codes for plakophilin 2, a protein involved in synthesis of desmosomes, structures that act as connection points between adjacent mammalian cells.
  - SYB1, specifying a vesicle-associated membrane protein whose role is to ensure that vesicles fuse with their correct target membranes within the cell.
  - A gene whose function has not yet been identified, called FLJ10143.
  - CD27, coding for a member of the tumor necrosis factor receptor superfamily, a group of proteins that regulate signal transduction pathways involved in **apoptosis** (programmed cell death) and cell differentiation.

  Note that each of these four genes is discontinuous, the number of introns ranging from two for SYB1 to eight for PKP2.

- 88 genome-wide repeat sequences. These are sequences that recur at many places in the genome. There are four main types of genome-wide repeat, called **LINEs** (long interspersed nuclear elements), **SINEs** (short interspersed nuclear elements), **LTR** (long terminal repeat) **elements**, and **DNA transposons**. Examples of each type are seen in this short segment of

Figure 7.12   A segment of the human genome. This map shows the location of genes, gene segments, genome-wide repeats, and microsatellites in a 50 kb segment of human chromosome 12.

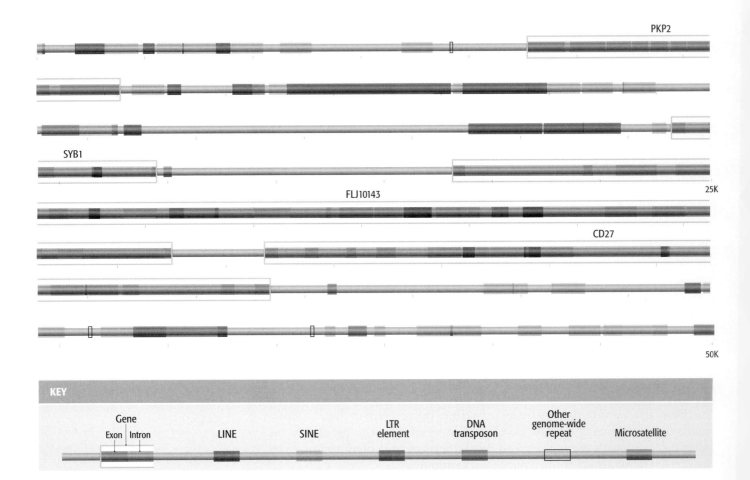

the genome. Most of the genome-wide repeats are located in the intergenic regions but several lie within introns.

- Seven microsatellites, which as described in Section 3.2.2, are sequences in which a short motif is repeated in tandem. One of the microsatellites seen here has the motif CA repeated 12 times, giving the sequence:

5′–CACACACACACACACACACACACA–3′
3′–GTGTGTGTGTGTGTGTGTGTGTGT–5′

The other six microsatellites comprise repeats of CAAA, CCTG, CTGGGG, CAAAA, TG, and TTTG, respectively. Four of the seven microsatellites are located within introns.

- Finally, approximately 30% of our 50 kb segment of the human genome is made up of stretches of nongenic, nonrepetitive, single-copy DNA of no known function or significance.

The most striking feature of this 50 kb segment of the human genome is the relatively small amount of space taken up by the genes. When added together, the total length of the exons—the parts of the four genes that contain the biological information—is 4745 bp, equivalent to 9.5% of the 50 kb segment. In fact, this segment is rather rich in genes: all the exons in the human genome make up only 48 Mb, just 1.5% of the total. In contrast, 44% of the genome is taken up by genome-wide repeats (Figure 7.13).

The yeast genome is very compact

How extensive are the differences in gene organization among eukaryotes? There are certainly very substantial differences in genome size, with the smallest eukaryotic genomes less than 10 Mb in length, and the largest over 100,000 Mb. As can be seen in Figure 7.14 and Table 7.2, this size range coincides to a certain extent with the complexity of the organism, the simplest eukaryotes such as fungi having the smallest genomes, and higher eukaryotes such as vertebrates and flowering plants having the largest ones. This might appear to make sense, as one would expect the complexity of an organism to be related to the number of genes in its genome—higher eukaryotes need larger

Figure 7.13 The composition of the human genome. Abbreviation: UTRs, untranslated regions.

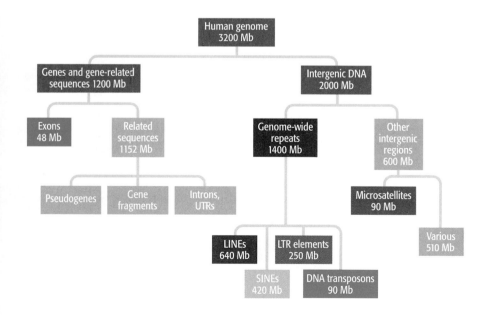

Figure 7.14 Size ranges of genomes in different groups of eukaryotes.

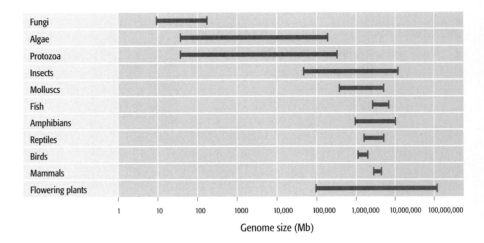

genomes to accommodate the extra genes. However, the correlation is far from precise: if it was, then the nuclear genome of the yeast *Saccharomyces cerevisiae*, which at 12 Mb is 0.004 times the size of the human nuclear genome, would be expected to contain $0.004 \times 35,000$ genes, which is just 140. In fact the *S. cerevisiae* genome contains about 6000 genes.

Table 7.2 Sizes of eukaryotic genomes

Species	Genome size (Mb)
Fungi	
Saccharomyces cerevisiae	12.1
Aspergillus nidulans	25.4
Protozoa	
Tetrahymena pyriformis	190
Invertebrates	
Caenorhabditis elegans	97
Drosophila melanogaster	180
Bombyx mori (silkworm)	490
Strongylocentrotus purpuratus (sea urchin)	845
Locusta migratoria (locust)	5000
Vertebrates	
Takifugu rubripes (pufferfish)	400
Homo sapiens	3200
Mus musculus (mouse)	3300
Plants	
Arabidopsis thaliana (vetch)	125
Oryza sativa (rice)	466
Zea mays (maize)	2500
Pisum sativum (pea)	4800
Triticum aestivum (wheat)	16,000
Fritillaria assyriaca (fritillary)	120,000

For many years the lack of precise correlation between the complexity of an organism and the size of its genome was looked on as a bit of a puzzle, the so-called **C-value paradox**. In fact the answer is quite simple: space is saved in the genomes of less-complex organisms because the genes are more closely packed together. The *S. cerevisiae* genome illustrates this point, as we can see from the top two parts of Figure 7.15, where the 50 kb segment of the human genome that we have just examined is compared with a 50 kb segment of the yeast genome. The yeast genome segment, which comes from chromosome III (the first eukaryotic chromosome to be sequenced), has the following distinctive features:

- It contains more genes than the human segment. This region of yeast chromosome III contains 26 genes thought to code for proteins and two that code for transfer RNAs.

- Relatively few of the yeast genes are discontinuous. In this segment of chromosome III none of the genes are discontinuous. In the entire yeast genome there are only 239 introns, compared with over 300,000 in the human genome.

- There are fewer genome-wide repeats. This part of chromosome III contains a single LTR element, called *Ty2*, and four truncated LTR elements called delta sequences. These five genome-wide repeats make up 13.5% of the 50 kb segment, but this figure is not typical of the yeast genome as a whole. When all 16 yeast chromosomes are considered, the total amount of sequence taken up by genome-wide repeats is only 3.4% of the total.

The picture that emerges is that the genetic organization of the yeast genome is much more economical than that of the human version. The genes themselves

Figure 7.15 Comparison of the genomes of humans, yeast, fruit flies, and maize. (A) The 50 kb segment of human chromosome 12 presented previously. This is compared with 50 kb segments from the genomes of (B) *Saccharomyces cerevisiae*, (C) *Drosophila melanogaster*, and (D) maize.

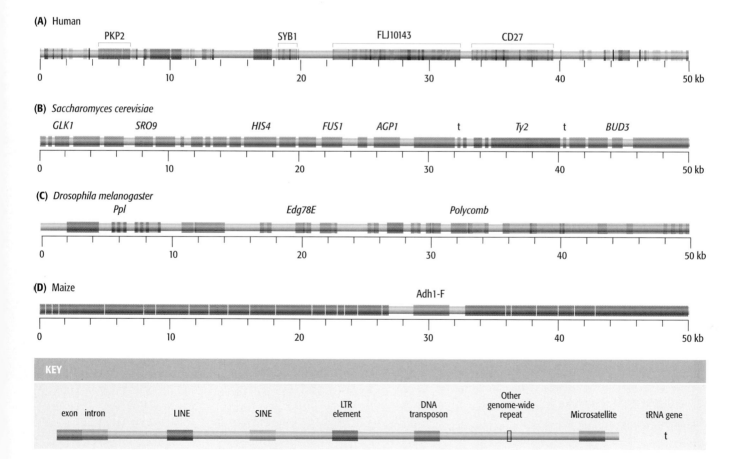

are more compact, having fewer introns, and the spaces between the genes are relatively short, with much less space taken up by genome-wide repeats and other noncoding sequences.

Gene organization in other eukaryotes

The hypothesis that more complex eukaryotes have less compact genomes holds when other species are examined. The third part of Figure 7.15 shows a 50 kb segment of the fruit-fly genome. If we agree that a fruit fly is more complex than a yeast cell but less complex than a human then we would expect the organization of the fruit-fly genome to be intermediate between that of yeast and humans. This is what we see in Figure 7.15C, this 50 kb segment of the fruit-fly genome having 11 genes, more than in the human segment but fewer than in the yeast sequence. All of these genes are discontinuous, but seven have just one intron each. The picture is similar when the entire genome sequences of the three organisms are compared (Table 7.3). The gene density in the fruit-fly genome is intermediate between that of yeast and humans, and the average fruit-fly gene has many more introns than the average yeast gene but still three times fewer than the average human gene.

The comparison between the yeast, fruit-fly, and human genomes also holds true when we consider the genome-wide repeats (see Table 7.3). These make up 3.4% of the yeast genome, about 12% of the fruit-fly genome, and 44% of the human genome. It is beginning to become clear that the genome-wide repeats play an intriguing role in dictating the compactness or otherwise of a genome. This is strikingly illustrated by the maize genome, which at 2500 Mb is relatively small for a flowering plant. Only a few regions of the maize genome have been sequenced, but some remarkable results have been obtained, revealing a genome dominated by repetitive elements. Figure 7.15D shows a 50 kb segment of this genome, either side of one member of a family of genes coding for the alcohol dehydrogenase enzymes. This is the only gene in this 50 kb region, although there is a second one, of unknown function, approximately 100 kb beyond the right-hand end of the sequence shown here. Instead of genes, the dominant feature of this genome segment is the genome-wide repeats, which have been described as forming a sea within which islands of genes are located. The genome-wide repeats are of the LTR element type, which comprise virtually all of the noncoding part of the segment, and on their own are estimated to make up approximately 50% of the maize genome. It is becoming clear that one or more families of genome-wide repeats have undergone a massive proliferation in the genomes of certain species. This may provide an explanation for the most puzzling aspect of the C-value paradox, which is not the general increase in genome size that is seen in increasingly complex organisms, but the fact that similar organisms

Table 7.3 Compactness of the yeast, fruit fly, and human genomes

Feature	Yeast	Fruit fly	Human
Gene density (average number per Mb)	496	76	11
Introns per gene (average)	0.04	3	9
Amount of the genome that is taken up by genome-wide repeats	3.4%	12%	44%

can differ greatly in genome size. A good example is provided by *Amoeba dubia* which, being a protozoan, might be expected to have a genome of 100–500 kb, similar to other protozoa such as *Tetrahymena pyriformis* (see Table 7.2). In fact the *Amoeba* genome is over 200,000 Mb. Similarly, we might guess that the genomes of crickets would be similar in size to those of other insects, but these bugs have genomes of approximately 2000 Mb, 11 times that of the fruit fly.

7.2.3 How many genes are there and what are their functions?

The most detailed annotations of the finished human chromosome sequences suggest that the human genome contains 30,000–40,000 genes, the uncertainty arising because of the difficulty, outlined in Section 5.1.1, in recognizing which sequences are genes and which are not. The number is much lower than originally expected, as a "best guess" of 80,000–100,000 was still in vogue up to a few months before the draft sequences were completed in 2000. These early estimates were high because they were based on the supposition that, in most cases, a single gene specifies a single mRNA and a single protein. According to this model, the number of genes in the human genome should be similar to the number of proteins in human cells, leading to the estimates of 80,000–100,000. The discovery that the number of genes is much lower than this indicates that alternative splicing, the process by which exons from a pre-mRNA are assembled in different combinations so that more than one protein can be coded by a single gene (see Figure 6.5), is more prevalent than was originally appreciated. Gene numbers for a variety of eukaryotic nuclear genomes are given in Table 7.4, but you should bear in mind that because of alternative splicing the question "How many genes are there?" has no real biological significance, as the number of genes does not indicate the number of proteins that can be synthesized and hence is not a measure of the biological complexity of a genome.

Although alternative splicing enables one gene to specify several proteins, those proteins will have at least some parts of their amino acid sequences in common and so will usually have similar or related functions. Categorizing genes according to their functions can therefore give meaningful information

Table 7.4 Genome sizes and gene numbers for various eukaryotes

Species	Size of genome (Mb)	Approximate number of genes
Saccharomyces cerevisiae (budding yeast)	12.1	6100
Schizosaccharomyces pombe (fission yeast)	12.5	4900
Caenorhabditis elegans (nematode worm)	97	19,000
Arabidopsis thaliana (plant)	125	25,500
Drosophila melanogaster (fruit fly)	180	13,600
Oryza sativa (rice)	466	40,000
Gallus gallus (chicken)	1200	20,000–23,000
Homo sapiens (human)	3200	30,000–40,000

on the range of biochemical activities specified by a genome, even if for many genes the splicing variants have either not been identified or not assigned individual functions. With gene catalogs the problem is lack of completeness, because of the difficulties in identifying functions, even for a relatively simple organism such as *Saccharomyces cerevisae*. It is quite probable that certain categories of gene are underrepresented in the existing catalogs, because those genes have functions that are particularly difficult to identify. With these qualifications in mind we will first look at the human gene catalog.

The human gene catalog

The functions of over half of the 30,000–40,000 human genes are known or can be inferred with a reasonable degree of certainty. The vast majority code for proteins; less than 2500 specify the various types of functional RNA. Almost one-quarter of the protein-coding genes are involved in expression, replication, and maintenance of the genome (Figure 7.16) and another 21% specify components of the **signal transduction** pathways that regulate genome expression and other cellular activities in response to signals received from outside the cell (Section 14.1.2). All of these genes can be looked on as having a function that is involved in one way or another with the activity of the genome. Enzymes responsible for the general biochemical functions of the cell account for another 17.5% of the known genes, and the remainder are involved in activities such as transport of compounds into and out of cells, the folding of proteins into their correct three-dimensional structures, the immune response, and synthesis of structural proteins such as those found in the cytoskeleton and in muscles. It is possible that as the human gene catalog is made more complete, the relative proportions of the genes in the three major categories in Figure 7.16 will decrease. This is because these major categories represent the most studied areas of cell biology, which means that many of the relevant genes can be recognized because their protein products are known. Genes whose products have not yet been identified are more likely to be involved in the less-well-studied areas of cellular activity.

One thing that the gene catalog cannot tell us, and will not be able to tell us even when it is complete, is what makes a human being. The minimalist approach to molecular biology, whereby the study of individual genes or groups of genes is expected to lead eventually to a full, biomolecular description of how a human being is constructed and functions, has been dealt a severe blow by the human genome sequence. There are no amazing revelations about what makes humans different from apes. Even though the chimpanzee genome has been completely sequenced it is still not possible simply from genome comparisons to determine what makes us human (Section 18.4). On the basis of gene number we are only three times more complex than a fruit fly and only twice as complex as the microscopic worm *Caenorhabditis elegans*. More detailed studies of how the human genome functions may reveal key features that underlie some of the special attributes of human beings, but genomics will never explain humanity.

Gene catalogs reveal the distinctive features of different organisms

There are various ways to categorize the genes in a eukaryotic genome. One possibility is to classify the genes according to their function, as shown in Figure 7.16 for the human genome. This system has the advantage that the fairly broad functional categories used in Figure 7.16 can be further subdivided

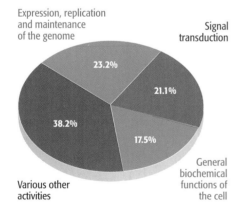

Expression, replication and maintenance of the genome

Signal transduction

23.2%

21.1%

38.2%

17.5%

Various other activities

General biochemical functions of the cell

Figure 7.16 Categorization of the human gene catalog. The pie chart shows a categorization of the identified human protein-coding genes. It omits approximately 13,000 genes whose functions are not yet known. The segment labeled "various other activities" includes, among others, proteins involved in biochemical transport processes and protein folding, immunological proteins, and structural proteins.

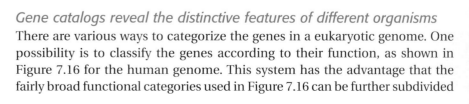

to produce a hierarchy of increasingly specific descriptions for smaller and smaller sets of genes. The weakness with this approach is that functions have not yet been assigned to many eukaryotic genes, so this type of classification leaves out a proportion of the total gene set. A more powerful method is to base the classification not on the functions of genes but on the structures of the proteins that they specify. A protein molecule is constructed from a series of **domains**, each of which has a particular biochemical function. Examples are the **zinc finger**, which is one of several domains that enable a protein to bind to a DNA molecule (Section 11.1.1), and the "death domain," which is present in many proteins involved in apoptosis. Each domain has a characteristic amino acid sequence, perhaps not exactly the same sequence in every example of that domain, but close enough for the presence of a particular domain to be recognizable by examining the amino acid sequence of the protein. The amino acid sequence of a protein is specified by the nucleotide sequence of its gene, so the domains present in a protein can be determined from the nucleotide sequence of the gene that codes for that protein. The genes in a genome can therefore be categorized according to the protein domains that they specify. This method has the advantage that it can be applied to genes whose functions are not known and hence can encompass a larger proportion of the set of genes in a genome.

Classification schemes that use domains to infer gene function suggest that all eukaryotes possess the same basic set of genes, but that more complex species have a greater number of genes in each category. For example, humans have the greatest number of genes in all but one of the categories used in Figure 7.17, the exception being "metabolism" where *Arabidopsis* comes out on top as a result of its photosynthetic capability, which requires a large set of genes not present in the other four genomes included in this comparison. This functional classification reveals other interesting features, notably that *C. elegans* has a relatively high number of genes involved in cell–cell signaling, which is surprising given that this organism has just 959 cells. Humans, who have 10^{13} cells, have only 250 more genes for cell–cell

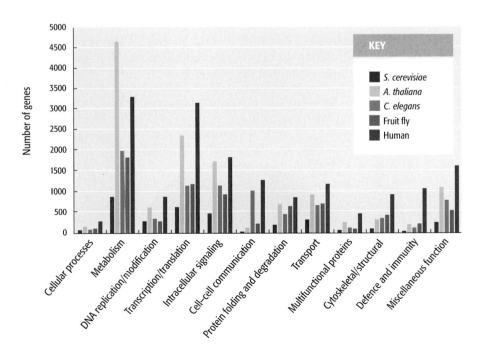

Figure 7.17 **Comparison of the gene catalogs of** *Saccharomyces cerevisiae*, *Arabidopsis thaliana*, *Caenorhabditis elegans*, **fruit fly, and humans.** Genes are categorized according to their function, as deduced from the protein domains specified by each gene.

Table 7.5 Examples of protein domains specified by different genomes

| Domain | Function | Number of genes in the genome containing the domain | | | | |
		Human	Fruit fly	*Caenorhabditis*	*Arabidopsis*	Yeast
Zinc finger, C₂H₂ type	DNA binding	564	234	68	21	34
Zinc finger, GATA type	DNA binding	11	5	8	26	9
Homeobox	Gene regulation during development	160	100	82	66	6
Death	Programmed cell death	16	5	7	0	0
Connexin	Electrical coupling between cells	14	0	0	0	0
Ephrin	Nerve cell growth	7	2	4	0	0

For more information on zinc fingers and the homeobox domain see Section 11.1.

signaling. In general, this type of analysis emphasizes the similarities between genomes, but does not reveal the genetic basis of the vastly different types of biological information contained in the genomes of, for example, fruit flies and humans. The domain approach is, however, promising in this respect because it shows that the human genome specifies a number of protein domains that are absent from the genomes of the other organisms, these domains including several involved in activities such as cell adhesion, electrical coupling between cells, and growth of nerve cells (Table 7.5). These functions are interesting because they are ones that we look on as conferring the distinctive features of vertebrates compared with other types of eukaryote.

Is it possible to identify a set of genes that are present in vertebrates but not in other eukaryotes? This analysis can only be done in an approximate way at present because only a few genome sequences are available. It currently appears that approximately one-fifth to one-quarter of the genes in the human genome are unique to vertebrates, and a further one-quarter are found only in vertebrates and other animals (Figure 7.18).

Figure 7.18 **Relationship between the human gene catalog and the catalogs of other groups of organisms.** The pie chart categorizes the human gene catalog according to the distribution of individual genes in other organisms. The chart shows, for example, that 22% of the human gene catalog is made up of genes that are specific to vertebrates, and that another 24% comprises genes specific to vertebrates and other animals.

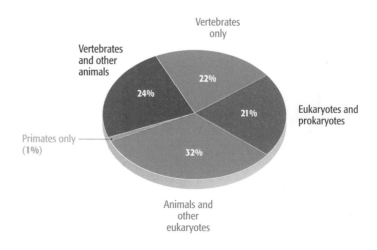

Families of genes

Since the earliest days of DNA sequencing it has been known that **multigene families**—groups of genes of identical or similar sequence—are common features of many genomes. For example, every eukaryote that has been studied (as well as all but the simplest bacteria) has multiple copies of the genes for the ribosomal RNAs. This is illustrated by the human genome, which contains approximately 2000 genes for the 5S rRNA (so-called because it has a sedimentation coefficient of 5S; see Technical Note 7.1), all located in a single cluster on chromosome 1. There are also about 280 copies of a repeat unit containing the 28S, 5.8S, and 18S rRNA genes, grouped into five clusters of 50–70 repeats, one on each of chromosomes 13, 14, 15, 21, and 22 (see Figure 7.5). Ribosomal RNAs are components of the protein-synthesizing particles called ribosomes, and it is presumed that their genes are present in multiple copies because there is a heavy demand for rRNA synthesis during cell division, when several tens of thousands of new ribosomes must be assembled.

The rRNA genes are examples of "simple" or "classical" multigene families, in which all the members have identical or nearly identical sequences. These families are believed to have arisen by gene duplication, with the sequences of the individual members kept identical by an evolutionary process that, as yet, has not been fully described (Section 18.2.1). Other multigene families, more common in higher eukaryotes than in lower eukaryotes, are called "complex" because the individual members, although similar in sequence, are sufficiently different for the gene products to have distinctive properties. One of the best examples of this type of multigene family are the mammalian globin genes. The globins are the blood proteins that combine to make hemoglobin, each molecule of hemoglobin being made up of two α-type and two β-type globins. In humans the α-type globins are coded by a small multigene family on chromosome 16, and the β-type globins by a second family on chromosome 11 (Figure 7.19). These genes were among the first to be sequenced, back in the late 1970s. The sequence data showed that the genes in each family are similar to one another, but by no means identical. In fact the nucleotide sequences of the two most different genes in the β-type cluster, coding for the β- and ε-globins, display only 79.1% identity. Although this is similar enough for both proteins to be β-type globins, it is sufficiently different for them to have distinctive biochemical properties. Similar variations are seen in the α-cluster.

Why are the members of the globin gene families so different from one another? The answer was revealed when the expression patterns of the individual genes were studied. It was discovered that the genes are expressed at different stages in human development: for example, in the β-type cluster, ε is expressed in the early embryo, G_γ and A_γ (whose protein products differ by just one amino acid) in the fetus, and δ and β in the adult (Figure 7.19). The different biochemical properties of the resulting globin proteins are thought

Figure 7.19 The human α- and β-globin gene clusters. The α-globin cluster is located on chromosome 16 and the β-cluster on chromosome 11. Both clusters contain genes that are expressed at different developmental stages and each includes at least one pseudogene. Note that expression of the α-type gene ξ_2 begins in the embryo and continues during the fetal stage; there is no fetal-specific α-type globin. The θ pseudogene is expressed but its protein product is inactive. None of the other pseudogenes is expressed. For more information on developmental regulation of the β-globin genes, see Section 10.1.2.

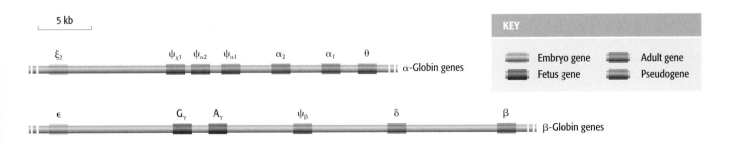

KEY

Embryo gene	Adult gene
Fetus gene	Pseudogene

to reflect slight changes in the physiological role that hemoglobin plays during the course of human development.

In some multigene families, the individual members are clustered, as with the globin genes, but in others the genes are dispersed around the genome. An example of a dispersed family is the five human genes for aldolase, an enzyme involved in energy generation, which are located on chromosomes 3, 9, 10, 16, and 17. The important point is that, even though dispersed, the members of the multigene family have sequence similarities that point to a common evolutionary origin. When these sequence comparisons are made it is sometimes possible to see relationships not only within a single gene family but also between different families. All of the genes in the α- and β-globin families, for example, have some sequence similarity and are thought to have evolved from a single ancestral globin gene. We therefore refer to these two multigene families as comprising a single globin **gene superfamily**, and from the similarities between the individual genes we can chart the duplication events that have given rise to the series of genes that we see today (Section 18.2.1).

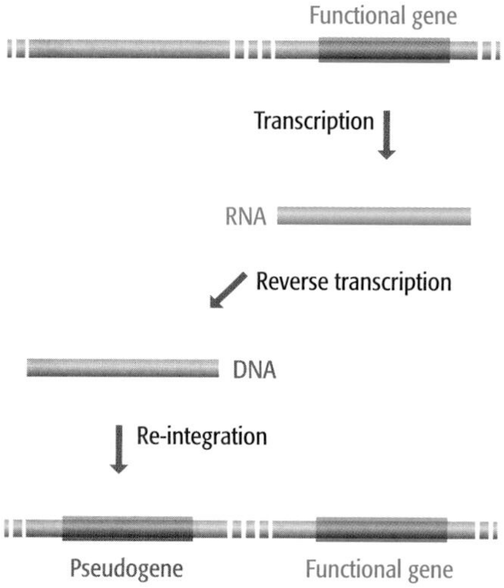

Pseudogenes and other evolutionary relics

The human globin gene clusters contain five genes that are no longer active. These are **pseudogenes**, nonfunctional gene copies. Pseudogenes are a type of evolutionary relic, an indication that genomes are continually undergoing change. There are two main types of pseudogene:

- A **conventional pseudogene** is a gene that has been inactivated because its nucleotide sequence has changed by **mutation** (Chapter 16). Many mutations have only minor effects on the activity of a gene but some are more important, and it is quite possible for a single nucleotide change to result in a gene becoming completely nonfunctional. Once a pseudogene has become nonfunctional it will degrade through accumulation of more mutations, and eventually will no longer be recognizable as a gene relic. The globin pseudogenes are examples of conventional pseudogenes.

- A **processed pseudogene** arises not by evolutionary decay but by an abnormal adjunct to gene expression. A processed pseudogene is derived from the mRNA copy of a gene by synthesis of a cDNA copy which subsequently reinserts into the genome (Figure 7.20). Because a processed pseudogene is a copy of an mRNA molecule, it does not contain any introns that were present in its parent gene. It also lacks the nucleotide sequences immediately upstream of the parent gene, which is the region in which the signals used to switch on expression of the parent gene are located. The absence of these signals means that a processed pseudogene is inactive.

As well as pseudogenes, genomes also contain other evolutionary relics in the form of **truncated genes**, which lack a greater or lesser stretch from one end of the complete gene, and **gene fragments**, which are short, isolated regions from within a gene (Figure 7.21).

7.2.4 The repetitive DNA content of eukaryotic nuclear genomes

The human genome sequence revealed that approximately 62% of the human genome comprises **intergenic regions**, the parts of the genome that lie between genes and which have no known function. These sequences used to be called **junk DNA** but the term is falling out of favor, partly because the number of surprises resulting from genome research over the last few years

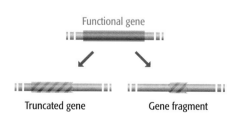

Figure 7.20 The origin of a processed pseudogene. A processed pseudogene is thought to arise by integration into the genome of a copy of the mRNA transcribed from a functional gene. The mRNA is reverse transcribed into a cDNA copy, which might integrate into the same chromosome as its functional parent, or possibly into a different chromosome.

Figure 7.21 A truncated gene and a gene fragment.

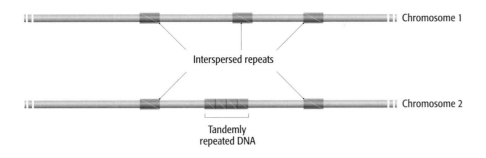

Figure 7.22 The two types of repetitive DNA: interspersed repeats and tandemly repeated DNA.

has meant that molecular biologists have become less confident in asserting that any part of the genome is unimportant simply because we do not currently know what its function might be. As we have seen, in most organisms the bulk of the intergenic DNA is made up of repeated sequences of one type or another. Repetitive DNA can be divided into two categories (Figure 7.22): genome-wide or **interspersed repeats**, whose individual repeat units are distributed around the genome in an apparently random fashion, and **tandemly repeated DNA**, whose repeat units are placed next to each other in an array.

Tandemly repeated DNA is found at centromeres and elsewhere in eukaryotic chromosomes

Tandemly repeated DNA is also called **satellite DNA** because DNA fragments containing tandemly repeated sequences form "satellite" bands when genomic DNA is fractionated by density gradient centrifugation (see Technical Note 7.1). For example, when broken into fragments 50–100 kb in length, human DNA forms a main band (buoyant density 1.701 g cm^{-3}) and three satellite bands (1.687, 1.693, and 1.697 g cm^{-3}). The main band contains DNA fragments made up mostly of single-copy sequences with GC compositions close to 40.3%, the average value for the human genome. The satellite bands contain fragments of repetitive DNA, and hence have GC contents and buoyant densities that are atypical of the genome as a whole (Figure 7.23). This repetitive DNA is made up of long series of tandem repeats, possibly hundreds of kilobases in length. A single genome can contain several different types of satellite DNA, each with a different repeat unit, these units being anything from less than 5 to more than 200 bp in length. The three satellite bands in human DNA include at least four different repeat types.

We have already encountered one type of human satellite DNA, the alphoid DNA repeats found in the centromere regions of chromosomes (Section 7.1.2). Although some satellite DNA is scattered around the genome, most is located in the centromeres, where it may play a structural role, possibly as binding sites for one or more of the special centromeric proteins.

Minisatellites and microsatellites

Although not appearing in satellite bands on density gradients, two other types of tandemly repeated DNA are also classed as "satellite" DNA. These are **minisatellites** and **microsatellites**. Minisatellites form clusters up to 20 kb in length, with repeat units up to 25 bp in length; microsatellite clusters are shorter, usually less than 150 bp, and the repeat unit is usually 13 bp or less.

Minisatellite DNA is a second type of repetitive DNA that we are already familiar with because of its association with structural features of chromosomes. Telomeric DNA, which in humans comprises hundreds of copies of the motif 5′–TTAGGG–3′ (see Figure 7.10), is an example of a minisatellite. We

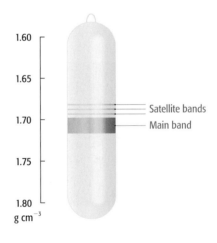

Figure 7.23 Satellite DNA from the human genome. Human DNA has an average GC content of 40.3% and average buoyant density of 1.701 g cm^{-3}. Fragments made up mainly of single-copy DNA have a GC content close to this average and are contained in the main band in the density gradient. The satellite bands at 1.687, 1.693, and 1.697 g cm^{-3} consist of fragments containing repetitive DNA. The GC contents of these fragments depend on their repeat motif sequences and are different from the genome average, meaning that these fragments have different buoyant densities to single-copy DNA and migrate to different positions in the density gradient.

know a certain amount about how telomeric DNA is formed, and we know that it has an important function in DNA replication (Section 15.2.4). In addition to telomeric minisatellites, some eukaryotic genomes contain various other clusters of minisatellite DNA, many, although not all, near the ends of chromosomes. The functions of these other minisatellite sequences have not been identified.

Microsatellites are also examples of tandemly repeated DNA. In a microsatellite the repeat unit is short—up to 13 bp in length. The commonest type of human microsatellite are dinucleotide repeats, with approximately 140,000 copies in the genome as a whole, about half of these being repeats of the motif "CA." Single-nucleotide repeats (e.g., AAAAA) are the next most common (about 120,000 copies in total). As with genome-wide repeats, it is not clear if microsatellites have a function. It is known that they arise through an error in the process responsible for copying of the genome during cell division (Section 16.1.1), and they might simply be unavoidable products of genome replication.

Although their function, if any, is unknown, microsatellites have proved very useful to geneticists. Many microsatellites are variable, meaning that the number of repeat units in the array is different in different members of a species. This is because "slippage" sometimes occurs when a microsatellite is copied during DNA replication, leading to insertion or, less frequently, deletion of one or more of the repeat units (see Figure 16.5). No two humans alive today have exactly the same combination of microsatellite length variants: if enough microsatellites are examined then a unique **genetic profile** can be established for every person. The only exceptions are genetically identical twins. Genetic profiling is well known as a tool in forensic science (Figure 7.24), but identification of criminals is a fairly trivial application of microsatellite variability. More sophisticated methodology makes use of the fact that a person's genetic profile is inherited partly from the mother and partly from the father. This means that microsatellites can be used to establish kinship relationships and population affinities, not only for humans but also for other animals, and for plants.

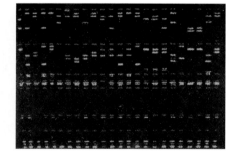

Figure 7.24 The use of microsatellite analysis in genetic profiling. In this example, microsatellites located on the short arm of chromosome 6 have been amplified by PCR. The PCR products are labeled with a blue or green fluorescent marker and run in a polyacrylamide gel, each lane showing the genetic profile of a different individual. No two individuals have the same genetic profile because each person has a different set of microsatellite length variants, the variants giving rise to bands of different sizes after PCR. The red bands are DNA size markers. Image courtesy of Applied Biosystems, Warrington, UK.

Interspersed repeats

Tandemly repeated DNA sequences are thought to have arisen by expansion of a progenitor sequence, either by replication slippage, as described for microsatellites, or by DNA recombination processes (Chapter 17). Both of these events are likely to result in a series of linked repeats, rather than individual repeat units scattered around the genome. Interspersed repeats must therefore have arisen by a different mechanism, one that can result in a copy of a repeat unit appearing in the genome at a position distant from the location of the original sequence. The most frequent way in which this occurs is by **transposition**, and most interspersed repeats have inherent transpositional activity. Transposition is also a feature of some viral genomes, which are able to insert into the genome of the infected cell and then move from place to place within that genome. Some interspersed repeats are clearly descended from transposable viruses, and because of this relationship we will postpone discussion of these and the other types of genome-wide repeat until Chapter 9, after we have looked in detail at the features of virus genomes.

Summary

The eukaryotic nuclear genome is split into a set of linear DNA molecules, each of which is contained in a chromosome. Within a chromosome, the DNA is packaged by association with histone proteins to form nucleosomes, which interact with one another to give the 30 nm fiber and higher orders of chromatin structure. The most compact organization results in the metaphase chromosomes that can be observed by light microscopy of dividing cells and which take up characteristic banding patterns after staining. The centromeres, which are visible in metaphase chromosomes, contain special proteins that make up the kinetochore, the attachment point for the microtubules which draw the divided chromosomes into the daughter nuclei. In *S. cerevisiae*, the centromeric DNA which acts as the binding site for these proteins is approximately 125 bp in length, but in most other eukaryotes this DNA region is much longer and is made up of repetitive DNA. Telomeres, the structures which maintain the chromosome ends, also contain repetitive DNA and special binding proteins. Genes are not evenly spread along chromosomes, the density in human chromosomes ranging from 0 to 64 genes per 100 kb. The coding parts of genes comprise only a small part of the human genome, less than 1.5%, with 44% of the genome made up of various types of repetitive DNA sequence. In contrast, the *S. cerevisiae* genome is much more compact, with only 3.4% taken up by repeat sequences. In general, larger genomes are less compact, explaining why organisms with similar numbers of genes can have genomes of very different sizes. Humans have 30,000–40,000 genes, about twice as many as the nematode worm *Caenorhabditis elegans* and about the same number as rice. Comparisons of gene catalogs listing the functions of the genes in a genome suggest that all eukaryotes possess the same basic set of genes, but that more complex species have a greater number of genes in each functional category. Many genes are organized into multigene families whose members have similar or identical sequences, and in some families, such as the vertebrate globin genes, the members are expressed at different developmental stages. Eukaryotic nuclear genomes also contain evolutionary relics, such as non-functional pseudogenes and gene fragments. The repetitive DNA content can be divided into interspersed DNA, much of which has transpositional activity, and tandemly repeated DNA, which includes the satellite DNA found at centromeres, minisatellites such as telomeric DNA, and microsatellites which are used by forensic scientists in genetic profiling.

Multiple Choice Questions

7.1.* The proteins that bind to DNA in the nucleosome and form a core octamer are called:

 a. Histidines.

 b. Histones.

 c. Chromatin.

 d. Chromatosome.

7.2. How is DNA thought to be packed during interphase?

 a. In single nucleosomes as seen in the "beads-on-a-string" images.

 b. In the 30 nm fiber.

 c. In a highly condensed state, visible using a light microscope.

 d. DNA is not packaged or associated with nucleosomes during interphase.

7.3.* What is the centromere of a chromosome?

 a. It is the end of the chromosome.

 b. It is the uncondensed region of a chromosome that contains active genes.

 c. It is the constricted region of a chromosome where the two copies are held together.

 d. It is the condensed, transcriptionally silent regions of chromosomes.

7.4. Holocentric chromosomes are which of the following?

 a. Chromosomes with multiple centromeres.

 b. Additional chromosomes possessed by some individuals in a population.

 c. Short chromosomes with many genes, as found in chicken.

 d. Circular chromosomes found in some lower eukaryotes.

7.5.* Why are centromeres often not included in a draft genome sequence?

 a. It is extremely difficult to clone this DNA because it is very condensed.

 b. Researchers are not interested in sequencing DNA regions that lack genes.

 c. Centromeres have the same sequences in all organisms.

 d. It is difficult to get an accurate sequence for these long regions of repetitive DNA.

7.6. What have scientists observed about the distribution of genes in eukaryotic genomes?

 a. Genes are evenly distributed throughout eukaryotic genomes.

 b. Genes are distributed at specific locations in eukaryotic genomes.

 c. There are always at least 10 genes per 100 kb in eukaryotic genomes.

 d. Genes appear to be randomly distributed throughout genomes and their density varies greatly.

7.7.* The yeast genome is 0.004 times the size of the human genome and yet it contains approximately 0.2 times fewer genes. The explanation for this is:

 a. The genes in yeast contain far fewer codons compared to human genes.

 b. Yeast chromosomes contain much smaller centromeres and telomeres.

 c. The yeast genome contains much less intergenic DNA and fewer introns.

 d. The yeast genome contains many overlapping genes.

7.8. What is the C-value paradox?

 a. The lack of correlation between the complexity of an organism and its genome size.

 b. The lack of correlation between the complexity of an organism and its number of chromosomes.

 c. The lack of correlation between the complexity of an organism and its number of genes.

 d. The lack of correlation between the number of genes and the number of chromosomes in organisms.

7.9.* Which of the following is an example of a protein domain?

 a. β-sheet.

 b. Zinc finger.

 c. Exon.

 d. Globin protein.

7.10. What do classification methods based on gene function tell researchers about different eukaryotic organisms?

 a. All eukaryotic organisms have the same numbers of genes in each functional category; complex organisms have a larger number of unknown genes.

 b. Complex organisms have greater numbers of genes in each functional category.

 c. Simpler organisms contain many fewer types of gene when compared to complex organisms.

 d. All eukaryotic organisms have approximately the same number of genes.

7.11.* The classification of genes based on protein domains reveals what about the human genome?

 a. The human genome contains no protein domains that are unique to humans.

 b. The human genome contains a small number of protein domains that are unique to vertebrates.

Multiple Choice Questions (continued)

    c. The human genome contains many protein domains that are unique to humans.

    d. The protein domains for the human genome are unique to humans and not present in other organisms.

7.12. Which of the following are NOT characteristics of the ribosomal RNA multigene families in the human genome?

    a. The gene families for each ribosomal subunit are present throughout the genome on every chromosome.

    b. The different members of the gene families all have identical or nearly identical sequences.

    c. These gene families are thought to have arisen by gene duplication.

    d. There are thought to be large numbers of these genes due to the need for new ribosomes during cell division.

7.13. What is a pseudogene?

    a. A gene that is only expressed at certain developmental stages.

    b. A nonfunctional gene.

    c. A gene that contains a mutation but is still functional.

    d. A sequence of DNA that is slowly evolving to become an active gene.

7.14. What region of a eukaryotic chromosome contains the highest density of genes?

    a. Centromere.

    b. Condensed heterochromatin.

    c. Euchromatin.

    d. Telomere.

Short Answer Questions

7.1. What does the treatment of eukaryotic chromatin with nucleases reveal about the packaging of eukaryotic DNA?

7.2. What is known about the 30 nm fiber form of chromatin? What is known about the packing of nucleosomes in this fiber?

7.3. How do minichromosomes differ from macrochromosomes?

7.4. What did researchers find when they sequenced the centromeres of *Arabidopsis*? Why was this finding surprising?

7.5. Why is it important that chromosomes have telomeres at their ends?

7.6. Prior to completion of the sequence, what were the two observations that led researchers to conclude that genes are distributed unevenly in the human genome?

7.7. What differences in gene distribution and repetitive DNA content are seen when yeast and human chromosomes are compared?

7.8. The human genome contains about 50,000 fewer genes than was predicted by many researchers. Why were these initial predictions so high?

7.9. What are the different methods used to catalog genes? What are the advantages or disadvantages of these methods?

7.10. What is the function of the different genes in the human globin gene families?

7.11. What is the difference between a conventional pseudogene and a processed pseudogene?

7.12. What types of repetitive DNA are present in the human genome?

continued …

In-depth Problems

*Guidance to odd-numbered questions can be found in the Appendix

7.1.* What impact is DNA packaging likely to have on the expression of individual genes?

7.2. Defend or attack the isochore model.

7.3.* Discuss possible functions for the intergenic component of the human genome.

7.4. To what extent is it possible to describe the "typical" features of a eukaryotic genome?

Figure Tests

*Answers to odd-numbered questions can be found in the Appendix

7.1.* What does this picture represent? How are the different chromosomes distinguished?

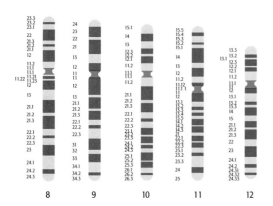

7.3.* What type of pseudogene is shown in this diagram? Why is the newly integrated copy of the gene not functional?

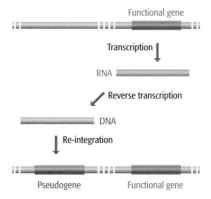

7.2. This is a diagram of the yeast centromere and associated proteins. What is the function of these sequences and proteins?

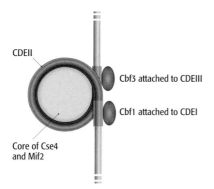

7.4. This figure shows the results of a genetic profiling study, such as might be used in a criminal or paternity case. What types of sequences are examined in DNA profiling and why are they useful for this purpose?

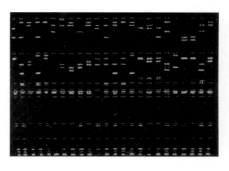

Further Reading

Key primary genome sequence publications, including initial gene catalogs

Adams, M.A., Celniker, S.E., Holt, R.A., *et al.* (2000) The genome sequence of *Drosophila melanogaster*. *Science* **287:** 2185–2195.

AGI (The Arabidopsis Genome Initiative) (2000) Analysis of the genome sequence of the flowering plant *Arabidopsis thaliana*. *Nature* **408:** 796–815.

CESC (The *C. elegans* Sequencing Consortium) (1998) Genome sequence of the nematode *C. elegans*: a platform for investigating biology. *Science* **282:** 2012–2018.

ICGSC (International Chicken Genome Sequencing Consortium) (2004) Sequence and comparative analysis of the chicken genome provide unique perspectives on vertebrate evolution. *Nature* **432:** 695–716.

IHGSC (International Human Genome Sequencing Consortium) (2001) Initial sequencing and analysis of the human genome. *Nature* **409:** 860–921.

Venter, J.C., Adams, M.D., Myers, E.W., *et al.* (2001) The sequence of the human genome. *Science* **291:** 1304–1351.

Chromosome structure

Cleveland, D.W., Mao, Y. and Sullivan, K.F. (2003) Centromeres and kinetochores: from epigenetics to mitotic checkpoint signalling. *Cell* **112:** 407–423. *Describes the DNA–protein interactions in the yeast and mammalian centromeres.*

Copenhaver, G.P., Nickel, K., Kuromori, T., *et al.* (1999) Genetic definition and sequence analysis of *Arabidopsis* centromeres. *Science* **286:** 2468–2474.

Dorigo, B., Schalch, T., Kulangara, A., Duda, S., Schroeder, R.R. and Richmond, T.J. (2004) Nucleosome arrays reveal the two-start organization of the chromatin fiber. *Science* **306:** 1571–1573. *New models of the 30 nm fiber.*

Ramakrishnan, V. (1997) Histone H1 and chromatin higher-order structure. *Crit. Rev. Eukaryot. Gene Expr.* **7:** 215–230. *Detailed descriptions of models for the 30 nm chromatin fiber.*

Schueler, M.G., Higgins, A.W., Rudd, M.K., Gustashaw, K. and Willard, H.W. (2001) Genomic and genetic definition of a functional human centromere. *Science* **294:** 109–115. *Details of the sequence features of human centromeres.*

Travers, A. (1999) The location of the linker histone on the nucleosome. *Trends Biochem. Sci.* **24:** 4–7.

van Steensel, B., Smogorzewska, A. and de Lange, T. (1998) TRF2 protects human telomeres from end-to-end fusions. *Cell* **92:** 401–413.

Genetic features

Balakirev, E.S. and Ayala, F.J. (2003) Pseudogenes: are they "junk" or functional DNA? *Annu. Rev. Biochem.* **37:** 123–151.

Csink, A.K. and Henikoff, S. (1998) Something from nothing: the evolution and utility of satellite repeats. *Trends Genet.* **14:** 200–204.

Fritsch, E.F., Lawn, R.M. and Maniatis, T. (1980) Molecular cloning and characterization of the human α-like globin gene cluster. *Cell* **19:** 959–972.

Gardiner, K. (1996) Base composition and gene distribution: critical patterns in mammalian genome organization. *Trends Genet.* **12:** 519–524. *The isochore model.*

Petrov, D.A. (2001) Evolution of genome size: new approaches to an old problem. *Trends Genet.* **17:** 23–28. *Reviews the C-value paradox and the genetic processes that might result in differences in genome size.*

Genomes of Prokaryotes and Eukaryotic Organelles

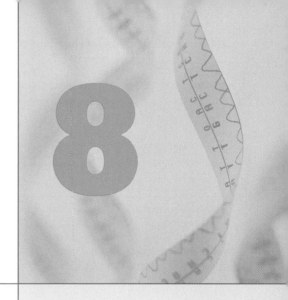

8

When you have read Chapter 8, you should be able to:

Describe how bacterial DNA is packaged into a nucleoid, and give the experimental evidence for the domain model of the *Escherichia coli* nucleoid.

Give examples of prokaryotic genomes that are linear and/or multipartite, and explain why the presence of plasmids in some prokaryotic cells complicates definition of what constitutes the "genome."

Outline the important features of gene organization in prokaryotic genomes.

Define, with examples, the term "operon."

Discuss the relationship between gene number and genome size in prokaryotes, and speculate on the content of the minimal prokaryotic genome and on the identity of distinctiveness genes.

Discuss the endosymbiont hypothesis for the origins of organelle genomes.

Describe the physical features and gene contents of mitochondrial and chloroplast genomes.

Prokaryotes are organisms whose cells lack extensive internal compartments. There are two very different groups of prokaryotes, distinguished from one another by characteristic genetic and biochemical features:

- The **bacteria**, which include most of the commonly encountered prokaryotes such as the gram-negatives (e.g., *Escherichia coli*), the gram-positives (e.g., *Bacillus subtilis*), the cyanobacteria (e.g., *Anabaena*), and many more.

- The **archaea**, which are less-well studied, and have mostly been found in extreme environments such as hot springs, brine pools, and anaerobic lake bottoms.

In this chapter we will examine the genomes of prokaryotes, and also of eukaryotic mitochondria and chloroplasts which, as they are descended from bacteria, have genomes that display many prokaryotic features. Because of the relatively small sizes of prokaryotic genomes, several hundred complete genome sequences for various bacteria and archaea have been published

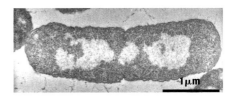

Figure 8.1 The *Escherichia coli* nucleoid. This transmission electron micrograph shows the cross section of a dividing *E. coli* cell. The nucleoid is the lightly staining area in the center of the cell. Image courtesy of Conrad Woldringh.

over the last few years. As a result, we are beginning to understand a great deal about the anatomies of prokaryotic genomes, and in some respects we know more about these organisms than we do about eukaryotes. The picture that is emerging is one of immense variability among the prokaryotes as a whole and in some cases even between closely related species.

8.1 The Physical Features of Prokaryotic Genomes

Prokaryotic genomes are very different from eukaryotic ones, in particular with regard to the physical organization of the genome within the cell. Although the word "chromosome" is used to describe the DNA–protein structures present in prokaryotic cells, this is a misnomer as this structure has few similarities with a eukaryotic chromosome.

8.1.1 The chromosomes of prokaryotes

The traditional view has been that in a typical prokaryote the genome is contained in a single, circular DNA molecule, localized within the **nucleoid**—the lightly staining region of the otherwise featureless prokaryotic cell (Figure 8.1). This is certainly true for *E. coli* and many of the other commonly studied bacteria. However, as we will see, our growing knowledge of prokaryotic genomes is leading us to question several of the preconceptions that became established during the pregenome era of microbiology. These preconceptions relate both to the physical structure of the prokaryotic genome and its genetic organization.

The traditional view of the prokaryotic chromosome

As with eukaryotic chromosomes, a prokaryotic genome has to squeeze into a relatively tiny space (the circular *E. coli* chromosome has a circumference of 1.6 mm whereas an *E. coli* cell is just 1.0×2.0 μm) and, as with eukaryotes, this is achieved with the help of DNA-binding proteins that package the genome in an organized fashion.

Most of what we know about the organization of DNA in the nucleoid comes from studies of *E. coli*. The first feature to be recognized was that the circular *E. coli* genome is **supercoiled**. Supercoiling occurs when additional turns are introduced into the DNA double helix (positive supercoiling) or if turns are removed (negative supercoiling). With a linear molecule, the torsional stress introduced by over- or underwinding is immediately released by rotation of the ends of the DNA molecule, but a circular molecule, having no ends, cannot reduce the strain in this way. Instead the circular molecule responds by

Figure 8.2 Supercoiling. The diagram shows how underwinding a circular, double-stranded DNA molecule results in negative supercoiling.

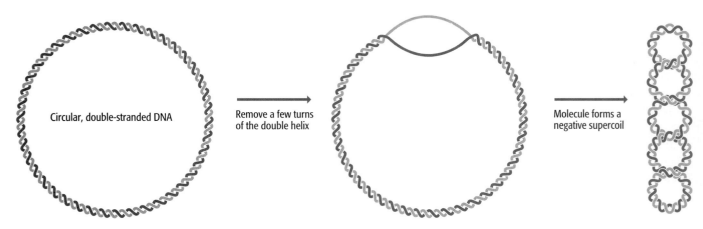

Circular, double-stranded DNA

Remove a few turns of the double helix

Molecule forms a negative supercoil

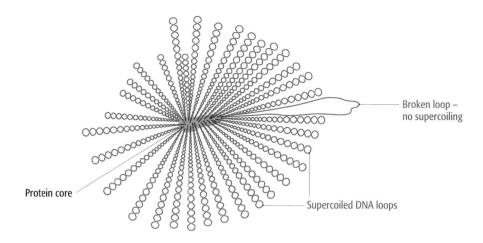

Protein core

Broken loop –
no supercoiling

Supercoiled DNA loops

Figure 8.3 A model for the structure of the *Escherichia coli* nucleoid. Between 40 and 50 supercoiled loops of DNA radiate from the central protein core. One of the loops is shown in circular form, indicating that a break has occurred in this segment of DNA, resulting in a loss of the supercoiling.

winding around itself to form a more compact structure (Figure 8.2). Supercoiling is therefore an ideal way to package a circular molecule into a small space. Evidence that supercoiling is involved in packaging the circular *E. coli* genome was first obtained in the 1970s from examination of isolated nucleoids, and subsequently confirmed as a feature of DNA in living cells in 1981. In *E. coli*, the supercoiling is thought to be generated and controlled by two enzymes, DNA gyrase and DNA topoisomerase I, which we will look at in more detail in Section 15.1.2 when we examine the roles of these enzymes in DNA replication.

Studies of isolated nucleoids suggest that the *E. coli* DNA molecule does not have unlimited freedom to rotate once a break is introduced. The most likely explanation is that the bacterial DNA is attached to proteins that restrict its ability to relax, so that rotation at a break site results in loss of supercoiling from only a small segment of the molecule (Figure 8.3). The strongest evidence for this domain model has come from experiments that exploit the ability of trimethylpsoralen to distinguish between supercoiled and relaxed DNA. When photoactivated by a pulse of light of wavelength 360 nm, trimethylpsoralen binds to double-stranded DNA at a rate that is directly proportional to the degree of torsional stress possessed by the molecule. The degree of supercoiling can therefore be assayed by measuring the amount of trimethylpsoralen that binds to a molecule in unit time. After *E. coli* cells have been irradiated to introduce single-strand breaks into their DNA molecules, the amount of trimethylpsoralen binding is proportional to the radiation dose (Figure 8.4). This is the response predicted by the domain model, in which the overall supercoiling of the molecule is gradually relaxed as greater doses of radiation cause breaks within an increasing number of domains. In contrast, if the *E. coli* nucleoid was not organized into domains then a single break in the DNA molecule would lead to complete loss of supercoiling: irradiation would therefore have an all-or-nothing effect on trimethylpsoralen binding.

The current model has the *E. coli* DNA attached to a protein core from which 40–50 supercoiled loops radiate out into the cell. Each loop contains approximately 100 kb of supercoiled DNA, the amount of DNA that becomes unwound after a single break. The protein component of the nucleoid includes DNA gyrase and DNA topoisomerase I, the two enzymes that are primarily responsible for maintaining the supercoiled state, as well as a set of at least four proteins believed to have a more specific role in packaging the bacterial DNA. The most abundant of these packaging proteins is HU, which

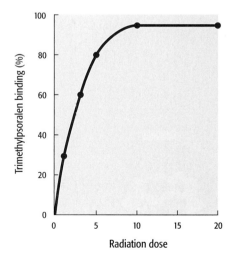

Figure 8.4 Graph showing the relationship between radiation dose and trimethylpsoralen binding.

is structurally very different to eukaryotic histones but acts in a similar way, forming a tetramer around which approximately 60 bp of DNA becomes wound. There are some 60,000 HU proteins per *E. coli* cell, enough to cover about one-fifth of the DNA molecule, but it is not known if the tetramers are evenly spaced along the DNA or restricted to the core region of the nucleoid.

The above discussion refers specifically to the *E. coli* chromosome, which we look on as typical of bacterial chromosomes in general. But we must be careful to make a distinction between the bacterial chromosome and that of the second group of prokaryotes, the archaea. One reason why the archaea are looked upon as a distinct group of organisms, different from the bacteria, is that archaea do not possess packaging proteins such as HU but instead have proteins that are much more similar to histones. These form a tetramer that associates with approximately 80 bp of DNA to form a structure similar to a eukaryotic nucleosome (see Figure 7.2). Currently we have very little information on the archaeal nucleoid, but the assumption is that these histone-like proteins play a central role in DNA packaging.

Some bacteria have linear or multipartite genomes

The *E. coli* genome, as described above, is a single, circular DNA molecule. This is also the case with the vast majority of bacterial and archaeal chromosomes that have been studied, but an increasing number of linear versions are being found. The first of these, for *Borrelia burgdorferi*, the organism that causes Lyme disease, was described in 1989, and during the following years similar discoveries were made for *Streptomyces coelicolor* and *Agrobacterium tumefaciens*. Linear molecules have free ends, which must be distinguishable from DNA breaks, so these chromosomes require terminal structures equivalent to the telomeres of eukaryotic chromosomes (Section 7.1.2). In *Borrelia* and *Agrobacterium*, the real chromosome ends are distinguishable because a covalent linkage is formed between the 5′ and 3′ ends of the polynucleotides in the DNA double helix, and in *Streptomyces* the ends appear to be marked by special binding proteins.

A second and more widespread variation on the *E. coli* theme is the presence in some prokaryotes of multipartite genomes—genomes that are divided into two or more DNA molecules. With these multipartite genomes a problem often arises in distinguishing a genuine part of the genome from a plasmid. A plasmid is a small piece of DNA, often but not always circular, that coexists with the main chromosome in a bacterial cell (Figure 8.5). Some types of plasmid are able to integrate into the main genome, but others are thought to be permanently independent. Plasmids carry genes that are not usually present in the main chromosome, but in many cases these genes are nonessential to the bacterium, coding for characteristics such as antibiotic resistance, which the bacterium does not need if the environmental conditions are amenable (Table 8.1). As well as this apparent dispensability, many plasmids are able to transfer from one cell to another, and the same plasmids are sometimes found in bacteria that belong to different species. These various features of plasmids suggest that they are independent entities and that in most cases the plasmid content of a prokaryotic cell should not be included in the definition of its genome.

With a bacterium such as *E. coli* K12, which has a 4.6 Mb chromosome and can harbor various combinations of plasmids, none of which is more than a few kilobases in size and all of which are dispensable, it is acceptable to define the main chromosome as the "genome." With other prokaryotes it is

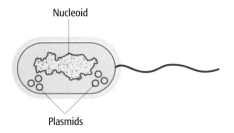

Nucleoid

Plasmids

Figure 8.5 Plasmids are small, circular DNA molecules that are found inside some prokaryotic cells.

Table 8.1 Features of typical plasmids

Type of plasmid	Gene functions	Examples
Resistance	Antibiotic resistance	Rbk of *Escherichia coli* and other bacteria
Fertility	Conjugation and DNA transfer between bacteria	F of *E. coli*
Killer	Synthesis of toxins that kill other bacteria	Col of *E. coli*, for colicin production
Degradative	Enzymes for metabolism of unusual molecules	TOL of *Pseudomonas putida*, for toluene metabolism
Virulence	Pathogenicity	Ti of *Agrobacterium tumefaciens*, conferring the ability to cause crown gall disease on dicotyledonous plants

Table 8.2 Examples of genome organization in prokaryotes

Species	DNA molecules	Genome organization Size (Mb)	Number of genes
Escherichia coli K12	One circular molecule	4.639	4405
Vibrio cholerae El Tor N16961	Two circular molecules		
	Main chromosome	2.961	2770
	Megaplasmid	1.073	1115
Deinococcus radiodurans R1	Four circular molecules		
	Chromosome 1	2.649	2633
	Chromosome 2	0.412	369
	Megaplasmid	0.177	145
	Plasmid	0.046	40
Borrelia burgdorferi B31	Seven or eight circular molecules, eleven linear molecules		
	Linear chromosome	0.911	853
	Circular plasmid cp9	0.009	12
	Circular plasmid cp26	0.026	29
	Circular plasmid cp32*	0.032	Not known
	Linear plasmid lp17	0.017	25
	Linear plasmid lp25	0.024	32
	Linear plasmid lp28-1	0.027	32
	Linear plasmid lp28-2	0.030	34
	Linear plasmid lp28-3	0.029	41
	Linear plasmid lp28-4	0.027	43
	Linear plasmid lp36	0.037	54
	Linear plasmid lp38	0.039	52
	Linear plasmid lp54	0.054	76
	Linear plasmid lp56	0.056	Not known

* There are five or six similar versions of plasmid cp32 per bacterium.

not so easy (Table 8.2). *Vibrio cholerae*, the pathogenic bacterium that causes cholera, has two circular DNA molecules, one of 2.96 Mb and the other of 1.07 Mb, with 71% of the organism's 3885 genes on the larger of these. It would appear obvious that these two DNA molecules together constitute the *Vibrio* genome, but closer examination reveals that most of the genes for the central cellular activities such as genome expression and energy generation, as well as the genes that confer pathogenicity, are located on the larger molecule. The smaller molecule contains many essential genes but also has certain features that are considered characteristic of plasmids, notably the presence of an **integron**—a set of genes and other DNA sequences that enable plasmids to capture genes from bacteriophages and other plasmids. It therefore appears possible that the smaller genome is a "megaplasmid" that was acquired by the ancestor to *Vibrio* at some period in the bacterium's evolutionary past. *Deinococcus radiodurans* R1, whose genome is of particular interest because it contains many genes that help this bacterium resist the harmful effects of radiation, is constructed on similar lines, with essential genes distributed among two circular chromosomes and two plasmids. However, the *Vibrio* and *Deinococcus* genomes are relatively noncomplex compared with *Borrelia burgdorferi* B31, whose linear chromosome of 911 kb, carrying 853 genes, is accompanied by 17 or 18 linear and circular plasmids, which together contribute another 533 kb and at least 430 genes. Although the functions of most of these genes are unknown, those that have been identified include several that would not normally be considered dispensable, such as genes for membrane proteins and purine biosynthesis. The implication is that at least some of the *Borrelia* plasmids are essential components of the genome, leading to the possibility that some prokaryotes have highly multipartite genomes, comprising a number of separate DNA molecules, more akin to what we see in the eukaryotic nucleus rather than the "typical" prokaryotic arrangement. This interpretation of the *Borrelia* genome is still controversial, and is complicated by the fact that the related bacterium *Treponema pallidum*, whose genome is a single, circular DNA molecule of 1138 kb containing 1041 genes, does not contain any of the genes present on the *Borrelia* plasmids.

8.2 The Genetic Features of Prokaryotic Genomes

Gene location by sequence inspection is much easier for prokaryotes compared with eukaryotes (Section 5.1.1), and for most of the prokaryotic genomes that have been sequenced we have reasonably accurate estimates of gene number and fairly comprehensive lists of gene functions. The results of these studies have been surprising, and have forced microbiologists to reconsider the meaning of "species" when applied to prokaryotes. We will examine these evolutionary issues in Section 8.2.3. First, we must look at the way in which the genes are organized in a prokaryotic genome.

8.2.1 How are the genes organized in a prokaryotic genome?

We are already familiar with the notion that bacterial genomes have compact genetic organizations with very little space between genes, as this was an important part of our discussion of the strengths and weaknesses of ORF scanning as a means of identifying the genes in a genome sequence (see Figure 5.3). To reemphasize this point, the complete circular gene map of the *E. coli* K12 genome is shown in Figure 8.6. There *is* noncoding DNA in the *E. coli* genome, but it accounts for only 11% of the total and it is distributed

Origin of replication

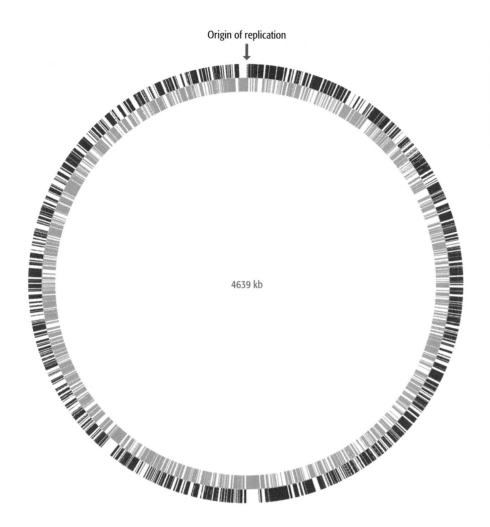

4639 kb

Figure 8.6 The genome of *Escherichia coli* **K12.** The map is shown with the origin of replication (Section 15.2.1) positioned at the top. Genes on the outside of the circle are transcribed in the clockwise direction and those on the inside are transcribed in the anticlockwise direction. Image courtesy of Dr F. R. Blattner.

around the genome in small segments that do not show up when the map is drawn at this scale. In this regard, *E. coli* is typical of all prokaryotes whose genomes have so far been sequenced—prokaryotic genomes have very little wasted space. There are theories that this compact organization is beneficial to prokaryotes, for example by enabling the genome to be replicated relatively quickly, but these ideas have never been supported by hard experimental evidence.

Gene organization in the E. coli genome

Let us look more closely at the *E. coli* genome. A typical 50 kb segment is shown in Figure 8.7. When we compare this segment with a typical part of the human genome (see Figure 7.12), it is immediately obvious that in the *E. coli* segment there are more genes and much less space between them, with 43 genes taking up 85.9% of the segment. Some genes have virtually no space between them: *thrA* and *thrB*, for example, are separated by a single nucleotide, and *thrC* begins at the nucleotide immediately following the last nucleotide of *thrB*. These three genes are an example of an **operon**, a group of genes involved in a single biochemical pathway (in this case, synthesis of the amino acid threonine) and expressed in conjunction with one another. In general, prokaryotic genes are shorter than their eukaryotic counterparts, the average length of a bacterial gene being about two-thirds that of a eukaryotic gene, even after the introns have been removed from the latter. Bacterial genes appear to be slightly longer than archaeal ones.

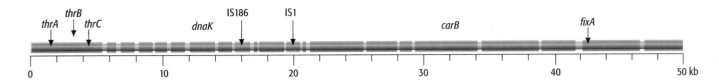

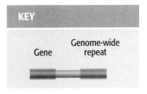

Figure 8.7 A 50 kb segment of the *Escherichia coli* genome.

Two other features of prokaryotic genomes can be deduced from Figure 8.7. First, there are no introns in the genes present in this segment of the *E. coli* genome. In fact *E. coli* has no discontinuous genes at all, and it is generally believed that discontinuous genes are virtually absent in prokaryotes, the few exceptions occurring mainly among the archaea. The second feature is the infrequency of repetitive sequences. Most prokaryotic genomes do not have anything equivalent to the high-copy-number genome-wide repeat families found in eukaryotic genomes. They do, however, possess certain sequences that might be repeated elsewhere in the genome, examples being the **insertion sequences** IS1 and IS186 that can be seen in the 50 kb segment shown in Figure 8.7. These are examples of transposable elements, sequences that have the ability to move around the genome and, in the case of insertion elements, to transfer from one organism to another, even sometimes between two different species (Section 9.2.2). The positions of the IS1 and IS186 elements shown in Figure 8.7 refer only to the particular *E. coli* isolate from which this sequence was obtained: if a different isolate is examined then the IS sequences could well be in different positions or might be entirely absent from the genome. Most other prokaryotic genomes have very few repeat sequences—there are virtually none in the 1.64 Mb genome of *Campylobacter jejuni* NCTC11168—but there are exceptions, notably the meningitis bacterium *Neisseria meningitidis* Z2491, which has over 3700 copies of 15 different types of repeat sequence, collectively making up almost 11% of the 2.18 Mb genome.

Operons are characteristic features of prokaryotic genomes

One characteristic feature of prokaryotic genomes illustrated by *E. coli* is the presence of operons. In the years before genome sequences were known, it was thought that we understood operons very well; now we are not so sure.

An operon is a group of genes that are located adjacent to one another in the genome, with perhaps just one or two nucleotides between the end of one gene and the start of the next. All the genes in an operon are expressed as a single unit. This type of arrangement is common in prokaryotic genomes. A typical *E. coli* example is the **lactose operon**, the first operon to be discovered, which contains three genes involved in conversion of the disaccharide sugar lactose into its monosaccharide units—glucose and galactose (Figure 8.8A). The monosaccharides are substrates for the energy-generating glycolytic pathway, so the function of the genes in the lactose operon is to convert lactose into a form that can be utilized by *E. coli* as an energy source. Lactose is not a common component of *E. coli*'s natural environment, so most of the time the operon is not expressed and the enzymes for lactose utilization are not made by the bacterium. When lactose becomes available, it switches on the operon; all three genes are expressed together, resulting in coordinated synthesis of the lactose-utilizing enzymes. This is the classic example of gene regulation in bacteria, and is examined in detail in Section 11.3.1.

(A) Lactose operon

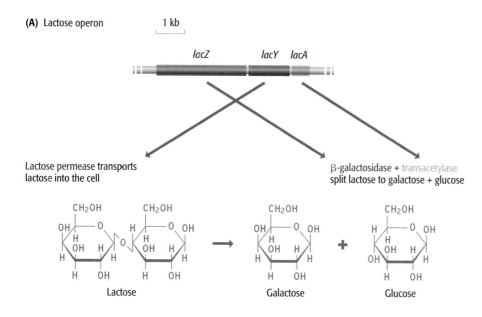

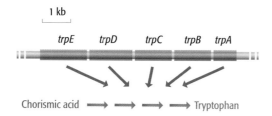

(B) Tryptophan operon

Figure 8.8 Two operons of *Escherichia coli*. (A) The lactose operon. The three genes are called *lacZ*, *lacY*, and *lacA*, the first two separated by 52 bp and the second two by 64 bp. All three genes are expressed together, *lacY* coding for the lactose permease that transports lactose into the cell, and *lacZ* and *lacA* coding for enzymes that split lactose into its component sugars—galactose and glucose. (B) The tryptophan operon, which contains five genes coding for enzymes involved in the multistep biochemical pathway that converts chorismic acid into the amino acid tryptophan. The genes in the tryptophan operon are closer together than those in the lactose operon: *trpE* and *trpD* overlap by 1 bp, as do *trpB* and *trpA*; *trpD* and *trpC* are separated by 4 bp, and *trpC* and *trpB* by 12 bp. For more details on the regulation of these operons, see Sections 11.3.1 and 14.1.1.

Altogether there are almost 600 operons in the *E. coli* K12 genome, each containing two or more genes, and a similar number are present in *Bacillus subtilis*. In most cases the genes in an operon are functionally related, coding for a set of proteins that are involved in a single biochemical activity such as utilization of a sugar as an energy source or synthesis of an amino acid. An example of the latter is the tryptophan operon of *E. coli* (Figure 8.8B). Microbial geneticists are attracted to the simplicity of this system whereby a bacterium is able to control its various biochemical activities by regulating the expression of groups of related genes linked together in operons. This may be a correct interpretation of the function of operons in *E. coli*, *Bacillus subtilis*, and many other prokaryotes, but in at least some species the picture is less straightforward. Both the archaeon *Methanococcus jannaschii* and the bacterium *Aquifex aeolicus* have operons, but the genes in an individual operon rarely have any biochemical relationship. For example, one of the operons in the *A. aeolicus* genome contains six linked genes, these genes coding for two proteins involved in DNA recombination, an enzyme used in protein synthesis, a protein required for motility, an enzyme involved in nucleotide synthesis, and an enzyme for lipid synthesis (Figure 8.9). This is typical of the operon structure in the *A. aeolicus* and *M. jannaschii* genomes. In other words, the notion

Figure 8.9 A typical operon in the genome of *Aquifex aeolicus*. The genes code for the following proteins: *gatC*, glutamyl-tRNA aminotransferase subunit C, which plays a role in protein synthesis; *recA*, recombination protein RecA; *pilU*, twitching mobility protein; *cmk*, cytidylate kinase, required for synthesis of cytidine nucleotides; *pgsA*, phosphotidylglycerophosphate synthase, an enzyme involved in lipid biosynthesis; and *recJ*, single-strand-specific endonuclease RecJ, which is another recombination protein.

that expression of an operon leads to the coordinated synthesis of enzymes required for a single biochemical pathway does not hold for these species.

Genome projects have therefore confused our understanding of operons. It is certainly too early to abandon the belief that operons play a central role in biochemical regulation in many bacteria, but we need to explain the unexpected features of the operons in *A. aeolicus* and *M. jannaschii*. It has been pointed out that both *A. aeolicus* and *M. jannaschii* are autotrophs, which means that, unlike many prokaryotes, they are able to synthesize organic compounds from carbon dioxide, but how this similarity between the species might be used to explain their operon structures is not clear.

8.2.2 How many genes are there and what are their functions?

There is some overlap in size between the largest prokaryotic and smallest eukaryotic genomes, but on the whole prokaryotic genomes are much smaller (Table 8.3). For example, the *E. coli* K12 genome is just 4639 kb, two-fifths the size of the yeast genome, and has only 4405 genes. Most prokaryotic genomes are less than 5 Mb in size, but the overall range among sequenced genomes is from 491 kb for *Nanoarchaeum equitans* to 9.1 Mb for *Bradyrhizobium japonicum*, and a few unsequenced genomes are substantially larger than this: *Bacillus megaterium*, for example, has a huge genome of 30 Mb.

Most of these genomes are organized along similar lines to *E. coli*, which means that genome size is proportional to gene number, with an average of approximately 950 genes per 1 Mb of DNA. Gene numbers therefore vary over an extensive range, with these numbers reflecting the nature of the ecological

Table 8.3 Genome sizes and gene numbers for various prokaryotes

Species	Size of genome (Mb)	Approximate number of genes
Bacteria		
Mycoplasma genitalium	0.58	500
Streptococcus pneumoniae	2.16	2300
Vibrio cholerae El Tor N16961	4.03	4000
Mycobacterium tuberculosis H37Rv	4.41	4000
Escherichia coli K12	4.64	4400
Yersinia pestis CO92	4.65	4100
Pseudomonas aeruginosa PA01	6.26	5700
Archaea		
Methanococcus jannaschii	1.66	1750
Archaeoglobus fulgidus	2.18	2500

For bacteria species, the strain designation (e.g., "K12") is given if specified by the group who sequenced the genome. With many bacterial species, different strains have different genome sizes and gene contents (Section 8.2.3).

niches within which different species of prokaryote live. The largest genomes tend to belong to free-living species that are found in the soil, the environment which is generally looked on as providing the broadest range of physical and biological conditions, to which the genomes of these species must be able to respond. At the other end of the scale, many of the smallest genomes belong to species that are obligate parasites, such as *Mycoplasma genitalium*, which has just 470 genes in a 0.58 Mb genome. The limited coding capacity of these small genomes means that these species are unable to synthesize many nutrients and hence must obtain these from their hosts. This is illustrated by Table 8.4, which compares the gene catalog of *M. genitalium* with that of *E. coli*. We see that, for instance, *E. coli* has 131 genes for biosynthesis of amino acids whereas *M. genitalium* has just one, that *E. coli* has 103 genes for cofactor synthesis compared with five in *M. genitalium*, and so on.

These comparisons have led to speculation about the smallest number of genes needed to specify a free-living cell. Theoretical considerations initially led to the suggestion that 256 genes are the minimum required, but experiments in which increasing numbers of *Mycoplasma* genes have been mutated suggest that 265–350 are needed. There has been similar interest in searching for "distinctiveness" genes—ones that distinguish one species from another. Of the 470 genes in the *M. genitalium* genome, 350 are also present in the distantly related bacterium *Bacillus subtilis*, which suggests that the biochemical and structural features that distinguish a *Mycoplasma* from a *Bacillus* are encoded in the 120 or so genes that are unique to the former. Unfortunately, the identities of these supposed distinctiveness genes do

Table 8.4 Partial gene catalogs for *Escherichia coli* K12 and *Mycoplasma genitalium*

Category	Number of genes in	
	E. coli K12	*M. genitalium*
Total protein-coding genes	4288	470
Biosynthesis of amino acids	131	1
Biosynthesis of cofactors	103	5
Biosynthesis of nucleotides	58	19
Cell envelope proteins	237	17
Energy metabolism	243	31
Intermediary metabolism	188	6
Lipid metabolism	48	6
DNA replication, recombination, and repair	115	32
Protein folding	9	7
Regulatory proteins	178	7
Transcription	55	12
Translation	182	101
Uptake of molecules from the environment	427	34

The numbers given refer only to those genes whose functions could be assigned when the genomes were sequenced. Additional gene assignments since then have not changed the overall picture.

not provide any obvious clues about what makes a bacterium a *Mycoplasma* rather than anything else.

8.2.3 Prokaryotic genomes and the species concept

Genome projects have confused our understanding of what constitutes a "species" in the prokaryotic world. This has always been a problem in microbiology because the standard biological definitions of species have been difficult to apply to microorganisms. The early taxonomists such as Linneaus described species in morphological terms, all members of one species having the same or very similar structural features. This form of classification was in vogue until the early twentieth century and was first applied to microorganisms in the 1880s by Robert Koch and others, who used staining and biochemical tests to distinguish between bacterial species. However, it was recognized that this type of classification was imprecise because many of the resulting species were made up of a variety of types with quite different properties. An example is provided by *E. coli* which, like many bacterial species, includes strains with distinctive pathogenic characteristics, ranging from harmless through to lethal. During the twentieth century, biologists redefined the species concept in evolutionary terms, and we now look on a species as a group of organisms that can interbreed with one another. If anything, this is more problematic with microorganisms because there are a variety of methods by which genes can be exchanged between prokaryotes that, according to their biochemical and physiological properties, are different species (see Figure 3.23). The barrier to **gene flow** that is central to the species concept therefore does not hold with prokaryotes.

Genome sequencing has emphasized the difficulties in applying the species concept to prokaryotes. It has become clear that different strains of a single species can have very different genome sequences, and may even have individual sets of strain-specific genes. This was first shown by a comparison between two strains of *Helicobacter pylori*, which causes gastric ulcers and other diseases of the human digestive tract. The two strains were isolated in the United Kingdom and the United States and have genomes of 1.67 Mb and 1.64 Mb, respectively. The larger genome contains 1552 genes and the smaller one 1495 genes, 1406 of these genes being present in both strains. In other words, some 6%–9% of the gene content of each genome is unique to that strain. A much more extreme distinction between strains was revealed when the genome sequence of the common laboratory strain of *E. coli*, K12, was compared with that of one of the most pathogenic strains, O157:H7. The lengths of the two genomes are significantly different—4.64 Mb for K12 and 5.53 Mb for O157:H7—with the extra DNA in the pathogenic strain scattered around the genome at almost 200 separate positions. These "O-islands" contain 1387 genes not present in *E. coli* K12, many of these genes coding for toxins and other proteins that are clearly involved in the pathogenic properties of O157:H7. But it is not simply a case of O157:H7 containing extra genes that make it pathogenic. K12 also has 234 segments of its own unique DNA, and although these "K-islands" are, on average, smaller than the O-islands, they still contain 528 genes that are absent from O157:H7. The situation, therefore, is that *E. coli* O157:H7 and *E. coli* K12 each has a set of strain-specific genes, which make up 26% and 12% of the gene catalogs, respectively. This is substantially more variation than can be tolerated by the species concept as

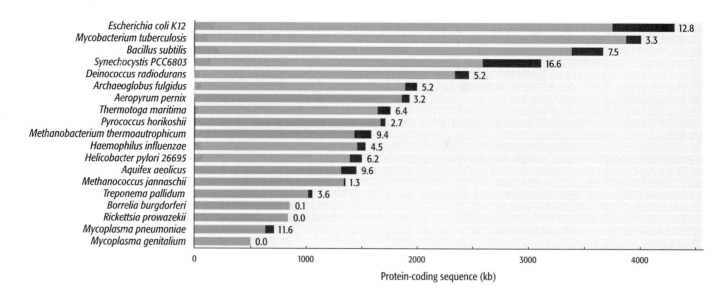

Figure 8.10 The impact of lateral gene transfer on the content of prokaryotic genomes. The chart shows the DNA that is unique to a particular species in blue and the DNA that has been acquired by lateral gene transfer in red. The number at the end of each bar indicates the percentage of the genome that derives from lateral gene transfer. Note that intergenic regions are omitted from this analysis.

applied to higher organisms, and is difficult to reconcile with any definition of species yet devised for microorganisms.

The difficulties become even more acute when other bacterial and archaeal genomes are examined. Because of the ease with which genes can flow between different prokaryotic species, it was anticipated that the same genes would occasionally be found in different species, but the extent of **lateral gene transfer** revealed by sequencing has taken everybody by surprise. Most genomes contain a few hundred kilobases of DNA acquired directly from a different species, and in some cases the figure is higher: 12.8% of the *E. coli* K12 genome, corresponding to 0.59 Mb, has been obtained in this way (Figure 8.10). A second surprise is that transfer has occurred between very different species, even between bacteria and archaea. For example, the thermophilic bacterium *Thermotoga maritima* has 1877 genes, 451 of which appear to have been obtained from archaeons. Transfer in the other direction, from bacteria to archaea, is equally prevalent. The picture that is emerging is one in which prokaryotes living in similar ecological niches exchange genes with one another in order to increase their individual fitness for survival in their particular environment. Many of the *Thermotoga* genes that have been obtained from archaeons have probably helped this bacterium acquire its ability to tolerate high temperatures.

Genome sequencing has also revealed that many bacterial species remain unidentified. Microbiologists have suspected this for many years, realizing that the artificial culture conditions used to isolate bacteria from their natural habitats do not suit all species, and that many will not grow under these conditions and hence will remain undetected. Practitioners of the new field of **metagenomics** are addressing this problem by obtaining DNA sequences from all the genomes in a particular habitat, for example from seawater or from an acidic soil. In one study, over a megabase of sequence was obtained from bacterial DNA from 1500 liters of surface water from the Sargasso Sea. The sequence included segments of the genomes of over 1800 species, of which 148 were totally new. Metagenomics raises the intriguing possibility that in the future we may have genome sequences for organisms that nobody has ever seen.

8.3 Eukaryotic Organelle Genomes

Now we return to the eukaryotic world to examine the genomes present in mitochondria and chloroplasts. The possibility that some genes might be located outside the nucleus—**extrachromosomal genes** as they were initially called—was first raised in the 1950s as a means of explaining the unusual inheritance patterns of certain genes in the fungus *Neurospora crassa*, the yeast *Saccharomyces cerevisiae*, and the photosynthetic alga *Chlamydomonas reinhardtii*. Electron microscopy and biochemical studies at about the same time provided hints that DNA molecules might be present in mitochondria and chloroplasts. Eventually, in the early 1960s, these various lines of evidence were brought together and the existence of mitochondrial and chloroplast genomes, independent of and distinct from the eukaryotic nuclear genome, was accepted.

8.3.1 The origins of organelle genomes

The discovery of organelle genomes led to many speculations about their origins. Today most biologists accept that the **endosymbiont theory** is correct, at least in outline, even though it was considered quite unorthodox when first proposed in the 1960s. The endosymbiont theory is based on the observation that the gene expression processes occurring in organelles are similar in many respects to equivalent processes in bacteria. In addition, when nucleotide sequences are compared, organelle genes are found to be more similar to equivalent genes from bacteria than they are to eukaryotic nuclear genes. The endosymbiont theory therefore holds that mitochondria and chloroplasts are the relics of free-living bacteria that formed a symbiotic association with the precursor of the eukaryotic cell, way back at the very earliest stages of evolution.

Support for the endosymbiont theory has come from the discovery of organisms which appear to exhibit stages of endosymbiosis that are less advanced than seen with mitochondria and chloroplasts. For example, an early stage in endosymbiosis is displayed by the protozoan *Cyanophora paradoxa*, whose photosynthetic structures, called **cyanelles**, are different from chloroplasts and instead resemble ingested cyanobacteria. Similarly, the *Rickettsia*, which live inside eukaryotic cells, might be modern versions of the bacteria that gave rise to mitochondria. It has also been suggested that the hydrogenosomes of trichomonads (unicellular microbes, many of which are parasites) represent an advanced type of mitochondrial endosymbiosis, as some hydrogenosomes possess a genome but most do not.

If mitochondria and chloroplasts were once free-living bacteria, then since the endosymbiosis was set up, there must have been a transfer of genes from the organelle into the nucleus. We do not understand how this occurred, or indeed whether there was a mass transfer of many genes at once, or a gradual trickle from one site to the other. But we do know that DNA transfer from organelle to nucleus, and indeed between organelles, still occurs. This was discovered in the early 1980s, when the first partial sequences of chloroplast genomes were obtained. It was found that in some plants the chloroplast genome contains segments of DNA, often including entire genes, that are copies of parts of the mitochondrial genome. The implication is that this so-called **promiscuous DNA** has been transferred from one organelle to the other. We now know that this is not the only type of transfer that can occur. The *Arabidopsis* mitochondrial genome contains various segments of nuclear DNA as well as 16 fragments of the chloroplast genome, including six tRNA genes that have retained their activity after transfer to the mitochondrion. The nuclear genome of this plant includes several short segments of the chloroplast and mitochondrial genomes as well as a 270 kb piece of mitochondrial DNA located within the centromeric region of chromosome 2. The transfer of mitochondrial DNA to vertebrate nuclear genomes has also been documented.

8.3.2 Physical features of organelle genomes

Almost all eukaryotes have mitochondrial genomes, and all photosynthetic eukaryotes have chloroplast genomes. Initially, it was thought that virtually all organelle genomes were circular DNA molecules. Electron microscopy had revealed both circular and linear DNA in some organelles, but it was assumed that the linear molecules were simply fragments of circular genomes that had become broken during preparation for electron microscopy. We still believe that most mitochondrial and chloroplast genomes are circular, but we now recognize that there is a great deal of variability in different organisms. In many eukaryotes, the circular genomes coexist in the organelles with linear versions and, in the case of chloroplasts, with smaller circles that contain subcomponents of the genome as a whole. The latter pattern reaches its extreme in the marine algae called dinoflagellates, whose chloroplast genomes are split into many small circles, each containing just a single gene. We also now realize that the mitochondrial genomes of some microbial eukaryotes (e.g., *Paramecium*, *Chlamydomonas*, and several yeasts) are always linear.

Copy numbers for organelle genomes are not particularly well understood. Each human mitochondrion contains about 10 identical molecules, which means that there are about 8000 per cell, but in *S. cerevisiae* the total number is probably smaller (less than 6500) even though there may be over 100 genomes per mitochondrion. Photosynthetic microorganisms such as *Chlamydomonas* have approximately 1000 chloroplast genomes per cell, about one-fifth the number present in a higher plant cell. One mystery, which dates back to the 1950s and has never been satisfactorily solved, is that when organelle genes are studied in genetic crosses the results suggest that there is just one copy of a mitochondrial or chloroplast genome per cell. This is clearly not the case but indicates that our understanding of the transmission of organelle genomes from parent to offspring is less than perfect.

Mitochondrial genome sizes are variable (Table 8.5) and are unrelated to the complexity of the organism. Most multicellular animals have small mitochondrial genomes with a compact genetic organization, the genes being close together with little space between them. The human mitochondrial genome (Figure 8.11), at 16,569 bp, is typical of this type. Most lower eukaryotes such as *S. cerevisiae* (Figure 8.12), as well as flowering plants, have larger and less compact mitochondrial genomes, with a number of the genes containing introns. Chloroplast genomes have less variable sizes (Table 8.5) and most have a structure similar to that shown in Figure 8.13 for the rice chloroplast genome.

8.3.3 The genetic content of organelle genomes

Organelle genomes are much smaller than their nuclear counterparts and we therefore anticipate that their gene contents are much more limited, which is indeed the case. Again, mitochondrial genomes display the greater

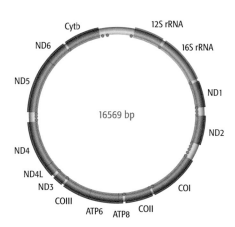

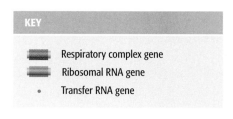

Figure 8.11 The human mitochondrial genome. The human mitochondrial genome is small and compact, with little wasted space—so much so that the ATP6 and ATP8 genes overlap. Abbreviations: ATP6, ATP8, genes for ATPase subunits 6 and 8; COI, COII, COIII, genes for cytochrome *c* oxidase subunits I, II, and III; Cytb, gene for apocytochrome *b*; ND1–ND6, genes for NADH hydrogenase subunits 1–6.

Table 8.5 Sizes of mitochondrial and chloroplast genomes

Species	Type of organism	Genome size (kb)
Mitochondrial genomes		
Plasmodium falciparum	Protozoan (malaria parasite)	6
Chlamydomonas reinhardtii	Green alga	16
Mus musculus	Vertebrate (mouse)	16
Homo sapiens	Vertebrate (human)	17
Metridium senile	Invertebrate (sea anemone)	17
Drosophila melanogaster	Invertebrate (fruit fly)	19
Chondrus crispus	Red alga	26
Aspergillus nidulans	Ascomycete fungus	33
Reclinomonas americana	Protozoa	69
Saccharomyces cerevisiae	Yeast	75
Suillus grisellus	Basidiomycete fungus	121
Brassica oleracea	Flowering plant (cabbage)	160
Arabidopsis thaliana	Flowering plant (vetch)	367
Zea mays	Flowering plant (maize)	570
Cucumis melo	Flowering plant (melon)	2500
Chloroplast genomes		
Pisum sativum	Flowering plant (pea)	120
Marchantia polymorpha	Liverwort	121
Oryza sativa	Flowering plant (rice)	136
Nicotiana tabacum	Flowering plant (tobacco)	156
Chlamydomonas reinhardtii	Green alga	195

variability, gene contents ranging from five for the malaria parasite *P. falciparum* to 92 for the protozoan *Reclinomonas americana* (Table 8.6). All mitochondrial genomes contain genes for the noncoding rRNAs and at least some of the protein components of the respiratory chain, the latter being the main biochemical feature of the mitochondrion. The more gene-rich genomes also code for tRNAs, ribosomal proteins, and proteins involved in transcription, translation, and transport of other proteins into the mitochondrion from the surrounding cytoplasm (Table 8.6). Most chloroplast genomes appear to possess the same set of 200 or so genes, again coding for rRNAs and tRNAs, as well as ribosomal proteins and proteins involved in photosynthesis (see Figure 8.13).

A general feature of organelle genomes emerges from Table 8.6. These genomes specify some of the proteins found in the organelle, but not all of them. The other proteins are coded by nuclear genes, synthesized in the cytoplasm, and transported into the organelle. If the cell has mechanisms for transporting proteins into mitochondria and chloroplasts, then why not have all the organelle proteins specified by the nuclear genome? We do not yet have a convincing answer to this question, although it has been suggested that at least some of the proteins coded by organelle genomes are extremely hydrophobic and cannot be transported through the membranes

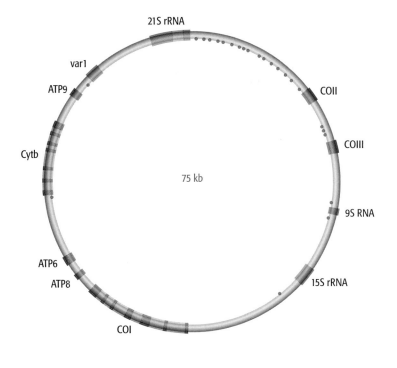

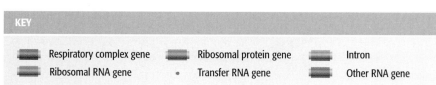

KEY

Respiratory complex gene	Ribosomal protein gene	Intron
Ribosomal RNA gene	· Transfer RNA gene	Other RNA gene

Figure 8.12 The *Saccharomyces cerevisiae* mitochondrial genome. Because of their relatively small sizes, many mitochondrial genomes have been completely sequenced. In the yeast mitochondrial genome, the genes are more widely spaced than in the human mitochondrial genome and some of the genes have introns. This type of organization is typical of many lower eukaryotes and plants. The yeast mitochondrial genome contains five additional open reading frames (not shown on this map) that have not yet been shown to code for functional gene products, and there are also several genes located within the introns of the discontinuous genes. Most of the latter code for maturase proteins involved in splicing the introns from the transcripts of these genes. Abbreviations: ATP6, ATP8, ATP9, genes for ATPase subunits 6, 8, and 9; COI, COII, COIII, genes for cytochrome *c* oxidase subunits I, II, and III; Cytb, gene for apocytochrome *b*; var1, gene for a ribosome-associated protein. The 9S RNA gene specifies the RNA component of the enzyme ribonuclease P (Section 12.1.3).

Table 8.6 Features of mitochondrial genomes

Feature	*Plasmodium falciparum*	*Chlamydomonas reinhardtii*	*Homo sapiens*	*Saccharomyces cerevisiae*	*Arabidopsis thaliana*	*Reclinomonas americana*
Total number of genes	5	12	37	35	52	92
Types of genes						
Protein-coding genes	3	7	13	8	27	62
Respiratory complex	3	7	13	7	17	24
Ribosomal proteins	0	0	0	1	7	27
Transport proteins	0	0	0	0	3	6
RNA polymerase	0	0	0	0	0	4
Translation factor	0	0	0	0	0	1
Functional RNA genes	2	5	24	27	25	30
Ribosomal RNA genes	2	2	2	2	3	3
Transfer RNA genes	0	3	22	24	22	26
Other RNA genes	0	0	0	1	0	1
Number of introns	0	1	0	8	23	1
Genome size (kb)	6	16	17	75	367	69

Figure 8.13 The rice chloroplast genome. Only those genes with known functions are shown. A number of the genes contain introns, which are not indicated on this map. These discontinuous genes include several of those for tRNAs, which is why the tRNA genes are of different lengths even though the tRNAs that they specify are all of similar size.

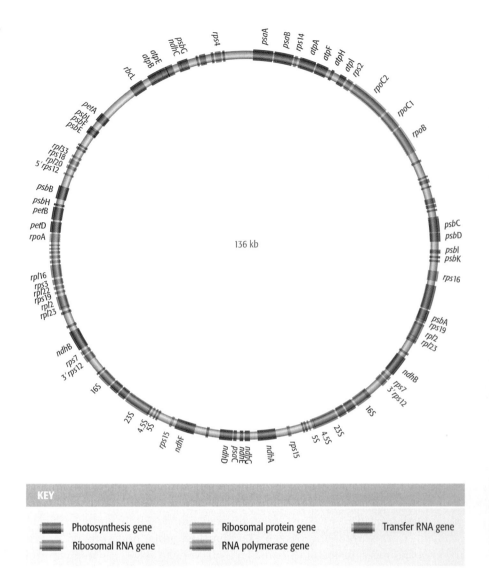

KEY

▬ Photosynthesis gene   ▬ Ribosomal protein gene   ▬ Transfer RNA gene
▬ Ribosomal RNA gene   ▬ RNA polymerase gene

that surround mitochondria and chloroplasts, and so simply cannot be moved into the organelle from the cytoplasm. The only way in which the cell can get them into the organelle is to make them there in the first place.

Summary

Prokaryotes comprise two distinct types of organism, the bacteria and the archaea. The bacterial genome is localized within the nucleoid—the lightly staining region of the otherwise featureless prokaryotic cell. The DNA is attached to a core of binding proteins from which 40–50 supercoiled loops of DNA radiate out into the cell. We know much less about the equivalent structures in archaea, but they are likely to be quite different as archaea possess proteins similar to eukaryotic histones rather than the bacterial nucleoid proteins. The *E. coli* genome is a single, circular DNA molecule but some prokaryotes have linear genomes and some have multipartite genomes made up of two or more circular and/or linear molecules. In the more complex cases it can be difficult to distinguish which molecules are genuine parts of the genome and which are dispensable plasmids. Prokaryotic genomes are very compact, with little repetitive DNA. Many genes are organized into operons, the members of which are expressed together and which may have a functional relationship. Gene number is related to genome size. The largest genomes belong to free-living species found in the soil, an environment that provides a broad range of physical and biological conditions to which these species must be able to respond. The smallest genomes belong to species that are obligate parasites, such as *Mycoplasma genitalium*, which has just 470 genes. Between 250 and 350 genes are thought to constitute the minimal set required to specify a free-living cell. Studies of prokaryotic genomes have complicated the species concept for these organisms, as the genomes of different strains of the same species often have very different gene contents, and lateral gene transfer can occur between different species. Metagenomics, the study of all the genomes in a habitat (such as seawater), is showing that a substantial proportion of the species that are present in a particular habitat have never been identified. The genomes in the mitochondria and chloroplasts of eukaryotic cells have prokaryotic features because they are descended from free-living bacteria that formed a symbiotic association with the precursor of the eukaryotic cell. Most mitochondrial and chloroplast genomes are circular, possibly multipartite, with a copy number of 1000–10,000 per cell. Mitochondrial genomes vary in size from 6 to 2500 kb and contain 5–100 genes, including genes for mitochondrial rRNAs, tRNAs, and proteins such as components of the respiratory complex. Chloroplast genomes are less variable, most being 100–200 kb in size with a similar set of some 200 genes, the majority coding for functional RNAs and photosynthetic proteins.

Multiple Choice Questions

8.1.* How is DNA in bacteria, such as *E. coli,* packaged in the cell?

   a. It is packaged into nucleosome complexes containing histone proteins.

   b. It is packaged into nucleoid structures containing histone proteins.

   c. It is packaged into nucleosome complexes containing DNA gyrase and DNA topoisomerase.

   d. It is supercoiled by DNA gyrase and DNA topoisomerase.

8.2. What is the bacterial nucleoid?

   a. It is a membrane-bound organelle that contains the genomic DNA.

   b. It is a lightly staining region of the bacterial cell that contains the genomic DNA.

   c. It is the protein complex of a bacterial cell that binds the genomic DNA.

   d. It is a membrane-bound complex that contains the bacterium's ribosomes.

8.3.* What is a plasmid?

   a. A small, usually circular DNA molecule that is independent of the main chromosome.

   b. A small, usually circular DNA molecule that contains essential genes.

   c. A small, usually circular DNA molecule that stabilizes the bacterial chromosome.

   d. A prokaryotic virus that can infect bacterial cells.

8.4. What is an integron?

   a. A plasmid that can integrate into the bacterial chromosome.

   b. A plasmid that can transfer to other bacteria.

   c. A set of genes and DNA sequences that allow a plasmid to capture genes from bacteriophages and other plasmids.

   d. A cloning vector that contains plasmid and bacteriophage sequences.

8.5.* What is a bacterial operon?

   a. A group of genes that have related biochemical functions.

   b. A group of genes that are evolutionarily related.

   c. A group of genes that are involved in a single biochemical pathway and are expressed together.

   d. A group of genes that are expressed from different promoters, but are regulated by the same repressor proteins.

8.6. What types of repeated sequences are present in bacterial genomes?

   a. Both microsatellites and minisatellites.

   b. Transposable elements.

   c. LINEs (long interspersed nuclear elements).

   d. Prokaryotic genomes do not possess any repeated sequences.

8.7.* Which of the following is NOT a feature of a typical bacterial operon?

   a. The genes are translated into a single polypeptide.

   b. The genes in the operon are transcribed into a single mRNA molecule.

   c. The genes frequently encode proteins that are involved in a single biochemical pathway.

   d. The genes are under the control of a single promoter.

8.8. How were prokaryotic organisms first classified into species?

   a. Staining and biochemical tests.

   b. Genetic tests.

   c. Microscopic studies.

   d. DNA sequence analyses.

8.9.* Lateral gene transfer includes all of the following DNA exchanges EXCEPT:

   a. The transfer of genes from bacteria to archaea.

   b. The transfer of genes from archaea to bacteria.

   c. The fusion of two bacterial species to produce diploid offspring.

   d. The transfer of a gene from one species to another.

8.10. Which of the following is NOT evidence that supports the endosymbiont theory?

   a. Mitochondria and chloroplasts have exterior structures similar to bacterial cell walls.

   b. Genes in these organelles are similar to bacterial genes.

   c. The gene expression processes in these organelles are similar to the bacterial processes.

   d. Organelle ribosomes resemble bacterial ribosomes.

8.11.* The smallest bacterial genome is several hundred thousand base pairs in length while the human mitochondrial genome is less than 17,000 base pairs. The smaller size of the mitochondrial genome is due to which of the following?

   a. The human mitochondrial genome has lost its protein-coding genes.

Multiple Choice Questions (continued)

    b. The human mitochondrial genome has lost its functional RNA genes.

    c. The human mitochondrial genome is nonfunctional and is an evolutionary relic.

    d. Genes from the human mitochondrial genome have been transferred to the nucleus.

8.12. A typical human mitochondrion contains how many identical copies of its DNA molecule?

    a. One.

    b. Ten.

    c. One hundred.

    d. Eight thousand.

8.13.* Which of the following types of gene are not known in any mitochondrial genome?

    a. tRNA genes.

    b. Respiratory chain genes.

    c. Glycolytic genes.

    d. rRNA genes.

8.14. What is the likely reason for the presence of some genes in the mitochondrion, rather than having all genes transferred to the nucleus?

    a. Some proteins are too large to be transported into the mitochondria.

    b. Some proteins have multiple subunits.

    c. Some proteins would be degraded before transport into the mitochondria.

    d. Some proteins are too hydrophobic to be transported into the mitochondria.

Short Answer Questions

8.1.* What are the differences between a eukaryotic chromosome and the *E. coli* chromosome?

8.2. What experimental evidence suggests that the *E. coli* chromosome is organized into supercoiled domains, and is attached to proteins that restrict its ability to relax?

8.3.* What similarities are there between *E. coli* HU proteins and eukaryotic histone proteins?

8.4. The *E. coli* genome is a single, circular DNA molecule. What other types of genome structure are found amongst prokaryotes?

8.5.* How are genes and other sequence features organized in a typical prokaryotic genome? When prokaryotic and mammalian genomes are compared, what differences are seen in gene density, number of introns, and repetitive DNA content?

8.6. How do the operons of *Methanococcus jannaschii* and *Aquifex aeolicus* differ from those of *E. coli*?

8.7.* The obligate intracellular parasitic bacterium *Mycoplasma genitalium* has just 470 genes. Why does this organism require so few genes?

8.8. What is the basis to the relationship between genome size and gene number in prokaryotes?

8.9.* Why does the species concept used for eukaryotes—that is, a group of organisms that can interbreed with one another—not apply to prokaryotes?

8.10. How is it possible to sequence the genome of an organism that has never been isolated?

continued …

In-depth Problems

8.1.* Should the traditional view of the prokaryotic genome as a single, circular DNA molecule be abandoned? If so, what new definition of "prokaryotic genome" should be adopted?

8.2. Speculate on the identities of the 250–350 genes that constitute the minimum set for a free-living cell.

8.3.* Can the concept of bacterial species survive genome sequencing?

8.4. Why do organelle genomes exist?

Figure Tests

8.1.* This represents a model of the *E. coli* nucleoid. How is the *E. coli* genome packaged within the nucleoid?

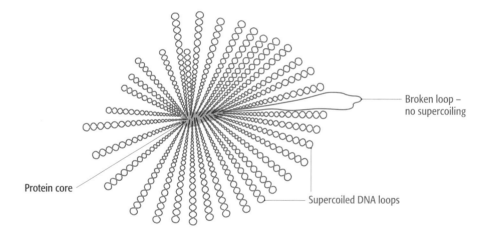

Broken loop – no supercoiling

Protein core

Supercoiled DNA loops

8.2. This figure shows a 50 kb segment of the *E. coli* genome. What are the key features of this sequence, and how does this compare with a typical 50 kb segment of the human genome?

thrB
thrA thrC
dnaK
IS186
IS1
carB
fixA

0   10   20   30   40   50 kb

Figure Tests (continued)

8.3.* As these genes are all involved in the biosynthesis of tryptophan, how would you expect their expression to be coordinated in *E. coli*?

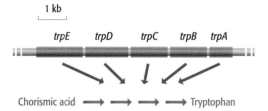

8.4. This figure shows the human mitochondrial genome. What are the key features of this genome? Where are the other genes coding for mitochondrial proteins located?

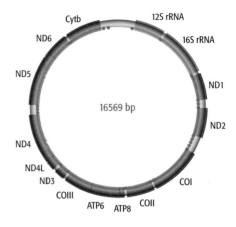

Further Reading

Prokaryotic nucleoids

Drlica, K. and Riley, M. (1990) *The Bacterial Chromosome.* American Society for Microbiology, Washington, DC. *A source of information on all aspects of bacterial DNA.*

Sinden, R.R. and Pettijohn, D.E. (1981) Chromosomes in living *Escherichia coli* cells are segregated into domains of supercoiling. *Proc. Natl Acad. Sci. USA* **78:** 224–228.

White, M.F. and Bell, S.D. (2002) Holding it all together: chromatin in the Archaea. *Trends Genet.* **18:** 621–626.

Multipartite genomes

Bentley, S.D. and Parkhill, J. (2004) Comparative genome structure of prokaryotes. *Annu. Rev. Genet.* **38:** 771–791. *Review of many of the aspects of prokaryotic genomes discussed in this chapter.*

Fraser, C.M., Casjens, S., Huang, W.M., *et al.* (1997) Genomic sequence of a Lyme disease spirochaete, *Borrelia burgdorferi. Nature* **390:** 580–586.

Heidelberg, J.F., Eisen, J.A., Nelson, W.C., *et al.* (2000) DNA sequence of both chromosomes of the cholera pathogen *Vibrio cholerae. Nature* **406:** 477–483.

White, O., Eisen, J.A., Heidelberg, J.F., *et al.* (1999) Genome sequence of the radioresistant bacterium *Deinococcus radiodurans* R1. *Science* **286:** 1571–1577.

Examples of gene organization

Blattner, F.R., Plunkett, G., Bloch, C.A., *et al.* (1997) The complete genome sequence of *Escherichia coli* K-12. *Science* **277:** 1453–1462.

Bult, C.J., White, O., Olsen, G.J., *et al.* (1996) Complete genome sequence of the methanogenic archaeon *Methanococcus jannaschii. Science* **273:** 1058–1073.

Deckert, G., Warren, P.V., Gaasterland, T., *et al.* (1998) The complete genome of the hyperthermophile bacterium *Aquifex aeolicus. Nature* **392:** 353–358.

Parkhill, J., Achtman, M., James, K.D., *et al.* (2000) Complete genome sequence of a serogroup A strain of *Neisseria meningitidis* Z2491. *Nature* **404:** 502–506.

Parkhill, J., Wren, B.W., Mungall, K., *et al.* (2000) The genome sequence of the food-borne pathogen *Campylobacter jejuni* reveals hypervariable sequences. *Nature* **403:** 665–668.

Minimal genome content

Koonin, E.V. (2000) How many genes can make a cell: the minimal-gene-set concept. *Annu. Rev. Genomics Hum.* *Genet.* **1:** 99–116. *Describes the experimental and theoretical work that is being done on the minimal genome.*

Problems with the species concept

Alm, R.A., Ling, L.-S.L., Moir, D.T., *et al.* (1999) Genomic-sequence comparison of two unrelated isolates of the human gastric pathogen *Helicobacter pylori. Nature* **397:** 176–180.

Boucher, Y., Douady, C.J., Papke, R.T., Walsh, D.A., Boudreau, M.E., Nesbo, C.L., Case, R.J. and Doolittle, W.F. (2003) Lateral gene transfer and the origins of prokaryotic groups. *Annu. Rev. Genet.* **37:** 283–328.

Nelson, K.E., Clayton, R.A., Gill, S.R., *et al.* (1999) Evidence for lateral gene transfer between Archaea and bacteria from genome sequence of *Thermotoga maritima. Nature* **399:** 323–329.

Ochman, H., Lawrence, J.G. and Groisman, E.A. (2000) Lateral gene transfer and the nature of bacterial innovation. *Nature* **405:** 299–304.

Perna, N.T., Plunkett, G., Burland, V., *et al.* (2001) Genome sequence of enterohaemorrhagic *Escherichia coli* O157:H7. *Nature* **409:** 529–532.

Metagenomics

Riesenfeld, C.S., Schloss, P.D. and Handelsman, J. (2004) Metagenomics: genomic analysis of microbial communities. *Annu. Rev. Genet.* **38:** 525–552.

Venter, J.C., Remington, K., Heidelberg, J.F., *et al.* (2004) Environmental genome shotgun sequencing of the Sargasso Sea. *Science* **304:** 66–74.

Organelle genomes

Lang, B.V., Gray, M.W. and Burger, G. (1999) Mitochondrial genome evolution and the origin of eukaryotes. *Annu. Rev. Genet.* **33:** 351–397.

Margulis, L. (1970) *Origin of Eukaryotic Cells.* Yale University Press, New Haven, Connecticut. *The first description of the endosymbiont theory for the origin of mitochondria and chloroplasts.*

Palmer, J.D. (1985) Comparative organization of chloroplast genomes. *Annu. Rev. Genet.* **32:** 437–459.

Virus Genomes and Mobile Genetic Elements

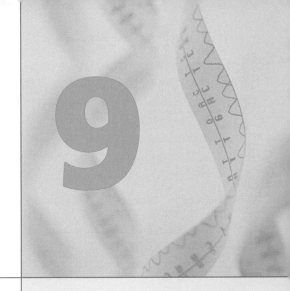

9

When you have read Chapter 9, you should be able to:

Describe the range of capsid structures displayed by bacteriophages and other types of virus.

Give an overview of the diverse ways in which bacteriophage genomes are organized.

Distinguish between the lytic and lysogenic infection pathways and give details of the key steps in each one.

Describe the major replication strategies used by eukaryotic viruses, in particular the viral retroelements.

Discuss the features of satellite RNAs, virusoids, viroids, and prions.

Distinguish between conservative and replicative transposition.

Give a detailed account of the structures of and relationships between different types of LTR retroelement.

Discuss, with examples, the key features of LINEs and SINEs.

Describe the range of DNA transposons found in prokaryotes.

Explain why DNA transposons of eukaryotes were important in the development of our understanding of transposition.

Give examples of DNA transposons in plants and *Drosophila melanogaster*.

The viruses are the last and simplest form of life whose genomes we will investigate. In fact, viruses are so simple in biological terms that we have to ask ourselves if they can really be thought of as living organisms. Doubts arise partly because viruses are constructed along lines different from all other forms of life—viruses are not cells—and partly because of the nature of the virus life cycle. Viruses are obligate parasites of the most extreme kind: they reproduce only within a host cell, and in order to replicate and express their genomes they must subvert at least part of the host's genetic machinery to

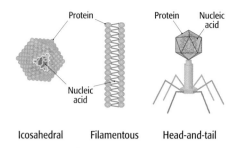

Figure 9.1 The three types of capsid structures commonly displayed by bacteriophages.

their own ends. Some viruses possess genes coding for their own DNA polymerase and RNA polymerase enzymes, but many depend on the host enzymes for genome replication and transcription. All viruses make use of the host's ribosomes and translation apparatus for synthesis of the polypeptides that make up the protein coats of their progeny. This means that virus genes must be matched to the host genetic system. Viruses are therefore quite specific for particular organisms, and individual types cannot infect a broad spectrum of species.

In this chapter we will also consider the mobile genetic elements that make up a substantial part of the repetitive component of eukaryotic and prokaryotic genomes. We link these elements with virus genomes because it has become clear in recent years that at least some of these repetitive sequences are derived from viruses, and are, in effect, viral genomes that have lost the ability to escape from their host cell.

9.1 The Genomes of Bacteriophages and Eukaryotic Viruses

There is a multitude of different types of virus but the ones that have received most attention from geneticists are those that infect bacteria. These are called bacteriophages and have been studied in great detail since the 1930s, when the early molecular biologists, notably Max Delbrück, chose phages as convenient model organisms with which to study genes. We will follow the lead taken by Delbrück and use bacteriophages as the starting point for our investigation of viral genomes.

9.1.1 Bacteriophage genomes

Bacteriophages are constructed from two basic components: protein and nucleic acid. The protein forms a coat, or **capsid**, within which the nucleic acid genome is contained. There are three basic capsid structures (Figure 9.1):

- Icosahedral, in which the individual polypeptide subunits (**protomers**) are arranged into a three-dimensional geometric structure that surrounds the nucleic acid. Examples are MS2 phage, which infects *Escherichia coli*, and PM2, which infects *Pseudomonas aeruginosa*.

- Filamentous, or helical, in which the protomers are arranged in a helix, producing a rod-shaped structure. The *E. coli* phage called M13 is an example.

- Head-and-tail, a combination of an icosahedral head, containing the nucleic acid, attached to a filamentous tail and possibly additional structures that facilitate entry of the nucleic acid into the host cell. This is a common structure possessed by, for example, the *E. coli* phages T4 and λ, and phage SPO1 of *Bacillus subtilis*.

Bacteriophage genomes have diverse structures and organizations

The term "nucleic acid" has to be used when referring to phage genomes because in some cases these molecules are made of RNA. Viruses are the one form of "life" that contradict the conclusion of Avery and his colleagues and of Hershey and Chase that the genetic material is DNA (Section 1.1.1). Phages and other viruses also break another rule: their genomes, whether of DNA or RNA, can be single stranded as well as double stranded. A whole

Table 9.1 Features of some typical bacteriophages and their genomes

Phage	Host	Capsid structure	Genome structure	Genome size (kb)	Number of genes
λ	*Escherichia coli*	Head-and-tail	Double-stranded linear DNA	49.5	48
φX174	*E. coli*	Icosahedral	Single-stranded circular DNA	5.4	11
f6	*Pseudomonas phaseolicola*	Icosahedral	Double-stranded segmented linear RNA	2.9, 4.0, 6.4	13
M13	*E. coli*	Filamentous	Single-stranded circular DNA	6.4	10
MS2	*E. coli*	Icosahedral	Single-stranded linear RNA	3.6	3
PM2	*Pseudomonas aeruginosa*	Icosahedral	Double-stranded linear DNA	10.0	approx. 21
SPO1	*Bacillus subtilis*	Head-and-tail	Double-stranded linear DNA	150	100+
T2, T4, T6	*E. coli*	Head-and-tail	Double-stranded linear DNA	166	150+
T7	*E. coli*	Head-and-tail	Double-stranded linear DNA	39.9	55+

The genome structure given is that in the phage capsid; some genomes exist in different forms within the host cell.

range of different genome structures is known among the phages, as summarized in Table 9.1. With most types of phage there is a single DNA or RNA molecule that comprises the entire genome. However, this is not always the case and a few RNA phages have **segmented genomes**, meaning that their genes are carried by a number of different RNA molecules. The sizes of phage genomes vary enormously, from about 1.6 kb for the smallest phages to over 150 kb for large ones such as T2, T4, and T6.

Bacteriophage genomes, being relatively small, were among the first to be studied comprehensively by the rapid and efficient DNA sequencing methods that were developed in the late 1970s. Gene numbers vary from just three in the case of MS2, to over 200 for the more complex head-and-tail phages (see Table 9.1). The smaller phage genomes of course contain relatively few genes, but these can be organized in a very complex manner. Phage φX174, for example, manages to pack into its genome "extra" biological information, as several of its genes overlap (Figure 9.2). These **overlapping genes** share nucleotide sequences (gene *B*, for example, is contained entirely within gene *A*) but code for different gene products, as the transcripts are translated from different start positions and in different reading frames. Overlapping genes are not uncommon in viruses. The larger phage genomes contain more genes, reflecting the more complex capsid structures of these phages and a dependence on a greater number of phage-encoded enzymes during the infection cycle. The T4 genome, for example, includes some 50 genes involved solely in construction of the phage capsid (Figure 9.3). Despite their complexity, even these large phages still require at least some host-encoded proteins and RNAs in order to carry through their infection cycles.

Replication strategies for bacteriophage genomes

Bacteriophages are classified into two groups according to their life cycle: lytic and lysogenic. The fundamental difference between these groups is that a lytic phage kills its host bacterium very soon after the initial infection,

Figure 9.2 The φX174 genome contains overlapping genes. The genome is made of single-stranded DNA. The expanded region shows the start and end of the overlap between genes *E* and *D*.

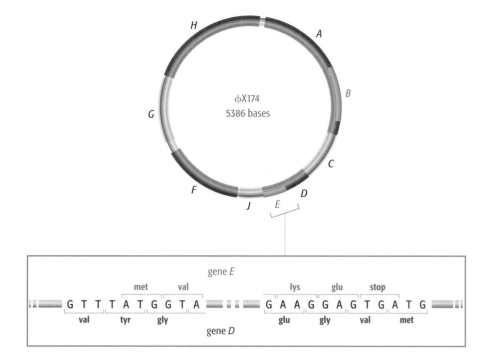

Figure 9.3 The T4 genome. The genome is made of double-stranded DNA and is shown in the circular form found in the host cell. Only those genes coding for components of the phage capsid are shown; about 100 additional genes involved in other aspects of the phage life cycle are omitted.

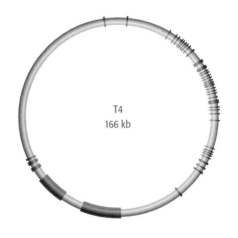

whereas a lysogenic phage can remain quiescent within its host for a substantial period of time, even throughout numerous generations of the host cell. These two life cycles are typified by two *E. coli* phages: the lytic (or **virulent**) T4 and the lysogenic (or **temperate**) λ.

The T series of *E. coli* phages (T1 to T7) were the first to become available to molecular geneticists and have been the subject of much study. Their lytic infection cycle was first investigated in 1939 by Emory Ellis and Max Delbrück, who added T4 phages to a culture of *E. coli*, waited 3 minutes for the phages to attach to the bacteria, and then measured the number of infected cells over a period of 60 minutes. Their results (Figure 9.4A) showed that there is no change in the number of infected cells during the first 22 minutes of infection, this **latent period** being the time needed for the phages to reproduce within their hosts. After 22 minutes the number of infected cells started to increase, showing that lysis of the original hosts had occurred and that the new phages that had been produced were now infecting other cells in the culture. The molecular events occurring at the different stages of this **one-step growth curve** are shown in Figure 9.4B. The initial event is attachment of the phage particle to a receptor protein on the outside of the bacterium. Different types of phage have different receptors: for example, for T4 the receptor is a protein called OmpC ("Omp" stands for "outer membrane protein"), which is a type of porin, the proteins that form channels through the outer cell membrane and facilitate the uptake of nutrients. After attachment, the phage injects its DNA genome into the cell through its tail structure. Immediately after entry of the phage DNA, the synthesis of host DNA, RNA, and protein stops and transcription of the phage genome begins. Within 5 minutes the bacterial DNA molecule has depolymerized and the resulting nucleotides are being utilized in replication of the T4 genome. After 12 minutes, new phage capsid proteins start to appear and the first complete phage particles are assembled. Finally, at the end of the latent period, the cell bursts and the new phages are released. A typical infection cycle produces 200 to 300 T4 phages per cell, all of which can go on to infect other bacteria.

(A) The one-step growth curve

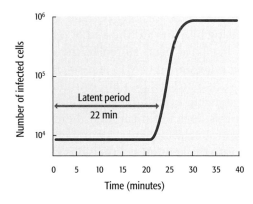

(B) The lytic infection cycle

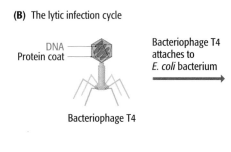

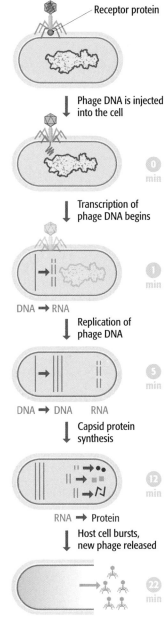

Receptor protein

Phage DNA is injected into the cell

Transcription of phage DNA begins

DNA → RNA

Replication of phage DNA

DNA → DNA   RNA

Capsid protein synthesis

RNA → Protein

Host cell bursts, new phage released

Most phages can follow the lytic infection cycle but some, such as λ, can also pursue a lysogenic cycle. In Section 2.2.1, when we looked at the use of λ phages as cloning vectors, we discovered that during a lysogenic cycle the phage genome becomes integrated into the host DNA. This occurs immediately after entry of the phage DNA into the cell, and results in a quiescent form of the bacteriophage, called the **prophage** (Figure 9.5A). Integration occurs by **site-specific recombination** (Section 17.2) between identical 15 bp sequences present in the λ and *E. coli* genomes. Note that this means that the λ genome always integrates at the same position within the *E. coli* DNA molecule. The integrated prophage can be retained in the host DNA molecule for many cell generations, being replicated along with the bacterial genome and passed with it to the daughter cells. However, the switch to the lytic mode of infection occurs if the prophage is **induced** by any one of several chemical or physical stimuli. Each of these appears to be linked to DNA damage, and possibly therefore signals the imminent death of the host by natural causes. In response to these stimuli, a second recombination event excises the phage genome from the host DNA, phage DNA replication begins, and phage coat proteins are synthesized (Figure 9.5B). Eventually, the cell bursts and new λ phages are released. Lysogeny adds an additional level of complexity to the phage life cycle, and ensures that the phage is able to adopt the particular infection strategy best suited to the prevailing conditions.

9.1.2 The genomes of eukaryotic viruses

The capsids of eukaryotic viruses are either icosahedral or filamentous: the head-and-tail structure is unique to bacteriophages. One distinct feature of eukaryotic viruses, especially those with animal hosts, is that the capsid may be surrounded by a lipid membrane, forming an additional component to the virus structure (Figure 9.6). This membrane is derived from the host when the new virus particle leaves the cell, and may subsequently be modified by insertion of virus-specific proteins.

Structures and replication strategies for eukaryotic viral genomes
Eukaryotic viral genomes display a great variety of structures (Table 9.2). They may be DNA or RNA, single or double stranded (or partly double stranded with single-stranded regions), linear or circular, segmented or nonsegmented. For reasons that no one has ever understood, the vast majority of plant viruses have RNA genomes. Genome sizes cover approximately the same range as seen with phages, although the largest viral genomes (e.g., vaccinia virus at 240 kb) are rather bigger than the largest phage genomes.

Figure 9.4 The lytic infection cycle of bacteriophage T4. (A) The one-step growth curve, as revealed by the experiment conducted by Ellis and Delbrück. (B) The molecular events occurring during the lytic infection cycle.

(A) Integration into the host DNA

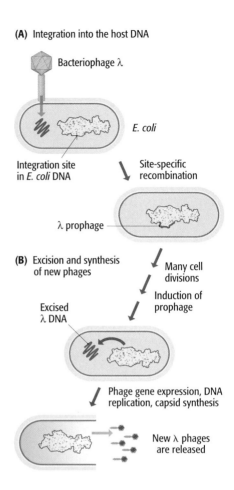

Figure 9.5 The lysogenic infection cycle, as followed by bacteriophage λ. After induction, the infection cycle is similar to the lytic mode. See Section 14.3.1 for a description of how λ genome expression is regulated during lysogeny.

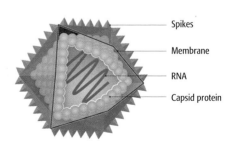

Figure 9.6 The structure of a eukaryotic retrovirus. The capsid is surrounded by a lipid membrane to which additional virus proteins are attached.

Although most eukaryotic viruses follow only the lytic infection cycle, few take over the host cell's genetic machinery to the extent that a bacteriophage does. Many viruses coexist with their host cells for long periods, possibly years, with the host cell functions ceasing only toward the end of the infection cycle, when the virus progeny that have been stored in the cell are released. Other viruses have infection cycles similar to M13 in *E. coli* (Section 4.1.1), continually synthesizing new virus particles that are extruded from the cell. These long-term infections can occur even if the viral genome does not integrate into the host DNA, but this does not mean that there are no eukaryotic viruses equivalent to lysogenic bacteriophages. A number of DNA and RNA viruses are able to integrate into the genomes of their hosts, sometimes with drastic effects on the host cell. The **viral retroelements** are examples of integrative eukaryotic viruses. Their replication pathways include a novel step in which an RNA version of the genome is converted into DNA. There are two kinds of viral retroelement: the **retroviruses**, whose capsids contain the RNA version of the genome, and the **pararetroviruses**, whose encapsidated genome is made of DNA. The ability of viral retroelements to convert RNA into DNA was confirmed independently in 1970 by Howard Temin and by David Baltimore. Working with cells infected with retroviruses, both Temin and Baltimore isolated the enzyme, now called **reverse transcriptase**, which is capable of making a DNA copy of an RNA template (and is of immense utility in the experimental study of genomes—Section 2.1.1). The typical retroviral genome is a single-stranded RNA molecule, 6000–9000 nucleotides in length. After entry into the cell, the genome is copied into double-stranded DNA by a few molecules of reverse transcriptase that the virus carries in its capsid. The double-stranded version of the genome then integrates into the host DNA (Figure 9.7). Unlike λ, the retroviral genome has no sequence similarity with its insertion site in the host DNA. Integration of the viral genome into the host DNA is a prerequisite for expression of the retrovirus genes. There are three of these, called *gag*, *pol*, and *env* (Figure 9.8). Each codes for a **polyprotein** that is cleaved, after translation, into two or more functional gene products. These products include the viral coat proteins (from *env*) and the reverse transcriptase (from *pol*). The protein products combine with full-length RNA transcripts of the retroviral genome to produce new virus particles.

The causative agents of AIDS (acquired immune deficiency syndrome) were shown to be retroviruses in 1983–84. The first AIDS virus was isolated independently by two groups, led by Luc Montagnier and Robert Gallo. This virus is called human immunodeficiency virus, or HIV-1, and is responsible for the most prevalent and pathogenic form of AIDS. A related virus, HIV-2, discovered by Montagnier in 1985, is less widespread and causes a milder form of the disease. The HIVs attack certain types of lymphocyte in the bloodstream, thereby depressing the immune response of the host. These lymphocytes carry on their surfaces multiple copies of a protein called CD4, which acts as a receptor for the virus. An HIV particle binds to a CD4 protein and then enters the lymphocyte after fusion between its lipid envelope and the cell membrane.

Genomes at the edge of life

Viruses occupy the boundary between the living and nonliving worlds. At the very edge of this boundary—or perhaps beyond it—reside a variety of nucleic acid molecules that might or might not be classified as genomes. The **satellite RNAs** or **virusoids** are examples. These are RNA molecules, some 320–400

Table 9.2 Features of some typical eukaryotic viruses and their genomes

Virus	Host	Genome structure	Genome size (kb)	Number of genes
Adenovirus	Mammals	Double-stranded linear DNA	36.0	30
Hepatitis B	Mammals	Partly double-stranded circular DNA	3.2	4
Influenza virus	Mammals	Single-stranded segmented linear RNA	22.0	12
Parvovirus	Mammals	Single-stranded linear DNA	1.6	5
Poliovirus	Mammals	Single-stranded linear RNA	7.6	8
Reovirus	Mammals	Double-stranded segmented linear RNA	22.5	22
Retroviruses	Mammals, birds	Single-stranded linear RNA	6.0–9.0	3
SV40	Monkeys	Double-stranded circular DNA	5.0	5
Tobacco mosaic virus	Plants	Single-stranded linear RNA	6.4	6
Vaccinia virus	Mammals	Double-stranded circular DNA	240	240

The genome structure given is that in the virus capsid; some genomes exist in different forms within the host cell.

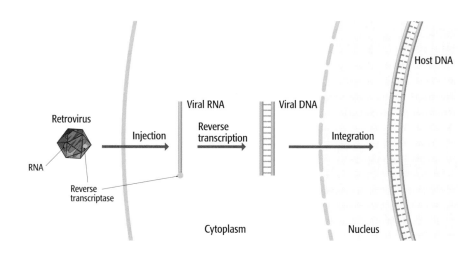

Figure 9.7 Insertion of a retroviral genome into a host chromosome.

nucleotides in length, which do not encode their own capsid proteins, instead moving from cell to cell within the capsids of helper viruses. The distinction between the two groups is that a satellite virus shares the capsid with the genome of the helper virus whereas a virusoid RNA molecule becomes encapsidated on its own. They are generally looked on as parasites of their helper viruses, although there appear to be at least a few cases where the helper cannot replicate without the satellite RNA or virusoid, suggesting that at least some of the relationships are symbiotic. Satellite RNAs and virusoids are both found predominantly in plants, as is a more extreme group called the **viroids.** These are RNA molecules, 240–375 nucleotides in length, which contain no genes and never become encapsidated, spreading from cell to cell as naked RNA. They include some economically important pathogens, such as the citrus exocortis viroid which reduces the growth of citrus fruit trees. Viroid and virusoid molecules are circular and single stranded and are replicated by enzymes coded by the host or helper virus genomes. The replication process results in a series of RNAs joined head to tail and, with some viroids and virusoids, these are cleaved

Figure 9.8 A retrovirus genome. Each LTR is a long terminal repeat of 250–1400 bp, which plays an important role in replication of the genome (see Section 17.3.2).

(A) Self-catalyzed cleavage of viroid and virusoid RNAs

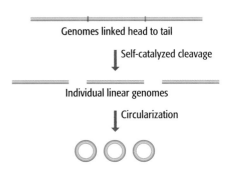

Genomes linked head to tail

↓ Self-catalyzed cleavage

Individual linear genomes

↓ Circularization

(B) The cleavage structure

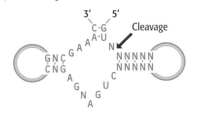

Figure 9.9 Self-catalyzed cleavage of linked genomes during replication of viroids and virusoids. (A) The replication pathway. (B) The "hammerhead" structure, which forms at each cleavage site and which has enzymatic activity. N, any nucleotide.

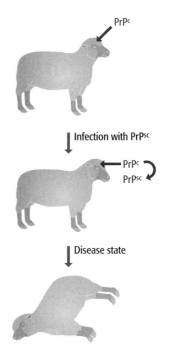

Figure 9.10 The mode of action of a prion. A normal, healthy sheep has PrP^C proteins in its brain. Infection with PrP^{SC} molecules leads to conversion of newly synthesized PrP^C proteins into PrP^{SC}, leading to the disease state—scrapie in sheep.

by a self-catalyzed reaction in which the RNA molecule acts as an enzyme (Figure 9.9). We will study these RNA enzymes in greater detail in Section 12.2.4.

Nucleic acid molecules that replicate within plant cells can perhaps be looked on as genomes even if they contain no genes. The same cannot be said for **prions**, as these infectious, disease-causing particles contain no nucleic acid. Prions are responsible for scrapie in sheep and goats and their transmission to cattle has led to the new disease called BSE—bovine spongiform encephalopathy. Whether their further transmission to humans causes a variant form of Creutzfeldt–Jakob disease (CJD) is controversial but accepted by many biologists. At first, prions were thought to be viruses but it is now clear that they are made solely of protein. The normal version of the prion protein, called PrP^C, is coded by a mammalian nuclear gene and synthesized in the brain, although its function is unknown. PrP^C is easily digested by proteases whereas the infectious version, PrP^{SC}, has a more highly β-sheeted structure that is resistant to proteases and forms fibrillar aggregates that are seen in infected tissues. Once inside a cell, PrP^{SC} molecules are able to convert newly synthesized PrP^C proteins into the infectious form, by a mechanism that is not yet understood, resulting in the disease state. Transfer of one or more of these PrP^{SC} proteins to a new animal results in accumulation of new PrP^{SC} proteins in the brain of that animal, transmitting the disease (Figure 9.10). Infectious proteins with similar properties are known in lower eukaryotes, examples being the Ure3 and Psi⁺ prions of *Saccharomyces cerevisiae*. It is clear, however, that prions are *gene products* rather than genetic material and despite their infectious properties, which led to the initial confusion regarding their status, they are unrelated to viruses or to subviral particles such as viroids and virusoids.

9.2 Mobile Genetic Elements

In Chapters 7 and 8 we learnt that eukaryotic genomes, and to a lesser extent those of prokaryotes, contain genome-wide or interspersed repeats, some with copy numbers of several thousand per genome, with the individual repeat units distributed in an apparently random fashion (Section 7.2.4). For many interspersed repeats, the genome-wide distribution pattern is set up by **transposition**, the process by which a segment of DNA can move from one position to another in a genome. These movable segments are called transposable elements, or **transposons**. Some types move by a **conservative** process, which involves the excision of the sequence from its original position followed by its reinsertion elsewhere. Conservative transposition therefore results in the transposon simply changing its position in the genome without increasing its copy number (Figure 9.11). **Replicative transposition**, on the other hand, results in an increase in copy number, because during this process the original element remains in place while a copy is inserted at the new position. This replicative process can therefore lead to a proliferation of the transposon at interspersed positions around the genome.

Both types of transposition involve recombination, and we will therefore deal with the details of the processes when we study recombination and related types of genome rearrangement in Chapter 17. What interests us here is the variety of structures displayed by the transposable elements found in eukaryotic and prokaryotic genomes, and the link that exists between these elements and viral genomes.

Figure 9.11 Conservative and replicative transposition.

9.2.1 Transposition via an RNA intermediate

Replicative transposons can be further subdivided into those that transpose via an RNA intermediate and those that do not. The process that involves an RNA intermediate, which is called **retrotransposition**, begins with synthesis of an RNA copy of the transposon by the normal process of transcription (Figure 9.12). The transcript is then copied into double-stranded DNA, which initially exists as an independent molecule outside of the genome. Finally, the DNA copy of the transposon integrates into the genome, possibly back into the same chromosome occupied by the original unit, or possibly into a different chromosome. The end result is that there are now two copies of the transposon, at different points in the genome.

If we compare the mechanism for retrotransposition with that for replication of a viral retroelement, as shown in Figure 9.7, then we see that the two processes are very similar, the one significant difference being that the RNA molecule that initiates the process is transcribed from an endogenous genomic sequence during retrotransposition, and an exogenous viral genome during replication of a viral retroelement. This close similarity alerts us to the relationships that exist between these two types of element.

RNA transposons with long terminal repeats are related to viral retroelements

RNA transposons, or **retroelements**, are features of eukaryotic genomes but have not so far been discovered in prokaryotes. They can be broadly classified into two types: those that possess **long terminal repeats** (**LTRs**) and those that do not. Long terminal repeats, which play a central role in the process by which the RNA copy of an LTR element is reverse transcribed into double-stranded DNA (Section 17.3.2), are also possessed by viral retroelements (see Figure 9.8). It is now clear that these viruses are one member of a superfamily of elements that also includes endogenous LTR transposons. The first of the endogenous elements to be discovered was the *Ty* sequence of yeast, which is 6.3 kb in length and has a copy number of 25–35 in most *Saccharomyces cerevisiae* genomes—recall that one such element was present in the 50 kb segment of the yeast genome that we examined in Section 7.2.2 (see Figure 7.15B). Yeast genomes also contain 100 or so additional copies of the 330 bp LTRs of *Ty* elements, these solo "delta" sequences probably arising by homologous recombination between the two LTRs of a *Ty* element, which could excise the bulk of the element leaving a single LTR (Figure 9.13). This excision event is probably unrelated to transposition of a *Ty* element, which occurs by the RNA-mediated process shown in Figure 9.12.

There are several types of *Ty* element in yeast genomes. The most abundant of these, *Ty1*, is similar to the *copia* retroelement of the fruit fly. These elements are therefore now called the *Ty1/copia* family. If we compare the structure of a *Ty1/copia* retroelement with that of a viral retroelement, then we

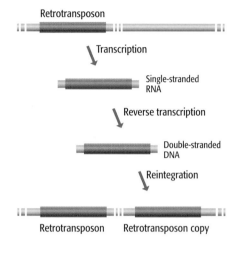

Figure 9.12 **Retrotransposition.** Compare with Figure 7.20 and note that the events are essentially the same as those that result in a processed pseudogene.

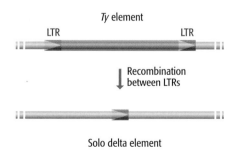

Figure 9.13 Homologous recombination between the LTRs at either end of a *Ty* element could give rise to a delta sequence.

(A) Viral retroelement

~7 kb

(B) *Ty1/copia* retroelement

~7 kb

(C) *Ty3/gypsy* retroelement

~7 kb

Figure 9.14 Genome structures for LTR retroelements.

see clear family relationships (Figure 9.14A and B). Each *Ty1/copia* element contains two genes, called *TyA* and *TyB* in yeast, which are similar to the *gag* and *pol* genes of a viral retroelement. In particular, *TyB* codes for a polyprotein that includes the reverse transcriptase that plays the central role in transposition of a *Ty1/copia* element. Note, however, that the *Ty1/copia* element lacks an equivalent of the viral *env* gene, the one that codes for the viral coat proteins. This means that *Ty1/copia* retroelements cannot form infectious virus particles and therefore cannot escape from their host cell. They do, however, form virus-like particles (VLPs) consisting of the RNA and DNA copies of the retroelements attached to core proteins derived from the TyA polyprotein. In contrast, the members of a second family of LTR retroelements, called *Ty3/gypsy* (again after the yeast and fruit fly versions), do have an equivalent of the *env* gene (Figure 9.14C) and at least some of these can form infectious viruses. Although classed as endogenous transposons, these infectious versions should be looked upon as viral retroelements.

LTR retroelements make up substantial parts of many eukaryotic genomes, and are particularly abundant in the larger plant genomes, especially those of grasses such as maize (see Figure 7.15D). They also make up an important component of invertebrate and some vertebrate genomes, but in the genomes of humans and other mammals all the LTR elements appear to be decayed viral retroelements rather than true transposons. These sequences are called **endogenous retroviruses (ERVs)** and with a copy number of approximately 240,000 they make up 4.7% of the human genome (Table 9.3). Human ERVs are 6–11 kb in length and have copies of the *gag*, *pol*, and *env* genes. Although most contain mutations or deletions that inactivate one or more of these genes, a few members of the human ERV group HERV-K have functional sequences. By comparing the positions of the HERV-K elements in the genomes of different individuals, it has been inferred that at least some of these are active transposons. The majority of human ERVs are, however, inactive sequences that are not capable of additional proliferation.

Table 9.3 Transposable elements in the human genome

Class	Family	Approximate number of copies	Fraction of genome (%)
SINE	Alu	1,200,000	10.7
	MIR	450,000	2.5
	MIR3	85,000	0.4
LINE	LINE-1	600,000	17.3
	LINE-2	370,000	3.3
	LINE-3	44,000	0.3
LTR retroelements	ERV	240,000	4.7
	MaLR	285,000	3.8
DNA transposons	MER-1	213,000	1.4
	MER-2	68,000	1.0
	Others	60,000	0.4

RNA transposons that lack LTRs

Not all types of RNA transposon have LTR elements. In mammals the most important types of nonLTR retroelements, or **retroposons**, are the **LINEs** (**long interspersed nuclear elements**) and **SINEs** (**short interspersed nuclear elements**). SINEs have the highest copy number for any type of interspersed repetitive DNA in the human genome, with over 1.7 million copies comprising 14% of the genome as a whole (Table 9.3). LINEs are less frequent, with just over 1 million copies, but as they are longer they make up a larger fraction of the genome—over 20%. The abundance of LINEs and SINEs in the human genome is underlined by their frequency in the 50 kb segment that we looked at in Section 7.2.2 (see Figure 7.12).

There are three families of LINEs in the human genome, of which one group, LINE-1, is both the most frequent and the only type that is able to transpose, the LINE-2 and LINE-3 families being made up of inactive relics. A full-length LINE-1 element is 6.1 kb and has two genes, one of which codes for a polyprotein similar to the product of the viral *pol* gene (Figure 9.15A). There are no LTRs, but the 3′ end of the LINE is marked by a series of A–T base pairs, giving what is usually referred to as a poly(A) sequence (though of course it is a poly(T) sequence on the other strand of the DNA). Not all copies of LINE-1 are full length, because the reverse transcriptase coded by LINEs does not always make a complete DNA copy of the initial RNA transcript, meaning that part of the 3′ end of the LINE may be lost. This truncation event is so common that only 1% of the LINE-1 elements in the human genome are full-length versions, with the average size of all the copies being just 900 bp. Although LINE-1 transposition is a rare event, it has been observed in cultured cells and appears to be responsible for hemophilia in some patients, due to movement of a LINE-1 sequence into the factor VIII gene, disrupting the gene and hence preventing synthesis of this important blood clotting protein.

SINEs are much shorter than LINEs, being just 100–400 bp and not containing any genes, which means that SINEs do not make their own reverse transcriptase enzymes (Figure 9.15B). Instead they "borrow" reverse transcriptases that have been synthesized by LINEs. The commonest SINE in primate genomes is **Alu**, which has a copy number of approximately 1.2 million in humans (Table 9.3). An Alu element comprises two halves, each half made up of a similar 120 bp sequence, with a 31–32 bp insertion in the right half (Figure 9.16). The mouse genome has a related element, called B1, which is 130 bp in length and equivalent to one half of an Alu sequence. Some Alu elements are actively copied into RNA, providing the opportunity for proliferation of the element.

Alu is derived from the gene for the 7SL RNA, a noncoding RNA involved in movement of proteins around the cell. The first Alu element may have arisen by the accidental reverse transcription of a 7SL RNA molecule and integration of the DNA copy into the human genome. Other SINEs are derived from tRNA genes which, like the gene for the 7SL RNA, are transcribed by RNA polymerase III in eukaryotic cells (Section 11.2.1), suggesting that some feature of the transcripts synthesized by this polymerase make these molecules prone to occasional conversion into retroposons.

9.2.2 DNA transposons

Not all transposons require an RNA intermediate. Many are able to transpose in a more direct DNA-to-DNA manner. In eukaryotes, these DNA transposons

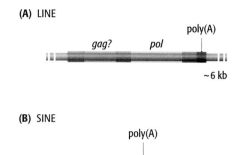

Figure 9.15 NonLTR retroelements. Both LINEs and SINEs have poly(A) sequences at their 3′ ends.

Figure 9.16 The structure of an Alu element. The element consists of two halves, each of 120 bp, with a 31–32 bp insertion in the right half, and a poly(A) tail at the 3′ end. The two halves (excluding the insertion) have about 85% sequence identity.

Figure 9.17 DNA transposons of prokaryotes. Four types are shown. Insertion sequences, Tn3-type transposons, and transposable phages are flanked by short (<50 bp) inverted terminal repeat (ITR) sequences. The resolvase gene of the Tn3-type transposon codes for a protein involved in the transposition process.

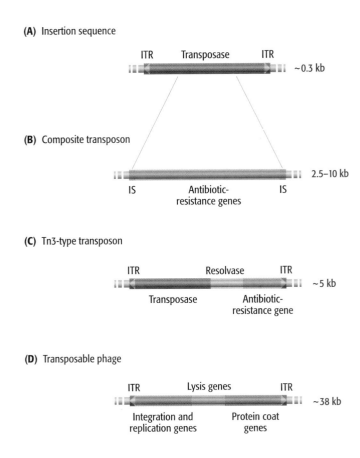

(A) Insertion sequence

ITR   Transposase   ITR

~0.3 kb

(B) Composite transposon

2.5–10 kb

IS   Antibiotic-resistance genes   IS

(C) Tn3-type transposon

ITR   Resolvase   ITR

~5 kb

Transposase   Antibiotic-resistance gene

(D) Transposable phage

ITR   Lysis genes   ITR

~38 kb

Integration and replication genes   Protein coat genes

are less common than retrotransposons, but they have a special place in genetics because a family of plant DNA transposons—the Ac/Ds elements of maize—were the first transposable elements to be discovered, by Barbara McClintock in the 1950s. Her conclusions—that some genes are mobile and can move from one position to another in a chromosome—were based on exquisite genetic experiments, the molecular basis of transposition not being understood until the late 1970s.

DNA transposons are common in prokaryotic genomes

DNA transposons are an important component of many prokaryotic genomes. The insertion sequences, IS1 and IS186, present in the 50 kb segment of *E. coli* DNA that we examined in Section 8.2.1 (see Figure 8.7), are examples of DNA transposons, and a single *E. coli* genome may contain as many as 20 of these of various types. Most of the sequence of an IS is taken up by one or two genes that specify the **transposase** enzyme that catalyzes its transposition (Figure 9.17A). There are a pair of inverted repeats at either end of each IS element, between 9 bp and 41 bp in length depending on the type of IS, and insertion of the element into the target DNA creates a pair of short (4–13 bp) *direct* repeats in the host genome. IS elements can transpose either replicatively or conservatively.

IS elements are also components of a second type of DNA transposon first characterized in *E. coli* and now known to be common in many prokaryotes. These **composite transposons** are made up of a pair of IS elements flanking a segment of DNA, usually containing one or more genes—often ones coding for antibiotic resistance (Figure 9.17B). Tn10, for example, carries a gene for tetracycline resistance, and Tn5 and Tn903 both carry a gene for resistance to kanamycin. Some composite transposons have identical IS elements at either end, and others have one element of one type and one of another. In some cases the IS elements are orientated as direct repeats and sometimes as inverted repeats. These variations do not appear to affect the transposition mechanism for a composite transposon, which is conservative in nature and catalyzed by the transposase coded by one or both of the IS elements.

Various other classes of DNA transposon are known in prokaryotes. Two additional important types from *E. coli* are:

- **Tn3-type transposons**, which have their own transposase gene and so do not require flanking IS elements in order to transpose (Figure 9.17C). Tn3 elements transpose replicatively.

- **Transposable phages**, which are bacterial viruses that transpose replicatively as part of their normal infection cycle (Figure 9.17D).

DNA transposons are less common in eukaryotic genomes

The human genome contains approximately 350,000 DNA transposons of various types (Table 9.3), all with terminal inverted repeats and all containing a gene for a transposase enzyme that catalyzes the transposition event. However, the vast majority of these elements are inactive, either because the transposase gene is nonfunctional or because sequences at the ends of the transposon, which are essential for active transposition, are missing or mutated.

Active DNA transposons are more common in plants, and include the Ac/Ds transposon, the first one to be discovered by McClintock, and the Spm element, both of which are found in maize. An interesting feature of these plant transposons is that they work together in family groups. For example, the Ac element codes for an active transposase that recognizes both Ac elements and Ds sequences. The latter are versions of Ac that have internal deletions that remove part of the transposase gene, meaning that a Ds element cannot

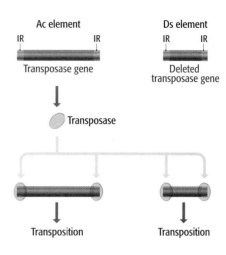

Figure 9.18 The Ac/Ds transposon family of maize. The full-length Ac element is 4.2 kb and contains a functional transposase gene. The transposase recognizes the 11 bp inverted repeats (IRs) at either end of the Ac sequence and catalyzes its transposition. The Ds element has an internal deletion and so does not synthesize its own transposase. But it still has the IR sequences, which are recognized by the transposase made by the Ac element. Hence the Ds element is also able to transpose. There are approximately ten different types of Ds element in the maize genome, with deletions ranging in size from 194 bp to several kilobases.

Figure 9.19 Variegated pigmentation in maize kernels caused by transposition in somatic cells. The highly colored forms of *Zea mays* are popularly known as "Indian corn." Image courtesy of Lena Struwe.

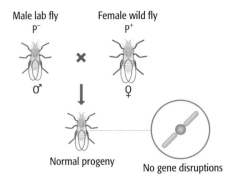

Male lab fly
P⁻

Female wild fly
P⁺

×

↓

Normal progeny   No gene disruptions

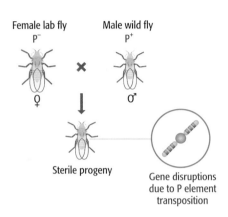

Female lab fly
P⁻

Male wild fly
P⁺

×

↓

Sterile progeny   Gene disruptions
                         due to P element
                         transposition

Figure 9.20 Hybrid dysgenesis. Crosses between male lab flies and female wild flies give normal progeny, but when the male partner is a wild fly the offspring are sterile. One possible explanation of hybrid dysgenesis is that the cytoplasm of flies with P elements (P⁺ in this diagram) contains a repressor that prevents P element transposition. The fertilized egg resulting from a cross between a female P⁺ fly and male P⁻ fly will contain this repressor and so the progeny are normal. However, the repressor will not be carried in the sperm from a male P⁺ fly, so the fertilized egg from a cross between a male P⁺ and a female P⁻ fly will lack the repressor, allowing P element transposition to occur and resulting in progeny displaying hybrid dysgenesis.

make its own transposase and can move only through the activity of the transposase synthesized by a full-length Ac element (Figure 9.18). Similarly, full-length Spm elements are accompanied by deleted versions which transpose through use of the transposase enzymes coded by the intact elements. The activity of Ac elements is apparent during the normal life cycle of a maize plant, transposition in somatic cells resulting in changes in gene expression which are manifested in, for example, variegated pigmentation in maize kernels (Figure 9.19).

McClintock's realization that the maize genome contains transposable elements resulted from her studies into the genetic basis of the different color patterns displayed by kernels. The P element, a DNA transposon in *Drosophila melanogaster*, was similarly discovered from studies of an unusual genetic event which, as it turns out, arises from transposition. This event is called **hybrid dysgenesis** and occurs when females from laboratory strains of *D. melanogaster* are crossed with males from wild populations. The offspring resulting from such crosses are sterile and have chromosomal abnormalities along with a variety of other genetic malfunctions. The explanation is that the genomes of wild fruit flies contain inactive versions of P elements—typical DNA transposons comprising a transposase gene flanked by inverted terminal repeats—but that laboratory strains lack these elements. After crossing, the elements inherited from the wild flies become active in the fertilized eggs, transposing into various new positions and causing the gene disruptions that characterize hybrid dysgenesis (Figure 9.20). Exactly why this activation occurs is not known, but a more interesting question is why the genomes of wild populations of *D. melanogaster* contain P elements whereas laboratory strains do not. Most of the laboratory strains are descended from flies collected by Thomas Hunt Morgan some 90 years ago, and used by Morgan and his colleagues in the first gene mapping experiments (Section 3.2.3). It appears that wild populations at that time lacked P elements, which have somehow proliferated in wild genomes during the last 90 years. The inability of wild and laboratory flies to produce viable offspring means that these two populations fail one of the main criteria used to identify biological species—the ability of all individuals to mate productively. This raises the intriguing possibility that speciation might, at least in some organisms, be driven by differential proliferation of transposable elements within the genomes of members of different populations.

Summary

Early studies of viruses focused largely on the bacteriophages—viruses that infect bacteria. Bacteriophages are constructed of protein and nucleic acid, the protein forming a capsid that encloses the genome. There are three basic types of capsid structure and many types of genome organization, different phages having single- or double-stranded DNA or RNA genomes, some with the entire genome contained in a single molecule, and some with segmented genomes. Bacteriophages follow two distinct infection cycles. All phages can infect via the lytic cycle, which results in the immediate synthesis of new bacteriophages, usually accompanied by death of the host cell. Some can also follow the lysogenic cycle, during which a copy of the phage genome becomes inserted into the host DNA, where it may remain in quiescent form for many cell generations. Eukaryotic viruses are equally diverse in terms in genome organization but display just two capsid structures. Most eukaryotic viruses follow a lytic infection cycle but this does not always result in the immediate death of the host cell. A number of DNA and RNA viruses can integrate their genomes into eukaryotic chromosomes in a manner similar to a lysogenic bacteriophage. The viral retroelements, which include HIV, the causative agent of AIDS, are examples of integrative RNA viruses. Satellite RNAs and virusoids are different types of infective RNA molecule that contain no genes and depend on other viruses for their transmission. Viroids are small, infective RNA molecules that never become encapsidated, and prions are infective proteins. Some mobile genetic elements, which are DNA sequences that can transpose within a genome but cannot escape from the cell, are related to RNA viruses. These elements transpose via an RNA intermediate in a pathway similar to the infection process of viral retroelements. The *Ty1/copia* and *Ty3/gypsy* retroelements, and the endogenous retroviruses of mammals, are the mobile elements most closely related to RNA viruses. Mammalian genomes also contain other types of RNA transposon, called LINEs and SINEs, most of which have lost their ability to transpose. DNA transposons do not make use of an RNA intermediate in their transposition pathway. These transposons are common in bacteria, within which they are responsible for the spread of genes coding for antibiotic resistance. DNA transposons are less widespread in eukaryotes but include some important examples, such as the Ac/Ds transposon of maize, the first transposon of any kind to be studied in detail, and the P element of *Drosophila melanogaster*, which is responsible for the hybrid dysgenesis that occurs when female laboratory fruit flies are crossed with male wild flies.

Multiple Choice Questions

*Answers to odd-numbered questions can be found in the Appendix

9.1.* Which type of bacteriophage capsid structure comprises polypeptide subunits arranged in a specific structure that surrounds a nucleic acid core, and a filamentous tail that facilitates entry into cells?

a. Icosahedral.
b. Filamentous.
c. Head-and-tail.
d. Segmented.

9.2. Which type of bacteriophage capsid structure comprises polypeptide subunits arranged in a helix resulting in a rod-like structure?

a. Icosahedral.
b. Filamentous.
c. Head-and-tail.
d. Segmented.

9.3.* In which type of bacteriophage life cycle is the host cell killed shortly after the initial infection?

a. Lytic.
b. Lysogenic.
c. Temperate.
d. Prophage.

9.4. A prophage is defined as:

a. A new phage particle that is assembled inside a host cell during infection.
b. An RNA molecule that does not encode its own capsid proteins.
c. A phage with an RNA genome that is converted to DNA by the enzyme reverse transcriptase.
d. A quiescent form of a bacteriophage that is integrated into the host cell genome.

9.5.* How do eukaryotic viruses acquire lipid membranes?

a. The lipids are synthesized by proteins coded by viral genes.
b. The viral capsid acquires the membrane when it leaves the host cell.
c. The viral capsid acquires the membrane when it is assembled inside the host cell.
d. The viral capsid acquires the membrane when it first binds to a host cell.

9.6. The enzyme reverse transcriptase is present in which type of viruses?

a. Prions.
b. Prophages.
c. Retroviruses.
d. Virusoids.

9.7.* Which of the following are RNA molecules that do not encode their own capsid proteins and move from cell to cell with the assistance of helper viruses?

a. Prions.
b. Prophages.
c. Retroviruses.
d. Virusoids.

9.8. How can viroids replicate and move from cell to cell if they contain no genes and never become encapsidated?

a. They are replicated and transferred from cell to cell with the assistance of a helper virus.
b. They are replicated by host cell enzymes and move from cell to cell with the assistance of a helper virus.
c. They are replicated by host cell or helper virus enzymes and move from cell to cell as naked RNA.
d. They are replicated with the assistance of a helper virus and move from cell to cell as naked DNA.

9.9.* Prions are defined as infectious, disease-causing particles that:

a. Contain only RNA.
b. Contain only DNA.
c. Contain only proteins (no nucleic acids).
d. Contain only lipids (no nucleic acids).

9.10. Conservative transposition is characterized by which of the following?

a. Excision of a transposon from one location and its subsequent insertion at a different location.
b. Replication of a transposon such that the original sequence remains in place and the new sequence is inserted at a different location.
c. Movement of a transposon from one cell to another.
d. Replication of repeated DNA sequences due to slippage of DNA polymerase.

9.11.* Which of the following enzymes is specified by a gene present in RNA transposons?

a. DNA polymerase.
b. RNA polymerase.
c. Reverse transcriptase.
d. Telomerase.

9.12. Which of the following are RNA transposons that lack long terminal repeats (LTRs) and are unable to synthesize their own reverse transcriptases?

a. Retroelements.
b. Endogenous retroviruses (ERVs).
c. Long interspersed nuclear elements (LINEs).
d. Short interspersed nuclear elements (SINEs).

Multiple Choice Questions (continued) *Answers to odd-numbered questions can be found in the Appendix

9.13.* What is thought to be the origin of the Alu RNA transposon?

    a. It is thought to be derived from a retrovirus.

    b. It is thought to be derived from a protein-coding gene.

    c. It is thought to be derived from a cellular noncoding RNA molecule.

    d. It is thought to be derived from a DNA virus.

9.14. Which enzyme is specified by a gene present in DNA transposons?

    a. DNA polymerase.

    b. RNA polymerase.

    c. Reverse transcriptase.

    d. Transposase.

9.15.* Name the researcher who first identified transposons and the organism he or she studied.

    a. David Baltimore and retroviruses.

    b. Barbara McClintock and maize.

    c. Thomas Hunt Morgan and fruit flies.

    d. Craig Venter and humans.

Short Answer Questions   *Answers to odd-numbered questions can be found in the Appendix

9.1.* How are viruses different from cells? Is it appropriate to look on viruses as living organisms?

9.2. How do the genomes of viruses differ from cellular genomes?

9.3.* What are overlapping genes, as found in some viral genomes?

9.4. How long does it take a lytic bacteriophage to lyse a host cell following the initial infection? What is the time line for the lytic infection cycle of T4 phage?

9.5.* Discuss the differences between the capsids of bacteriophages and eukaryotic viruses.

9.6. Discuss the life cycle of retroviruses.

9.7.* What is a transposon?

9.8. What are the characteristics of the LTR retroelements present in the human genome?

9.9.* Discuss the properties and types of retroposons present in the human genome.

9.10. What are the general properties of composite transposons?

9.11.* What are the important features of the DNA transposons found in plants?

9.12. Describe the basis to hybrid dysgenesis in fruit flies.

In-depth Problems   *Guidance to odd-numbered questions can be found in the Appendix

9.1.* To what extent can viruses be considered a form of life?

9.2. Bacteriophages with small genomes (for example, φX1s74) are able to replicate very successfully in their hosts. Why then should other bacteriophages, such as T4, have large and complicated genomes?

9.3.* Genetic elements that reproduce within or along with a host genome, but confer no benefit on the host, are sometimes called "selfish" DNA. Discuss this concept, in particular as it applies to transposons.

9.4. Some bacteriophages, such as T4, modify the host RNA polymerase after infection, so that this polymerase no longer recognizes *E. coli* genes, but transcribes bacteriophage genes instead. How might this modification be carried out?

9.5.* Why do LTR retroelements have long terminal repeats?

continued …

Figure Tests

9.1.* Identify the three types of bacteriophage capsid structure.

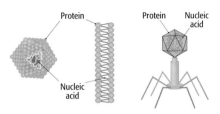

9.2. What type of bacteriophage life cycle is represented in the figure?

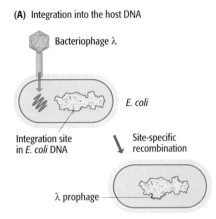

9.3.* What type of viral infection is shown in the figure?

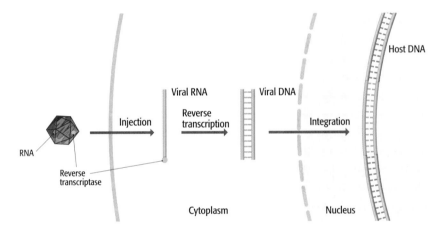

9.4. This figure shows the genome of what type of virus? What are the functions of the LTR sequences?

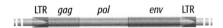

9.5.* Name the researcher who first described the Ac and Ds elements. What is the difference between these elements?

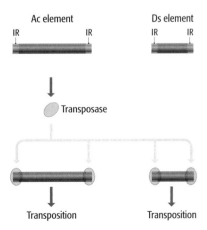

Further Reading

Classic papers on bacteriophage genetics

Delbrück, M. (1940) The growth of bacteriophage and lysis of the host. *J. Gen. Physiol.* **23:** 643–660.

Doermann, A.H. (1952) The intracellular growth of bacteriophage. *J. Gen. Physiol.* **35:** 645–656.

Ellis, E.L. and Delbrück, M. (1939) The growth of bacteriophage. *J. Gen. Physiol.* **22:** 365–383.

Lwoff, A. (1953) Lysogeny. *Bacteriol. Rev.* **17:** 269–337.

Bacteriophage genome sequences

Dunn, J.J. and Studier, F.W. (1983) Complete nucleotide sequence of bacteriophage T7 DNA and the locations of T7 genetic elements. *J. Mol. Biol.* **166:** 477–535.

Sanger, F., Air, G.M., Barrell, B.G., Brown, N.L., Coulson, A.R., Fiddes, C.A., Hutchison, C.A., Slocombe, P.M. and Smith, M. (1977) Nucleotide sequence of bacteriophage φX174 DNA. *Nature* **265:** 687–695.

Sanger, F., Coulson, A.R., Hong, G.F., Hill, D.F. and Petersen, G.B. (1982) Nucleotide sequence of bacteriophage λ DNA. *J. Mol. Biol.* **162:** 729–773.

Eukaryotic viruses

Baltimore, D. (1970) RNA-dependent DNA polymerase in virions of RNA tumour viruses. *Nature* **226:** 1209–1211.

Dimmock, N.J., Easton, A.J. and Leppard, K.N. (2001) *An Introduction to Modern Virology*, 5th Edn. Blackwell Scientific Publishers, Oxford. *The best general text on viruses.*

Temin, H.M. and Mizutani, S. (1970) RNA-dependent DNA polymerase in virions of Rous sarcoma virus. *Nature* **226:** 1211–1213.

Varmus, H. and Brown, P. (1989) Retroviruses. In: *Mobile DNA* (eds D.E. Berg and M. Howe). American Society for Microbiology, Washington, DC, pp. 3–108.

Prions

Prusiner, S.B. (1996) Molecular biology and pathogenesis of prion diseases. *Trends Biochem. Sci.* **21:** 482–487.

RNA transposons

Kumar, A. and Bennetzen, J.L. (1999) Plant retrotransposons. *Annu. Rev. Genet.* **33:** 479–532. *Detailed review of this subject.*

Ostertag, E.M. and Kazazian, H.H. (2005) LINEs in mind. *Nature* **435:** 890–891. *Brief review of recent research into LINEs.*

Patience, C., Wilkinson, D.A. and Weiss, R.A. (1997) Our retroviral heritage. *Trends Genet.* **13:** 116–120. *ERVs.*

Peterson-Burch, B.D., Wright, D.A., Laten, H.M. and Voytas, D.F. (2000) Retroviruses in plants? *Trends Genet.* **16:** 151–152.

Song, S.U., Gerasimova, T., Kurkulos, M., Boeke, J.D. and Corces, V.G. (1994) An env-like protein encoded by a *Drosophila* retroelement: evidence that *gypsy* is an infectious retrovirus. *Genes Dev.* **8:** 2046–2057.

Volff, J.-N., Bouneau, L., Ozouf-Costaz, C. and Fischer, C. (2003) Diversity of retrotransposable elements in compact pufferfish genomes. *Trends Genet.* **19:** 674–678.

DNA transposons

Comfort, N.C. (2001) *The Tangled Field: Barbara McClintock's Search for the Patterns of Genetic Control.* Harvard University Press, Cambridge, MA. *A biography of the geneticist who discovered transposable elements; for a highly condensed version, see* Trends Genet. **17:** 475–478.

Engels, W.R. (1983) The P family of transposable elements in *Drosophila. Annu. Rev. Genet.* **17:** 315–344.

Gierl, A., Saedler, H. and Peterson, P.A. (1989) Maize transposable elements. *Annu. Rev. Genet.* **23:** 71–85.

Kleckner, N. (1981) Transposable elements in prokaryotes. *Annu. Rev. Genet.* **15:** 341–404.

PART 3

How Genomes Function

Part 3 – How Genomes Function examines the events that result in the transfer of biological information from genome to transcriptome to proteome. We begin with the genome itself and the way in which chromatin structure influences the overall pattern of genome expression (Chapter 10). We then examine the structures of the transcription initiation complexes of eukaryotes and prokaryotes, and investigate the ways in which these structures are assembled (Chapter 11), before moving on to the events responsible for synthesis and processing of the components of the transcriptome (Chapter 12) and proteome (Chapter 13). Finally, in Chapter 14, we survey the various strategies that are used by cells to regulate expression of their genomes, these strategies enabling genome activity to be responsive to extracellular signals and enabling multicellular organisms to follow complex developmental pathways.

Accessing the Genome

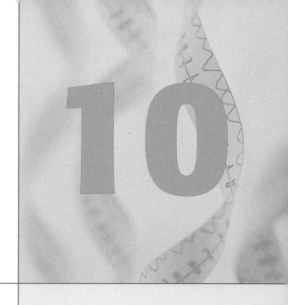

10

When you have read Chapter 10, you should be able to:

Explain how chromatin structure influences genome expression.

Describe the internal architecture of the eukaryotic nucleus.

Distinguish between the terms "constitutive heterochromatin," "facultative heterochromatin," and "euchromatin."

Discuss the key features of functional domains, insulators, and locus control regions, and describe the experimental evidence supporting our current knowledge of these structures.

Give details of how histone acetylation and deacetylation are carried out and how these modifications influence genome expression.

Describe the other types of chemical modification that can be made to histone proteins, and link this information to the concept of the "histone code."

State why nucleosome positioning is important in genome expression, and give details of protein complexes involved in nucleosome remodeling.

Explain how DNA methylation is carried out and describe the importance of methylation in silencing the genome.

Give details of the involvement of DNA methylation in genomic imprinting and X inactivation.

In order for the cell to utilize the biological information contained within its genome, groups of genes, each gene representing a single unit of information, have to be expressed in a coordinated manner. This coordinated gene expression determines the makeup of the transcriptome, which in turn specifies the nature of the proteome and defines the activities that the cell is able to carry out. In Part 3 of *Genomes* we examine the events that result in the transfer of biological information from genome to proteome. Our knowledge of these events was initially gained through studies of individual genes, often as "naked" DNA in test-tube experiments. These experiments provided an interpretation of gene expression that in recent years has been embellished by more sophisticated studies that have taken greater account of the fact that, in reality, it is the genome that is expressed, not individual genes, and this expression occurs in living cells rather than in a test tube.

We begin our investigation of genome expression, here in Chapter 10, by examining the substantial and important impact that the nuclear environment has on the utilization of the biological information contained in the genomes of eukaryotes, the accessibility of that information being dependent on the way in which the DNA is packaged into chromatin, and being responsive to processes that can silence or inactivate part or all of a chromosome. Chapter 11 then describes the events involved in initiation of transcription, and emphasizes further the critical role that DNA-binding proteins play during the early stages of genome expression. The synthesis of transcripts and their subsequent processing into functional RNAs is dealt with in Chapter 12, and Chapter 13 covers the equivalent events that lead to synthesis of the proteome. As you read Chapters 10–13, you will discover that control over the composition of the transcriptome and of the proteome can be exerted at various stages during the overall chain of events that make up genome expression. These regulatory threads will be drawn together in Chapter 14, where we examine how genome activity changes in response to extracellular signals and during differentiation and development.

10.1 Inside the Nucleus

When one looks at a genome sequence written out as a series of As, Cs, Gs, and Ts, or drawn as a map (as in Figure 7.12, for example), there is a tendency to imagine that all parts of the genome are readily accessible to the DNA-binding proteins that are responsible for its expression. In reality, the situation is very different. The DNA in the nucleus of a eukaryotic cell or the nucleoid of a prokaryote is attached to a variety of proteins that are not directly involved in genome expression and which must be displaced in order for the RNA polymerase and other expression proteins to gain access to the genes. We know very little about these events in prokaryotes, a reflection of our generally poor knowledge about the physical organization of the prokaryotic genome (Section 8.1.1), but we are beginning to understand how the packaging of DNA into chromatin (Section 7.1.1) influences genome expression in eukaryotes. This is an exciting area of molecular biology, with recent research indicating that histones and other packaging proteins are not simply inert structures around which the DNA is wound, but instead are active participants in the processes that determine which parts of the genome are expressed in an individual cell. Many of the discoveries in this area have been driven by new insights into the substructure of the nucleus, and it is with this topic that we begin the chapter.

10.1.1 The internal architecture of the eukaryotic nucleus

The internal architecture of the nucleus was first examined by light and electron microscopy. The apparent lack of structure that emerged from these studies led to the view that the inside of the nucleus is relatively homogeneous, a typical "black box" in common parlance. In recent years this interpretation has been overthrown, and we now appreciate that the nucleus has a complex internal structure that is related to the variety of biochemical activities that it must carry out. Indeed, the inside of the nucleus is just as complex as the cytoplasm of the cell, the only difference being that, in contrast to the

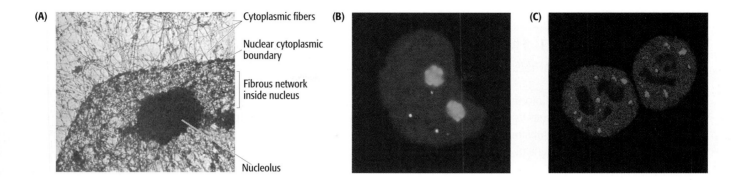

cytoplasm, the functional compartments within the nucleus are not individually enclosed by membranes, and so are not visible when the cell is observed using conventional light or electron microscopy techniques.

The nucleus has a highly ordered internal structure

This revised picture of nuclear structure has emerged from two novel types of microscopic analysis. First, conventional electron microscopy has been supplemented by examination of mammalian cells that have been prepared in a special way. After dissolution of membranes by soaking in a mild, nonionic detergent such as one of the Tween compounds, followed by treatment with a deoxyribonuclease to degrade the nuclear DNA, and salt extraction to remove the chemically basic histone proteins, the nuclear substructure has been revealed as a complex network of protein and RNA fibrils, called the **nuclear matrix** (Figure 10.1A). The matrix permeates the entire nucleus and includes regions defined as the **chromosome scaffold**, which changes its structure during cell division, resulting in condensation of the chromosomes into their metaphase forms (see Figure 7.4).

A second novel type of microscopy has involved the use of fluorescent labeling, designed specifically to reveal areas within the nucleus where particular biochemical activities are occurring. The **nucleolus** (Figure 10.1B), which is the center for synthesis and processing of rRNA molecules, has been recognized for many years as it is the one structure within the nucleus that can be seen by conventional electron microscopy. Fluorescent labeling directed at the proteins involved in RNA splicing (Section 12.2.2) has shown that this activity is also localized into distinct regions (Figure 10.1C), although these are more widely distributed and less-well defined than the nucleoli. Other structures, such as Cajal bodies (visible in Figure 10.1B), which are probably involved in synthesis of small nuclear RNAs (Section 12.2.2), are also seen after fluorescent labeling.

The complexity of the nuclear matrix, as shown by Figure 10.1A, could be taken as an indication that the nucleus has a static internal environment, with limited movement of molecules from one site to another. Another new microscopy technique, called **fluorescence recovery after photobleaching** (**FRAP**; Technical Note 10.1), which enables the movement of proteins within the nucleus to be visualized, shows that this is not the case. The migration of nuclear proteins does not occur as rapidly as would be expected if their movement were totally unhindered, which is entirely expected in view of the large amounts of DNA and RNA in the nucleus, but it is still possible for a protein to traverse the entire diameter of a nucleus in a matter of minutes. Proteins involved in genome expression therefore have the freedom needed

Figure 10.1 The internal architecture of the eukaryotic nucleus. (A) Transmission electron micrograph showing the nuclear matrix of a cultured human HeLa cell. Cells were treated with a nonionic detergent to remove membranes, digested with a deoxyribonuclease to degrade most of the DNA, and extracted with ammonium sulfate to remove histones and other chromatin-associated proteins. (B) and (C) Images of living nuclei containing fluorescently labeled proteins (see Technical Note 10.1). In (B), the nucleolus is shown in blue and Cajal bodies in yellow. The purple areas in (C) indicate the positions of proteins involved in RNA splicing. Image (A) reprinted with permission from Penman *et al.*, *Symp. Quant. Biol.*, **46**, 1013. Copyright 1982 Cold Spring Harbor Laboratory Press. Images (B) and (C) reprinted from Misteli (2001), *Science*, **291**, 843–847.

Technical Note 10.1 Fluorescence recovery after photobleaching (FRAP)

Visualization of protein mobility in living nuclei

FRAP is perhaps the most informative of the various innovative microscopy techniques that have opened up our understanding of nuclear substructure. It has enabled, for the first time, the movement of proteins to be visualized inside living nuclei, the resulting data allowing biophysical models of protein dynamism to be tested.

The starting point for a FRAP experiment is a nucleus in which every copy of the protein of interest carries a fluorescent tag. Labeling the protein molecules *in vitro* and then reintroducing them into the nucleus is not possible, so the host organism has to be genetically engineered so that the fluorescent tag is an integral part of the protein that is synthesized *in vivo*. This is achieved by ligating the coding sequence for the **green fluorescent protein** to the gene for the protein being studied. Standard cloning techniques are then used to insert the modified gene into the host genome, leading to a recombinant cell that synthesizes a fluorescent version of the protein. Observation of the cell using a fluorescence microscope now reveals the distribution of the labeled protein within the nucleus.

To study the mobility of the protein, a small area of the nucleus is **photobleached** by exposure to a tightly focused pulse from a high-energy laser. The laser pulse inactivates the fluorescent signal in the exposed area, leaving a region that appears bleached in the microscopic image. This bleached area gradually retrieves its fluorescent signal, not by a reversal of the bleaching effect, but by migration into the bleached region of fluorescent proteins from the unexposed area of the nucleus. Rapid reappearance of the fluorescent signal in the bleached area therefore indicates that the tagged proteins are highly mobile, whereas a slow recovery indicates that the proteins are relatively static. The kinetics of signal recovery can be used to test theoretical models of protein dynamism derived from biophysical parameters such as binding constants and flux rates.

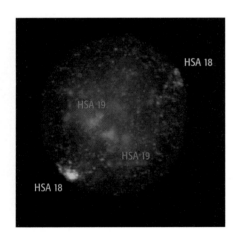

Figure 10.2 Chromosome territories.
Human chromosomes 18 (HSA 18) and 19 (HSA 19) have been painted green and red, respectively. Each occupies its own discrete territory within the nucleus. Image courtesy of Wendy Bickmore.

to move from one activity site to another, as dictated by the changing requirements of the cell. In particular, the linker histones (Section 7.1.1) continually detach and reattach to their binding sites on the genome. This discovery is important because it emphasizes that the DNA–protein complexes that make up chromatin are dynamic, an observation that has considerable relevance to genome expression, as we will discover later in this chapter.

Each chromosome has its own territory within the nucleus

Initially it was thought that chromosomes are distributed randomly within a eukaryotic nucleus. We now know that this view is incorrect and that each chromosome occupies its own space, or **territory**. These can be visualized by **chromosome painting**, which is a version of fluorescent *in situ* hybridization (FISH; Section 3.3.2) in which the hybridization probe is a mixture of DNA molecules, each specific for different regions of a single chromosome. When applied to interphase nuclei, chromosome painting reveals territories occupied by individual chromosomes (Figure 10.2). These territories take up the majority of the space within the nucleus, but are separated from one another by **nonchromatin regions**, within which the enzymes and other proteins involved in expression of the genome are located.

Chromosome territories appear to be fairly static within an individual nucleus. This has been concluded from experiments in which CENP-B proteins, components of the centromeres (Section 7.1.2), are labeled with green fluorescent protein (see Technical Note 10.1) and the locations of these proteins, and hence of the centromeres, observed over a period of time. On the whole, individual centromeres remain stationary throughout the cell cycle, though there are occasional bursts of relatively slow movement. Although

Figure 10.3 Products of translocation between human chromosomes 9 and 22. The normal human chromosomes 9 and 22 are shown on the left and the translocation products are on the right. The Philadelphia chromosome is the smaller of the two translocation products. Chromosomes 9 and 22 commonly break at the positions indicated. Often the breaks are correctly repaired, but occasionally misrepair creates the hybrid products. It is thought that the relatively high frequency with which the Philadelphia chromosome arises indicates that chromosomes 9 and 22 occupy adjacent territories in the human nucleus. The chromosome 9 breakpoint lies within the *ABL* gene, the product of which is involved in cell signaling (Section 14.1.2). The translocation attaches a new coding sequence to the start of this gene, resulting in an abnormal protein which causes cell transformation and gives rise to chronic myeloid leukemia.

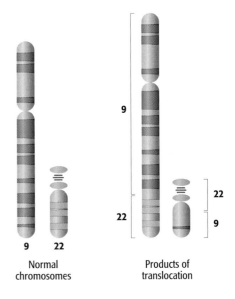

Normal
chromosomes

Products of
translocation

fairly static during the lifetime of a cell, most studies suggest that the relative positioning of territories is not retained after cell division, as different patterns are observed in the nuclei of the daughter cells. There may, however, be certain constraints on territory locations, as it has been known for several years that chromosome **translocations**, which result in a segment of one chromosome becoming attached to another chromosome, are more frequent between certain pairs than others. For example, a translocation between human chromosomes 9 and 22, resulting in the abnormal product called the **Philadelphia chromosome,** is a common cause of chronic myeloid leukemia (Figure 10.3). The repeated occurrence of the same translocation suggests that the territories of the interacting pair of chromosomes are frequently close to one another in the nucleus. There is also evidence that, at least in some organisms, certain chromosomes preferentially occupy territories close to the periphery of the nucleus. Relatively little genome expression occurs in this region, and it is often here that those chromosomes that contain few active genes are found, examples being the macrochromosomes of the chicken genome (Section 7.1.2).

The positioning of active genes within individual chromosome territories is a further topic of debate. At one time it was thought that the active genes were located on the surface of a territory, adjacent to the nonchromatin region and hence within easy reach of the enzymes and proteins involved in gene transcription (Figure 10.4). This view is now being questioned, partly as a result of experiments that have shown that RNA transcripts are distributed within territories as well as on their surfaces. More refined microscopic examination has shown that channels run through chromosome territories, linking different parts of the nonchromatin regions, and providing a means by which the transcription machinery can penetrate into the internal parts of these territories.

10.1.2 Chromatin domains

In Section 7.1.1 we learnt that chromatin is the complex of genomic DNA and chromosomal proteins present in the eukaryotic nucleus. Chromatin structure is hierarchic, ranging from the two lowest levels of DNA packaging—the nucleosome and the 30 nm chromatin fiber (see Figures 7.2 and 7.3)—to the metaphase chromosomes, which represent the most compact form of chromatin in eukaryotes and occur only during nuclear division. After division, the chromosomes become less compact and cannot be distinguished as individual structures unless specialized techniques such as chromosome painting are used. When nondividing nuclei are examined by light microscopy all that can be seen is a mixture of light- and dark-staining areas within the nucleus. The dark areas are called **heterochromatin** and contain DNA that is still in a relatively compact organization, although less compact than in the metaphase structure. Two types of heterochromatin are recognized:

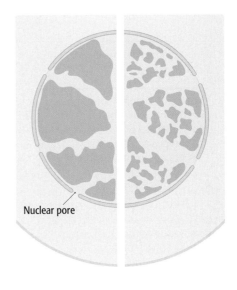

Nuclear pore

Figure 10.4 Chromosome territories. The view on the left shows the original model with each territory forming a block, implying that active genes are located on the surface of a territory. The view on the right shows the revised model, with channels running through the territories.

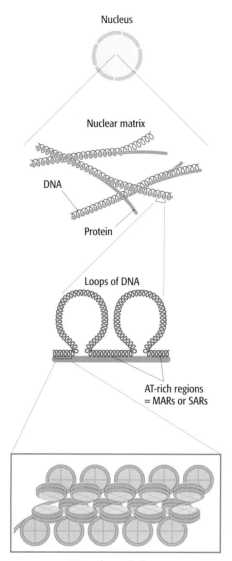

Figure 10.5 **A scheme for organization of DNA in the nucleus.** The nuclear matrix is a fibrous, protein-based structure whose precise composition and arrangement in the nucleus has not been described. Euchromatin, predominantly in the form of the 30 nm chromatin fiber (see Figure 7.3), is thought to be attached to the matrix by AT-rich sequences called matrix-associated regions or scaffold attachment regions (MARs or SARs).

- **Constitutive heterochromatin** is a permanent feature of all cells and represents DNA that contains no genes and so can always be retained in a compact organization. This fraction includes centromeric and telomeric DNA, as well as certain regions of some other chromosomes. For example, most of the human Y chromosome is made of constitutive heterochromatin (see Figure 7.5).

- **Facultative heterochromatin** is not a permanent feature but is seen in some cells some of the time. Facultative heterochromatin is thought to contain genes that are inactive in some cells or at some periods of the cell cycle. When these genes are inactive, their DNA regions are compacted into heterochromatin.

It is assumed that the organization of heterochromatin is so compact that proteins involved in genome expression simply cannot access this type of DNA. In contrast, the parts of the chromosomal DNA where the active genes are located are less compact and permit entry of the expression proteins. These regions are called **euchromatin**. The exact organization of the DNA within euchromatin is not known, but with the electron microscope it is possible to see loops of DNA within the euchromatin regions, each loop between 40 kb and 100 kb in length and predominantly in the form of the 30 nm chromatin fiber. The loops are attached to the nuclear matrix via AT-rich DNA segments called **matrix-associated regions** (**MARs**), or **scaffold attachment regions** (**SARs**) (Figure 10.5).

The loops of DNA between the nuclear matrix attachment points are called **structural domains**. An intriguing question is the precise relationship between these and the **functional domains** that can be discerned when the region of DNA around an expressed gene or set of genes is examined. A functional domain is delineated by treating a region of purified chromatin with deoxyribonuclease I (DNase I) which, being a DNA-binding protein, cannot gain access to the more compacted regions of DNA (Figure 10.6). Regions sensitive to DNase I extend to either side of a gene or set of genes that is being expressed, indicating that in this area the chromatin has a more open organization, although it is not clear whether this organization is the 30 nm fiber or the "beads-on-a-string" structure (Figure 7.2A). Intuition suggests that there should be a correspondence between structural and functional domains, and this view is supported by the location of some MARs, which mark the limits of a structural domain, at the boundary of a functional domain. But the correspondence does not seem to be complete because some structural domains contain genes that are not expressed at the same time, and the boundaries of some structural domains lie within genes.

Functional domains are defined by insulators

The boundaries of functional domains are marked by sequences, 1–2 kb in length, called **insulators**. Insulator sequences were first discovered in *Drosophila* and have now been identified in a range of eukaryotes. The best

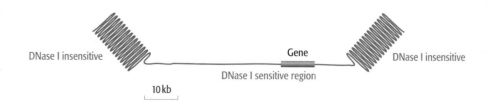

DNase I insensitive   Gene   DNase I insensitive
DNase I sensitive region
10 kb

Figure 10.6 **A functional domain in a DNase I-sensitive region.**

studied are the pair of sequences called scs and scs′ (scs stands for "specialized chromatin structure"), which are located either side of the two *hsp70* genes in the fruit-fly genome (Figure 10.7).

Insulators display two special properties related to their role as the delimiters of functional domains. The first is their ability to overcome the **positional effect** that occurs during a gene-cloning experiment with a eukaryotic host. The positional effect refers to the variability in gene expression that occurs after a new gene has been inserted into a eukaryotic chromosome. It is thought to result from the random nature of the insertion event, which could deliver the gene to a region of highly packaged chromatin, where it will be inactive, or into an area of open chromatin, where it will be expressed (Figure 10.8A). The ability of scs and scs′ to overcome the positional effect was demonstrated by placing them either side of a fruit-fly gene for eye color. When flanked by the insulators, this gene was always highly expressed when inserted back into the *Drosophila* genome, in contrast to the variable expression that was seen when the gene was cloned without the insulators (Figure 10.8B). The deduction from this and related experiments is that insulators can bring about modifications to chromatin packaging and hence establish a functional domain when inserted into a new site in the genome.

Insulators also maintain the independence of each functional domain, preventing "cross-talk" between adjacent domains. If scs or scs′ is excised from its normal location and reinserted between a gene and the upstream regulatory modules that control expression of that gene, then the gene no longer responds to its regulatory modules: it becomes "insulated" from their effects (Figure 10.9A). This observation suggests that, in their normal positions, insulators prevent the genes within a domain from being influenced by the regulatory modules present in an adjacent domain (Figure 10.9B).

How insulators carry out their roles is not yet known but it is presumed that the functional component is not the insulating sequence itself but the DNA-binding proteins, such as Su(Hw) in *Drosophila*, that attach specifically to insulators. As well as binding to insulators, these proteins form associations with the nuclear matrix, possibly indicating that the functional domains that they define are also structural domains within the chromatin. This is an attractive hypothesis that can be tied in with the ability of insulators to establish open chromatin regions and to prevent cross-talk between functional domains, but it implies that insulators contain MAR sequences, which has not been proven. An equivalence between functional and structural domains therefore remains elusive.

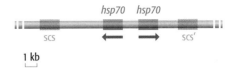

Figure 10.7 Insulator sequences in the fruit-fly genome. The diagram shows the region of the *Drosophila* genome containing the two *hsp70* genes. The insulator sequences scs and scs′ are either side of the gene pair. The arrows below the two genes indicate that they lie on different strands of the double helix and so are transcribed in opposite directions.

Figure 10.8 The positional effect. (A) A cloned gene that is inserted into a region of highly packaged chromatin will be inactive, but one inserted into open chromatin will be expressed. (B) The results of cloning experiments without (red) and with (blue) insulator sequences. When insulators are absent, the expression level of the cloned gene is variable, depending on whether it is inserted into packaged or open chromatin. When flanked by insulators, the expression level is consistently high because the insulators establish a functional domain at the insertion site.

(A) The positional effect

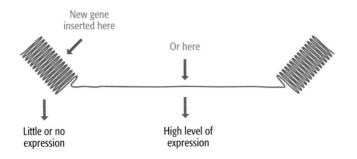

(B) Insulators overcome the positional effect

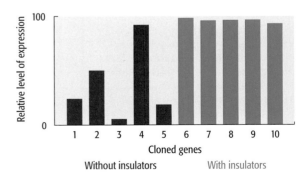

See Figure 7.19 for more information on developmental regulation of expession of the β-globin gene cluster.

Figure 10.9 Insulators maintain the independence of a functional domain. (A) When placed between a gene and its upstream regulatory modules, an insulator sequence prevents the regulatory signals from reaching the gene. (B) In their normal positions, insulators prevent cross-talk between functional domains, so the regulatory modules of one gene do not influence expression of a gene in a different domain.

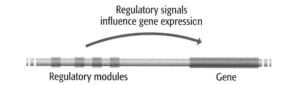

(A) Insulators block the regulatory signals that control gene expression

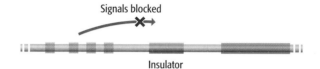

(B) Insulators prevent cross-talk between functional domains

Some functional domains contain locus control regions

The formation and maintenance of an open functional domain, at least for some domains, is the job of a DNA sequence called the **locus control region**, or **LCR**. Like insulators, an LCR can overcome the positional effect when linked to a new gene that is inserted into a eukaryotic chromosome. Unlike insulators, an LCR also stimulates the expression of genes contained within its functional domain.

LCRs were first discovered during a study of the human β-globin genes (Section 7.2.3) and are now thought to be involved in expression of many genes that are active in only some tissues or during certain developmental stages. The globin LCR is contained in a stretch of DNA some 12 kb in length, positioned upstream of the genes in the 60 kb β-globin functional domain (Figure 10.10). The LCR was initially identified during studies of individuals with thalassemia, a blood disease that results from defects in the α- or β-globin proteins. Many thalassemias result from mutations in the coding regions of the globin genes, but a few were shown to map to a 12 kb region upstream of the β-globin gene cluster, the region now called the LCR. The ability of mutations in the LCR to cause thalassemia is a clear indication that disruption of the LCR results in a loss of globin gene expression.

Figure 10.10 DNase I hypersensitive sites indicate the position of the locus control region for the human β-globin gene cluster. A series of hypersensitive sites are located in the 20 kb of DNA upstream of the start of the β-globin gene cluster. These sites mark the position of the locus control region. Additional hypersensitive sites are seen immediately upstream of each gene, at the position where RNA polymerase attaches to the DNA. These hypersensitive sites are specific to different developmental stages, being seen only during the phase of development when the adjacent gene is active. The 60 kb region shown here represents the entire β-globin functional domain. See Figure 7.19 for more information on developmental regulation of expession of the β-globin gene cluster.

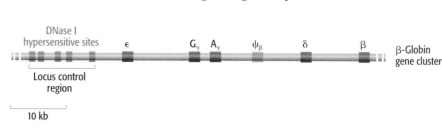

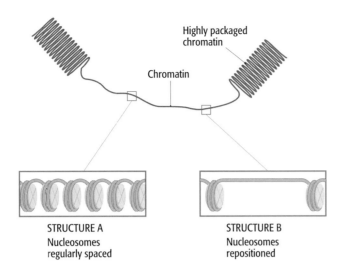

STRUCTURE A
Nucleosomes
regularly spaced

STRUCTURE B
Nucleosomes
repositioned

Figure 10.11 Two ways in which chromatin structure can influence gene expression. A region of unpackaged chromatin in which the genes are accessible is flanked by two more-compact segments. Within the unpackaged region, the positioning of the nucleosomes influences gene expression. On the left, the nucleosomes have regular spacing, as displayed by the typical "beads-on-a-string" structure. On the right, the nucleosome positioning has changed and a short stretch of DNA, approximately 300 bp in length, is exposed.

More detailed study of the β-globin LCR has shown that it contains five separate **DNase I hypersensitive sites**, short regions of DNA that are cleaved by DNase I more easily than other parts of the functional domain. These sites are thought to coincide with positions where nucleosomes have been modified or are absent and which are therefore accessible to binding proteins that attach to the DNA. It is these proteins, not the DNA sequence of the LCR, that control the chromatin structure within the functional domain. Exactly how, and in response to what biochemical signals, is not known.

DNase I hypersensitive sites also occur immediately upstream of each of the genes in the β-globin functional domain (see Figure 10.10) at the positions where the transcription initiation complex is assembled on the DNA (Section 11.2.2). These assembly positions illustrate an interesting feature of DNase I hypersensitive sites: they are not invariant components of a functional domain. Recall that the different β-type globin genes are expressed at different stages of the human developmental cycle, ε being active in the early embryo, G_γ and A_γ in the fetus, and δ and β in the adult (see Figure 7.19). Only when the gene is active is its assembly position for the transcription initiation complex marked by a hypersensitive site. Initially it was thought that this was an *effect* of the differential expression of these genes; in other words, that in the absence of gene activity it was possible for nucleosomes to cover the assembly site, presumably to be pushed to one side when it became time to express the gene. Now it is thought that the presence or absence of nucleosomes is a *cause* of gene expression, the gene being switched off if nucleosomes cover the assembly site, or switched on if access to the site is open.

10.2 Chromatin Modifications and Genome Expression

The previous sections have introduced us to two ways in which chromatin structure can influence genome expression (Figure 10.11):

- The degree of chromatin packaging displayed by a segment of a chromosome determines whether or not genes within that segment are expressed.

- If a gene is accessible, then its transcription is influenced by the precise nature and positioning of the nucleosomes in the region where the transcription initiation complex will be assembled.

Ac Ac
H2A S G R G K Q G G K A R A K A K T R S S R
10 20
Ac Ac Ac Ac
H2B P E P S K S A P A P K K G S K K A I T K A
10 20
Ac Ac Ac Ac
H3 A R T K Q T A R K S T G G K A P R K Q L A T K A R K S A P
10 20
Ac Ac Ac Ac
H4 S G R G K G G K G L G K G G A K R H R K
10 20

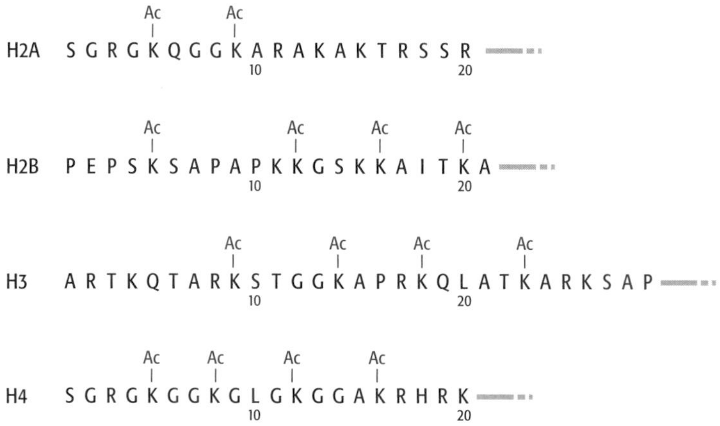

Figure 10.12 The positions at which acetyl groups are attached within the N-terminal regions of the four core histones. Each sequence begins with the N-terminal amino acid.

Significant advances in understanding both types of chromatin modification have been made in recent years. We will begin with the processes that influence chromatin packaging.

10.2.1 Chemical modification of histones

Nucleosomes appear to be the primary determinants of genome activity in eukaryotes, not only by virtue of their positioning on a strand of DNA, but also because the precise chemical structure of the histone proteins contained within nucleosomes is the major factor determining the degree of packaging displayed by a segment of chromatin.

Acetylation of histones influences many nuclear activities including genome expression

Histone proteins can undergo various types of modification, the best studied of these being **histone acetylation**—the attachment of acetyl groups to lysine amino acids in the N-terminal regions of each of the core molecules (Figure 10.12). These N termini form tails that protrude from the nucleosome core octamer (Figure 10.13) and their acetylation reduces the affinity of the histones for DNA and possibly also reduces the interaction between individual

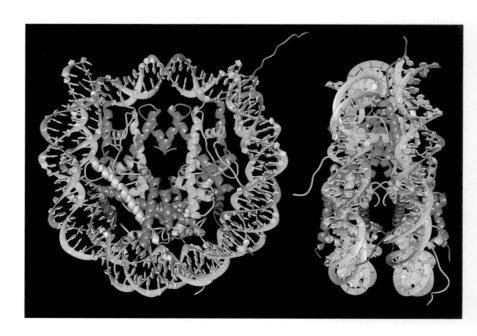

Figure 10.13 Two views of the nucleosome core octamer. The view on the left is downward from the top of the barrel-shaped octamer; the view on the right is from the side. The two strands of the DNA double helix wrapped around the octamer are shown in brown and green. The octamer comprises a central tetramer of two histone H3 (blue) and two histone H4 (bright green) subunits plus a pair of H2A (yellow)–H2B (red) dimers, one above and one below the central tetramer. Note the N-terminal tails of the histone proteins protruding from the core octamer. Reprinted with permission from Luger *et al.*, (1997), *Nature*, **389**, 251–260.

nucleosomes, destabilizing the 30 nm chromatin fiber. The histones in heterochromatin are generally unacetylated whereas those in functional domains are acetylated, a clear indication that this type of modification is linked to DNA packaging.

The relevance of histone acetylation to genome expression was underlined in 1996 when, after several years of trying, the first examples of **histone acetyltransferases** (**HATs**)—the enzymes that add acetyl groups to histones—were identified. It was realized that some proteins that had already been shown to have important influences on genome expression had HAT activity. For example, one of the first HATs to be discovered, the *Tetrahymena* protein called p55, was shown to be a homolog of a yeast protein, GCN5, which was known to activate assembly of the transcription initiation complex (Section 11.3.2). Similarly, the mammalian protein called p300/CBP, which had been ascribed a clearly defined role in activation of a variety of genes, was found to be a HAT. These observations, plus the demonstration that different types of cell display different patterns of histone acetylation, underline the prominent role that histone acetylation plays in regulating genome expression.

Individual HATs can acetylate histones in the test tube but most have negligible activity on intact nucleosomes, indicating that, in the nucleus, HATs almost certainly do not work independently, but instead form multiprotein complexes, such as the SAGA and ADA complexes of yeast and the TFTC complex of humans. These complexes are typical of the large multiprotein structures that catalyze and regulate the various steps in genome expression, many examples of which we will meet as we progress through the next few chapters. SAGA, for example, comprises at least 15 proteins with a combined molecular mass of 1.8 million. The complex is 18×28 nm in size, which means that it is larger than the nucleosome core octamer, which with its associated DNA is 11×13 nm, and comparable in one dimension with the 30 nm chromatin fiber. As well as GCN5—the protein with the HAT activity—the SAGA complex contains a set of proteins related to the TATA-binding protein (TBP), which initiates the process by which a gene is transcribed (Section 11.2.3), as well as five of the TBP-associated factors (TAFs) which help TBP fulfill its role. The complexity of SAGA and the other HAT complexes, and the presence within these complexes of proteins with distinct roles in the initiation of gene expression, indicates that the individual events that result in a gene becoming active are intimately linked, with histone acetylation being an integral part, but just one part, of the overall process.

There are at least five different families of HAT proteins. The GCN5-related acetyltransferases, or GNATs, which are components of SAGA, ADA, and TFTC, are clearly associated with activation of gene transcription, but also are involved in the repair of some types of damaged DNA, in particular double-strand breaks and lesions resulting from ultraviolet irradiation (Section 16.2.4). A second family of HATs, called MYST after the initial letters of four of the proteins in this family, is similarly involved in transcription activation and DNA repair, and has also been implicated in control of the cell cycle, though this may simply be another aspect of the DNA repair function, as the cell cycle stalls if the genome is extensively damaged (Section 15.3.2). Different complexes appear to acetylate different histones and some can also acetylate other proteins involved in genome expression, such as the general transcription factors TFIIE and TFIIF, which we will meet in Section 11.2.3. HATs are therefore emerging as versatile proteins that may have diverse functions in expression, replication, and maintenance of the genome.

Histone deacetylation represses active regions of the genome

Gene activation must be reversible otherwise genes that become switched on will remain permanently active. Hence it is no surprise that there is a set of enzymes that can remove acetyl groups from histone tails, overturning the transcription-activating effects of the HATs described above. This is the role of the **histone deacetylases** (**HDACs**). The link between HDAC activity and gene silencing was established in 1996, when mammalian HDAC1, the first of these enzymes to be discovered, was shown to be related to the yeast protein called Rpd3, which was known to be a repressor of transcription. The link between histone deacetylation and repression of transcription was therefore established in the same way as the link between acetylation and activation—by showing that two proteins that were initially thought to have different activities are in fact related. These are good examples of the value of homology analysis in studies of gene and protein function (Section 5.2.1).

HDACs, like HATs, are contained in multiprotein complexes. One of these is the mammalian Sin3 complex, which comprises at least seven proteins, including HDAC1 and HDAC2 along with others that do not have deacetylase activity but which provide ancillary functions essential to the process. Examples of ancillary proteins are RbAp46 and RbAp48, which are members of the Sin3 complex and are thought to contribute the histone-binding capability. RbAp46 and RbAp48 were first recognized through their association with the retinoblastoma protein, which controls cell proliferation by inhibiting expression of various genes until their activities are required and which, when mutated, leads to cancer. This link between Sin3 and a protein implicated in cancer provides a powerful argument for the importance of histone deacetylation in gene silencing. Other deacetylation complexes include NuRD in mammals, which combines HDAC1 and HDAC2 with a different set of ancillary proteins, and yeast Sir2, which is different from other HDACs in that it has an energy requirement. The distinctive features of Sir2 show that HDACs are more diverse than originally realized, possibly indicating that novel roles for histone deacetylation are waiting to be discovered.

Studies of HDAC complexes are beginning to reveal links between the different mechanisms for genome activation and silencing. Both Sin3 and NuRD contain proteins that bind to methylated DNA (Section 10.3.1), and NuRD contains proteins that are very similar to components of the nucleosome remodeling complex Swi/Snf (Section 10.2.2). NuRD does in fact act as a typical nucleosome remodeling machine *in vitro*. Further research will almost certainly unveil additional links between what we currently look on as different types of chromatin modification system, but which in reality may simply be different facets of a single grand design.

Acetylation is not the only type of histone modification

Lysine acetylation/deacetylation is the best-studied form of histone modification but it is by no means the only type. Three other kinds of covalent modification are known to occur:

- Methylation of lysine and arginine residues in the N-terminal regions of histones H3 and H4. Methylation was originally thought to be irreversible and hence responsible for permanent changes to chromatin structure.

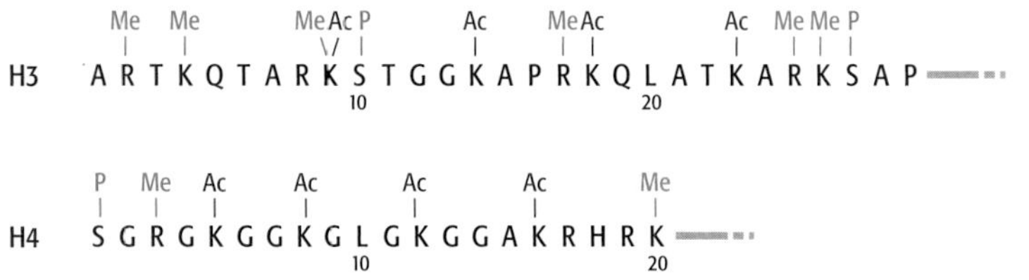

```
       Me   Me        MeAc P       Ac      MeAc            Ac   Me Me P
        |    |          | \/ |      |       | |            |    |  |  |
H3    A R T K Q T A R K S T G G K A P R K Q L A T K A R K S A P ----- ·
                       10                   20
```

```
       P    Me  Ac      Ac      Ac      Ac      Me
       |    |   |       |       |       |       |
H4    S G R G K G G K G L G K G G A K R H R K ----- ·
                  10                    20
```

Figure 10.14 Modifications of the N-terminal regions of mammalian histones H3 and H4. All of the known modifications occurring in these regions are shown. Abbreviations: Ac, acetylation; Me, methylation; P, phosphorylation.

This view has been challenged by the discovery of enzymes that demethylate lysine and arginine residues, but it is still accepted that the effects of methylation are relatively long term.

- Phosphorylation of serine residues in the N-terminal regions of H2A, H2B, H3, and H4.

- Ubiquitination of lysine residues at the C termini of H2A and H2B. This modification involves addition of the small, common ("ubiquitous") protein called **ubiquitin**, or a related protein rather unhelpfully called **SUMO**.

As with acetylation, these other types of modification influence chromatin structure and have a significant impact on cellular activity. For example, phosphorylation of histone H3 and of the linker histone has been associated with formation of metaphase chromosomes, and ubiquitination of histone H2B is part of the general role that ubiquitin plays in control of the cell cycle. The effects of methylation of a pair of lysine amino acids at the fourth and ninth positions from the N terminus of histone H3 are particularly interesting. Methylation of lysine-9 forms a binding site for the HP1 protein, which induces chromatin packaging and silences gene expression, but this event is blocked by the presence of two or three methyl groups attached to lysine-4. Methylation of lysine-4 therefore promotes an open chromatin structure and is associated with active genes. Within the β-globin functional domain, and probably elsewhere, lysine-4 methylation also prevents binding of the NuRD deacetylase to histone H3, ensuring that this histone remains acetylated. Lysine-4 methylation may therefore work hand-in-hand with histone acetylation to activate regions of chromatin.

Altogether, 29 sites in the N- and C-terminal regions of the four core histones are known to be subject to covalent modification (Figure 10.14). Our growing awareness of the variety of histone modifications that occur, and of the way in which different modifications work together, has led to the suggestion that there is a **histone code**, by which the pattern of chemical modifications specifies which regions of the genome are expressed at a particular time and dictates other aspects of genome biology, such as the repair of damaged sites and coordination of genome replication with the cell cycle. This idea is still unproven, but it is clear that the pattern of specific histone modifications within the genome is linked closely to gene activity. Studies of human chromosomes 21 and 22, for example, have shown that regions within these chromosomes where lysine-4 of histone H3 is trimethylated and lysine-9 and lysine-14 are acetylated correspond to the transcription start points for active genes, and that dimethylated lysine-4 is also sometimes found in these regions (Figure 10.15). As with all aspects of chromatin modification, the key question is to distinguish cause and effect: are these patterns of histone modification the reason why these particular genes are active, or merely a by-product of the processes responsible for their activation?

Figure 10.15 The pattern of histone modifications is linked to gene activity. Segments of human chromosomes 21 and 22 are shown, each segment 100 kb in length. Regions that are enriched for dimethylated lysine-4, trimethylated lysine-4, and acetylated lysine-9 and lysine-14 in lung fibroblasts are shown relative to the locations of known genes. The arrows indicate the directions in which the genes are transcribed.

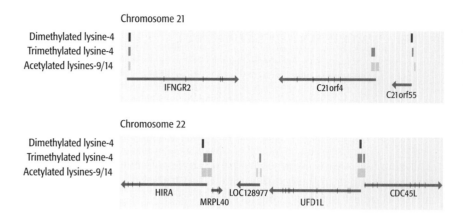

10.2.2 The influence of nucleosome remodeling on genome expression

The second type of chromatin modification that can influence genome expression is **nucleosome remodeling**. This term refers to the modification or repositioning of nucleosomes within a short region of the genome, so that DNA-binding proteins can gain access to their attachment sites. This does not appear to be an essential requirement for transcription of all genes, and in at least a few cases it is possible for a protein that switches on gene expression to achieve its effect either by binding to surfaces of nucleosomes or somehow interacting with the linker DNA without affecting the nucleosome positions. In other examples, repositioning of nucleosomes has been clearly shown to be a prerequisite for gene activation. For example, transcription of the *hsp70* gene of *Drosophila melanogaster*, which codes for a protein involved in folding other proteins (Section 13.3.1), is activated by the GAGA protein in response to a heat shock. Activation is associated with the creation of a DNase I hypersensitive region (Section 10.1.2) upstream of *hsp70*, a clear indication that the nucleosomes in this area have been moved and a segment of naked DNA exposed (Figure 10.16).

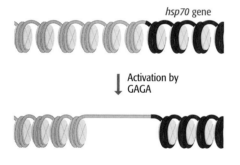

Figure 10.16 Activation of the *hsp70* gene is associated with the creation of a DNase I hypersensitive region. The diagram shows nucleosomes, immediately upstream of the start of the gene, being repositioned when the gene is activated.

Unlike acetylation and the other chemical modifications described in the previous section, nucleosome remodeling does not involve covalent alterations to histone molecules. Instead, remodeling is induced by an energy-dependent process that weakens the contact between the nucleosome and the DNA with which it is associated. Three distinct types of change can occur (Figure 10.17):

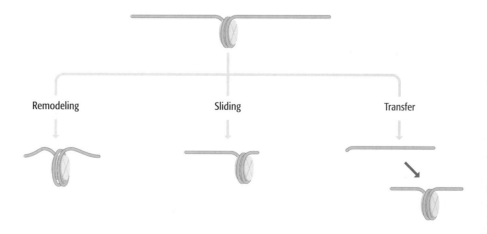

Figure 10.17 Nucleosome remodeling, sliding, and transfer.

- Remodeling, in the strict sense, involves a change in the structure of the nucleosome, but no change in its position. The nature of the structural change is not known, but when induced *in vitro* the outcome is a doubling in size of the nucleosome and an increased DNase sensitivity of the attached DNA.

- Sliding, or *cis*-**displacement**, physically moves the nucleosome along the DNA.

- Transfer, or *trans*-**displacement**, results in the nucleosome being transferred to a second DNA molecule, or to a nonadjacent part of the same molecule.

As with histone acetyltransferases, the proteins responsible for nucleosome remodeling work together in large complexes. One of these is Swi/Snf, which is made up of at least 11 proteins and is present in many eukaryotes. Little is currently known about the way in which Swi/Snf, or any other nucleosome remodeling complex, carries out its role in increasing access to the genome. None of the components of Swi/Snf appears to have a DNA-binding capability, so the complex must be recruited to its target site by additional proteins. Interactions have been detected between Swi/Snf and HATs, suggesting that nucleosome remodeling might occur in conjunction with histone acetylation. This is an attractive hypothesis because it links the two activities that are currently looked on as central to genome activation. But there are problems with this hypothesis because Swi/Snf does not appear to have a global effect across an entire genome, but instead influences gene expression at only a limited number of positions: in the case of yeast, perhaps at no more than 6% of all the genes in the genome. This observation suggests that the more important interactions set up by Swi/Snf might not be with HATs, which work throughout the genome, but with other proteins that target a limited set of genes. The most likely candidates for these other proteins are transcription activators (Section 11.3.2), each of which is specific for a small number of genes and some of which form associations with Swi/Snf *in vitro*.

10.3 DNA Modification and Genome Expression

Important alterations in genome activity can also be achieved by making chemical changes to the DNA itself. These changes are associated with the semipermanent silencing of regions of the genome, possibly entire chromosomes, and often the modified state is inherited by the progeny arising from cell division. The modifications are brought about by **DNA methylation**.

10.3.1 Genome silencing by DNA methylation

In eukaryotes, cytosine bases in chromosomal DNA molecules are sometimes changed to 5-methylcytosine by the addition of methyl groups by enzymes called **DNA methyltransferases** (see Figure 3.29). Cytosine methylation is relatively rare in lower eukaryotes but in vertebrates up to 10% of the total number of cytosine bases in a genome are methylated, and in plants the figure can be as high as 30%. The methylation pattern is not random, instead being limited to the cytosine in some copies of the sequences 5′–CG–3′ and, in plants, 5′–CNG–3′. Two types of methylation activity have been distinguished (Figure 10.18). The first is **maintenance methylation** which, following genome replication, is responsible for adding methyl groups to the newly synthesized strand of DNA at positions opposite methylated sites on the parent strand

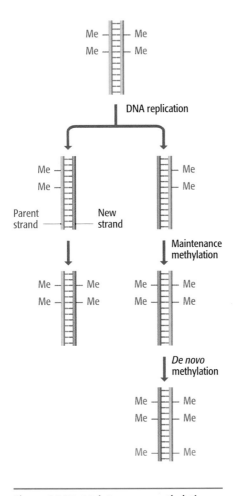

Figure 10.18 Maintenance methylation and *de novo* methylation.

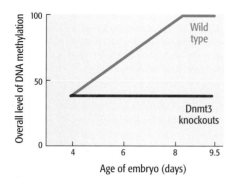

Figure 10.19 Experimental evidence that DNA methyltransferases 3a and 3b are de novo methylases. The graph shows the overall level of DNA methylation in normal (wild type) mice embryos and in Dnmt3 knockouts. In normal embryos, the level of DNA methylation increases due to de novo methylation carried out by DNA methyltransferases 3a and 3b, but in the knockout embryos the level of DNA methylation stays at its original level.

(Section 16.2.3). The maintenance activity therefore ensures that the two daughter DNA molecules retain the methylation pattern of the parent molecule, which means that the pattern can be inherited after cell division. The second activity is ***de novo* methylation**, which adds methyl groups at totally new positions and so changes the pattern of methylation in a localized region of the genome.

DNA methyltransferases and the repression of genome activity

DNA methyltransferases have been extensively studied and are similar in all organisms, from bacteria (which methylate their DNA to protect it from being degraded by their restriction endonucleases, enabling these enzymes to be directed at invading bacteriophage DNA) through to mammals such as humans. Although a lot of work has been done on these enzymes, for several years the apparent presence of only one DNA methyltransferase in mammalian cells was something of a puzzle. This enzyme, now called DNA methyltransferase 1 (Dnmt1), is responsible for maintenance methylation but not *de novo* methylation, as knockout mice with inactivated Dnmt1 genes are still capable of carrying out *de novo* methylation, as shown by their ability to add methyl groups to the DNA genome of an infecting retrovirus. During the late 1990s, by which time most of the genes in the human and mouse genomes had become available as expressed sequence tags (Section 3.3.3), searches of the relevant databases for homologs of Dnmt1 revealed two genes—Dnmt3a and Dnmt3b—that are now known to code for the de novo methyltransferases. Mice with these genes knocked out are unable to complete their full developmental pathways, those with an inactivated Dnmt3b gene dying only a few days after birth, and those lacking Dnmt3a surviving only 1–2 weeks longer. Analysis of methylation patterns in the DNA of embryos during the period before death shows that in the knockouts the extent of DNA methylation is only half that seen in normal mice (Figure 10.19), indicating that maintenance methylation, resulting from Dnmt1 activity, is occurring, but that *de novo* methylation is absent, and hence there is no increase in the overall level of DNA methylation in these mice over time.

Both maintenance and *de novo* methylation result in repression of gene activity. This has been shown by experiments in which methylated or unmethylated genes have been introduced into cells by cloning and their expression levels measured: expression does not occur if the DNA sequence is methylated. The link with gene expression is also apparent when the methylation patterns in chromosomal DNAs are examined, these showing that active genes are located in unmethylated regions. For example, in humans, 40%–50% of all genes are located close to CpG islands (Section 5.1.1), with the methylation status of the CpG island reflecting the expression pattern of the adjacent gene. Housekeeping genes—those that are expressed in all tissues—have unmethylated CpG islands, whereas the CpG islands associated with tissue-specific genes are unmethylated only in those tissues in which the gene is expressed. Note that because the methylation pattern is maintained after cell division, information specifying which genes should be expressed is inherited by the daughter cells, ensuring that in a differentiated tissue the appropriate pattern of gene expression is retained even though the cells in the tissue are being replaced and/or added to by new cells.

The importance of DNA methylation is underlined by studies of human diseases. The syndrome called ICF (immunodeficiency, centromere instability,

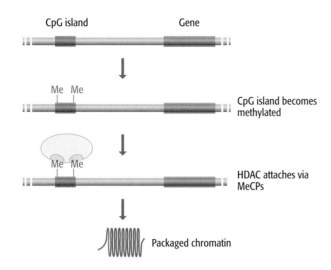

Figure 10.20 A model for the link between DNA methylation and genome expression. Methylation of the CpG island upstream of a gene provides recognition signals for the methyl-CpG-binding protein (MeCP) components of a histone deacetylase complex (HDAC). The HDAC modifies the chromatin in the region of the CpG island and hence inactivates the gene. Note that the relative positions and sizes of the CpG island and the gene are not drawn to scale.

and facial anomalies) which, as the name suggests, has wide-ranging phenotypic effects, is associated with undermethylation of various genomic regions, and is caused by a mutation in the gene for Dnmt3b. The opposing situation—hypermethylation—is seen within the CpG islands of genes that exhibit altered expression patterns in certain types of cancer, although in these cases the abnormal methylation could equally well be a *result* rather than the *cause* of the disease state.

How methylation influences genome expression was a puzzle for many years. Now it is known that **methyl-CpG-binding proteins** (**MeCPs**) are components of both the Sin3 and NuRD histone deacetylase complexes. This discovery has led to a model in which methylated CpG islands are the target sites for attachment of HDAC complexes that modify the surrounding chromatin in order to silence the adjacent genes (Figure 10.20).

Methylation is involved in genomic imprinting and X inactivation
Further evidence, if it is needed, of the link between DNA methylation and genome silencing is provided by two intriguing phenomena called **genomic imprinting** and **X inactivation**.

Genomic imprinting is a relatively uncommon but important feature of mammalian genomes in which only one of a pair of genes, present on homologous chromosomes in a diploid nucleus, is expressed, the second being silenced by methylation. It also occurs in some insects (though apparently not *Drosophila melanogaster*) and some plants. It is always the same member of a pair of genes that is imprinted and hence inactive: for some genes this is the version inherited from the mother, and for other genes it is the paternal version. Just over 60 genes in humans and mice have been shown to display imprinting, these including both protein-coding and functional RNA genes. Imprinted genes are distributed around the genome but tend to occur in clusters. For example, in humans there is a 2.2 Mb segment of chromosome 15 within which there are at least ten imprinted genes, and a smaller, 1 Mb region of chromosome 11 which contains eight imprinted genes.

An example of an imprinted gene in humans is *Igf2*, which codes for a growth factor, a protein involved in signaling between cells (Section 14.1). Only the

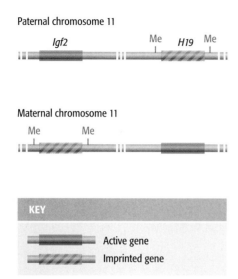

Paternal chromosome 11

Maternal chromosome 11

KEY

Active gene

Imprinted gene

Figure 10.21 A pair of imprinted genes on human chromosome 11. *Igf2* is imprinted on the chromosome inherited from the mother, and *H19* is imprinted on the paternal chromosome. The drawing is not to scale: the two genes are approximately 90 kb apart.

(A) Inactivation of unusual karyotypes

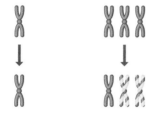

(B) Inactivation involves chromosome counting

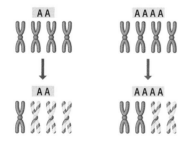

Figure 10.22 X inactivation. (A) If a single X chromosome is present then no inactivation occurs; if there are three X chromosomes then two are inactivated. (B) Three X chromosomes are inactivated in a cell that has a diploid complement of autosomes (AA) but four X chromosomes. In contrast, only two X chromosomes are inactivated if the cell is tetraploid (AAAA).

paternal gene is active (Figure 10.21), because on the chromosome inherited from the mother, various segments of DNA in the region of *Igf2* are methylated, preventing expression of this copy of the gene. A second imprinted gene, *H19*, is located some 90 kb away from *Igf2*, but the imprinting is the other way round: the maternal version of *H19* is active and the paternal version is silent. Imprinting is controlled by **imprint control elements**, DNA sequences that are found within a few kilobases of clusters of imprinted genes. These centers mediate the methylation of the imprinted regions, but the mechanism by which they do this has not yet been described in detail. There is also uncertainty regarding the function of imprinting. One possibility is that it has a role in development, because artificially created parthenogenetic mice, which have two copies of the maternal genome, fail to develop properly. More subtle explanations based on the evolutionary conflicts between the males and females of a species have also been proposed.

X inactivation is less enigmatic. This is a special form of imprinting that leads to almost total inactivation of one of the X chromosomes in a female mammalian cell. It occurs because females have two X chromosomes whereas males have only one. If both of the female X chromosomes were active then proteins coded by genes on the X chromosome might be synthesized at twice the rate in females compared with males. To avoid this undesirable state of affairs, one of the female X chromosomes is silenced and is seen in the nucleus as a condensed structure called the **Barr body**, which is comprised entirely of heterochromatin. Most of the genes on the inactivated X chromosome become silenced but, for reasons that are unknown, some 20% escape the process and remain functional.

Silencing occurs early in embryo development and is controlled by the X inactivation center (*Xic*), a discrete region present on each X chromosome. In a cell undergoing X inactivation, the inactivation center on one of the X chromosomes initiates the formation of heterochromatin, which spreads out from the nucleation point until the entire chromosome is affected, with the exception of a few short segments containing those genes that remain active. The process takes several days to complete. The exact mechanism is not understood but it involves, although is not entirely dependent upon, a gene called *Xist*, located in the inactivation center, which is transcribed into a 25 kb noncoding RNA, copies of which coat the chromosome as heterochromatin is formed. At the same time various histone modifications occur. Lysine-9 of histone H3 becomes methylated (recall that this modification is associated with genome inactivation—Section 10.2.1), histone H4 becomes deacetylated (as usually occurs in heterochromatin), and the histone H2A molecules are replaced with a special histone, macroH2A1. Certain DNA sequences become hypermethylated by DNA methyltransferase 3a, although this appears to occur after the inactive state has been set up. X inactivation is heritable and is displayed by all cells descended from the initial one within which the inactivation took place.

In a normal diploid female, one X chromosome is inactivated and the other remains active. Remarkably, in diploid females with unusual sex chromosome constitutions, the process still results in just a single X chromosome remaining active. For example, in those rare individuals that possess just a single X chromosome no inactivation occurs, and in those individuals with an XXX karyotype, two of the three X chromosomes are inactivated (Figure 10.22A). This means that there must be a mechanism by which the X chromosomes in the nucleus are counted and the appropriate number inactivated. In fact, this mechanism does not simply count the X chromosomes; it also counts the autosomes and compares the two numbers. This is evident because if the cell has four X chromosomes but is otherwise diploid then three X chromosomes are inactivated, but if it is tetraploid (i.e., has four X chromosomes and also four copies of each autosome) then two X chromosomes are inactivated (Figure 10.22B). How the cell counts its chromosomes has puzzled cytogeneticists for many years, and continues to puzzle us, but the most recent research suggests that two genes within the X inactivation center, called *Tsix* and *Xite*, control the process, deletion or overexpression of one or both of these genes causing an incorrect number of chromosomes to be inactivated.

Summary

The nuclear environment has a substantial and important impact on expression of the genome. A eukaryotic nucleus has a highly ordered internal structure which includes a complex network of protein and RNA fibrils called the nuclear matrix. Each chromosome has its own territory within the nucleus, these territories separated from one another by nonchromatin regions within which the enzymes and other proteins involved in genome expression are located. The most compact form of chromatin is heterochromatin, within which genes are inaccessible and cannot be expressed. Constitutive heterochromatin is a permanent feature of all cells and represents DNA that contains no genes, whereas facultative heterochromatin is not permanent and is thought to contain genes that are inactive in some tissues or at some stages of the cell cycle. The more open type of chromatin, called euchromatin, is organized into loops attached to the nuclear matrix, the loops possibly corresponding to the functional domains into which the genes present in the genome are organized. Each functional domain is delimited by a pair of insulators, and some contain locus control regions that are involved in regulating expression of the genes contained in the domain. Nucleosomes appear to be the primary determinants of genome activity in eukaryotes, not only by virtue of their positioning on a strand of DNA, but also because the precise chemical structure of the histone proteins contained within nucleosomes is the major factor determining the degree of packaging displayed by a segment of chromatin. Acetylation of lysine amino acids in the N-terminal regions of each of the core histones is associated with activation of a region of the genome, and deacetylation leads to genome silencing. Histones can also be modified by methylation, phosphorylation, and ubiquitination, each of these events having different, specific effects on the activity of adjacent genes. There may be a histone code by which combinations of these various modifications are interpreted by the genome. Nucleosome repositioning is required for the expression of some, but not all, genes. Regions of the genome can be also silenced by DNA methylation, the relevant enzymes possibly working in conjunction with histone deacetylases. Methylation is responsible for genomic imprinting, which results in one of a pair of genes on homologous chromosomes becoming silenced, and X inactivation, which leads to the almost complete inactivation of one of the X chromosomes in a female nucleus.

Multiple Choice Questions

10.1.* What is the nuclear matrix?

    a. A complex of histone proteins and DNA that provides a structural network throughout the nucleus.

    b. A homogenous mixture of DNA, RNA, and proteins that makes up the nucleus.

    c. The microtubules that provide the structural foundation for the nucleus.

    d. A complex network of protein and RNA fibrils that make up the nuclear substructure.

10.2. What is the function of the nucleolus?

    a. It is the site for expression of genes coding for proteins.

    b. It is the chromosomal scaffold that changes its structure to condense the chromosomes during cell division.

    c. It is the site for synthesis and processing of rRNA molecules.

    d. It is the site for processing of mRNA molecules.

10.3.* Which of the following techniques is useful in determining the movement of proteins within a nucleus?

    a. Electron microscopy.

    b. Fluorescence recovery after photobleaching (FRAP).

    c. Fluorescent *in situ* hybridization (FISH).

    d. Confocal light microscopy.

10.4. Heterochromatin is defined as:

    a. Chromatin that is composed of heterogeneous nucleotide sequences.

    b. Chromatin that contains heterogeneous proteins.

    c. Chromatin that is relatively condensed and contains inactive genes.

    d. Chromatin that is relatively relaxed and contains active genes.

10.5.* Which type of chromatin contains genes that are being expressed?

    a. Euchromatin.

    b. Facultative heterochromatin.

    c. Constitutive heterochromatin.

    d. All of the above.

10.6. Which of the following is a region of eukaryotic DNA, containing one or more active genes, that can be delineated by treatment with DNase I?

    a. Euchromatin.

    b. Heterochromatin.

    c. Functional domain.

    d. Structural domain.

10.7.* Which of the following is able to prevent gene expression when inserted between a gene and its regulatory sequences?

    a. Functional domain.

    b. Structural domain.

    c. Insulator sequence.

    d. Locus control region.

10.8. What is the role of locus control regions (LCRs) in regulation of gene expression?

    a. DNA-binding proteins attach to the LCRs and modify chromatin structure.

    b. Transcription factors bind to the LCRs and promote gene expression.

    c. DNA-binding proteins attach to the LCRs and stimulate DNA methylation.

    d. LCRs make attachments with the nuclear matrix.

10.9.* Which type of amino acid is acetylated within the N-terminal regions of histone proteins?

    a. Arginine.

    b. Lysine.

    c. Serine.

    d. Tyrosine.

10.10. Which of the following is NOT a type of histone modification?

    a. Acetylation.

    b. ADP-ribosylation.

    c. Methylation.

    d. Phosphorylation.

10.11.* Which of the following remodeling events results in a nucleosome being moved to a second DNA molecule?

    a. Acetylation.

    b. Remodeling.

    c. Sliding.

    d. Transfer.

10.12. Which of the following types of DNA modification results in silencing of a region of the genome in such a way that the silencing can be passed on to offspring?

    a. Acetylation.

    b. Methylation.

    c. Phosphorylation.

    d. Ubiquitination.

10.13.* *De novo* methylation of DNA is defined as which of the following?

    a. The addition of methyl groups to DNA at new positions to change the methylation pattern of the genome.

    b. The addition of methyl groups to newly synthesized strands of DNA to ensure that the daughter strands possess the same methylation patterns as the parent strands.

    c. The addition of methyl groups to gene promoters to activate gene expression.

    d. The addition of methyl groups to insulator regions to repress gene expression.

10.14. What is the methylation state of the CpG islands of housekeeping genes?

    a. They have hypermethylated CpG islands.

    b. They are methylated in some tissues but not all.

    c. They have unmethylated CpG islands.

    d. These genes lack CpG islands.

10.15.* Genomic imprinting occurs when:

    a. DNA methylation patterns in a genome are passed on to the offspring.

    b. Genes are incorrectly silenced due to DNA methylation, resulting in altered phenotypes.

    c. Only one of a pair of genes is expressed, the other being methylated and silenced.

    d. DNA methylation is removed allowing for genes that should be silenced to be expressed.

10.16. If a diploid individual has three X chromosomes, how many of the X chromosomes are inactivated?

    a. One.

    b. Two.

    c. Three.

    d. It varies and can be either one or two.

Short Answer Questions

10.1.* Which types of microscopic analysis have led to advances in our understanding of the structural organization of the nucleus?

10.2. What has chromosome painting revealed about the location of chromosomes within the nucleus?

10.3.* Translocations occur at a higher frequency between certain pairs of chromosomes. What does this tell us about the distribution of chromosomes in the nucleus?

10.4. What is the difference between constitutive heterochromatin and facultative heterochromatin?

10.5.* What is the explanation of the positional effect that sometimes occurs when a gene is cloned in a eukaryotic host?

10.6. What are insulator sequences and what unique properties do they possess?

10.7.* Discuss the similarities and differences between insulator sequences and locus control regions (LCRs).

10.8. What can we conclude from the finding that histone acetyltransferases (HATs) have low activity on intact nucleosomes?

10.9.* What is the role of histone deacetylases (HDACs) in regulation of genome expression?

10.10. What is the "histone code"?

10.11.*Why is DNase I used to study changes in chromatin structure? What does the susceptibility of DNA to cleavage by DNase I tell us about gene expression?

10.12. What changes occur within nucleosomes during X inactivation?

continued ...

In-depth Problems

*Guidance to odd-numbered questions can be found in the Appendix

10.1.* To what extent can it be assumed that the picture of nuclear architecture built up by modern electron microscopy is an accurate depiction of the actual structure of the nucleus, as opposed to an artifact of the methods used to prepare cells for examination?

10.2. In many areas of biology it is difficult to distinguish between *cause* and *effect*. Evaluate this issue with regard to nucleosome remodeling and genome expression—does nucleosome remodeling *cause* changes in genome expression or is it the *effect* of these expression changes?

10.3.* Explore and assess the histone code hypothesis.

10.4. Maintenance methylation ensures that the pattern of DNA methylation displayed by two daughter DNA molecules is the same as the pattern on the parent molecule. In other words, the methylation pattern, and the information on gene expression that it conveys, is inherited. Other aspects of chromatin structure might also be inherited in a similar way. How do these phenomena affect the Mendelian view that inheritance is specified by genes?

10.5.* What might be the means by which the numbers of X chromosomes and autosomes in a nucleus are counted so that the appropriate number of X chromosomes can be inactivated?

Figure Tests

*Answers to odd-numbered questions can be found in the Appendix

10.1.* The figure shows two sites where a gene could be inserted into a genome. What level of gene expression is expected after insertion into the region of highly packaged chromatin and into the region of open chromatin?

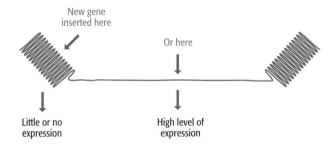

10.2. Discuss how insulator sequences increase the level of gene expression if they are attached to a cloned gene that is inserted into a genome.

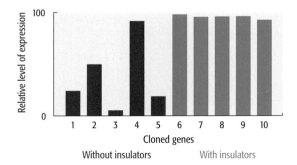

10.3.* How does the methylation of the CpG island affect the expression of the gene?

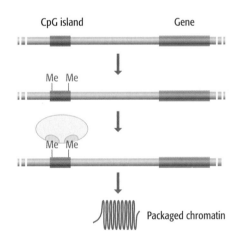

10.4. Discuss the differences in the expression patterns of the genes inherited from each parent.

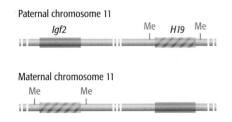

Further Reading

The internal structure of the nucleus

Gerlich, D., Beaudouin, J., Kalbfuss, B., Daigle, N., Eils, R. and Ellenberg, J. (2003) Global chromosome positions are transmitted through mitosis in mammalian cells. *Cell* **112:** 751–764.

Misteli, T. (2001) Protein dynamics: implications for nuclear architecture and gene expression. *Science* **291:** 843–847.

Williams, R.R.E. (2003) Transcription and the territory: the ins and outs of gene positioning. *Trends Genet.* **19:** 298–302. *Chromosome territories.*

Chromatin domains

Bell, A.C., West, A.G. and Felsenfeld, G. (2001) Insulators and boundaries: versatile regulatory elements in the eukaryotic genome. *Science* **291:** 447–450.

Gerasimova, T.I., Byrd, K. and Corces, V.G. (2000) A chromatin insulator determines the nuclear localization of DNA. *Mol. Cell* **6:** 1025–1035.

Li, Q., Harju, S. and Peterson, K.R. (1999) Locus control regions: coming of age at a decade plus. *Trends Genet.* **15:** 403–408.

Covalent modification of histones

Ahringer, J. (2000) NuRD and SIN3: histone deacetylase complexes in development. *Trends Genet.* **16:** 351–356.

Bannister, A.J. and Kouzarides, T. (2005) Reversing histone methylation. *Nature* **436:** 1103–1106.

Bernstein, B.E., Kamal, M., Lindblad-Toh, K., *et al.* (2005) Genomic maps and comparative analysis of histone modifications in human and mouse. *Cell* **120:** 169–181. *Correlates the positions of histone modifications in chromosomes 21 and 22 with gene activity.*

Carrozza, M.J., Utley, R.T., Workman, J.L. and Côté, J. (2003) The diverse functions of histone acetyltransferase complexes. *Trends Genet.* **19:** 321–329.

Imai, S., Armstrong, C.M., Kaeberlein, M. and Guarente, L. (2000) Transcriptional silencing and longevity protein Sir2 is an NAD-dependent histone deacetylase. *Nature* **403:** 795–800.

Jenuwein, T. and Allis, C.D. (2001) Translating the histone code. *Science* **293:** 1074–1080.

Khorasanizedeh, S. (2004) The nucleosome: from genomic organization to genomic regulation. *Cell* **116:** 259–272. *Review of histone modification, nucleosome remodeling, and DNA methylation.*

Lachner, M., O'Carroll, D., Rea, S., Mechtler, K. and Jenuwein, T. (2001) Methylation of histone H3 lysine 9 creates a binding site for HP1 proteins. *Nature* **410:** 116–120.

Sims, R.J., Nishioka, K. and Reinberg, D. (2003) Histone lysine methylation: a signature for chromatin function. *Trends Genet.* **19:** 629–639.

Strahl, B.D. and Allis, D. (2000) The language of covalent histone modifications. *Nature* **403:** 41–45.

Taunton, J., Hassig, C.A. and Schreiber, S.L. (1996) A mammalian histone deacetylase related to the yeast transcriptional regulator Rpd3p. *Science* **272:** 408–411.

Timmers, H.T. and Tora, L. (2005) SAGA unveiled. *Trends Biochem. Sci.* **30:** 7–10.

Verdin, E., Dequiedt, F. and Kasler, H.G. (2003) Class II histone deacetylases: versatile regulators. *Trends Genet.* **19:** 286–293.

Nucleosome remodeling

Aalfs, J.D. and Kingston, R.E. (2000) What does 'chromatin remodelling' mean? *Trends Biochem. Sci.* **25:** 548–555. *Stimulating discussion of histone modification and nucleosome remodeling.*

Sudarsanam, P. and Winston, F. (2000) The Swi/Snf family: nucleosome-remodeling complexes and transcriptional control. *Trends Genet.* **16:** 345–351.

Discovery of the mammalian DNA methyltransferases

Bird, A. (1999) DNA methylation *de novo*. *Science* **286:** 2287–2288.

Okano, M., Bell, D.W., Haber, D.A. and Li, E. (1999) DNA methyltransferases Dnmt3a and Dnmt3b are essential for *de novo* methylation and mammalian development. *Cell* **99:** 247–257.

Xu, G.-L., Bestor, T.H., Bourc´his, D., *et al.* (1999) Chromosome instability and immunodeficiency syndrome caused by mutations in a DNA methyltransferase gene. *Nature* **402:** 187–191.

Imprinting

Feil, R. and Khosia, S. (1999) Genomic imprinting in mammals: an interplay between chromatin and DNA methylation? *Trends Genet.* **15:** 431–434.

Jeppesen, P. and Turner, B.M. (1993) The inactive X chromosome in female mammals is distinguished by a lack of histone H4 acetylation, a cytogenetic marker for gene expression. *Cell* **74:** 281–289.

continued …

X inactivation

Ballabio, A. and Willard, H.F. (1992) Mammalian X-chromosome inactivation and the XIST gene. *Curr. Opin. Genet. Devel.* **2:** 439–448.

Brown, C.J. and Greally, J.M. (2003) A stain upon the silence: genes escaping X inactivation. *Trends Genet.* **19:** 432–438.

Costanzi, C. and Pehrson, J.R. (1998) Histone macroH2A1 is concentrated in the inactive X chromosome of female mammals. *Nature* **393:** 599–601.

Heard, E., Clerc, P. and Avner, P. (1997) X-chromosome inactivation in mammals. *Annu. Rev. Genet.* **31:** 571–610.

Lee, J.T. (2005) Regulation of X-chromosome counting by *Tsix* and *Xite* sequences. *Science* **309:** 768–771.

Assembly of the Transcription Initiation Complex

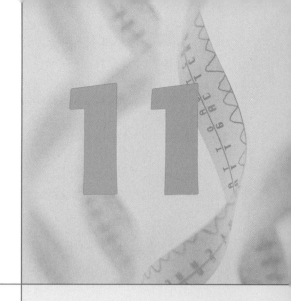

When you have read Chapter 11, you should be able to:

Describe the key structural motifs that enable proteins to make sequence-specific attachments to DNA molecules.

Outline the various techniques that are used to locate the position at which a DNA-binding protein attaches to a DNA molecule.

Discuss the features of the double helix that are important in interactions between DNA and its binding proteins, and give details of the chemical events that underlie the interactions.

Identify the key features of the various prokaryotic and eukaryotic RNA polymerases and describe the structures of the promoter sequences that they recognize.

Give a detailed description of how the *Escherichia coli* transcription initiation complex is assembled and discuss the various ways in which this process can be controlled, in particular distinguishing between constitutive and regulatory control mechanisms.

Give a detailed description of the assembly of the RNA polymerase II transcription initiation complex in eukaryotes, and outline the equivalent processes for RNA polymerases I and III.

Describe, with examples, the modular structure of eukaryotic promoters.

Explain how activators and repressors influence assembly of transcription initiation complexes in eukaryotes.

Discuss our current understanding of the structures of mediator complexes and their roles in transcription initiation in eukaryotes.

The initial product of genome expression is the transcriptome, the collection of RNA molecules derived from those protein-coding genes whose biological information is required by the cell at a particular time (see Figure 1.2). The transcriptome is maintained by the process called transcription, in which individual genes are copied into RNA molecules. Transcription was once looked on simply as "DNA makes RNA," and in essence this is what happens,

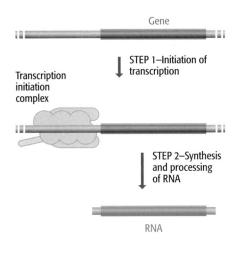

Gene

STEP 1—Initiation of transcription

Transcription initiation complex

STEP 2—Synthesis and processing of RNA

RNA

Figure 11.1 The two stages in the process that leads from genome to transcriptome.

but we now realize that the process that leads from the genome to the transcriptome is much more complex than suggested by this trite statement. This part of genome expression is now divided into two key stages (Figure 11.1):

● Initiation of transcription, which results in assembly upstream of the gene of the complex of proteins, including the RNA polymerase enzyme and its various accessory proteins, that will subsequently copy the gene into an RNA transcript. Inherent in this step are the events that determine whether or not the gene is actually transcribed.

● Synthesis and processing of RNA, which begins when the RNA polymerase leaves the initiation region and starts to make an RNA copy of the gene, and ends after completion of the processing and modification events that convert the initial transcript into a mature RNA molecule, able to carry out its function in the cell.

This chapter deals with the initiation of transcription, and Chapter 12 covers RNA synthesis and processing. But before we move on to these topics we must do a little groundwork. The central players in transcription are **DNA-binding**

Table 11.1 Functions of DNA- and RNA-binding proteins

Function	Examples
DNA-binding proteins	
Genome expression	
Transcription initiation	Eukaryotic TATA-binding protein (Section 11.2.3)
	σ subunit of bacterial RNA polymerase (Section 11.2.3)
RNA synthesis	RNA polymerases (Section 11.2.1)
Regulation of transcription	Eukaryotic activators and repressors (Section 11.3.2)
	Bacterial repressors (Section 11.3.1)
DNA packaging	Eukaryotic histones (Section (7.1.1)
	Bacterial nucleoid proteins (Section 8.1.1)
DNA recombination	RecA (Section 17.1.2)
DNA repair	DNA glycosylases, nucleases (Section 16.2.2)
DNA replication	Origin recognition proteins (Section 15.2.1)
	DNA polymerases and ligases (Sections 2.1.1, 2.1.3, and 15.2.2)
	Single-strand binding proteins (Section 15.2.2)
	DNA topoisomerases (Section 15.1.2)
Others	Prokaryotic restriction endonucleases (Section 2.1.2)
RNA-binding proteins	
Genome expression	
Intron splicing	snRNP proteins (Section 12.2.2)
mRNA polyadenylation	CPSF, CstF (Section 12.2.1)
mRNA editing	Adenosine deaminases (Section 12.2.5)
rRNA and tRNA processing	Ribonucleases (Sections 12.1.3 and 12.2.4)
Translation	Aminoacyl-tRNA synthetases (Section 13.1.1)
	Translation factors (Sections 13.2.2, 13.2.3, and 13.2.4)
RNA degradation	Ribonucleases (Sections 12.1.4 and 12.2.6)
Ribosome structure	Ribosomal proteins (Section 13.2.1)

proteins that attach to the genome in order to perform their biochemical functions (see Table 11.1). Histones are examples of DNA-binding proteins, and we will encounter many others later in this chapter when we look at assembly of the initiation complexes of prokaryotes and eukaryotes. There are also DNA-binding proteins that are involved in DNA replication, repair, and recombination, as well as a large group of related proteins that bind to RNA rather than DNA (see Table 11.1). Many DNA-binding proteins recognize specific nucleotide sequences and bind predominantly to these target sites, whereas others bind nonspecifically at various positions in the genome.

The mode of action of DNA-binding proteins is central to the initiation of transcription, and without a knowledge of how they function we can never hope to understand how the information in the genome is utilized. We will therefore spend some time examining what is known about DNA-binding proteins and how they interact with the genome.

11.1 DNA-binding Proteins and Their Attachment Sites

Our main interest lies with those proteins that are able to target a specific nucleotide sequence and hence bind to a limited number of positions on a DNA molecule, this being the type of interaction that is most important in expression of the genome. To bind in this specific fashion, a protein must make contact with the double helix in such a way that the nucleotide sequence can be recognized, which generally requires that part of the protein penetrates into the major and/or minor grooves of the helix (see Figures 1.8A and 1.9) in order to achieve **direct readout** of the sequence (Section 11.1.3). This is usually accompanied by more general interactions with the surface of the DNA molecule, which may simply stabilize the DNA–protein complex or which may access the indirect information on nucleotide sequence that is provided by the conformation of the helix.

11.1.1 The special features of DNA-binding proteins

The structures of many proteins, including over 100 that bind to DNA or RNA, have been determined by methods such as **X-ray crystallography** and **nuclear magnetic resonance (NMR) spectroscopy** (Technical Note 11.1). When the structures of sequence-specific DNA-binding proteins are compared, it is immediately evident that the family as a whole can be divided into a limited number of different groups on the basis of the structure of the segment of the protein that interacts with the DNA molecule (Table 11.2). Each of these **DNA-binding motifs** is present in a range of proteins, often from very different organisms, and at least some of them probably evolved more than once. We will look at two in detail—the **helix–turn–helix (HTH) motif** and the **zinc finger**—and then briefly survey the others.

The helix–turn–helix motif is present in prokaryotic and eukaryotic proteins

The HTH motif was the first DNA-binding structure to be identified. As the name suggests, the motif is made up of two α-helices separated by a turn (Figure 11.2). The latter is not a random conformation but a specific structure, referred to as a **β-turn**, made up of four amino acids, the second of which is usually glycine. This turn, in conjunction with the first α-helix, positions the

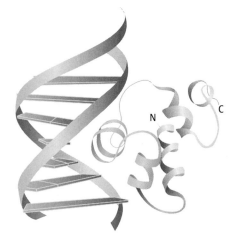

Figure 11.2 The helix–turn–helix motif. The drawing shows the orientation of the helix–turn–helix motif (in purple) of the *Escherichia coli* bacteriophage 434 repressor in the major groove of the DNA double helix. "N" and "C" indicate the N- and C-termini of the motif, respectively.

Technical Note 11.1 X-ray crystallography and nuclear magnetic resonance spectroscopy

Methods for studying the structures of proteins and protein–nucleic acid complexes

Once a DNA- or RNA-binding protein has been purified, attempts can be made to determine its structure, in isolation or attached to its binding site. This enables the precise structure of the nucleic acid–binding part of the protein to be determined, and allows the identity and nature of the contacts with the DNA or RNA molecule to be elucidated. Two techniques—X-ray crystallography and nuclear magnetic resonance (NMR) spectroscopy—are central to this area of research.

X-ray crystallography, which is a long-established technique whose pedigree stretches back to the late nineteenth century, is based on **X-ray diffraction**. X-rays have very short wavelengths—between 0.01 nm and 10 nm—which is 4000 times shorter than visible light and comparable with the spacings between atoms in chemical structures. When

a beam of X-rays is directed onto a crystal, some of the X-rays pass straight through, but others are diffracted and emerge from the crystal at a different angle from which they entered (Figure T11.1A). If the crystal is made up of many copies of the same molecule, all positioned in a regular array, then different X-rays are diffracted in similar ways, resulting in overlapping circles of diffracted waves which interfere with one another. An X-ray-sensitive photographic

Figure T11.1 X-ray crystallography. (A) An X-ray diffraction pattern is obtained by passing a beam of X-rays through a crystal of the molecule being studied. (B) The diffraction pattern obtained with crystals of ribonuclease. (C) Part of the electron-density map derived from this diffraction pattern. (D) If the electron-density map has sufficient resolution then it is possible to identify the R groups of individual amino acids, as shown here for tyrosine.

(A) Production of a diffraction pattern

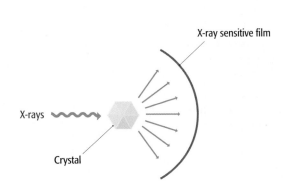

(B) X-ray diffraction pattern for ribonuclease

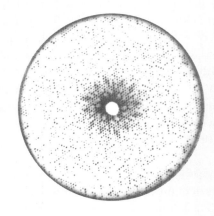

(C) Part of the ribonuclease electron-density map

(D) Interpretation of an electron-density map—a 2Å resolution electron-density map revealing a tyrosine R group

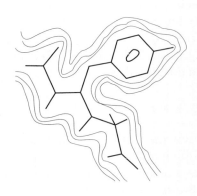

film or electronic detector placed across the beam reveals a series of spots (Figure T11.1B), an **X-ray diffraction pattern**, from which the structure of the molecule in the crystal can be deduced. This is because the relative positioning of the spots indicates the arrangement of the molecules in the crystal, and their relative intensities provide information on the structure of the molecule. The more complex the molecule, the greater the number of spots and the larger the number of comparisons that must be made between them. Computational help is therefore required for all but the simplest molecules. If successful, the result is an electron-density map (Figure T1C and D) which, with a protein, provides a chart of the folded polypeptide from which the positioning of structural features such as α-helices and β-sheets can be determined. If sufficiently detailed, the R groups of the individual amino acids in the polypeptide can be identified and their orientations relative to one another established, allowing deductions to be made about the hydrogen bonding and other chemical interactions occurring within the protein structure. With luck, these deductions lead to a detailed, three-dimensional model of the protein.

Like X-ray crystallography, NMR is a long-established technique that traces its origins to the early part of the twentieth century, first being described in 1936. The principle of the technique is that rotation of a charged chemical nucleus generates a magnetic moment. When placed in an applied electromagnetic field, the spinning nucleus orientates in one of two ways, called α and β (Figure T11.2), the α-orientation (which is aligned with the magnetic field) having a slightly lower energy. In NMR spectroscopy the magnitude of this energy separation is determined by measuring the frequency of the electromagnetic radiation needed to induce the transition from α to β, the value being described as the resonance frequency of the nucleus being studied. The critical point is that although each type of nucleus (e.g., ^{1}H, ^{13}C, ^{15}N) has its own specific resonance frequency, the measured frequency is often slightly different from the standard value (typically by less than 10 parts per million) because electrons in the vicinity of the rotating nucleus shield it to a certain extent from the applied magnetic field. This **chemical shift** (the difference between the observed resonance energy and the standard value for the nucleus being studied) enables the chemical environment of the nucleus to be inferred, and hence provides structural information. Particular types of analysis (called COSY and TOCSY) enable atoms linked by chemical bonds to the spinning nucleus to be identified; other analyses (e.g., NOESY) identify atoms that are close to the spinning nucleus in space but not directly connected to it. Not all

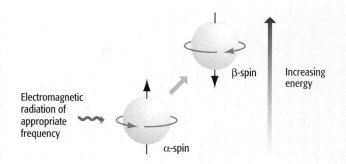

Figure T11.2 The basis of NMR. A rotating nucleus can take up either of two orientations in an applied electromagnetic field. The energy separation between the α and β spin states is determined by measuring the frequency of electromagnetic radiation needed to induce an $\alpha \rightarrow \beta$ transition.

chemical nuclei are suitable for NMR. Most protein NMR projects are ^{1}H studies, the aim being to identify the chemical environments and covalent linkages of every hydrogen atom, and from this information to infer the overall structure of the protein. These studies are frequently supplemented by analyses of substituted proteins in which at least some of the carbon and/or nitrogen atoms have been replaced with the rare isotopes ^{13}C and ^{15}N, these also giving good results with NMR.

When successful, NMR results in the same level of resolution as X-ray crystallography and so provides very detailed information on protein structure. The main advantage of NMR is that it works with molecules in solution and so avoids the problems that sometimes occur when attempting to obtain crystals of a protein for X-ray analysis. Solution studies also offer greater flexibility if the aim is to examine changes in protein structure, as occur during protein folding or in response to addition of a substrate. The disadvantage of NMR is that it is only suitable for relatively small proteins. There are several reasons for this, one being the need to identify the resonance frequencies for each, or as many as possible, of the ^{1}H or other nuclei being studied. This depends on the various nuclei having different chemical shifts so that their frequencies do not overlap. The larger the protein, the greater the number of nuclei and the greater the chances that frequencies overlap and structural information is lost. Although this limits the applicability of NMR, the technique is still very valuable. There are many interesting proteins that are small enough to be studied by NMR, and important information can also be obtained by structural analysis of peptides which, although not complete proteins, can act as models for aspects of protein activity such as nucleic acid binding.

Table 11.2 DNA-binding motifs

Motif	Examples of proteins with this motif
Sequence-specific DNA-binding motifs	
Helix–turn–helix family	
Standard helix–turn–helix	*Escherichia coli* lactose repressor, tryptophan repressor
Homeodomain	*Drosophila* Antennapedia protein
Paired homeodomain	Vertebrate Pax transcription factor
POU domain	Vertebrate regulatory proteins Pit-1, Oct-1, and Oct-2
Winged helix–turn–helix	GABP regulatory protein of higher eukaryotes
High mobility group (HMG) domain	Mammalian sex determination protein SRY
Zinc-finger family	
Cys_2His_2 finger	Transcription factor TFIIIA of eukaryotes
Multicysteine zinc finger	Steroid receptor family of higher eukaryotes
Zinc binuclear cluster	Yeast GAL4 transcription factor
Basic domain	Yeast GCN4 transcription factor
Ribbon–helix–helix	Bacterial MetJ, Arc, and Mnt repressors
TBP domain	Eukaryotic TATA-binding protein
β-Barrel dimer	Papillomavirus E2 protein
Rel homology domain (RHB)	Mammalian transcription factor NF-κB
Non-sequence-specific DNA-binding motifs	
Histone fold	Eukaryotic histones
HU/IHF motif*	Bacterial HU and IHF proteins
Polymerase cleft	DNA and RNA polymerases

*The HU/IHF motif is a non-sequence-specific DNA-binding motif in bacterial HU proteins (the nucleoid packaging proteins; Section 8.1.1) but directs sequence-specific binding of the IHF (integration host factor) protein (Section 17.2.1).

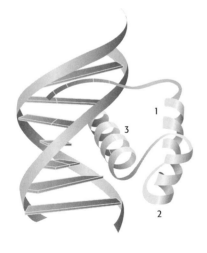

Figure 11.3 The homeodomain motif. The first three helices of a typical homeodomain are shown with helix 3 orientated in the major groove and helix 1 making contacts in the minor groove. Helices 1–3 run in the N→C direction along the motif.

second α-helix on the surface of the protein in an orientation that enables it to fit inside the major groove of a DNA molecule. This second α-helix is therefore the **recognition helix** that makes the vital contacts which enable the DNA sequence to be read. The HTH structure is usually 20 or so amino acids in length and so is just a small part of the protein as a whole. Some of the other parts of the protein form attachments with the surface of the DNA molecule, primarily to aid the correct positioning of the recognition helix within the major groove.

Many prokaryotic and eukaryotic DNA-binding proteins utilize an HTH motif. In bacteria, HTH motifs are present in some of the best-studied regulatory proteins, which switch on and off the expression of individual genes. An example is the **lactose repressor**, which regulates expression of the lactose operon (Section 11.3.1). The various eukaryotic HTH proteins include many whose DNA-binding properties are important in the developmental regulation of genome expression, such as the **homeodomain** proteins, whose roles we will examine in Section 14.3.4. The homeodomain is an extended HTH motif possessed by each of these proteins. It is made up of 60 amino acids which form four α-helices, numbers 2 and 3 separated by a β-turn, with number 3 acting as the recognition helix and number 1 making contacts within the minor groove (Figure 11.3). Other versions of the HTH motif found in eukaryotes include:

- The **POU domain**, which is usually found in proteins that also have a homeodomain, the two motifs probably working together by binding different regions of a double helix. The name "POU" comes from the initial letters of the names of the first proteins found to contain this motif.

- The **winged helix–turn–helix** motif, which is another extended version of the basic HTH structure, this one with a third α-helix on one side of the HTH motif and a β-sheet on the other side.

Many proteins, prokaryotic and eukaryotic, possess an HTH motif, but the details of the interaction of the recognition helix with the major groove are not exactly the same in all cases. The length of the recognition helix varies, generally being longer in eukaryotic proteins, the orientation of the helix in the major groove is not always the same, and the position within the recognition helix of those amino acids that make contacts with nucleotides is different.

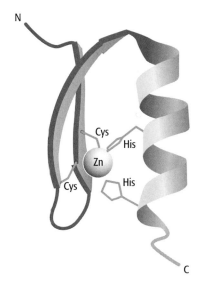

Figure 11.4 The Cys$_2$His$_2$ zinc finger. This particular zinc finger is from the yeast SWI5 protein. The zinc atom is held between two cysteines within the β-sheet of the motif and two histidines in the α-helix. The orange lines indicate the R groups of these amino acids. "N" and "C" indicate the N- and C-termini of the motif, respectively.

Zinc fingers are common in eukaryotic proteins

The second type of DNA-binding motif that we will look at in detail is the zinc finger, which is rare in prokaryotic proteins but very common in eukaryotes. There appear to be more than 500 different zinc-finger proteins in the worm *Caenorhabditis elegans*, out of a total 19,000 proteins, and it is estimated that 1% of all mammalian genes code for zinc-finger proteins.

There are at least six different versions of the zinc finger. The first to be studied in detail was the **Cys$_2$His$_2$ finger**, which comprises a series of 12 or so amino acids, including two cysteines and two histidines, which form a segment of β-sheet followed by an α-helix. These two structures, which form the "finger" projecting from the surface of the protein, hold between them a bound zinc atom, coordinated to the two cysteines and two histidines (Figure 11.4). The α-helix is the part of the motif that makes the critical contacts within the major groove, its positioning within the groove being determined by the β-sheet, which interacts with the sugar–phosphate backbone of the DNA, and the zinc atom, which holds the β-sheet and α-helix in the appropriate positions relative to one another. Other versions of the zinc finger differ in the structure of the finger, some lacking the β-sheet component and consisting simply of one or more α-helices, and the precise way in which the zinc atom is held in place also varies. For example, the **multicysteine zinc fingers** lack histidines, the zinc atom being coordinated between four cysteines.

An interesting feature of the zinc finger is that multiple copies of the finger are sometimes found on a single protein. Several proteins have two, three, or four fingers, but there are examples with many more than this—37 for one toad protein. In most cases, the individual zinc fingers are thought to make independent contacts with the DNA molecule, but in some cases the relationship between different fingers is more complex. In one particular group of proteins—the nuclear, or steroid, receptor family—two α-helices containing six cysteines combine to coordinate two zinc atoms in a single DNA-binding domain, larger than a standard zinc finger (Figure 11.5). Within this motif it appears that one of the α-helices enters the major groove whereas the second makes contacts with other proteins.

Other nucleic acid–binding motifs

The various other DNA-binding motifs that have been discovered in different proteins include:

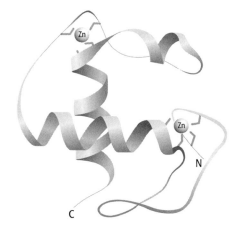

Figure 11.5 The steroid receptor zinc finger. The R groups of the amino acids involved in the interactions with the zinc atoms are shown as orange lines. "N" and "C" indicate the N- and C-termini of the motif, respectively.

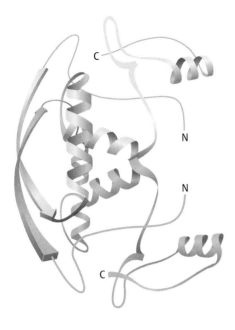

Figure 11.6 The ribbon–helix–helix motif. The drawing is of the ribbon–helix–helix motif of the *Escherichia coli* MetJ repressor, which consists of a dimer of two identical proteins, one shown in gray and the other in purple. The β-strands at the left of the structure make contact with the major groove of the double helix. "N" and "C" indicate the N- and C-termini of the motif, respectively.

DNA-binding proteins

Gene

200 bp

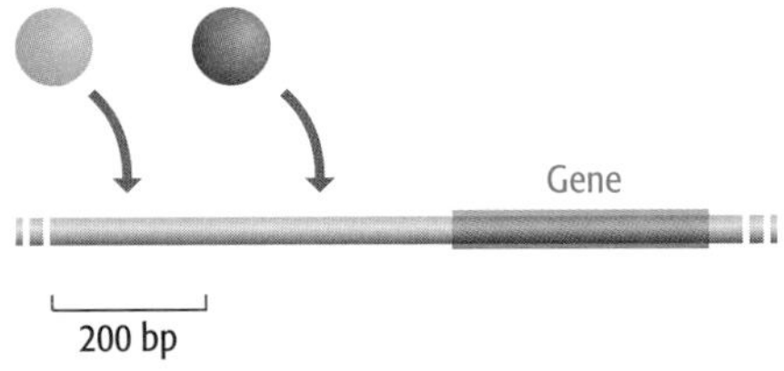

Figure 11.7 Attachment sites for DNA-binding proteins are located immediately upstream of a gene.

- The **basic domain**, in which the DNA recognition structure is an α-helix that contains a high number of basic amino acids (e.g., arginine, serine, and threonine). A peculiarity of this motif is that the α-helix only forms when the protein interacts with DNA: in the unbound state the helix has a disorganized structure. Basic domains are found in a number of eukaryotic proteins involved in transcription of DNA into RNA.

- The **ribbon–helix–helix** motif, which is one of the few motifs that achieves sequence-specific DNA binding without making use of an α-helix as the recognition structure. Instead, the ribbon (i.e., two strands of a β-sheet) makes contact with the major groove (Figure 11.6). Ribbon–helix–helix motifs are found in some gene-regulation proteins in bacteria.

- The **TBP domain**, which has so far only been discovered in the **TATA-binding protein** (Section 11.2.3), after which it is named. As with the ribbon–helix–helix motif, the recognition structure is a β-sheet, but in this case the main contacts are with the minor, not major, groove of the DNA molecule.

RNA-binding proteins also have specific motifs that form the attachment with the RNA molecule. The most important of these are as follows:

- The **ribonucleoprotein (RNP) domain** comprises four β-strands and two α-helices in the order β–α–β–β–α–β. The two central β-strands make the critical attachments with the RNA molecule. The RNP domain is the commonest RNA-binding motif and has been found in more than 250 proteins.

- The **double-stranded RNA binding domain (dsRBD)** is similar to the RNP domain but has the structure α-β-β-β-α. The RNA-binding function lies between the β-strand and α-helix at the end of the structure. As the name implies, the motif is found in proteins that bind double-stranded RNA.

- The **κ-homology domain** has the structure β-α-α-β-β-α, with the binding function between the pair of α-helices. It is relatively uncommon but present in at least one nuclear RNA-binding protein.

Additionally, the DNA-binding homeodomain may also have RNA-binding activity in some proteins. One ribosomal protein uses a structure similar to a homeodomain to attach to rRNA, and some homeodomain proteins such as Bicoid of *Drosophila melanogaster* (Section 14.3.4) can bind both DNA and RNA.

11.1.2 Locating the positions of DNA-binding sites in a genome

Often the first thing that is discovered about a DNA-binding protein is not the identity of the protein itself but the features of the DNA sequence that the protein recognizes. This is because genetic and molecular biology experiments, which we will deal with later in this chapter, have shown that many of the proteins that are involved in genome expression bind to short DNA sequences immediately upstream of the genes on which they act (Figure 11.7). This means that the sequence of a newly discovered gene, assuming that it includes both the coding DNA and the regions upstream of it, provides immediate access to the binding sites of at least some of the proteins responsible for expression of that gene. Because of this, a number of methods have been developed for locating protein-binding sites within DNA fragments up to several kilobases in length, these methods working perfectly well even if the relevant DNA-binding proteins have not been identified.

Gel retardation identifies DNA fragments that bind to proteins

The first of these methods makes use of the substantial difference between the electrophoretic properties of a "naked" DNA fragment and one that carries a bound protein. Recall that DNA fragments are separated by agarose gel electrophoresis because smaller fragments migrate through the pore-like structure of the gel more quickly than larger fragments (see Technical Note 2.2). If a DNA fragment has a protein bound to it then its mobility through the gel will be impeded: the DNA–protein complex therefore forms a band at a position nearer to the starting point (Figure 11.8). This is called **gel retardation**. In practice the technique is carried out with a collection of restriction fragments that span the region thought to contain a protein-binding site. The digest is mixed with an extract of nuclear proteins (assuming that a eukaryote is being studied) and retarded fragments are identified by comparing the banding pattern obtained after electrophoresis with the pattern for restricted fragments that have not been mixed with proteins. A nuclear extract is used because at this stage of the project the DNA-binding protein has not usually been purified. If, however, the protein is available then the experiment can be carried out just as easily with the pure protein as with a mixed extract.

Protection assays pinpoint binding sites with greater accuracy

Gel retardation gives a general indication of the location of a protein-binding site in a DNA sequence, but does not pinpoint the site with great accuracy. Often the retarded fragment is several hundred base pairs in length, compared with the expected length of the binding site of a few tens of base pairs at most, and there is no indication of where in the retarded fragment the binding site lies. Also, if the retarded fragment is long then it might contain separate binding sites for several proteins, or if it is quite small then there is the possibility that the binding site also includes nucleotides on adjacent fragments, ones that on their own do not form a stable complex with the protein and so do not lead to gel retardation. Gel retardation studies are therefore a starting point but other techniques are needed to provide more accurate information.

Modification protection assays can take over where gel retardation leaves off. The basis of these techniques is that if a DNA molecule carries a bound protein then part of its nucleotide sequence will be protected from modification. There are two ways of carrying out the modification:

- By treatment with a nuclease, which cleaves all phosphodiester bonds except those protected by the bound protein.

- By exposure to a methylating agent, such as dimethyl sulfate which adds methyl groups to G nucleotides. Any Gs protected by the bound protein will not be methylated.

The practical details of these two techniques are shown in Figures 11.9 and 11.10. Both utilize an experimental approach called **footprinting**. In nuclease footprinting, the DNA fragment being examined is labeled at one end, complexed with binding protein (as a nuclear extract or as pure protein), and treated with deoxyribonuclease I (DNase I). Normally, DNase I cleaves every

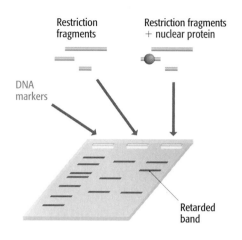

Figure 11.8 Gel retardation analysis. A nuclear extract has been mixed with a DNA restriction digest, and a DNA-binding protein in the extract has attached to one of the restriction fragments. The DNA–protein complex has a larger molecular mass than the "naked" DNA and so runs more slowly during gel electrophoresis. As a result, the band for this fragment is retarded and can be recognized by comparison with the banding pattern produced by restriction fragments that have not been mixed with the nuclear extract.

Figure 11.9 DNase I footprinting. The restriction fragments used at the start of the procedure must be labeled at just one end. This is usually achieved by treating a set of longer restriction fragments with an enzyme that attaches labels at *both* ends, then cutting these labeled molecules with a second restriction enzyme, and purifying one of the sets of end fragments. The DNase I treatment is carried out in the presence of a manganese salt, which induces the enzyme to make random, double-stranded cuts in the target molecules, leaving blunt-ended fragments.

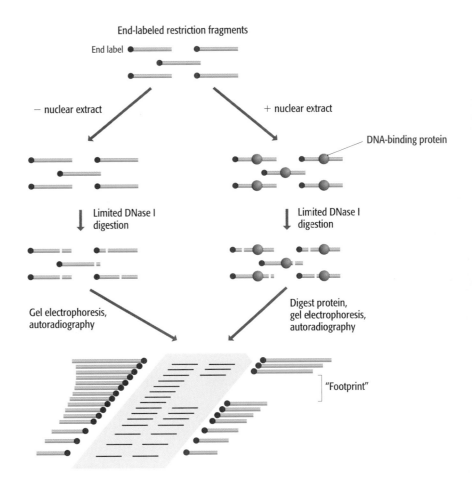

phosphodiester bond, leaving only the DNA segment protected by the binding protein. This is not very useful because it can be difficult to sequence such a small fragment. It is quicker to use the more subtle approach shown in Figure 11.9. The nuclease treatment is carried out under limiting conditions, such as a low temperature and/or very little enzyme, so that on average each copy of the DNA fragment suffers a single "hit"—meaning that it is cleaved at just one position along its length. Although each fragment is cut just once, in the entire population of fragments all bonds are cleaved except those protected by the bound protein. The protein is now removed, the mixture electrophoresed, and the labeled fragments visualized. Each of these fragments has the label at one end and a cleavage site at the other. The result is a ladder of bands corresponding to fragments that differ in length by one nucleotide, with the ladder broken by a blank area in which no labeled bands occur. This blank area, or "footprint," corresponds to the positions of the protected phosphodiester bonds, and hence of the bound protein, in the starting DNA.

Modification interference identifies nucleotides central to protein binding

Modification protection should not be confused with **modification interference**, a different technique that provides an extra dimension to the study of protein binding. Modification interference works on the basis that if a nucleotide critical for protein binding is altered, for example by addition of a methyl group, then binding may be prevented. One of this family of techniques is illustrated in Figure 11.11. The DNA fragment, labeled at one end, is treated with the modification reagent, in this case dimethyl sulfate, under

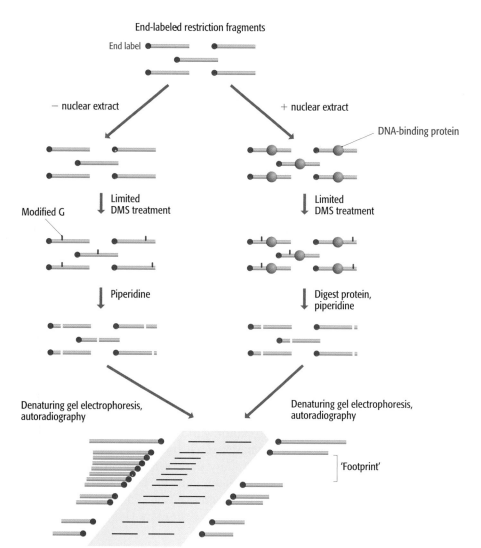

End-labeled restriction fragments

End label

− nuclear extract

+ nuclear extract

DNA-binding protein

Modified G

Limited
DMS treatment

Limited
DMS treatment

Piperidine

Digest protein,
piperidine

Denaturing gel electrophoresis,
autoradiography

Denaturing gel electrophoresis,
autoradiography

'Footprint'

Figure 11.10 The dimethyl sulfate (DMS) modification protection assay. The technique is similar to DNase I footprinting. Instead of DNase I digestion, the fragments are treated with limited amounts of DMS so that a single guanine base is methylated in each fragment. Guanines that are protected by the bound protein cannot be modified. After removal of the protein, the DNA is treated with piperidine, which cuts at the modified nucleotide positions. For simplicity, the diagram shows the double-stranded molecules being cleaved at this stage. In fact, they are only nicked, as piperidine only cuts the strand that is modified, rather than making a double-stranded cut across the entire molecule. The samples are therefore examined by *denaturing* gel electrophoresis so that the two strands are separated. The resulting autoradiograph shows the sizes of the strands that have one labeled end and one end created by piperidine nicking. The banding pattern for the control DNA strands—those not incubated with the nuclear extract—indicates the positions of guanines (Gs) in the restriction fragment, and the footprint seen in the banding pattern for the test sample shows which Gs were protected.

limiting conditions so that just one guanine per fragment is methylated. Now the binding protein or nuclear extract is added, and the fragments electrophoresed. Two bands are seen, one corresponding to the DNA–protein complex and one containing DNA without bound protein. The latter contains molecules that have been prevented from attaching to the protein because the methylation treatment has modified one or more Gs that are crucial for the binding. To identify which Gs are modified, the fragment is purified from the gel and treated with piperidine, the compound that cleaves DNA at methylguanine nucleotides. The result of this treatment is that each fragment is cut into two segments, one of which carries the label. The length(s) of the labeled segment(s), determined by a second round of electrophoresis, tells us which nucleotide(s) in the original fragment were methylated and hence identifies the positions in the DNA sequence of Gs that participate in the binding reaction. Equivalent techniques can be used to identify the A, C, and T nucleotides involved in binding.

11.1.3 The interaction between DNA and its binding proteins

In recent years, our understanding of the part played by the DNA molecule in the interaction with a binding protein has begun to change. It has always been accepted that proteins that recognize a specific sequence as their binding site

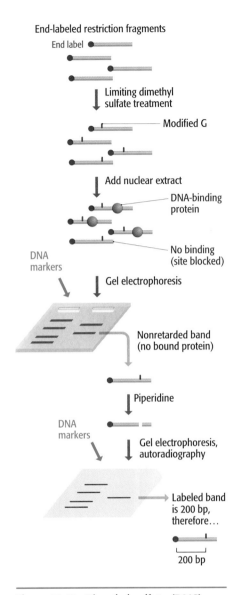

Figure 11.11 Dimethyl sulfate (DMS) modification interference assay. See the legend to Figure 11.9 for a description of the procedure used to obtain DNA fragments labeled at just one end.

can locate this site by forming contacts with chemical groups attached to the bases that are exposed within the major and minor grooves that spiral around the double helix (see Figure 1.8). It is now recognized that the nucleotide sequence also influences the precise conformation of each region of the helix, and that these conformational features represent a second, less direct way in which the DNA sequence can influence protein binding.

Direct readout of the nucleotide sequence

It was clear from the double helix structure described by Watson and Crick (Section 1.1.2) that although the nucleotide bases are on the inside of the DNA molecule, they are not entirely buried, and some of the chemical groups attached to the purine and pyrimidine bases are accessible from outside the helix. **Direct readout** of the nucleotide sequence should therefore be possible without breaking the base pairs and opening up the molecule.

In order to form chemical bonds with groups attached to the nucleotide bases, a binding protein must make contacts within one or both of the grooves on the surface of the helix. With the B-form of DNA, the identity and orientation of the exposed parts of the bases within the major groove is such that most sequences can be read unambiguously, whereas within the minor groove it is possible to identify if each base pair is A–T or G–C but difficult to know which nucleotide of the pair is in which strand of the helix (Figure 11.12). Direct readout of the B-form therefore predominantly involves contacts in the major groove. With other DNA types there is much less information on the contacts formed with binding proteins, but the picture is likely to be quite different. In the A-form, for example, the major groove is deep and narrow and less easily penetrated by any part of a protein molecule (see Table 1.1). The shallower minor groove is therefore likely to play the main part in direct readout. With Z-DNA, the major groove is virtually nonexistent and direct readout is possible to a certain extent without moving beyond the surface of the helix.

The nucleotide sequence has a number of indirect effects on helix structure

Originally it was thought that cellular DNA molecules have fairly uniform structures, made up mainly of the B-form of the double helix. Some short segments might be in the A-form, and there might be some Z-DNA tracts, especially near the ends of a molecule, but the majority of the length of a double helix would be unvarying B-DNA. We now recognize that DNA is much more polymorphic, and that it is possible for the A-, B-, and Z-DNA configurations, and intermediates between them, to coexist within a single DNA molecule, different parts of the molecule having different structures. These conformational variations are sequence dependent, being largely the result of the base-stacking interactions that occur between adjacent base pairs. As well as being responsible, along with base pairing, for the stability of the helix, the base stacking also influences the amount of rotation that occurs around the covalent bonds within individual nucleotides and hence determines the conformation of the helix at a particular position. The rotational possibilities in one base pair are influenced, via the base-stacking interactions, by the identities of the neighboring base pairs. This means that the nucleotide sequence indirectly affects the overall conformation of the helix, possibly providing structural information that a binding protein can use to help it locate its appropriate attachment site on a DNA molecule. At present this is just a theoretical possibility as no protein that specifically recognizes a non-B-form of the helix has been identified, but many researchers believe that helix conformation is likely to play some role in the interaction between DNA and protein.

A second type of conformational change is **DNA bending**. This does not refer to the natural flexibility of DNA, which enables it to form circles and supercoils, but instead to localized positions where the nucleotide sequence causes the DNA to bend. Like other conformational variations, DNA bending is sequence dependent. In particular, a DNA molecule in which one polynucleotide contains two or more groups of repeated adenines—each group comprising 3–5 As, with the individual groups separated by 10 or 11 nucleotides—will bend at the 3′ end of the adenine-rich region. As with helix conformation, it is not yet known to what extent DNA bending influences protein binding, although protein-induced bending at flexible sites has a clearly demonstrated function in the regulation of some genes (Section 11.3.2).

Contacts between DNA and proteins

The contacts formed between DNA and its binding proteins are noncovalent. Within the major groove, hydrogen bonds form between the nucleotide bases and the R groups of amino acids in the recognition structure of the protein, whereas in the minor groove hydrophobic interactions are more important. On the surface of the DNA helix, the major interactions are electrostatic, between the negative charges on the phosphate component of each nucleotide and the positive charges on the R groups of amino acids such as lysine and arginine, although some hydrogen bonding also occurs. In some cases, hydrogen bonding on the surface of the helix or in the major groove is direct between DNA and protein; in others it is mediated by water molecules. Few generalizations can be made: at this level of DNA–protein interaction each example has its own unique features and the details of the bonding have to be worked out by structural studies rather than by comparisons with other proteins.

Most proteins that recognize specific nucleotide sequences are also able to bind nonspecifically to other parts of a DNA molecule. In fact, it has been suggested that the amount of DNA in a cell is so large, and the numbers of each DNA-binding protein so small, that the proteins spend most, if not all, of their time attached nonspecifically to DNA. The distinction between the nonspecific and specific forms of binding is that the latter is more favorable in thermodynamic terms. As a result, a protein is able to bind to its specific site even though there are literally millions of other sites to which it could attach nonspecifically. To achieve this thermodynamic favorability, the specific binding process must involve the greatest possible number of DNA–protein contacts, which explains in part why the recognition structures of many DNA-binding motifs have evolved to fit snugly into the major groove of the helix, where the opportunity for DNA–protein contacts is greatest. It also explains why some DNA–protein interactions result in conformational changes to one or other partner, increasing still further the complementarity of the interacting surfaces, and allowing additional bonding to occur.

The need to maximize contacts in order to ensure specificity is also one of the reasons why many DNA-binding proteins are dimers, consisting of two proteins attached to one another. This is the case for most HTH proteins and

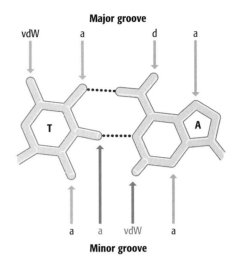

Figure 11.12 Recognition of an A–T base pair in the B-form double helix. An A–T base pair is shown in outline (Figure 1.8B), with arrows indicating the chemical features that can be recognized by accessing the base pair via the major groove (above) and minor groove (below). In the major groove, the chemical features are asymmetric and the orientation of the A–T pair can be identified by a binding protein. For some time it was believed that this is not possible in the minor groove, because only the two features shown in gray were thought to be present, and these are symmetric. Using these two features, the binding protein could recognize the A–T base pair but not know which nucleotide is in which strand of the helix. The asymmetric features shown in green have recently been discovered, suggesting that the orientation of the pair might, in fact, be discernible via the minor groove. Abbreviations: a, hydrogen bond acceptor; d, hydrogen bond donor; vdW, van der Waals interaction.

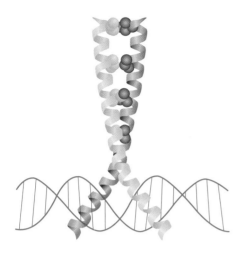

Figure 11.13 A leucine zipper. This is a bZIP type of leucine zipper. The red and orange structures are parts of different proteins. Each set of spheres represents the R group of a leucine amino acid. Leucines in the two helices associate with one another via hydrophobic interactions to hold the two proteins together in a dimer. In this example, the dimerization helices are extended to form a pair of basic domain DNA-binding motifs, shown making contacts in the major groove.

many of the zinc-finger type. Dimerization occurs in such a way that the DNA-binding motifs of the two proteins are both able to access the helix, possibly with some degree of cooperativity between them, so that the resulting number of contacts is greater than twice the number achievable by a monomer. As well as their DNA-binding motifs, many proteins contain additional characteristic domains that participate in the protein–protein contacts that result in dimer formation. One of these is the **leucine zipper**, which is an α-helix that coils more tightly than normal and presents a series of leucines on one of its faces. These can form contacts with the leucines of the zipper on a second protein, forming the dimer (Figure 11.13). A second dimerization domain is, rather unfortunately, called the **helix–loop–helix** motif, which is distinct from, and should not be confused with, the helix–turn–helix DNA-binding motif.

An intriguing question is whether the specificity of DNA binding can be understood in sufficient detail for the sequence of a protein's target site to be predicted from examination of the structure of the recognition helix of a DNA-binding motif. To date, this objective has largely eluded us, but it has been possible to deduce some rules for the interaction involving certain types of zinc finger. In these proteins, four amino acids—three in the recognition helix and one immediately adjacent to it—form critical attachments with the nucleotide bases of the target site. Some of these attachments involve a single amino acid with a single base, others involve two amino acids with one base. By comparing the sequences of amino acids in the recognition helices of different zinc fingers with the sequences of nucleotides at the binding sites, it has been possible to identify a set of rules governing the interaction. These enable the nucleotide sequence specificity of a new zinc-finger protein to be predicted, admittedly with the possibility of some ambiguity, once the amino acid composition of its recognition helix is known.

11.2 DNA–Protein Interactions During Transcription Initiation

Now that we have established that DNA–protein interactions are the key to understanding the initiation of transcription, we can move on to begin our examination of the events involved in the assembly of the initiation complex. We will do this in two stages. First, we will study the DNA–protein interactions that are involved in transcription initiation. Then, in Section 11.3, we will investigate how assembly of the initiation complex, and its ability to initiate transcription, can be controlled by various additional proteins that respond to stimuli from inside or outside the cell and ensure that the correct genes are transcribed at the appropriate times.

11.2.1 RNA polymerases

In Section 1.2.1 we learnt that the enzymes responsible for transcription of DNA into RNA are called DNA-dependent RNA polymerases. Transcription of eukaryotic nuclear genes requires three different RNA polymerases: **RNA polymerase I**, **RNA polymerase II**, and **RNA polymerase III**. Each is a multi-subunit protein (8–12 subunits) with a molecular mass in excess of 500 kDa. Structurally, these polymerases are quite similar to one another, the three largest subunits being closely related and some of the smaller ones being found in more than one enzyme; functionally, however, they are quite distinct. Each works on a different set of genes, with no interchangeability

Table 11.3 Functions of the three eukaryotic nuclear RNA polymerases

Polymerase	Genes transcribed
RNA polymerase I	28S, 5.8S, and 18S ribosomal RNA (rRNA) genes
RNA polymerase II	Protein-coding genes, most small nuclear RNA (snRNA) genes, microRNA (miRNA) genes
RNA polymerase III	Genes for transfer RNAs (tRNA), 5S rRNA, U6 snRNA, small nucleolar (sno) RNAs

(Table 11.3). Most research attention has been directed at RNA polymerase II, as this is the enzyme that transcribes genes that code for proteins. It also works on a set of genes specifying the snRNAs that are involved in RNA processing, and with the genes for the miRNAs. RNA polymerase III transcribes other genes for small RNAs, including those for tRNAs. RNA polymerase I transcribes the multicopy repeat units containing the 28S, 5.8S, and 18S rRNA genes. The functions of all these RNAs were summarized in Section 1.2.2 and are described in detail in Chapters 12 and 13.

Archaea possess a single RNA polymerase that is structurally very similar to the eukaryotic enzymes. But this is not typical of the prokaryotes in general because the bacterial RNA polymerase is very different, consisting of just five subunits, with its composition described as $\alpha_2\beta\beta'\sigma$ (two α subunits, one each of β and the related β', and one of σ). The α, β, and β' subunits are equivalent to the three largest subunits of the eukaryotic RNA polymerases, but the σ subunit has its own special properties, both in terms of its structure and, as we will see in the next section, its function. An RNA polymerase very similar to the bacterial enzyme is also found in chloroplasts, reflecting the bacterial origins of these organelles (Section 8.3.1). Interestingly, however, the mitochondrial RNA polymerase, which consists of a single subunit with a molecular mass of 140 kDa, is more closely related to the RNA polymerases of certain bacteriophages than it is to the standard bacterial version.

11.2.2 Recognition sequences for transcription initiation

It is essential that transcription initiation complexes are constructed at the correct positions on DNA molecules. These positions are marked by target sequences that are recognized either by the RNA polymerase itself or by a DNA-binding protein which, once attached to the DNA, forms a platform onto which the RNA polymerase binds (see Figure 11.14).

Bacterial RNA polymerases bind to promoter sequences

In bacteria, the target sequence for RNA polymerase attachment is called the **promoter**. This term was first used by geneticists in 1964 to describe the function of a locus immediately upstream of the three genes in the lactose operon (Figure 11.15). When this locus was inactivated by mutation, the genes in the operon were not expressed; the locus therefore appeared to *promote* expression of the genes. We now know that this is because the locus is the binding site for the RNA polymerase that transcribes the operon.

The sequences that make up the *Escherichia coli* promoter were first identified by comparing the regions upstream of over 100 genes. It was assumed that promoter sequences would be very similar for all genes and so should be recognizable when the upstream regions were compared. These analyses

(A) Direct attachment of RNA polymerase

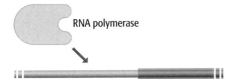

(B) Indirect attachment of RNA polymerase

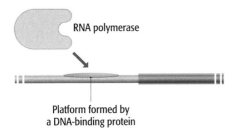

Platform formed by
a DNA-binding protein

Figure 11.14 Two ways in which RNA polymerases bind to their promoters. (A) The direct recognition of the promoter by the RNA polymerase, as occurs in bacteria. (B) Recognition of the promoter by a DNA-binding protein, which forms a platform onto which the RNA polymerase binds. This indirect mechanism occurs with eukaryotic and archaeal RNA polymerases

Figure 11.15 The promoter for the lactose operon of *Escherichia coli*. The promoter is located immediately upstream of *lacZ*, the first gene in the operon. The DNA sequence shows the positions of the −35 and −10 boxes, the two distinct sequence components of the promoter. Compare these sequences with the consensus sequences described in the text. For more information on the lactose operon, see Figure 8.8A.

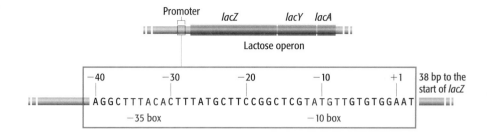

showed that the *E. coli* promoter consists of two segments, both of six nucleotides, described as follows (see Figure 11.15):

−35 box   5′–TTGACA–3′

−10 box   5′–TATAAT–3′

These are consensus sequences and so describe the "average" of all promoter sequences in *E. coli*: the actual sequences upstream of any particular gene might be slightly different (Table 11.4). The names of the boxes indicate their positions relative to the point at which transcription begins. The nucleotide at this point is labeled "+1" and is anything between 20 and 600 nucleotides upstream of the start of the coding region of the gene. The spacing between the two boxes is important because it places the two motifs on the same face of the double helix, facilitating their interaction with the DNA-binding component of the RNA polymerase (Section 11.2.3).

Eukaryotic promoters are more complex

In eukaryotes, the term "promoter" is used to describe all the sequences that are important in initiation of transcription of a gene. For some genes these sequences can be numerous and diverse in their functions, including not only the **core promoter**, sometimes called the **basal promoter**, which is the site at which the initiation complex is assembled, but also one or more **upstream promoter elements** which, as their name implies, lie upstream of the core promoter. Assembly of the initiation complex on the core promoter can usually occur in the absence of the upstream elements, but only in an inefficient way. This indicates that the proteins that bind to the upstream elements include at least some that are activators of transcription, and which therefore "promote" gene expression. Inclusion of these sequences in the "promoter" is therefore justified.

Each of the three eukaryotic RNA polymerases recognizes a different type of promoter sequence; indeed, it is the difference between the promoters that defines which genes are transcribed by which polymerases. The details for vertebrates are as follows (Figure 11.16):

Table 11.4 Sequences of *Escherichia coli* promoters

Promoter	Sequence	
	−35 box	−10 box
Consensus	5′–TTGACA–3′	5′–TATAAT–3′
Lactose operon	5′–TTTACA–3′	5′–TATGTT–3′
Tryptophan operon	5′–TTGACA–3′	5′–TTAACT–3′

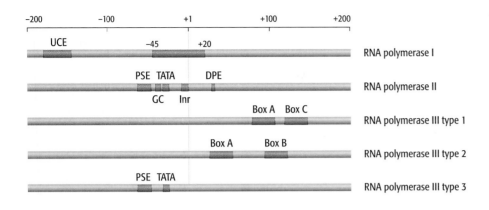

Figure 11.16 **Structures of eukaryotic promoters.** Abbreviations are defined in the text.

- RNA polymerase I promoters consist of a core promoter spanning the transcription start point, between nucleotides –45 and +20, and an **upstream control element** (UCE) about 100 bp further upstream.

- RNA polymerase II promoters are variable and can stretch for several kilobases upstream of the transcription start site. The core promoter consists of two main segments: the –25 or **TATA box** (consensus 5′–TATAWAAR–3′, where W is A or T, and R is A or G) and the **initiator (Inr) sequence** (mammalian consensus 5′–YCANTYY–3′, where Y is C or T, and N is any nucleotide) located around nucleotide +1. Some genes transcribed by RNA polymerase II have only one of these two components of the core promoter, and some, surprisingly, have neither. The latter are called "null" genes. They are still transcribed, although the start position for transcription is more variable than for a gene with a TATA box and/or Inr sequence. A few genes have additional sequences that are sometimes looked on as part of the core promoter. Examples are:

  - The downstream promoter element (DPE; located at positions +28 to +32), which has a variable sequence but has been identified through its ability to bind TFIID, a protein complex that plays a central role in the preinitiation complex (Section 11.2.3).

  - A 7 bp GC-rich motif immediately upstream of the TATA box, which is recognized by TFIIB, another component of the preinitiation complex.

  - The proximal sequence element (PSE), which is located between positions –45 and –60 upstream of those snRNA genes that are transcribed by RNA polymerase II.

  As well as the components of the core promoter, genes transcribed by RNA polymerase II have various upstream promoter elements, the functions of which are described in Section 11.3.2.

- RNA polymerase III promoters are variable, falling into at least three categories. Two of these categories are unusual in that the important sequences are located within the genes whose transcription they promote. Usually these sequences span 50–100 bp and comprise two conserved boxes separated by a variable region. The third category of RNA polymerase III promoter is similar to those for RNA polymerase II, having a TATA box and a range of additional promoter elements (sometimes including the PSE mentioned above) located upstream of the target gene. Interestingly, this arrangement is seen with the U6 gene, which is the single snRNA gene transcribed by RNA polymerase III, all the other snRNA genes being transcribed by RNA polymerase II.

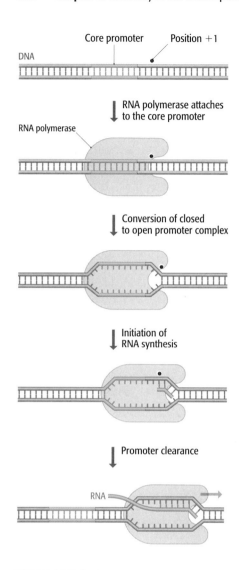

Figure 11.17 Generalized scheme for the events occurring during initiation of transcription. The core promoter is shown in green and the transcription initiation site is indicated by a red dot. After RNA polymerase attachment, the closed complex is converted into the open complex by breakage of base pairs within a short region of the DNA double helix. RNA synthesis begins but successful initiation is not achieved until the polymerase moves away from the promoter region.

11.2.3 Assembly of the transcription initiation complex

In a general sense, initiation of transcription operates along the same lines with each of the four types of RNA polymerase that we have been considering (Figure 11.17). The bacterial polymerase and the three eukaryotic enzymes all begin by attaching, directly or via accessory proteins, to their promoter or core promoter sequences. Next, this **closed promoter complex** is converted into an **open promoter complex** by breakage of a limited number of base pairs around the transcription initiation site. Finally, the RNA polymerase moves away from the promoter. This last step is more complicated than it might appear because some attempts by the polymerase to achieve **promoter clearance** are unsuccessful and lead to truncated transcripts that are degraded soon after they are synthesized. The true completion of the initiation stage of transcription is therefore the establishment of a stable transcription complex that is actively transcribing the gene to which it is attached.

Although the scheme shown in Figure 11.17 is correct in outline for all four RNA polymerases, the details are different for each one. We will begin with the more straightforward events occurring in *E. coli* and other bacteria, and then move on to the ramifications of transcription initiation in eukaryotes.

Transcription initiation in E. coli

In *E. coli*, a direct contact is formed between the promoter and RNA polymerase. The sequence specificity of the polymerase resides in its σ subunit: the "core enzyme," which lacks this component, can only make loose and nonspecific attachments to DNA.

Mutational studies of *E. coli* promoters have shown that changes to the sequence of the –35 box affect the ability of RNA polymerase to bind, whereas changes to the –10 box affect the conversion of the closed promoter complex into the open form. These results led to the model for *E. coli* transcription initiation shown in Figure 11.18, where recognition of the promoter occurs by an interaction between the σ subunit and the –35 box, forming a closed promoter complex in which the RNA polymerase spans some 80 bp from upstream of the –35 box to downstream of the –10 box. The closed promoter complex is converted to the open form by the combined action of the β′ and σ subunits, which break the base pairs within the –10 box. The model is consistent with the fact that the –10 boxes of different promoters are comprised mainly or entirely of A–T base pairs, which are weaker than G–C pairs, being linked by just two hydrogen bonds as opposed to three (see Figure 1.8B).

Opening up of the helix involves contacts between the polymerase and the non-template strand (i.e., the one that is not copied into RNA), again with the σ subunit playing a central role. However, the σ subunit is not all-important because it usually (but not always) dissociates soon after initiation is complete, converting the holoenzyme ($\alpha_2\beta\beta'\sigma$) to the core enzyme ($\alpha_2\beta\beta'$) which carries out the elongation phase of transcription (Section 12.1.1). Initially the core enzyme covers some 60 bp of the DNA, but soon after the start of elongation the polymerase undergoes a second conformational change, reducing its footprint to just 30–40 bp.

Transcription initiation with RNA polymerase II

How does the easily understandable series of events occurring in *E. coli* compare with the equivalent processes in eukaryotes? Study of RNA polymerase II will show us that eukaryotic transcription initiation involves more proteins and has added complexities.

The first difference between initiation of transcription in *E. coli* and eukaryotes is that eukaryotic RNA polymerases do not directly recognize their core promoter sequences. For genes transcribed by RNA polymerase II, the initial contact is made by the **general transcription factor** (**GTF**) TFIID, which is a complex made up of the **TATA-binding protein** (**TBP**) and at least 12 **TBP-associated factors**, or **TAFs**. TBP is a sequence-specific protein that binds to DNA via its unusual TBP domain (Section 11.1.1), which makes contact with the minor groove in the region of the TATA box. X-ray crystallography studies of TBP show that it has a saddle-like shape that wraps partially around the double helix, forming a platform onto which the remainder of the initiation complex can be assembled (Figure 11.19).

The TAFs assist in attachment of TBP to the TATA box and, in conjunction with other proteins called **TAF- and initiator-dependent cofactors** (**TICs**), possibly also participate in recognition of the Inr sequence, especially at those promoters that lack a TATA box. TAFs are intriguing proteins that appear to play a variety of roles during initiation of transcription and also during other events that involve assembly of multiprotein complexes onto the genome. Five of the yeast TAFs are also present in SAGA, one of the histone acetyltransferase complexes that we met in Section 10.2.1, and TAF1 of *Drosophila melanogaster* possesses a kinase activity that enables it to phosphorylate serine-33 of histone H2B, activating expression of adjacent genes (Section 10.2.1). TAFs have also been implicated in control of the cell cycle in various eukaryotes and in regulation of the developmental changes that result in formation of gametes in animals. A clue as to how TAFs carry out their multifarious roles has been provided by structural studies which have shown that at least three of them contain a histone fold—a non-sequence-specific DNA-binding motif, comprising a long α-helix flanked on either side by two shorter α-helices, that is a characteristic feature of histones (see Table 11.2). In the complex formed between TAF$_{II}$42 and TAF$_{II}$62, the two histone folds are oriented in almost exactly the same way as in the histone H3/H4 dimer found in the central tetramer of the nucleosome core particle (Figure 11.20). It has been proposed that these TAFs might be able to form a DNA-binding structure resembling a nucleosome, this pseudonucleosome acting as the platform for assembly of the initiation complex. The idea is attractive but more research is needed to determine if the similarities between TAFs and histones extend to the details of the attachment between the initiation complex and its DNA target. TAFs lack certain amino acids that are looked on as essential for stabilizing the contacts between real nucleosomes and DNA, and the similarities could simply reflect an equivalence in the protein–protein interactions within the initiation complex and the nucleosome, rather than being directly relevant to DNA attachment.

After TBP has attached to the core promoter, the **preinitiation complex** (**PIC**) is formed by recruitment of additional GTFs, whose roles are summarized in Table 11.5. TBP binding induces formation of a bend of approximately 80° in the DNA, widening the minor groove in the region of the TATA box (Figure 11.21). TFIIB now attaches to the complex, this attachment involving contacts with the TATA box via the widened minor groove and, via the major groove, with the TFIIB recognition motif immediately upstream of the TATA

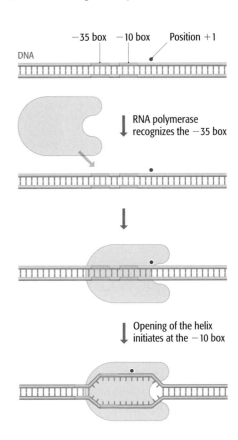

Figure 11.18 Initiation of transcription in *Escherichia coli*. The *E. coli* RNA polymerase recognizes the −35 box as its binding sequence. After attachment to the DNA, the transition from closed to open complex is initiated by breakage of base pairs in the AT-rich −10 box. Note that although the polymerase is shown as a single structure it is the σ subunit that possesses sequence-specific DNA-binding activity and which therefore recognizes the −35 sequence. For subsequent events leading to promoter clearance, refer to Figure 11.17.

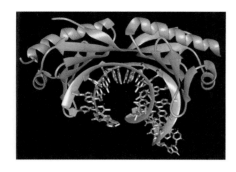

Figure 11.19 TBP attachment to the TATA box forms a platform onto which the initiation complex can be assembled. The dimer of TBP proteins is shown in brown and the DNA is in silver. Image courtesy of Song Tan, Penn State University.

box (Section 11.2.2). These attachments ensure the correct positioning, relative to the transcription start site, of RNA polymerase II, which is brought into the complex by TFIIF. The preinitiation complex is completed by addition of TFIIE and TFIIH, the latter of which has a helicase activity and is thought to disrupt the DNA base pairing and hence to convert the promoter into the open form. TFIIH is an interesting protein as it also has a role in DNA repair, some types of which are coupled with transcription, and we will encounter it again when we study DNA repair mechanisms in Section 16.2.2.

Activation of the initiation complex requires the addition of phosphate groups to the **C-terminal domain** (**CTD**) of the largest subunit of RNA polymerase II. In mammals, this domain consists of 52 repeats of the seven-amino-acid sequence Tyr–Ser–Pro–Thr–Ser–Pro–Ser. Two of the three serines in each repeat unit can be modified by addition of a phosphate group, causing a substantial change in the ionic properties of the polymerase. Once phosphorylated, the polymerase is able to leave the initiation complex and begin

Figure 11.20 The dimers formed between TAF$_{II}$42/TAF$_{II}$62 and histones H3/H4. The drawings show the orientations of the histone folds of the pair of proteins in each complex.

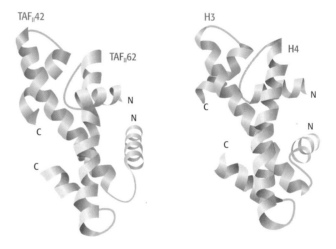

Table 11.5 Functions of the human general transcription factors (GTFs)

GTF	Function
TFIID (TBP component)	Recognition of the TATA box and possibly Inr sequence; forms a platform for TFIIB binding
TFIID (TAFs)	Recognition of the core promoter; regulation of TBP binding
TFIIA	Stabilizes TBP and TAF binding
TFIIB	Intermediate in recruitment of RNA polymerase II; influences selection of the start point for transcription
TFIIF	Recruitment of RNA polymerase II; interaction with the non-template strand
TFIIE	Intermediate in recruitment of TFIIH; modulates the various activities of TFIIH
TFIIH	Helicase activity responsible for the transition from the closed to open promoter complex; possibly influences promoter clearance by phosphorylation of the C-terminal domain of the largest subunit of RNA polymerase II

synthesizing RNA. Phosphorylation might be carried out by TFIIH, which has the appropriate protein kinase capability, or it might be the function of the **mediator** (Section 11.3.2), which transduces signals from activator proteins that regulate expression of individual genes. After departure of the polymerase, at least some of the GTFs detach from the core promoter, but TFIID, TFIIA, and TFIIH remain, enabling reinitiation to occur without the need to rebuild the entire assembly from the beginning. Reinitiation is therefore a more rapid process than primary initiation, which means that once a gene is switched on, transcripts can be initiated from its promoter with relative ease until such a time as a new set of signals switches the gene off.

Transcription initiation with RNA polymerases I and III

Initiation of transcription at RNA polymerase I and III promoters involves similar events to those seen with RNA polymerase II, but the details are different. One of the most striking similarities is that TBP, first identified as the key sequence-specific DNA-binding component of the RNA polymerase II preinitiation complex, is also involved in initiation of transcription by the two other eukaryotic RNA polymerases.

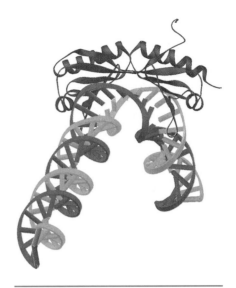

Figure 11.21 **Attachment of TBP induces a bend in the DNA molecule.** TBP is shown in purple and the DNA in green. The bend in the DNA opens up the minor groove, facilitating attachment of TFIIB. Photograph kindly provided by Stephen K. Burley.

The RNA polymerase I initiation complex involves four protein complexes in addition to the polymerase itself. One of these, UBF, is a dimer of identical proteins that interacts with both the core promoter and the upstream control element (see Figure 11.16). UBF is another protein which, like some of the RNA polymerase II TAFs, resembles a histone and may form a nucleosome-like structure in the promoter region. A second protein complex, called SL1 in humans and TIFIB in mice, contains TBP and, together with UBF, directs RNA polymerase I and the last two complexes, TIFIA and TIFIC, to the promoter. Originally it was thought that the initiation complex was built up in a stepwise fashion, but recent results suggest that RNA polymerase I binds the four protein complexes before promoter recognition, the entire assembly attaching to the DNA in a single step.

RNA polymerase III promoters are variable in structure (see Figure 11.16) and this is reflected by a nonuniformity of the processes through which they are recognized. Initiation at the different categories of RNA polymerase III promoter requires different sets of GTFs, but each type of initiation process involves TFIIIB, one of whose subunits is TBP. With promoters of the type seen with the U6 snRNA gene, which contain a TATA sequence, TBP probably binds directly to the DNA. At those RNA polymerase III promoters that lie within genes, and which have no TATA sequence, binding of TBP is via a pair of assembly factors called TFIIIA and TFIIIC. The name "assembly factor" indicates that these two proteins are needed only for attachment of TBP to the promoter and are not essential for the subsequent binding of RNA polymerase III.

11.3 Regulation of Transcription Initiation

As we progress through the next few chapters we will encounter a number of strategies that organisms use to regulate expression of individual genes. We will discover that virtually every step in the pathway from genome to proteome is subject to some degree of control. Of all these regulatory systems, it appears that transcription initiation is the stage at which the critical controls over the expression of individual genes (i.e., those controls that have greatest

Figure 11.22 Primary and secondary levels of genome regulation. According to this scheme, "primary" regulation of genome expression occurs at the level of transcription initiation, this step determining which genes are expressed in a particular cell at a particular time and setting the relative rates of expression of those genes that are switched on. "Secondary" regulation involves all steps in the genome expression pathway after transcription initiation, and serves to modulate the amount of protein that is synthesized or to change the nature of the protein in some way, for example by chemical modification.

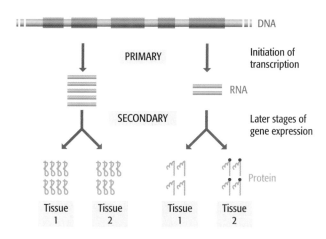

impact on the biochemical properties of the cell) are exerted. This is perfectly understandable. It makes sense that transcription initiation, being the first step in genome expression, should be the stage at which "primary" regulation occurs, this being the level of regulation that determines which genes are expressed. Later steps in the pathway might be expected to respond to "secondary" regulation, the function of which is not to switch genes on or off but to modulate expression by making small changes to the rate at which the protein product is synthesized, or possibly by changing the nature of the product in some way (Figure 11.22).

In Chapter 10 we looked at how chromatin structure can influence gene expression by controlling the accessibility of promoter sequences to RNA polymerase and its associated proteins. This is just one way in which initiation of transcription can be regulated. To obtain a broader picture, we will establish some general principles with bacteria, and then examine the events in eukaryotes.

11.3.1 Strategies for controlling transcription initiation in bacteria

In bacteria such as *E. coli*, we recognize two distinct ways in which transcription initiation is controlled:

- **Constitutive control**, which depends on the structure of the promoter.
- **Regulatory control**, which depends on the influence of regulatory proteins.

Promoter structure determines the basal level of transcription initiation

The consensus sequence for the *E. coli* promoter (Section 11.2.2) is quite variable, with a range of different motifs being permissible at both the –35 and –10 boxes (see Table 11.4). These variations, together with less-well-defined sequence features around the transcription start site and in the first 50 or so nucleotides of the transcription unit, affect the efficiency of the promoter. Efficiency is defined as the number of productive initiations that are promoted per second, a productive initiation being one that results in the RNA polymerase clearing the promoter and beginning synthesis of a full-length transcript. The exact way in which the sequence of the promoter affects initiation is not known, but from our discussion of the events involved in transcription initiation (Section 11.2.3) we might, intuitively, expect that the precise sequence of the –35 box would influence recognition by the σ subunit and hence the rate of attachment of RNA polymerase, that the transition from the closed to open promoter complex might be dependent on the sequence of the –10 box, and that the frequency of abortive initiations (ones that terminate before they progress very far into the transcription unit) might be influenced by the sequence at, and immediately downstream of, nucleotide +1. All this is speculation but it is a sound "working hypothesis." What is clear is that different promoters vary 1000-fold in their efficiencies, the most efficient promoters (called **strong promoters**) directing 1000 times as many productive initiations as the weakest promoters. We refer to these as differences in the **basal rate** of transcription initiation.

Note that the basal rate of transcription initiation for a gene is preprogrammed by the sequence of its promoter and so, under normal circumstances, cannot be changed. It could be changed by a mutation that alters a critical nucleotide in the promoter, and undoubtedly this happens from time to time, but it is not something that the bacterium has control over. The bacterium can, however, determine which promoter sequences are favored by changing the σ subunit of its RNA polymerase. The σ subunit is the part of the polymerase that has the sequence-specific DNA-binding capability (Section 11.2.3), so replacing one version of this subunit with a different version with a slightly different DNA-binding motif, and hence an altered sequence specificity, would result in a different set of promoters being recognized. In *E. coli*, the standard σ subunit, which recognizes the consensus promoter sequence and hence directs transcription of most genes, is called σ^{70} (its molecular mass is approximately 70 kDa). *E. coli* also has a second σ subunit, σ^{32}, which is made when the bacterium is exposed to a heat shock. During a heat shock, *E. coli*, in common with other organisms, switches on a set of genes coding for special proteins that help the bacterium withstand the stress (Figure 11.23). These genes have special promoter sequences, ones specifically recognized by the σ^{32} subunit. The bacterium is therefore able to switch on a whole range of different genes by making one simple alteration to the structure of its RNA polymerase. This system is common in bacteria: for example, *Klebsiella pneumoniae* uses it to control expression of genes involved in nitrogen fixation, this time with the σ^{54} subunit, and *Bacillus* species use a whole range of different σ subunits to switch on and off groups of genes during the changeover from normal growth to formation of spores (Section 14.3.2).

Regulatory control over bacterial transcription initiation

Promoter structure determines the basal rate of transcription initiation for a bacterial gene but, with the exception of recognition by alternative σ subunits, does not provide any general means by which the expression of the gene can respond to changes in the environment or to the biochemical requirements of the cell. Other types of regulatory control are needed.

The foundation of our understanding of regulatory control over transcription initiation in bacteria was laid in the early 1960s by François Jacob, Jacques Monod, and other geneticists who studied the lactose operon and other model systems. We have already seen how this work led to discovery of the promoter for the lactose operon (Section 11.2.2). It also resulted in identification of the **operator**, a region adjacent to the promoter and which regulates initiation of transcription of the operon (Figure 11.24A). The original model envisaged that a DNA-binding protein—the **lactose repressor**—attached to

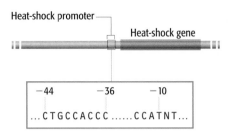

(A) An *E. coli* heat shock gene

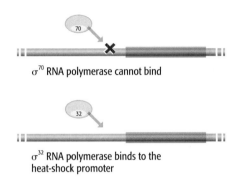

(B) Recognition by the σ^{32} subunit

σ^{70} RNA polymerase cannot bind

σ^{32} RNA polymerase binds to the heat-shock promoter

Figure 11.23 Recognition of an *Escherichia coli* **heat-shock gene by the** σ^{32} **subunit.** (A) The sequence of the heat-shock promoter is different from that of the normal *E. coli* promoter (compare with Table 11.4). (B) The heat-shock promoter is not recognized by the normal *E. coli* RNA polymerase containing the σ^{70} subunit, but is recognized by the σ^{32} RNA polymerase that is active during heat shock. Abbreviation: N, any nucleotide. For more details of the use of novel σ factors by bacteria, see Section 14.3.2.

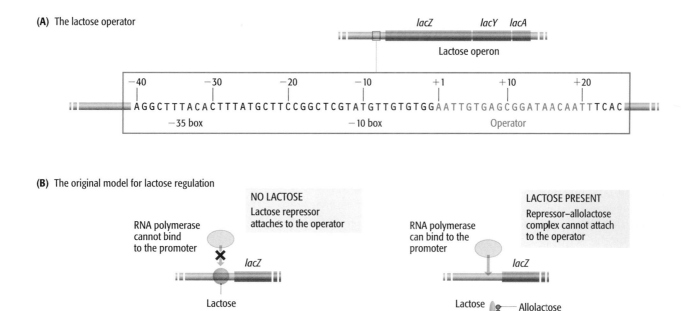

Figure 11.24 Regulation of the lactose operon of *Escherichia coli*. (A) The operator sequence lies immediately downstream of the promoter for the lactose operon. Note that this sequence has inverted symmetry: when read in the 5′→3′ direction, the sequence is the same in both strands. This enables two subunits of the tetrameric repressor protein to make contact with a single operator sequence. (B) In the original model for lactose regulation, the lactose repressor is looked on as a simple blocking device that binds to the operator and prevents the RNA polymerase gaining access to the promoter. The three genes in the operon are therefore switched off. This is the situation in the absence of lactose, although transcription is not completely blocked because the repressor occasionally detaches, allowing a few transcripts to be made. Because of this basal level of transcription, the bacterium always possesses a few copies of each of the three enzymes coded by the operon (see Figure 8.8A), probably less than five of each. This means that when the bacterium encounters a source of lactose it is able to transport a few molecules into the cell and split these into glucose and galactose. An intermediate in this reaction is allolactose, an isomer of lactose, which induces expression of the lactose operon by binding to the repressor, causing a change in the conformation of the latter so it is no longer able to attach to the operator. This allows the RNA polymerase to bind to the promoter and transcribe the three genes. When fully induced, approximately 5000 copies of each protein product are present in the cell. When the lactose supply is used up and allolactose is no longer present, the repressor reattaches to the operator and the operon is switched off. The transcripts of the operon, which have a half-life of less than 3 minutes, decay and the enzymes are no longer made. Note that the shapes of the repressor and polymerase structures shown here are purely schematic.

the operator and prevented the RNA polymerase from binding to the promoter, simply by denying it access to the relevant segment of DNA (Figure 11.24B). Whether the repressor binds depends on the presence in the cell of allolactose, an isomer of lactose, the latter being the substrate for the biochemical pathway carried out by the enzymes coded by the three genes in the operon. Allolactose is an **inducer** of the lactose operon. When allolactose is present it binds to the lactose repressor, causing a slight structural change which prevents the HTH motifs of the repressor from recognizing the operator as a DNA binding site. The allolactose–repressor complex therefore cannot bind to the operator, enabling the RNA polymerase to gain access to the promoter. When the supply of lactose is used up and there is no allolactose left to bind to the repressor, the repressor reattaches to the operator and prevents transcription. The operon is therefore expressed only when the enzymes coded by the operon are needed.

Most of the original scheme for regulation of the lactose operon has been confirmed by DNA sequencing of the control region and by structural studies of

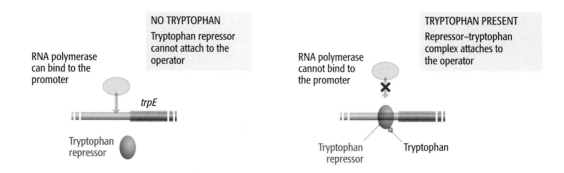

Figure 11.25 Regulation of the tryptophan operon of *Escherichia coli.*
Regulation occurs via a repressor–operator system in a similar way to that described for the lactose operon, but with the difference that the operon is repressed by the regulatory molecule, tryptophan, which is the product of the biochemical pathway specified by the genes in the operon (see Figure 8.8B). When tryptophan is present, and so does not need to be synthesized, the operon is switched off because the repressor–tryptophan complex binds to the operator. In the absence of tryptophan, the repressor cannot attach to the operator, and the operon is expressed.

the repressor bound to its operator. The one complication has been the discovery that the repressor has three potential binding sites, centered on nucleotide positions –82, +11, and +412. The operator defined by genetic studies is the sequence located at +11 (see Figure 11.24A), and this is the only one of the three sites whose occupancy by the repressor would be expected to prevent access of the RNA polymerase to the promoter. But the other two sites also play some role in repression as their removal, individually or together, significantly impairs the ability of the repressor to switch off gene expression. The repressor is a tetramer of four identical subunits that work in pairs to attach to a single operator, so the repressor has the capacity to bind to two of the three operator sites at once. One possibility is that binding of one pair of subunits to the +11 site is enhanced or stabilized by attachment of the other pair of subunits to the –82 or +412 site. It is also possible that the repressor can bind to a pair of operator sequences in such a way that it does not block attachment of the polymerase to the promoter, but does prevent a later step in initiation, such as formation of the open promoter complex.

The basic principle underlying the regulatory control of transcription initiation, as illustrated by the lactose operon, is that attachment of a DNA-binding protein to its specific recognition site can influence the events involved in assembly of the transcription initiation complex and/or initiation of productive RNA synthesis by an RNA polymerase. Several variations on this theme are seen with other bacterial genes:

- Some repressors respond not to an inducer but to a **co-repressor**. An example is provided by the tryptophan operon of *E. coli*, which codes for a set of genes involved in synthesis of tryptophan (see Figure 8.8B). In contrast to the lactose operon, the regulatory molecule for the tryptophan operon is not a substrate for the relevant biochemical pathway, but the product, tryptophan itself (Figure 11.25). Only when tryptophan is attached to the tryptophan repressor can the latter bind to the operator. The tryptophan operon is therefore switched off in the presence of tryptophan, and switched on when tryptophan is needed.

- Some DNA-binding proteins are **activators** rather than repressors of transcription initiation. The best-studied example in *E. coli* is the **catabolite activator protein**, which binds to sites upstream of several operons, including the lactose operon, and increases the efficiency of transcription initiation, probably by forming a direct contact with the RNA polymerase. The biological role of the catabolite activator protein is described in Section 14.1.1.

Figure 11.26 The multiple targets of the tryptophan repressor.

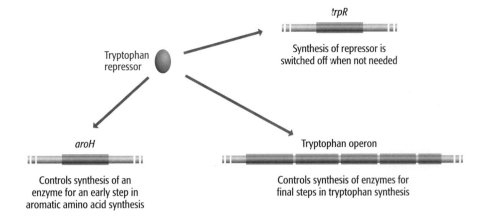

- The same repressor or activator can control two or more promoters. For example, in *E. coli* the tryptophan repressor controls the tryptophan operon, the *aroH* gene (which specifies an enzyme for an early step in the biochemical pathway leading to tryptophan), and the *trpR* gene (which is the tryptophan repressor's own gene, meaning that the repressor protein is synthesized only when it is needed) (Figure 11.26).

- Recognition sequences for DNA-binding proteins can work singly or together to increase or repress transcription of genes to which they are not closely linked. These **enhancers** and **silencers** are not common in bacteria but a few examples are known, including an enhancer that acts on the *E. coli* heat-shock genes whose promoters are recognized by the σ^{32} version of the RNA polymerase. Because they are so far from the genes that they control, they can only form a contact with the RNA polymerase if the DNA forms a loop. A characteristic feature is that a single enhancer or silencer can control expression of more than one gene.

All of these basic principles of gene regulation apply not only to bacteria but, as we will see in the next section, also to eukaryotes.

11.3.2 Control of transcription initiation in eukaryotes

The key lesson that we have learnt from our examination of transcriptional control in bacteria is that transcription initiation can be influenced by DNA-binding proteins that recognize specific sequences located near the attachment site for the RNA polymerase. This is also the basis to transcriptional control in eukaryotes, with two differences, the first of which concerns the basal rate of transcription initiation. The bacterial RNA polymerase has a strong affinity for its promoter and the basal rate of transcription initiation is relatively high for all but the weakest promoters. With most eukaryotic genes, the reverse is true. The RNA polymerase II and III preinitiation complexes do not assemble efficiently and the basal rate of transcription initiation is therefore very low, regardless of how "strong" the promoter is. In order to achieve effective initiation, formation of the complex must be activated by additional proteins. This means that, compared with bacteria, eukaryotes use different strategies to control transcription initiation, with activators playing a much more prominent role than repressor proteins.

The second difference is that, as with all aspects of genome expression, the processes that regulate transcription initiation in eukaryotes are substantially more complex than those occurring in bacteria.

Eukaryotic promoters contain regulatory modules

This greater complexity is apparent when we examine eukaryotic promoters. Transcription initiation for a typical protein-coding gene is influenced by a variety of different biochemical signals which act together to ensure that the gene is expressed at precisely the level that is appropriate for the prevailing conditions within and external to the cell in which that gene is located. An RNA polymerase II promoter can be looked on as a series of modules, each comprising a short sequence of nucleotides and each acting as the binding site for a protein that influences assembly of the transcription initiation complex. Many different genes are transcribed by RNA polymerase II (over 30,000 in humans) but there are only a limited number of promoter modules for this polymerase. The expression pattern for a gene is therefore determined not by an individual module but by the combination of modules within its promoter, and possibly by their relative positions. The amount of transcription initiation that occurs is dependent on which modules are occupied by their binding proteins at a particular time.

The modules for an RNA polymerase II promoter can be categorized in various ways. One scheme is as follows:

- The **core promoter** modules (Section 11.2.2), the most important of these being the TATA box and the Inr sequence.

- **Basal promoter elements** are modules that are present in many RNA polymerase II promoters and set the basal level of transcription initiation, without responding to any tissue-specific or developmental signals. These include: the **CAAT box** (consensus 5′–GGCCAATCT–3′), recognized by the activators NF-1 and NF-Y; the **GC box** (consensus 5′–GGGCGG–3′), recognized by the Sp1 activator; and the **octamer** module (consensus 5′–ATG-CAAAT–3′), recognized by Oct-1.

- **Response modules** are found upstream of various genes and enable transcription initiation to respond to general signals from outside of the cell. Examples are: the cyclic AMP response module CRE (consensus 5′–WCGTCA–3′, where W is A or T), recognized by the CREB activator; the heat-shock module (consensus 5′–CTNGAATNTTCTAGA–3′, where N is any nucleotide), recognized by Hsp70 and other activators; and the serum response module (consensus 5′–CCWWWWWWGG–3′), recognized by the serum response factor.

- **Cell-specific modules** are located in the promoters of genes that are expressed in just one type of tissue. Examples include: the erythroid module (consensus 5′–WGATAR–3′, where R is A or G), which is the binding site for the GATA-1 activator; the pituitary cell module (consensus 5′–ATATTCAT–3′), recognized by Pit-1; the myoblast module (consensus 5′–CAACTGAC–3′), recognized by MyoD; and the lymphoid cell module, or κB site (consensus 5′–GGGACTTTCC–3′), recognized by NF-κB. Note that in lymphoid cells, the octamer module is recognized by the tissue-specific Oct-2 activator.

- Modules for **developmental regulators** mediate expression of genes that are active at specific developmental stages. Two examples in *Drosophila* are the Bicoid module (consensus 5′–TCCTAATCCC–3′) and the Antennapedia module (consensus 5′–TAATAATAATAATAA–3′) (Section 14.3.4).

The modular structure of a typical RNA polymerase II promoter is shown in Figure 11.27. As well as the modules in the region immediately upstream of the

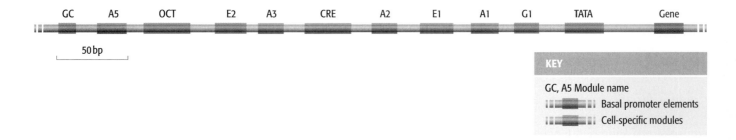

Figure 11.27 The modular structure of the promoter for the human insulin gene.

gene, the same and other modules can also be contained within enhancers, which are 200–300 bp in length and can be located some distance upstream or downstream of their target gene. Silencers are similar to enhancers but, as their name suggests, their modules have a negative rather than enhancing influence on transcription initiation.

An additional level of complexity is seen with some genes that have **alternative promoters** that give rise to different versions of the transcript specified by the gene. An example is provided by the human dystrophin gene, which has been extensively studied because defects in this gene result in the genetic disease called Duchenne muscular dystrophy. The dystrophin gene is one of the largest known in the human genome, stretching over 2.4 Mb and containing 78 introns. It has at least seven alternative promoters (Figure 11.28). Three of these lie upstream of the gene and give rise to full-length transcripts that differ only in the identity of the first exon; these three promoters are active in cortical tissue, muscles, and the cerebellum, respectively. The other four promoters are located within the gene and give shortened transcripts that again are synthesized in a tissue-specific manner. Each promoter has its own modular structure though each is influenced by the same enhancer and silencer sequences. Alternative promoters are also used to generate related versions of some proteins at different stages in development, and to enable a single cell to synthesize similar proteins with slightly different biochemical properties. The last point indicates that although usually referred to as "alternative" promoters these are, more correctly, "multiple" promoters as more than one may be active at a single time. Indeed, this may be the normal situation for many genes. For example, a genome-wide survey has revealed that 10,500 promoters are active in human fibroblast cells, but that these promoters are driving expression of less than 8000 genes, indicating that a substantial number of genes in these cells are being expressed from two or more promoters simultaneously.

Activators and coactivators of eukaryotic transcription initiation

A protein that stimulates transcription initiation is called an **activator** if it is a sequence-specific DNA-binding protein, or a **coactivator** if it binds nonspecifically to DNA or works via protein–protein interactions. Some activators recognize upstream promoter elements and influence transcription initiation only

Figure 11.28 Alternative promoters. The positions of seven alternative promoters for the human dystrophin gene are shown. Abbreviations indicate the tissue within which each promoter is active: C, cortical tissue; M, muscle; Ce, cerebellum; R, retinal tissue (and also brain and cardiac tissue); CNS, central nervous system (and also kidney); S, Schwann cells; G, general (most tissues other than muscle).

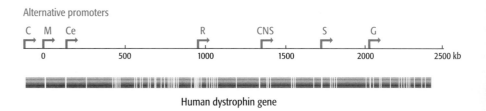

Activator

Enhancer

Figure 11.29 Activators of eukaryotic transcription initiation. The blue activator is attached to a regulatory module upstream of a gene, and influences transcription initiation only at that single gene. The green activator is attached to a site within an enhancer and is influencing transcription of all three genes.

at the promoter to which these elements are attached, and other activators target sites within enhancers and influence transcription of several genes at once (Figure 11.29). As with bacteria, eukaryotic enhancers can be some distance from their genes, their target specificity being ensured by the presence of insulators at either side of each functional domain, preventing the enhancers within that domain from influencing gene expression in adjacent domains (Section 10.1.2). Whether bound to an upstream promoter element or to a more distant enhancer, the activator stabilizes the preinitiation complex.

Coactivators are much more diverse. They include histone acetylation complexes such as SAGA (Section 10.2.1) and nucleosome remodeling complexes such as Swi/Snf (Section 10.2.2). Other proteins classed as coactivators influence gene expression by introducing bends and other distortions into DNA, possibly as a prelude to chromatin modification, or possibly to bring together proteins attached to nonadjacent sites, enabling the bound factors to work together in a structure that has been called an **enhanceosome**. An example of a coactivator that works in this way is SRY, which is the primary protein responsible for determining sex of mammals.

Activators have been looked upon as important in initiation by RNA polymerases II and III, but their role at RNA polymerase I promoters has been less well defined. RNA polymerase I is unusual in that it transcribes just a single set of genes: the multiple copies of the transcription unit containing the 28S, 5.8S, and 18S rRNA sequences (Section 7.2.3). These genes are expressed continuously in most cells, but the rate of transcription varies during the cell cycle and is subject to a certain amount of tissue-specific regulation. The regulatory mechanism has not been described in detail but recent research has suggested a role for the RNA polymerase I **termination factor**. This factor, called TTF-I in mice and Reb1p in *Saccharomyces cerevisiae*, was first identified as an activator of RNA polymerase II transcription. It appears that the termination factor may also activate RNA polymerase I transcription, a binding site for it having been located immediately upstream of the promoter for the rRNA transcription unit.

The mediator forms the contact between an activator and the RNA polymerase II preinitiation complex

A critical feature of the "traditional" type of activator—those that bind to upstream promoter elements or to enhancers—is the contact that is formed with the preinitiation complex. The part of the activator involved in this contact is called the **activation domain**. Structural studies have shown that although activation domains are variable, most of them fall into one of three categories:

- **Acidic domains** are ones that are relatively rich in acidic amino acids (aspartic acid and glutamic acid). This is the commonest category of activation domain.

- **Glutamine-rich domains** are often found in activators whose DNA-binding motifs are of the homeodomain or POU type (Section 11.1.1).

- **Proline-rich domains** are less common.

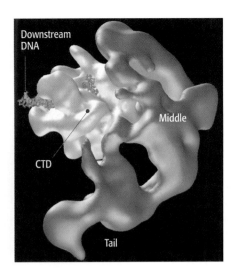

Downstream
DNA

Middle

CTD

Tail

Figure 11.30 Interaction between the yeast mediator and the RNA polymerase II preinitiation complex. The preinitiation complex is shown in white and the mediator in orange. Reprinted from *Molecular Cell*, Vol. 10, Davis *et al.*, 'Structure of the Yeast RNA Polymerase II Holoenzyme', 409–415, 2002, with permission fom Elsevier.

Details of the interaction between activators and the RNA polymerase II preinitiation complex were obscure for several years, with apparently conflicting evidence coming from work with different organisms. A number of protein–protein interaction studies had suggested that direct contacts could be made between different activators and various parts of the complex, with TBP, various TAFs, TFIIB, TFIIH, and RNA polymerase II all implicated as partners in different interactions. The solution to this conundrum was suggested when a large protein complex called the **mediator** was identified in yeast. The yeast mediator comprises 21 subunits that form a structure with distinct head, middle, and tail domains. The tail forms a physical contact with an activator protein attached to its DNA recognition sequence, and the middle and head sections interact with the preinitiation complex (Figure 11.30). Therefore, rather than the activator associating directly with the preinitiation complex, the association is indirect with the activation signal transduced by the mediator. Initially it was thought that the mediator activated transcription initiation by directly phosphorylating the CTD of RNA polymerase II, stimulating promoter clearance (Section 11.2.3), but it is now known that this kinase activity is provided by Kin28, a subunit of TFIIH. There is evidence that the mediator is able to stimulate Kin28 activity, but this is not its only interaction with the preinitiation complex, and the precise way in which it regulates transcription initiation is not yet clear. The mediator is present when TBP attaches to the TATA box, and could form part of the platform onto which the remainder of the preinitiation complex is constructed. A complicating factor is that the mediator transduces signals not only from activators but also from repressors, which means that its influence on the preinitiation complex has to be both positive and negative.

Mediators in higher eukaryotes are larger than the yeast complex, with 30 or more proteins making up the human version. One feature of the mammalian mediator is that its protein composition is variable, raising the possibility that there are several versions, each one responding to a different, although possibly overlapping, set of activators. Current opinion tends to the view that a mediator is an obligatory component of the RNA polymerase II preinitiation complex, and that the stimulatory effects of all activators pass through the mediator. The possibility that some activators bypass the mediator and have a direct effect on one or other part of the preinitiation complex cannot, however, be discounted.

Repressors of eukaryotic transcription initiation

Most of the research on regulation of transcription initiation in eukaryotes has concentrated on activation, partly because the low level of basal initiation occurring at RNA polymerase II and III promoters suggests that the repression of initiation, which is so important in bacteria (Section 11.3.1), is unlikely to play a major part in control of eukaryotic transcription. This view is probably incorrect because a growing number of DNA-binding proteins that repress transcription initiation are being discovered, these proteins binding to upstream promoter elements or to more distant sites in silencers. Some influence genome expression in a general way through histone deacetylation (Section 10.2.1) or DNA methylation (Section 10.3.1), but others have more specific effects at individual promoters. The yeast repressors called Mot1 and NC2, for example, inhibit assembly of the preinitiation complex by binding directly to TBP and disrupting its activity. Mot1 causes TBP to dissociate from the DNA, and NC2 prevents further assembly of the complex

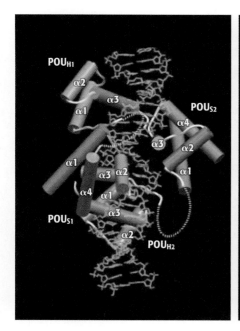 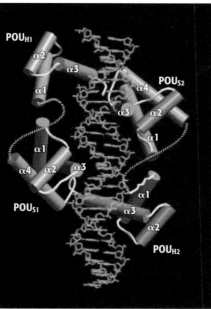

Figure 11.31 Conformation of the POU domains of the Pit-1 activator bound to its target sites upstream of the prolactin (left) and growth hormone (right) genes. Pit-1 is a dimer, and each monomer has two POU domains. The two domains of one monomer are shown in red and the two domains of the other monomer are shown in blue. The barrels are α-helices, with α3 being the recognition helix of each domain. Note the difference between the conformations of the domains when bound to the two binding sites. The more open structure adopted at the growth hormone site enables the Pit-1 dimer to interact with N-CoR and other proteins to repress transcription of the growth hormone gene. Pit-1 therefore activates the prolactin gene but represses the growth hormone gene. Reprinted with permission from Scully *et al.*, *Science*, **290**, 1127–1131. © 2000 AAAS.

on the bound TBP. Both of these repressors have a broad spectrum of activity, inactivating a large set of genes, as does the Ssn6-Tup1 repressor, which is one of the main gene silencers in the yeast *Schizosaccharomyces pombe*, and which has homologs in many other eukaryotes.

Another indication of the importance of repression in eukaryotic transcription comes from the demonstration that some proteins can exert both activating and repressing effects, depending on the circumstances. NC2, for example, represses initiation of transcription from promoters with a TATA box but has an activating effect on promoters that lack the TATA sequence. Pit-1, which is the first of the three proteins after which the POU domain is named (Section 11.1.1), activates some genes and represses others, depending on the sequence of the DNA binding site. The presence in this site of two additional nucleotides induces a change in the conformation of Pit-1, enabling it to interact with a second protein called N-CoR and repress transcription of the target gene (Figure 11.31).

Controlling the activities of activators and repressors

The operation of individual activators and repressors must be controlled in order to ensure that the appropriate set of genes is expressed by a cell. We will return to this topic in Chapter 14, when it will form the central theme of our study of the ways in which genome activity is regulated in response to extracellular signals and during differentiation and development.

There are several ways in which an activator or repressor could be regulated. One possibility is to control its synthesis, but this does not permit rapid changes in genome expression because it takes time to accumulate an activator or repressor in the cell, or to destroy it when it is not needed. This type of control is therefore associated with activators and repressors responsible for maintaining stable patterns of genome expression, for example those underlying cellular differentiation and some aspects of development. An alternative way of controlling an activator or repressor is by chemical modification, for example by phosphorylation, or by inducing a change in its conformation. These changes are much more rapid than *de novo* synthesis, and enable the cell to respond to extracellular signaling compounds that induce transient changes in genome expression. We will examine the details of these various regulatory mechanisms in Chapter 14.

Summary

The central players in transcription, and other aspects of genome activity, are DNA-binding proteins that attach to the genome in order to perform their biochemical functions. Many of these proteins are able to attach to specific DNA sequences by virtue of DNA-binding motifs such as the helix–turn–helix structure or the zinc finger. Their binding positions on DNA molecules can be identified by gel retardation analysis and delineated in greater detail by modification protection and modification interference assays. Some proteins recognize their binding sites by direct readout of the DNA sequence, which is possible by making contacts within the major groove of the double helix, as here the identity of nucleotides can be determined from the positions of the chemical groups attached to the purine and pyrimidine nucleotides. Direct readout can be influenced by various indirect effects that the nucleotide sequence has on the conformation of the helix, including the formation of bends in adenine-rich sequences. Many DNA-binding proteins act as dimers, contacting the helix at two positions simultaneously. Special structures on the protein surface, such as leucine zippers, aid dimerization. Bacteria have a single RNA polymerase that transcribes all the genes in the genome, but eukaryotes have three different nuclear RNA polymerases, and a different enzyme that works in mitochondria. Promoter sequences mark the positions at which the transcription initiation complex must be assembled. The bacterial promoter comprises two sequence segments but eukaryotic promoters are much more complex, having a modular structure that includes sequences that are recognized by the initiation complex and others that act as binding sites for regulatory proteins. The bacterial RNA polymerase attaches directly to its promoter, but the initiation complexes for each type of eukaryotic RNA polymerase contain ancillary proteins and must be built up in an ordered fashion. Successful initiation results in a polymerase complex that is competent to achieve promoter clearance and begin the transcription process. Transcription initiation is a key step at which genome expression is regulated. Some bacteria are able to alter the structure of their RNA polymerase so the enzyme recognizes a different set of promoters and hence transcribes a different set of genes. Expression of individual genes in bacteria can also be controlled by repressor and activator proteins. The same is true with eukaryotes, though activator proteins appear to be more important than repressors. The regulatory proteins attach to sites adjacent to or some distance away from the promoters they control. In eukaryotes, activation of the transcription initiation complex occurs via the mediator, a multisubunit protein which forms a physical bridge between an activator and the RNA polymerase.

Multiple Choice Questions

*Answers to odd-numbered questions can be found in the Appendix

11.1.* How are proteins able to bind to DNA at specific sequences?

a. By interacting with the sugar–phosphate backbone.

b. By opening up the double helix and forming bonds with the bases.

c. By interacting with the bases through the histone proteins.

d. By interacting with the bases in the major and minor grooves of the double helix.

11.2. Which of the following DNA binding domains makes its major contact with the nucleotide bases via the minor groove of the double helix?

a. Helix–turn–helix.

b. Zinc finger.

c. Basic domain.

d. TBP domain.

11.3.* Which of the following techniques, based on migration of DNA fragments in a gel in the presence or absence of proteins, is used to identify proteins that bind to DNA?

a. Nuclear magnetic resonance spectroscopy.

b. Gel retardation.

c. Nuclease protection.

d. DNA footprinting.

11.4. Modification interference assays use which of the following techniques to identify nucleotides central to protein binding?

a. The DNA–protein complex is treated with nucleases to degrade the unprotected phosphodiester bonds.

b. The DNA–protein complex is treated with methylating agents to demarcate the binding site.

c. The DNA is treated with methylating agents prior to protein binding.

d. The protein is treated with methylating agents prior to DNA binding.

11.5.* Which of the following RNA polymerases is responsible for the transcription of protein-coding genes in eukaryotes?

a. RNA polymerase I.

b. RNA polymerase II.

c. RNA polymerase III.

d. RNA polymerase IV.

11.6. The attachment site for RNA polymerase in bacteria is called the:

a. Initiator.

b. Operator.

c. Promoter.

d. Start codon.

11.7.* The specificity of bacterial RNA polymerases for their promoters is due to which subunit?

a. α.

b. β.

c. γ.

d. σ.

11.8. The first protein complex to bind to the core promoter for a protein-coding gene in eukaryotes is:

a. RNA polymerase II.

b. General transcription factor TFIIB.

c. General transcription factor TFIID.

d. General transcription factor TFIIE.

11.9.* Which type of modification must be made to RNA polymerase II in order to activate the preinitiation complex?

a. Acetylation.

b. Methylation.

c. Phosphorylation.

d. Ubiquitination.

11.10. Why is the reinitiation of transcription at an RNA polymerase II promoter a more rapid process than the primary initiation event?

a. All the general transcription factors remain at the promoter to facilitate reinitiation.

b. Some of the general transcription factors remain at the promoter to facilitate reinitiation.

c. All the general transcription factors dissociate from the promoter, but the promoter is still exposed and this allows rapid reassembly of the initiation complex.

d. None of the above.

11.11.* What is the name of the DNA sequence that is located near the promoter of the lactose operon, and which regulates expression of the operon in *E. coli*?

a. Activator.

b. Inducer.

c. Operator.

d. Repressor.

11.12. Allolactose acts as which of the following in regulating the expression of the lactose operon?

a. Activator.

b. Inducer.

c. Operator.

d. Repressor.

continued …

Multiple Choice Questions (continued) *Answers to odd-numbered questions can be found in the Appendix

11.13.* Which of the following sequence modules is NOT a basal promoter element?

 a. CAAT box.

 b. GC box.

 c. Octamer module.

 d. TATA box.

11.14. Which of the following types of sequence module enables transcription to respond to general signals from outside of the cell?

 a. Cell-specific modules.

 b. Developmental regulator modules.

 c. Repressor modules.

 d. Response modules.

11.15.* Which of the following DNA sequences can increase the rate of transcription initiation and may be located hundreds of base pairs upstream or downstream from the genes they regulate?

 a. Activators.

 b. Enhancers.

 c. Silencers.

 d. Terminators.

11.16. Which of the following is NOT a type of activation domain?

 a. Acidic domains.

 b. Glutamine-rich domains.

 c. Leucine-zipper domains.

 d. Proline-rich domains.

Short Answer Questions   *Answers to odd-numbered questions can be found in the Appendix

11.1.* How does the homeodomain motif bind to specific DNA sequences?

11.2. What are the general properties of the Cys_2His_2 zinc finger motif and how does this motif bind to DNA?

11.3.* What are the two versions of the modification protection assay?

11.4. Discuss the different ways in which proteins can contact the bases in A-, B-, and Z-DNA.

11.5.* Describe the types of bonds and interactions that occur between proteins and DNA molecules.

11.6. What would happen if the −10 and −35 components of an *E. coli* promoter were moved closer together or further apart? Explain the basis to your answer.

11.7.* Distinguish between the roles of the core and the upstream elements of a eukaryotic promoter.

11.8. What factors control the basal rate of transcription initiation at a bacterial promoter?

11.9.* How does allolactose interact with the repressor protein to regulate transcription of the lactose operon?

11.10. Name two fundamental differences between the control of transcription in bacteria and equivalent events in eukaryotes.

11.11.* Why do some genes have alternative or multiple promoters?

11.12. What are the differences between activator and coactivator proteins?

In-depth Problems

*Guidance to odd-numbered questions can be found in the Appendix

11.1.* Use your knowledge of DNA chip and microarray technologies (Technical Note 3.1) to devise a method for identifying the attachment sites for a DNA-binding protein across an entire chromosome, as opposed to just within the region upstream of a single gene.

11.2. Construct a hypothesis to explain why eukaryotes have three RNA polymerases. Can your hypothesis be tested?

11.3.* To what extent is *E. coli* a good model for the regulation of transcription initiation in eukaryotes? Justify your opinion by providing specific examples of how extrapolations from *E. coli* have been helpful

and/or unhelpful in the development of our understanding of equivalent events in eukaryotes.

11.4. A model for control of transcription of the lactose operon in *E. coli* was first proposed by François Jacob and Jacques Monod in 1961 (see Further Reading). Explain the extent to which their work, which was based almost entirely on genetic analysis, provided an accurate description of the molecular events that are now known to occur.

11.5.* Assess the accuracy and usefulness of the module concept for the structure of an RNA polymerase II promoter.

Figure Tests

*Answers to odd-numbered questions can be found in the Appendix

11.1.* This figure shows the interaction between the *E. coli* bacteriophage 434 repressor and the DNA sequence to which it binds. Name the type of DNA binding motif present in this protein and describe how the domain interacts with DNA.

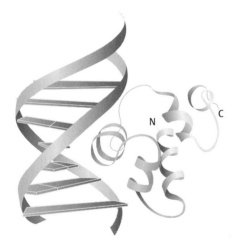

11.2. Discuss the type of experiment shown in the figure. How is the experiment set up, and what is the objective of the experiment?

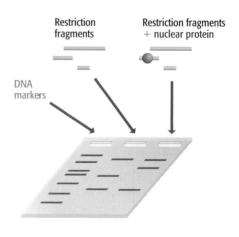

continued ...

Figure Tests (continued)

11.3.* What is the significance of the contact sites shown in the figure for the A–T base pair?

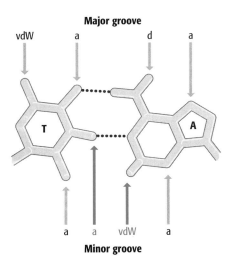

11.4. The figure shows two mechanisms by which an RNA polymerase can attach to its promoter. Identify which mechanism is used by the bacterial RNA polymerase and which is used by RNA polymerase II of eukaryotes.

(A) Direct attachment of RNA polymerase

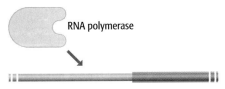

(B) Indirect attachment of RNA polymerase

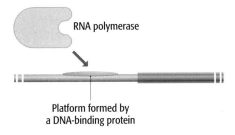

11.5.* Discuss the binding of a bacterial RNA polymerase to a promoter, as is shown in the figure.

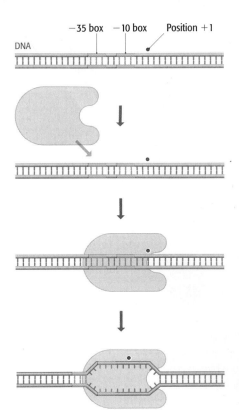

Further Reading

DNA- and RNA-binding motifs

Fierro-Monti, I. and Mathews, M.B. (2000) Proteins binding to duplexed RNA: one motif, multiple functions. *Trends Biochem. Sci.* **25**: 241–246.

Gangloff, Y.G., Romier, C., Thuault, S., Werten, S. and Davidson, I. (2001) The histone fold is a key structural motif of transcription factor TFIID. *Trends Biochem. Sci.* **26**: 250–257.

Harrison, S.C. and Aggarwal, A.K. (1990) DNA recognition by proteins with the helix-turn-helix motif. *Annu. Rev. Biochem.* **59**: 933–969.

Herr, W., Sturm, R.A., Clerc, R.G., *et al.* (1988) The POU domain: a large conserved region in the mammalian *pit*-1, *oct*-1, *oct*-2, and *Caenorhabditis elegans unc*-86 gene products. *Genes Dev.* **2**: 1513–1516.

Mackay, J.P. and Crossley, M. (1998) Zinc fingers are sticking together. *Trends Biochem. Sci.* **23**: 1–4.

Methods for studying DNA-binding proteins

Galas, D. and Schmitz, A. (1978) DNase footprinting: a simple method for the detection of protein-DNA binding specificity. *Nucleic Acids Res.* **5**: 3157–3170.

Garner, M.M. and Revzin, A. (1981) A gel electrophoretic method for quantifying the binding of proteins to specific DNA regions: application to components of the *Escherichia coli* lactose operon regulatory system. *Nucleic Acids Res.* **9**: 3047–3060. *Gel retardation.*

Interactions between DNA and DNA-binding proteins

Kielkopf, C.L., White, S., Szewczyk, J.W., Turner, J.M., Baird, E.E., Dervan, P.B. and Rees, D.C. (1998) A structural basis for recognition of A•T and T•A base pairs in the minor groove of B-DNA. *Science* **282**: 111–115.

Stormo, G.D. and Fields, D.S. (1998) Specificity, free energy and information content in protein–DNA interactions. *Trends Biochem. Sci.* **23**: 109–113.

RNA polymerases and their promoters

Geiduschek, E.P. and Kassavetis, G.A. (2001) The RNA polymerase III transcription apparatus. *J. Mol. Biol.* **310**: 1–26.

Kim, T.H., Barrera, L.O., Zheng, M., Qu, C., Singer, M.A., Richmond, T.A., Wu, Y., Green, R.D. and Ren, B. (2005) A high-resolution map of active promoters in the human genome. *Nature* **436**: 876–880. *Reveals the extent to which alternative promoters are used in the human genome.*

Russell, J. and Zomerdijk, J.C.B.M. (2005) RNA-polymerase-I-directed rDNA transcription, life and works. *Trends Biochem. Sci.* **30**: 87–96.

Seither, P., Iben, S. and Grummt, I. (1998) Mammalian RNA polymerase I exists as a holoenzyme with associated basal transcription factors. *J. Mol. Biol.* **275**: 43–53.

Smale, S.T. and Kadonaga, J.T. (2003) The RNA polymerase II core promoter. *Annu. Rev. Biochem.* **72**: 449–479.

Young, B.A., Gruber, T.M. and Gross, C.A. (2004) Minimal machinery of RNA polymerase holoenzyme sufficient for promoter melting. *Science* **303**: 1382–1384. *Probes the fine structure of the bacterial RNA polymerase.*

Assembly of the transcription initiation complex

Dieci, G. and Sentenac, A. (2003) Detours and shortcuts to transcription reinitiation. *Trends Biochem. Sci.* **28**: 202–209.

Green, M.R. (2000) TBP-associated factors (TAF$_{II}$s): multiple, selective transcriptional mediators in common complexes. *Trends Biochem. Sci.* **25**: 59–63.

Kadonaga, J.T. (2004) Regulation of RNA polymerase II transcription by sequence-specific DNA binding factors. *Cell* **116**: 247–257.

Kim, T.-K., Ebright, R.H. and Reinberg, D. (2000) Mechanism of ATP-dependent promoter melting by transcription factor IIH. *Science* **288**: 1418–1421.

Verrijzer, C.P. (2001) Transcription factor IID – not so basal after all. *Science* **293**: 2010–2011.

Xie, X., Kokubo, T., Cohen, S.L., Mirza, U.A., Hoffmann, A., Chait, B.T., Roeder, R.G., Nakatani, Y. and Burley, S.K. (1996) Structural similarity between TAFs and the heterotetrameric core of the histone octamer. *Nature* **380**: 316–322.

Control of transcription initiation in bacteria

Jacob, F. and Monod, J. (1961) Genetic regulatory mechanisms in the synthesis of proteins. *J. Mol. Biol.* **3**: 318–389. *The original proposal of the operon theory for control of bacterial gene expression.*

Oehler, S., Eismann, E.R., Krämer, H. and Müller-Hill, B. (1990) The three operators of the lac operon cooperate in repression. *EMBO J.* **9**: 973–979.

Schleif, R. (2000) Regulation of the L-arabinose operon of *Escherichia coli*. *Trends Genet.* **16**: 559–565. *Gives details of one example of bacterial gene regulation.*

continued …

Control of transcription initiation in eukaryotes

Hanna-Rose, W. and Hansen, U. (1996) Active repression mechanisms of eukaryotic transcription repressors. *Trends Genet.* **12:** 229–234.

Kim, Y.-J. and Lis, J.T. (2005) Interactions between subunits of *Drosophila* Mediator and activator proteins. *Trends Biochem. Sci.* **30:** 245–249.

Latchman, D.S. (2001) Transcription factors: bound to activate or repress. *Trends Biochem. Sci.* **26**: 211–213. *Short review of proteins that combine activation with repression.*

Scully, K.M., Jacobson, E.M., Jepsen, K., *et al.* (2000) Allosteric effects of Pit-1 DNA sites on long-term repression in cell type specification. *Science* **290:** 1127–1131. *Describes how Pit-1 acts as an activator of some genes and a repressor of others.*

Wolffe, A.P. (1994) Architectural transcription factors. *Science* **264:** 1100–1101. *Proteins such as SRY which introduce bends into DNA.*

Synthesis and Processing of RNA

12

When you have read Chapter 12, you should be able to:

Give details of the elongation and termination phases of transcription in *Escherichia coli*, and explain how these are regulated by antitermination, attenuation, and transcript cleavage proteins.

Describe the cutting events and chemical modifications involved in processing of functional RNA in bacteria.

Summarize our current knowledge of RNA degradation in bacteria.

Give details of elongation and termination of eukaryotic transcripts, including the processes responsible for capping and polyadenylation of eukaryotic mRNAs.

Distinguish between the splicing pathways of different types of intron, and in particular give a detailed description of splicing of GU–AG introns, including examples of alternative splicing and trans-splicing.

Describe the synthesis and processing of functional RNA in eukaryotes.

Define the term "ribozyme," and give examples of ribozymes.

Explain how eukaryotic rRNAs are chemically modified at specific nucleotide positions.

Give examples of RNA editing in mammals and outline the more complex types of RNA editing that occur in various other eukaryotes.

Describe the RNA degradation mechanisms of eukaryotes with emphasis on the roles of siRNAs and miRNAs in RNA silencing.

Outline the events involved in transport of eukaryotic RNAs from the nucleus to the cytoplasm.

Initiation of transcription, culminating with the RNA polymerase leaving the promoter and beginning synthesis of an RNA molecule, is simply the first step in the genome expression pathway. In this chapter and the next, we will follow the process onward and examine how transcription and translation eventually result in synthesis of the proteome. We begin by looking at the synthesis and processing of RNAs, including the mRNAs that make up the transcriptome and which specify the protein content of the cell, and the functional

RNAs that play essential roles in genome expression and other aspects of cell biology (Section 1.2.2). The underlying events are similar in prokaryotes and eukaryotes, but there are substantial differences in some of the details and, as with other aspects of genome expression, there is a greater complexity to the processes occurring in eukaryotes. We will therefore begin by examining RNA synthesis and processing in bacteria.

12.1 Synthesis and Processing of Bacterial RNAs

Because there is just one bacterial RNA polymerase (Section 11.2.1), the general mechanism of RNA synthesis is the same for all bacterial genes. Distinctions do, however, arise when we consider the mechanisms by which synthesis of transcripts from individual genes is regulated.

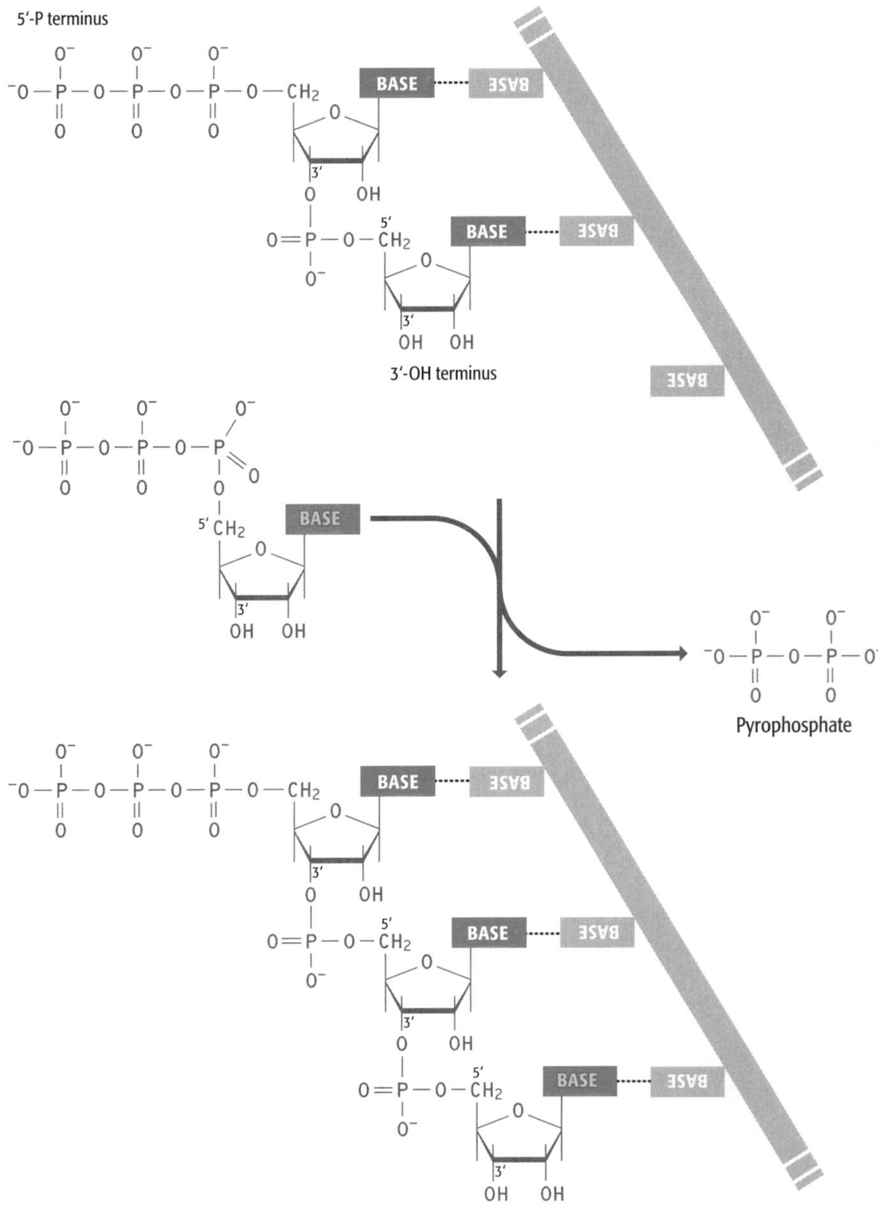

Figure 12.1 The chemical basis of RNA synthesis. Compare this reaction with polymerization of DNA, as illustrated in Figure 1.6.

12.1.1 Synthesis of bacterial transcripts

The chemical basis of the template-dependent synthesis of RNA is shown in Figure 12.1. Ribonucleotides are added one after another to the growing 3′ end of the RNA transcript, the identity of each nucleotide specified by the base-pairing rules: A base-pairs with T or U; G base-pairs with C. During each nucleotide addition, the β- and γ-phosphates are removed from the incoming nucleotide, and the hydroxyl group is removed from the 3′-carbon of the nucleotide present at the end of the chain.

Elongation of a transcript by the bacterial RNA polymerase

During the elongation stage of transcription, the bacterial RNA polymerase is in its core enzyme form, comprising four proteins: two relatively small (approximately 35 kDa) α subunits, and one each of the related subunits β and β′ (both approximately 150 kDa); the σ subunit that has the key role in transcription initiation has usually left the complex at this stage (Section 11.2.3). The RNA polymerase covers about 30 bp of the template DNA, including the **transcription bubble** of 12–14 bp, within which the growing transcript is held to the template strand of the DNA by approximately eight RNA–DNA base pairs (Figure 12.2).

The RNA polymerase has to keep a tight grip on both the DNA template and the RNA that it is making in order to prevent the transcription complex from falling apart before the end of the gene is reached. However, this grip must not be so tight as to prevent the polymerase from moving along the DNA. To understand how these apparently contradictory requirements are met, the interactions between the polymerase, the DNA template, and the RNA transcript have been examined by X-ray crystallography studies (see Technical

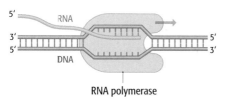

Figure 12.2 Schematic representation of the *Escherichia coli* transcription elongation complex. The RNA polymerase covers approximately 30 bp of DNA, including a transcription bubble of 12–14 bp, with the RNA attached to the template strand of the DNA by eight or so RNA–DNA base pairs. The arrow shows the direction in which the RNA polymerase moves along the DNA.

Figure 12.3 Interactions within the bacterial RNA polymerase. The β and β′ subunits of the RNA polymerase are shown in orange, the numbers indicating the positions of the amino acids which, according to cross-linking studies, lie close to one or more of the DNA/RNA polynucleotides within the complex. The interactions revealed by cross-links are shown by the thin pink lines.

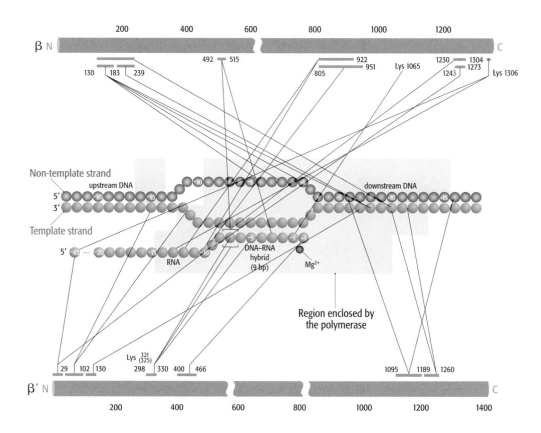

Note 11.1), combined with cross-linking experiments in which covalent bonds are formed between the DNA or RNA and the polymerase, these bonds enabling the amino acids that are closest to the DNA and RNA to be identified. The cross-linking experiments make use of a variety of photoreactive compounds that can be attached to synthetic DNA and RNA molecules. After assembly of the complex, the labels are activated by a pulse of light so that they form cross-links between the nucleic acid and any amino acids within the polymerase that are located close to the position of the label. Some labels form cross-links with any amino acid and others are more discriminatory, forming links only with lysines, for example. After cross-linking, the RNA polymerase is disassembled into subunits and each subunit treated with cyanogen bromide, which cuts polypeptides specifically at methionine residues, yielding a characteristic set of fragments. Fragments that are cross-linked to the nucleic acid are then identified. These fragments must contain the amino acids that were adjacent to the label in the intact transcription elongation complex. By identifying as many cross-links as possible, a detailed map can be built up (Figure 12.3). This information can then be combined with the structural data from X-ray crystallography to construct a model of the transcription elongation complex, showing the precise positioning of the different parts of the DNA double helix and of the RNA transcript within the polymerase (Figure 12.4). These experiments have shown that the double helix lies between the β and β′ subunits, within a trough on the enclosed surface of β′. The active site for RNA synthesis also lies between these two subunits, with the non-template strand of DNA looping away from the active site and held within the β subunit. The RNA transcript extrudes from the complex via a channel formed partly by the β and partly by the β′ subunit (at the bottom right of the structures in Figure 12.4).

The polymerase does not synthesize its transcript at a constant rate. Instead, synthesis is discontinuous, with periods of rapid elongation interspersed by brief pauses during which the active site of the polymerase undergoes a slight structural rearrangement. A pause rarely lasts longer than 6 milliseconds, and might be accompanied by the polymerase moving in reverse (**backtracking**) along the template. Pauses occur randomly rather than being caused by any particular feature of the template DNA. Pausing plays an important role in transcript termination, as described below, but whether or not this is its only function is currently unknown.

Figure 12.4 A model of the bacterial transcription elongation complex. The β and β′ subunits of the RNA polymerase are depicted in blue/green and pink, respectively, the double helix is colored red (template strand) and yellow (non-template strand), and the RNA transcript is gold. The two views show that the double helix lies between the β and β′ subunits, within a trough on the enclosed surface of β′. The arrows indicate the direction in which the DNA moves through the polymerase. Reprinted with permission from Korzheva *et al., Science,* **289**, 619–625. © 2000 AAAS.

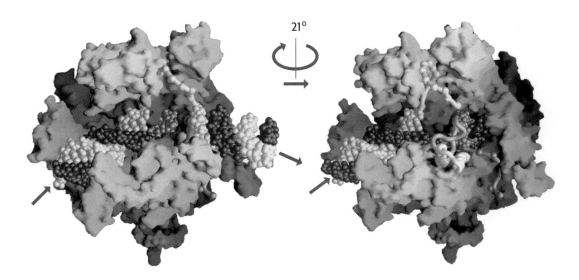

Termination of a bacterial transcript

Current thinking views transcription as a discontinuous process, with the polymerase pausing regularly and making a "choice" between continuing elongation by adding more ribonucleotides to the transcript, or terminating by dissociating from the template. Which choice is selected depends on which alternative is more favorable in thermodynamic terms. This model emphasizes that, in order for termination to occur, the polymerase has to reach a position on the template where dissociation is more favorable than continued RNA synthesis.

Bacteria appear to use two distinct strategies for transcription termination. About half the positions in *Escherichia coli* at which transcription terminates correspond to DNA sequences where the template strand contains an inverted palindrome followed by a run of deoxyadenosine nucleotides (Figure 12.5). These **intrinsic terminators** are thought to utilize two processes to promote dissociation of the polymerase and to destabilize the attachment of the growing transcript to the template. First, when the inverted palindrome is transcribed, the RNA sequence folds into a stable hairpin, this RNA–RNA base pairing being favored over the DNA–RNA pairing that normally occurs within the transcription bubble. This reduces the number of contacts made between the template and transcript, weakening the overall interaction and favoring dissociation. Second, the interaction is further weakened when the run of As in the template is transcribed, because the resulting A–U base pairs have only two hydrogen bonds each, compared with three for each G–C pair. The net result is that termination is favored over continued elongation. This model is easy to rationalize with the known properties of DNA–RNA hybrids, but an alternative hypothesis has been prompted by the result of cross-linking experiments, which have shown that the RNA hairpin makes contact with a flap structure on the outer surface of the RNA polymerase β subunit, adjacent to the exit point of the channel through which the RNA emerges from the complex (Figure 12.6). Although the flap structure is quite distant (some 6.5 nm) from the active site of the polymerase, a direct connection is made between the two by a segment of β-sheet within the β subunit. Movement of the flap could therefore affect the positioning of amino acids within the active site, possibly leading to breakage of the DNA–RNA base pairs and termination of transcription. Additional evidence in support of this model comes from the demonstration that the

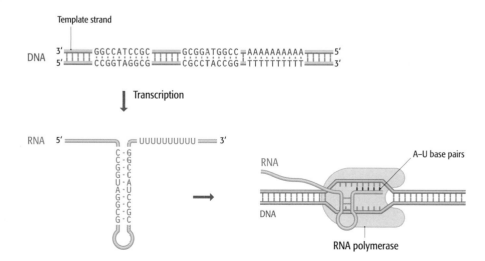

Figure 12.5 Termination at an intrinsic terminator. The presence of an inverted palindrome in the DNA sequence results in formation of a hairpin loop in the transcript.

Figure 12.6 A flap structure on the surface of the RNA polymerase could mediate termination. The hairpin that forms in the RNA transcript when the termination region is reached makes contact with a flap structure on the outer surface of the RNA polymerase β subunit, adjacent to the exit point of the channel through which the RNA emerges from the complex. The part of the β subunit polypeptide that forms the flap is colored dark blue, with the remainder of the β subunit in light blue, the β' subunit in pink, and the α subunit in white. Note that although the flap structure is located on the surface of the polymerase, the region of the β polypeptide that forms the flap is connected directly to the active site, indicated by the magnesium ion in magenta and the nucleoside 5'-triphosphate substrate in green. An interaction between the RNA hairpin and the flap could therefore affect the positioning of amino acids within the active site, possibly causing the DNA–RNA base pairs to break, leading to termination of transcription. Reprinted with permission from Toulokhonov *et al.*, *Science,* **292**, 730–733. © 2001 AAAs.

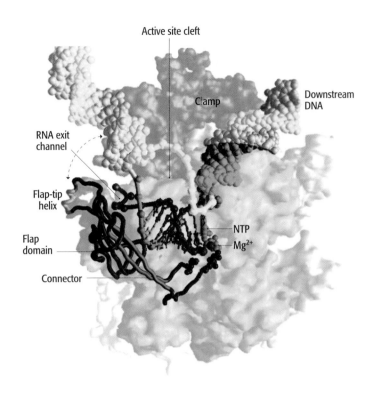

protein called NusA, which enhances termination at intrinsic terminators, interacts with the hairpin loop and flap structure and may stabilize the contact between the two.

The second type of bacterial transcription termination signal is **Rho dependent**. These signals usually retain the hairpin feature of intrinsic terminators, although the hairpin is less stable and there is no run of As in the template. Termination requires the activity of a protein called Rho, which attaches to the transcript and moves along the RNA toward the polymerase. If the polymerase continues to synthesize RNA then it keeps ahead of the pursuing Rho, but at the termination signal the polymerase stalls (see Figure 12.7). Exactly why has not been explained—presumably the hairpin loop that forms in the RNA is responsible in some way—but the result is clear: Rho is able to catch up. Rho is a **helicase**, which means that it actively breaks base pairs, in this case between the template and transcript, resulting in termination of transcription.

12.1.2 Control over the choice between elongation and termination

Can the choice that a paused polymerase makes between continuing elongation or terminating transcription by dissociating from the template be influenced in any way? The answer is yes, this being the primary means by which the synthesis (as opposed to initiation) of bacterial transcripts is regulated.

Antitermination results in termination signals being ignored

The first of these regulatory processes is called **antitermination**. This occurs when the RNA polymerase ignores a termination signal and continues elongating its transcript until a second signal is reached (Figure 12.8). It provides

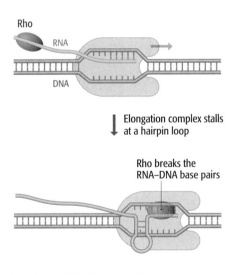

Figure 12.7 Rho-dependent termination. Rho is a helicase that follows the RNA polymerase along the transcript. When the polymerase stalls at a hairpin, Rho catches up and breaks the RNA–DNA base pairs, releasing the transcript. Note that the diagram is schematic and does not reflect the relative sizes of Rho and the RNA polymerase.

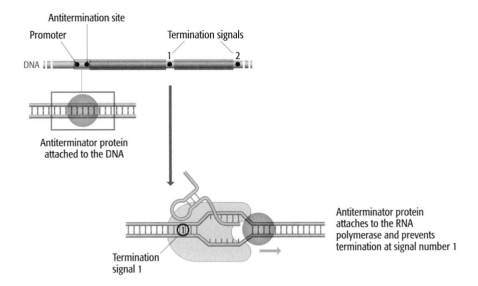

Figure 12.8 Antitermination. The antiterminator protein attaches to the DNA and transfers to the RNA polymerase as it moves past, subsequently enabling the polymerase to continue transcription through termination signal number 1, so the second of the pair of genes in this operon is transcribed.

a mechanism whereby one or more of the genes at the end of an operon can be switched off or on by the polymerase recognizing or not recognizing a termination signal located upstream of those genes. Antitermination is controlled by an **antiterminator protein**, which attaches to the DNA near the beginning of the operon and then transfers to the RNA polymerase as it moves past en route to the first termination signal. The presence of the antiterminator protein causes the enzyme to ignore the termination signal, presumably by countering the destabilizing properties of an intrinsic terminator or by preventing stalling at a Rho-dependent terminator.

Although the mechanics of antitermination are unclear, the impact that it can have on gene expression has been described in detail, especially during the infection cycle of bacteriophage λ. Immediately after entering an *E. coli* cell, transcription of the λ genome is initiated by the bacterial RNA polymerase attaching to two promoters, P_L and P_R, and synthesizing two "immediate-early" mRNAs, these terminating at positions t_{L1} and t_{R1} (Figure 12.9A). The mRNA transcribed from P_R to t_{R1} codes for a protein called Cro, one of the major regulatory proteins involved in the λ infection cycle. The second mRNA specifies the N protein, which is an antiterminator. The N protein attaches to the λ genome at sites *nutL* and *nutR* and transfers to the RNA polymerase as it passes. Now the RNA polymerase ignores the t_{L1} and t_{R1} terminators and continues transcription downstream of these points. The resulting mRNAs encode the "delayed-early" proteins (Figure 12.9B). Antitermination controlled by the N protein therefore ensures that the immediate-early and delayed-early proteins are synthesized at the appropriate times during the λ infection cycle. One of the delayed-early proteins, Q, is a second antiterminator that controls the switch to the late stage of the infection cycle.

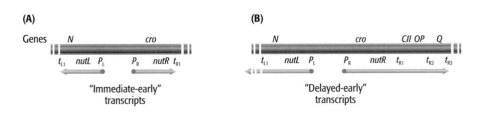

(A) "Immediate-early" transcripts

(B) "Delayed-early" transcripts

Figure 12.9 Antitermination during the infection cycle of bacteriophage λ.

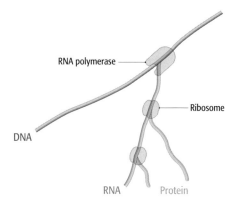

RNA polymerase

Ribosome

DNA

RNA   Protein

Figure 12.10 In bacteria, transcription and translation are often coupled.

Attenuation results in premature termination

Bacterial mRNAs do not undergo any significant forms of processing: the primary transcript that is synthesized by the RNA polymerase is itself the mature mRNA, and its translation usually begins before transcription is complete (Figure 12.10). This coupling of transcription and translation is important in that it allows a special type of control, called **attenuation**, to be applied to the regulation of bacterial mRNA synthesis.

Attenuation operates primarily with operons that code for enzymes involved in amino acid biosynthesis, but a few other examples are also known. The tryptophan operon of *E. coli* (Section 11.3.1) illustrates how it works. In this operon, two hairpin loops can form in the region between the start of the transcript and the beginning of *trpE*. The smaller of these loops acts as a termination signal, but the larger hairpin loop, which is closer to the start of the transcript, is more stable. The larger loop overlaps with the termination hairpin, so only one of the two hairpins can form at any one time. Which loop forms depends on the relative positioning between the RNA polymerase and a ribosome which attaches to the 5′ end of the transcript as soon as it is synthesized in order to translate the genes into protein (Figure 12.11). If the ribosome stalls so that it does not keep up with the polymerase, then the larger hairpin forms and transcription continues. However, if the

Figure 12.11 Attenuation at the tryptophan operon.

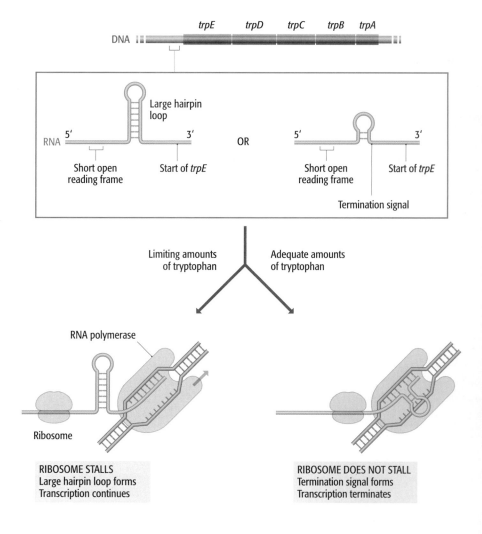

ribosome keeps pace with the RNA polymerase then it disrupts the larger hairpin by attaching to the RNA that forms part of the stem of this hairpin. When this happens, the termination hairpin is able to form, and transcription stops. Ribosome stalling can occur because upstream of the termination signal is a short open reading frame (ORF) coding for a 14-amino-acid peptide that includes two tryptophans. If the amount of free tryptophan is limiting, then the ribosome stalls as it attempts to synthesize this peptide, while the polymerase continues to make its transcript. Because this transcript contains copies of the genes coding for the biosynthesis of tryptophan, its continued elongation addresses the requirement that the cell has for this amino acid. When the amount of tryptophan in the cell reaches a satisfactory level, the attenuation system prevents further transcription of the tryptophan operon, because now the ribosome does not stall while making the short peptide, and instead keeps pace with the polymerase, allowing the termination signal to form.

The *E. coli* tryptophan operon is controlled not only by attenuation but also by a repressor (Section 11.3.1). Exactly how attenuation and repression work together to regulate expression of the operon is not known, but it is thought that repression provides the basic on–off switch and attenuation modulates the precise level of gene expression that occurs. Other *E. coli* operons, such as those for biosynthesis of histidine, leucine, and threonine, are controlled solely by attenuation. Interestingly, in some bacteria, including *Bacillus subtilis*, the tryptophan operon is one of those that does not have a repressor system and so is regulated entirely by attenuation. In these bacteria, attenuation is mediated not by the speed at which the ribosome tracks along the mRNA, but by an RNA-binding protein called ***trp* RNA-binding attenuation protein (TRAP)** which, in the presence of tryptophan, attaches to the mRNA in the region equivalent to the short ORF of the *E. coli* transcript (Figure 12.12). Attachment of TRAP leads to formation of the termination signal and cessation of transcription.

Transcript cleavage proteins can prevent stalling of a backtracked polymerase

Backtracking occurs when a paused polymerase enzyme reverses a short distance along the DNA template strand (Section 12.1.1). This displaces the newly synthesized RNA molecule, whose 3′ end becomes detached from the template. To prevent the polymerase becoming stalled at this point, the detached portion of the RNA molecule must be snipped off (Figure 12.13). The polymerase possesses the necessary RNA cleavage activity, but normally this activity is disfavored and hence stalling is likely to occur. Stalling is prevented by a pair of transcript cleavage factors, called GreA and GreB which, despite their name, do not actually cut the RNA, instead stimulating the polymerase to do this itself. The two proteins probably work by repositioning one of two magnesium ions that are present within the active site of the polymerase.

The role of GreA and GreB in preventing polymerase arrest has only recently become clear, and no link has yet been made between these factors and any specific regulatory process. There are, however, intriguing aspects to their mode of action which suggest that they might act as mediators for regulatory signals of one kind or another. The major parts of the GreA and GreB proteins are two α-helices separated by a short turn containing one aspartic acid and one glutamic acid. These two helices form a needle-like structure

Figure 12.12 Regulation of the tryptophan operon of *Bacillus subtilis*.
(A) Regulation centers on the protein called TRAP, which attaches to the leader region of the transcript when tryptophan levels are adequate. When attached, TRAP *prevents* formation of the large hairpin loop and so *allows* formation of the termination signal. Compare with Figure 12.12. (B) Structure of the TRAP–RNA complex. TRAP comprises 11 identical subunits, each one made up mainly of β-sheets, which associate together to form a circular structure with a diameter of 8 nm. The 11 TRAP subunits are shown in different colors, each one with an attached tryptophan molecule indicated by the red–yellow–blue spherical structures. The RNA molecule is shown as a ball-and-stick structure wound around the TRAP complex. Part (B) reprinted with permission from Antson *et al.*, (1999), *Nature*, **401**, 235–242.

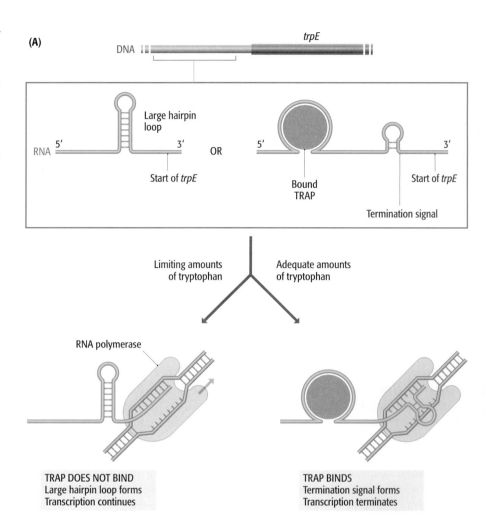

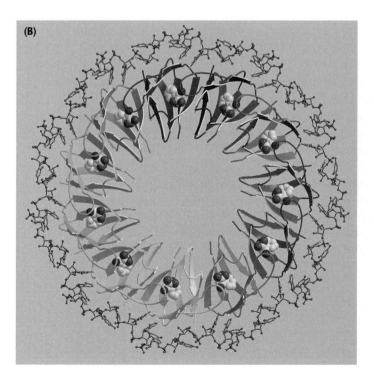

that penetrates into the RNA polymerase via the **secondary channel**, which leads from the surface of the enzyme to the active site within the protein complex (Figure 12.14). The two acidic amino acids at the point of the needle are thought to interact with the pair of magnesium ions within the active site, promoting cleavage of the detached RNA segment. The structure of GreA and GreB is remarkably similar to that of a third protein, called DksA, which mediates the **stringent response** in *E. coli* and other bacteria. This response is activated when a bacterium encounters poor growth conditions such as low levels of essential amino acids. To conserve resources, the bacterium reduces its rate of transcription, in particular the synthesis of rRNA and tRNA, to about 5% of the normal level. Activation of the stringent response involves ppGpp and pppGpp (Figure 12.15), two unusual nucleotides also called **alarmones**, which are synthesized by the RelA protein in response to amino acid starvation. The alarmones work in conjunction with DksA to shut down transcription. How this happens is still speculative, but the similarity between the structure of DksA and that of GreA and GreB suggests that DksA, like the transcript cleavage factors, inserts its needle-like structure into the secondary channel of the polymerase and, by virtue of the pair of acidic amino acids at its tip (in this case, two aspartic acids), influences polymerase activity by interacting with the essential magnesium ions within the active site. A recent model gives the alarmones a direct role in this process by envisaging that ppGpp enters the channel at the tip of the DksA needle and participates in the contact with the magnesium ions.

Exactly which step in transcription is inhibited during the stringent response is not known, some evidence pointing toward events occurring during the initiation process and other results suggesting that polymerase pausing is the target. The important point, however, is that the role of DksA in this regulatory process indicates that GreA and GreB, which act in a manner almost indistinguishable from DksA, may also be involved in other, as yet undiscovered, mechanisms for regulating bacterial RNA synthesis.

12.1.3 Processing of bacterial RNAs

In Section 1.2.3 we learnt that most RNA molecules are synthesized as precursors which must be processed before they can carry out their functions in the cell. Bacterial mRNA is the one exception to this rule, the initial transcripts of most bacterial protein-coding genes being functional mRNAs that can immediately be translated (see Figure 12.10). Bacterial tRNAs and rRNAs, on the other hand, are first transcribed as pre-RNAs which become functional only after a series of cutting events and chemical modifications.

Cutting events release mature rRNAs and tRNAs from their precursor molecules

Bacteria synthesize three different rRNAs, called 5S rRNA, 16S rRNA, and 23S rRNA, the names indicating the sizes of the molecules as measured by **sedimentation analysis** (see Technical Note 7.1). The three genes for these rRNAs are linked into a single transcription unit (which is usually present in multiple copies, seven for *E. coli*) and so the pre-rRNA contains copies of all three rRNAs. Cutting events are therefore needed to release the mature rRNAs. These cuts are made by ribonucleases III, P, and F at positions specified by double-stranded regions formed by base pairing between different

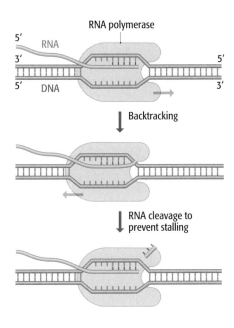

Figure 12.13 Preventing a backtracked RNA polymerase from becoming stalled. Backtracking detaches a short piece of newly synthesized RNA. Cleavage of this detached segment is necessary to prevent the polymerase becoming stalled.

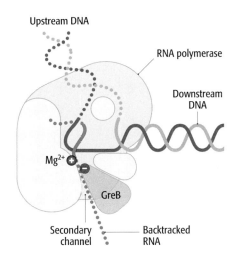

Figure 12.14 The interaction between GreB and the bacterial RNA polymerase.

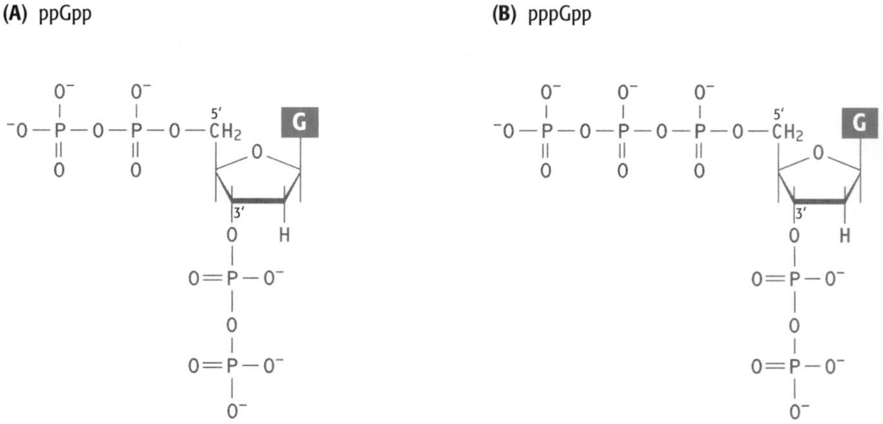

Figure 12.15 The structures of the alarmones ppGpp and pppGpp. Compare with the standard nucleotide structures shown in Figure 1.4.

(A) ppGpp

(B) pppGpp

parts of the pre-rRNA (Figure 12.16). The cut ends are subsequently trimmed by the exonuclease activities of ribonucleases M16, M23, and M5 to give the mature rRNAs.

Genes for tRNAs are distributed around bacterial genomes, some on their own and some as tandem arrays in multi-tRNA transcription units. In some bacteria, tRNA genes also occur as infiltrators in the rRNA transcription units, as is the case with *E. coli*, which has either one or two tRNA genes between the 16S and 23S genes in each of its seven rRNA transcription units (see Figure 12.16). All pre-tRNAs are processed in similar though not identical ways. The example shown in Figure 12.17 refers to the precursor for the tRNA^{Tyr} molecule of *E. coli*. The tRNA sequence within the precursor molecule adopts its base-paired cloverleaf structure (see Figure 13.2) and two additional hairpin structures form, one on either side of the tRNA. Processing begins with the cut by ribonuclease E or F, forming a new 3′ end just upstream of one of the hairpins. Ribonuclease D, which is an exonuclease, trims seven nucleotides from this new 3′ end and then pauses while ribonuclease P makes a cut at the start of the cloverleaf, forming the 5′ end of the mature tRNA. Ribonuclease D then removes two more nucleotides, creating the 3′ end of the mature molecule. All

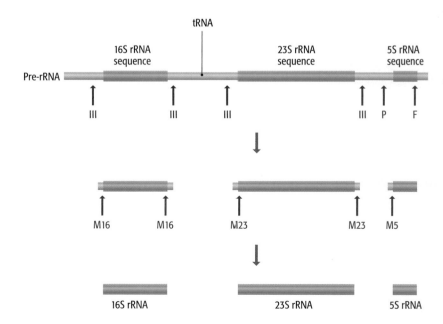

Figure 12.16 Pre-rRNA processing in *Escherichia coli*. Note that a tRNA is located between the 16S and 23S sequences in this *E. coli* pre-rRNA. This tRNA is processed as shown in Figure 12.17.

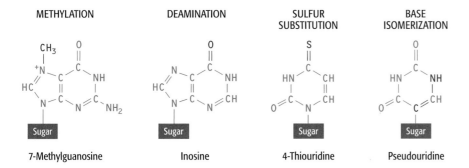

Figure 12.17 Processing of an *Escherichia coli* pre-tRNA. The example shown results in synthesis of tRNA^Tyr. The cut made with some other pre-tRNAs by RNase Z is shown in red. Abbreviation: RNase, ribonuclease.

mature tRNAs must end with the trinucleotide 5′–CCA–3′. With tRNA^Tyr the terminal CCA is present in the pre-RNA and is not removed by ribonuclease D, but with some other pre-tRNAs this sequence is absent, or is removed by the processing ribonucleases. This occurs with most of those pre-tRNAs whose 3′ ends are created by an endonuclease called ribonuclease Z, which makes a cut adjacent to the first base pair in the tRNA cloverleaf (see Figure 12.17) and hence removes the region that would contain the terminal CCA. When the CCA is absent, it has to be added by one or more *template-independent* RNA polymerases such as **tRNA nucleotidyltransferase**.

Of the various nucleases mentioned above, ribonuclease P is particularly interesting because the various subunits of this enzyme include one that is made of RNA rather than protein. RNA subunits are present in several enzymes involved in RNA processing, including ribonuclease MRP, which works with ribonuclease P in processing of eukaryotic pre-rRNAs (Section 12.2.4). These hybrid protein–RNA enzymes may be relics of the **RNA world**, the early period of evolution when all biological reactions centered around RNA (Section 18.1.1).

Figure 12.18 Examples of chemical modifications occurring with nucleotides in rRNA and tRNA. Methylation is the addition of one or more –CH₃ groups to the base or sugar. Deamination is the removal of an amino (–NH₂) group from the base: inosine is the deaminated form of adenosine. Sulfur substitution involves replacement of oxygen with sulfur. Base isomerization occurs when the positions of atoms in the ring component of the base are changed: isomerization of uridine gives pseudouridine. Double-bond saturation converts a double bond to a single bond: for example, to convert uridine to dihydrouridine. Nucleotide replacement is the substitution of an existing nucleotide with a new one, such as queosine.

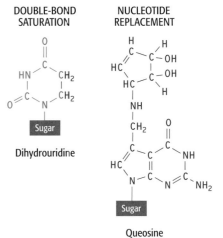

METHYLATION	DEAMINATION	SULFUR SUBSTITUTION	BASE ISOMERIZATION	DOUBLE-BOND SATURATION	NUCLEOTIDE REPLACEMENT
7-Methylguanosine	Inosine	4-Thiouridine	Pseudouridine	Dihydrouridine	Queosine

Nucleotide modifications broaden the chemical properties of tRNAs and rRNAs

The final type of processing that occurs with pre-RNAs is the chemical modification of nucleotides within the transcript. This occurs with both pre-rRNAs and pre-tRNAs. A broad spectrum of chemical changes has been identified with different pre-RNAs: over 50 different modifications are known in total (Figure 12.18). Most of these are carried out directly on an existing nucleotide within the transcript but two modified nucleotides, queosine and wyosine, are put in place by cutting out an entire nucleotide and replacing it with the modified version.

Many of these unusual nucleotides were first identified in tRNAs, within which approximately one in ten nucleotides becomes altered. These modifications are thought to mediate the recognition of individual tRNAs by the enzymes that attach amino acids to these molecules (Section 13.1.1), and to increase the range of the interactions that can occur between tRNAs and codons during translation, enabling a single tRNA to recognize more than one codon (Section 13.1.2). We know relatively little about how tRNA modifications are carried out, beyond the fact that there are a number of enzymes that catalyze these changes. It is presumed that these enzymes use specific features of the base-paired structure of the tRNA to identify the appropriate nucleotides to modify.

Ribosomal RNAs are modified in two ways: by addition of methyl groups to, mainly, the 2′–OH group on nucleotide sugars, and by conversion of uridine to pseudouridine (see Figure 12.18). The same modification occurs at the same position on all copies of an rRNA, and these modified positions are, to a certain extent, the same in different species. Some similarities in modification patterns are even seen when bacteria and eukaryotes are compared, although bacterial rRNAs are less heavily modified than eukaryotic ones. Functions for the modifications have not been identified, although most occur within those parts of rRNAs thought to be most critical to the activity of these molecules in ribosomes (Section 13.2.1). Modified nucleotides might, for example, be involved in rRNA-catalyzed reactions such as synthesis of peptide bonds. In eukaryotes a complex machinery exists for modifying rRNA molecules (Section 12.2.5), but this machinery is absent from bacteria, whose rRNAs are modified by enzymes that directly recognize the sequence and/or structures of the regions of RNA that contain the nucleotides to be modified. Often two or more nucleotides in the same region are modified at once. Bacterial rRNA modification is therefore similar to the systems for modifying tRNAs in both bacteria and eukaryotes.

12.1.4 Degradation of bacterial RNAs

So far this chapter has concentrated on synthesis of RNAs. Their degradation is equally important, especially with regard to mRNAs whose presence or absence in the cell determines which proteins will be synthesized. Degradation of specific mRNAs could be a powerful way of regulating genome expression.

The rate of degradation of an mRNA can be estimated by determining its half-life in the cell. The estimates show that there are considerable variations between and within organisms. Bacterial mRNAs are generally turned over very rapidly, their half-lives rarely being longer than a few minutes, a reflection of the rapid changes in protein synthesis patterns that can occur in an actively growing bacterium with a generation time of 20 minutes or so. Eukaryotic mRNAs are longer lived, with half-lives of, on average, 10–20 minutes for yeast and several hours for mammals.

Bacterial mRNAs are degraded in the 3′→5′ direction

Studies of mutant bacteria whose mRNAs have extended half-lives have identified a range of ribonucleases and other RNA-degrading enzymes that are thought to be involved in mRNA degradation. These include:

- RNase E and RNase III, which are endonucleases that make internal cuts in RNA molecules.

- RNase II, which is an exonuclease that removes nucleotides in the 3′→5′ direction.

- Polynucleotide phosphorylase (PNPase), which also removes nucleotides sequentially from the 3′ end of an mRNA but, unlike true nucleases, requires inorganic phosphate as a cosubstrate.

No enzyme capable of degrading RNA in the 5′→3′ direction has yet been isolated from bacteria. This absence leads to the assumption that the main degradative process for bacterial mRNAs is removal of nucleotides from the 3′ end. This is not possible under normal circumstances because most mRNAs have a hairpin structure near the 3′ end, the same hairpin that induced termination of transcription (see Figures 12.5 and 12.7). This structure blocks the progress of RNase II and PNPase, preventing them from gaining access to the coding part of the transcript (Figure 12.19). The model for mRNA degradation therefore begins with removal of the 3′ terminal region, including the hairpin, by one of the endonucleases, exposing a new end from which RNase II and PNPase can enter the coding region, destroying the functional activity of the mRNA. Polyadenylation may also have a role. Although looked on primarily as a feature of eukaryotic mRNAs (Section 12.2.1), it has been known since 1975 that many bacterial transcripts have poly(A) tails at some stage in their existence, but that these tails are rapidly degraded. At present it is not clear whether polyadenylation precedes degradation of an mRNA, or whether it occurs at various intermediate stages after degradation has begun.

In the cell, RNase E and PNPase are located within a multiprotein complex called the **degradosome**. Other components of the degradosome include an RNA helicase, which is thought to aid degradation by unwinding the double-helix structure of the stems of RNA stem-loops. Fragments of rRNA occasionally copurify with the degradosome, suggesting that the complex might be involved in both rRNA and mRNA degradation. But the exact role of the degradosome is still not clear and a few researchers are sceptical about its actual existence, pointing out that proteins not obviously involved in mRNA degradation, such as the glycolysis enzyme enolase, appear to be components of the degradosome, possibly indicating that the complex is an artifact that is produced during extraction of proteins from bacterial cells. A more significant gap in our knowledge concerns the way in which degradation is specifically targeted at individual mRNAs. We know that specific degradation occurs because mRNA degradation has been implicated in the regulation of several sets of bacterial genes, such as the *pap* operon of *E. coli*, which codes for proteins involved in synthesis of the cell surface pili. Unfortunately, the process by which such control is exerted remains a mystery.

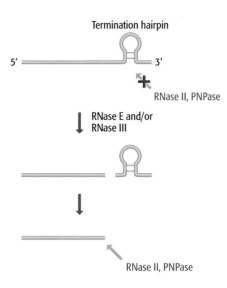

Figure 12.19 Degradation of bacterial RNA. The termination hairpin blocks the exonuclease activities of RNase II and PNPase, and so must be removed by endonuclease action (RNase E and/or RNase III) before degradation can proceed.

12.2 Synthesis and Processing of Eukaryotic RNA

At the most fundamental level, transcription is similar in bacteria and eukaryotes. The chemistry of RNA polymerization is identical in all types of organism, and the three eukaryotic nuclear RNA polymerases are all structurally related to the *E. coli* RNA polymerase, their three largest subunits being equivalent to the α, β, and β′ subunits of the bacterial enzyme. The contacts between the eukaryotic RNA polymerase II, the template DNA, and the RNA transcript, as revealed by X-ray crystallography and cross-linking studies, are similar to the interactions described for bacterial transcription (Section 12.1.1), and the basic principle that transcription is a step-by-step competition between elongation and termination also holds.

12.2.1 Synthesis of eukaryotic mRNAs by RNA polymerase II

Despite the underlying similarities, the overall processes for mRNA synthesis in bacteria and eukaryotes are quite different. The most striking dissimilarity is the extent to which eukaryotic mRNAs are processed during transcription. In bacteria, the transcripts of protein-coding genes are not processed at all: the primary transcripts are mature mRNAs. In contrast, all eukaryotic mRNAs have a cap added to the 5′ end, most are also polyadenylated by addition of a series of adenosines to the 3′ end, many contain introns and so undergo splicing, and a few are subject to RNA editing. A function has been assigned to capping, but the reason for polyadenylation largely remains a mystery. With splicing and editing we can appreciate why the events occur—the former removes introns that block translation of the mRNA, the latter changes the coding properties of the mRNA—but we do not understand why these mechanisms have evolved. Why do genes have introns in the first place? Why edit an mRNA rather than encoding the desired sequences in the DNA?

Eukaryotic mRNAs are processed while they are being synthesized. The cap is added as soon as transcription has been initiated, splicing and editing begin while the transcript is still being made, and polyadenylation is an inherent part of the termination mechanism for RNA polymerase II. To deal with all of these events together would be confusing, with too many different things being described at once. We will therefore postpone study of RNA editing until later in the chapter, which means it can be dealt with in tandem with similar forms of chemical modification that occur during rRNA and tRNA processing, and we will devote a separate section to splicing after we have studied capping, elongation, and polyadenylation.

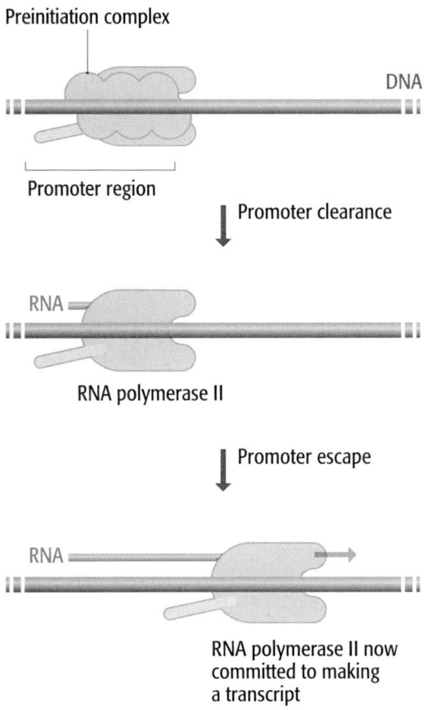

Capping of RNA polymerase II transcripts occurs immediately after initiation

Although phosphorylation of the C-terminal domain (CTD) of the largest subunit of RNA polymerase II is the final step in initiation of transcription of mRNA-encoding genes in eukaryotes (Section 11.2.3), it is not immediately followed by the onset of elongation. A somewhat gray area exists in our understanding of the events that distinguish **promoter clearance**, which refers to the transition from the preinitiation complex to a complex that has begun to synthesize RNA, and **promoter escape**, during which the polymerase moves away from the promoter region and becomes committed to making a transcript (Figure 12.20). The opposing effects of negative and positive elongation factors influence the ability of the polymerase to begin productive RNA synthesis, and if the negative factors predominate then transcription halts before

Preinitiation complex

DNA

Promoter region

↓ Promoter clearance

RNA

RNA polymerase II

↓ Promoter escape

RNA

RNA polymerase II now committed to making a transcript

Figure 12.20 Promoter clearance and promoter escape. Promoter clearance is the transition from the preinitiation complex to a complex that has begun to synthesize RNA. Promoter escape occurs when the polymerase moves away from the promoter region and becomes committed to making a transcript. Note that the drawing is schematic and is not intended to indicate the shape or subunit composition of the RNA polymerase II complex that synthesizes the transcript.

the polymerase has moved more than 30 nucleotides from the initiation point. Promoter escape could therefore be an important control point, but how regulation is applied at this stage is not yet known.

Successful promoter escape could be linked with capping, this processing event being completed before the transcript reaches 30 nucleotides in length. The first step in capping is addition of an extra guanosine to the extreme 5′ end of the RNA. Rather than occurring by normal RNA polymerization, capping involves a reaction between the 5′ triphosphate of the terminal nucleotide and the triphosphate of a GTP nucleotide. The γ-phosphate of the terminal nucleotide (the outermost phosphate) is removed, as are the β and γ phosphates of the GTP, resulting in a 5′–5′ bond (Figure 12.21). The reaction is carried out by the enzyme **guanylyl transferase**. The second step of the capping reaction converts the new terminal guanosine into 7-methylguanosine by attachment of a methyl group to nitrogen number 7 of the purine ring, this modification catalyzed by **guanine methyltransferase**. The two capping enzymes make attachments with the CTD and it is possible that they are intrinsic components of the RNA polymerase II complex during promoter clearance.

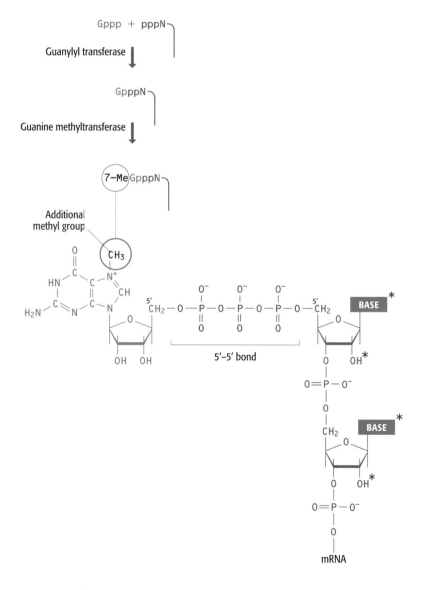

Figure 12.21 **Capping of eukaryotic mRNA.** The top part of the diagram shows the capping reaction in outline. A GTP molecule (drawn as Gppp) reacts with the 5′ end of the mRNA to give a triphosphate linkage. In the second step of the process, the terminal G is methylated at nitrogen number 7. The bottom part of the diagram shows the chemical structure of the type 0 cap, with asterisks indicating the positions where additional methylations might occur to produce type 1 and type 2 cap structures.

The 7-methylguanosine structure is called a **type 0 cap** and is the commonest form in yeast. In higher eukaryotes, additional modifications occur (see Figure 12.21):

● A second methylation replaces the hydrogen of the 2′–OH group of what is now the second nucleotide in the transcript. This results in a **type 1 cap**.

● If this second nucleotide is an adenosine, then the amino group attached to carbon number 6 of the purine ring might also be methylated.

● Another 2′–OH methylation might occur at the third nucleotide position, resulting in a **type 2 cap**.

All RNAs synthesized by RNA polymerase II are capped in one way or another. This means that as well as mRNAs, the snRNAs that are transcribed by this enzyme are also capped (see Table 11.3). The cap may be important for export of mRNAs and snRNAs from the nucleus (Section 12.2.7), but its best-defined role is in translation of mRNAs, which is covered in Section 13.2.2.

Elongation of eukaryotic mRNAs

As mentioned above, the fundamental aspects of transcript elongation are the same in bacteria and eukaryotes. The one major distinction concerns the length of transcript that must be synthesized. The longest bacterial genes are only a few kilobases in length and can be transcribed in a matter of minutes by the bacterial RNA polymerase, which has a polymerization rate of several hundred nucleotides per minute. In contrast, RNA polymerase II can take hours to transcribe a single gene, even though it can work at a rate of up to 2000 nucleotides per minute. This is because the presence of multiple introns in many eukaryotic genes (Section 12.2.2) means that considerable lengths of DNA must be copied. For example, the pre-mRNA for the human dystrophin gene is 2400 kb in length and takes about 20 hours to synthesize.

The extreme length of eukaryotic genes places demands on the stability of the transcription complex. RNA polymerase II on its own is not able to meet these demands: when the purified enzyme is studied *in vitro* its polymerization rate is less than 300 nucleotides per minute because the enzyme pauses frequently on the template and sometimes stops altogether. In the nucleus, pausing and stopping are reduced because of the action of a series of **elongation factors**, proteins that associate with the polymerase after it has cleared the promoter and left behind the transcription factors involved in initiation. Thirteen elongation factors are currently known in mammalian cells, displaying a variety of functions (Table 12.1). Their importance is shown by the effects of mutations that disrupt the activity of one or other of the factors.

Table 12.1 Examples of elongation factors for mammalian RNA polymerase II

Elongation factor	Function
TFIIF, CSB, ELL, Elongin	These factors suppress pausing of RNA polymerase II, which can occur when the enzyme transcribes through a region where intrastrand base pairs (e.g., a hairpin loop) can form
TFIIS	Prevents arrest (complete cessation of elongation)
FACT	Thought to modify chromatin in order to assist elongation

Inactivation of CSB, for example, results in Cockayne syndrome, a disease characterized by developmental defects such as mental retardation, and disruption of ELL causes acute myeloid leukemia.

A second difference between bacterial and eukaryotic transcript elongation is that RNA polymerase II, as well as the other eukaryotic nuclear polymerases, has to negotiate the nucleosomes that are attached to the template DNA that is being transcribed. At first glance it is difficult to imagine how the polymerase can elongate its transcript through a region of DNA wound around a nucleosome (see Figure 7.2). The solution to this problem is probably provided by elongation factors that are able to modify the chromatin structure in some way. In mammals, the elongation factor FACT has been shown to interact with histones H2A and H2B, possibly influencing nucleosome positioning, and less-well-defined interactions have been demonstrated for other factors. Yeast possesses a factor called **elongator**, which has been assigned a role in chromatin modification because it contains a subunit that has histone acetyltransferase activity, but so far a homolog of this complex has not been identified in mammals. An intriguing question is whether the first polymerase to transcribe a particular gene is a "pioneer" with a special elongation factor complement that opens up the chromatin structure, with subsequent rounds of transcription being performed by standard polymerase complexes that take advantage of the changes induced by the pioneer.

Termination of synthesis of most mRNAs is combined with polyadenylation

Most eukaryotic mRNAs have a series of up to 250 adenosines at their 3′ ends. These As are not specified by the DNA and are added to the transcript by a template-independent RNA polymerase called **poly(A) polymerase**. This polymerase does not act at the extreme 3′ end of the transcript, but at an internal site which is cleaved to create a new 3′ end to which the poly(A) tail is added.

The basic features of polyadenylation have been understood for some time. In mammals, polyadenylation is directed by a signal sequence in the mRNA, almost invariably 5′–AAUAAA–3′. This sequence is located between 10 and 30 nucleotides upstream of the polyadenylation site, which is often immediately after the dinucleotide 5′–CA–3′ and is followed 10–20 nucleotides later by a GU-rich region. Both the poly(A) signal sequence and the GU-rich region are binding sites for multisubunit protein complexes, which are, respectively, the **cleavage and polyadenylation specificity factor** (**CPSF**) and the **cleavage stimulation factor** (**CstF**). Poly(A) polymerase and at least two other protein factors must associate with bound CPSF and CstF in order for polyadenylation to occur (Figure 12.22). These additional factors include **polyadenylate-binding protein** (**PADP**), which helps the polymerase to add the adenosines, possibly influences the length of the poly(A) tail that is synthesized, and appears to play a role in maintenance of the tail after synthesis. In yeast, the signal sequences in the transcript are slightly different, but the protein complexes are similar to those in mammals and polyadenylation is thought to occur by more or less the same mechanism.

Polyadenylation was once looked on as a "posttranscriptional" event but it is now recognized that the process is an inherent part of the mechanism for termination of transcription by RNA polymerase II. CPSF is known to interact

Figure 12.22 Polyadenylation of eukaryotic mRNA. Note that the diagram is schematic and is not intended to indicate the relative sizes and shapes of the various protein complexes, nor their precise positioning, although CPSF and CstF are thought to bind to the 5'–AAUAAA–3' and GU-rich sequences, respectively, as shown. Note that "GU" indicates a GU-rich sequence rather than the dinucleotide 5'–GU–3'.

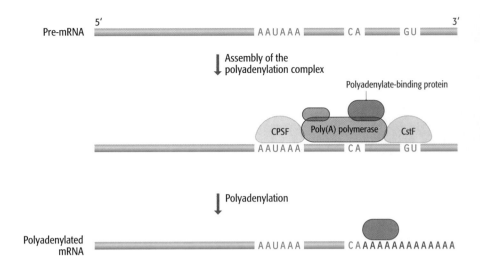

with TFIID and is recruited into the polymerase complex during the transcription initiation stage. By riding along the template with RNA polymerase II, CPSF is able to bind to the poly(A) signal sequence as soon as it is transcribed, initiating the polyadenylation reaction (Figure 12.23). Both CPSF and CstF form contacts with the CTD of the polymerase. It has been suggested that the nature of these contacts changes when the poly(A) signal sequence is located, and that this change alters the properties of the elongation complex so that termination becomes favored over continued RNA synthesis. As a result, transcription stops soon after the poly(A) signal sequence has been transcribed.

Even though polyadenylation can be identified as an inherent part of the termination process, this does not explain why it is necessary to add a poly(A) tail to the transcript. A role for the poly(A) tail has been sought for several years, but no convincing evidence has been found for any of the various suggestions that have been made. These suggestions include an influence on mRNA stability, which seems unlikely as some stable transcripts have very short poly(A) tails, and a role in initiation of translation. The latter proposal is supported by research showing that poly(A) polymerase activity is repressed during those periods of the cell cycle when relatively little protein synthesis occurs.

Figure 12.23 The link between polyadenylation and termination of transcription by RNA polymerase II. CPSF is shown attached to the RNA polymerase II elongation complex that is synthesizing RNA. CPSF binds to the polyadenylation signal sequence as soon as it is transcribed. This changes the interaction between CPSF and the CTD of RNA polymerase II so that termination of transcription is now favored over continued elongation. Note that this is a schematic representation and ignores the possibility that CstF may also be a component of the elongation complex. This representation also shows CPSF leaving the complex in order to bind to the polyadenylation signal, when in reality it may maintain its attachment to RNA polymerase II during the polyadenylation process.

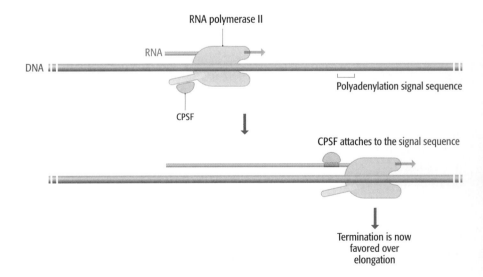

The role that is eventually ascribed to polyadenylation must take account of the fact that not all eukaryotic mRNAs have a poly(A) tail. The nonpolyadenylated mRNAs are a small group but they include several important members, most notably the mRNAs that specify the histone proteins. The 3' termini of these nonpolyadenylated mRNAs are, like the polyadenylated versions, created by cleavage of the primary transcript at a specific position, but the cleavage signals and the proteins involved in the process are quite different. There appear to be two cleavage signals within the transcript, the first of these a hairpin loop that forms downstream of the coding region (Figure 12.24A). The hairpin loop always comprises 6 bp in the stem and 4 nucleotides in the loop, this particular configuration being essential for successful cleavage, although the exact sequence of nucleotides in the structure can vary. The second signal is a 9-nucleotide sequence (consensus 5'–CAAGAAAGA–3') positioned about 12 nucleotides downstream of the hairpin. This sequence base-pairs with a part of the U7-snRNA (Figure 12.24B), one of the family of small nuclear RNAs whose members are involved in various aspects of RNA processing, including intron splicing, as we will see in the next section. The pairing is stabilized by a hairpin loop–binding protein and the cut is made four or five nucleotides downstream of the hairpin. Exactly how this cut is made is not known: no protein with the necessary activity has yet been identified.

Regulation of mRNA synthesis in eukaryotes

When we examined synthesis of bacterial mRNAs we encountered a number of processes by which control could be exerted over genome expression by regulating the elongation–termination stage of transcription (Section 12.1.2). Equivalent regulatory mechanisms in eukaryotes have proved elusive and, from our current standpoint, transcriptional control processes appear to operate almost exclusively at the initiation stage, as described in Section 11.3.2, rather than during the elongation and termination phases of mRNA synthesis.

Three potential control mechanisms should, however, be mentioned. First, the RNA polymerase II elongation factor called TFIIS, although structurally distinct from the bacterial Gre proteins that we met in Section 12.1.2, acts in a similar fashion, accessing the active site of RNA polymerase II by insertion

(A) Cleavage signals

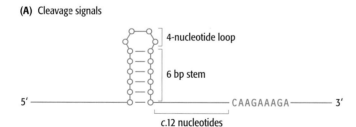

(B) Cleavage involves U7-snRNA

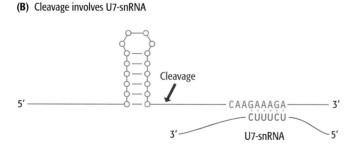

Figure 12.24 Processing of the 3' ends of histone mRNAs. (A) The structure of the 3' region of a histone mRNA, showing the hairpin loop and consensus 9-nucleotide sequence. (B) The consensus sequence base-pairs with a short region of the U7-snRNA. The attachment is stabilized by a hairpin loop–binding protein (not shown in this diagram) and the cut made by an unidentified protein at a position four or five nucleotides downstream of the hairpin loop.

of a needle-like structure through a channel that runs from the surface of the enzyme complex. In this manner, TFIIS, like GreA and GreB, is able to restart a stalled RNA polymerase by stimulating cleavage of the detached 3′ segment of the RNA transcript. We decided in Section 12.1.2 that although no specific regulatory process has yet been ascribed to the Gre proteins, such a role is possible, and by extrapolation that conclusion also applies to TFIIS.

The second potential mechanism for regulation of elongation and/or termination of eukaryotic mRNA synthesis centers on the phosphorylation status of the CTD of RNA polymerase II. Recall that during the transcription initiation process, phosphorylation of serines within the repeat component of the CTD activates the polymerase and stimulates promoter clearance (Section 11.2.3). This implies that a phosphorylated CTD might be essential for efficient RNA synthesis. The observation that the phosphorylation status of the CTD is not constant throughout the transcription process therefore suggests that changes to this status might play some role in regulating transcription. Several kinase proteins have been identified which phosphorylate the CTD, and at least three phosphatases are known which can carry out the reverse reaction, removing phosphates from the structure. Hence the machinery exists for modifying the phosphorylation status of the CTD during elongation and termination, but as yet no link with the regulation of transcription has been uncovered.

Finally, there is growing evidence that control of mRNA polyadenylation might be an important regulatory mechanism in eukaryotes. Many eukaryotic genes have more than one polyadenylation signal sequence, which means that termination can occur at different positions, resulting in mRNAs with identical coding properties but with distinctive 3′ ends. The different polyadenylation signals appear to be used in different tissues, suggesting that **alternative polyadenylation** could be an important mechanism for establishing tissue-specific patterns of genome expression.

12.2.2 Removal of introns from nuclear pre-mRNA

The existence of introns was not suspected until 1977 when DNA sequencing was first applied to eukaryotic genes and it was realized that many of these contain "intervening sequences" that separate different segments of the coding DNA from one another (Figure 12.25). We now recognize seven distinct types of intron in eukaryotes, and additional forms in the archaea (Table 12.2). Two of these types—the GU–AG and AU–AC introns—are found in eukaryotic protein-coding genes and are dealt with in this section; the other types will be covered later in the chapter.

Few rules can be established for the distribution of introns in protein-coding genes, beyond the fact that introns are less common in lower eukaryotes: the 6000 genes in the yeast genome contain only 239 introns in total, whereas many individual mammalian genes contain 50 or more introns. When the same gene is compared in related species, we usually find that some of the introns are in identical positions but that each species has one or more unique introns. This implies that some introns remain in place for millions of years, retaining their positions while species diversify, whereas others appear or disappear during this same period. These observations have implications for theories regarding genome evolution (Section 18.3.2). The important point, however, is that a eukaryotic pre-mRNA may contain many introns, perhaps over 100, taking up a considerable length of the transcript (Table 12.3), and that

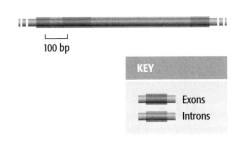

100 bp

KEY	
	Exons
	Introns

Figure 12.25 Introns. The structure of the human β-globin gene is shown. This gene is 1423 bp in length and contains two introns, one of 131 bp and one of 851 bp, which together make up 69% of the length of the gene.

Table 12.2 Types of intron

Intron type	Where found
GU–AG introns	Eukaryotic nuclear pre-mRNA
AU–AC introns	Eukaryotic nuclear pre-mRNA
Group I	Eukaryotic nuclear pre-rRNA, organelle RNAs, few bacterial RNAs
Group II	Organelle RNAs, some prokaryotic RNAs
Group III	Organelle RNAs
Twintrons	Organelle RNAs
Pre-tRNA introns	Eukaryotic nuclear pre-tRNA
Archaeal introns	Various RNAs

these introns must be excised and the exons joined together in the correct order before the transcript can function as a mature mRNA.

Conserved sequence motifs indicate the key sites in GU–AG introns

With the vast bulk of pre-mRNA introns, the first two nucleotides of the intron sequence are 5′–GU–3′ and the last two are 5′–AG–3′. They are therefore called "GU–AG" introns and all members of this class are spliced in the same way. These conserved motifs were recognized soon after introns were discovered and it was immediately assumed that they must be important in the splicing process. As intron sequences started to accumulate in the databases, it was realized that the GU–AG motifs are merely parts of longer consensus sequences that span the 5′ and 3′ splice sites. These consensus sequences vary in different types of eukaryote; in vertebrates they can be described as:

5′ splice site 5′–AG↓GUAAGU–3′

3′ splice site 5′–PyPyPyPyPyPyNCAG↓–3′

In these designations, "Py" is one of the two pyrimidine nucleotides (U or C), "N" is any nucleotide, and the arrow indicates the exon–intron boundary. The 5′ splice site is also known as the **donor site** and the 3′ splice site as the **acceptor site**.

Table 12.3 Introns in human genes

Gene	Length (kb)	Number of introns	Amount of the gene taken up by the introns (%)
Insulin	1.4	2	69
β-globin	1.6	2	61
Serum albumin	18	13	79
Type VII collagen	31	117	72
Factor VIII	186	25	95
Dystrophin	2400	78	98

Figure 12.26 Conserved sequences in vertebrate introns. The longer consensus sequences around the splice sites are given in the text. Abbreviation: Py, pyrimidine nucleotide (U or C).

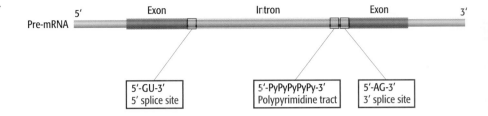

Other conserved sequences are present in some but not all eukaryotes. Introns in higher eukaryotes usually have a **polypyrimidine tract**, a pyrimidine-rich region located just upstream of the 3′ end of the intron sequence (Figure 12.26). This tract is less frequently seen in yeast introns, but these have an invariant 5′–UACUAAC–3′ sequence, located between 18 and 140 nucleotides upstream of the 3′ splice site, which is not present in higher eukaryotes. The polypyrimidine tract and the 5′–UACUAAC–3′ sequence are not functionally equivalent, as described in the next two sections.

Outline of the splicing pathway for GU–AG introns

The conserved sequence motifs indicate important regions of GU–AG introns, regions that we would anticipate either acting as recognition sequences for RNA-binding proteins involved in splicing, or playing some other central role in the process. Early attempts to understand splicing were hindered by technical problems (in particular, difficulties in developing a cell-free splicing system with which the process could be probed in detail), but during the 1990s there was an explosion of information. This work showed that the splicing pathway can be divided into two steps (Figure 12.27):

- Cleavage of the 5′ splice site occurs by a transesterification reaction promoted by the hydroxyl group attached to the 2′ carbon of an adenosine nucleotide located within the intron sequence. In yeast, this adenosine is the last one in the conserved 5′–UACUAAC–3′ sequence. The result of the hydroxyl attack is cleavage of the phosphodiester bond at the 5′ splice site, accompanied by formation of a new 5′–2′ phosphodiester bond linking the first nucleotide of the intron (the G of the 5′–GU–3′ motif) with the internal adenosine. This means that the intron has now been looped back on itself to create a **lariat** structure.

- Cleavage of the 3′ splice site and joining of the exons result from a second transesterification reaction, this one promoted by the 3′–OH group attached to the end of the upstream exon. This group attacks the phosphodiester

Figure 12.27 Splicing in outline. Cleavage of the 5′ splice site is promoted by the hydroxyl (OH) attached to the 2′–carbon of an adenosine nucleotide within the intron sequence. This results in the lariat structure and is followed by the 3′–OH group of the upstream exon inducing cleavage of the 3′ splice site. This enables the two exons to be ligated, with the released intron being debranched and degraded.

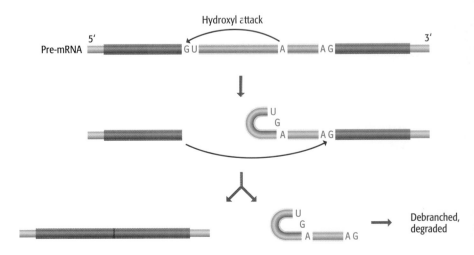

bond at the 3' splice site, cleaving it and so releasing the intron as the lariat structure, which is subsequently converted back to a linear RNA and degraded. At the same time, the 3' end of the upstream exon joins to the newly formed 5' end of the downstream exon, completing the splicing process.

In a chemical sense, intron splicing is not a great challenge for the cell. It is simply a double transesterification reaction, no more complicated than many other biochemical reactions that are dealt with by individual enzymes. But a complex machinery has evolved to deal with it. The difficulty lies with the topological problems. The first of these is the substantial distance that might lie between splice sites, possibly a few tens of kilobases, representing 100 nm or more if the mRNA is in the form of a linear chain. A means is therefore needed of bringing the splice sites into proximity. The second topological problem concerns selection of the correct splice site. All splice sites are similar, so if a pre-mRNA contains two or more introns then there is the possibility that the wrong splice sites could be joined, resulting in **exon skipping**—the loss of an exon from the mature mRNA (Figure 12.28A). Equally unfortunate would be selection of a **cryptic splice site**, a site within an intron or exon that has sequence similarity with the consensus motifs of real splice sites (Figure 12.28B). Cryptic sites are present in most pre-mRNAs and must be ignored by the splicing apparatus.

snRNAs and their associated proteins are the central components of the splicing apparatus

The central components of the splicing apparatus for GU–AG introns are the snRNAs called U1, U2, U4, U5, and U6. These are short molecules (between 106 nucleotides [U6] and 185 nucleotides [U2] in vertebrates) that associate with proteins to form **small nuclear ribonucleoproteins (snRNPs)** (Figure 12.29). The snRNPs, together with other accessory proteins, attach to the transcript and form a series of complexes, the last one of which is the **spliceosome**, the structure within which the actual splicing reactions occur. The process operates as follows (Figure 12.30):

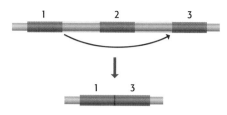

(A) Exon skipping

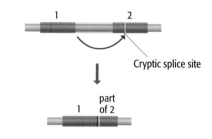

(B) Cryptic splice site selection

Cryptic splice site

part
1   of 2

Figure 12.28 Two aberrant forms of splicing. (A) In exon skipping the aberrant splicing results in an exon being lost from the mRNA. (B) When a cryptic splice site is selected, part of an exon might be lost from the mRNA, as shown here, or if the cryptic site lies within an intron then a segment of that intron will be retained in the mRNA.

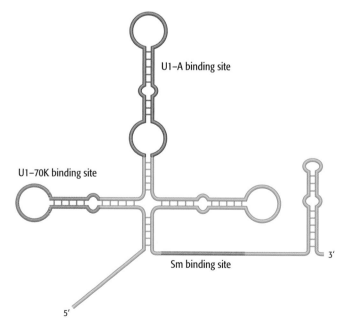

U1–A binding site

U1–70K binding site

Sm binding site

3'

5'

Figure 12.29 Structure of U1–snRNP. The mammalian U1–snRNP comprises the 165-nucleotide U1–snRNA plus ten proteins. Three of these proteins (U1–70K, U1–A, and U1–C) are specific to this snRNP, the other seven are Sm proteins that are found in all the snRNPs involved in splicing. The U1 snRNA forms a base-paired structure as shown. The U1–70K and U1–A proteins attach to two of the major stem-loops of this base-paired structure, and U1–C attaches via a protein–protein interaction. The Sm proteins attach to the Sm site.

Figure 12.30 The roles of snRNPs and associated proteins during splicing. There are several unanswered questions about the series of events occurring during splicing and it is unlikely that the scheme shown here is entirely accurate. The key point is that associations between the snRNPs are thought to bring the three critical parts of the intron—the two splice sites and the branch point—into close proximity.

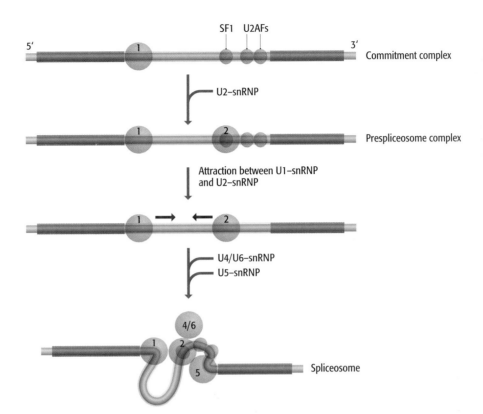

- The **commitment complex** initiates a splicing activity. This complex comprises U1–snRNP, which binds to the 5′ splice site, partly by RNA–RNA base pairing, and the protein factors SF1, U2AF^{35}, and U2AF^{65}, which make protein–RNA contacts with the branch site, the polypyrimidine tract, and the 3′ splice site, respectively.

- The **prespliceosome complex** comprises the commitment complex plus U2–snRNP, the latter attached to the branch site. At this stage, an association between U1–snRNP and U2–snRNP brings the 5′ splice site into close proximity with the branch point.

- The **spliceosome** is formed when U4/U6–snRNP (a single snRNP containing two snRNAs) and U5–snRNP attach to the prespliceosome complex. This results in additional interactions that bring the 3′ splice site close to the 5′ site and the branch point. All three key positions in the intron are now in proximity and the two transesterifications occur as a linked reaction, possibly catalyzed by U6–snRNP, completing the splicing process.

The series of events shown in Figure 12.30 provides no clues about how the correct splice sites are selected so that exons are not lost during splicing, and cryptic sites are ignored. This aspect of splicing is still poorly understood but it has become clear that a set of splicing factors called **SR proteins** are important in splice-site selection. The SR proteins—so-called because their C-terminal domains contain a region rich in serine (abbreviation S) and arginine (R)—were first implicated in splicing when it was discovered that they are components of the spliceosome. They appear to have several functions, including the establishment of a connection between bound U1–snRNP and the bound U2AF proteins in the commitment complex. This is perhaps the clue to their role in splice-site selection, formation of the commitment complex being the critical stage of the splicing process, as this is the event that identifies which sites will be linked.

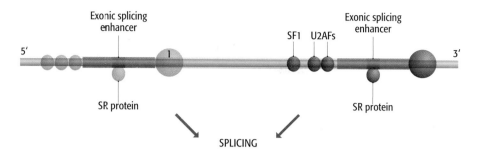

Figure 12.31 An alternative model for assembly of the commitment complex. In this model, each individual commitment complex (one shown in orange and one in red) is built up across an exon, bringing the complex into close association with an exonic splicing enhancer or silencer and its attached SR proteins. Once complexes have been built up over adjacent exons, splicing follows the pathway illustrated in Figure 12.30, the only difference being that the resulting spliceosome is made up of components of adjacent commitment complexes, rather than being derived from the single commitment complex shown in Figure 12.30.

SR proteins also interact with **exonic splicing enhancers** (**ESEs**), which are purine-rich sequences located in the exon regions of a transcript. We are still at an early stage in our understanding of ESEs and their counterparts, the **exonic splicing silencers** (ESSs), but their importance in controlling splicing is clear from the discovery that several human diseases, including one type of muscular dystrophy, are caused by mutations in ESE sequences. The location of ESEs and ESSs indicates that assembly of the spliceosome is driven not simply by contacts within the intron but also by interactions with adjacent exons. In fact, it is possible that an individual commitment complex is not assembled within an intron as shown in Figure 12.30, but initially bridges an exon (Figure 12.31). This model is attractive not only because it provides a means by which contact between an ESE or ESS and an SR protein could influence splicing, but also because it takes account of the large disparity between the lengths of exons and introns in vertebrate genes. In the human genome, for example, the exons have an average length of 145 nucleotides compared with 3365 nucleotides for introns. Initial assembly of a commitment complex across an exon might therefore be a less difficult task than assembly across a much longer intron.

There is one final aspect of SR proteins that we should address. This is the possibility that a subset of these SR proteins, called **CASPs** (**CTD-associated SR-like proteins**) or **SCAFs** (**SR-like CTD-associated factors**), form a physical connection between the spliceosome and the CTD of the RNA polymerase II transcription complex, and hence provide a link between transcript elongation and processing. As with some of the polyadenylation proteins (Section 12.2.1), it is probable that these splicing factors ride with the polymerase as it synthesizes the transcript, and are deposited at their appropriate positions at intron splice sites as soon as these are transcribed. Electron microscopy studies have shown that transcription and splicing occur together, and the discovery of splicing factors that have an affinity for RNA polymerase II provides a biochemical basis for this observation.

Alternative splicing is common in many eukaryotes

When introns were first discovered it was imagined that each gene always gives rise to the same mRNA: in other words, that there is a single **splicing pathway** for each primary transcript (Figure 12.32A). This assumption was found to be incorrect in the 1980s, when it was shown that the primary transcripts of some genes can follow two or more **alternative splicing** pathways, enabling a single transcript to be processed into related but different mRNAs and hence to direct synthesis of a range of proteins (Figure 12.32B). In some organisms alternative splicing is uncommon, only three examples being known in *Saccharomyces cerevisiae*, but in higher eukaryotes it is much more prevalent. This first became apparent when the draft *Drosophila melanogaster* genome sequence was examined, and it was realized that fruit flies have fewer

Figure 12.32 The assumption that each pre-mRNA follows a single splicing pathway was shown to be incorrect when alternative splicing was discovered.

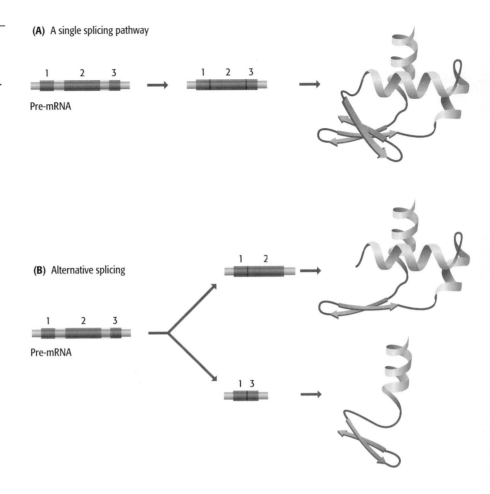

(A) A single splicing pathway

Pre-mRNA

(B) Alternative splicing

Pre-mRNA

genes than the microscopic worm *Caenorhabditis elegans* (see Table 7.4), despite the obviously greater physical complexity of *Drosophila*, which should be reflected in a more diverse proteome. The most likely explanation for the lack of congruence between the number of genes in the *Drosophila* genome and the number of proteins in its proteome is that a substantial number of the genes give rise to multiple proteins via alternative splicing. At about the same time as these observations, the first human chromosome sequences were obtained and it was recognized that rather than having 80,000–100,000 genes, as suggested by the size of the human proteome, humans have only 35,000 or so genes. It is now believed that at least 35% of the genes in the human genome undergo alternative splicing: the principle "one gene, one protein,"—biological dogma since the 1940s—has been completely overthrown.

Alternative splicing is now looked on as a crucial innovation in the genome expression pathway. Two examples will suffice to illustrate its importance. The first of these concerns sex determination, a fundamental aspect of the biology of any organism, and which in *Drosophila* is determined by an alternative splicing cascade. The first gene in this cascade is *sxl*, whose transcript contains an optional exon which, when spliced to the one preceding it, results in an inactive version of protein SXL. In females the splicing pathway is such that this exon is skipped so that functional SXL is made (Figure 12.33). SXL promotes selection of a cryptic splice site in a second transcript, *tra*, by directing U2AF^{65} away from its normal 3′ splice site to a second site further downstream. The resulting female-specific TRA protein is again involved in alternative splicing, this time by interacting with SR proteins to form a multi-

(A) Sex-specific alternative splicing of *sxl* pre-mRNA

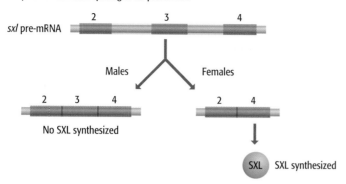

(B) SXL induces cryptic splice site selection in the *tra* pre-mRNA

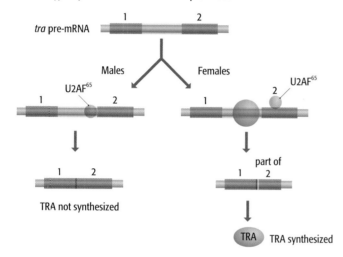

(C) TRA induces alternative splicing of the *dsx* pre-mRNA

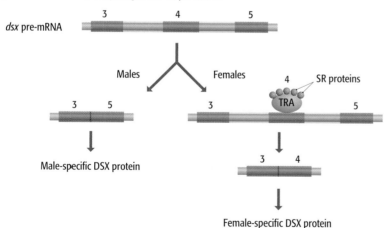

Figure 12.33 Regulation of splicing during expression of genes involved in sex determination in *Drosophila*. (A) The cascade begins with sex-specific alternative splicing of the *sxl* pre-mRNA. In males, all exons are present in the mRNA, but this means that a truncated protein is produced because exon 3 contains a termination codon. In females, exon 3 is skipped, leading to a full-length, functional SXL protein. (B) In females, SXL blocks the 3′ splice site in the first intron of the *tra* pre-mRNA. U2AF^{65} is unable to locate this site and instead directs splicing to a cryptic site in exon 2. This results in an mRNA that codes for a functional TRA protein. In males, there is no SXL so the 3′ splice site is not blocked and a dysfunctional mRNA is produced. (C) In males, exon 4 of the *dsx* pre-mRNA is skipped. The resulting mRNA codes for a male-specific DSX protein. In females, TRA stabilizes the attachment of SR proteins to an exonic splicing enhancer located within exon 4, so this exon is not skipped, resulting in the mRNA that codes for the female-specific DSX protein. The two versions of DSX are the primary determinants of the male and female physiologies. The female *dsx* mRNA ends with exon 4 because the intron between exons 4 and 5 has no 5′ splice site, meaning that exon 5 cannot be ligated to the end of exon 4. Instead a polyadenylation site at the end of exon 4 is recognized in females. Note that the diagram is schematic and that the introns are not drawn to scale.

factor complex that attaches to an ESE within an exon of a third pre-mRNA, *dsx*, promoting selection of a secondary, female-specific splice site in this transcript. The male and female versions of the DSX proteins are the primary determinants of *Drosophila* sex.

The second example of alternative splicing illustrates the multiplicity of mRNAs synthesized from some primary transcripts. The human *slo* gene codes

Figure 12.34 The human *slo* gene. The gene comprises 35 exons, shown as boxes, eight of which (in green) are optional and appear in different combinations in different *slo* mRNAs. There are 8! = 40,320 possible splicing pathways and hence 40,320 possible mRNAs, but only some 500 of these are thought to be synthesized in the human cochlea.

for a membrane protein that regulates the entry and exit of potassium ions into and out of cells. The gene has 35 exons, eight of which are involved in alternative splicing events (Figure 12.34). The alternative splicing pathways involve different combinations of the eight optional exons, leading to over 500 distinct mRNAs, each specifying a membrane protein with slightly different functional properties. The human *slo* genes are active in the inner ear and determine the auditory properties of the hair cells on the basilar membrane of the cochlea. Different hair cells respond to different sound frequencies between 20 Hz and 20,000 Hz, their individual capabilities determined in part by the properties of their Slo proteins. Alternative splicing of *slo* genes in cochlear hair cells therefore determines the auditory range of humans.

At present we do not understand how alternative splicing is regulated and cannot describe the process that determines which of several splicing pathways is followed by a particular transcript. The players are thought to be the SR proteins in conjunction with ESEs and ESSs, but the way in which they control splice-site selection is not known.

Trans-splicing links exons from different transcription units

In the examples of splicing that we have considered so far, the two exons that are joined together are both located within the same transcript. In a few organisms, splicing also takes place between exons that are contained within different RNA molecules. This is called **trans-splicing** and it occurs in the chloroplasts of some plants, with some genes in *C. elegans*, and in trypanosomes, the protozoan parasites of vertebrates that cause sleeping sickness in humans.

All of the examples of trans-splicing that have been studied so far are similar in that they result in the same short leader segment becoming attached to the 5′ ends of each member of a set of mRNAs (Figure 12.35). The transcript which donates this leader segment is called the **spliced leader RNA** (**SL RNA**). In *C. elegans*, this SL RNA is approximately 100 nucleotides in length and contains a 22-nucleotide sequence that is attached to the 5′ ends of the target mRNAs. The splicing reaction proceeds in a manner very similar to the standard scheme shown in Figure 12.27, although as the splicing partners are different molecules a forked structure is formed instead of the lariat. The only complication is that the SL RNA is able to fold into a base-paired structure similar to an snRNA and, according to some models of trans-splicing, the SL RNA replaces U1–snRNP in the splicing process.

Trans-splicing in *C. elegans* has one other interesting aspect. Some of the mRNAs that participate in trans-splicing contain two genes which are transcribed together, head to tail, from a single promoter. If one of these two-gene

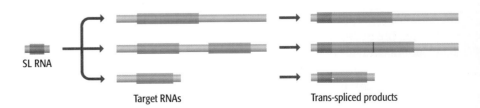

Figure 12.35 Trans-splicing. The leader exon from a single SL RNA is attached by splicing to a variety of target RNAs.

SL RNA

Target RNAs

Trans-spliced products

mRNAs in *C. elegans* is not trans-spliced then only the upstream gene can be translated because, as we will see in Chapter 13, the ribosome that translates a eukaryotic mRNA attaches to the extreme 5′ end of the molecule and usually dissociates from the transcript when it reaches a termination codon. Hence, with an unspliced mRNA, the downstream gene is inaccessible to the translation apparatus. With these mRNAs, trans-splicing is therefore the process that activates translation of the downstream gene by creating a new 5′ end to which a ribosome can bind (Figure 12.36).

AU–AC introns are similar to GU–AG introns but require a different splicing apparatus

One of the more surprising events of recent years has been the discovery of a few introns in eukaryotic pre-mRNAs that do not fall into the GU–AG category, having different consensus sequences at their splice sites. These are the **AU–AC introns** which, to date, have been found in approximately 20 genes in organisms as diverse as humans, plants, and *Drosophila*.

As well as the sequence motifs at their splice sites, AU–AC introns have a conserved (though not invariant) branch site sequence with the consensus 5′–UCCUUAAC–3′, the last adenosine in this motif being the one that participates in the first transesterification reaction. This points us toward the remarkable feature of AU–AC introns: their splicing pathway is very similar to that for GU–AG introns, but involves a different set of splicing factors. Only the U5–snRNP is involved in the splicing mechanisms of both types of intron. For AU–AC introns, the roles of U1–snRNP and U2–snRNP are taken by U11/U12–snRNP, a previously discovered complex that had never been assigned a function, and an entirely new U4atac/U6atac–snRNP has subsequently been isolated to complete the picture.

The splicing pathways for the "major" and "minor" types of intron are not identical but many of the interactions between the transcript and the snRNPs and other splicing proteins are remarkably similar. This means that AU–AC introns, rather than simply being a curiosity, are proving useful in testing models for interactions occurring during GU–AG intron splicing. The argument is that a predicted interaction between two components of the GU–AG spliceosome can be checked by seeing if the same interaction is possible with the equivalent AU–AC components. This has already been informative in helping to define a base-paired structure formed between the U2- and U6-snRNAs in the GU–AG spliceosome.

12.2.3 Synthesis of functional RNAs in eukaryotes

In general, we know less about transcript elongation and termination by RNA polymerases I and III than we do about equivalent processes for RNA polymerase II. The interaction of the polymerase with the template and transcript during elongation appears to be similar with all three enzymes, a reflection of the structural relatedness of the three largest subunits in each RNA polymerase. One difference is the rate of transcription—RNA polymerase I, for example, being much slower than RNA polymerase II, managing a polymerization rate of only 20 nucleotides per minute, compared with up to 2000 per minute for mRNA synthesis. A second difference is that neither RNA polymerase I transcripts nor RNA polymerase III transcripts are capped. Various proteins that might act as elongation factors for RNA polymerase I or III have been isolated, including SGS1 and SRS2 of yeast, which code for two related

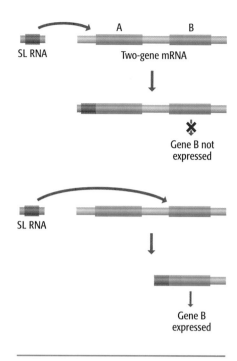

Figure 12.36 Gene regulation by trans-splicing. In the upper drawing, the leader exon is trans-spliced to Gene A. As a result, Gene B is not expressed because the ribosome cannot traverse the gap between the end of Gene A and the start of Gene B. In the lower drawing, trans-splicing is to Gene B, which is now expressed.

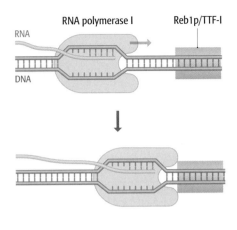

RNA polymerase I Reb1p/TTF-I
RNA
DNA

Figure 12.37 A possible scheme for termination of transcription by RNA polymerase I.

DNA helicases. Mutations in the genes for SGS1 and SRS2 cause a reduction in RNA polymerase I transcription as well as DNA replication. SGS1 is interesting because it is a homolog of a pair of human proteins that are defective in the growth disorders Bloom's and Werner's syndromes (Section 5.2.1) but the exact involvement of SGS1 and SRS2, and other putative elongation factors, in transcription by RNA polymerases I and III is not known.

The major differences between the three RNA polymerases are seen when the termination processes are compared. The polyadenylation system for RNA polymerase II termination (Section 12.2.1) is unique to that enzyme and no equivalent has been described for the other two RNA polymerases. Termination of transcription by RNA polymerase I involves a DNA-binding protein, called Reb1p in *Saccharomyces cerevisiae* and TTF-I in mice, which attaches to the DNA at a recognition sequence located 12–20 bp downstream of the point at which transcription terminates (Figure 12.37). Exactly how the bound protein causes termination is not known, but a model in which the polymerase becomes stalled because of the blocking effect of Reb1p/TTF-I has been proposed. A second protein, PTRF (polymerase I and transcript release factor), is thought to induce dissociation of the polymerase and the transcript from the DNA template. Even less is known about RNA polymerase III termination: a run of adenosines in the template is implicated but the process does not involve a hairpin loop and so is not analogous to termination in bacteria.

12.2.4 Splicing of eukaryotic pre-rRNA and pre-tRNA

In eukaryotes there are four rRNAs. One of these, the 5S rRNA, is transcribed by RNA polymerase III and does not undergo processing. The remaining three (the 5.8S, 18S, and 28S rRNAs) are transcribed by RNA polymerase I from a single unit, producing a pre-rRNA which, as with the bacterial pre-rRNAs, is processed by cutting and end-trimming. Several nucleases are required, including the multifunctional **ribonuclease MRP** which, as well as 5.8S rRNA processing, is involved in replication of mitochondrial DNA and control of the cell cycle. Genes for tRNAs occur singly and as multigene transcription units, and are processed in a manner very similar to that seen in bacteria (see Figure 12.17). The main distinction between functional RNA processing in bacteria and eukaryotes is that the primary transcripts for some eukaryotic rRNAs and tRNAs contain introns. Neither type of intron is similar to the GU–AG and AU–AC introns of pre-mRNA, and we must therefore spend some time examining them.

Introns in eukaryotic pre-rRNAs are autocatalytic
Introns are quite uncommon in eukaryotic pre-rRNAs but a few are known in microbial eukaryotes such as *Tetrahymena*. These introns are members of the Group I family (see Table 12.2) and are also found in mitochondrial and chloroplast genomes, where they occur in pre-mRNA as well as pre-rRNA. A few isolated examples are known in bacteria, for instance in a tRNA gene of the cyanobacterium *Anabaena* and in the thymidylate synthase gene of the *E. coli* bacteriophage T4.

The splicing pathway for Group I introns is similar to that for pre-mRNA introns in that two transesterifications are involved. The first is induced not by a nucleotide within the intron but by a free nucleoside or nucleotide, any one of guanosine or guanosine mono-, di-, or triphosphate (Figure 12.38).

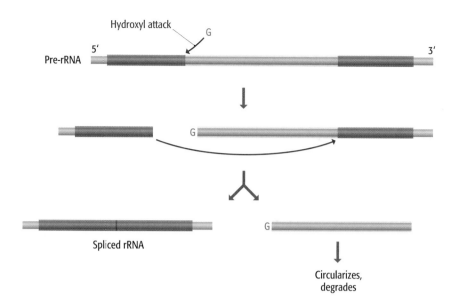

Figure 12.38 The splicing pathway for the *Tetrahymena* rRNA intron.

The 3′–OH of this cofactor attacks the phosphodiester bond at the 5′ splice site, cleaving it, with transfer of the G to the 5′ end of the intron. The second transesterification involves the 3′–OH at the end of the exon, which attacks the phosphodiester bond at the 3′ splice site, causing cleavage, joining of the two exons, and release of the intron. The released intron is linear, rather than the lariat seen with pre-mRNA introns, but may undergo additional transesterifications, leading to circular products, as part of its degradation process.

The remarkable feature of the Group I intron splicing pathway is that it proceeds in the absence of proteins and hence is autocatalytic, the RNA itself possessing enzymatic activity. This was the first example of an RNA enzyme, or **ribozyme**, to be discovered, back in the early 1980s. Initially this caused quite a stir but it is now realized that, although uncommon, there are several examples of ribozymes (Table 12.4). The self-splicing activity of Group I introns resides in the base-paired structure taken up by the RNA. This structure was first described in two-dimensional terms by comparing the sequences of different Group I introns and working out a common base-paired arrangement that could be adopted by all versions. This resulted in a model comprising nine major base-paired regions (Figure 12.39). More recently, the three-dimensional structure has been solved by X-ray crystallography. The ribozyme consists of a catalytic core made up of two domains, each one comprising two of the base-paired regions, with the splice sites brought into proximity by interactions between two other parts of the secondary structure. Although this RNA structure is sufficient for splicing, it is possible that with some introns the stability of the ribozyme is enhanced by noncatalytic protein factors that bind to it. This has long been suspected with the Group I introns in organelle genes, many of these containing an open reading frame coding for a protein called a **maturase** that appears to play a role in splicing.

Removal of introns from eukaryotic pre-tRNAs

Transfer RNA introns are relatively common in lower eukaryotes but less frequent in vertebrates—introns are present in only 6% of all human tRNA genes. Introns in eukaryotic pre-tRNAs are 14–60 nucleotides in length and are usually found at the same position in the transcript, within the anticodon

Table 12.4 Examples of ribozymes

Ribozyme	Description
Self-splicing introns	Some introns of Groups I, II, and III splice themselves by an autocatalytic process. There is also growing evidence that the splicing pathway of GU–AG introns includes at least some steps that are catalyzed by snRNAs
Ribonuclease P	The enzyme that creates the 5′ ends of bacterial tRNAs (see Section 12.1.3) consists of an RNA subunit and a protein subunit, with the catalytic activity residing in the RNA
Ribosomal RNA	The peptidyl transferase activity required for peptide bond formation during protein synthesis (Section 13.2.3) is associated with the 23S rRNA of the large subunit of the ribosome
tRNA^{Phe}	Undergoes self-catalyzed cleavage in the presence of divalent lead ions
Virus genomes	Replication of the RNA genomes of some viruses involves self-catalyzed cleavage of chains of newly synthesized genomes linked head to tail. Examples are the plant viroids and virusoids (Section 9.1.2) and the animal hepatitis delta virus. These viruses form a diverse group with the self-cleaving activity specified by a variety of different base-paired structures, including a well-studied one that resembles a hammerhead (see Figure 9.9)

Figure 12.39 The base-paired structure of the *Tetrahymena* rRNA intron. The sequence of the intron is shown in capital letters, with the exons in lower case. Additional interactions fold the intron into a three-dimensional structure that brings the two splice sites close together.

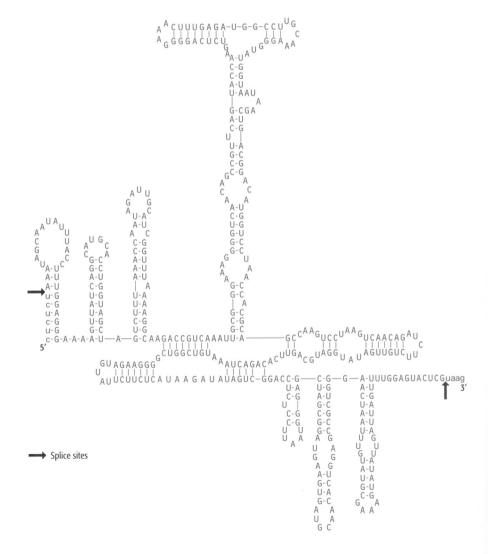

loop, one nucleotide downstream of the anticodon itself. The intron sequence is variable, but includes a short region complementary to the anticodon and possibly one or two adjacent nucleotides. Base pairing between the complementary sequences forms a short stem between two loops in the unspliced pre-tRNA (Figure 12.40).

Unlike all other types of introns in eukaryotes, splicing of pre-tRNA introns does not involve transesterifications. Instead, the two splice sites are cut by an endonuclease. This enzyme contains four nonidentical subunits, one of which uses the structure of the base-paired intron as a guide to identify the correct positions at which the RNA should be cut. The upstream and downstream cuts are then made by two of the other enzyme subunits. Cleavage leaves a cyclic phosphate structure attached to the 3′ end of the upstream exon, and a hydroxyl group at the 5′ end of the downstream exon (Figure 12.40). The cyclic phosphate is converted to a 3′–OH end by a phosphodiesterase, and the 5′–OH terminus is converted to 5′–P by a kinase. These two ends are held in proximity by the natural base pairing adopted by the tRNA sequence and are ligated by an RNA ligase. The phosphodiesterase, kinase, and ligase activities are provided by a single protein.

Other types of intron

There are eight different types of intron (see Table 12.2). Four of these have already been described in this chapter: the nuclear pre-mRNA introns of the GU–AG and AU–AC classes, the self-splicing Group I introns, and the introns in eukaryotic pre-tRNA genes. For completeness, the details of the four other categories are as follows:

- **Group II introns** are found in the organelle genomes of fungi and plants, in both pre-mRNA and pre-rRNA, and a few are known in prokaryotes. Group II introns take up a characteristic secondary structure and they are able to self-splice in the test tube, but they are distinct from Group I introns. The secondary structure is different, and the splicing mechanism is more closely allied to that of pre-mRNA introns, the initial transesterification being promoted by the hydroxyl group of an internal adenosine nucleotide and the intron being converted into a lariat structure. These similarities have prompted suggestions that Group II and pre-mRNA introns may have a common evolutionary origin (see Section 18.3.2). Some Group II introns are mobile elements that transpose by a remarkable process called **retrohoming**, during which the excised intron, which of course is a single-stranded RNA molecule, inserts directly into the organelle genome prior to being copied into the DNA version.

- **Group III introns** are also found in organelle genomes and they self-splice via a mechanism very similar to that of Group II introns, but Group III introns are smaller and have their own distinctive secondary structure. The resemblance to Group II introns again suggests an evolutionary relationship.

- **Twintrons** are composite structures made up of two or more Group II and/or Group III introns. The simplest twintrons consist of one intron embedded in another, but more complex ones contain multiple embedded introns. The individual introns that make up a twintron are usually spliced in a defined sequence.

- **Archaeal introns** are present in tRNA and rRNA genes. They are cleaved by a ribonuclease similar to the one involved in eukaryotic pre-tRNA splicing.

12.2.5 Chemical modification of eukaryotic RNAs

Eukaryotic tRNAs and rRNAs undergo the same types of chemical modification as the bacterial molecules (Section 12.1.3). With tRNAs, the enzymes that carry out the chemical modifications appear to be directed to the correct

Figure 12.40 Splicing of the *Saccharomyces cerevisiae* pre-tRNA^Tyr.

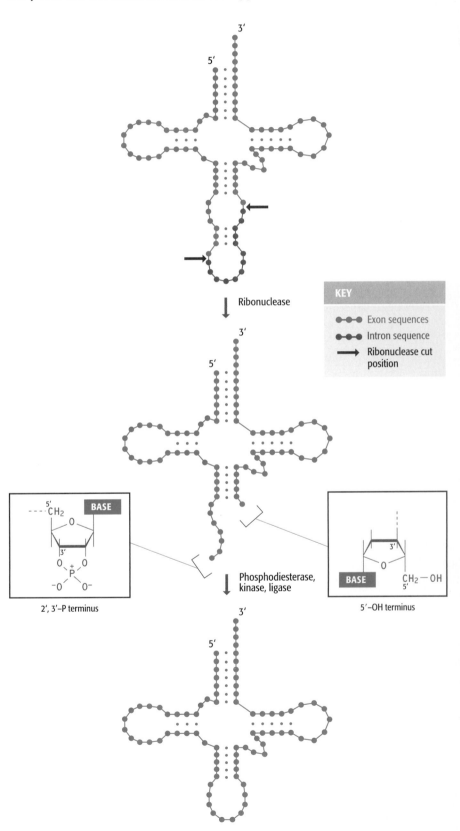

KEY

●—●—● Exon sequences

●—●—● Intron sequence

→ Ribonuclease cut position

nucleotides by the base-paired structure of the tRNA molecule, similar to the way in which the endonucleases that splice tRNA introns use the base-paired structure as a guide. With eukaryotic rRNAs, the situation is rather different.

Small nucleolar RNAs act as guides for chemical modification of eukaryotic rRNAs

It is not easy, simply by intuition, to imagine how specificity of rRNA modification can be ensured. Human pre-rRNA, for example, undergoes 106 methylations and 95 pseudouridinylations, each alteration at a specified position, with no obvious sequence similarities that can be inferred as target motifs for the modifying enzymes. Not surprisingly, progress in understanding rRNA modification was slow to begin with. The breakthrough came when it was shown that in eukaryotes the short RNAs called snoRNAs are involved in the modification process. These molecules are 70–100 nucleotides in length and are located in the nucleolus, the region within the nucleus where rRNA processing takes place. The initial discovery was that by base-pairing to the relevant region, snoRNAs pinpoint positions at which the pre-rRNA must be methylated. The base pairing involves only a few nucleotides, not the entire length of the snoRNA, but these nucleotides are always located immediately upstream of a conserved sequence called the D box (Figure 12.41A). The base pair involving the nucleotide that will be modified is five positions away from the D box. The hypothesis is that the D box is the recognition signal for the methylating enzyme, which is therefore directed toward the appropriate nucleotide. After these initial discoveries with regard to methylation, it was shown that a different family of snoRNAs carries out the same guiding role in conversion of uridines to pseudouridines. These snoRNAs do not have D boxes but still have conserved motifs that could be recognized by the modifying enzyme, and each is able to form a specific base-paired interaction with its target site, specifying the nucleotide to be modified.

The implication is that there is a different snoRNA for each modified position in a pre-rRNA, except possibly for a few sites that are close enough together to be dealt with by a single snoRNA. This means that there must be a few hundred snoRNAs per cell. At one time this seemed unlikely because very few snoRNA genes could be located, but now it appears that only a fraction of all the snoRNAs are transcribed from these standard genes, most being specified by sequences within the introns of other genes and released by cutting up the intron after splicing (Figure 12.41B).

RNA editing

Because rRNAs and tRNAs are noncoding, chemical modifications to their nucleotides affect only the structural features and, possibly, catalytic activities of the molecules. With mRNAs the situation is very different: chemical modification has the potential to change the coding properties of the transcript, resulting in an equivalent alteration in the amino acid sequence of the protein that is specified. This is called **RNA editing**. A notable example of RNA editing occurs with the human mRNA for apolipoprotein B. The gene for this protein codes for a 4563-amino-acid polypeptide, called apolipoprotein B100, which is synthesized in liver cells and secreted into the bloodstream where it transports lipids around the body. A related protein, apolipoprotein B48, is made by intestinal cells. This protein is only 2153 amino acids in length and is synthesized from an edited version of the mRNA for the full-length protein (Figure 12.42). In intestinal cells this mRNA is modified by

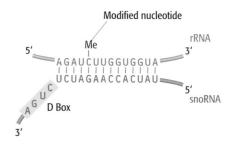

(A) Methylation by yeast U24-snoRNA

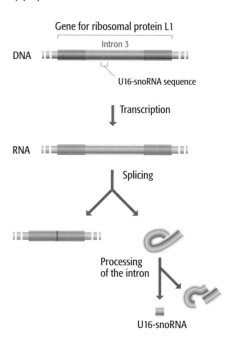

(B) Synthesis of human U16-snoRNA

Figure 12.41 Methylation of rRNA by a snoRNA. (A) This example shows methylation of the C at position 1436 in the *Saccharomyces cerevisiae* 25S rRNA (equivalent to the 28S rRNA of vertebrates), directed by U24-snoRNA. The D box of the snoRNA is highlighted. Modification always occurs at the base pair five positions away from the D box. Note that the interaction between rRNA and snoRNA involves an unusual G–U base pair, which is permissible between RNA polynucleotides. (B) Many snoRNAs are synthesized from intron RNA, as shown here for human U16-snoRNA, which is specified by a sequence in intron 3 of the gene for ribosomal protein L1.

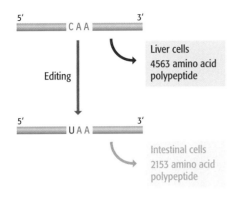

Figure 12.42 Editing of the human apolipoprotein B mRNA. Conversion of a C to a U creates a termination codon, resulting in a shortened form of apolipoprotein B being synthesized in intestinal cells.

deamination of a cytosine, converting this into a uracil. This changes a CAA codon, specifying glutamine, into a UAA codon, which causes translation to stop, resulting in the truncated protein. The deamination is carried out by an RNA-binding enzyme which, in conjunction with a set of auxiliary protein factors, binds to a sequence immediately downstream of the modification position within the mRNA.

Although not common, RNA editing occurs in a number of different organisms and includes a variety of different nucleotide changes (Table 12.5). Some editing events have a significant impact on the organism: in humans, RNA editing is partly responsible for the generation of antibody diversity (Section 14.2.1) and has also been implicated in control of the HIV-1 infection cycle. One particularly interesting type of RNA editing is the deamination of adenosine to inosine, which is carried out by enzymes called **adenosine deaminases acting on RNA** (**ADARs**). Some of the target mRNAs for these enzymes are selectively edited at a limited number of positions. These positions are apparently specified by double-stranded segments of the pre-mRNA, formed by base pairing between the modification site and sequences from adjacent introns. This type of editing occurs, for example, during processing of the mRNAs for mammalian glutamate receptors. There is evidence that ADAR editing is closely linked with RNA synthesis, as some nucleotides within introns are edited (indicating that editing occurs before intron splicing) and editing efficiency is reduced if changes are artificially made to the CTD of RNA polymerase II.

Selective editing contrasts with the second type of modification carried out by ADARs, in which the target molecules become extensively deaminated, over 50% of the adenosines in the RNA becoming converted to inosines. This hyperediting has so far been observed mainly, but not exclusively, with viral RNAs and is thought to occur by chance, these RNAs adopting base-paired structures that fortuitously act as substrates for ADAR. It may, however, have physiological importance in the etiology of diseases caused by the edited viruses. This possibility is raised by the discovery that viral RNAs associated with persistent measles infections (as opposed to the more usual transient version of the disease) are hyperedited.

The above examples of RNA editing are relatively straightforward events which, with the exception of hyperediting, lead to nucleotide changes at a

Table 12.5 Examples of RNA editing in mammals

Tissue	Target RNA	Change	Comments
Intestine	Apolipoprotein B mRNA	C→U	Converts a glutamine codon to a stop codon
Muscle	α-galactosidase mRNA	U→A	Converts a phenylalanine codon into a tyrosine codon
Testis, tumors	Wilms tumor-1 mRNA	U→C	Converts a leucine codon into a proline codon
Tumors	Neurofibromatosis type-1 mRNA	C→U	Converts an arginine codon into a stop codon
B lymphocytes	Immunoglobulin mRNA	Various	Contributes to the generation of antibody diversity
HIV-infected cells	HIV-1 transcript	G→A, C→U	Involved in regulation of the HIV-1 infection cycle
Brain	Glutamate receptor mRNA	A→inosine	Multiple positions leading to various codon changes

single or limited number of positions in selected mRNAs. More complex types of RNA editing are also known:

- **Pan-editing** involves the extensive insertion of nucleotides into abbreviated RNAs in order to produce functional molecules. It is particularly common in the mitochondria of trypanosomes. Many of the RNAs transcribed in trypanosome mitochondria are specified by **cryptogenes**— sequences lacking some of the nucleotides present in the mature RNAs. The pre-RNAs transcribed from these cryptogenes are processed by multiple insertions of U nucleotides, at positions defined by short **guide RNAs**. These are short RNAs that can base-pair to the pre-RNA and which contain As at the positions where Us must be inserted (Figure 12.43).

- Less extensive **insertional editing** occurs with some viral RNAs. For example, the paramyxovirus P gene gives rise to at least two different proteins because of the insertion of Gs at specific positions in the mRNA. These insertions are not specified by guide RNAs: instead they are added by the RNA polymerase as the mRNA is being synthesized.

- **Polyadenylation editing** is seen with many animal mitochondrial mRNAs. Five of the mRNAs transcribed from the human mitochondrial genome end with just a U or UA, rather than with one of the termination codons (UAA and UAG in the human mitochondrial genetic code). Polyadenylation converts the terminal U or UA into UAAAA..., and so creates a termination codon. This is just one of several features that appear to have evolved in order to make vertebrate mitochondrial genomes as small as possible.

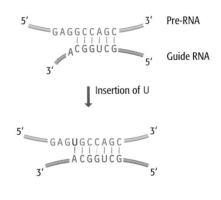

Figure 12.43 The role of a guide RNA in pan-editing.

12.2.6 Degradation of eukaryotic RNAs

Eukaryotic mRNAs are longer lived than their bacterial counterparts, with half-lives of, on average, 10–20 minutes for yeast mRNA and several hours for mammalian mRNA. Within individual cells the variations are almost equally striking: some yeast mRNAs have half-lives of only 1 minute whereas for others the figure is more like 35 minutes. These observations raise two questions: What are the processes for mRNA degradation? And how are these processes controlled?

Eukaryotes have diverse mechanisms for RNA degradation

Among eukaryotes, most progress in understanding mRNA degradation has been made with yeast. At least four pathways have been identified. One of these involves a multiprotein complex called the **exosome**, which degrades transcripts in the 3′→5′ direction and contains nucleases related to the enzymes of the bacterial degradosome. Exosomes are probably also present in mammalian cells and are clearly important, but they are not particularly well studied. Their role may not be in mRNA degradation *per se*, but in monitoring polyadenylation and ensuring that transcripts that are about to leave the nucleus have an appropriate poly(A) tail.

Rather more is known about two other eukaryotic mRNA degradation processes. The first of these is **deadenylation-dependent decapping** (Figure 12.44), which is triggered by removal of the poly(A) tail, possibly by exonuclease-mediated cleavage or possibly by loss of the polyadenylate-binding protein which stabilizes the tail (Section 12.2.1). Poly(A) tail removal is followed by cleavage of the 5′ cap by the decapping enzyme Dcp1p. Decapping prevents the mRNA from being translated (Section 13.2.2) and so ends its

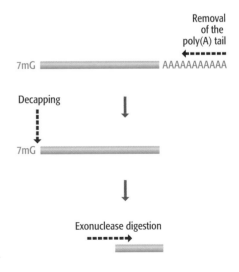

Figure 12.44 The deadenylation-dependent decapping pathway for degradation of an mRNA.

functional life. The mRNA then undergoes rapid exonuclease digestion from its 5′ end. Whether or not an individual mRNA is degraded is probably determined by the ability of Dcp1p to gain access to the cap structure, which in turn depends on the association between the cap and the proteins that bind to it in order to initiate translation (Section 13.2.2). Degradation is also influenced, at least with some yeast mRNAs, by sequences called **instability elements**, located within the transcript. The importance of these sequences has been demonstrated by experiments in which an element is artificially deleted, which leads to increased translation and reduced degradation of the mRNA.

The second well-studied system for degradation of eukaryotic mRNAs is called **nonsense-mediated RNA decay** (**NMD**), or **mRNA surveillance**. The first of these names gives a clue to its function, because in molecular biology jargon a "nonsense" sequence is a termination codon. NMD results in the specific degradation of mRNAs that have a termination codon at an incorrect position, either because the gene has undergone a mutation or as a result of incorrect splicing. The incorrect codon is thought to be detected by a "surveillance" mechanism that involves a complex of proteins which scans the mRNA and somehow is able to distinguish between the correct termination codon, located at the end of the coding region of the transcript, and one that is in the wrong place (Figure 12.45A). There are a number of conceptual difficulties with this model because it is not easy to imagine how the surveillance complex could discriminate between correct and incorrect termination codons. Current hypotheses are based on the demonstration that the correct termination codon is recognized as aberrant if the transcript is engineered so that an exon–intron boundary is placed downstream of this termination codon (Figure 12.45B). The surveillance enzymes may therefore use exon–intron boundaries as orientation positions in order to distinguish the correct termination codon, which is usually downstream from the last intron. Alternative schemes have also been proposed, in which importance is placed not on the position of the termination codon but on the precise nature of the events involved in termination of translation at a premature

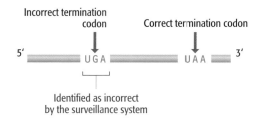

(A) mRNA surveillance can locate incorrect termination codons

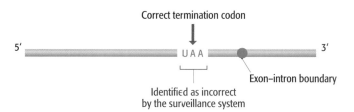

(B) The influence of an exon–intron boundary

Figure 12.45 mRNA surveillance.

stop codon compared with one that is at its correct position. Whatever the mechanism, identification of an incorrect termination codon induces cap cleavage and 5′→3′ exonuclease degradation, without prior removal of the poly(A) tail, by proteins different to those involved in deadenylation-dependent decapping. Although NMD is designed primarily to degrade mRNAs that have become altered by mutation or have been incorrectly spliced, there is evidence that the pathway is also responsible for degradation of normal mRNAs, but probably not in a way that leads to control over expression of any individual gene.

RNA silencing was first identified as a means of destroying invading viral RNA

The systems described above are used by eukaryotic cells to degrade endogenous mRNAs. For several years it has been known that eukaryotes also possess other RNA degradation mechanisms that protect their cells from attack by foreign RNAs such as the genomes of viruses. Originally called **RNA silencing**, this process is already familiar to us under its alternative name of **RNA interference**, as its underlying mechanism has been utilized by genome researchers as a means of inactivating selected genes in order to study their function (Section 5.2.2).

The target for RNA silencing must be double stranded, which excludes cellular mRNAs but encompasses viral genomes, many of which are either double-stranded RNA in their native state or replicate via a double-stranded RNA intermediate (Section 9.1.2). The double-stranded RNA is recognized by binding proteins that form an attachment site for a ribonuclease called **Dicer**, which cuts the molecule into **short interfering RNAs** (**siRNAs**) of 21–28 nucleotides in length (Figure 12.46). This inactivates the virus genome, but what if the virus genes have already been transcribed? If this has occurred then the harmful effects of the virus will already have been initiated and RNA silencing would appear to have failed in its attempt to protect the cell from damage. One of the more remarkable discoveries of recent years has revealed a second stage of the interference process that is directed specifically at the viral mRNAs. The siRNAs produced by cleavage of the viral genome are separated into individual strands, one strand of each siRNA subsequently base-pairing to any viral mRNAs that are present in the cell. The double-stranded regions that are formed are target sites for assembly of the **RNA induced silencing complex** (**RISC**), which includes an RNA-binding protein of the Argonaut family and a nuclease (which may or may not be Argonaut itself), which cleaves and hence silences the mRNA.

The work which resulted in the initial description of the molecular process underlying RNA interference was carried out in the late 1990s with *C. elegans*. Since then, RNA interference has been shown to occur in all eukaryotes, with a few exceptions including *S. cerevisiae*, and interference has been linked to various events that involve RNA degradation but which were previously thought to be unrelated. For example, the movement of some types of transposable element involves a double-stranded RNA intermediate which can be degraded by a process now known to be RNA interference. This is one way in which eukaryotes prevent the wholesale proliferation of transposons within their genomes. Genetic engineers had also been puzzled by the ability of some organisms, especially plants, to silence new genes that had been inserted into their genomes by cloning techniques. We now know

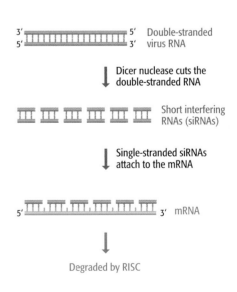

Figure 12.46 The RNA interference pathway.

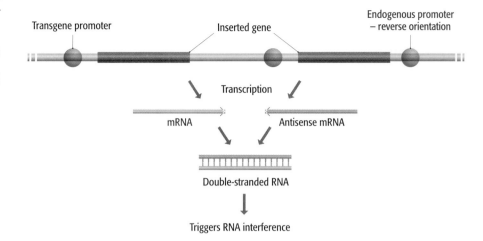

Figure 12.47 **RNA interference explains why transgenes are sometimes inactive.** For clarity, the mRNA and antisense RNA are shown being transcribed from different copies of the inserted transgene. They could also come from a single transgene that is transcribed from both its own promoter and an endogenous promoter.

that this type of silencing can occur if the transgene is inserted, by chance, upstream of a promoter that directs synthesis of an antisense RNA copy of all or part of the gene, this RNA then base-pairing with the sense mRNA produced from the transgene's own promoter to form a double-stranded RNA that triggers the RNA interference pathway (Figure 12.47). Other phenomena in diverse organisms, variously known as quelling, cosuppression, and posttranscriptional gene silencing, are all now known to be different guises of RNA interference.

MicroRNAs regulate genome expression by causing specific target mRNAs to be degraded

In many organisms, more than one type of Dicer protein has been identified. *Drosophila melanogaster*, for example, has two related Dicer enzymes, and *Arabidopsis thaliana* has four. The multiplicity of Dicer proteins with slightly different properties alerts us to the possibility that there might be additional RNA degradation processes related to, but perhaps different from, the form of RNA interference described above. It turns out that the second type of Dicer in *Drosophila* works not with the double-stranded RNAs produced by the pathway illustrated in Figure 12.46, but with **microRNAs** (miRNAs), which are coded by the fruit-fly genome and synthesized by RNA polymerase II. MicroRNAs are initially synthesized as precursor molecules called foldback RNAs, this name indicating that these RNAs can form intrastrand base pairs resulting in one or more internal hairpin structures (Figure 12.48). Within the nucleus these foldback RNAs are cut by the enzyme Drosha into individual hairpins that are transported into the cytoplasm. The double-stranded RNA component of the stem then stimulates the RNA interference pathway, with the second of the two *Drosophila* Dicer enzymes cleaving the molecule into miRNAs approximately 21 nucleotides in length. Each miRNA is complementary to part of a cellular mRNA and hence base-pairs with this target, stimulating assembly of a microribonucleoprotein (miRNP) complex which is functionally identical to the RISC and contains many of the same proteins. This leads to cleavage of the mRNA. Often the miRNA annealing site is present in the 3′ untranslated region of the target mRNA, sometimes in multiple copies (Figure 12.49). Cleavage by the miRNP therefore does not disrupt the coding region of the mRNA, but will lead to detachment of the poly(A) tail. This might interfere with the process for translation initiation, which involves the poly(A) tail (Section 13.2.2), or it might target the mRNA for degradation by the deadenylation-dependent decapping pathway. Whatever the precise mechanism, cleavage by the miRNP leads to the mRNA being silenced.

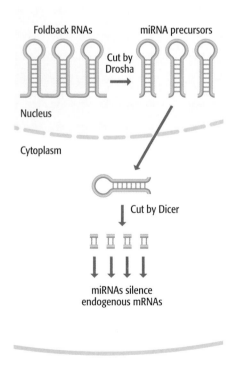

Foldback RNAs   miRNA precursors

Cut by Drosha →

Nucleus

Cytoplasm

Cut by Dicer

miRNAs silence endogenous mRNAs

Figure 12.48 **The microRNA interference pathway.**

The first miRNA silencing system to be characterized involved the *C. elegans* genes called *lin-4* and *let-7*, both of which code for foldback RNAs that generate miRNAs after cleavage by Dicer. A mutation in either of these two genes causes defects in the worm's development pathway, indicating that this type of RNA degradation is not simply a means of getting rid of unwanted or potentially harmful mRNAs, but instead plays a fundamental role in regulation of genome expression. Further support for this notion was provided by other studies of *C. elegans* miRNAs, which revealed that these molecules are involved in biological events as diverse as cell death, specification of neuron cell types, and control of fat storage. Genome analysis shows that most animals have the capacity to synthesize at least 100 to 200 different miRNAs, and possibly many more. Although as yet very few of the targets for these miRNAs have been identified, the miRNA system is clearly emerging as a wide-ranging and extremely important aspect of genome regulation. In the past, the focus has largely been on the way in which genome expression is regulated by proteins, and the discovery that RNA molecules might be equally important in this regard has resulted in a major shift in our perception of how control is exerted over the composition of a cell's proteome.

12.2.7 Transport of RNA within the eukaryotic cell

In a typical mammalian cell, about 14% of the total RNA is present in the nucleus. About 80% of this nuclear fraction is RNA that is being processed before leaving for the cytoplasm. The other 20% is snRNAs and snoRNAs, playing an active role in the processing events, at least some of these molecules having already been to the cytoplasm where they were coated with protein molecules before being transported back into the nucleus. In other words, eukaryotic RNAs are continually being moved from nucleus to cytoplasm and possibly back to the nucleus again.

The only way for RNAs to leave or enter the nucleus is via one of the many **nuclear pore complexes** that cover the nuclear membrane (Figure 12.50). Initially looked upon as little more than a hole in the membrane, pore complexes are now regarded as complex structures that play an active role in movement of molecules into and out of the nucleus. Small molecules can move unimpeded through a pore complex but RNAs and most proteins are too large to diffuse through unaided and so have to be transported across by an energy-dependent process. As in many biochemical systems, the energy is obtained by hydrolysis of one of the high-energy phosphate–phosphate bonds in a ribonucleotide triphosphate, in this case by converting GTP to GDP (other processes use ATP→ADP). Energy generation is carried out by a protein called Ran, and transport requires receptor proteins called **karyopherins**, or **exportins** and **importins** depending on the direction of their transport activity. There are at least 20 different human karyopherins, each responsible for the transport of a different class of molecule—mRNA, rRNA, and so on. Examples are exportin-t, which has been identified as the karyopherin for export of tRNAs in yeasts and mammals. Transfer RNAs are directly recognized by exportin-t, but other types of RNA are probably exported by protein-specific karyopherins which recognize the proteins bound to the RNA, rather than the RNA itself. This also appears to be the case for import of snRNA from cytoplasm to nucleus, which makes use of importin β, a component of one of the protein-transport pathways.

Export of mRNAs is triggered by completion of the splicing pathway, possibly through the action of the protein called Yra1p in yeast and Aly in animals.

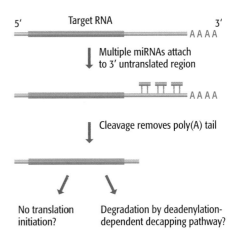

Figure 12.49 MicroRNA target sites are often in the 3′ untranslated region of the target mRNA.

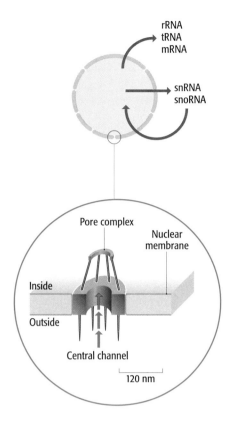

rRNA
tRNA
mRNA

snRNA
snoRNA

Pore complex

Nuclear
membrane

Inside

Outside

Central channel

120 nm

Figure 12.50 Eukaryotic RNAs must be transported through the nuclear pore complexes. In eukaryotes, rRNAs, tRNAs, and mRNAs are transported from the nucleus to the cytoplasm, where these molecules carry out their cellular functions. At least some of the snRNAs and snoRNAs are also transported to the cytoplasm, where they are coated with proteins before returning to the nucleus to carry out their roles in RNA processing. The nuclear pore is not simply a hole in the nuclear membrane. It contains a protein assembly comprising a ring embedded in the pore, with structures radiating into both the nucleus and the cytoplasm. Not shown in this diagram is the central channel complex, a 12 kDa protein that is thought to reside in the channel that connects the cytoplasm to the nucleus. There are thought to be approximately 3000 pores on the surface of a nucleus in an animal cell.

Once outside the nucleus, there are mechanisms that ensure that mRNAs are transported to their appropriate places in the cell. It is not known to what extent protein localization within the cell is due to translation of an mRNA at a specific position or to movement of the protein after it has been synthesized, but it is clear that at least some mRNAs are translated at defined places. For example, those mRNAs coding for proteins that are to be transferred into a mitochondrion are translated by ribosomes located on the surface of the organelle. It is assumed that protein "address tags" are attached to mRNAs in order to direct them to their correct locations after they are transported out of the nucleus, but very little is known about this process.

Summary

Structural studies are beginning to reveal the precise nature of the contacts formed between an RNA polymerase, the template DNA, and the RNA transcript during the elongation stage of transcription. The RNA polymerase does not synthesize its transcript at a constant rate. Instead, synthesis is discontinuous, with periods of rapid elongation interspersed by brief pauses during which the active site of the polymerase undergoes a slight structural rearrangement. Termination of bacterial transcripts can occur by either of two methods, one requiring an ancillary protein called Rho. Bacteria possess various mechanisms for regulating termination, either by reading through termination signals so downstream sequences are transcribed, which is central to expression of the λ genome, or by stopping transcription before a gene or operon is transcribed if the gene products are not required. Ribosomal and transfer RNAs in bacteria are initially synthesized as precursor molecules which are processed by cutting and trimming events to release the functional RNAs. These RNAs are also chemically modified at various nucleotide positions. A variety of enzymes are involved in the controlled degradation of bacterial RNAs. In eukaryotes, the mRNAs made by RNA polymerase II are capped by addition of 7-methylguanosine to the 5′ end, and polyadenylated at the 3′ end by addition of a series of adenine nucleotides. Many eukaryotic pre-mRNAs contain introns which are spliced from the transcripts by a complex pathway involving the small nuclear ribonucleoproteins acting in a structure called the spliceosome. Alternative splicing pathways enable more than one protein to be synthesized from a single gene, and are important in various physiological processes including sex determination in *Drosophila melanogaster*. Eukaryotic pre-rRNAs and pre-tRNAs may also contain introns. Those in pre-rRNAs are self-splicing and hence are examples of ribozymes. Eukaryotic rRNA is chemically modified by a process in which small nucleolar RNAs act as guides to indicate the positions at which modifications must be made. Chemical modification of mRNA is less common but results in alteration of the coding specification, as occurs during synthesis of the liver version and intestinal version of apolipoprotein B in mammals. Eukaryotes have diverse mechanisms for RNA degradation, including the process called RNA silencing, or RNA interference, in which small interfering RNAs and microRNAs degrade and hence silence invading viral RNAs and mRNAs transcribed from cellular genes whose products are no longer required.

Multiple Choice Questions

12.1.* Approximately how many base pairs form the attachment between the DNA template and RNA transcript during transcription in prokaryotes?

a. 8.

b. 12–14.

c. 30.

d. The entire RNA molecule remains base-paired to the template until transcription is finished.

12.2. Which factor is thought to be most important in determining whether a bacterial RNA polymerase continues or terminates transcription?

a. Nucleotide concentration.

b. Structure of the polymerase.

c. Methylation of terminator sequences.

d. Thermodynamic events.

12.3.* What is the role of the Rho protein in termination of transcription?

a. It is a helicase that actively breaks base pairs between the template and transcript.

b. It is a DNA-binding protein that blocks the movement of RNA polymerase down the template.

c. It is a subunit of RNA polymerase that binds to RNA hairpins and stalls transcription.

d. It is a nuclease that degrades the 3′ ends of RNA transcripts.

12.4. Antitermination is involved in regulation of which of the following?

a. Operons encoding enzymes involved in the biosynthesis of amino acids, with regulation dependent on the concentration of the amino acids.

b. Operons encoding enzymes involved in the degradation of metabolites, regulation dependent on the presence of the metabolite.

c. Genes present in the upstream region of the operon.

d. Genes present in the downstream region of the operon.

12.5.* What is the major transcriptional change that occurs during the stringent response in *E. coli*?

a. Transcription rates are increased for most genes.

b. Transcription rates are increased only for the amino acid biosynthesis operons.

c. Transcription rates are decreased for most genes.

d. Transcription rates are decreased only for the amino acid biosynthesis operons.

12.6. Which of the following is NOT thought to be a reason why tRNA nucleotides are modified?

a. To achieve stronger base pairing between ribonucleotides.

b. To assist the recognition of different tRNA molecules by aminoacyl-tRNA synthetases.

c. To increase the range of interactions that can occur between tRNAs and codons.

d. To enable a single tRNA molecule to recognize more than one codon.

12.7.* Promoter escape in eukaryotes is associated with:

a. Transition of RNA polymerase from the preinitiation complex to a complex synthesizing RNA.

b. Movement of RNA polymerase away from the promoter region and its commitment to making the RNA transcript.

c. Release of RNA polymerase from the preinitiation complex so that no transcript is synthesized.

d. Termination of transcription caused by the dissociation of RNA polymerase from the template DNA.

12.8. How is the lariat structure formed during splicing of a GU–AG intron?

a. After cleavage of the 5′ splice site, a new phosphodiester bond is formed between the 5′ nucleotide and the 2′ carbon of the nucleotide at the 3′ splice site.

b. After cleavage of the 5′ splice site, a new phosphodiester bond is formed between the 5′ nucleotide and the 2′ carbon of an internal adenosine.

c. After cleavage of the 3′ splice site, a new phosphodiester bond is formed between the 5′ nucleotide and the 2′ carbon of the nucleotide at the 5′ splice site.

d. After cleavage of the 3′ splice site, a new phosphodiester bond is formed between the 5′ nucleotide and the 2′ carbon of an internal adenosine.

12.9.* What are cryptic splice sites?

a. These are splice sites that are used in some cells, but not in others.

b. These are splice sites that are always used.

c. These are sites that are involved in alternative splicing, resulting in the removal of exons from some mRNA molecules.

d. These are sequences within exons or introns that resemble consensus splicing signals, but are not true splice sites.

continued …

Multiple Choice Questions (continued)

12.10. Which statement correctly describes trans-splicing?

  a. The order of exons within an RNA transcript is rearranged to yield a different mRNA sequence.

  b. Exons are deleted from some RNA transcripts but not others.

  c. Intron sequences are not removed from RNA transcripts and are translated into proteins.

  d. Exons from different RNA transcripts are joined together.

12.11.* Group I introns are remarkable because:

  a. They are spliced by external RNA molecules without protein involvement.

  b. They are spliced by protein molecules in the absence of external RNA molecules.

  c. They are autocatalytic.

  d. They are only present in mitochondrial and chloroplast genomes.

12.12. The chemical modification of eukaryotic rRNA molecules takes place in the:

  a. Cytoplasm.

  b. Endoplasmic reticulum.

  c. Nuclear envelope.

  d. Nucleolus.

12.13.* Which of the following is an example of RNA editing?

  a. Removal of introns from an RNA transcript.

  b. Degradation of an RNA molecule by nucleases.

  c. Alteration of the nucleotide sequence of an RNA molecule.

  d. Capping of the 5′ end of an RNA transcript.

12.14. Nonsense-mediated RNA decay (NMD) is a system for the degradation of eukaryotic mRNA molecules with what features?

  a. NMD degrades mRNA molecules with stop codons at incorrect positions.

  b. NMD degrades mRNA molecules that encode nonfunctional proteins.

  c. NMD degrades mRNA molecules that lack a start codon.

  d. NMD degrades mRNA molecules that lack a stop codon.

12.15.* Which of the following describes RNA interference?

  a. Antisense RNA molecules block the translation of mRNA molecules.

  b. Double-stranded RNA molecules are bound by proteins that block their translation.

  c. Double-stranded RNA molecules are cleaved by a nuclease into short interfering RNA molecules.

  d. Short interfering RNA molecules bind to the ribosome to prevent the translation of viral mRNAs.

12.16. How are RNA molecules transported out of the nucleus?

  a. Passive diffusion through the membrane.

  b. Through membrane pores in an energy-independent process.

  c. Through membrane pores in an energy-dependent process.

  d. Through a channel in the membrane that leads to the endoplasmic reticulum.

Short Answer Questions

12.1.* Describe the process of Rho-dependent termination of transcript synthesis in *E. coli*.

12.2. How are antiterminator proteins thought to prevent dissociation of RNA polymerase at termination signals?

12.3.* Why is attenuation absent in eukaryotic organisms?

12.4. Describe the factors that determine which of two hairpin loops forms during transcription of the region upstream of *trpE* in the tryptophan operon. How do these hairpins function to regulate the expression of the tryptophan operon?

12.5.* How are mature tRNA molecules processed from pre-tRNAs in *E. coli*? What enzymes are involved in this process?

Short Answer Questions (continued)

*Answers to odd-numbered questions can be found in the Appendix

12.6. Why does RNA degradation play an important role in the regulation of genome expression?

12.7. * How do RNA exonucleases degrade a bacterial mRNA from its 3′ end if the hairpin structure that induced termination of transcription is present, thereby blocking the activity of these enzymes?

12.8. What are the common modifications made to transcripts of protein-coding genes in eukaryotes?

12.9. * How is the type 0 cap structure attached to a eukaryotic mRNA?

12.10. Discuss how several hundred distinct mRNA molecules can be produced from a single eukaryotic gene such as the human *slo* gene.

12.11. * What is the role of the small nucleolar RNA (snoRNA) molecules in the modification of eukaryotic pre-rRNA molecules?

12.12. How is it possible for microRNAs to regulate eukaryotic genome expression by binding to the 3′ untranslated end of an mRNA molecule?

In-depth Problems

*Guidance to odd-numbered questions can be found in the Appendix

12.1. * "Current thinking views transcription as a discontinuous process, with the polymerase pausing regularly and making a 'choice' between continuing elongation by adding more ribonucleotides to the transcript, or terminating by dissociating from the template. Which choice is selected depends on which alternative is more favorable in thermodynamic terms" (page 337). Evaluate this view of transcription.

12.2. To what extent has the study of AU–AC introns provided insights into the details of GU–AG intron splicing?

12.3. * Discontinuous genes are common in higher organisms but virtually absent in bacteria. Discuss the possible reasons for this.

12.4. Discuss the issues raised by the discovery of RNA editing.

12.5. * The existence of ribozymes is looked upon as evidence that RNA evolved before proteins and therefore at one time, during the earliest stages of evolution, all enzymes were made of RNA. Assuming that this hypothesis is correct, explain why some ribozymes persist to the present day.

continued ...

Figure Tests

12.1.* Discuss the mechanism of transcription termination at an intrinsic terminator in bacteria.

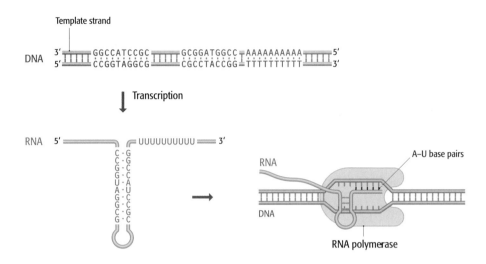

12.2. Discuss the mechanism for the polyadenylation of mRNA molecules in eukaryotes.

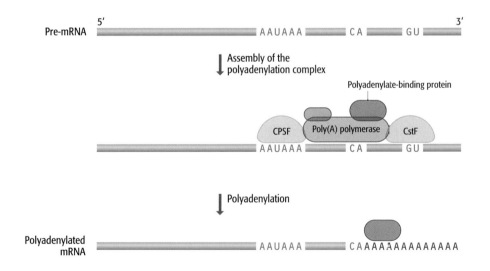

Figure Tests (continued)

12.3.* Discuss the steps in the removal of the intron shown in the figure.

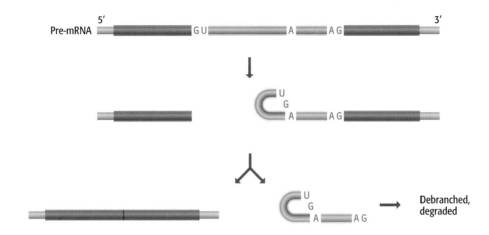

12.4. Discuss the deadenylation pathway for the degradation of eukaryotic mRNAs. At what point are eukaryotic mRNA molecules no longer translated?

12.5.* Name the pathway and discuss the processes shown in the figure.

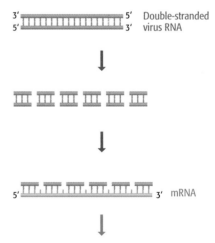

Further Reading

RNA synthesis by the bacterial RNA polymerase

Klug, A. (2001) A marvellous machine for making messages. *Science* **292:** 1844–1846.

Korzheva, N., Mustaev, A., Kozlov, M., Malhotra, A., Nikiforov, V., Goldfarb, A. and Darst, S.A. (2000) A structural model of transcription elongation. *Science* **289:** 619–625.

Toulokhonov, I., Artsimovitch, I. and Landick, R. (2001) Allosteric control of RNA polymerase by a site that contacts nascent RNA hairpins. *Science* **292:** 730–733. *A model for termination of transcription.*

Control of mRNA synthesis in bacteria

Henkin, T.M. (1996) Control of transcription termination in prokaryotes. *Annu. Rev. Genet.* **30:** 35–57. *A detailed account of antitermination and attenuation.*

Losick, R.L. and Sonenshein, A.L. (2001) Turning gene regulation on its head. *Science* **293:** 2018–2019. *Describes the attenuation systems at the tryptophan operons of* E. coli *and* B. subtilis.

Nickels, B.E. and Hochschild, A. (2004) Regulation of RNA polymerase through the secondary channel. *Cell* **118:** 281–284. *The mode of action of transcript cleavage proteins.*

Synthesis and processing of eukaryotic mRNA

Arndt, K.M. and Kane, C.M. (2003) Running with RNA polymerase: eukaryotic transcript elongation. *Trends Genet.* **19:** 543–550. *Includes details of the roles of elongation factors.*

Conaway, J.W., Shilatifard, A., Dvir, A. and Conaway, R.C. (2000) Control of elongation by RNA polymerase II. *Trends Biochem. Sci.* **25:** 375–380.

Conaway, R.C., Kong, S.E. and Conaway, J.W. (2003) TFIIS and GreB: two like-minded transcription elongation factors with sticky fingers. *Cell* **114:** 272–273. *A eukaryotic transcript cleavage protein.*

Cougot, N., van Dijk, E., Babajko, S. and Seeraphin, B. (2004) 'Cap-tabolism'. *Trends Biochem. Sci.* **29:** 436–444. *mRNA capping.*

Manley, J.L. and Takagaki, Y. (1996) The end of the message – another link between yeast and mammals. *Science* **274:** 1481–1482. *Polyadenylation.*

Proudfoot, N. (2000) Connecting transcription to messenger RNA processing. *Trends Biochem. Sci.* **25:** 290–293.

Shilatifard, A., Conaway, R.C. and Conaway, J.W. (2003) The RNA polymerase II elongation complex. *Annu. Rev. Biochem.* **72:** 693–716. *Includes details of elongation factors.*

Studitsky, V.M., Walter, W., Kireeva, M., Kashlev, M. and Felsenfeld, G. (2004) Chromatin remodeling by RNA polymerases. *Trends Biochem. Sci.* **29:** 127–135. *Possible ways by which RNA polymerases deal with nucleosomes attached to the DNA being transcribed.*

Splicing of pre-mRNA

Black, D.L. (2003) Mechanisms of alternative pre-messenger RNA splicing. *Annu. Rev. Biochem.* **72:** 291–336.

Blencowe, B.J. (2000) Exonic splicing enhancers: mechanism of action, diversity and role in human genetic diseases. *Trends Biochem. Sci.* **25:** 106–110.

Corden, J.L. and Patturajan, M. (1997) A CTD function linking transcription to splicing. *Trends Biochem. Sci.* **22:** 413–416.

Graveley, B.R. (2001) Alternative splicing: increasing diversity in the proteomic world. *Trends Genet.* **17:** 100–107.

Stetefeld, J. and Ruegg, M.A. (2005) Structural and functional diversity generated by alternative mRNA splicing. *Trends Biochem. Sci.* **30:** 515–521.

Tarn, W.-Y. and Steitz, J.A. (1997) Pre-mRNA splicing: the discovery of a new spliceosome doubles the challenge. *Trends Biochem. Sci.* **22:** 132–137. *AU–AC introns.*

Valcärcel, J. and Green, M.R. (1996) The SR protein family: pleiotropic functions in pre-mRNA splicing. *Trends Biochem. Sci.* **21:** 296–301.

Other types of intron

Burke, J.M., Belfort, M., Cech, T.R., Davies, R.W., Schweyen, R.J., Shub, D.A., Szostak, J.W. and Tabak, H.F. (1987) Structural conventions for Group I introns. *Nucleic Acids Res.* **15:** 7217–7221. *The nomenclature for the two-dimensional representation of the Group I intron structure.*

Cech, T.R. (1990) Self-splicing of group I introns. *Annu. Rev. Biochem.* **59:** 543–568. *Written by one of the discoverers of autocatalytic RNA.*

Copertino, D.W. and Hallick, R.B. (1993) Group II and Group III introns of twintrons: potential relationships with nuclear pre-mRNA introns. *Trends Biochem. Sci.* **18:** 467–471.

Lambowitx, A.M. and Zimmerly, S. (2004) Mobile Group II introns. *Annu. Rev. Genet.* **38:** 1–35.

Lykke-Andersen, J., Aagaard, C., Semionenkov, M. and Garrett, R.A. (1997) Archaeal introns: splicing, intercellular mobility and evolution. *Trends Biochem. Sci.* **22:** 326–331.

Transcription by RNA polymerases I and III

Geiduschek, E.P. and Kassavetis, G.A. (2001) The RNA polymerase III transcription apparatus. *J. Mol. Biol.* **310:** 1–26.

Reeder, R.H. and Lang, W.H. (1997) Terminating transcription in eukaryotes: lessons learned from RNA polymerase I. *Trends Biochem. Sci.* **22:** 473–477.

Russell, J. and Zomerdijk, J.C.B.M. (2005) RNA-polymerase-I-directed rDNA transcription, life and works. *Trends Biochem. Sci.* **30:** 87–96.

Processing of functional RNA in bacteria and eukaryotes

Tollervey, D. (1996) Small nucleolar RNAs guide ribosomal RNA methylation. *Science* **273:** 1056–1057.

Venema, J. and Tollervey, D. (1999) Ribosome synthesis in *Saccharomyces cerevisiae. Annu. Rev. Genet.* **33:** 261–311. *Extensive details on rRNA processing.*

RNA editing

Bass, B.L. (1997) RNA editing and hypermutation by adenosine deamination. *Trends Biochem. Sci.* **22:** 157–162.

Bourara, K., Litvak, S. and Araya, A. (2000) Generation of G-to-A and C-to-U changes in HIV-1 transcripts by RNA editing. *Science* **289:** 1564–1566.

Gott, J.M. and Emeson, R.B. (2000) Functions and mechanisms of RNA editing. *Annu. Rev. Genet.* **34:** 499–531.

Stuart, K.D., Schnaufer, A., Ernst, N.L. and Panigrahi, A.K. (2005) Complex management: RNA editing in trypanosomes. *Trends Biochem. Sci.* **30:** 97–105.

RNA degradation in bacteria and eukaryotes

Carpousis, A.J., Vanzo, N.F. and Raynal, L.C. (1999) mRNA degradation: a tale of poly(A) and multiprotein machines. *Trends Genet.* **15:** 24–28.

Coller, J. and Parker, R. (2004) Eukaryotic mRNA decapping. *Annu. Rev. Biochem.* **73:** 861–890.

Hilleren, P., McCarthy, T., Rosbach, M., Parker, R. and Jensen, T.H. (2001) Quality control of mRNA 3′-end processing is linked to the nuclear exosome. *Nature* **413:** 538–542.

Singh, G. and Lykke-Andersen, J. (2003) New insights into the formation of active nonsense-mediated decay complexes. *Trends Biochem. Sci.* **28:** 464–466.

RNA silencing

Mello, C.C. and Conte, D. (2004) Revealing the world of RNA interference. *Nature* **431:** 338–342.

Sontheimer, E.J. and Carthew, R.W. (2005) Silence from within: endogenous siRNAs and miRNAs. *Cell* **122:** 9–12.

Zamore, P.D. and Haley, B. (2005) Ribo-genome: the big world of small RNAs. *Science* **309:** 1519–1524.

RNA transport

Fahrenkrog, B., Köser, J. and Aebi, U. (2004) The nuclear pore complex: a jack of all trades. *Trends Biochem. Sci.* **29:** 175–182.

Nigg, E.A. (1997) Nucleocytoplasmic transport: signals, mechanisms and regulation. *Nature* **386:** 779–787.

Weis, K. (1998) Importins and exportins: how to get in and out of the nucleus. *Trends Biochem. Sci.* **23:** 185–189.

Synthesis and Processing of the Proteome

13

When you have read Chapter 13, you should be able to:

Draw the general structure of a tRNA and explain how this structure enables the tRNA to play both a physical and an informational role during protein synthesis.

Describe how an amino acid becomes attached to a tRNA, and outline the processes that ensure that combinations are formed between the correct pairs of amino acids and tRNAs.

Explain how codons and anticodons interact, and discuss the influence of wobble on this interaction.

Outline the techniques that have been used to study the structure of the ribosome, and summarize the information that has resulted from these studies.

Give a detailed description of the process of translation in bacteria and eukaryotes, with emphasis on the roles of the various translation factors.

Describe the experimental evidence that has led to the conclusion that peptidyl transferase is a ribozyme.

Explain how translation is regulated and give an outline of the unusual events, such as frameshifting, that can occur during the elongation phase.

Explain why posttranslational processing of proteins is an important component of the genome expression pathway, and describe the key features of protein folding, protein processing by proteolytic cleavage and chemical modification, and intein splicing.

Describe the major processes responsible for protein degradation in bacteria and eukaryotes.

The end result of genome expression is the proteome, the collection of functioning proteins synthesized by a living cell. The identity and relative abundance of the individual proteins in a proteome represents a balance between the synthesis of new proteins and the degradation of existing ones. The biochemical capabilities of the proteome can also be changed by chemical modification and other processing events. The combination of synthesis, degradation, and modification/processing enables the proteome to meet the changing requirements of the cell and to respond to external stimuli.

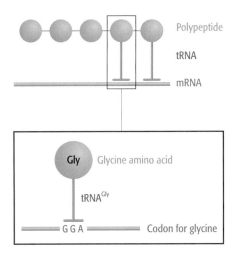

Figure 13.1 The adaptor role of tRNA in translation. The top drawing shows the physical role of tRNA, forming an attachment between the polypeptide and the mRNA. The lower drawing shows the informational link, the tRNA carrying the amino acid specified by the codon to which it attaches.

In this chapter we will study the synthesis, processing, and degradation of the components of the proteome. To understand protein synthesis we will first examine the role of tRNAs in decoding the genetic code and then investigate the events, occurring at the ribosome, that result in polymerization of amino acids into polypeptides. The ribosomal events are sometimes looked upon as the final stage in expression of the genome but the polypeptide that is initially synthesized is inactive until it has been folded, and may also have to undergo cutting and chemical modification before it becomes functional. We will study these processing events in Section 13.3. At the end of the chapter we will investigate how the cell degrades proteins that it no longer requires.

13.1 The Role of tRNA in Protein Synthesis

Transfer RNAs play the central role in translation. They are the adaptor molecules, whose existence was predicted by Francis Crick in 1956, which form the link between the mRNA and the polypeptide that is being synthesized. This is both a *physical* link, tRNAs binding to both the mRNA and the growing polypeptide, and an *informational* link, tRNAs ensuring that the polypeptide being synthesized has the amino acid sequence that is denoted, via the genetic code, by the sequence of nucleotides in the mRNA (Figure 13.1). To understand how tRNAs play this dual role we must examine **aminoacylation**, the process by which the correct amino acid is attached to each tRNA, and **codon–anticodon recognition**, the interaction between tRNA and mRNA.

13.1.1 Aminoacylation: the attachment of amino acids to tRNAs

Bacteria contain 30–45 different tRNAs and eukaryotes have up to 50 tRNAs. As only 20 amino acids are designated by the genetic code, this means that all organisms have at least some **isoaccepting tRNAs**, different tRNAs that are specific for the same amino acid. The terminology used when describing tRNAs is to indicate the amino acid specificity with a superscript suffix, using the numbers 1, 2, and so on, to distinguish different isoacceptors: for example, two tRNAs specific for glycine would be written as tRNA^{Gly1} and tRNA^{Gly2}.

All tRNAs have a similar structure

The smallest tRNAs are only 74 nucleotides in length, and the largest are rarely more than 90 nucleotides. Because of their small size, and because it is possible to purify individual tRNAs, they were among the first nucleic acids to be sequenced, way back in 1965 by Robert Holley and colleagues. The sequences revealed one unexpected feature, that as well as the standard RNA nucleotides (A, C, G, and U), tRNAs contain a number of modified nucleotides, 5–10 in any particular tRNA, with over 50 different modifications known altogether (Section 12.1.3).

Examination of the first tRNA sequence, for tRNA^{Ala} of *Saccharomyces cerevisiae*, showed that the molecule could adopt various base-paired secondary structures. After more tRNAs had been sequenced, it became clear that one particular structure could be taken up by all of them. This is the **cloverleaf** (Figure 13.2), and has the following features:

- The **acceptor arm** is formed by seven base pairs between the 5′ and 3′ ends of the molecule. The amino acid is attached to the extreme 3′ end of

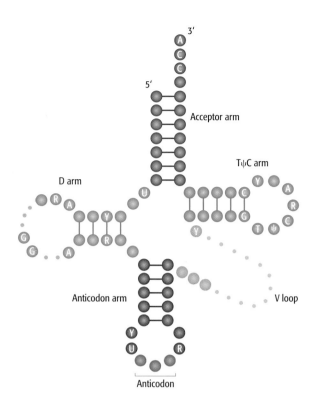

Figure 13.2 The cloverleaf structure of a tRNA. The tRNA is drawn in the conventional cloverleaf structure, with the different components labeled. Invariant nucleotides (A, C, G, T, U, Ψ, where Ψ = pseudouridine) and semi-invariant nucleotides (abbreviations: R, purine; Y, pyrimidine) are indicated. Optional nucleotides not present in all tRNAs are shown as smaller dots. The standard numbering system places position 1 at the 5′ end and position 76 at the 3′ end; it includes some but not all of the optional nucleotides. The invariant and semi-invariant nucleotides are at positions 8, 11, 14, 15, 18, 19, 21, 24, 32, 33, 37, 48, 53, 54, 55, 56, 57, 58, 60, 61, 74, 75, and 76. The nucleotides of the anticodon are at positions 34, 35, and 36.

the tRNA, to the adenosine of the invariant CCA terminal sequence (Section 12.1.3).

- The **D arm**, named after the modified nucleoside dihydrouridine (see Figure 12.18), which is always present in this structure.

- The **anticodon arm** contains the triplet of nucleotides called the **anticodon** which base pair with the mRNA during translation.

- The **V loop** contains 3–5 nucleotides in Class 1 tRNAs or 13–21 nucleotides in Class 2 tRNAs.

- The **TΨC arm**, named after the sequence thymidine–pseudouridine–cytidine, which is always present.

The cloverleaf structure can be formed by virtually all tRNAs, the main exceptions being the tRNAs used in vertebrate mitochondria, which are coded by the mitochondrial genome and which sometimes lack parts of the structure. An example is the human mitochondrial tRNA^Ser, which has no D arm. As well as the conserved secondary structure, the identities of nucleotides at some positions are completely invariant (always the same nucleotide) or semi-invariant (always a purine or always a pyrimidine), and the positions of the modified nucleotides are almost always the same.

Many of the invariant nucleotide positions are important in the tertiary structure of tRNA. X-ray crystallography studies have shown that nucleotides in the D and TΨC loops form base pairs that fold the tRNA into a compact L-shaped structure (Figure 13.3). Each arm of the L-shape is approximately 7 nm long and 2 nm in diameter, with the amino acid binding site at the end of one arm and the anticodon at the end of the other. The additional base pairing means that the base stacking (see Section 1.1.2) is almost continuous from one end of the tRNA to the other, providing stability to the structure.

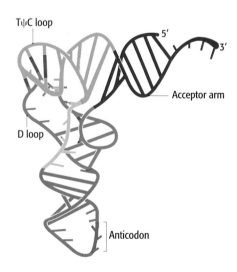

Figure 13.3 The three-dimensional structure of a tRNA. Additional base pairs, shown in black and mainly between the D and TΨC loops, fold the cloverleaf structure into this L-shaped configuration. Depending on its sequence, the V loop might also form interactions with the D arm, as indicated by the black lines. The color scheme is the same as in Figure 13.2.

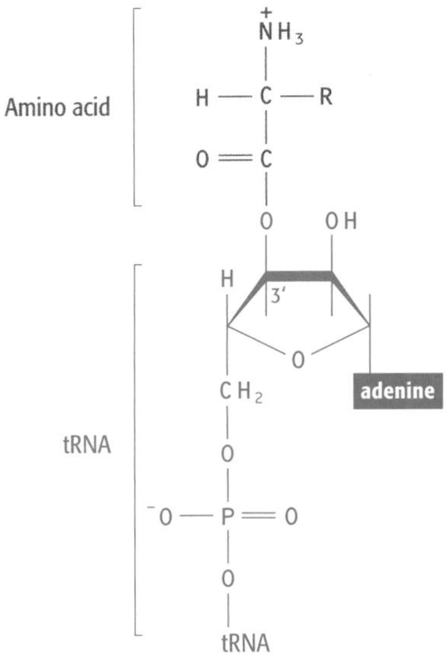

Figure 13.4 Aminoacylation of a tRNA.
The result of aminoacylation by a Class II aminoacyl-tRNA synthetase is shown, the amino acid being attached via its –COOH group to the 3′–OH of the terminal nucleotide of the tRNA. A Class I aminoacyl-tRNA synthetase attaches the amino acid to the 2′–OH group.

Aminoacyl-tRNA synthetases attach amino acids to tRNAs

The attachment of amino acids to tRNAs—"charging" in molecular biology jargon—is the function of the group of enzymes called **aminoacyl-tRNA synthetases**. The chemical reaction that results in aminoacylation occurs in two steps. An activated amino acid intermediate is first formed by reaction between the amino acid and ATP, and then the amino acid is transferred to the 3′ end of the tRNA, the link being formed between the –COOH group of the amino acid and the –OH group attached to either the 2′ or 3′ carbon on the sugar of the last nucleotide, which is always an A (Figure 13.4).

With a few exceptions, organisms have 20 aminoacyl-tRNA synthetases, one for each amino acid. This means that groups of isoaccepting tRNAs are aminoacylated by a single enzyme. Although the basic chemical reaction is the same for each amino acid, the 20 aminoacyl-tRNA synthetases fall into two distinct groups, Class I and Class II, with several important differences between them (Table 13.1). In particular, Class I enzymes attach the amino acid to the 2′–OH group of the terminal nucleotide of the tRNA, whereas Class II enzymes attach the amino acid to the 3′–OH group.

Aminoacylation must be carried out accurately: the correct amino acid must be attached to the correct tRNA if the rules of the genetic code are to be followed during protein synthesis. It appears that an aminoacyl-tRNA synthetase has high fidelity for its tRNA, the result of an extensive interaction between the two, covering some 25 nm^2 of surface area and involving the acceptor arm and anticodon loop of the tRNA, as well as individual nucleotides in the D and TΨC arms. The interaction between enzyme and amino acid is, of necessity, less extensive, amino acids being much smaller than tRNAs, and presents greater problems with regard to specificity because several pairs of amino acids are structurally similar. Errors do therefore occur, at a very low rate for most amino acids but possibly as frequently as one aminoacylation in 80 for difficult pairs such as isoleucine and valine. Most errors are corrected by the aminoacyl-tRNA synthetase itself, by an editing process that is distinct from aminoacylation, involving different contacts with the tRNA.

In most organisms, aminoacylation is carried out by the process just described, but a few unusual events have been documented. These include a

Table 13.1 Features of aminoacyl-tRNA synthetases

Feature	Class I enzymes	Class II enzymes
Structure of the enzyme active site	Parallel β-sheet	Antiparallel β-sheet
Interaction with the tRNA	Minor groove of the acceptor arm	Major groove of the acceptor arm
Orientation of the bound tRNA	V loop faces away from the enzyme	V loop faces the enzyme
Amino acid attachment	To the 2′–OH of the terminal nucleotide of the tRNA	To the 3′–OH of the terminal nucleotide of the tRNA
Enzymes for	Arg, Cys, Gln, Glu, Ile, Leu, Lys*, Met, Trp, Tyr, Val	Ala, Asn, Asp, Gly, His, Lys*, Phe, Pro, Thr, Ser

* The aminoacyl-tRNA synthetase for lysine is a Class I enzyme in some archaea and bacteria and a Class II enzyme in all other organisms.

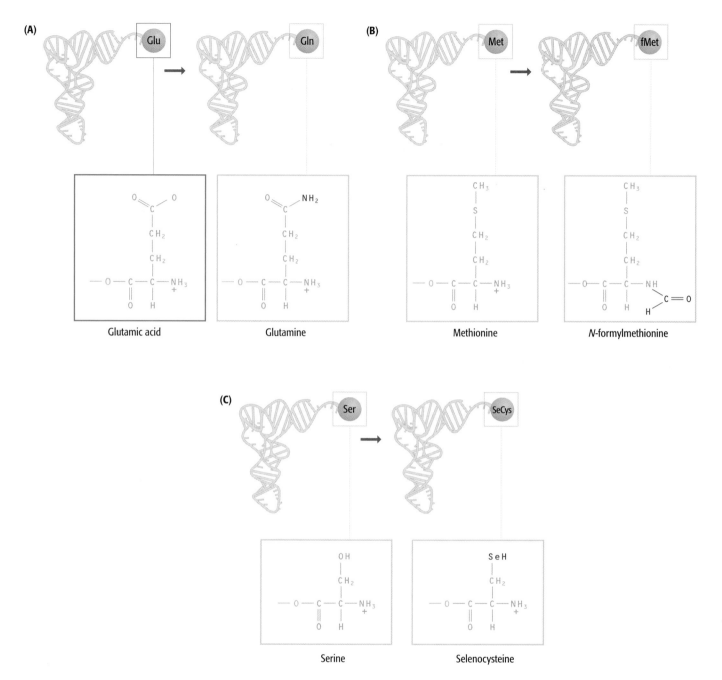

Figure 13.5 Unusual types of aminoacylation. (A) In some bacteria, tRNA^{Gln} is aminoacylated with glutamic acid, which is then converted to glutamine by transamidation. (B) The special tRNA used in initiation of translation in bacteria is aminoacylated with methionine, which is then converted to *N*-formylmethionine. (C) tRNA^{SeCys} in various organisms is initially aminoacylated with serine.

number of instances where the aminoacyl-tRNA synthetase attaches the incorrect amino acid to a tRNA, this amino acid subsequently being transformed into the correct one by a second, separate chemical reaction. This was first discovered in the bacterium *Bacillus megaterium* for synthesis of glutamine–tRNA^{Gln} (i.e., glutamine attached to its tRNA). This aminoacylation is carried out by the enzyme responsible for synthesis of glutamic acid–tRNA^{Glu}, and initially results in attachment of a glutamic acid to the tRNA^{Gln} (Figure 13.5A). This glutamic acid is then converted to glutamine by transamidation catalyzed by a second enzyme. The same process is used by various other bacteria (although not *Escherichia coli*) and by the archaea. Some archaea also use transamidation to synthesize asparagine–tRNA^{Asn} from aspartic acid–tRNA^{Asn}. In both of these cases, the amino acid that is synthesized by the modification process is one of the 20 that are specified by the genetic code. There are also two examples where the modification results in an unusual

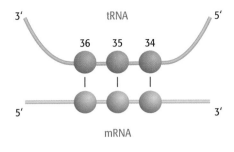

Figure 13.6 The interaction between a codon and an anticodon. The numbers indicate the nucleotide positions in the tRNA (see Figure 13.2).

amino acid. The first example is the conversion of methionine to *N*-formyl-methionine (Figure 13.5B), producing the special aminoacyl-tRNA used in initiation of bacterial translation (Section 13.2.2). The second example occurs in both prokaryotes and eukaryotes and results in synthesis of selenocysteine, which is specified in a context-dependent manner by some 5′–UGA–3′ codons (Section 1.3.2). These codons are recognized by a special tRNA^SeCys, but there is no aminoacyl-tRNA synthetase that is able to attach selenocysteine to this tRNA. Instead, the tRNA is aminoacylated with a serine by the seryl-tRNA synthetase, and then modified by replacement of the –OH group of the serine with an –SeH, to give selenocysteine (Figure 13.5C). The second context-dependent codon reassignment that has been discovered is the occasional use of 5′–UAG–3′ to code for pyrrolysine in archaea (Section 1.3.2). This does not involve modification of a precharged tRNA. Instead, there is a specific aminoacyl-tRNA synthetase that directly attaches pyrrolysine to the tRNA^pLys.

13.1.2 Codon–anticodon interactions: the attachment of tRNAs to mRNA

Aminoacylation represents the first level of specificity displayed by a tRNA. The second level is the specificity of the interaction between the anticodon of the tRNA and the mRNA being translated. This specificity ensures that protein synthesis follows the rules of the genetic code (see Figure 1.20).

In principle, codon–anticodon recognition is a straightforward process involving base pairing between the anticodon of the tRNA and a codon in the mRNA (Figure 13.6). The specificity of aminoacylation ensures that the tRNA carries the amino acid denoted by the codon that it pairs with, and the ribosome controls the topology of the interaction in such a way that only a single triplet of nucleotides is available for pairing. Because base-paired polynucleotides are always antiparallel, and because the mRNA is read in the 5′→3′ direction, the first nucleotide of the codon pairs with nucleotide 36 of the tRNA, the second with nucleotide 35, and the third with nucleotide 34.

In practice, codon recognition is complicated by the possibility of **wobble**. This is another of the principles of gene expression originally proposed by Crick and subsequently shown to be correct. Because the anticodon is in a loop of RNA, the triplet of nucleotides is slightly curved (see Figures 13.2 and 13.3) and so cannot make an entirely uniform alignment with the codon. As a result, a nonstandard base pair can form between the third nucleotide of the codon and the first nucleotide (number 34) of the anticodon. This is called "wobble." A variety of pairings is possible, especially if the nucleotide at position 34 is modified. In bacteria, the two main features of wobble are:

- **G–U base pairs** are permitted. This means that an anticodon with the sequence 3′–♦♦G–5′ can base-pair with both 5′–♦♦C–3′ and 5′–♦♦U–3′. Similarly, the anticodon 3′–♦♦U–5′ can base-pair with both 5′–♦♦A–3′ and 5′–♦♦G–3′. The consequence is that rather than needing a different tRNA for each codon, the four members of a codon family (e.g., 5′–GCN–3′, all coding for alanine) can be decoded by just two tRNAs (Figure 13.7A).

- **Inosine**, abbreviated to I, is a modified purine (see Figure 12.18) that can base-pair with A, C, and U. Inosine can only occur in the tRNA because the mRNA is not modified in this way. The triplet 3′–UAI–5′ is sometimes used as the anticodon in a tRNA^Ile molecule because it pairs with 5′–AUA–3′, 5′–AUC–3′, and 5′–AUU–3′ (Figure 13.7B), which form the three-codon family for this amino acid in the standard genetic code.

(A) G–U base-pairing

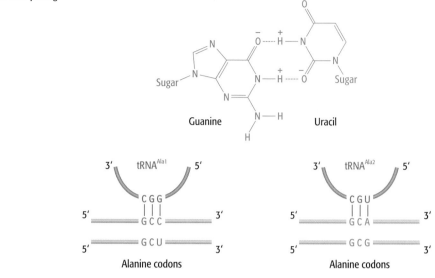

Guanine

Uracil

tRNA^{Ala1}

Alanine codons

tRNA^{Ala2}

Alanine codons

(B) Inosine base-pairs with A, C, and U

Inosine   Adenine

Inosine   Cytosine

Inosine   Uracil

Wobble reduces the number of tRNAs needed in a cell by enabling one tRNA to read two or possibly three codons. Hence bacteria can decode their mRNAs with as few as 30 tRNAs. Eukaryotes also make use of wobble but in a restricted way. The human genome, which in this regard is fairly typical of higher eukaryotes, has 48 tRNAs. Of these, 16 are predicted to use wobble to decode two codons each, with the remaining 32 being specific for just a single triplet (Figure 13.8). The distinctive features compared with wobble in bacteria are:

- G–U wobble is used with eight tRNAs but in every case the wobble involves an anticodon with the sequence 3′–◆◆G–5′. The alternative version of G–U wobble, where the anticodon sequence is 3′–◆◆U–5′, appears not to be used in eukaryotes, possibly because this could result in a tRNA^{Ile} with the anticodon 3′–UAU–5′ reading the methionine codon 5′–AUG–3′ (Figure 13.9). Eukaryotes therefore have a means of preventing this type of wobble from occurring.

Figure 13.7 Two examples of wobble in bacteria. (A) Wobble involving a G–U base pair enables the four-codon family for alanine to be decoded by just two tRNAs. Note that wobble involving G–U also enables accurate decoding of a four-codon family that specifies two amino acids. For example, the anticodon 3′–AAG–5′ can decode 5′–UUC–3′ and 5′–UUU–3′, both coding for phenylalanine (see Figure 1.20), and the anticodon 3′–AAU–5′ can decode the other two members of this family, 5′–UUA–3′ and 5′–UUG–3′, which code for leucine. (B) Inosine can base-pair with A, C, or U, meaning that a single tRNA can decode all three codons for isoleucine. Dotted lines indicate hydrogen bonds. Abbreviation: I, inosine.

Figure 13.8 The predicted usage of wobble in decoding the human genome. Pairs of codons that are predicted to be decoded by a single tRNA using G–U wobble are highlighted in pink, and those pairs predicted to be decoded by wobble involving inosine are highlighted in yellow. Codons that are not highlighted have their own individual tRNAs. The predictions are based largely on examination of the anticodon sequences of the tRNAs that have been located in the human genome sequence. The analysis shown here implies that there are 45 tRNAs in human cells—the 16 for the wobble pairs and 29 singletons. In fact, there are 48 tRNAs. This is because three codons thought to be decoded as part of a wobble pair (5′–AAU–3′, 5′–AUC–3′, and 5′–UAU–3′) also have their own individual tRNAs, although these are present in low abundance.

UUU	phe	UCU		UAU	tyr	UGU	cys
UUC		UCC	ser	UAC		UGC	
UUA	leu	UCA		UAA	stop	UGA	stop
UUG		UCG		UAG		UGG	trp
CUU		CCU		CAU	his	CGU	
CUC	leu	CCC	pro	CAC		CGC	arg
CUA		CCA		CAA	gln	CGA	
CUG		CCG		CAG		CGG	
AUU		ACU		AAU	asn	AGU	ser
AUC	ile	ACC	thr	AAC		AGC	
AUA		ACA		AAA	lys	AGA	arg
AUG	met	ACG		AAG		AGG	
GUU		GCU		GAU	asp	GGU	
GUC	val	GCC	ala	GAC		GGC	gly
GUA		GCA		GAA	glu	GGA	
GUG		GCG		GAG		GGG	

KEY

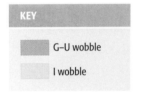

⬛ G–U wobble

⬜ I wobble

- Eight other human tRNAs have anticodons containing inosine (3′–♦♦I–5′) but these decode only 5′–♦♦C–3′ and 5′–♦♦U–3′. The base pairing between I and A is weak, which means that 5′–♦♦A–3′ codons are only inefficiently recognized by a 3′–♦♦I–5′ anticodon. To avoid this inefficiency, in every example of wobble involving inosine in the human tRNA set, the 5′–♦♦A–3′ codon is recognized by a separate tRNA. Note, however, that recognition by a separate tRNA does not preclude the 5′–♦♦A–3′ codon from also being decoded by the tRNA containing 3′–♦♦I–5′, albeit inefficiently. This does not compromise the specificity of the genetic code, because wobble involving inosine is limited to codon families in which all three triplets that can be decoded by 3′–♦♦I–5′ specify the same amino acid (see Figure 13.8).

Other genetic systems use more extreme forms of wobble. Human mitochondria, for example, use only 22 tRNAs. With some of these tRNAs the nucleotide in the wobble position of the anticodon is virtually redundant because it can base-pair with any nucleotide, enabling all four codons of a family to be recognized by the same tRNA. This phenomenon has been called **superwobble**.

13.2 The Role of the Ribosome in Protein Synthesis

An *E. coli* cell contains approximately 20,000 ribosomes, distributed throughout its cytoplasm. The average human cell contains rather more (nobody has ever counted them all), some free in the cytoplasm and some attached to the outer surface of the endoplasmic reticulum, the membranous network of tubes and vesicles that permeates the cell. Originally, ribosomes were looked

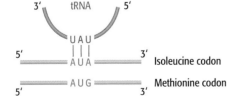

Figure 13.9 A tRNA with the anticodon 3′–UAU–5′ could read the isoleucine codon 5′–AUA–3′ as well as the methionine codon.

on as passive partners in protein synthesis, merely the structures on which translation occurs. This view has changed over the years and ribosomes are now considered to play two active roles in protein synthesis:

- Ribosomes *coordinate* protein synthesis by placing the mRNA, aminoacyl-tRNAs, and associated protein factors in their correct positions relative to one another.

- Components of ribosomes, including the rRNAs, *catalyze* at least some of the chemical reactions occurring during translation.

To understand how ribosomes play these roles we will first survey the structural features of ribosomes in bacteria and eukaryotes, and then examine the detailed mechanism for protein synthesis in these two types of organism.

13.2.1 Ribosome structure

Our understanding of ribosome structure has gradually developed over the last 50 years as more and more powerful techniques have been applied to the problem. Originally called "microsomes," ribosomes were first observed in the early decades of the twentieth century as tiny particles almost beyond the resolving power of light microscopy. In the 1940s and 1950s, the first electron micrographs showed that bacterial ribosomes are oval shaped, with dimensions of 29×21 nm, rather smaller than eukaryotic ribosomes, the latter varying a little in size depending on species but averaging about 32×22 nm. In the mid-1950s, the discovery that ribosomes are the sites of protein synthesis stimulated attempts to define the structures of these particles in greater detail.

Ultracentrifugation was used to measure the sizes of ribosomes and their components

The initial progress in understanding the detailed structure of the ribosome came not from observing them with the electron microscope but by analyzing their components by ultracentrifugation (Technical Note 7.1). Intact ribosomes have sedimentation coefficients of 80S for eukaryotes and 70S for bacteria, and each can be broken down into smaller components (Figure 13.10):

- Each ribosome comprises two subunits. In eukaryotes these subunits have sedimentation coefficients of 60S and 40S; in bacteria they are 50S and 30S. Note that sedimentation coefficients are not additive because they depend on shape as well as mass: it is perfectly acceptable for the intact ribosome to have an S value less than the sum of its two subunits.

- The large subunit contains three rRNAs in eukaryotes (the 28S, 5.8S, and 5S rRNAs) but only two in bacteria (23S and 5S rRNAs). In bacteria the equivalent of the eukaryotic 5.8S rRNA is contained within the 23S rRNA.

- The small subunit contains a single rRNA in both types of organism: an 18S rRNA in eukaryotes and a 16S rRNA in bacteria.

- Both subunits contain a variety of **ribosomal proteins**, the numbers of these detailed in Figure 13.10. The ribosomal proteins of the small subunit are called S1, S2, and so on; those of the large subunit are L1, L2, and so on. There is just one of each protein per ribosome, except for L7 and L12, which are present as dimers.

Probing the fine structure of the ribosome

Once the basic composition of eukaryotic and bacterial ribosomes had been worked out, attention was focused on the way in which the various rRNAs

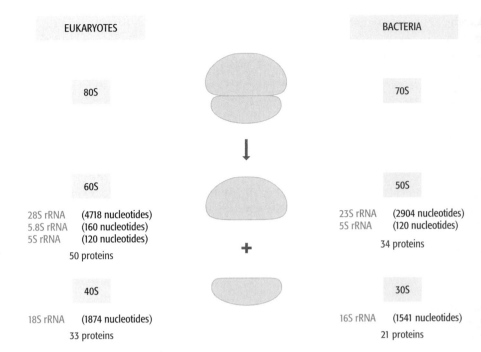

EUKARYOTES

BACTERIA

80S

70S

60S

50S

28S rRNA	(4718 nucleotides)
5.8S rRNA	(160 nucleotides)
5S rRNA	(120 nucleotides)

50 proteins

| 23S rRNA | (2904 nucleotides) |
| 5S rRNA | (120 nucleotides) |

34 proteins

+

40S

30S

| 18S rRNA | (1874 nucleotides) |

33 proteins

| 16S rRNA | (1541 nucleotides) |

21 proteins

Figure 13.10 The composition of eukaryotic and bacterial ribosomes. The details refer to a "typical" eukaryotic ribosome and the *Escherichia coli* ribosome. Variations between different species mainly concern the numbers of ribosomal proteins.

and proteins fit together. Important information was provided by the first rRNA sequences, comparisons between these identifying conserved regions that can base-pair to form complex two-dimensional structures (Figure 13.11). This suggested that the rRNAs provide a scaffolding within the ribosome, to which the proteins are attached, an interpretation that underemphasizes the active role that rRNAs play in protein synthesis but which nonetheless was a useful foundation on which to base subsequent research.

Much of that subsequent research has concentrated on the bacterial ribosome, which is smaller than the eukaryotic version and available in large amounts from extracts of cells grown to high density in liquid cultures. A number of technical approaches have been used to study the bacterial ribosome:

- **Nuclease protection experiments** (Section 7.1.1) enable contacts between rRNAs and proteins to be identified.

- **Protein–protein cross-linking** identifies pairs or groups of proteins that are located close to one another in the ribosome.

- **Electron microscopy** has gradually become more sophisticated, enabling the overall structure of the ribosome to be resolved in greater detail. For example, innovations such as **immunoelectron microscopy**, in which ribosomes are labeled with antibodies specific for individual ribosomal proteins before examination, have been used to locate the positions of these proteins on the surface of the ribosome.

- **Site-directed hydroxyl radical probing** makes use of the ability of Fe(II) ions to generate hydroxyl radicals that cleave RNA phosphodiester bonds located within 1 nm of the site of radical production. This technique has been used to determine the exact positioning of ribosomal proteins in the *E. coli* ribosome. For example, to determine the position of S5, different amino acids within this protein were labeled with Fe(II) and hydroxyl radicals induced in reconstituted ribosomes. The positions at which the 16S rRNA was cleaved were then used to infer the topology of the rRNA in the vicinity of S5 protein (Figure 13.12).

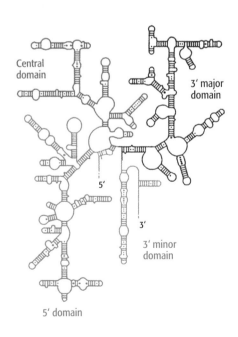

Central domain

3' major domain

5'

3'

3' minor domain

5' domain

Figure 13.11 The base-paired structure of the *Escherichia coli* 16S rRNA. In this representation, standard base pairs (G–C, A–U) are shown as bars; nonstandard base pairs (e.g., G–U) are shown as dots.

In recent years these techniques have been increasingly supplemented by X-ray crystallography (Technical Note 11.1), which has been responsible for the most exciting insights into ribosome structure. Analyzing the massive amounts of X-ray diffraction data that are produced by crystals of an object as large as a ribosome is a huge task, particularly at the level needed to obtain a structure that is detailed enough to be informative about the way in which the ribosome works. This challenge has been met, and structures have been deduced for ribosomal proteins bound to their segments of rRNA, for the large and small subunits, and for the entire bacterial ribosome attached to mRNA and tRNAs. As well as revealing the structure of the ribosome (Figure 13.13), this recent explosion of information has had an important impact on our understanding of the translation process.

13.2.2 Initiation of translation

Although ribosomal architecture is similar in bacteria and eukaryotes, there are distinctions in the way in which translation is carried out in the two types of organism. The most important of these differences occurs during the first stage of translation, when the ribosome is assembled on the mRNA at a position upstream of the initiation codon.

Initiation in bacteria requires an internal ribosome binding site

The main difference between initiation of translation in bacteria and eukaryotes is that in bacteria the translation initiation complex is built up directly at the initiation codon, the point at which protein synthesis will begin, whereas

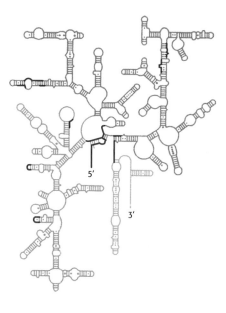

Figure 13.12 Positions within the *Escherichia coli* 16S rRNA that form contacts with ribosomal protein S5. The distribution of the contact positions (shown in red) for this single ribosomal protein emphasizes the extent to which the base-paired secondary structure of the rRNA is further folded within the three-dimensional structure of the ribosome.

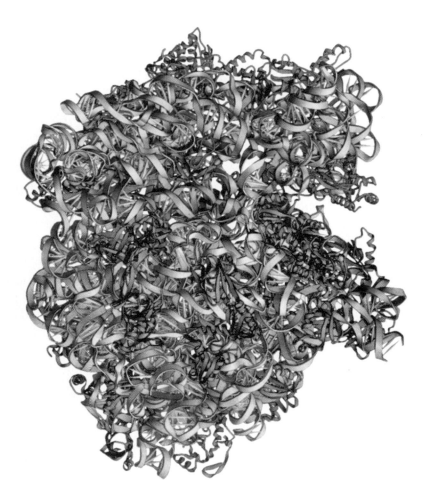

Figure 13.13 The bacterial ribosome. The picture shows the ribosome of the bacterium *Thermus thermophilus*. The small subunit is at the top, with the 16S rRNA in light blue and the small subunit ribosomal proteins in dark blue. The large subunit rRNAs are in grey and the proteins in purple. The gold area is the A site (Section 13.2.3)—the point at which aminoacylated tRNAs enter the ribosome during protein synthesis. This site, and most of the region within which protein synthesis actually occurs, is located in the cleft between the two subunits. Reprinted from *Trends Biochem. Sci.*, **26**, Mathews and Pe'ery, The machine that decodes the Genome. 585–587, 2001, with permission from Elsevier.

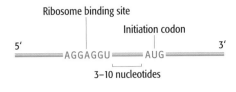

Figure 13.14 The ribosome binding site for bacterial translation. In *Escherichia coli*, the ribosome binding site has the consensus sequence 5'–AGGAGGU–3' and is located between 3 and 10 nucleotides upstream of the initiation codon.

eukaryotes use a more indirect process for locating the initiation point, as we will see in the next section.

When not actively participating in protein synthesis, ribosomes dissociate into their subunits, which remain in the cytoplasm waiting to be used for a new round of translation. In bacteria, the process initiates when a small subunit, in conjunction with the translation **initiation factor** IF-3 (Table 13.2), attaches to the **ribosome binding site** (also called the **Shine–Dalgarno sequence**). This is a short target site, consensus 5'–AGGAGGU–3' in *E. coli* (Table 13.3), located about 3–10 nucleotides upstream of the initiation codon, the point at which translation will begin (Figure 13.14). The ribosome binding site is complementary to a region at the 3' end of the 16S rRNA, the one present in the small subunit, and it is thought that base pairing between the two is involved in the attachment of the small subunit to the mRNA.

Table 13.2 Functions of the initiation factors in bacteria and eukaryotes

Factor	Function
Bacteria	
IF-1	Unclear; X-ray crystallography studies show that binding of IF-1 blocks the A site, so its function may be to prevent premature entry of tRNAs into the A site. Alternatively IF-1 may cause conformational changes that prepare the small subunit for attachment to the large subunit
IF-2	Directs the initiator tRNA^{Met} to its correct position in the initiation complex
IF-3	Prevents premature reassociation of the large and small subunits of the ribosome
Eukaryotes	
eIF-1	Component of the preinitiation complex
eIF-1A	Component of the preinitiation complex
eIF-2	Binds to the initiator tRNA^{Met} within the ternary complex component of the preinitiation complex; phosphorylation of eIF-2 results in a global repression of translation
eIF-2B	Regenerates the eIF2–GTP complex
eIF-3	Component of the preinitiation complex; makes direct contact with eIF-4G and so forms the link with the cap binding complex
eIF-4A	Component of the cap binding complex; a helicase that aids scanning by breaking intramolecular base pairs in the mRNA
eIF-4B	Aids scanning, possibly by acting as a helicase that breaks intramolecular base pairs in the mRNA
eIF-4E	Component of the cap binding complex, possibly the component that makes direct contact with the cap structure at the 5' end of the mRNA
eIF-4F	The cap binding complex, comprising eIF-4A, eIF-4E, and eIF-4G, which makes the primary contact with the cap structure at the 5' end of the mRNA
eIF-4G	Component of the cap binding complex; forms a bridge between the cap binding complex and eIF-3 in the preinitiation complex. In at least some organisms, eIF-4G also forms an association with the poly(A) tail, via the polyadenylate-binding protein
eIF-4H	In mammals, aids scanning in a manner similar to eIF-4B
eIF-5	Aids release of the other initiation factors at the completion of initiation
eIF-6	Associated with the large subunit of the ribosome; prevents large subunits from attaching to small subunits in the cytoplasm

Table 13.3 Examples of ribosome binding site sequences in *Escherichia coli*

Gene	Codes for	Ribosome binding sequence	Nucleotides to the initiation codon
E. coli consensus	–	5′–AGGAGGU–3′	3–10
Lactose operon	Lactose utilization enzymes	5′–AGGA–3′	7
galE	Hexose-1-phosphate uridyltransferase	5′–GGAG–3′	6
rplJ	Ribosomal protein L10	5′–AGGAG–3′	8

Attachment to the ribosome binding site positions the small subunit of the ribosome over the initiation codon (Figure 13.15). This codon is usually 5′–AUG–3′, which codes for methionine, although 5′–GUG–3′ and 5′–UUG–3′ are sometimes used. All three codons can be recognized by the same initiator tRNA, the last two by wobble. This initiator tRNA is the one that was amino-acylated with methionine and subsequently modified by conversion of the methionine to *N*-formylmethionine (see Figure 13.5B). The modification attaches a formyl group (–COH) to the amino group, which means that only the carboxyl group of the initiator methionine is free to participate in peptide bond formation. This ensures that polypeptide synthesis can take place only in the N→C direction. The initiator tRNA^{Met} is brought to the small subunit of the ribosome by a second initiation factor, IF-2, along with a molecule of GTP, the latter acting as a source of energy for the final step of initiation. The formyl group remains attached until translation has proceeded into the elongation phase but it is then removed from the growing polypeptide, either on its own or along with the rest of the initial methionine. Note that the initiator tRNA^{Met} is only able to decode the initiation codon; it cannot enter the complete ribosome during the elongation phase of translation during which internal 5′–AUG–3′ codons are recognized by a different tRNA^{Met} carrying an unmodified methionine. The discrimination between these tRNAs appears to center on two unusual features of the initiator molecule. First, unlike all other bacterial tRNAs that have been sequenced, the initiator tRNA has a run of three G–C base pairs in its anticodon stem that aid the attachment of this tRNA to the small subunit. Second, the 5′ nucleotide, which in most tRNAs participates in the first base pair of the acceptor stem (see Figure 13.2) is unpaired in the initiator tRNA. This unusual feature acts as a signal for the enzyme that converts the attached methionine to *N*-formylmethionine, and possibly also prevents the initiator tRNA from entering the ribosome during the elongation phase.

Completion of the initiation phase occurs when IF-1 binds to the initiation complex. The precise role of IF-1 is unclear (see Table 13.2), but it may induce a conformational change in the initiation complex, enabling the large subunit of the ribosome to attach. Attachment of the large subunit requires energy, which is generated by hydrolysis of the bound GTP, and results in release of the initiation factors.

Initiation in eukaryotes is mediated by the cap structure and poly(A) tail

Only a small number of eukaryotic mRNAs have internal ribosome binding sites. Instead, with most mRNAs the small subunit of the ribosome makes its initial attachment at the 5′ end of the molecule and then **scans** along the

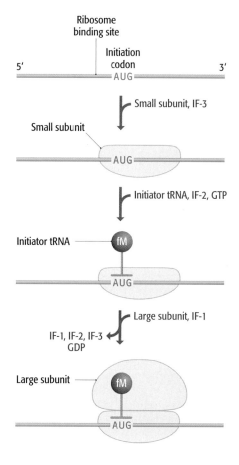

Figure 13.15 Initiation of translation in *Escherichia coli*. Note that the different components of the initiation complex are not drawn to scale. Abbreviation: fM, *N*-formylmethionine.

(A) Attachment of the preinitiation complex to the mRNA

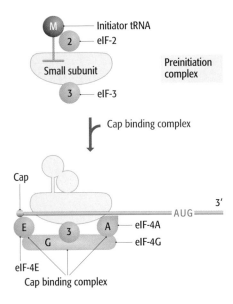

(B) Scanning

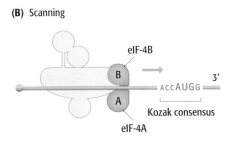

Figure 13.16 Initiation of translation in eukaryotes. (A) Assembly of the preinitiation complex and its attachment to the mRNA. For clarity, several proteins whose precise roles are not understood have been omitted. The overall configuration of the complex is not known. (B) The preinitiation complex scans along the mRNA until it reaches the initiation codon, which is recognizable because it is located within the Kozak consensus sequence. Scanning is aided by eIF-4A and eIF-4B, which are thought to have helicase activity. It is probable that eIF-3 remains attached to the preinitiation complex during scanning, as shown here. It is not clear whether eIF-4E and eIF-4G also remain attached at this stage. Note that scanning is an energy-dependent process that requires hydrolysis of ATP. Abbreviation: M, methionine.

sequence until it locates the initiation codon. The process requires a plethora of initiation factors and there is still some confusion over the functions of all of these (Table 13.2). The details are as follows (Figure 13.16).

The first step involves assembly of the **preinitiation complex**. This structure comprises the 40S subunit of the ribosome, a "ternary complex" made up of the initiation factor eIF-2 bound to the initiator tRNA^{Met} and a molecule of GTP, and three additional initiation factors, eIF-1, eIF-1A, and eIF-3. As in bacteria, the initiator tRNA is distinct from the normal tRNA^{Met} that recognizes internal 5'–AUG–3' codons but, unlike bacteria, it is aminoacylated with normal methionine, not the formylated version.

After assembly, the preinitiation complex associates with the 5' end of the mRNA. This step requires the **cap binding complex** (sometimes called eIF-4F), which comprises the initiation factors eIF-4A, eIF-4E, and eIF-4G. The contact with the cap might be made by eIF-4E alone (as shown in Figure 13.16) or might involve a more general interaction with the cap binding complex. The factor eIF-4G acts as a bridge between eIF-4E, bound to the cap, and eIF-3, attached to the preinitiation complex. The result is that the preinitiation complex becomes attached to the 5' region of the mRNA. Attachment of the preinitiation complex to the mRNA is also influenced by the poly(A) tail, at the distant 3' end of the mRNA. This interaction is thought to be mediated by the polyadenylate-binding protein (PADP), which is attached to the poly(A) tail (Section 12.2.1). In yeast and plants it has been shown that PADP can form an association with eIF-4G, this association requiring that the mRNA bends back on itself. With artificially uncapped mRNAs, the PADP interaction is sufficient to load the preinitiation complex onto the 5' end of the mRNA, but under normal circumstances the cap structure and poly(A) tail probably work together. The poly(A) tail could have an important regulatory role, as the length of the tail appears to be correlated with the extent of initiation that occurs with a particular mRNA.

After becoming attached to the 5' end of the mRNA, the **initiation complex**, as it is now called, has to scan along the mRNA molecule and find the initiation codon. The leader regions of eukaryotic mRNAs can be several tens, or even hundreds, of nucleotides in length and often contain regions that form hairpins and other base-paired structures. These are probably removed by a combination of eIF-4A and eIF-4B. eIF-4A, and possibly also eIF-4B, has a helicase activity and so is able to break intramolecular base pairs in the mRNA, freeing the passage for the initiation complex (Figure 13.16B). The initiation codon, which is usually 5'–AUG–3' in eukaryotes, is recognizable because it is contained in a short consensus sequence, 5'–ACCAUGG–3', referred to as the **Kozak consensus**.

Once the initiation complex is positioned over the initiation codon, the large subunit of the ribosome attaches. As in bacteria, this requires hydrolysis of GTP and leads to release of the initiation factors. Two final initiation factors are involved at this stage: eIF-5, which aids release of the other factors, and eIF-6, which is associated with the unbound large subunit and prevents it from attaching to a small subunit in the cytoplasm.

Initiation of eukaryotic translation without scanning

The scanning system for initiation of translation does not apply to every eukaryotic mRNA. This was first recognized with the picornaviruses, a group of viruses with RNA genomes which includes the human poliovirus and rhinovirus, the latter being responsible for the common cold. Transcripts from these viruses are not capped but instead have an **internal ribosome entry site (IRES)** which is similar in function to the ribosome binding site of bacteria, although the sequences of IRESs and their positions relative to the initiation codon are more variable than the bacterial versions. The presence of IRESs on their transcripts means that picornaviruses can block protein synthesis in the host cell by inactivating the cap binding complex, without affecting translation of their own transcripts, although this is not a normal part of the infection strategy of all picornaviruses.

Remarkably, no virus proteins are required for recognition of an IRES by a host ribosome. In other words, the normal eukaryotic cell possesses proteins and/or other factors that enable it to initiate translation by the IRES method. Because of their variability, IRESs are difficult to identify by inspection of DNA sequences, but it is becoming clear that a few nuclear gene transcripts possess them and that these are translated, at least under some circumstances, via their IRES rather than by scanning. Examples are the mRNAs for the mammalian immunoglobulin heavy-chain binding protein and the *Drosophila* Antennapedia protein (Section 14.3.4). IRESs are also found on several mRNAs whose protein products are translated when the cell is put under stress, for example by exposure to heat, irradiation, or low oxygen conditions. Under these circumstances, cap-dependent translation is globally suppressed (as described in the next section). The presence of IRESs on the "survival" mRNAs therefore enables these to undergo preferential translation at the time when their products are needed.

Regulation of translation initiation

The initiation of translation is an important control point in protein synthesis, at which two different types of regulation can be exerted. The first of these is **global regulation**, which involves a general alteration in the amount of protein synthesis occurring, with all mRNAs translated by the cap mechanism being affected to a similar extent. This is commonly achieved by phosphorylation of eIF-2, which results in repression of translation initiation by preventing eIF-2 from binding the molecule of GTP that it needs before it can transport the initiator tRNA to the small subunit of the ribosome. Phosphorylation of eIF-2 occurs during stresses such as heat shock, when the overall level of protein synthesis is decreased and IRES-mediated translation takes over.

Transcript-specific regulation involves mechanisms that act on a single transcript or a small group of transcripts coding for related proteins. The most frequently cited example of transcript-specific regulation involves the operons for the ribosomal protein genes of *E. coli* (Figure 13.17A). The leader region of the mRNA transcribed from each operon contains a sequence that acts as a binding site for one of the proteins coded by the operon. When this protein is synthesized it can either attach to its position on the ribosomal RNA, or bind to the leader region of the mRNA. The rRNA attachment is favored and occurs if there are free rRNAs in the cell. Once all the free rRNAs have been assembled into ribosomes, the ribosomal protein binds to its mRNA, blocking translation initiation and hence switching off further synthesis of the ribosomal proteins coded by that particular mRNA. Similar events involving other mRNAs ensure that synthesis of each ribosomal protein is coordinated with the amount of free rRNA in the cell. Other proteins with RNA-binding capability, such as

(A) Autoregulation of ribosomal protein synthesis

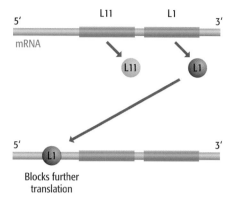

(B) Regulation by iron-response elements

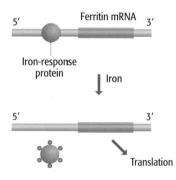

Figure 13.17 Transcript-specific regulation of translation initiation. (A) Regulation of ribosomal protein synthesis in bacteria. The L11 operon of *Escherichia coli* is transcribed into an mRNA carrying copies of the genes for the L11 and L1 ribosomal proteins. When the L1 binding sites on the available 23S rRNA molecules have been filled, L1 binds to the 5′ untranslated region of the mRNA, blocking further initiation of translation. (B) Regulation of ferritin protein synthesis in mammals. The iron-response protein binds to the 5′ untranslated region of the ferritin mRNA when iron is absent, preventing the synthesis of ferritin.

some aminoacyl-tRNA synthetases, also employ transcript-specific regulation in a manner similar to ribosomal proteins.

A second example of transcript-specific regulation, one occurring in mammals, involves the mRNA for ferritin, an iron-storage protein (Figure 13.17B). In the absence of iron, ferritin synthesis is inhibited by proteins that bind to sequences called **iron-response elements** located in the leader region of the ferritin mRNA. The bound proteins block the ribosome as it attempts to scan along the mRNA in search of the initiation codon. When iron is present, the binding proteins detach and the mRNA is translated. Interestingly, the mRNA for a related protein—the transferrin receptor involved in the uptake of iron—also has iron-response elements, but in this case detachment of the binding proteins in the presence of iron results not in translation of the mRNA but in its degradation. This is logical because when iron is present in the cell, there is less requirement for transferrin receptor activity because there is less need to import iron from outside.

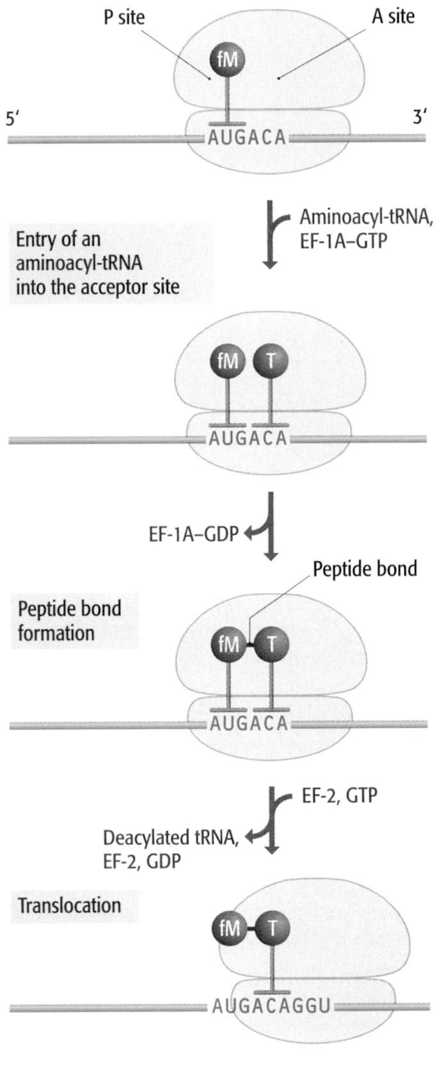

Initiation of translation of some bacterial mRNAs can also be regulated by short RNAs which attach to recognition sequences within the mRNAs. This does not always result in translation being prevented, as some short RNAs can also activate translation of one or more of their target mRNAs. An example is the *E. coli* RNA called OxyS, which is 109 nucleotides in length and regulates translation of 40 or so mRNAs. Synthesis of OxyS is activated by hydrogen peroxide and other reactive oxygen compounds which can cause oxidative damage to the cell. Once synthesized, OxyS switches on translation of some mRNAs whose products help protect the bacterium from oxidative damage, and switches off translation of other mRNAs whose products would be deleterious under these circumstances. The structures formed when OxyS and other short, regulatory RNAs bind to their target mRNAs have been described, but these do not give us any substantial insights into how the regulatory process is mediated, especially as the structures formed on mRNAs that are silenced, and which presumably act by blocking access of the small subunit of the ribosome to the mRNA, are not obviously different from the structures formed with mRNAs whose translation is activated.

13.2.3 The elongation phase of translation

The main differences between translation in bacteria and in eukaryotes occur during the initiation phase: the events occurring after the large subunit of the ribosome becomes associated with the initiation complex are similar in both types of organism. We can therefore deal with them together, by looking at what happens in bacteria and referring to the distinctive features of eukaryotic translation where appropriate.

Elongation in bacteria and eukaryotes

Attachment of the large subunit results in two sites at which aminoacyl-tRNAs can bind. The first of these, the **P** or **peptidyl site**, is already occupied by the initiator tRNA^{Met}, charged with *N*-formylmethionine or methionine, and base-paired with the initiation codon. The second site, the **A** or **aminoacyl site**, covers the second codon in the open reading frame (Figure 13.18). The structures revealed by X-ray crystallography show that these sites are located in the cavity between the large and small subunits of the ribosome, the codon–anticodon interaction being associated with the small subunit, and the aminoacyl end of the tRNA with the large subunit (Figure 13.19).

Figure 13.18 Elongation of translation. The diagram shows the events occurring during a single elongation cycle in *Escherichia coli*. See text for details regarding eukaryotic translation. Abbreviations: fM, *N*-formylmethionine; T, threonine.

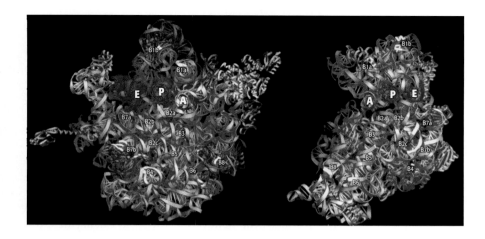

Figure 13.19 The important sites in the ribosome. The structure on the left is the large subunit of the *Thermus thermophilus* ribosome; that on the right is the small subunit. The views look down onto the two surfaces that contact one another when the subunits are placed together to make the intact ribosome. The A, P, and E sites are labeled, and each one is occupied by a tRNA shown in red or orange. The main part of each tRNA is embedded within the large subunit, with just the anticodon arms and loops associated with the small subunit. Those parts of the ribosome that make the important bridging contacts between the two subunits are labeled as Bla, etc. Reprinted with permission from Yuspov *et al.*, *Science*, **292**, 883–896. © 2001 AAAS.

The A site becomes filled with the appropriate aminoacyl-tRNA, which in *E. coli* is brought into position by the **elongation factor** EF-1A, which ensures that only tRNAs that carry the correct amino acid are able to enter the ribosome, mischarged tRNAs being rejected at this point. EF-1A is an example of a G protein, meaning that it binds a molecule of GTP which it can hydrolyze to release energy. In eukaryotes the equivalent factor is called eEF-1, which is a complex of four subunits: eEF-1a, eEF-1b, eEF-1d, and eEF-1g (Table 13.4). The first of these exists in at least two forms, eEF-1a1 and eEF-1a2, which are highly similar proteins that probably have equivalent functions in different tissues. Specific contacts between the tRNA, mRNA, and the small subunit rRNA within the A site ensure that only the correct tRNA is accepted. These contacts are able to discriminate between a codon–anticodon interaction in which all three base pairs have formed and one in which one or more mis-pairs are present, the latter signaling that the wrong tRNA is present. This is probably just one part of a series of safeguards that ensure the accuracy of the translation process.

When the aminoacyl-tRNA has entered the A site, a peptide bond is formed between the two amino acids. This involves a **peptidyl transferase** enzyme,

Table 13.4 Elongation factors for bacterial and eukaryotic translation

Factor	Function
Bacteria	
EF-1A	Directs the next tRNA to its correct position in the ribosome
EF-1B	Regenerates EF-1A after the latter has yielded the energy contained in its attached GTP molecule
EF-2	Mediates translocation
Eukaryotes	
eEF-1	Complex of four subunits (eEF-1a, eEF-1b, eEF-1d, and eEF-1g); directs the next tRNA to its correct position in the ribosome
eEF-2	Mediates translocation

Note that the bacterial elongation factors have recently been renamed. The older designations were EF-Tu, EF-Ts, and EF-G for EF-1A, EF-1B, and EF-2, respectively.

which releases the amino acid from the initiator tRNA^{Met} and then forms a peptide bond between this amino acid and the one attached to the second tRNA. The reaction is energy dependent and requires hydrolysis of the GTP attached to EF-1A (eEF-1 in eukaryotes). This inactivates EF-1A, which is ejected from the ribosome and regenerated by EF-1B. A eukaryotic equivalent of EF-1B has not been identified, and it is possible that one of the subunits of eEF-1 has the regenerative activity.

Now the dipeptide corresponding to the first two codons in the open reading frame is attached to the tRNA in the A site. The next step is **translocation** (see Figure 13.18), during which the ribosome moves three nucleotides along the mRNA, so a new codon enters the A site. This moves the dipeptide-tRNA to the P site, which in turn displaces the deacylated tRNA. In eukaryotes, the deacylated tRNA is simply ejected from the ribosome, but in bacteria the deacylated tRNA departs via a third position, the **E** or **exit site**. This site was originally looked on as a simple exit point from the ribosome, but it is now known to have an important role in ensuring that translocation moves the ribosome along the mRNA by precisely three nucleotides, thereby ensuring that the ribosome keeps to the correct reading frame.

Translocation requires hydrolysis of a molecule of GTP and is mediated by EF-2 in bacteria and by eEF-2 in eukaryotes. Electron microscopy of ribosomes at different intermediate stages in translocation shows that, during translocation, the ribosome adopts a less compact structure, with the two subunits rotating slightly in opposite directions, opening up the space between them and enabling the ribosome to slide along the mRNA. Translocation results in the A site becoming vacant, allowing a new aminoacyl-tRNA to enter. The elongation cycle is now repeated, and continues until the end of the open reading frame is reached.

Peptidyl transferase is a ribozyme

A ribosome-associated protein that has the peptidyl transferase activity needed to synthesize peptide bonds during translation has never been isolated. The reason for this lack of success is now known, at least for bacteria: the enzyme activity is specified by part of the 23S rRNA and hence is another example of a ribozyme (Section 12.2.4).

When the base-paired structures of rRNAs (see Figure 13.11) were first determined in the early 1980s, the possibility that an RNA molecule could have enzymatic activity was unheard of, the breakthrough discoveries with regard to ribozymes not being made until the period 1982–86. Ribosomal RNAs were therefore initially assigned purely structural roles in the ribosome, their base-paired conformations being looked upon as scaffolds to which the important components of the ribosome—the proteins—were attached. Problems with this interpretation began to arise in the late 1980s when difficulties were encountered in identifying the protein or proteins responsible for the central catalytic activity of the ribosome—the formation of peptide bonds. By now the existence of ribozymes had been established and molecular biologists began to take seriously the possibility that rRNAs might have an enzymatic role in protein synthesis. To test this hypothesis it was necessary to locate with precision the site of peptidyl transferase activity in the ribosome. Over the years, antibiotics and other inhibitors of protein synthesis have played an important role in studies of ribosome function. In 1995, a new inhibitor called CCdA-phosphate-puromycin was synthesized, this compound being an analog of the intermediate structure formed when two amino acids are joined by formation of a peptide bond during protein synthesis. CCdA-phosphate-puromycin binds tightly to the bacterial ribosome and, because of its structure, this binding site must be at precisely the position where peptide bonds are formed in the functioning ribosome. X-ray crystallography was therefore used to reveal exactly where CCdA-phosphate-puromycin binds within the large subunit. Its position is deep down within the body of the subunit, closely associated with the 23S rRNA of the large subunit, but 1.84 nm away from the nearest protein, L3, and slightly more distant from L2, L4, and L10 (Figure 13.20). In atomic terms, 1–2 nm is a massive distance, making it inconceivable that peptide bond synthesis could be catalyzed by any one of these four proteins. The positioning of CCdA-phosphate-puromycin therefore provides convincing evidence that peptidyl transferase is a ribozyme.

Now that this evidence has been obtained, researchers are moving on to determine exactly how the rRNA backbone acts as a ribozyme in peptide bond formation. Attention was initially concentrated on an adenine nucleotide at position 2451 in the *E. coli* 23S rRNA, because this adenine has unusual charge properties compared with other nucleotides. The hypothesis was that an interaction between this adenine and a nearby guanine, at position 2447, is the key to protein synthesis. But this model has been questioned by mutational studies, which have shown that although replacing A2451 with a uracil reduces the rate of peptide bond synthesis by 99%, both A2451 and G2447 can be replaced by other nucleotides without detectable effect. Attention is therefore turning to the other parts of the 23S rRNA that are located in the vicinity of the active site.

Frameshifting and other unusual events during elongation

The straightforward codon-by-codon translation of an mRNA is looked upon as the standard way in which proteins are synthesized. But an increasing number of unusual elongation events are being discovered. One of these is **frameshifting**, which occurs when a ribosome pauses in the middle of an mRNA, moves back one nucleotide or, less frequently, forward one nucleotide, and then continues translation. The result is that the codons that are read after the pause are not contiguous with the preceding set of codons: they lie in a different reading frame (Figure 13.21A).

Spontaneous frameshifts occur randomly and are deleterious because the polypeptide synthesized after the frameshift has the incorrect amino acid sequence. But not all frameshifts are spontaneous: a few mRNAs utilize **programmed frameshifting** to induce the ribosome to change frame at a specific point within the transcript. Programmed frameshifting occurs in all types of organism, from bacteria through to humans, as well as during expression of a number of viral genomes. An example occurs during synthesis of DNA polymerase III in *E. coli*, the main enzyme involved in replication of DNA (Section 15.2.2). Two of the DNA polymerase III subunits, γ and τ, are coded by a single gene, *dnaX*. Subunit τ is the full-length translation product of the *dnaX* mRNA, and subunit γ is a shortened version. Synthesis of γ involves a frameshift in the middle of the *dnaX* mRNA, the ribosome encountering a termination codon immediately after the frameshift and so producing the

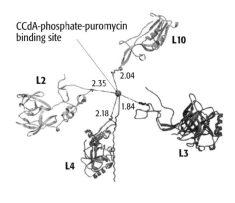

Figure 13.20 The location of a CCdA-phosphate-puromycin molecule within the large subunit of the bacterial ribosome. Parts of ribosomal proteins L2, L3, L4, and L10 are shown. Distances in nanometers between the CCdA-phosphate-puromycin molecule (shown as a red dot) and each protein are indicated.

(A) Programmed frameshifting in the *dnaX* mRNA

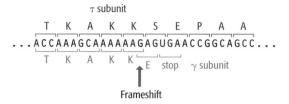

(B) Translational slippage in the lactose operon mRNA

(C) Translational bypassing in the T4 gene *60* mRNA

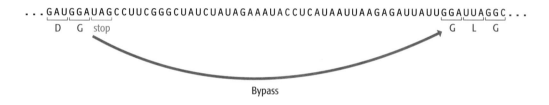

Figure 13.21 Three unusual translation elongation events occurring in *Escherichia coli.* (A) Programmed frameshifting during translation of the *dnaX* mRNA. During synthesis of the γ subunit the ribosome shifts back one nucleotide, immediately after a series of As. The ribosome inserts a glutamic acid into the polypeptide and then encounters a termination codon. (B) Slippage between the *lacZ* and *lacY* genes of the lactose operon mRNA. (C) Bypassing during translation of the bacteriophage T4 gene *60* mRNA involves a jump between two glycine codons. See Table 1.2 for the one-letter abbreviations of the amino acids.

truncated γ version of the translation product. It is thought that the frameshift is induced by three features of the *dnaX* mRNA:

- A hairpin loop, located immediately after the frameshift position, which stalls the ribosome.

- A sequence similar to a ribosome binding site immediately upstream of the frameshift position, which is thought to base-pair with the 16S rRNA (as does an authentic ribosome binding site), again causing the ribosome to stall.

- The codon 5′–AAG–3′ at the frameshift position. The presence of a modified nucleotide at the wobble position of the tRNA^{Lys} that decodes 5′–AAG–3′ means that the codon–anticodon interaction is relatively weak at this position, enabling the frameshift to occur.

A similar phenomenon—translational **slippage**—enables a single ribosome to translate an mRNA that contains copies of two or more genes (Figure 13.21B). This means that, for example, a single ribosome can synthesize each of the five proteins coded by the mRNA transcribed from the tryptophan operon of *E. coli* (see Figure 8.8B). When the ribosome reaches the end of one series of codons, it releases the protein it has just made, slips to the next initiation codon, and begins synthesizing the next protein. A more extreme form of slippage is **translational bypassing**, in which a larger part of the transcript, possibly a few tens of base pairs, is skipped, and elongation of the original protein continues after the bypassing event (Figure 13.21C). The bypass starts and ends either at two identical codons or at two codons that can be translated by the same tRNA by wobble. This suggests that the jump is controlled by the tRNA attached to the growing polypeptide, which scans the mRNA as the ribosome tracks along, and halts the bypass when a new codon to which it can base-pair is reached. Translational bypassing of 44 nucleotides occurs in *E. coli* during translation of the mRNA for gene *60* of T4 bacteriophage, which codes for a DNA topoisomerase subunit. Similar events have also been identified in a variety of other bacteria. Bypassing could result in two different proteins being synthesized from one mRNA—one protein from normal translation and one from bypassing—but whether this is its general function is not yet known.

13.2.4 Termination of translation

Protein synthesis ends when one of the three termination codons is reached (Figure 13.22). The A site is now entered not by a tRNA but by a protein **release factor** (Table 13.5). Bacteria have three of these: RF-1, which recognizes the termination codons 5′–UAA–3′ and 5′–UAG–3′; RF-2, which recognizes the termination codons 5′–UAA–3′ and 5′–UGA–3′; and RF-3, which stimulates release of RF-1 and RF-2 from the ribosome after termination, in a reaction requiring energy from the hydrolysis of GTP. Eukaryotes have just two release factors: eRF-1, which recognizes all three termination codons, and eRF-3, which might play the same role as RF-3 although this has not been proven. The structure of eRF-1 has been solved by X-ray crystallography, showing that the shape of this protein is very similar to that of a tRNA (Figure 13.23). This leads to a model in which a release factor mimics the structure of a tRNA and hence is able to enter the A site when the termination codon is reached. This model is attractive, but other studies suggest that the release factor adopts a different conformation when associated with a ribosome, one that is less similar to the shape of a tRNA.

The release factors terminate translation but they do not appear to be responsible for disassociation of the ribosomal subunits, at least not in bacteria. This is the function of an additional protein called **ribosome recycling factor** (**RRF**) which, like eRF-1, has a tRNA-like structure. RRF probably enters the P or A site and "unlocks" the ribosome (see Figure 13.22). Disassociation requires energy, which is released from GTP by EF-2, one of the elongation factors, and also requires the initiation factor IF-3 to prevent the subunits from attaching together again. A eukaryotic equivalent of RRF has not been identified, and this may be one of the functions of eRF-3. The disassociated ribosome subunits enter the cytoplasmic pool, where they remain until used again in another round of translation.

13.2.5 Translation in the archaea

The descriptions above refer to the events occurring during translation in bacteria and eukaryotes. We should not ignore the second group of prokaryotes, the archaea, and so before moving on we will briefly review what is know about translation in these organisms.

In most respects, translation in the archaea more closely resembles the equivalent events in the eukaryotes rather than in bacteria. The one apparent exception is that the archaeal ribosome, with a sedimentation coefficient of 70S, is comparable in size to the bacterial ribosome and, like bacterial ribosomes, contains 23S, 16S, and 5S rRNAs. This apparent similarity is illusory because the archaeal rRNAs form base-paired secondary structures

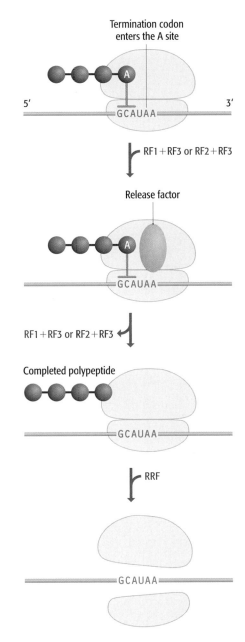

Figure 13.22 Termination of translation. Termination in *Escherichia coli* is illustrated. For differences in eukaryotes, see the text. The amino acid labeled with an "A" is an alanine. Abbreviations: RF, release factor; RRF, ribosome recycling factor.

Table 13.5 Release and ribosome recycling factors in bacteria and eukaryotes

Factor	Function
Bacteria	
RF-1	Recognizes the termination codons 5'–UAA–3' and 5'–UAG–3'
RF-2	Recognizes the termination codons 5'–UAA–3' and 5'–UGA–3'
RF-3	Stimulates dissociation of RF-1 and RF-2 from the ribosome after termination
RRF	Ribosome release factor, responsible for disassociating the ribosome subunits after translation has terminated
Eukaryotes	
eRF-1	Recognizes the termination codon
eRF-3	Possibly stimulates dissociation of eRF-1 from the ribosome after termination; possibly causes the ribosome subunits to disassociate after termination of translation

Figure 13.23 The structure of the eukaryotic release factor eRF-1 is similar to that of a tRNA. The left panel shows eRF-1 and the right panel shows a tRNA. The part of eRF-1 that resembles the tRNA is highlighted in white. The purple segment of eRF-1 interacts with the second eukaryotic release factor, eRF-3. Reprinted from *Trends Biochem. Sci.*, **25**, Kisselev and Buckingham, Transitional termination comes of age, 561–566, 2000, with permission from Elsevier.

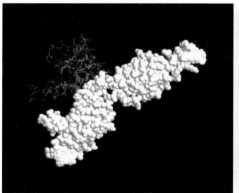

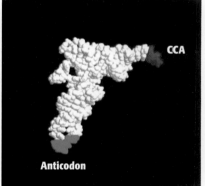

that are significantly different from the equivalent bacterial structures. The archaeal rRNA structures are also different from the eukaryotic versions, but the ribosomal proteins that attach to the rRNAs are homologs of the eukaryotic proteins.

Archaeal mRNAs are capped and polyadenylated, and translation initiation is thought to involve a scanning process similar to that described for eukaryotic mRNAs. Archaeal tRNAs display a few unique features, including the absence of thymidine in the so-called TΨC arm of the cloverleaf, and the presence at various positions of modified nucleotides not seen in either bacteria or eukaryotes. The methionine carried by the initiator tRNA is not N-formylated and the initiation and elongation factors resemble the eukaryotic molecules.

13.3 Posttranslational Processing of Proteins

Translation is not the end of the genome expression pathway. The polypeptide that emerges from the ribosome is inactive, and before taking on its functional

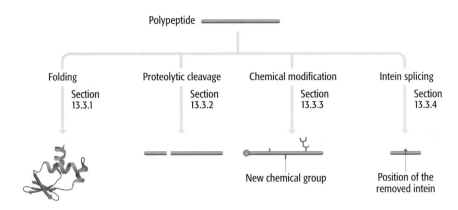

Figure 13.24 Schematic representation of the four types of posttranslational processing event. Not all events occur in all organisms.

role in the cell must undergo at least the first of the following four types of posttranslational processing (Figure 13.24):

- **Protein folding.** The polypeptide is inactive until it is folded into its correct tertiary structure.

- **Proteolytic cleavage.** Some proteins are processed by cutting events carried out by enzymes called **proteases**. These cutting events may remove segments from one or both ends of the polypeptide, resulting in a shortened form of the protein, or they may cut the polypeptide into a number of different segments, all or some of which are active.

- **Chemical modification.** Individual amino acids in the polypeptide might be modified by attachment of new chemical groups.

- **Intein splicing. Inteins** are intervening sequences in some proteins, similar in a way to introns in mRNAs. They have to be removed and the **exteins** ligated in order for the protein to become active.

Often these different types of processing occur together, the polypeptide being cut and modified at the same time that it is folded. If this is the case then the cutting, modification, and/or splicing events may be necessary for the polypeptide to take up its correct three-dimensional conformation, because this is dependent in part on the relative positioning of the various chemical groups along the molecule. Alternatively, a cutting event or a chemical modification may occur after the protein has been folded, possibly as part of a regulatory mechanism that converts a folded but inactive protein into an active form.

13.3.1 Protein folding

Recall that in Section 1.3.1 we examined the four levels of protein structure (primary, secondary, tertiary, and quaternary) and learnt that all of the information that a polypeptide needs in order to fold into its correct three-dimensional structure is contained within its amino acid sequence. This is one of the central principles of molecular biology. We must therefore examine its experimental basis and consider how the information contained in the amino acid sequence is utilized during the folding process for a newly translated polypeptide.

Not all proteins fold spontaneously in the test tube

The notion that the amino acid sequence contains all the information needed to fold the polypeptide into its correct tertiary structure derives from experiments carried out with ribonuclease in the 1960s. Ribonuclease is a small protein, just 124 amino acids in length, containing four disulfide bridges and with

Figure 13.25 Denaturation and spontaneous renaturation of a small protein. As the urea concentration increases to 8 M, the protein becomes denatured by unfolding: its activity decreases and the viscosity of the solution increases. When the urea is removed by dialysis, this small protein readopts its folded conformation. The activity of the protein increases back to the original level and the viscosity of the solution decreases.

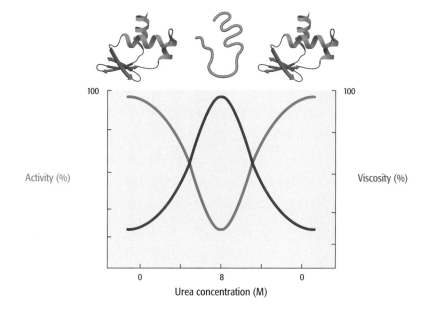

Figure 13.26 An incorrectly folded protein might be able to refold into its correct conformation. The black arrow represents the correct folding pathway, leading from the unfolded protein on the left to the active protein on the right. The red arrow leads to an incorrectly folded conformation, but this conformation is unstable and the protein is able to unfold partially, return to its correct folding pathway and, eventually, reach its active conformation.

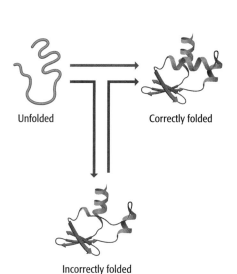

a tertiary structure that is made up predominantly of β-sheet, with very little α-helix. Studies of its folding were carried out with ribonuclease that had been purified from cow pancreas and resuspended in buffer. Addition of urea, a compound that disrupts hydrogen bonding, resulted in a decrease in the activity of the enzyme (measured by testing its ability to cut RNA) and an increase in the viscosity of the solution (Figure 13.25), indicating that the protein was being **denatured** by unfolding into an unstructured polypeptide chain. The critical observation was that when the urea was removed by dialysis, the viscosity decreased and the enzyme activity reappeared. The conclusion is that the protein refolds spontaneously when the denaturant (in this case, urea) is removed. In these initial experiments, the four disulfide bonds remained intact because they were not disrupted by urea, but the same result occurred when the urea treatment was combined with addition of a reducing agent to break the disulfide bonds: the activity was still regained on renaturation. This shows that the disulfide bonds are not critical to the protein's ability to refold, they merely stabilize the tertiary structure once it has been adopted.

More detailed study of the spontaneous folding pathways for ribonuclease and other small proteins has led to the following general two-step description of the process:

- The secondary structural motifs along the polypeptide chain form within a few milliseconds of the denaturant being removed. This step is accompanied by the protein collapsing into a compact, but not folded, organization, with its hydrophobic groups on the inside, shielded from water.

- During the next few seconds or minutes, the secondary structural motifs interact with one another and the tertiary structure gradually takes shape, often via a series of intermediate conformations. In other words, the protein follows a **folding pathway**. There may, however, be more than one possible pathway that a protein can follow to reach its correctly folded structure. The pathways may also have side-branches into which the protein can be diverted, leading to an incorrect structure. If an incorrect structure is sufficiently unstable then partial or complete unfolding may occur, allowing the protein a second opportunity to pursue a productive route towards its correct conformation (Figure 13.26).

For several years it was more or less assumed that all proteins would fold spontaneously in the test tube, but experiments have shown that only smaller proteins with less complex structures possess this ability. Two factors seem to prevent larger proteins from folding spontaneously. The first of these is their tendency to form insoluble aggregates when the denaturant is removed: the polypeptides may collapse into interlocked networks when they attempt to protect their hydrophobic groups from water in step one of the general folding pathway. Experimentally, this can be avoided by using a low dilution of the protein, but this is not an option that the cell can take to prevent its unfolded proteins from aggregating. The second factor that prevents folding is that a large protein tends to get stuck in nonproductive side-branches of its folding pathway, taking on an intermediate form that is incorrectly folded but which is too stable to unfold to any significant extent. Concerns have also been raised about the relevance of *in vitro* folding, as studied with ribonuclease, to the folding of proteins in the cell, because a cellular protein might begin to fold before it has been fully synthesized. If the initial folding occurs when only part of the polypeptide is available, then there might be an increased possibility of incorrect branches of the folding pathway being followed. These various considerations prompted research into protein folding in living cells.

In cells, folding is aided by molecular chaperones

Most of our current understanding of protein folding in the cell is founded on the discovery of proteins that help other proteins to fold. These are called **molecular chaperones** and have been studied in most detail in *E. coli*. It is clear that both eukaryotes and archaea possess equivalent proteins, although some of the details of the way they work are different.

The molecular chaperones can be divided into two groups:

- The **Hsp70 chaperones**, which include the proteins called Hsp70 (coded by the *dnaK* gene in *E. coli* and sometimes called DnaK protein), Hsp40 (coded by *dnaJ*), and GrpE.

- The **chaperonins**, the main version of which is the GroEL/GroES complex present in bacteria and eukaryotes, and TRiC found only in eukaryotes.

Molecular chaperones do not specify the tertiary structure of a protein, they merely help the protein find that correct structure. The two types of chaperone do this in different ways. The Hsp70 family bind to hydrophobic regions of unfolded proteins, including proteins that are still being translated (Figure 13.27). They hold the protein in an open conformation and aid folding, presumably by modulating the association between those parts of the polypeptide which form interactions in the folded protein. Exactly how this is achieved is not understood but it involves repeated binding and detachment of the Hsp70 protein, each cycle of which requires energy provided by the hydrolysis of ATP. The Hsp70 protein, as well as binding to the target polypeptide, has an ATPase activity and hence can release this energy, but it can function efficiently only with the help of Hsp40 and GrpE. Hsp40 stimulates the ATPase activity of Hsp70, and GrpE removes from the complex the ADP molecule (into which ATP is converted after energy release) enabling the cycle to begin again. As well as protein folding, the Hsp70 chaperones are also involved in other processes that require shielding of hydrophobic regions in proteins, such as transport through membranes, association of proteins into multisubunit complexes, and disaggregation of proteins that have been damaged by heat stress.

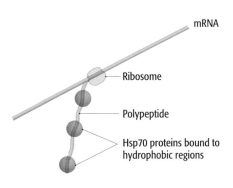

Figure 13.27 The Hsp70 chaperone system. Hsp70 chaperones bind to hydrophobic regions in unfolded polypeptides, including those that are still being translated, and hold the protein in an open conformation to aid its folding.

Figure 13.28 The GroEL/GroES chaperonin. On the left is a view from the top, and on the right a view from the side. 1 Å is equal to 0.1 nm. The GroES part of the structure is made up of seven identical protein subunits and is shown in gold. The GroEL components consist of 14 identical proteins arranged into two rings (shown in red and green), each containing seven subunits. The main entrance into the central cavity is through the bottom of the structure shown on the right. Reprinted with permission from Xu *et al.*, (1997), *Nature*, **388**, 741–750.

The chaperonins work in a quite different way. GroEL and GroES form a multisubunit structure that looks like a hollowed-out bullet with a central cavity (Figure 13.28). A single, unfolded protein enters the cavity and emerges folded. The mechanism for this is not known but it is postulated that the GroEL/GroES complex acts as a cage that prevents the unfolded protein from aggregating with other proteins, and that the inside surface of the cavity changes from hydrophobic to hydrophilic in such a way as to promote the burial of hydrophobic amino acids within the protein. This is not the only hypothesis: other researchers hold that the cavity unfolds proteins that have folded incorrectly, passing these unfolded proteins back to the cytoplasm so they can have a second attempt at adopting their correct tertiary structure.

Although both the Hsp70 family of chaperones and the GroEL/GroES chaperonins are present in eukaryotes, it seems that in these organisms protein folding depends mainly on the action of the Hsp70 proteins. This is probably true also of bacteria, even though the GroEL/GroES chaperonins play a major role in the folding of metabolic enzymes and proteins involved in transcription and translation.

13.3.2 Processing by proteolytic cleavage

Proteolytic cleavage has two functions in posttranslational processing of proteins (Figure 13.29):

- It is used to remove short pieces from the N- and/or C-terminal regions of polypeptides, leaving a single, shortened molecule that folds into the active protein.

- It is used to cut **polyproteins** into segments, all or some of which are active proteins.

These events are relatively common in eukaryotes but less frequent in bacteria.

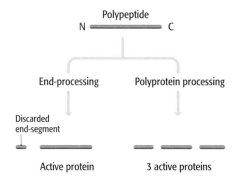

Figure 13.29 Protein processing by proteolytic cleavage. On the left, the protein is processed by removal of the N-terminal segment. C-terminal processing also occurs with some proteins. On the right, a polyprotein is processed to give three different proteins. Not all proteins undergo proteolytic cleavage.

Cleavage of the ends of polypeptides

Processing by cleavage is common with secreted polypeptides whose biochemical activities might be deleterious to the cell producing the protein. An example is provided by melittin, the most abundant protein in bee venom and the one responsible for causing cell lysis after injection of the bee sting into the person or animal being stung. Melittin lyses cells in bees as well as animals and so must initially be synthesized as an inactive precursor. This precursor, promelittin, has 22 additional amino acids at its N terminus. The presequence is removed by an extracellular protease that cuts it at 11 positions, releasing

Cut sites

A P E P E P A P E P E A E A D A E A D P E A G I G A V L K V L T T G L P A L I S W I K R K R Q Q G

Figure 13.30 Processing of promelittin, the bee-sting venom. Arrows indicate the cut sites.

the active venom protein. The protease does not cleave within the active sequence because its mode of action is to release dipeptides with the sequence X–Y, where X is alanine, aspartic acid, or glutamic acid, and Y is alanine or proline; these motifs do not occur in the active sequence (Figure 13.30).

A similar type of processing occurs with insulin, the protein made in the islets of Langerhans in the vertebrate pancreas and responsible for controlling blood sugar levels. Insulin is synthesized as preproinsulin, which is 105 amino acids in length (Figure 13.31). The processing pathway involves the removal of the first 24 amino acids to give proinsulin, followed by two additional cuts which excise a central segment, leaving two active parts of the protein, the A and B chains, which link together by formation of two disulfide bonds to form mature insulin. The first segment to be removed, the 24 amino acids from the N terminus, is a **signal peptide**, a highly hydrophobic stretch of amino acids that attaches the precursor protein to a membrane prior to transport across that membrane and out of the cell. Signal peptides are commonly found on proteins that bind to and/or cross membranes, in both eukaryotes and prokaryotes.

Proteolytic processing of polyproteins

In the examples shown in Figures 13.30 and 13.31, proteolytic processing results in a single mature protein. This is not always the case. Some proteins

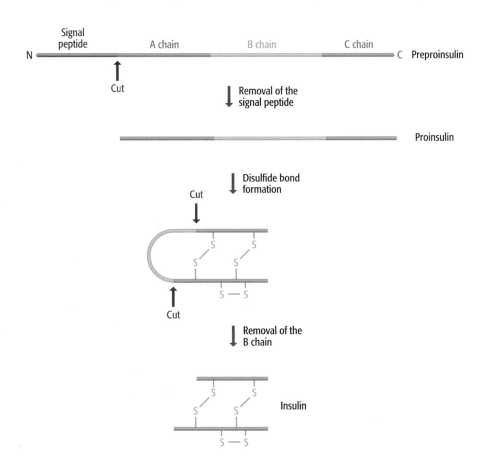

Figure 13.31 Processing of preproinsulin.

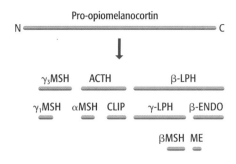

Figure 13.32 Processing of the pro-opiomelanocortin polyprotein. Abbreviations: ACTH, adrenocorticotropic hormone; CLIP, corticotropin-like intermediate lobe protein; ENDO, endorphin; LPH, lipotropin; ME, met-enkephalin; MSH, melanotropin.

are initially synthesized as polyproteins, long polypeptides that contain a series of mature proteins linked together in head-to-tail fashion. Cleavage of the polyprotein releases the individual proteins, which may have very different functions from one another.

Polyproteins are not uncommon in eukaryotes. Several types of virus that infect eukaryotic cells use them as a way of reducing the sizes of their genomes, a single polyprotein gene with one promoter and one terminator taking up less space than a series of individual genes. Polyproteins are also involved in the synthesis of peptide hormones in vertebrates. For example, the polyprotein called pro-opiomelanocortin, made in the pituitary gland, contains at least ten different peptide hormones. These are released by proteolytic cleavage of the polyprotein (Figure 13.32), but not all can be produced at once because of overlaps between individual peptide sequences. Instead, the exact cleavage pattern is different in different cells.

13.3.3 Processing by chemical modification

Genomes have the capacity to code for 22 different amino acids: the 20 specified by the standard genetic code, selenocysteine, and (at least in archaea) pyrrolysine, the latter two being inserted into polypeptides by context-dependent reassignment of 5′–UGA–3′ and 5′–UAG–3′, respectively (Section 1.3.2). This repertoire is increased dramatically by posttranslational chemical modification of proteins, which results in a vast array of different amino acid types. The simpler types of modification occur in all organisms; the more complex ones, especially glycosylation, are rare in bacteria.

The simplest types of chemical modification involve addition of a small chemical group (e.g., an acetyl, methyl, or phosphate group; Table 13.6) to an amino acid side chain, or to the amino or carboxyl groups of the terminal amino acids in a polypeptide. Over 150 different modified amino acids have

Table 13.6 Examples of posttranslational chemical modifications

Modification	Amino acids that are modified	Examples of proteins
Addition of small chemical groups		
Acetylation	Lysine	Histones
Methylation	Lysine	Histones
Phosphorylation	Serine, threonine, tyrosine	Some proteins involved in signal transduction
Hydroxylation	Proline, lysine	Collagen
N-formylation	N-terminal glycine	Melittin
Addition of sugar side chains		
O-linked glycosylation	Serine, threonine	Many membrane proteins and secreted proteins
N-linked glycosylation	Asparagine	Many membrane proteins and secreted proteins
Addition of lipid side chains		
Acylation	Serine, threonine, cysteine	Many membrane proteins
N-myristoylation	N-terminal glycine	Some protein kinases involved in signal transduction
Addition of biotin		
Biotinylation	Lysine	Various carboxylase enzymes

```
         Me  Me        Me Ac P        Ac     Me Ac         Ac   Me Me P
          |   |         |  \ /         |      |  |          |    |  | |
H3   A R T K Q T A R K S T G G K A P R K Q L A T K A R K S A P
                      10                      20
```

Figure 13.33 Posttranslational modification of mammalian histone H3. All of the known modifications occurring in this region are shown. Abbreviations: Ac, acetylation; Me, methylation; P, phosphorylation.

been documented in different proteins, with each modification carried out in a highly specific manner, the same amino acids being modified in the same way in every copy of the protein. This is illustrated in Figure 13.33 for histone H3. The example reminds us that chemical modification often plays a central role in determining the precise biochemical activity of the target protein: we saw in Section 10.2.1 how acetylation, methylation, and phosphorylation of H3 and other histones have an important influence on chromatin structure and hence on genome expression. Chemical modification has several other regulatory roles, phosphorylation in particular being used to activate many proteins involved in signal transduction (Section 14.1.2).

A more complex type of modification is **glycosylation**, the attachment of large, carbohydrate side chains to polypeptides. There are two general types of glycosylation (Figure 13.34):

- **O-linked glycosylation** is the attachment of a sugar side chain via the hydroxyl group of a serine or threonine amino acid.

- **N-linked glycosylation** involves attachment of a sugar side chain through the amino group on the side chain of asparagine.

Glycosylation can result in attachment to the protein of grand structures comprising branched networks of 10–20 sugar units of various types. These side chains help to target proteins to particular sites in cells and determine the stability of proteins circulating in the bloodstream. Another type of large-scale modification involves attachment of long-chain lipids, often to serine or cysteine amino acids. This process is called **acylation** and occurs with many proteins that become associated with membranes. A less common modification is **biotinylation**, in which a molecule of biotin is attached to a small number of enzymes that catalyze the carboxylation of organic acids such as acetate and propionate.

13.3.4 Inteins

The final type of posttranslational processing that we must consider is intein splicing, a protein version of the more extensive intron splicing that occurs with pre-RNAs. Inteins are internal segments of proteins that are removed soon after translation, the two external segments, or exteins, becoming linked together (Figure 13.35). The first intein was discovered in *S. cerevisiae* in 1990, and there have been only 100 confirmed identifications so far. Despite their scarcity, inteins are widespread across different organisms. Most are known in bacteria and archaea but there are also examples in lower eukaryotes. In a few cases there is more than one intein in a single protein.

Most inteins are approximately 150 amino acids in length and, like pre-mRNA introns (Section 12.2.2), the sequences at the splice junctions of inteins have some similarity in most of the known examples. In particular, the first amino acid of the downstream extein is cysteine, serine, or threonine. A few other amino acids within the intein sequence are also conserved. These conserved amino acids are involved in the splicing process, which is self-catalyzed by the intein itself.

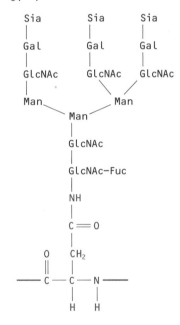

(A) O-linked glycosylation

(B) N-linked glycosylation

Figure 13.34 Glycosylation. (A) O-linked glycosylation. The structure shown is found in a number of glycoproteins. It is drawn here attached to a serine amino acid but it can also be linked to a threonine. (B) N-linked glycosylation usually results in larger sugar structures than are seen with O-linked glycosylation. The drawing shows a typical example of a complex glycan attached to an asparagine amino acid. Abbreviations: Fuc, fucose; Gal, galactose; GalNAc, *N*-acetylgalactosamine; GlcNAc, *N*-acetylglucosamine; Man, mannose; Sia, sialic acid.

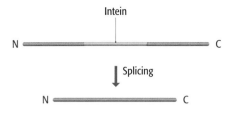

Figure 13.35 Intein splicing.

Two interesting features of inteins have recently come to light. The first of these was discovered when the structures of two inteins were determined by X-ray crystallography. These structures are similar in some respects to that of a *Drosophila* protein called Hedgehog, which is involved in development of the segmentation pattern of the fly embryo. Hedgehog is an autoprocessing protein that cuts itself in two. The structural similarity with inteins lies in the part of the Hedgehog protein that catalyzes its self-cleavage. Possibly the same protein structure has evolved twice, or possibly inteins and Hedgehog shared a common link at some stage in the evolutionary past.

The second interesting feature is that with some inteins the excised segment is a sequence-specific endonuclease. The intein cuts DNA at the sequence corresponding to its insertion site in a gene coding for an intein-free version of the protein from which it is derived (Figure 13.36). If the cell also contains a gene coding for the intein-containing protein, then the DNA sequence for the intein is able to jump into the cut site, converting the intein-minus gene into an intein-plus version, a process called **intein homing**. The same type of event occurs with some Group I introns (Section 12.2.4), which code for proteins that direct **intron homing**. It is possible that transfer of inteins and Group I introns might also occur between cells or even between species. This is thought to be a mechanism by which **selfish DNA** is able to propagate (Section 18.3).

13.4 Protein Degradation

The protein synthesis and processing events that we have studied so far in this chapter result in new, active proteins that take up their place in the cell's proteome. These proteins either replace existing ones that have reached the end of their working lives or provide new protein functions in response to the changing requirements of the cell. The concept that the proteome of a cell can change over time requires not only *de novo* protein synthesis but also the removal of proteins whose functions are no longer required. This removal must be highly selective so that only the correct proteins are degraded, and

Figure 13.36 Intein homing. The cell is heterozygous for the intein-containing gene, possessing one allele with the intein and one allele without the intein. After protein splicing, the intein cuts the intein-minus gene at the appropriate place, allowing a copy of the intein DNA sequence to jump into this gene, converting it into the intein-plus version.

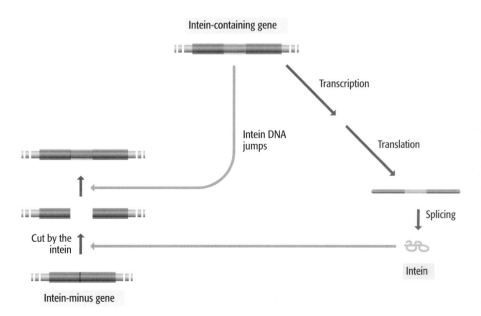

must also be rapid in order to account for the abrupt changes that occur under certain conditions, for example during key transitions in the cell cycle.

For many years, protein degradation was an unfashionable subject and it was not until the 1990s that real progress was made in understanding how specific proteolysis events are linked with processes such as the cell cycle and differentiation. Even now, our knowledge centers largely on descriptions of general protein breakdown pathways and less on the regulation of the pathways and the mechanisms used to target specific proteins. There appear to be a number of different types of breakdown pathway whose interconnectivities have not yet been traced. This is particularly true in bacteria, which seem to have a range of proteases that work together in controlled degradation of proteins. In eukaryotes, most breakdown involves a single system, involving **ubiquitin** and the **proteasome**.

A link between ubiquitin and protein degradation was first established in 1975 when it was shown that this abundant 76-amino-acid protein is involved in energy-dependent proteolysis reactions in rabbit cells. Subsequent research identified a series of three enzymes that attach ubiquitin molecules, singly or in chains, to lysine amino acids in proteins that are targeted for breakdown. There are also ubiquitin-like proteins, such as SUMO, which act in the same way as ubiquitin. Whether or not a protein becomes ubiquitinated depends on the presence or absence within it of amino acid motifs that act as degradation-susceptibility signals. These signals have not been completely characterized but there are thought to be at least ten different types in *S. cerevisiae*, including:

- The **N-degron**, a sequence element present at the N terminus of a protein.
- **PEST sequences**, internal sequences that are rich in proline (P), glutamic acid (E), serine (S), and threonine (T).

These sequences are permanent features of the proteins that contain them and so cannot be straightforward "degradation signals": if they were then these proteins would be broken down as soon as they are synthesized. Instead, they must determine susceptibility to degradation and hence the general stability of a protein in the cell. How this might be linked to the controlled breakdown of selected proteins at specific times, for instance during the cell cycle, is not yet clear.

The second component of the ubiquitin-dependent degradation pathway is the proteasome, the structure within which ubiquitinated proteins are broken down. In eukaryotes, the proteasome is a large, multisubunit structure with a sedimentation coefficient of 26S, comprising a hollow cylinder of 20S and two "caps" of 19S (Figure 13.37). Archaea also have proteasomes of about the same size but these are less complex, being composed of multiple copies of just two proteins: eukaryotic proteasomes contain 14 different types of protein subunit. The entrance into the cavity within the proteasome is narrow, and a protein must be unfolded before it can enter. This unfolding probably occurs through an energy-dependent process and may involve structures similar to chaperonins (Section 13.3.1) but with unfolding rather than folding activity. After unfolding, the protein can enter the proteasome within which it is cleaved into short peptides 4–10 amino acids in length. These are released back into the cytoplasm where they are broken down into individual amino acids which can be reutilized in protein synthesis.

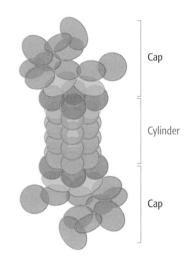

Cap

Cylinder

Cap

Figure 13.37 The eukaryotic proteasome. The protein components of the two caps are shown in orange and red and those forming the cylinder in blue.

Summary

The end result of genome expression is the proteome, the collection of functioning proteins synthesized by a living cell. The identity and relative abundance of the individual proteins in a proteome represents a balance between the synthesis of new proteins and the degradation of existing ones. Transfer RNAs play the central role in protein synthesis by acting as adaptor molecules which form the physical and informational link between the mRNA being translated and the polypeptide being synthesized. Virtually all tRNAs fold into the same base-paired structure, whose two-dimensional representation is called the cloverleaf. The correct amino acid is attached to the 3′ end of a tRNA by an aminoacyl-tRNA synthetase. Aminoacylation is very accurate because each aminoacyl-tRNA synthetase has a high fidelity for its tRNA and amino acid, and because there is also a proofreading mechanism which checks that the correct amino acid has been attached to the correct tRNA. Codon–anticodon interactions ensure that the rules of the genetic code are obeyed. Although the code has 61 triplets that specify amino acids, most cells have fewer than this number of tRNAs, because a single tRNA can read two or more codons by the process called wobble. The structure of the ribosome is being probed in ever-increasing detail by various techniques including X-ray crystallography. Initiation of translation in bacteria involves the small subunit of the ribosome recognizing an internal binding site on the mRNA, a few nucleotides upstream of the initiation codon. In eukaryotes only a few mRNAs have internal binding sites, the more common process involving attachment of the small subunit of the ribosome to the capped 5′ end of the mRNA, followed by scanning along the mRNA to find the initiation codon. Various mechanisms by which initiation of translation can be regulated, either globally or at specific transcripts, are known. Addition of the large subunit of the ribosome to the initiation complex enables the elongation phase of translation to commence. The central activity during elongation is synthesis of peptide bonds. Peptide bond formation is catalyzed by peptidyl transferase, an enzyme activity that resides within one of the rRNA molecules, which acts as a ribozyme. Unusual elongation events include programmed changes of reading frame and translational bypassing, the latter resulting in a substantial segment of the mRNA being skipped by the ribosome. Termination of translation involves special proteins which enter the ribosome when a termination codon is reached. The polypeptide that is initially synthesized must be folded into its correct tertiary structure and possibly processed by proteolytic cleavage and/or chemical modification. A very small number of proteins contain intervening sequences called inteins which must be removed by protein splicing. Protein degradation in eukaryotes involves ubiquitination of proteins targeted for destruction followed by their degradation within a proteasome.

Multiple Choice Questions

*Answers to odd-numbered questions can be found in the Appendix

13.1.* Which of the following is a definition of an isoaccepting tRNA molecule?

   a. A single tRNA molecule that can interact with different codons for the same amino acid.

   b. Different tRNA molecules that are specific for the same amino acid.

   c. Different tRNA molecules that recognize the same codon.

   d. A tRNA molecule that can be aminoacylated with different amino acids.

13.2. Which of the following statements about the specificity of aminoacyl-tRNA synthetases is TRUE?

   a. Each aminoacyl-tRNA synthetase catalyzes the addition of a single amino acid to a single tRNA molecule.

   b. Each aminoacyl-tRNA synthetase catalyzes the addition of a single amino acid to one or more tRNA molecules.

   c. Each aminoacyl-tRNA synthetase catalyzes the addition of one or more amino acids to a single tRNA molecule.

   d. Each aminoacyl-tRNA synthetase catalyzes the addition of one or more amino acids to one or more tRNA molecules.

13.3.* Codon–anticodon interactions occur by:

   a. Covalent bonds.

   b. Electrostatic interactions.

   c. Hydrogen bonds.

   d. Hydrophobic interactions.

13.4. Wobble between a codon and anticodon occurs between:

   a. The first nucleotide of the codon and the first nucleotide of the anticodon.

   b. The first nucleotide of the codon and the third nucleotide of the anticodon.

   c. The third nucleotide of the codon and the first nucleotide of the anticodon.

   d. The third nucleotide of the codon and the third nucleotide of the anticodon.

13.5.* Wobble occurs because of all of the following EXCEPT:

   a. The anticodon is in a loop of the tRNA molecule and does not align uniformly with the codon.

   b. An inosine nucleotide in the tRNA molecule can base-pair with A, C, and U in the mRNA.

   c. An inosine nucleotide in the mRNA molecule can base-pair with A, C, and U in the tRNA.

   d. Guanine can base-pair with uracil.

13.6. What is the first step in the initiation of translation in bacteria?

   a. The small subunit of the ribosome binds to the 5′ cap of the mRNA and scans the mRNA for the initiation codon.

   b. The large subunit of the ribosome binds to the ribosome binding site on the mRNA molecule.

   c. The ribosome binds to the initiation codon on the mRNA molecule.

   d. The small subunit of the ribosome binds to the ribosome binding site on the mRNA molecule.

13.7.* The formyl group attached to the initiator methionine of bacteria serves what function?

   a. It links the initiator tRNA to the large ribosomal subunit as translation is initiated.

   b. It binds the GTP molecule required for assembly of the initiation complex.

   c. It blocks the amino group of the methionine ensuring that protein synthesis occurs in the N→C direction.

   d. It blocks the side chain of the methionine so that it does not react with the initiation factor IF-3.

13.8. What is the general mechanism for the initiation of translation in eukaryotes?

   a. The small subunit of the ribosome binds to the 5′ cap of the mRNA and scans the mRNA for the initiation codon.

   b. The large subunit of the ribosome binds to the 5′ cap of the mRNA and scans the mRNA for the initiation codon.

   c. The ribosome binds to the initiation codon on the mRNA molecule.

   d. The small subunit of the ribosome binds to the ribosome binding site on the mRNA molecule.

13.9.* What is the function of the initiation factor eIF-6?

   a. It binds to the initiator tRNA^{Met} and GTP during assembly of the preinitiation complex.

   b. It functions as a bridge between the 5′ cap of the mRNA and the preinitiation complex.

   c. It releases the other initiation factors as the ribosome assembles at the initiation codon.

   d. It prevents the large ribosomal subunit from attaching to the small subunit in the cytoplasm.

13.10. What is the function of elongation factor EF-1A?

   a. It catalyzes the formation of peptide bonds.

continued …

Multiple Choice Questions (continued) *Answers to odd-numbered questions can be found in the Appendix

b. It ensures that the correct aminoacyl-tRNA enters the ribosome.

c. It prevents tRNA molecules from leaving the ribosome prior to the formation of peptide bonds.

d. It hydrolyzes GTP to assist the translocation of the ribosome along the mRNA.

13.11. *How does frameshifting occur during translation?

a. A ribosome translates an mRNA molecule that contains an extra or missing nucleotide.

b. A ribosome skips a codon during the translation of an mRNA molecule.

c. A ribosome pauses during translation and moves back or forward one nucleotide and then continues translation.

d. A ribosome terminates translation at a codon that usually specifies an amino acid.

13.12. How is protein synthesis terminated?

a. A release factor recognizes the termination codon and enters the A site.

b. A tRNA for the termination codon enters the A site.

c. A tRNA for the termination codon enters the P site.

d. The ribosome stalls at the termination codon and catalyzes the release of the protein from the tRNA.

13.13. *Which of the following is NOT a reason why proteins require assistance in folding during translation or following denaturation?

a. After denaturation, proteins may form insoluble aggregates caused by an inability to shield their hydrophobic groups from water.

b. After denaturation, proteins may take up an incorrectly folded but stable state.

c. During translation, the partially translated protein is a random coil, unable to fold into any specific conformation.

d. During translation, partially translated proteins might begin to fold incorrectly before the entire protein is synthesized.

13.14. Which of the following is NOT a function of molecular chaperones in protein folding?

a. Molecular chaperones assist proteins in finding their correct structure.

b. Molecular chaperones specify the tertiary structure of a protein.

c. Molecular chaperones can stabilize partially folded proteins and prevent them from aggregating with other proteins.

d. Molecular chaperones can shield and protect exposed hydrophobic regions of proteins.

13.15. *Which of the following is NOT an example of posttranslational chemical modification of proteins?

a. Glycosylation.

b. Methylation.

c. Phosphorylation.

d. Proteolysis.

13.16. What are inteins?

a. External or internal segments of proteins that are removed by proteolysis resulting in an active protein.

b. External segments of proteins that are added to other proteins by protein ligases.

c. Internal segments of proteins that are removed after translation with the external segments becoming linked together.

d. External segments of proteins that are covalently attached to lipids for membrane insertion.

Short Answer Questions

*Answers to odd-numbered questions can be found in the Appendix

13.1. * How do tRNA molecules function as both a physical and informational link between the mRNA molecule being translated and the protein being synthesized?

13.2. Describe briefly the two-step reaction that results in the attachment of an amino acid to a tRNA molecule.

13.3. * What would happen if an aminoacyl-tRNA synthetase incorrectly attached the wrong amino acid to a tRNA molecule (for example, aminoacylated the tRNA for valine with isoleucine)?

13.4. What are the two active roles of ribosomes during protein synthesis?

Short Answer Questions (continued)

13.5.* List the molecules present in the preinitiation complex that assembles during the first step of translation initiation in eukaryotes.

13.6. What role is the poly(A) tail thought to have during translation initiation in eukaryotes?

13.7.* How can eukaryotic cells quickly repress translation in response to a stress such as heat shock?

13.8. What features of the *dnax* mRNA are thought to induce programmed frameshifting when the mRNA is translated?

13.9.* What are the two steps that occurr during the refolding of a small protein?

13.10. Discuss the differences between the ways in which Hsp70 chaperones and the GroEL/GroES chaperonin function.

13.11.* How is an intein removed from a protein?

13.12. Describe two of the signals that are recognized by the enzymes that attach ubiquitin molecules to proteins targeted for degradation. What happens to proteins that have become ubiquitinylated in this way?

In-depth Problems

13.1.* Why are there two classes of aminoacyl-tRNA synthetases?

13.2. In human mitochondria, protein synthesis requires only 22 different tRNAs. What implications does this have for the rules governing codon–anticodon interactions in this system?

13.3.* How might the genetic code have originated?

13.4. Most organisms display a distinct codon bias in their genes. For instance, of the four codons for proline, only two—CCU and CCA—appear at all frequently in genes of *Saccharomyces cerevisiae*: CCC and CCG are less common. It has been suggested that a gene that contains a relatively high number of unfavored codons might be expressed at a relatively slow rate. Explain the thinking behind this hypothesis and discuss its ramifications.

13.5.* To what extent have studies of ribosome structure been of value in understanding the detailed process by which proteins are synthesized?

continued …

Figure Tests

13.1.* At which position in a tRNA anticodon is inosine sometimes located? When inosine is present, what is its function?

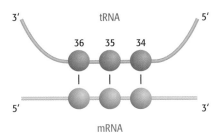

13.2. Describe the initiation of translation in *E. coli*.

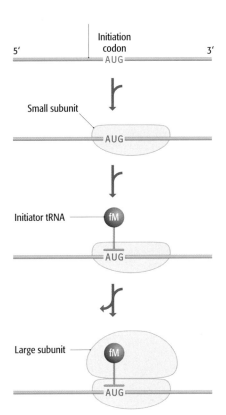

13.3.* The figure shows the distances between the CCdA-phosphate-puromycin molecule (red dot) and the nearest proteins in the large ribosomal subunit of *E. coli*. What does this figure reveal about the formation of peptide bonds in the ribosome?

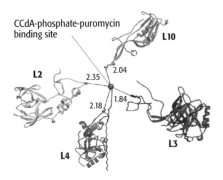

13.4. Discuss how the activity of a protein is dependent on events that occur after translation.

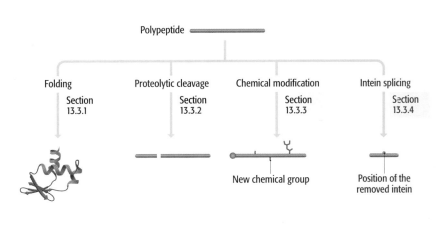

Further Reading

tRNA structure and function

Clark, B.F.C. (2001) The crystallization and structural determination of tRNA. *Trends Biochem. Sci.* **26:** 511–514.

Hale, S.P., Auld, D.S., Schmidt, E. and Schimmel, P, (1997) Discrete determinants in transfer RNA for editing and aminoacylation. *Science* **276:** 1250–1252. *Ensuring the accuracy of aminoacylation.*

Ibba, M. and Söll, D. (2000) Aminoacyl-tRNA synthetases. *Annu. Rev. Biochem.* **69:** 617–650.

Percudani, R. (2001) Restricted wobble rules for eukaryotic genomes. *Trends Genet.* **17:** 133–135.

Probing the structure of the ribosome

Ban, N., Nissen, P., Hansen, J., Moore, P.B. and Steitz, T.A. (2000) The complete atomic structure of the large ribosomal subunit at 2.4 Å resolution. *Science* **289:** 905–920.

Heilek, G.M. and Noller, H.F. (1996) Site-directed hydroxyl radical probing of the rRNA neighborhood of ribosomal protein S5. *Science* **272:** 1659–1662.

Moore, P.B. and Steitz, T.A. (2003) The structural basis of large ribosome subunit function. *Annu. Rev. Biochem.* **72:** 813–850.

Wimberly, B.T., Brodersen, D.E., Clemons, W.M., Morgan-Warren, R.J., Carter, A.P., Vonrhein, C., Hartsch, T. and Ramakrishnan, V. (2000) Structure of the 30S ribosomal subunit. *Nature* **407:** 327–339.

Yusupov, M.M., Yusupova, G.Z., Baucom, A., Lieberman, K., Earnest, T.N., Cate, J.H. and Noller, H.F. (2001) Crystal structure of the ribosome at 5.5 Å resolution. *Science* **292:** 883–896.

Mechanics of protein synthesis

Andersen, G.R., Nissen, P. and Nyborg, J. (2003) Elongation factors in protein biosynthesis. *Trends Biochem. Sci.* **28:** 434–441.

Frank, J. and Agarwal, R.K. (2000) A ratchet-like inter-subunit reorganization of the ribosome during translocation. *Nature* **406:** 318–322.

Ibba, M. and Söll, D. (1999) Quality control mechanisms during translation. *Science* **286:** 1893–1897.

Kapp, L.D. and Lorsch, J.R. (2004) The molecular mechanics of eukaryotic translation. *Annu. Rev. Biochem.* **73:** 657–704.

McCarthy, J.E.G. (1998) Posttranscriptional control of gene expression in yeast. *Microbiol. Mol. Biol. Rev.* **62:** 1492–1553. *Detailed review of translation and its control in yeast.*

Nakamura, Y. and Ito, K. (2003) Making sense of mimic in translation termination. *Trends Biochem. Sci.* **28:** 99–105. *The mode of action of release and ribosome recycling factors.*

Rodnina, M.V. and Wintermeyer, W. (2001) Ribosome fidelity: tRNA discrimination, proofreading and induced fit. *Trends Biochem. Sci.* **26:** 124–130.

Peptidyl transferase is a ribozyme

Nissen, P., Hansen, J., Ban, N., Moore, P.B. and Steitz, T.A. (2000) The structural basis of ribosome activity in peptide bond synthesis. *Science* **289:** 920–930.

Polacek, N., Gaynor, M., Yassin, A. and Mankin, A.S. (2001) Ribosomal peptidyl transferase can withstand mutations at the putative catalytic nucleotide. *Nature* **411:** 498–501.

Steitz, T.A. and Moore, P.B. (2003) RNA, the first macromolecular catalyst: the ribosome is a ribozyme. *Trends Biochem. Sci.* **28:** 411–418.

Unusual events in translation

Farabaugh, P.J. (1996) Programmed translational frameshifting. *Annu. Rev. Genet.* **30:** 507–528.

Herr, A.J., Atkins, J.F. and Gesteland, R.F. (2000) Coupling of open reading frames by translational bypassing. *Annu. Rev. Biochem.* **69:** 343–372.

Protein folding

Anfinsen, C.B. (1973) Principles that govern the folding of protein chains. *Science* **181:** 223–230. *The first experiments on protein folding.*

Daggett, V. and Fersht, A.R. (2003) Is there a unifying mechanism for protein folding? *Trends Biochem. Sci.* **28:** 18–25.

Frydman, J. (2001) Folding of newly translated proteins *in vivo*: the role of molecular chaperones. *Annu. Rev. Biochem.* **70:** 603–649.

Xu, Z., Horwich, A.L. and Sigler, P.B. (1997) The crystal structure of the asymmetric GroEL-GroES-(ADP)$_7$ chaperonin complex. *Nature* **388:** 741–750.

Protein processing and modification

Chapman-Smith, A. and Cronan, J.E. (1999) The enzymatic biotinylation of proteins: a post-translational modification of exceptional specificity. *Trends. Biochem. Sci.* **24:** 359–363.

Drickamer, K. and Taylor, M.E. (1998) Evolving views of protein glycosylation. *Trends Biochem. Sci.* **23:** 321–324.

Paulus, H. (2000) Protein splicing and related forms of protein autoprocessing. *Annu. Rev. Biochem.* **69:** 447–496.

Protein degradation

Varshavsky, A. (1997) The ubiquitin system. *Trends Biochem. Sci.* **22:** 383–387.

Voges, D., Zwickl, P. and Baumeister, W. (1999) The 26S proteasome: a molecular machine designed for controlled proteolysis. *Annu. Rev. Biochem.* **68:** 1015–1068.

Regulation of Genome Activity

14

When you have read Chapter 14, you should be able to:

Distinguish between differentiation and development, and outline how regulation of genome expression underlies these two processes.

Describe, with examples, the various ways in which imported signaling compounds, such as lactoferrin and steroid hormones, can bring about transient changes in genome activity.

Give a detailed account of catabolite repression in bacteria.

Discuss the various pathways by which signals from cell surface receptors are transmitted to the genome.

Describe, with examples, the various ways in which permanent and semipermanent changes in genome activity can be brought about, making clear distinction between those processes that involve rearrangement of the genome, those that involve changes in chromatin structure, and those that involve feedback loops.

Discuss how studies of the lysogenic infection cycle of bacteriophage λ, and of sporulation in *Bacillus subtilis*, provide basic information relevant to differentiation and development.

Explain why *Caenorhabditis elegans* is a useful model organism and describe how cell fate is determined during development of the *C. elegans* vulva.

Describe the genetic events that underlie embryogenesis in *Drosophila melanogaster*.

Discuss the roles of homeotic genes in *D. melanogaster*, vertebrates, and plants.

We have followed the pathway by which expression of the genome specifies the content of the proteome, which in turn determines the biochemical signature of the cell. In no organism is this biochemical signature entirely constant. Even the simplest unicellular organisms are able to alter their proteomes to take account of changes in the environment, so that their biochemical capabilities are continually in tune with the available nutrient supply and the prevailing physical and chemical conditions. Cells in multicellular organisms are equally responsive to changes in the extracellular environment, the only difference being that the major stimuli include hormones and growth factors as well as nutrients. The resulting *transient* changes in genome activity enable the proteome to be remodeled continuously to satisfy the demands that the outside world places on the cell (Figure 14.1). Other changes in genome activity are *permanent* or at least *semipermanent*, and result in the cell's biochemical signature becoming altered in a way that is not readily reversible. These

Figure 14.1 Two ways in which genome activity is regulated. The genes on the left are subject to transient regulation and are switched on and off in response to changes in the extracellular environment. The genes on the right have undergone a permanent or semipermanent change in their expression pattern, resulting in the same three genes being expressed continuously.

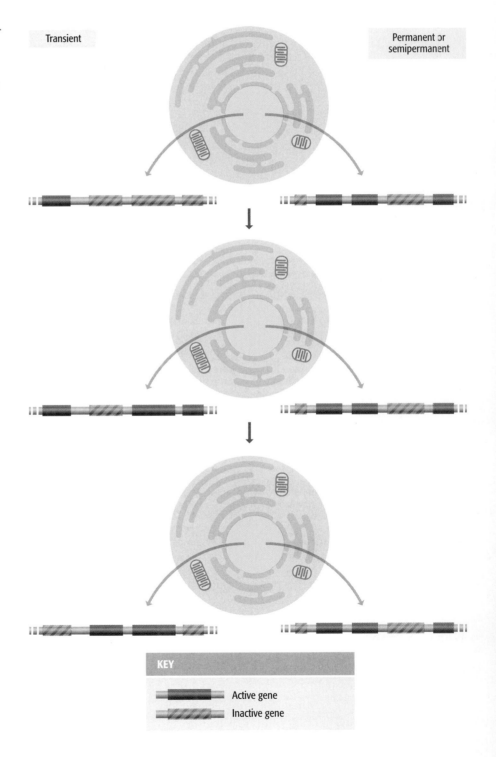

changes lead to cellular **differentiation**, the adoption by the cell of a specialized physiological role. Differentiation pathways are known in many unicellular organisms, an example being the production of spore cells by bacteria such as *Bacillus*, but we more frequently associate differentiation with multicellular organisms, in which a variety of specialized cell types (over 250 types in humans) are organized into tissues and organs. Assembly of these complex multicellular structures, and of the organism as a whole, requires coordination of the activities of the genomes in different cells. This coordination involves both transient and permanent changes, and must continue over a long period of time during the **development** of the organism.

There are many steps within the expression pathways for individual genes at which regulation can be exerted (Table 14.1). Examples of the biological roles of different regulatory mechanisms were given at the appropriate places in Chapters 10–13. The objective of this chapter is not to reiterate these gene-specific control systems, but to explain how the activity of the genome as a whole is regulated. In doing this we should bear in mind that the biosphere is so diverse, and the numbers of genes in individual genomes so large, that it is reasonable to assume that any mechanism that could have evolved to regulate genome expression is likely to have done so. It is therefore no surprise that we can nominate examples of regulation for every point in the genome expression pathway. But are all these control points of equal importance in regulating the activity of the genome as a whole? Our current perception is that they are not. Our understanding may be imperfect, based as it is on investigation of just a limited number of genes in a few organisms, but it appears that the critical controls over genome expression—the decisions about which genes are switched on and which are switched off—are exerted at the level of transcription initiation. For most genes, control that is exerted at later steps serves to modulate expression but does not act as the primary determinant of whether the gene is on or off (see Figure 11.22). Most, but not all, of what we will discuss in this chapter therefore concerns control of genome activity by mechanisms that specify which genes are transcribed and which are silent. We will address two issues: the ways in which transient and permanent changes in genome activity are brought about, and the ways in which these changes are linked in time and space in developmental pathways.

14.1 Transient Changes in Genome Activity

Transient changes in genome activity occur predominantly in response to external stimuli. For unicellular organisms, the most important external stimuli relate to nutrient availability, these cells living in variable environments in which the identities and relative amounts of the nutrients change over time. The genomes of unicellular organisms therefore include genes for uptake and utilization of a range of nutrients, and changes in nutrient availability are shadowed by changes in genome activity, so that at any one time only those genes needed to utilize the available nutrients are expressed. Most cells in multicellular organisms live in less variable environments, but an environment whose maintenance requires coordination between the activities of different cells. For these cells, the major external stimuli are therefore hormones, growth factors, and related compounds that convey signals within the organism and stimulate coordinated changes in genome activity.

Table 14.1 Examples of steps in the genome expression pathway at which regulation can be exerted

Step	Example of regulation
Transcription	
Gene accessibility	Locus control regions determine chromatin structure in areas that contain genes (Section 10.1.2)
	Histone modifications influence chromatin structure and determine which genes are accessible (Section 10.2.1)
	Nucleosome positioning controls access of RNA polymerase and transcription factors to the promoter region (Section 10.2.2)
	DNA methylation silences regions of the genome (Section 10.3.1)
Initiation of transcription	Productive initiation is influenced by activators, repressors, and other control systems (Section 11.3)
Synthesis of RNA	Prokaryotes use antitermination and attenuation to control the amount and nature of individual transcripts (Section 12.1.2)
Eukaryotic mRNA processing	
Capping	Some animals use capping as a means of regulating protein synthesis during egg maturation
Polyadenylation	Alternative polyadenylation sites control flowering in *Arabidopsis*
	Translation of *bicoid* mRNA in *Drosophila* eggs is activated after fertilization by extension of the poly(A) tail (Section 14.3.4)
Splicing	Alternative splice site selection controls sex determination in *Drosophila* (Section 12.2.2)
Chemical modification	RNA editing of apolipoprotein-B mRNA results in liver- and intestine-specific versions of this protein (Section 12.2.5)
mRNA degradation	MicroRNAs control cell death, specification of neuron cell types, and control of fat storage in *Caenorhabditis elegans*, as well as many diverse processes in other eukaryotes (Section 12.2.6)
	Iron controls degradation of transferrin receptor mRNA (Section 13.2.2)
Protein synthesis and processing	
Initiation of translation	Phosphorylation of eIF-2 results in a general reduction in translation initiation in eukaryotes (Section 13.2.2)
	Ribosomal proteins in bacteria control their own synthesis by modulating ribosome attachment to their mRNAs (Section 13.2.2)
	In some eukaryotes, iron controls ribosome scanning on ferritin mRNAs (Section 13.2.2)
	Small RNAs in bacteria regulate the response to oxidative stress by modulating initiation of translation of various mRNAs (Section 13.2.2)
Protein synthesis	Frameshifting enables two DNA polymerase III subunits to be translated from the *Escherichia coli dnaX* gene (Section 13.2.3)
Cutting events	Alternative cleavage pathways for polyproteins result in tissue-specific protein products (Section 13.3.2)
Chemical modification	Many proteins involved in signal transduction are activated by phosphorylation (Section 14.1.2)

To exert an effect on genome activity, the nutrient, hormone, growth factor, or other extracellular compound that represents the external stimulus must influence events within the cell. There are two ways in which it can do this (Figure 14.2):

- Directly, by acting as a signaling compound that is transported across the cell membrane and into the cell.

- Indirectly, by binding to a cell surface receptor which transmits a signal into the cell.

Signal transmission, by direct or indirect means, is one of the major research areas in cell biology, with attention focused in particular on its relevance to the abnormal biochemical activities that underlie cancer. Many examples of signal transmission have been discovered, some of general importance in a variety of organisms and others restricted to just a few species. In the first part of this chapter we will survey these signaling pathways.

14.1.1 Signal transmission by import of the extracellular signaling compound

In the direct method of signal transmission, the extracellular compound that represents the external stimulus crosses the cell membrane and enters the cell. After import into the cell, the signaling compound could influence genome activity by any one of three routes (Figure 14.3).

- If the signaling compound is a protein, then it could act in the same way as one of the various regulatory proteins that we met in Chapters 10–13, for example by activating or repressing assembly of the transcription initiation complex (Section 11.3), or by interacting with a splicing enhancer or silencer (Section 12.2.2).

- The signaling compound could directly influence the activity of an existing regulatory protein. Such a signaling compound need not itself be a protein: it could, theoretically, be any type of compound.

- The signaling compound could influence the activity of an existing regulatory protein via one or more intermediates, rather than by interacting with it directly.

Examples of each of these three modes of action are described below.

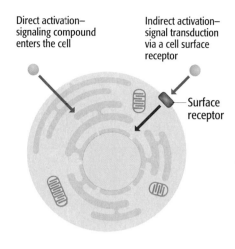

Figure 14.2 Two ways in which an extracellular signaling compound can influence events occurring within a cell.

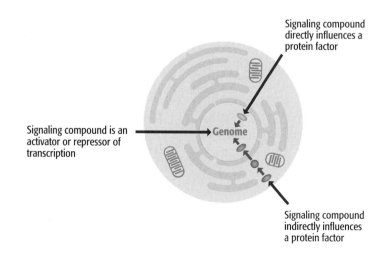

Figure 14.3 Three ways in which an extracellular signaling compound could influence genome activity after import into the cell.

Lactoferrin is an extracellular signaling protein which acts as a transcription activator

If the extracellular signaling compound that is imported into the cell is a protein with suitable properties then it could directly affect the activity of its target genes by acting as an activator or repressor of some stage in the genome expression pathway. This might appear to be an attractively straightforward way of regulating genome activity, but it is not a common mechanism. The reason for this is not clear but probably relates, at least partly, to the difficulty in designing a protein that combines the hydrophobic properties needed for effective transport across a membrane with the hydrophilic properties needed for migration through the aqueous cytoplasm to the protein's site of action in the nucleus or on a ribosome.

The one clear example of a signaling compound that can function in this way is provided by lactoferrin, a mammalian protein found mainly in milk and, to a lesser extent, in the bloodstream. Lactoferrin is a transcription activator (Section 11.3.2). Its specific function has been difficult to pin down, but it seems to play a role in the body's defenses against microbial attack. As its name suggests, lactoferrin is able to bind iron, and it is thought that at least part of its protective role arises from its ability to reduce free-iron levels in milk, thereby starving invading microbes of this essential cofactor. It might therefore appear unlikely that lactoferrin would have a role in genome expression, but it has been known since the early 1980s that the protein is multitalented and, among other things, can bind to DNA. This property was linked to a second function of lactoferrin—stimulation of the blood cells involved in the immune response—when in 1992 it was shown that the protein is taken up by immune cells, enters their nuclei, and attaches to the genome. Subsequently the DNA binding was shown to be sequence specific and to result in transcription being stimulated, confirming that lactoferrin is a true transcription activator.

Some imported signaling compounds directly influence the activity of preexisting regulatory proteins

Although only a few imported signaling compounds are able themselves to act as activators or repressors of genome expression, many have the ability to influence directly the activity of regulatory proteins that are already present in the cell. We encountered one example of this type of regulation in Section 11.3.1 when we studied the lactose operon of *Escherichia coli*. This operon responds to extracellular levels of lactose, this compound acting as a signaling molecule which enters the cell and, after conversion to its isomer allolactose, influences the DNA-binding properties of the lactose repressor and hence determines whether or not the lactose operon is transcribed (see Figure 11.24). Many other bacterial operons coding for genes involved in sugar utilization are controlled in this way.

Direct interaction between a signaling compound and a transcription activator or repressor is also a common means of regulating genome activity in eukaryotes. A good example is provided by the control system that maintains the intracellular metal-ion content at an appropriate level. Cells need metal ions such as copper and zinc as cofactors in biochemical reactions, but these metals are toxic if they accumulate in the cell above a certain level. Their uptake therefore has to be carefully controlled so that the cell contains sufficient metal ions when the environment is lacking in metal compounds, but does not overaccumulate metal ions when the environmental concentrations

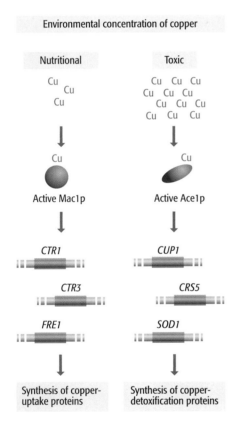

Figure 14.4 Copper-regulated gene expression in *Saccharomyces cerevisiae*. Yeast requires low amounts of copper because a few of its enzymes (e.g., cytochrome *c* oxidase and tyrosinase) are copper-containing metalloproteins, but too much copper is toxic for the cell. When copper levels are low, the Mac1p protein factor is activated by copper binding and switches on expression of genes for copper uptake. When the copper levels are too high, a second factor, Ace1p, is activated, switching on expression of a different set of genes, these coding for proteins involved in copper detoxification.

are high. The strategies used are illustrated by the copper-control system of *Saccharomyces cerevisiae*. This yeast has two copper-dependent transcription activators, Mac1p and Ace1p. Both of these activators bind copper ions, the binding inducing a conformational change that enables the factor to stimulate expression of its target genes (Figure 14.4). For Mac1p these target genes code for copper-uptake proteins, whereas for Ace1p they are genes coding for proteins such as superoxide dismutase that are involved in copper detoxification. The metal-controlled balance between the activities of Mac1p and Ace1p ensures that the copper content of the cell remains within acceptable levels.

Transcription activators are also the targets for **steroid hormones,** which are signaling compounds that coordinate a range of physiological activities in the cells of higher eukaryotes. They include the sex hormones (estrogens for female sex development, androgens for male sex development), and the glucocorticoid and mineralocorticoid hormones. Steroids are hydrophobic and so easily penetrate the cell membrane. Once inside the cell, each hormone binds to a specific **steroid receptor** protein, which is usually located in the cytoplasm. After binding, the activated receptor migrates into the nucleus, where it attaches to a **hormone response element** upstream of a target gene. Once bound, the receptor acts as a transcription activator. Response elements for each receptor are located upstream of 50–100 genes, often within enhancers, so a single steroid hormone can induce a large-scale change in the biochemical properties of the cell. All steroid receptors are structurally similar, not just with regard to their DNA-binding domains but also in other parts of their protein structures (Figure 14.5). Recognition of these similarities has led to the identification of a number of putative or orphan steroid receptors whose hormonal partners and cellular functions are not yet known. The structural similarities have also shown that a second set of receptor proteins, the **nuclear receptor superfamily,** belongs to the same general class as steroid receptors, although the hormones that they work with are not steroids. As their name suggests, these receptors are located in the nucleus rather than the cytoplasm. They include the receptors for vitamin D3, whose roles include control of bone development, and thyroxine, which stimulates the tadpole-to-frog metamorphosis.

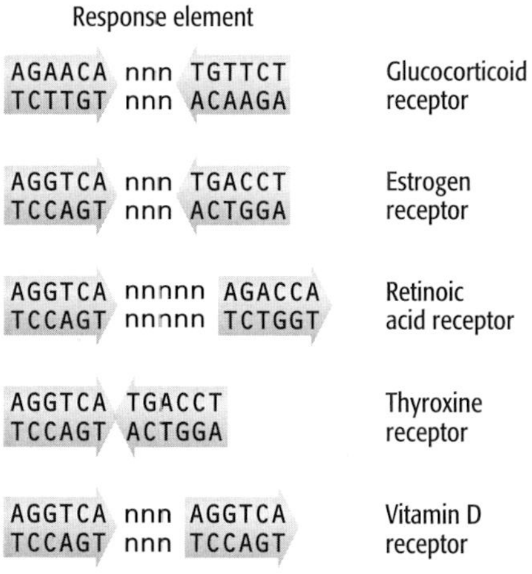

Steroid and nuclear receptors are dimers, each subunit of which possesses one of the special zinc fingers that are characteristic of this group of proteins (see Figure 11.5). Each of these zinc fingers recognizes and binds to a 6 bp sequence in its hormone response element. For most steroid receptors, the pair of 6 bp sequences are arranged as a direct or inverted repeat separated by a 0–4 bp spacer (Figure 14.6). The response elements for nuclear receptors are similar except that the recognition sequences are almost always direct repeats. The spacer serves simply to ensure that the distance between the recognition sequences is appropriate for the orientation of the zinc fingers in the receptor protein. This means that different receptor proteins can possess the same pair of zinc fingers but recognize different response elements, specificity being maintained by the orientation of the fingers and the spacing between the recognition sequences.

Some imported signaling compounds influence genome activity indirectly

The link between a signaling molecule and the regulatory proteins involved in genome expression does not have to be as direct as in the examples

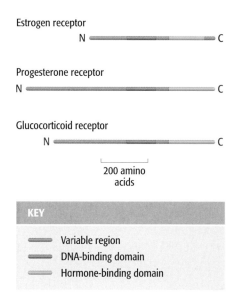

Figure 14.5 All steroid hormone receptor proteins have similar structures. Three receptor proteins are compared. Each one is shown as an unfolded polypeptide with the two conserved functional domains aligned. The DNA-binding domain is very similar in all steroid receptors, displaying 50%–90% amino acid sequence identity. The hormone-binding domain (Section 11.3.2) is less well conserved, with 20%–60% sequence identity. The activation domain lies between the N terminus and the DNA-binding domain, but this region displays little sequence similarity in different receptors.

Figure 14.6 The sequences of typical steroid and nuclear receptor response elements. The retinoic acid receptor is unusual as the 6 bp sequences are not exact repeats and they are separated by more than four nucleotides.

described in the previous section. Signaling molecules can also influence genome activity in an indirect manner via one or more intermediates. An example is provided by the **catabolite repression** system of bacteria. This is the means by which extracellular and intracellular glucose levels dictate whether or not operons for utilization of other sugars are switched on when those alternative sugars are present in the medium.

This phenomenon was discovered in 1941 by Jacques Monod, who showed that if *E. coli* or *Bacillus subtilis* are provided with a mixture of sugars, then one sugar will be metabolized first, the bacteria turning to the second sugar only when the first has been used up. Monod used a French word to describe this: **diauxie**. One combination of sugars that elicits a diauxic response is glucose plus lactose, glucose being used before the lactose (Figure 14.7A). When the details of the lactose operon were worked out some 20 years later (Section 11.3.1), it became clear that the diauxie between glucose and lactose must involve a mechanism whereby the presence of glucose can override the inductive effect that lactose usually has on its operon. In the presence of lactose plus glucose, the lactose operon is switched off, even though some of the lactose in the mixture is converted into allolactose which binds to the lactose repressor, so that under normal circumstances the operon would be transcribed (Figure 14.7B).

The explanation for the diauxic response is that glucose acts as a signaling compound that represses expression of the lactose operon, as well as other sugar utilization operons, through an indirect influence on the **catabolite activator protein** (also called the CRP activator). This protein binds to a recognition sequence at various sites in the bacterial genome and activates transcription initiation at downstream promoters, probably by interacting with the α subunit of the RNA polymerase. Inherent in this activation is the creation of a sharp, 90° bend in the double helix in the region of the binding site when the catabolite activator protein is attached. Productive initiation of transcription at these promoters is dependent on the presence of bound catabolite activator protein: if the protein is absent then the genes controlled by the promoter are not transcribed.

Glucose does not itself interact with the catabolite activator protein. Instead, glucose controls the level in the cell of the modified nucleotide **cyclic AMP** (**cAMP**; Figure 14.7C). It does this by inhibiting the activity of **adenylate cyclase**, the enzyme that synthesizes cAMP from ATP. Inhibition is mediated by a protein called IIA^{Glc}, a component of a multiprotein complex that transports sugars into the bacterium. When glucose is being transported into the cell, IIA^{Glc} becomes dephosphorylated (i.e., additional phosphate groups previously added to the protein by posttranslational modification are removed). The dephosphorylated version of IIA^{Glc} inhibits adenylate cyclase activity. This means that if glucose levels are high, the cAMP content of the cell is low. The catabolite activator protein can bind to its target sites only in the presence of cAMP, so when glucose is present the protein remains detached and the operons it controls are switched off. In the specific case of diauxie involving glucose plus lactose, the indirect effect of glucose on the catabolite activator

(A)

(B)

(C)

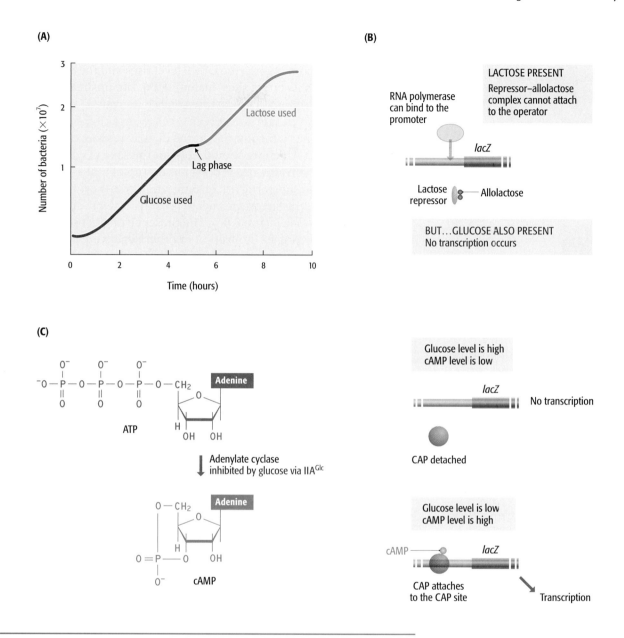

Figure 14.7 Catabolite repression. (A) A typical diauxic growth curve, as seen when *Escherichia coli* is grown in a medium containing a mixture of glucose and lactose. During the first few hours, the bacteria divide exponentially, using the glucose as the carbon and energy source. When the glucose is used up, there is a brief lag period while the *lac* genes are switched on before the bacteria return to exponential growth, now using up the lactose. (B) Glucose overrides the lactose repressor. If lactose is present, then the repressor detaches from the operator and the lactose operon should be transcribed, but it remains silent if glucose is also present. Refer to Figure 11.24B for details of how the lactose repressor controls expression of the lactose operon. (C) Glucose exerts its effect on the lactose operon and other target genes via IIA^{Glc}, which controls the activity of adenylate cyclase and hence regulates the amount of cAMP in the cell. The catabolite activator protein (CAP) can attach to its DNA binding site only in the presence of cAMP. If glucose is present, the cAMP level is low, so CAP does not bind to the DNA and does not activate the RNA polymerase. Once the glucose has been used up, the cAMP level rises, allowing CAP to bind to the DNA and activate transcription of the lactose operon and its other target genes.

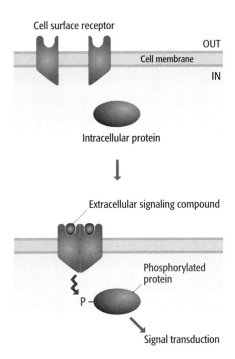

Figure 14.8 The role of a cell surface receptor in signal transduction. Binding of the extracellular signaling compound to the outer surface of the receptor protein causes a conformational change, often involving dimerization, that results in activation of an intracellular protein, for example by phosphorylation. The events occurring "downstream" of this initial protein activation are diverse, as described in the text. "P" indicates a phosphate group, PO_3^{2-}.

protein means that the lactose operon remains inactivated, even though the lactose repressor is not bound, and so the glucose in the medium is used up first. When the glucose is gone, the cAMP level rises and the catabolite activator protein binds to its target sites, including the site upstream of the lactose operon, and transcription of the lactose genes is activated.

Protein IIA^{Glc} also has a second role in the diauxic response, one not involving the genome but which we should note for the sake of completeness. The dephosphorylated form of IIA^{Glc} prevents the uptake of lactose and other sugars by inhibiting the permease enzymes that transport these sugars into the cell—recall that the lactose permease is coded by *lacY*, the second gene in the lactose operon (see Figure 8.8A). The presence of glucose therefore has a dual effect: the operons for utilization of other sugars are switched off, and the uptake of those sugars is prevented.

14.1.2 Signal transmission mediated by cell surface receptors

Many extracellular signaling compounds are unable to enter the cell because they are too hydrophilic to penetrate the lipid membrane and the cell lacks a specific transport mechanism for their uptake. In order to influence genome activity these signaling compounds must bind to cell surface receptors that carry their signals across the cell membrane. These receptors are proteins that span the membrane, with a site for binding the signaling compound on the outer surface. Binding of the signaling compound results in a conformational change in the receptor. Often this conformation change is dimerization, the liquid nature of the cell membrane permitting a limited amount of lateral movement by membrane proteins, enabling the subunits of dimers to associate and disassociate in response to the presence or absence of the extracellular signal (Figure 14.8). The conformation change induces a biochemical event within the cell. For example, the intracellular segments of many receptor proteins have kinase activity, so when the subunits are brought together in a dimer they phosphorylate each other. This biochemical event, whether

Table 14.2 Cell surface receptor proteins involved in signal transmission into eukaryotic cells

Receptor type	Description	Signals
G-protein-coupled	Activate intracellular G-proteins, which bind GTP and control biochemical activities by conversion of this GTP to GDP with the release of energy	Diverse: epinephrine, peptides (e.g., glucagon), protein hormones, odorants, light
Tyrosine kinases	Activate intracellular proteins by tyrosine phosphorylation	Hormones (e.g., insulin), various growth factors
Tyrosine-kinase-associated	Similar to tyrosine kinase receptors but activate intracellular proteins indirectly (e.g., see description of STATs in the text)	Hormones, growth factors
Serine–threonine kinases	Activate intracellular proteins by serine and/or threonine phosphorylation	Hormones, growth factors
Ion channels	Control intracellular activities by regulating the movement of ions and other small molecules into and out of cells	Chemical stimuli (e.g., glutamate), electrical charges

mutual phosphorylation or some other reaction, forms the first step in the intracellular stage of the **signal transduction** pathway.

Several types of cell surface receptor have been discovered (Table 14.2) and the intracellular events that they initiate are diverse, with many variations on each theme, not all of these specifically involved in regulating genome activity. Three examples will help us to appreciate the complexity of the system.

Signal transduction with one step between receptor and genome

With some signal transduction systems, stimulation of the cell surface receptor by attachment of the extracellular signaling compound results in the direct activation of a protein that influences genome activity. This is the simplest system by which an extracellular signal can be transduced into a genomic response.

The direct system is used by many cytokines such as interleukins and interferons, which are extracellular signaling polypeptides that control cell growth and division. Binding of these polypeptides to their cell surface receptors results in activation of a type of transcription factor called a **STAT** (signal transducer and activator of transcription). Activation is by phosphorylation of a single tyrosine amino acid at a position near to the C terminus of the STAT polypeptide. If the cell surface receptor is a member of the tyrosine kinase family (see Table 14.2) then it is able to activate the STAT directly (Figure 14.9A). If it is a tyrosine-kinase-associated receptor then it does not itself have the ability to phosphorylate a STAT, or any other intracellular protein, but acts through intermediaries called **Janus kinases** (**JAKs**). Binding of the signaling molecule to a tyrosine-kinase-associated receptor causes a change in the conformation of the receptor, often by inducing dimerization. This causes a JAK that is associated with the receptor to phosphorylate itself, this autoactivation being followed by phosphorylation of the STAT by the JAK (Figure 14.9B).

Seven STATs have so far been identified in mammals. Three of these—STATs 2, 4, and 6—are specific for just one or two extracellular cytokines, but the others are broad spectrum and can be activated by several different interleukins and interferons. Discrimination is provided by the cell surface receptors: a particular receptor binds just one type of cytokine, and most cells have only one or a few types of cytokine receptor. Different cells therefore respond in different ways to the presence of particular cytokines, even though the internal signaling process involves only a limited number of STATs.

The consensus sequence of the DNA binding sites for STATs has been defined as 5′–TTN$_{5-6}$AA–3′, largely by studies in which purified STATs have been tested against oligonucleotides of known sequence. The DNA-binding domain of the STAT protein is made up of three loops emerging from a barrel-shaped β-sheet structure. This is an unusual type of DNA-binding domain and has not been identified in precisely the same form in any other type of protein, although it has similarities with the DNA-binding domains of the NK-κB and Rel transcription activators. These similarities refer only to the tertiary structures of the DNA-binding domains because STATs, NK-κB, and Rel, as a whole, have very little amino acid sequence identity. Many target genes are activated by STATs but the overall genomic response is modulated by other proteins which interact with STATs and influence which genes are switched on under a particular set of circumstances. Complexity is entirely

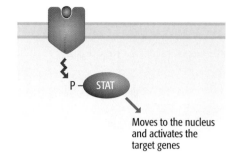

(A) Direct activation of a STAT

Moves to the nucleus and activates the target genes

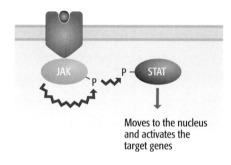

(B) Activation via a JAK

Moves to the nucleus and activates the target genes

Figure 14.9 Signal transduction involving STATs. (A) If the receptor is a member of the tyrosine kinase family then it can activate the STAT directly. (B) If the receptor is a tyrosine-kinase-associated type then it acts via a Janus kinase (JAK), which autophosphorylates when the extracellular signal binds and then activates the STAT. Note that activation of the JAK usually involves dimerization, the extracellular signal inducing two subunits to associate, resulting in the version of the JAK with phosphorylation activity. Dimerization is also central to activation of a STAT, phosphorylation causing two STATs, not necessarily of the same type, to form a dimer. This dimer is able to act as a transcription activator. "P" indicates a phosphate group, PO_3^{2-}.

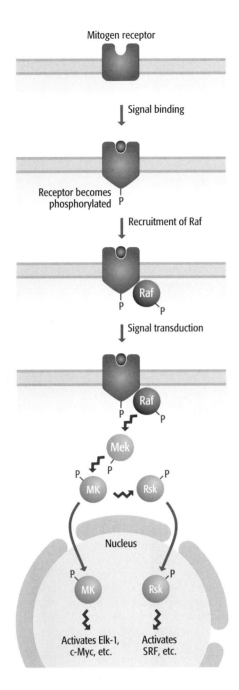

Figure 14.10 Signal transduction by the MAP kinase pathway. "MK" is the MAP kinase and "P" indicates a phosphate group, PO_3^{2-}. Elk-1, c-Myc, and SRF (serum response factor) are examples of transcription factors activated at the end of the pathway.

expected because the cellular processes that STATs mediate—growth and division—are themselves complex, and we anticipate that changes in these processes will require extensive remodeling of the proteome and hence large-scale alterations in genome activity.

Signal transduction with many steps between receptor and genome

The simplicity of the system whereby the cell surface receptor activates a STAT, directly or through a JAK associated with the receptor, contrasts with the more prevalent forms of signal transduction, in which the receptor represents just the first in a series of steps that lead eventually to one or more transcription activators or repressors being switched on or off. A number of these **cascade** pathways have been delineated in different organisms. The following are the important ones in mammals:

- The **MAP** (mitogen activated protein) **kinase system** (Figure 14.10) responds to many extracellular signals, including mitogens—compounds with similar effects to cytokines but that specifically stimulate cell division. Binding of the signaling compound results in dimerization of the mitogen receptor and mutual phosphorylation of the internal parts of the two subunits (see Figure 14.8). Phosphorylation stimulates attachment to the receptor, on the internal side of the membrane, of various cytoplasmic proteins, one of which is Raf, a protein kinase that is activated when it becomes membrane bound. Raf initiates a cascade of phosphorylation reactions. It phosphorylates Mek, activating this protein so that it, in turn, phosphorylates the MAP kinase. The activated MAP kinase now moves into the nucleus where it switches on, again by phosphorylation, a series of transcription activators. The MAP kinase also phosphorylates another protein kinase, this one called Rsk, which phosphorylates and activates a second set of factors. Additional flexibility is provided by the possibility of replacing one or more of the proteins in the MAP kinase pathway with related proteins, ones with slightly different specificities, and so switching on another suite of activators. The MAP kinase pathway is used by vertebrate cells; equivalent pathways, using intermediates similar to those identified in mammals, are known in other organisms (see Sections 14.2.1 and 14.3.3 for examples).

- The **Ras system** is centered around the Ras proteins, three of which are known in mammalian cells (H-, K-, and N-Ras), and similar proteins such as Rac and Rho. These proteins are involved in regulation of cell growth and differentiation and, as with many proteins in this category, when dysfunctional they can give rise to cancer. The Ras family proteins are not limited to mammals, examples being known in other eukaryotes such as the fruit fly. Ras proteins are intermediates in signal transduction pathways that initiate with autophosphorylation of a tyrosine kinase receptor in response to an extracellular signal. The phosphorylated version of the receptor forms protein–protein complexes with **GNRPs** (**guanine nucleotide releasing proteins**) and **GAPs** (**GTPase activating proteins**) which activate and inactivate Ras, respectively (Figure 14.11). The extracellular signals can therefore switch Ras-mediated signal transduction on or off, the choice between the two depending on the nature of the signal and the relative amounts of active GNRPs and GAPs in the cell. When activated, Ras stimulates Raf activity, so in effect Ras provides a second entry point into the MAP kinase pathway, although this is unlikely to be the only function of Ras and it probably also activates proteins involved in signal transduction by second messengers (as described in the next section).

- The **SAP** (stress activated protein) **kinase system** is induced by stress-related signals such as ultraviolet radiation, and growth factors associated with inflammation. The pathway has not been described in detail but is similar to the MAP kinase system although targeting a different set of transcription activators.

Each of the steps in these cascade pathways involves a physical interaction between two proteins, often resulting in the "downstream" member of the pair becoming phosphorylated. Phosphorylation activates the downstream protein, enabling it to form a connection with the next protein in the cascade. These interactions involve special protein–protein binding domains, such as the ones called SH2 and SH3, which bind to receptor domains on their partner proteins. The receptor domains contain one or more tyrosines which must be phosphorylated in order for docking to take place. Hence the upstream protein contains the receptor domain, whose phosphorylation status determines whether the protein can bind its downstream partner and hence propagate the signal (Figure 14.12).

Signal transduction via second messengers

Some signal transduction cascades do not involve the direct transfer of the external signal to the genome but instead utilize an indirect means of influencing transcription. The indirectness is provided by **second messengers**, which are less specific internal signaling compounds that transduce the signal from a cell surface receptor in several directions so that a variety of cellular activities, not just transcription, respond to the one signal.

In Section 14.1.1 we saw how glucose modulates the catabolite activator protein by influencing cAMP levels in bacteria (see Figure 14.7). Cyclic nucleotides are also important second messengers in eukaryotic cells. Some cell surface receptors have guanylate cyclase activity, and so convert GTP to cGMP, but most receptors in this family work indirectly by influencing the activity of cytoplasmic cyclases and decyclases. These cyclases and decyclases determine the cellular levels of cGMP and cAMP, which in turn control the activities of various target enzymes. The latter include protein kinase A, which is stimulated by cAMP. One of the functions of protein kinase A is to phosphorylate, and hence activate, a transcription activator called **CREB**. This is one of several proteins that influence the activity of a variety of genes by interacting with a second activator, p300/CBP, which is able to modify histone proteins and so affect chromatin structure and nucleosome positioning (Sections 10.2.1 and 10.2.2).

As well as being activated indirectly by cAMP, p300/CBP responds to another second messenger, calcium. The calcium ion concentration is substantially lower inside than outside the cell, so proteins that open calcium channels in the cell membrane allow calcium ions to enter. This can be induced by extracellular signals that activate tyrosine kinase receptors which in turn activate phospholipases that cleave phosphatidylinositol-4,5-bisphosphate, a lipid component of the inner cell membrane, into inositol-1,4,5-trisphosphate ($Ins(1,4,5)P_3$) and 1,2-diacylglycerol (DAG). $Ins(1,4,5)P_3$ opens calcium channels (Figure 14.13). $Ins(1,4,5)P_3$ and DAG are themselves second messengers that can initiate other signal transduction cascades. Both the calcium- and the lipid-induced cascades target transcription activators, but only indirectly: the primary targets are other proteins. Calcium, for example,

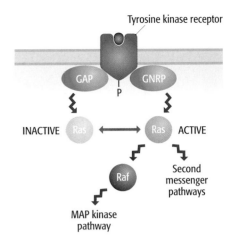

Figure 14.11 The Ras signal transduction system. Abbreviations: GAP, GTPase activating protein; GNRP, guanine nucleotide releasing protein. "P" indicates a phosphate group, PO_3^{2-}.

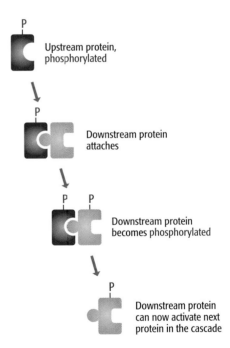

Figure 14.12 A scheme for interaction of proteins in a signaling cascade. The upstream protein is phosphorylated and hence able to bind its downstream partner. Binding leads to phosphorylation of the receptor domain in the downstream protein, propagating the signal. "P" indicates a phosphate group, PO_3^{2-}.

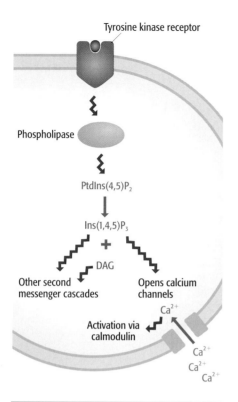

Tyrosine kinase receptor

Phospholipase

PtdIns(4,5)P₂

Ins(1,4,5)P₃

+

DAG

Other second
messenger cascades

Opens calcium
channels

Ca²⁺

Activation via
calmodulin

Ca²⁺
Ca²⁺
Ca²⁺

Figure 14.13 Induction of the calcium second messenger system. Abbreviations: DAG, 1,2-diacylglycerol; Ins(1,4,5)P₃, inositol-1,4,5-trisphosphate; PtdIns(4,5)P₂, phosphatidylinositol-4,5-bisphosphate.

Figure 14.14 Understanding the SMAD signaling pathway. (A) Outline of the signaling pathway. Smad1 is shown being activated by a TGF-β receptor. The same pathway applies to Smad2, Smad3, and Smad5. (B) Two models for the inhibitory effects of Smad6 (shown in this diagram) and Smad7.

binds to and activates the protein called calmodulin, which regulates a variety of enzyme types, including protein kinases, ATPases, phosphatases, and nucleotide cyclases.

Unraveling a signal transduction pathway

How do cell biologists unravel the complexities of signal transduction pathways? To answer this question, we will examine some of the recent research directed at the signaling pathway activated by transforming growth factor-β (TGF-β), a family of some 30 related polypeptides that control processes such as cell division and differentiation in vertebrates. The cell surface receptors for TGF-β are serine–threonine kinases (see Table 14.2), which activate a variety of target proteins within the cell. Part of the signal transduction process initiated by TGF-β binding involves a set of proteins called the SMAD family, the name being an abbreviation of "SMA/MAD related," referring to the proteins in *Drosophila melanogaster* and *Caenorhabditis elegans*, respectively, which were the original members of the family to be isolated.

Initially, five SMADs were discovered in vertebrate cells. Four of these— Smad1, Smad2, Smad3, and Smad5—are called receptor-regulated SMADs because they associate directly with the cell surface receptor. Each of these SMADs is specific for a different type of serine–threonine receptor and hence responds to different members of the TGF-β family of signaling compounds. Binding of the extracellular signal induces a receptor to phosphorylate its SMAD, which then associates with Smad4, moves to the nucleus, and via interactions with DNA-binding proteins, activates a set of target genes (Figure 14.14A). Smad4 is therefore a co-mediator that participates in the signaling pathway of each of the other four SMADs. The SMAD system is a second example of signal transduction with one step between receptor and genome, similar to the STAT pathway described above.

This interpretation of the SMAD pathway was complicated by the discovery of two additional SMADs—numbers 6 and 7—that do not fit into the scheme. These SMADs lack the amino acid sequence motif serine–serine–X–serine (where X is valine or methionine), present in the C-terminal region of Smad1, Smad2, Smad3, and Smad5, and which is phosphorylated by the receptor. Clearly, therefore, Smad6 and Smad7 do not respond directly to attachment of the extracellular signals to the receptor protein. Are they co-mediators similar to Smad4, or do they have some other role in TGF-β signal transduction?

(A) The SMAD signaling pathway

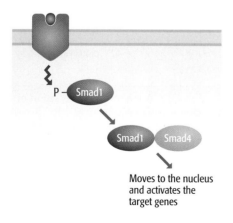

P — Smad1

Smad1   Smad4

Moves to the nucleus
and activates the
target genes

(B) Models for the inhibitory effect of Smad6 and Smad7

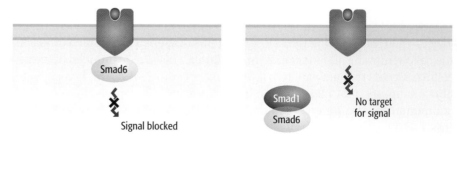

Smad6

Signal blocked

Smad1
Smad6

No target
for signal

The first step in understanding the functions of Smad6 and Smad7 was to determine the effect of overexpression of these proteins on TGF-β signal transduction. Overexpression was achieved by attaching the Smad6 or Smad7 gene to a strong promoter and then using cloning techniques to introduce the gene into cultured cells. The outcome was that nuclear genes normally switched on by TGF-β became nonresponsive to the extracellular signal in cells overexpressing Smad6 or Smad7. This result gave the first indication that Smad6 and Smad7 have inhibitory effects on the TGF-β pathway.

Two models have been proposed to explain how the inhibitory SMADs, as Smad6 and Smad7 are now called, repress the TGF-β pathway (Figure 14.14B). The first model is based on the observation that in cell extracts the Smad6 and Smad7 proteins are associated with the intracellular parts of the cell surface receptors. The hypothesis is that Smad6 and Smad7 inhibit signal transduction by preventing the activated receptors from phosphorylating the other SMADs. This model probably explains the inhibitory effects of overexpression of Smad6 and Smad7, but in normal cells there may not be enough copies of these proteins to block the cell surface receptors entirely. An alternative model has therefore been proposed, in which the inhibitory SMADs bind to one or more of the other SMADs, removing these from the pathway and hence stopping signal transduction. There is good evidence that this type of interaction is the explanation for the inhibitory effect that Smad6 has on Smad1. Yeast two-hybrid studies (Section 6.2.2) have shown that these two SMADs interact, and after binding to Smad6, Smad1 is unable to influence the transcriptional activators that it normally stimulates, even after it has been phosphorylated by the cell surface receptors.

Whatever the mechanism for Smad6 and Smad7 activity, the discovery of these inhibitory SMADs shows that TGF-β signaling via the SMAD pathway is more flexible than was originally envisaged. Rather than being an all-or-nothing response, the activity of the receptor-regulated SMADs can be modified by the inhibitory effects of Smad6 and Smad7, these proteins presumably responding to as-yet-unidentified intra- and/or intercellular signals in order to modulate the effects of TGF-β binding in an appropriate way.

14.2 Permanent and Semipermanent Changes in Genome Activity

Transient changes in genome activity are, by definition, readily reversible, the genome expression pattern reverting to its original state when the external stimulus is removed or replaced by a contradictory stimulus. In contrast, the permanent and semipermanent changes in genome activity that underlie cellular differentiation must persist for long periods, and ideally should be maintained even when the stimulus that originally induced them has disappeared. We therefore anticipate that the regulatory mechanisms bringing about these longer-term changes will involve systems additional to the modulation of transcription activators and repressors. This expectation is correct. We will look at three mechanisms:

- Changes resulting from physical rearrangement of the genome.

- Changes due to chromatin structure.

- Changes maintained by feedback loops.

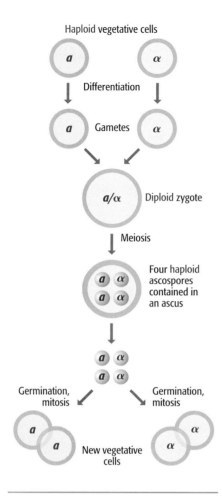

Haploid vegetative cells

Differentiation

Gametes

a/α   Diploid zygote

Meiosis

Four haploid ascospores contained in an ascus

Germination, mitosis

Germination, mitosis

New vegetative cells

Figure 14.15 The life cycle of the yeast *Saccharomyces cerevisiae.*

14.2.1 Genome rearrangements

Changing the physical structure of the genome is an obvious, although drastic, way to bring about a permanent change in genome expression. It is not a common regulatory mechanism, but several important examples are known.

Yeast mating types are determined by gene conversion events

Mating type is the equivalent of sex in yeasts and other eukaryotic microorganisms. Because these organisms reproduce mainly by vegetative cell division, there is the possibility that a population, being derived from just one or a few ancestral cells, will be largely or completely composed of a single mating type and so will not be able to reproduce sexually. In *Saccharomyces cerevisiae* and some other species this problem is avoided because cells are able to change sex by the process called **mating-type switching**.

The two *S. cerevisiae* mating types are called a and α. Each mating type secretes a short polypeptide pheromone (12 amino acids in length for a and 13 for α) that binds to receptors on the surfaces of cells of the opposite mating type. Binding of the pheromone initiates a MAP kinase signal transduction pathway (see Figure 14.10) that alters the genome expression profile within the cell, leading to subtle morphological and physiological changes that convert the cell into a gamete able to participate in sexual reproduction. Mixing two haploid strains of opposite mating type therefore stimulates formation of gametes which fuse to produce a diploid zygote. Meiosis occurs within the zygote, giving rise to a tetrad of four haploid ascospores, contained in a structure called an ascus. The ascus bursts open, releasing the ascospores, which then divide by mitosis to produce new haploid vegetative cells (Figure 14.15).

The mating type is specified by the *MAT* gene, located on chromosome III. This gene has two alleles, *MATa* and *MATα*, a haploid yeast cell displaying the mating type corresponding to whichever allele it possesses. Elsewhere on chromosome III are two additional *MAT*-like genes, called *HMRa* and *HMLα* (Figure 14.16). These have the same sequences as *MATa* and *MATα*, respectively, but neither gene is expressed because upstream of each one is a silencer that represses transcription initiation. These two genes are called "silent mating-type cassettes." Their silencing involves the Sir proteins, several of which have histone deacetylase activity (Section 10.2.1), indicating that silencing involves changes in the chromatin structure in the region of *HMRa* and *HMLα*.

Mating-type switching is initiated by the HO endonuclease, which makes a double-strand cut at a 24 bp sequence located within the *MAT* gene. This enables a **gene conversion** event to take place. We examine the details of gene conversion in Section 17.1.1; all that concerns us at the moment is that one of the free 3′ ends produced by the endonuclease can be extended by DNA synthesis, using one of the two silent cassettes as the template (see Figure 14.16). The newly synthesized DNA subsequently replaces the DNA currently at the *MAT* locus. The silent cassette chosen as the template is usually the one that is different to the allele originally at *MAT*, so replacement with the newly synthesized strand converts the *MAT* gene from *MATa* to *MATα*, or vice versa. This results in mating-type switching.

The *MAT* genes code for regulatory proteins (one in the case of *MATa* and two for *MATα*) that interact with a transcription activator, MCM1, thus determining

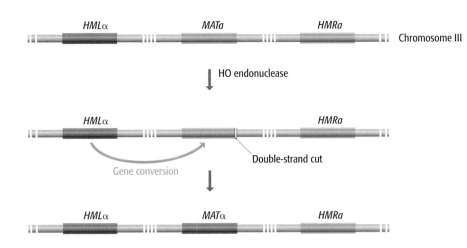

which set of genes are switched on by this factor. The *MATa* and *MATα* gene products have different effects on MCM1, and so specify different allele-specific genome expression patterns. These expression patterns are maintained in a semipermanent fashion until another *MAT* gene conversion occurs.

Genome rearrangements are responsible for immunoglobulin and T-cell receptor diversities

In vertebrates, there are two striking examples of the use of DNA rearrangements to achieve permanent changes in genome activity. These two examples, which are very similar, are responsible for the generation of immunoglobulin and T-cell receptor diversities.

Immunoglobulins and T-cell receptors are related proteins that are synthesized by B and T lymphocytes, respectively. Both types of protein become attached to the outer surfaces of their cells, and immunoglobulins are also released into the bloodstream. The proteins help to protect the body against invasion by bacteria, viruses, and other unwanted substances by binding to these **antigens**, as they are called. During its lifetime, an organism could be exposed to any number of a vast range of antigens, which means that the immune system must be able to synthesize an equally vast range of immunoglobulin and T-cell receptor proteins. In fact, humans can make approximately 10^8 different immunoglobulin and T-cell receptor proteins. But there are only 3.5×10^4 genes in the human genome, so where do all these proteins come from?

To understand the answer we will look at the structure of a typical immunoglobulin protein. Each immunoglobulin is a tetramer of four polypeptides linked by disulfide bonds (Figure 14.17). There are two long "heavy" chains and two short "light" chains. When the sequences of different heavy chains are compared it becomes clear that the variability between them lies mainly in the N-terminal regions of these polypeptides, the C-terminal parts being very similar, or "constant," in all heavy chains. The same is true for the light chains, except that two families, κ and λ, can be distinguished, differing in the sequences of their constant regions.

In the vertebrate genomes, there are no complete genes for the immunoglobulin heavy and light polypeptides. Instead, these proteins are specified by gene segments. The heavy-chain segments are on chromosome 14 and comprise 11

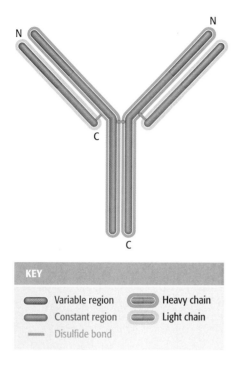

KEY

- Variable region
- Constant region
- Disulfide bond
- Heavy chain
- Light chain

Figure 14.17 Immunoglobulin structure. Each immunoglobulin protein is made up of two heavy and two light chains, linked by disulfide bonds. Each heavy chain is 446 amino acids in length and consists of a variable region (shown in red) spanning amino acids 1–108 followed by a constant region. Each light chain is 214 amino acids, again with an N-terminal variable region of 108 amino acids. Additional disulfide bonds form between different parts of individual chains: these and other interactions fold the protein into a more complex three-dimensional structure.

constant region (C_H) gene segments, preceded by 123–129 V_H gene segments, 27 D_H gene segments, and 9 J_H gene segments, these last three types coding for different versions of the V (variable), D (diverse), and J (joining) components of the variable part of the heavy chain (Table 14.3; Figure 14.18). The entire heavy-chain locus stretches over several megabase pairs. A similar arrangement is seen with the light-chain loci on chromosomes 2 (κ locus) and 22 (λ locus), the only difference being that the light chains do not have D segments (Table 14.3).

During the early stage of B lymphocyte development, the immunoglobulin loci in its genome undergo rearrangements. Within the heavy-chain locus, these rearrangements link one of the V_H gene segments with one of the D_H gene segments, and then link this V–D combination with a J_H gene segment (Figure 14.19). These rearrangements occur by an unusual type of recombination, catalyzed by a pair of proteins called RAG1 and RAG2, with the positions at which the breakage and reunion reactions must occur in order to link the gene segments marked by a series of 8 bp and 9 bp consensus sequences. The end result is an exon that contains the complete open reading frame specifying the V_H, D_H, and J_H segments of the immunoglobulin protein. This exon becomes linked to a C_H segment exon by splicing during the transcription process, creating a complete heavy-chain mRNA that can be translated into an immunoglobulin protein that is specific for just that one lymphocyte. A similar series of DNA rearrangements results in the lymphocyte's light-chain V–J exon being constructed at either the κ or λ locus, and once again splicing attaches a light-chain C segment exon when the mRNA is synthesized.

Despite its name, the constant region is not identical in every immunoglobulin protein. The small variations that occur result in five different classes of immunoglobulin—IgA, IgD, IgE, IgG, and IgM—each with its own specialized role in the immune system. Initially, each B lymphocyte synthesizes an IgM molecule, the C_H segment of which is specified by the $C\mu$ sequence that lies at the 5' end of the C_H segment cluster. As shown in Figure 14.19, later in its development the immature cell might also synthesize some IgD proteins, utilizing the second C_H sequence in the cluster ($C\delta$), the exon for this sequence becoming attached to the V–D–J segment by alternative splicing. Later in their lifetimes, when they have reached maturity, some B lymphocytes undergo a second type of **class switching**, which results in a complete change in the type of immunoglobulin that the lymphocyte synthesizes. This second class switching requires a further recombination event that deletes the $C\mu$ and $C\delta$ sequence along with the part of the chromosome between this region and the C_H segment specifying the class of immunoglobulin that the

Table 14.3 Immunoglobulin gene segments in the human genome

Component	Locus	Chromosome	V	D	J	C
			\multicolumn{4}{c}{Number of gene segments}
Heavy chain	*IGH*	14	123–129	27	9	11
Light κ chain	*IGK*	2	76	0	5	1
Light λ chain	*IGL*	22	70–71	0	7–11	7–11

Some numbers are variable because of differences between human genotypes. It is not known if all the gene segments are functional—some may be pseudogenes.

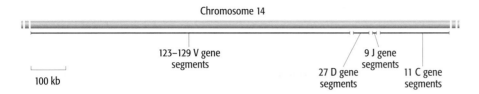

lymphocyte will now synthesize. For example, for the lymphocyte to switch to synthesis of IgG, the most prevalent type of immunoglobulin made by mature lymphocytes, then the deletion will place one of the Cγ segments, which specify the IgG heavy chain, at the 5′ end of the cluster (Figure 14.20). Class switching is therefore a second example of genome rearrangement occurring during B lymphocyte development. The mechanism is distinct from V–D–J joining and the recombination event does not involve the RAG proteins.

Diversity of T-cell receptors is based on similar rearrangements which link V, D, J, and C gene segments in different combinations to produce cell-specific genes. Each receptor comprises a pair of β molecules, which are similar to the immunoglobulin heavy chain, and two α molecules, which resemble the immunoglobulin κ light chains. As with immunoglobulins, the T-cell receptors become embedded in the cell membrane and enable the lymphocyte to recognize and respond to extracellular antigens.

14.2.2 Changes in chromatin structure

Some of the effects that chromatin structure can have on genome expression were described in Section 10.2. These range from the modulation of transcription initiation at an individual promoter by nucleosome positioning, through to the silencing of large segments of DNA locked-up in higher-order chromatin structure. The latter is an important means of bringing about long-term changes in genome activity and is implicated in a number of regulatory events. One of these concerns the yeast mating-type loci that we looked at earlier in this section, the silencing of the *HMRa* and *HMLα* cassettes resulting

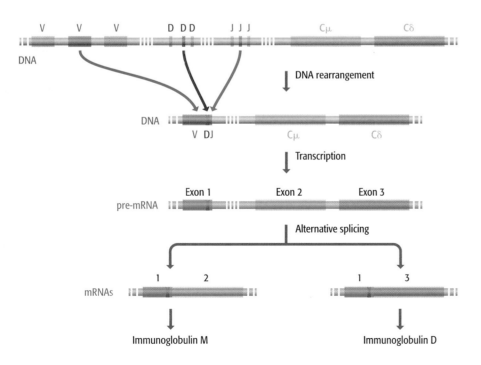

Figure 14.19 Synthesis of a specific immunoglobulin protein. DNA rearrangement within the heavy-chain locus links V, D, and J segments, which are then linked to a C segment by splicing of the mRNA. In immature B cells, the V–D–J exon always becomes linked to the Cμ exon (exon 2) to produce an mRNA specifying a class M immunoglobulin. At an early stage in development of the B cell, some immunoglobulin D proteins are also produced by alternative splicing that links the V–D–J exon to the Cδ exon. Both types of immunoglobulin become bound to the cell membrane.

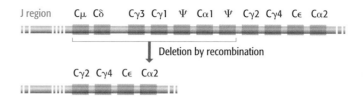

Figure 14.20 Immunoglobulin class switching. In this example, seven C$_H$ segments are deleted, placing the Cγ2 segment adjacent to the J region. This B cell will therefore synthesize an immunoglobulin G molecule, which is secreted by the cell. The two segments labeled ψ are pseudogenes.

mainly from these loci being buried in inaccessible chromatin in response to the influence of their upstream silencer sequences. X inactivation (Section 10.3.1) also involves the formation of inaccessible chromatin, in this case along virtually the entire length of one of the two X chromosomes in a female nucleus.

One other example of chromatin silencing merits attention. This is a system that we will meet again later in the chapter when we look at development processes in the fruit fly. It concerns the *Polycomb* gene family, which comprises some 30 genes that code for proteins that bind to DNA sequences, called Polycomb response elements, and induce formation of heterochromatin, the condensed form of chromatin that prevents transcription of the genes that it contains (Figure 14.21). Each response element is approximately 10 kb in length but does not appear to contain specific Polycomb binding sites, suggesting that additional proteins act as intermediaries in Polycomb attachment. A candidate for this intermediary role is the non-Polycomb protein called Dorsal switch protein 1 (DSP1), which binds to 5′–GAAAA–3′ sequences present in the core region of a Polycomb response element. Mutations to these sequences that prevent DSP1 binding also prevent recruitment of Polycomb proteins in experimental systems. Whatever the mechanism, the result of Polycomb attachment is nucleation of heterochromatin around the Polycomb proteins, the heterochromatin then propagating along the DNA for tens of kilobases in either direction.

The regions that become silenced contain homeotic genes which, as we will see in Section 14.3.4, specify the development of the individual body parts of the fly. As only one body part must be specified at a particular position in the fruit fly, it is important that a cell expresses only the correct homeotic gene. This is ensured by the action of Polycomb, which permanently silences the homeotic genes that must be switched off. Polycomb proteins do not, however, determine which genes will be silenced: expression of these genes is already repressed before the Polycomb proteins bind to their response elements. The role of Polycomb is therefore to *maintain* rather than *initiate* gene

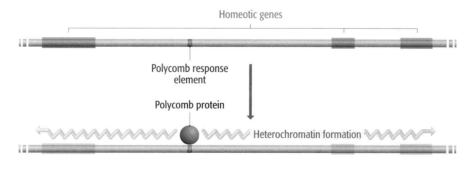

Figure 14.21 Polycomb maintains silencing in regions of the *Drosophila* genome by initiating heterochromatin formation. Note that attachment of the Polycomb protein to its response element is mediated by additional proteins not shown in this drawing.

silencing. An important point is that the heterochromatin induced by Polycomb is heritable: after division, the two new cells retain the heterochromatin established in the parent cell. This type of regulation of genome activity is therefore permanent not only in a single cell, but also in a cell lineage.

The trithorax proteins act in a similar manner to Polycomb, but have the opposite effect: they maintain an open chromatin state in the regions of active genes, the targets including the same homeotic genes as those that are silenced, in different body parts, by Polycomb proteins. The modes of action of trithorax and Polycomb may be closely linked, as there is evidence that trithorax proteins attach to their target sites via a mediator protein, called GAGA, which binds to sequences within Polycomb response elements. Some mutations abolish both Polycomb and trithorax activity, indicating that the two systems may share common components.

14.2.3 Genome regulation by feedback loops

The final mechanism that we will consider for bringing about long-term changes in genome activity involves the use of a feedback loop. In this system, a regulatory protein activates its own transcription so that once its gene has been switched on, it is expressed continuously (Figure 14.22). A number of examples of this type of feedback regulation are known:

- The **MyoD transcription activator**, which is involved in muscle development, is one of the best-understood examples of cellular differentiation in vertebrates. A cell becomes committed to becoming a muscle cell when it begins to express the *myoD* gene. The product of this gene is a transcription activator that targets a number of other genes coding for muscle-specific proteins, such as actin and myosin, and is also indirectly responsible for one of the key features of muscle cells—the absence of a normal cell cycle, these cells being stopped in the G1 phase (Section 15.3.1). The MyoD protein also binds upstream of *myoD*, ensuring that its own gene is continuously expressed. The result of this positive feedback loop is that the cell continues to synthesize the muscle-specific proteins and remains a muscle cell. The differentiated state is heritable because cell division is accompanied by transmission of MyoD to the daughter cells, ensuring that these are also muscle cells.

- **Deformed of *Drosophila*** is one of several proteins coded by homeotic selector genes and is responsible for specifying segment identity in the fruit fly (Section 14.3.4). The Deformed (Dfd) protein is responsible for the identity of the head segments. To perform this function, Dfd must be continuously expressed in the relevant cells. This is achieved by a feedback system, Dfd binding to an enhancer located upstream of the *Dfd* gene. Feedback autoregulation also controls the expression of at least some homeotic selector genes of vertebrates.

14.3 Regulation of Genome Activity During Development

The developmental pathway of a multicellular eukaryote begins with a fertilized egg cell and ends with an adult form of the organism. In between lies a complex series of genetic, cellular, and physiological events that must occur in the correct order, in the correct cells, and at the appropriate times if the

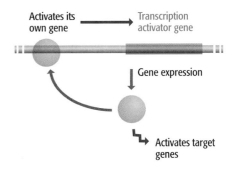

Figure 14.22 Feedback regulation of gene expression.

pathway is to reach a successful culmination. With humans, this developmental pathway results in an adult containing 10^{13} cells differentiated into approximately 250 specialized types, the activity of each individual cell coordinated with that of every other cell. Developmental processes of such complexity might appear intractable, even to the powerful investigative tools of modern molecular biology, but remarkably good progress toward understanding them has been made in recent years. The research that has underpinned this progress has been designed around three guiding principles:

● It should be possible to describe and comprehend the genetic and biochemical events that underlie differentiation of individual cell types. This in turn means that an understanding of how specialized tissues, and even complex body parts, are constructed should be within reach.

● The signaling processes that coordinate events in different cells should be amenable to study. We saw in Section 14.1 that a start is being made to describing these systems at the molecular level.

● There should be similarities and parallels between developmental processes in different organisms, reflecting common evolutionary origins. This means that information relevant to human development can be obtained from studies of model organisms chosen for the relative simplicity of their developmental pathways.

Developmental biology encompasses areas of genetics, molecular biology, cell biology, physiology, biochemistry, and systems biology. We are concerned only with the role of the genome in development and so will not attempt a wide-ranging overview of developmental research in all its guises. Instead we will concentrate on four model systems of increasing complexity in order to investigate the types of change in genome activity that occur during development.

14.3.1 The lysogenic cycle of bacteriophage λ

A bacteriophage that infects *Escherichia coli* might seem an odd place to begin a study of genome regulation during development. But this is exactly where molecular biologists began the lengthy program of research that today is revealing the underlying genomic basis to development in humans and other vertebrates. We will therefore follow this same progression from the relatively simple to the relatively complex.

In Section 9.1.1 we learnt that lysogenic bacteriophages such as λ can follow two alternative replication pathways after infection of a host cell. As well as the lytic pathway, during which new phages are assembled and released from the cell soon after the initial infection (after 45 minutes for λ), these phages can also pursue a lysogenic cycle characterized by insertion of the phage DNA into the host chromosome. The integrated prophage remains quiescent for many bacterial generations until a chemical or physical stimulus linked to DNA damage induces excision of the λ genome, rapid assembly of phages, and lysis of the host cell (see Figures 9.4 and 9.5).

Bacteriophage λ must make a choice between lysis and lysogeny
The ability of bacteriophages such as λ to follow a lysogenic infection cycle raises three questions: how does the phage "decide" whether to follow the lytic or lysogenic cycle, how is lysogeny maintained, and how is the prophage induced to break lysogeny? A considerable amount is known about genome expression during λ infection, so much so that very detailed and complex answers can be given for these questions.

The first step in the lytic infection cycle is expression of the two immediate-early λ genes, which are called *N* and *cro*. These are transcribed from two promoters, P_L and P_R, respectively (Figure 14.23A). Protein N is the antiterminator which enables the host RNA polymerase to ignore the termination signals that it encounters immediately downstream of the *N* and *cro* coding sequences and to transcribe the delayed-early genes (see Figure 12.9). These genes include *cII* and *cIII*, which together activate a third promoter P_{RM}, resulting in transcription of *cI*. This is an all-important gene as it codes for the λ repressor protein, the key master switch that shuts down the lytic cycle and maintains lysogeny. The repressor does this by binding to the operators O_L and O_R, adjacent to P_L and P_R, respectively (Figure 14.23B). As a result almost the entire λ genome is silenced, because P_L and P_R direct transcription not only of the immediate-early and delayed-early genes, but also of the late genes, which code for the proteins needed for assembly of new phages and host cell lysis. One of the few genes to remain active is *int*, which is transcribed from its own promoter. The integrase protein coded by this gene catalyzes the site-specific recombination which inserts the λ DNA into the host genome. Lysogeny is maintained for numerous cell divisions because the *cI* gene is continuously expressed, albeit at a low level, so that the amount of cI repressor present in the cell is always enough to keep P_L and P_R switched off. This continued expression of *cI* occurs because the cI repressor, when bound to O_R, not only blocks transcription from P_R, but also stimulates transcription from its own promoter P_{RM}. The dual role of the cI repressor is therefore the key to lysogeny.

Once *cI* is expressed, the repressor protein prevents entry into the lytic cycle and ensures that lysogeny is set up and maintained. But λ does not always enter the lysogenic cycle—on some occasions an infection proceeds immediately to host lysis. This is because of the activity of the second immediate-early gene, *cro*, which also codes for a repressor, but in this case one that prevents transcription of *cI* (Figure 14.23C). The decision between lysis and lysogeny is therefore determined by the outcome of a race between *cI* and *cro*. If the cI repressor is synthesized more quickly than the Cro repressor, then genome expression is blocked and lysogeny follows. However, if *cro* wins the race, then the Cro repressor blocks *cI* expression before enough cI repressor has been synthesized to silence the genome. As a result, the phage enters the lytic infection cycle. The decision appears to be random, depending on chance events that lead to either the cI or the Cro repressor accumulating the quickest in the cell, although environmental conditions can have an influence. Growth on a rich medium, for example, shifts the balance toward the lytic cycle, presumably because it is beneficial to produce new phages when the host cells are proliferating. This shift is brought about by activation of proteases that degrade the *cII* protein, reducing the ability of the *cII*–*cIII* combination to switch on transcription of the *cI* repressor gene.

If the bacteriophage enters the lysogenic cycle, then this state is maintained as long as the cI repressor is bound to the operators O_L and O_R. The prophage will therefore be induced if the level of active cI repressor declines below a certain point. This may happen by chance, leading to spontaneous induction, or may occur in response to physical or chemical stimuli. These stimuli activate a general protective mechanism in *E. coli*, the **SOS response**. Part of the SOS response is expression of an *E. coli* gene, *recA*, whose product inactivates the cI repressor by cleaving it in half. This switches on expression of the early genes, enabling the phage to enter the lytic cycle. Inactivation of the

(A) Synthesis of immediate-early gene transcripts

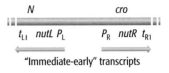

(B) The role of the cI repressor

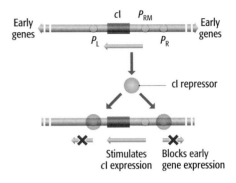

(C) The role of the Cro repressor

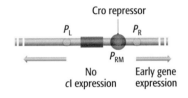

Figure 14.23 The genetic basis to the choice made by bacteriophage λ between lysis and lysogeny.

cI repressor also means that transcription of *cI* is no longer stimulated, avoiding the possibility of lysogeny being reestablished through the synthesis of more cI repressor. Inactivation of the cI repressor therefore leads to induction of the prophage.

What do we learn from this model system?

- A simple genetic switch can determine which of two developmental pathways is followed by a cell.

- Genetic switches can involve a combination of activation and repression of different promoters.

- It is possible to reprogram a developmental pathway, and transfer to an alternative pathway, in response to appropriate stimuli.

14.3.2 Sporulation in *Bacillus*

The second system that we will examine is formation of spores by the bacterium *Bacillus subtilis*. As with λ lysogeny this is not, strictly speaking, a developmental pathway, merely a type of cellular differentiation, but the process illustrates two of the fundamental issues that have to be addressed when genuine development in multicellular organisms is studied. These issues are how a series of changes in genome activity over time is controlled, and how signaling establishes coordination between events occurring in different cells. The advantages of *Bacillus* as a model system are that it is easy to grow in the laboratory and is amenable to study by genetic and molecular biology techniques such as analysis of mutants and sequencing of genes.

Sporulation involves coordinated activities in two distinct cell types
Bacillus is one of several genera of bacteria that produce endospores in response to unfavorable environmental conditions. These spores are highly resistant to physical and chemical abuse and can survive for decades or even centuries—the possibility of infection with anthrax spores produced by *B. anthracis* is taken very seriously by archaeologists excavating sites containing human and animal remains. Resistance is due to the specialized nature of the spore coat, which is impermeable to many chemicals, and to biochemical changes within the spore that retard the decay of DNA and other polymers and enable the spore to survive a prolonged period of dormancy.

In the laboratory, sporulation is usually induced by nutrient starvation. This causes the bacteria to abandon their normal vegetative mode of cell division, which involves synthesis of a septum (or cross-wall) in the center of the cell. Instead the cells construct an unusual septum, one that is thinner than normal, at one end of the cell (Figure 14.24). This produces two cellular compartments, the smaller of which is called the prespore and the larger the mother cell. As sporulation proceeds, the prespore becomes entirely engulfed by the mother cell. By now the two cells are committed to different but coordinated differentiation pathways, the prespore undergoing the biochemical changes that enable it to become dormant, and the mother cell constructing the resistant coat around the spore and eventually dying.

Special σ subunits control genome activity during sporulation
Changes in genome activity during sporulation are controlled largely by the synthesis of special σ subunits that change the promoter specificity of the *Bacillus* RNA polymerase. Recall that the σ subunit is the part of the RNA

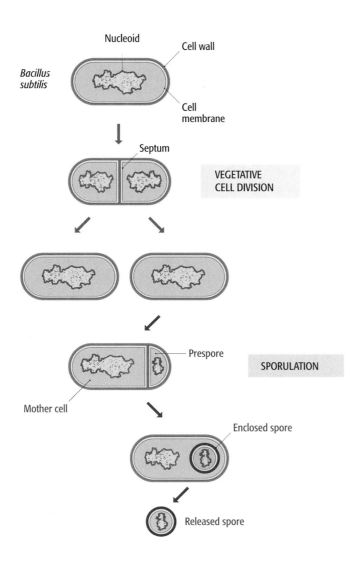

Nucleoid

Cell wall

Bacillus subtilis

Cell membrane

Septum

VEGETATIVE CELL DIVISION

Prespore

SPORULATION

Mother cell

Enclosed spore

Released spore

Figure 14.24 Sporulation in *Bacillus subtilis*. The top part of the diagram shows the normal vegetative mode of cell division, involving formation of a septum across the center of the bacterium and resulting in two identical daughter cells. The lower part of the diagram shows sporulation, in which the septum forms near one end of the cell, leading to a mother cell and prespore of different sizes. Eventually the mother cell completely engulfs the prespore. At the end of the process, the mature, resistant spore is released.

polymerase that recognizes the bacterial promoter sequence, and that replacement of one σ subunit with another with a different DNA-binding specificity can result in a different set of genes being transcribed (Section 11.3.1). We have seen how this simple control system is used by *E. coli* in response to heat stress (see Figure 11.23). It is also the key to the changes in genome activity that occur during sporulation.

The standard *B. subtilis* σ subunits are called σ^A and σ^H. These subunits are synthesized in vegetative cells and enable the RNA polymerase to recognize promoters for all the genes it needs to transcribe in order to maintain normal growth and cell division. In the prespore and mother cell these subunits are replaced by σ^F and σ^E, respectively, which recognize different promoter sequences and so give rise to large-scale changes in genome expression patterns. The master switch from vegetative growth to spore formation is provided by a protein called SpoOA, which is present in vegetative cells but in an inactive form. This protein is activated by phosphorylation, via a cascade of protein kinases that respond to various extracellular signals that indicate the presence of an environmental stress such as lack of nutrients. The initial response is provided by three kinases, called KinA, KinB, and KinC, which phosphorylate themselves and then pass the phosphate via SpoOF and

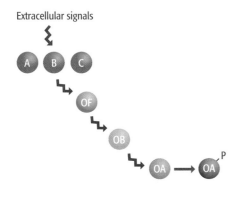

Extracellular signals

Figure 14.25 The phosphorylation cascade that leads to activation of SpoOA. Abbreviations: A, KinA; B, KinB; C, KinC; OF, SpoOF; OB, SpoOB; OA, SpoOA; P indicates a phosphate group, PO_3^{2-}.

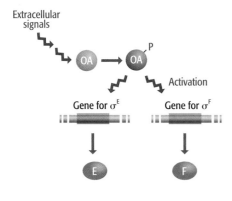

Extracellular signals

Activation

Gene for σ^E Gene for σ^F

E   F

Figure 14.26 The role of SpoOA in *Bacillus* **sporulation.** SpoOA is phosphorylated in response to extracellular signals derived from environmental stresses as shown in Figure 14.25. It is a transcription activator with roles that include activation of the genes for the σ^E and σ^F RNA polymerase subunits. Abbreviations: E, σ^E; F, σ^F; OA, SpoOA; P indicates a phosphate group, PO_3^{2-}.

SpoOB to SpoOA (Figure 14.25). Activated SpoOA is a transcription factor that modulates the expression of various genes transcribed by the vegetative RNA polymerase and hence recognized by the regular σ^A and σ^H subunits. The genes that are switched on include those for σ^F and σ^E, resulting in the switch to prespore and mother cell differentiation (Figure 14.26).

Initially, both σ^F and σ^E are present in each of the two differentiating cells. This is not exactly what is wanted because σ^F is the prespore-specific subunit and so should be active only in this cell, and σ^E is mother-cell specific. A means is therefore needed of activating or inactivating the appropriate subunit in the correct cell. This is thought to be achieved as follows (Figure 14.27):

- σ^F is activated by release from a complex with a second protein, SpoIIAB. This is controlled by a third protein, SpoIIAA, which, when unphosphorylated, can also attach to SpoIIAB and prevent the latter from binding to σ^F. If SpoIIAA is unphosphorylated then σ^F is released and is active; when SpoIIAA is phosphorylated then σ^F remains bound to SpoIIAB and so is inactive. In the mother cell, SpoIIAB phosphorylates SpoIIAA and so keeps σ^F in its bound, inactive state. But in the prespore, SpoIIAB's attempts to phosphorylate SpoIIAA are antagonized by yet another protein, SpoIIE, and so σ^F is released and becomes active. SpoIIE's ability to antagonize SpoIIAB in the prespore but not the mother cell derives from the fact that SpoIIE molecules are bound to the membrane on the surface of the septum. Because the prespore is much smaller than the mother cell, but the septum surface area is similar in both, the concentration of SpoIIE is greater in the prespore, and this enables it to antagonize SpoIIAB.

- σ^E is activated by proteolytic cleavage of a precursor protein. The protease that carries out this cleavage is the SpoIIGA protein, which spans the septum between the prespore and mother cell. The protease domain, which is on the mother-cell side of the septum, is activated by binding of SpoIIR to a receptor domain on the prespore side. It is a typical receptor-mediated signal transduction system (Section 14.1.2). The gene for SpoIIR is one of those whose promoter is recognized specifically by σ^F, so activation of the protease, and conversion of pre-σ^E to active σ^E, occurs once σ^F-directed transcription is underway in the prespore.

Activation of σ^F and σ^E is just the beginning of the story. In the prespore, about 1 hour after its activation, σ^F responds to an unknown signal (possibly from the mother cell) which results in a slight change in genome activity in the spore. This includes transcription of a gene for another σ subunit, σ^G, which recognizes promoters upstream of genes whose products are required during the later stages of spore differentiation. One of these proteins is SpoIVB, which activates another septum-bound protease, SpoIVF (Figure 14.28). This protease then activates a second mother-cell σ subunit, σ^K, which is coded by a σ^E-transcribed gene but retained in the mother cell in an inactive form until the signal for its activation is received from the prespore. σ^K directs transcription of the genes whose products are needed during the later stages of the mother-cell differentiation pathway.

To summarize, the key features of *Bacillus* sporulation are as follows:

- The master protein, SpoOA, responds to external stimuli via a cascade of phosphorylation events to determine if and when the switch to sporulation should occur.

- A succession of σ subunits in prespore and mother cell brings about time-dependent changes in genome activity in the two cells.

- Cell–cell signaling ensures that the events occurring in prespore and mother cell are coordinated.

14.3.3 Vulva development in *Caenorhabditis elegans*

B. subtilis is a unicellular organism and, although sporulation involves the coordinated differentiation of two cell types, it can hardly be looked upon as comparable to the developmental processes that occur in multicellular organisms. Sporulation provides pointers to the general ways in which genome activity might be regulated during the development of a multicellular organism, but it does not indicate the specific events to expect. We therefore need to examine development in a simple multicellular eukaryote.

C. elegans *is a model for multicellular eukaryotic development*

Research with the microscopic nematode worm *C. elegans* (Figure 14.29) was initiated by Sydney Brenner in the 1960s with the aim of utilizing it as a simple model for multicellular eukaryotic development. *C. elegans* is easy to grow in the laboratory and has a short generation time, measured in days but still convenient for genetic analysis. The worm is transparent at all stages of its life cycle, so internal examination is possible without killing the animal. This is an important point because it has enabled researchers to follow the entire developmental process of the worm at the cellular level. Every cell division in the pathway from fertilized egg to adult worm has been charted, and every point at which a cell adopts a specialized role has been identified. In addition, the complete connectivity of the 302 cells that comprise the nervous system of the worm has been mapped.

The genome of *C. elegans* is relatively small, just 97 Mb (see Table 7.2), and the entire sequence is known. Analysis of the sequence, using many of the techniques described in Chapter 5, is beginning to assign functions to the unknown genes and to establish links between genome activity and the developmental pathways. The objective is a complete genetic description of development in *C. elegans*, a goal that is attainable in the not-too-distant future.

Determination of cell fate during development of the C. elegans *vulva*

A critical feature that underpins the usefulness of *C. elegans* as a tool for research is the fact that its development is more or less invariant: the pattern of cell division and differentiation is virtually the same in every individual. This appears to be due in large part to cell–cell signaling, which induces each cell to follow its appropriate differentiation pathway. To illustrate this we will look at development of the *C. elegans* vulva.

Most *C. elegans* worms are hermaphrodites, meaning that they have both male and female sex organs. The vulva is part of the female sex apparatus, being the tube through which sperm enter and fertilized eggs are laid. The adult vulva comprises 22 cells, which are the progeny of three ancestral cells originally located in a row on the undersurface of the developing worm (Figure 14.30). Each of these ancestral cells becomes committed to the differentiation pathway that leads to production of vulva cells. The central cell, called P6.p, adopts the "primary vulva cell fate" and divides to produce eight new cells. The other two cells—P5.p and P7.p—take on the "secondary vulva

(A) Activation of σ^F in the prespore

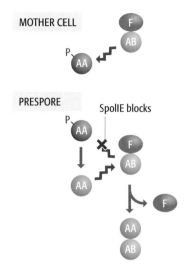

(B) Activation of σ^E in the mother cell

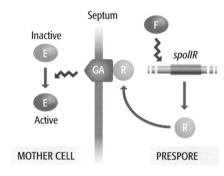

Figure 14.27 Activation of the prespore- and mother-cell-specific σ subunits during *Bacillus* sporulation. (A) In the mother cell, σ^F is inactive because it is bound to SpoIIAB, which phosphorylates SpoIIAA and prevents the latter releasing σ^F. Activation of σ^F in the prespore occurs by its release from its complex with SpoIIAB, which is indirectly influenced by the concentration of membrane-bound SpoIIE. (B) In the mother cell, σ^E is activated by proteolytic cleavage by SpoIIGA, which responds to the presence in the prespore of the σ^F-dependent protein, SpoIIR. Abbreviations: AA, SpoIIAA; AB, SpoIIAB; E, σ^E; F, σ^F; GA, SpoIIGA; R, SpoIIR.

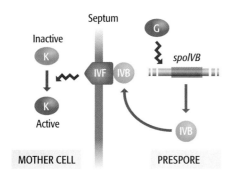

Figure 14.28 Activation of σ^K during *Bacillus* **sporulation.** Note that the scheme is very similar to the procedure used to activate σ^E (see Figure 14.27B). Abbreviations: G, σ^G; K, σ^K; IVB, SpoIVB; IVF, SpoIVF.

Figure 14.29 The nematode worm *Caenorhabditis elegans*. The micrograph shows an adult hermaphrodite worm, approximately 1 mm in length. The vulva is the small projection located on the underside of the animal, about halfway along. Egg cells can be seen inside the worm's body in the region either side of the vulva. Reprinted with permission from Kendrew, J. (Ed.), *Encylopaedia of Molecular Biology*. © 1994 Blackwell Publishing.

cell fate" and divide into seven cells each. These 22 cells then reorganize their positions to construct the vulva.

A critical aspect of vulva development is that it must occur in the correct position relative to the gonad, the structure containing the egg cells. If the vulva develops in the wrong place then the gonad will not receive sperm and the egg cells will never be fertilized. The positional information needed by the vulva progenitor cells is provided by a cell within the gonad called the anchor cell (Figure 14.31). The importance of the anchor cell has been demonstrated by experiments in which it is artificially destroyed in the embryonic worm: in the absence of the anchor cell, a vulva does not develop. The implication is that the anchor cell secretes an extracellular signaling compound that induces P5.p, P6.p, and P7.p to differentiate. This signaling compound is the protein called LIN-3, coded by the *lin-3* gene.

Why does P6.p adopt the primary cell fate whereas P5.p and P7.p take on secondary cell fates? There are two possibilities. The first is that LIN-3 forms a concentration gradient and therefore has different effects on P6.p, the cell which is closest to it, and the more distant P5.p and P7.p, as shown in Figure 14.31. Evidence in favor of this idea comes from studies showing that isolated cells adopt the secondary fate when exposed to low levels of LIN-3. Alternatively, the signal that commits P5.p and P7.p to their secondary fates might not come directly from the anchor cell but via P6.p in the form of a different extracellular signaling compound whose synthesis by P6.p is switched on by LIN-3 activation. This hypothesis is supported by the abnormal features displayed by certain mutants in which more than three cells become committed to vulva development. With these mutants there is more than one primary cell, but each one is invariably surrounded by two secondary cells, suggesting that in the living worm, adoption of the secondary cell fate is dependent on the presence of an adjacent primary cell.

There are other instructive features of vulva development in *C. elegans*. The first is that the signaling process that commits P6.p to its primary cell fate has many similarities with the MAP kinase signal transduction system of vertebrates (see Figure 14.10). The cell surface receptor for LIN-3 is a protein kinase called LET-23 which, when activated by binding LIN-3, initiates a series of intracellular reactions that leads to activation of a MAP-kinase-like protein, which in turn switches on a variety of transcription activators. Unfortunately, the target genes have not yet been delineated in either the primary or secondary vulva progenitor cells, but the system is open to study.

A second noteworthy feature is that as well as the activation signal provided by the anchor cell in the form of LIN-3, the vulva progenitor cells are also subject to the deactivating effects of a second signaling compound secreted by the hypodermal cell, a multinuclear sheath that surrounds most of the worm's body. This repressive signal is overcome by the positive signals that

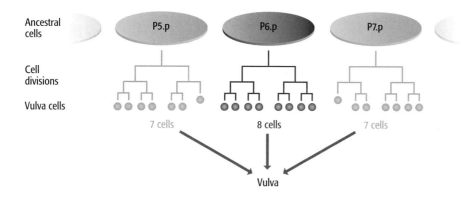

Figure 14.30 Cell divisions resulting in production of the vulva cells of *Caenorhabditis elegans.* Three ancestral cells divide in a programmed manner to produce 22 progeny cells, which reorganize their positions relative to one another to construct the vulva.

induce P5.p, P6.p, and P7.p to differentiate, but prevents the unwanted differentiation of three adjacent cells, P3.p, P4.p, and P8.p, each of which can become committed to vulva development if the repressive signal malfunctions, for example in a mutant worm.

In summary, the general concepts to emerge from the study of vulva development in *C. elegans* are as follows:

- In a multicellular organism, positional information is important: the correct structure must develop at the appropriate place.

- The commitment to differentiation of a small number of progenitor cells can lead to construction of a multicellular structure.

- Cell–cell signaling can utilize a concentration gradient to induce different responses in cells at different positions relative to the signaling cell.

- A cell might be subject to competitive signaling, where one signal tells it to do one thing and a second signal tells it to do the opposite.

14.3.4 Development in *Drosophila melanogaster*

The last organism whose development we will study is *Drosophila melanogaster.* The experimental history of the fruit fly dates back to 1910 when Thomas Hunt Morgan first used this organism as a model system in genetic research. For Morgan, the advantages of *Drosophila* were its small size, enabling large numbers to be studied in a single experiment, its minimal nutritional requirements (the flies like bananas), and the presence in

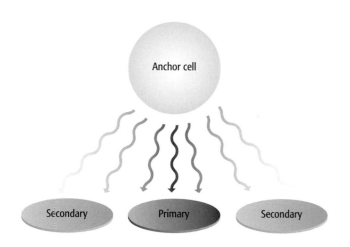

Figure 14.31 The postulated role of the anchor cell in determining cell fate during vulva development in *Caenorhabditis elegans.* It is thought that release of the signaling compound LIN-3 by the anchor cell commits P6.p (shown in red), the cell closest to the anchor cell, to the primary vulva cell fate. P5.p and P7.p (shown in yellow) are further away from the anchor cell and so are exposed to a lower concentration of LIN-3 and become secondary vulva cells. As described in the text, there is evidence that commitment of the secondary cells to their fates is also influenced by signals from the primary vulva cell.

natural populations of occasional variants with easily recognized genetic characteristics such as unusual eye colors. Morgan was not aware that other advantages are a small genome (180 Mb; see Table 7.2) and the fact that gene isolation is aided by the presence in the salivary glands of "giant" chromosomes. These are made up of multiple copies of the same DNA molecule laid side by side, displaying banding patterns that can be correlated with the physical map of each chromosome to pinpoint the positions of desired genes. But Morgan did foresee that *Drosophila* might become an important organism for developmental research, a topic that he was as interested in as we are today.

The major contribution that *Drosophila* has made to our understanding of development has been through the insights it has provided into how an undifferentiated embryo acquires positional information that eventually results in the construction of complex body parts at the correct places in the adult organism. Although in some respects *Drosophila* is quite unusual in its embryonic organization (as we will see in the next section), the genetic mechanisms that specify the fly's body plan are similar to those in other organisms, including humans. Knowledge gained from *Drosophila* has therefore directed research into areas of human development that for a long time were thought to be inaccessible. To explore this story we must start with the events that occur in the developing *Drosophila* embryo.

Maternal genes establish protein gradients in the Drosophila *embryo*
The unusual feature of the early *Drosophila* embryo is that it is not made up of lots of cells, as in most organisms, but instead is a single **syncytium** comprising a mass of cytoplasm and multiple nuclei (Figure 14.32). This structure persists until 13 successive rounds of nuclear division have produced some 1500 nuclei: only then do individual uninucleate cells start to appear around the outside of the syncytium, producing the structure called the blastoderm. Before the blastoderm stage has been reached, the positional information has begun to be established.

Initially, the positional information that the embryo needs is a definition of which end is the front (anterior) and which the back (posterior), as well as similar information relating to up (dorsal) and down (ventral). This information is provided by concentration gradients of proteins that become established in the syncytium. The majority of these proteins are not synthesized from genes in the embryo, but are translated from mRNAs injected into the embryo by the mother. To see how these **maternal-effect genes** work we will examine the synthesis of Bicoid, one of the four proteins involved in determining the anterior–posterior axis.

The *bicoid* gene is transcribed in the maternal nurse cells, which are in contact with the egg cells, and the mRNA is injected into the anterior end of the unfertilized egg. This position is defined by the orientation of the egg cell in the egg chamber. The *bicoid* mRNA remains in the anterior region of the egg cell, attached by its 3′ untranslated region to the cell's cytoskeleton. It is not translated immediately, probably because its poly(A) tail is too short. This is inferred because translation, which occurs after fertilization of the egg, is preceded by extension of the poly(A) tail through the combined efforts of the Cortex, Grauzone, and Staufen proteins, all of which are synthesized from genes in the egg. Bicoid protein then diffuses through the syncytium, setting

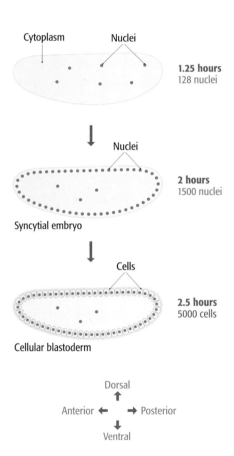

Cytoplasm · Nuclei

1.25 hours
128 nuclei

Nuclei

2 hours
1500 nuclei

Syncytial embryo

Cells

2.5 hours
5000 cells

Cellular blastoderm

Dorsal
Anterior ← → Posterior
Ventral

Figure 14.32 Early development of the *Drosophila* embryo. To begin with, the embryo is a single syncytium containing a gradually increasing number of nuclei. These nuclei migrate to the periphery of the embryo after about 2 hours, and within another 30 minutes cells begin to be constructed. The embryo is approximately 500 μm in length and 170 μm in diameter.

up a concentration gradient, highest at the anterior end and lowest at the posterior end (Figure 14.33).

Three other maternal-effect gene products are also involved in setting up the anterior–posterior gradient. These are the Hunchback, Nanos, and Caudal proteins. All are injected as mRNAs into the anterior region of the unfertilized egg. The *nanos* mRNA is transported to the posterior part of the egg and attached to the cytoskeleton while it awaits translation. The *hunchback* and *caudal* mRNAs become distributed evenly through the cytoplasm, but their proteins subsequently form gradients through the action of Bicoid and Nanos:

- Bicoid activates the *hunchback* gene in the embryonic nuclei, supplementing the *hunchback* mRNA in the anterior region, and represses translation of the maternal *caudal* mRNA. The result is an increase in the concentration of the Hunchback protein in the anterior region and a decrease in that of Caudal.

- Nanos represses translation of *hunchback* mRNA, contributing further to the anterior–posterior gradient of the Hunchback protein.

The net result is a gradient of Bicoid and Hunchback, greater at the anterior end, and of Nanos and Caudal, greater at the posterior end (see Figure 14.33). The gradient is supplemented with Torso protein, another maternal-effect gene product, which accumulates at the extreme anterior and posterior ends. Similar events result in a dorsal-to-ventral gradient, predominantly of the protein called Dorsal.

A cascade of gene expression converts positional information into a segmentation pattern

The body plan of the adult fly, as well as that of the larva, is built from a series of segments, each with a different structural role. This is clearest in the thorax, which has three segments, each carrying one pair of legs, and the abdomen, which is made up of eight segments, but is also true for the head, even though in the head the segmented structure is less visible (Figure 14.34). The objective of embryo development is therefore production of a young larva with the correct segmentation pattern.

The gradients established in the embryo by the maternal-effect gene products are the first stage in formation of the segmentation pattern. These gradients provide the interior of the embryo with a basic amount of positional information, each point in the syncytium now having its own unique chemical signature defined by the relative amounts of the various maternal-effect gene products. This positional information is made more precise by expression of the **gap genes**. Three of the anterior–posterior gradient proteins—Bicoid, Hunchback, and Caudal—are transcription activators that target the gap genes in the nuclei that now line the inside of the embryo (see Figure 14.32). The identities of the gap genes expressed in a particular nucleus depend on the relative concentrations of the gradient proteins and hence on the position of the nucleus along the anterior–posterior axis. Some gap genes are activated directly by Bicoid, Hunchback, and Caudal, examples being *buttonhead*, *empty spiracles*, and *orthodenticle* which are activated by Bicoid. Other gap genes are switched on indirectly, as is the case with *huckebein* and *tailless*, which respond to transcription activators that are switched on by Torso. There are also repressive effects (e.g., Bicoid represses expression of

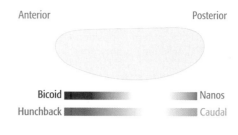

Figure 14.33 Establishment of the anterior–posterior axis in a *Drosophila* embryo. The anterior–posterior axis is established by gradients of Bicoid, Nanos, Caudal, and Hunchback proteins, as descrtibed in the text. In this diagram, the concentration gradients are indicated by the colored bars under the outline of the embryo.

Figure 14.34 The segmentation pattern of the adult *Drosophila melanogaster*. Note that the head is also segmented, but the pattern is not easily discernible from the morphology of the adult fly.

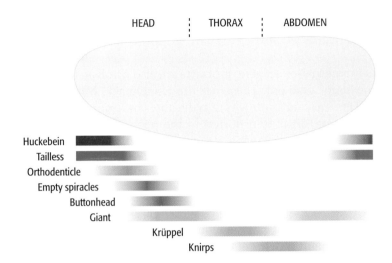

Figure 14.35 The role of the gap gene products in conferring positional information during embryo development in *Drosophila melanogaster*. The concentration gradient of each gap gene product is denoted by the colored bars. The parts of the embryo that give rise to the head, thorax, and abdomen regions of the adult fly are indicated.

knirps) and the gap gene products regulate their own expression in various ways. This complex interplay results in the positional information in the embryo, now carried by the relative concentrations of the gap gene products, becoming more detailed (Figure 14.35).

The next set of genes to be activated, the **pair-rule genes**, establish the basic segmentation pattern. Transcription of these genes responds to the relative concentrations of the gap gene products and occurs in nuclei that have become enclosed in cells. The pair-rule gene products therefore do not diffuse through the syncytium but remain localized within the cells that express them. The result is that the embryo can now be looked upon as comprising a series of stripes, each stripe consisting of a set of cells expressing a particular pair-rule gene. In a further round of gene activation, the **segment polarity genes** become switched on, providing greater definition to the stripes by setting the sizes and precise locations of what will eventually be the segments of the larval fly. Gradually we have converted the imprecise positional information of the maternal-effect gradients into a sharply defined segmentation pattern.

Segment identity is determined by the homeotic selector genes

The pair-rule and segment polarity genes establish the segmentation pattern of the embryo but do not determine the identities of the individual segments. This is the job of the **homeotic selector genes**, which were first discovered by virtue of the extravagant effects that mutations in these genes have on the appearance of the adult fly. The *antennapedia* mutation, for example, transforms the head segment that usually produces an antenna into one that makes a leg, so the mutant fly has a pair of legs where its antennae should be. The early geneticists were fascinated by these monstrous **homeotic mutants** and many were collected during the first few decades of the twentieth century.

Genetic mapping of homeotic mutations has revealed that the selector genes are clustered in two groups on chromosome 3. These clusters are called the Antennapedia complex (ANT-C), which contains genes involved in determination of the head and thorax segments, and the Bithorax complex (BX-C), which contains genes for the abdomen segments (Figure 14.36). Some additional non-selector development genes, such as *bicoid*, are also located in

ANT-C. One interesting feature of the ANT-C and BX-C clusters, which is still not understood, is that the order of genes corresponds to the order of the segments in the fly, the first gene in ANT-C being *labial palps*, which controls the most anterior segment of the fly, and the last gene in BX-C being *Abdominal B*, which specifies the most posterior segment.

The correct selector gene is expressed in each segment because the activation of each one is responsive to the positional information represented by the distributions of the gap gene and pair-rule gene products. The selector gene products are themselves transcription activators, each containing a homeodomain version of the helix–turn–helix DNA-binding structure (Section 11.1.1). Each selector gene product, possibly in conjunction with a coactivator such as Extradenticle, switches on the set of genes needed to initiate development of the specified segment. Maintenance of the differentiated state is ensured partly by the repressive effect that each selector gene product has on expression of the other selector genes, and partly by the work of Polycomb which, as we saw in Section 14.2.2, constructs inactive chromatin over the selector genes that are not expressed in a particular cell.

Homeotic selector genes are universal features of higher eukaryotic development

The homeodomains of the various *Drosophila* selector genes are strikingly similar. This observation led researchers in the 1980s to search for other homeotic genes by using the homeodomain as a probe in hybridization experiments. First, the *Drosophila* genome was searched, resulting in isolation of several previously unknown homeodomain-containing genes. These have turned out not to be selector genes but other types of genes coding for transcription activators involved in development. Examples include the pair-rule genes *even-skipped* and *fushi tarazu*, and the segment polarity gene *engrailed*.

The real excitement came when the genomes of other organisms were probed and it was realized that homeodomains are present in genes in a wide variety of animals, including humans. Even more unexpected was the discovery that some of the homeodomain genes in these other organisms are homeotic selector genes organized into clusters similar to ANT-C and BX-C, and that these genes have equivalent functions to the *Drosophila* versions, specifying construction of the body plan. For example, mutations in the HoxC8 gene of mouse results in an animal that has an extra pair of ribs, due to conversion of a lumbar vertebra (normally in the lower back) into a thoracic vertebra (from which the ribs emerge). Other Hox mutations in animals lead to limb deformations, such as absence of the lower arm, or extra digits on the hands or feet.

We now look on the ANT-C and BX-C clusters of selector genes in *Drosophila* as two parts of a single complex, usually referred to as the homeotic gene complex, or HOM-C. In vertebrates there are four homeotic gene clusters, called HoxA to HoxD. When these four clusters are aligned with one another and with HOM-C (Figure 14.37), similarities are seen between the genes at equivalent positions, such that the evolutionary history of the homeotic selector gene clusters can be traced from insects through to humans (see Section 18.2.1). As in *Drosophila*, the order of genes in the vertebrate clusters reflects the order of the structures specified by the genes in the adult body

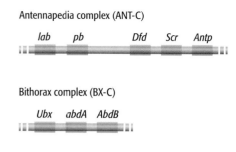

Figure 14.36 The Antennapedia and Bithorax gene complexes of *Drosophila melanogaster*. Both complexes are located on the fruit-fly chromosome 3, ANT-C upstream of BX-C. The genes are usually drawn in the order shown, although this means that they are transcribed from right to left. The diagram does not reflect the actual lengths of the genes. The full gene names are as follows: *lab*, labial palps; *pb*, proboscipedia; *Dfd*, Deformed; *Scr*, Sex combs reduced; *Antp*, Antennapedia; *Ubx*, Ultrabithorax; *abdA*, abdominal A; *AbdB*, Abdominal B. In ANT-C, the non-selector genes *zerknüllt* and *bicoid* occur between *pb* and *Dfd*, and *fushi tarazu* lies between *Scr* and *Antp*.

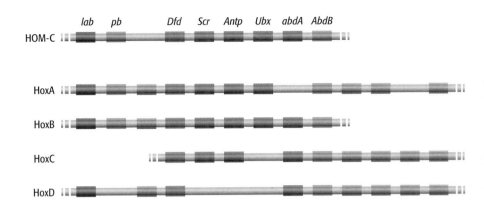

plan. This is clearly seen with the mouse HoxB cluster, which controls development of the nervous system (Figure 14.38). The remarkable conclusion is that, at this fundamental level, developmental processes in fruit flies and other "simple" eukaryotes are similar to the processes occurring in humans and other "complex" organisms. The discovery that studies of fruit flies are directly relevant to human development opens up vast vistas of future research possibilities.

Homeotic genes also underlie plant development

The power of *Drosophila* as a model system for development extends even beyond vertebrates. Developmental processes in plants are, in most respects, very different from those of fruit flies and other animals, but at the genetic level there are certain similarities, sufficient for the knowledge gained about *Drosophila* development to be of value in interpreting similar research carried out with plants. In particular, the recognition that a limited number of homeotic selector genes control the *Drosophila* body plan has led to a model for plant development which postulates that the structure of the flower is determined by a small number of homeotic genes.

All flowers are constructed along similar lines, made up of four concentric whorls, each comprising a different floral organ (Figure 14.39). The outer whorl, number 1, contains sepals, which are modified leaves that envelop and protect the bud during its early development. The next whorl, number 2, contains the distinctive petals, and within these are whorls 3 (stamens, the male reproductive organs) and 4 (carpels, the female reproductive organs).

Most of the research on plant development has been carried out with *Antirrhinum* (the snapdragon) and *Arabidopsis thaliana*, a small vetch that

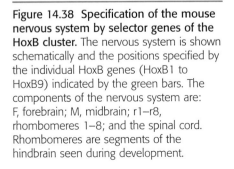

Figure 14.38 Specification of the mouse nervous system by selector genes of the HoxB cluster. The nervous system is shown schematically and the positions specified by the individual HoxB genes (HoxB1 to HoxB9) indicated by the green bars. The components of the nervous system are: F, forebrain; M, midbrain; r1–r8, rhombomeres 1–8; and the spinal cord. Rhombomeres are segments of the hindbrain seen during development.

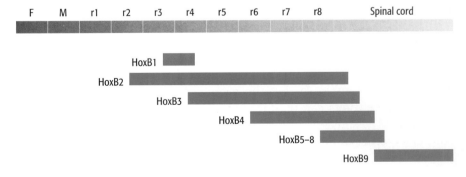

has been adopted as a model species, partly because it has a genome of only 125 Mb (see Table 7.2), one of the smallest known among flowering plants. Although these plants do not appear to contain homeodomain proteins, they do have genes which, when mutated, lead to homeotic changes in the floral architecture, such as replacement of sepals by carpels. Analysis of these mutants has led to the "ABC model," which states that there are three types of homeotic genes—A, B, and C—which control flower development as follows:

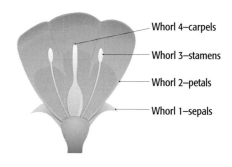

Figure 14.39 Flowers are constructed from four concentric whorls.

- Whorl 1 is specified by A-type genes: examples in *Arabidopsis* are *apetala1* and *apetala2*.

- Whorl 2 is specified by A genes acting in concert with B genes, examples of the latter including *apetala3* and *pistillata*.

- Whorl 3 is specified by the B genes plus the C gene, *agamous*.

- Whorl 4 is specified by the C gene acting on its own.

As anticipated from the work with *Drosophila*, the A, B, and C homeotic gene products are transcription activators. All except the APETALA2 protein contain the same DNA-binding domain, the **MADS box**, which is also found in other proteins involved in plant development, including SEPALLATA1, 2, and 3, which work with the A, B, and C proteins in defining the detailed structure of the flower. Other components of the flower development system include at least one master gene, called *floricaula* in *Antirrhinum* and *leafy* in *Arabidopsis*, which controls the switch from vegetative to reproductive growth, initiating flower development, and also has a role in establishing the pattern of homeotic gene expression. In *Arabidopsis* there is also a gene, called *curly leaf*, whose product acts like Polycomb of *Drosophila* (Section 14.2.2), maintaining the differentiated state of each cell by repressing those homeotic genes that are inactive in a particular whorl.

Summary

There are many steps within the expression pathways of individual genes at which regulation can be exerted, but the key control mechanisms act during the initiation of transcription. Transient changes in genome expression patterns occur predominantly in response to external stimuli which influence the transcription of individual genes. Some extracellular signaling compounds are imported into the cell and directly influence transcription, an example being lactoferrin in mammals. Steroid hormones also enter the cell but influence genome expression via receptor proteins which act as transcription activators. In catabolite repression in bacteria, glucose influences expression of a variety of genes involved in sugar utilization by indirectly controlling the level of cyclic AMP in the cell, which in turn influences the activity of a transcription activator called the catabolite activator protein. Other signaling pathways are mediated by cell surface receptors, many of which dimerize in response to the extracellular signal, initiating a signal transduction pathway that leads to the genome. The MAP kinase pathway is one such example, but there are several others, including ones involving transcription activators called STATs. Some signal transduction pathways make use of second messengers such as cyclic nucleotides and calcium ions, which influence a number of cellular activities including genome expression. Permanent and semipermanent changes in genome expression can be brought about by genome rearrangements, as occur during mating-type switching in yeast and generation of immunoglobulin and T-cell receptor diversity in vertebrates. Permanent and semipermanent changes can also be produced by proteins such as Polycomb of *Drosophila melanogaster*, which induces the formation of heterochromatin over regions of a chromosome that must be silenced. Our understanding of the genetic basis of development has been aided by the study of model systems in relatively simple organisms such as bacteria, *Caenorhabditis elegans*, and *Drosphila melanogaster*. The lysogenic infection cycle of bacteriophage λ indicates ways in which simple genetic switches can determine which of two developmental pathways are followed, and studies of sporulation in *Bacillus subtilis* have illustrated how time-dependent changes in genome expression can be brought about and how cell–cell signaling can regulate a differentiation pathway. Mechanisms for the determination of cell fate have been revealed by studies of vulva development in *C. elegans*. The most informative pathway for developmental genetics has been embryogenesis in the fruit fly, study of which has shown how a complex body plan can be specified by controlled patterns of genome expression. This work has also revealed the existence of homeotic selector genes which control development processes not only in flies, but also in vertebrates and in plants.

Multiple Choice Questions

14.1.* The term differentiation is defined as which of the following?

   a. Alterations in genome expression that do not change the proteome of the cell.

   b. Transient changes in the genome activity of a cell in response to extracellular factors.

   c. A coordinated series of changes that occurs during the life history of a cell.

   d. The adoption of a specialized physiological role by a cell.

14.2. What is the most important control point for regulation of genome expression?

   a. Transcription initiation.

   b. Processing of transcripts.

   c. Translation initiation.

   d. Degradation of proteins and RNA molecules.

14.3.* Which of the following is NOT a mechanism by which a signaling compound is known to influence genome expression after being imported into a cell?

   a. Some signaling molecules methylate DNA sequences to silence specific genes.

   b. Some signaling molecules are proteins that function as regulators of genome expression.

   c. Some signaling molecules directly influence the activity of regulatory proteins within the cell.

   d. Some signaling molecules influence the activity of regulatory proteins in the cell indirectly through intermediate molecules.

14.4. How do steroid hormones, such as estrogen, modulate genome expression in responsive cells?

   a. By binding to enhancer sequences.

   b. By binding to receptors in the cytoplasm which then migrate to the nucleus where they bind to DNA to regulate genome expression.

   c. By binding to receptors in the nucleus that are activated and then bind to DNA to regulate genome expression.

   d. By binding to receptors in the cell membrane, the signal is then transduced to the nucleus through a signaling pathway.

14.5.* What types of signal transduction pathways involve the STAT proteins?

   a. These pathways have a single step between the receptor and the genome.

   b. These pathways have several steps between the receptor and the genome.

   c. These pathways utilize second messengers to transduce the signal to the genome.

   d. These pathways activate a receptor that then moves to the nucleus to regulate genome expression.

14.6. What is the most common type of covalent modification that activates proteins in signaling pathways?

   a. Acetylation.

   b. Glycosylation.

   c. Methylation.

   d. Phosphorylation.

14.7.* What are second messenger molecules?

   a. They are hormones that initiate a signaling pathway.

   b. They are receptors that bind to hormones and activate a pathway.

   c. They are internal molecules that transduce a signal inside the cell.

   d. They are transcriptional activators that function at the end of a pathway.

14.8. The mating-type switching process that occurs in yeast is an example of which of the following?

   a. Alternative splicing.

   b. Change in a feedback loop.

   c. Change in DNA methylation patterns.

   d. Change due to physical rearrangement of the genome.

14.9.* What changes are made in B cells when they switch from producing IgM or IgD to IgG immunoglobulins?

   a. This change is accomplished by alternative splicing of the RNA transcripts.

   b. This change occurs in the proteome as the IgM/IgD constant regions are proteolytically removed from the IgG protein.

   c. This change occurs in the genome as the genes encoding the constant regions for IgM and IgD are deleted by the RAG1 and RAG2 proteins.

   d. This change occurs in the genome as the genes encoding the constant regions for IgM and IgD are deleted independently of the RAG proteins.

14.10. The Polycomb proteins in *Drosophila* function by:

   a. Condensing chromatin to induce gene silencing and this silencing is passed on to daughter cells.

   b. Condensing chromatin to maintain gene silencing and this silencing is passed on to daughter cells.

   c. Condensing chromatin to induce gene silencing and this silencing is not passed on to daughter cells.

   d. Condensing chromatin to maintain gene silencing and this silencing is not passed on to daughter cells.

continued …

Multiple Choice Questions (continued) *Answers to odd-numbered questions can be found in the Appendix

14.11. Feedback loops could bring about long-term changes in genome expression by what mechanism?

- **a.** A regulatory protein activates its own transcription so it is continuously expressed.
- **b.** A regulatory protein represses its own transcription so that it is permanently silenced.
- **c.** A regulatory protein activates another protein that stimulates expression of the regulatory protein.
- **d.** All of the above.

14.12. How is the sporulation pathway in *Bacillus* activated?

- **a.** A lack of nutrients signals the activation of the gene encoding the SpoOA protein.
- **b.** A lack of nutrients signals the activation of the SpoOA protein by proteolytic cleavage.
- **c.** A lack of nutrients signals the activation of the SpoOA protein by acetylation.
- **d.** A lack of nutrients signals the activation of the SpoOA protein by phosphorylation.

14.13. What prevents the cells adjacent to the progenitor vulva cells in *C. elegans* from differentiating into vulva cells?

- **a.** These cells are too distant from the anchor cell to receive the LIN-3 signal.
- **b.** These cells lack the ability to bind and respond to the LIN-3 signal.
- **c.** These cells receive a signaling compound from a hypodermal cell that deactivates the LIN-3 signal.

- **d.** These cells have condensed their chromatin in the regions responsive to the LIN-3 signal.

14.14. What is the syncytium structure of the *Drosophila* embryo?

- **a.** A highly compacted mass of undifferentiated cells.
- **b.** An oblong structure that contains a concentration gradient of developmental proteins.
- **c.** A mass of cytoplasm and multiple nuclei.
- **d.** A mixture of diploid and haploid cells generated by mitotic and meiotic cell divisions.

14.15. Which *Drosophila* genes determine the identification of the segments of the fruit-fly larva?

- **a.** The gap genes.
- **b.** The pair-rule genes.
- **c.** The segment polarity genes.
- **d.** The homeotic selector genes.

14.16. If a loss-of-function mutation occurs in the B-type genes of *Arabidopsis*, what will be the composition of the flower whorls (from whorl 1 to whorl 4)?

- **a.** Sepals–petals–stamens–carpels.
- **b.** Sepals–sepals–stamens–carpels.
- **c.** Sepals–sepals–carpels–carpels.
- **d.** Petals–petals–stamens–stamens.

Short Answer Questions   *Answers to odd-numbered questions can be found in the Appendix

14.1. Outline the differences between differentiation and development and describe the basis for these differences.

14.2. Why are changes in nutrient conditions likely to result in greater changes in genome activity in unicellular organisms compared to multicellular organisms?

14.3. Explain the relationship between glucose transport and cAMP levels in *E. coli*.

14.4. How are STAT proteins phosphorylated if the receptor is not a tyrosine kinase?

14.5. How does MAP kinase function to regulate genome expression?

14.6. How do the Ras proteins function in signaling pathways?

14.7. What is the basis to class switching, as occurs with some B lymphocytes?

14.8. How do muscle cells remain differentiated as muscle cells?

14.9. During sporulation in *Bacillus*, σ^E and σ^F are present in both the prespore and mother cells. How is σ^F activated in the prespore?

14.10. How does the anchor cell of *C. elegans* induce the vulva progenitor cells to differentiate into vulva cells? Why do the vulva progenitor cells follow different pathways upon receiving the signal from the anchor cell?

Short Answer Questions (continued) *Answers to odd-numbered questions can be found in the Appendix

14.11.* How is the concentration gradient for the Bicoid protein established in the syncytium of the *Drosophila* embryo?

14.12. How did mutations in the *Drosophila* ANT-C gene complex and mouse Hox genes provide researchers with information on the function of these genes?

In-depth Problems *Guidance to odd-numbered questions can be found in the Appendix

14.1.* Describe how studies of signal transduction have improved our understanding of the abnormal biochemical activities that underlie cancer.

14.2. Explore the influence of signal transduction by second messengers on the regulation of genome activity.

14.3.* Are *Caenorhabditis elegans* and *Drosophila* *melanogaster* good model organisms for development in higher eukaryotes?

14.4. Is there any value in having model organisms for development in higher eukaryotes?

14.5.* What would be the key features of an ideal model organism for development in higher eukaryotes?

Figure Tests *Answers to odd-numbered questions can be found in the Appendix

14.1.* The graph shows that glucose is utilized before lactose by *E. coli* when the cells are grown in a medium containing both sugars. What term is used to describe this phenomenon? Describe the mechanism that underlies the process.

14.2. The figure depicts a change of the *MAT* locus from a *MATa* genotype to a *MATα* genotype. What is the term used to describe this change and what is the mechanism by which the change occurs?

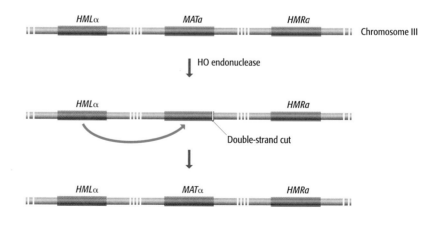

continued ...

Figure Tests (continued)

*Answers to odd-numbered questions can be found in the Appencix

14.3.* Discuss the processes by which the great diversity of immunoglobulins is produced by B cells.

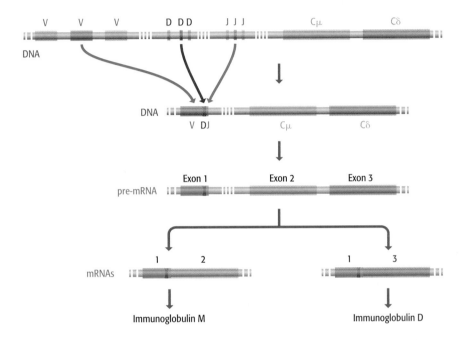

14.4. The figure shows the distribution of proteins during embryo development in *Drosophila*. These proteins are encoded by which types of genes? Which proteins regulate the expression of these genes?

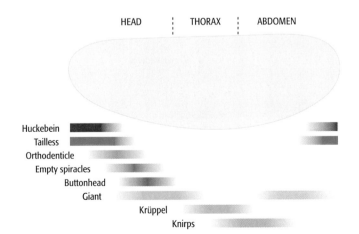

Further Reading

Imported extracellular signaling compounds

He, J. and Furmanski, P. (1995) Sequence specificity and transcriptional activation in the binding of lactoferrin to DNA. *Nature* **373:** 721–724.

Tsai, M.-J. and O'Malley, B.W. (1994) Molecular mechanisms of action of steroid/thyroid receptor superfamily members. *Annu. Rev. Biochem.* **63:** 451–486.

Winge, D.R., Jensen, L.T. and Srinivasan, C. (1998) Metal ion regulation of gene expression in yeast. *Curr. Opin. Chem. Biol.* **2:** 216–221.

Signal transduction via cell surface receptors

Horvath, C.M. (2000) STAT proteins and transcriptional responses to extracellular signals. *Trends Biochem. Sci.* **25:** 496–502.

Karin, M. and Hunter, T. (1995) Transcriptional control by protein phosphorylation: signal transmission from the cell surface to the nucleus. *Curr. Biol.* **5:** 747–757.

Maruta, H. and Burgess, A.W. (1994) Regulation of the Ras signaling network. *Bioessays* **16:** 489–496.

Robinson, M.J. and Cobb, M.H. (1997) Mitogen-activated kinase pathways. *Curr. Opin. Cell Biol.* **9:** 180–186.

Schlessinger, J. (1993) How receptor tyrosine kinases activate Ras. *Trends Biochem. Sci.* **18:** 273–275.

Spiegel, S., Foster, D. and Kolesnick, R. (1996) Signal transduction through lipid second messengers. *Curr. Opin. Cell Biol.* **8:** 159–167.

Whitman, M. (1998) Feedback from inhibitory SMADs. *Nature* **389:** 549–551.

Genome rearrangements

Alt, F.W., Blackwell, T.K. and Yancopoulos, G.D. (1987) Development of the primary antibody repertoire. *Science* **238:** 1079–1087. *Generation of immunoglobulin diversity.*

Nasmyth, K. and Shore, D. (1987) Transcriptional regulation in the yeast life cycle. *Science* **237:** 1162–1170. *Yeast mating–type switching.*

Polycomb

Chan, C.S., Rastelli, L. and Pirrotta, V. (1994) A Polycomb response element in the Ubx gene that determines an epigenetically inherited state of repression. *EMBO J.* **13:** 2553–2564.

Déjardin, J., Rappailles, A., Cuvier, O., Grimaud, C., Decoville, M., Locker, D. and Cavalli, G. (2005) Recruitment of *Drosophila* Polycomb group proteins to chromatin by DSP1. *Nature* **434:** 533–538.

Feedback loops

Popperl, H., Bienz, M., Studer, M., Chan, S.K., Aparicio, S., Brenner, S., Mann, R.S. and Krumlauf, R. (1995) Segmental expression of HoxB-1 is controlled by a highly conserved autoregulatory loop dependent upon exd/pbx. *Cell* **81:** 1031–1042.

Regulski, M., Dessain, S., McGinnis, N. and McGinnis, W. (1991) High affinity binding sites for the Deformed protein are required for the function of an autoregulatory enhancer of the *deformed* gene. *Genes Devel.* **5:** 278–286.

Sporulation in *B. subtilis*

Errington, J. (1996) Determination of cell fate in *Bacillus subtilis. Trends Genet.* **12:** 31–34.

Sonenshein, A.L. (2000) Control of sporulation initiation in *Bacillus subtilis. Curr. Opin. Microbiol.* **3:** 561–566.

Stragier, P. and Losick, R. (1996) Molecular genetics of sporulation in *Bacillus subtilis. Annu. Rev. Genet.* **30:** 297–341.

C. elegans vulva development

Aroian, R.V., Koga, M., Mendel, J.E., Ohshima, Y. and Sternberg, P.W. (1990) The *let-23* gene necessary for *Caenorhabditis elegans* vulval induction encodes a tyrosine kinase of the EGF receptor subfamily. *Nature* **348:** 693–699.

Katz, W.S., Hill, R.J., Clandinin, T.R. and Sternberg, P.W. (1995) Different levels of the *C. elegans* growth factor LIN-3 promote distinct vulval precursor fates. *Cell* **82:** 297–307.

Kornfeld, K. (1997) Vulval development in *Caenorhabditis elegans. Trends Genet.* **13:** 55–61.

Labouesse, M. and Mango, S.E. (1999) Patterning the *C. elegans* embryo: moving beyond the cell lineage. *Trends Genet.* **15:** 307–313. *Reviews the developmental pathways of C. elegans.*

Sharma-Kishore, R., White, J.G., Southgate, E. and Podbilewicz, B. (1999) Formation of the vulva in *Caenorhabditis elegans*: a paradigm for organogenesis. *Development* **126:** 691–699.

Embryogenesis in fruit flies and homeotic selector genes in vertebrates

Ingham, P.W. (1988) The molecular genetics of embryo pattern formation in *Drosophila. Nature* **335:** 25–34.

Krumlauf, R. (1994) Hox genes in vertebrate development. *Cell* **78:** 191–201.

Maconochie, M., Nonchev, S., Morrison, A. and Krumlauf, R. (1996) Paralogous Hox genes: function and regulation. *Annu. Rev. Genet.* **30:** 529–556. *Describes homeotic selector genes in vertebrates.*

Mahowald, A.P. and Hardy, P.A. (1985) Genetics of *Drosophila* embryogenesis. *Annu. Rev. Genet.* **19:** 149–177.

Plant development

Goodrich, J., Puangsomlee, P., Martin, M., Long, D., Meyerowitz, E.M. and Coupland, G. (1997) A Polycomb-group gene regulates homeotic gene expression in *Arabidopsis. Nature* **386:** 44–51.

Ma, H. (1998) To be, or not to be, a flower – control of floral meristem identity. *Trends Genet.* **14:** 26–32.

Parcy, F., Nilsson, O., Busch, M.A., Lee, I. and Weigel, D. (1998) A genetic framework for floral patterning. *Nature* **395:** 561–566.

How Genomes Replicate and Evolve

Part 4 – How Genomes Replicate and Evolve links replication, mutation and recombination with the gradual evolution of genomes over time. We begin with a detailed examination of the molecular processes underlying genome replication (Chapter 15), mutation and repair (Chapter 16), and recombination (Chapter 17) before exploring the ways in which these processes have shaped the structures and genetic contents of genomes over evolutionary time (Chapter 18). Finally, Chapter 19 describes how molecular phylogenetics makes use of the evolutionary information contained within genomes to investigate questions such as the relationships between humans and other primates, the origins of AIDS, and the migratory routes followed by humans as we spread across the planet from our birthplace in Africa.

Genome Replication

15

When you have read Chapter 15, you should be able to:

State what is meant by the topological problem and explain how DNA topoisomerases solve this problem.

Describe the key experiment that proved that DNA replication occurs by the semiconservative process.

Outline the displacement and rolling circle modes of genome replication.

Discuss how replication is initiated in bacteria, yeast, and mammals.

Describe the key features of bacterial and eukaryotic DNA polymerases.

Explain why the leading and lagging strands of a DNA molecule must be replicated by different processes.

Give a detailed description of the events occurring at the bacterial replication fork, and indicate how these events differ from those occurring in eukaryotes.

Describe what is currently known about termination of replication in bacteria and eukaryotes.

Explain how telomerase maintains the ends of a chromosomal DNA molecule in eukaryotes, and appraise the possible links between telomere length, cell senescence, and cancer.

Describe how genome replication is coordinated with the cell cycle.

The primary function of a genome is to specify the biochemical signature of the cell in which it resides. We have seen that the genome achieves this objective by the coordinated expression of genes and groups of genes, resulting in maintenance of a proteome whose individual protein components carry out and regulate the cell's biochemical activities. In order to continue carrying out this function, the genome must replicate every time that the cell divides. This means that the entire DNA content of the cell must be copied at the appropriate period in the cell cycle, and the resulting DNA molecules must be

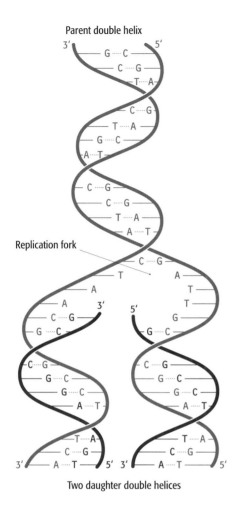

Parent double helix

Replication fork

Two daughter double helices

Figure 15.1 DNA replication, as predicted by Watson and Crick. The polynucleotides of the parent double helix are shown in black. Both act as templates for synthesis of new strands of DNA, shown in red. The sequences of these new strands are determined by base pairing with the template molecules. The topological problem arises because the two polynucleotides of the parent helix cannot simply be pulled apart; the helix has to be unwound in some way.

distributed to the daughter cells so that each one receives a complete copy of the genome. This elaborate process, which spans the interface between molecular biology, biochemistry, and cell biology, is described in this chapter.

Genome replication has been studied since Watson and Crick first discovered the double helix structure of DNA back in 1953. In the years since then, research has been driven by three related but distinct issues:

- The topological problem was the primary concern in the years from 1953 to 1958. This problem arises from the need to unwind the double helix in order to make copies of its two polynucleotides (see Figure 15.1). The issue assumed center stage in the mid-1950s because it was the main stumbling block to acceptance of the double helix as the correct structure for DNA, but moved into the background in 1958 when Matthew Meselson and Franklin Stahl demonstrated that, despite the perceived difficulties, DNA replication in *Escherichia coli* occurs by the method predicted by the double helix structure. The **Meselson–Stahl experiment** enabled research into genome replication to move forward, even though the topological problem itself was not solved until the early 1980s when the mode of action of **DNA topoisomerases** was first understood (Section 15.1.2).

- The replication process has been studied intensively since 1958. During the 1960s, the enzymes and proteins involved in replication in *E. coli* were identified and their functions delineated, and in the following years similar progress was made in understanding the details of eukaryotic genome replication. This work is ongoing, with research today centered on topics such as the initiation of replication and the precise modes of action of the proteins that are active at the replication fork.

- The regulation of genome replication, particularly in the context of the cell cycle, has become the predominant area of research in recent years. This work has shown that initiation is the key control point in genome replication and has begun to explain how replication is synchronized with the cell cycle so that daughter genomes are available when the cell divides.

Our study of genome replication will deal with each of these three topics in the order listed above.

15.1 The Topological Problem

In their paper in *Nature* announcing the discovery of the double helix structure of DNA, Watson and Crick made one of the most famous statements in molecular biology:

"It has not escaped our notice that the specific pairing we have postulated immediately suggests a possible copying mechanism for the genetic material."

The pairing process that they refer to is one in which each strand of the double helix acts as a template for synthesis of a second complementary strand, the end result being that both of the daughter double helices are identical to the parent molecule (Figure 15.1). The scheme is almost implicit in the double helix structure, but it presents problems, as admitted by Watson and Crick in a second paper published in *Nature* just a month after the report of the structure. This paper describes the postulated replication process in more detail, but points out the difficulties that arise from the need to unwind the double helix. The most trivial of these difficulties is the possibility of the daughter molecules getting tangled up. More critical is the rotation that would accompany the unwinding: with one turn occurring for every 10 bp of the double helix, complete replication of the DNA molecule in human chromosome 1, which is 250 Mb in length, would require 25 million rotations of the chromosomal DNA. It is difficult to imagine how this could occur within the constrained volume of the nucleus, but the unwinding of a linear chromosomal DNA molecule is not physically impossible. In contrast, a circular double-stranded molecule, for example a bacterial or bacteriophage genome, having no free ends, would not be able to rotate in the required manner and so, apparently, could not be replicated by the Watson–Crick scheme. Finding an answer to this dilemma was a major preoccupation of molecular biology during the 1950s.

15.1.1 Experimental proof for the Watson–Crick scheme for DNA replication

The topological problem was considered so serious by some molecular biologists, notably Max Delbrück, that there was initially some resistance to accepting the double helix as the correct structure of DNA. The difficulty relates to the **plectonemic** nature of the double helix, this being the topological arrangement that prevents the two strands of a coil being separated without unwinding. The problem would therefore be resolved if the double helix was in fact **paranemic**, because this would mean that the two strands could be separated simply by moving each one sideways without unwinding the molecule. It was suggested that the double helix could be converted into a paranemic structure by supercoiling in the direction opposite to the turn of the helix itself, or that within a DNA molecule the right-handed helix proposed by Watson and Crick might be "balanced" by equal lengths of a left-handed helical structure. The possibility that double-stranded DNA was not a helix at all, but a side-by-side ribbon structure, was also briefly considered, this idea surprisingly being revived in the late 1970s. Each of these proposed solutions to the topological problem were individually rejected for one reason or another, most of them because they required alterations to the double helix structure, alterations that were not compatible with the X-ray diffraction results and other experimental data pertaining to DNA structure.

The first real progress toward a solution of the topological problem came in 1954 when Delbrück proposed a "breakage-and-reunion" model for separating the strands of the double helix. In this model, the strands are separated not by unwinding the helix with accompanying rotation of the molecule, but by breaking one of the strands, passing the second strand though the gap, and rejoining the first strand. This scheme is in fact very close to the correct solution to the topological problem, being one of the ways in which DNA topoisomerases work (see Figure 15.4A), but unfortunately Delbrück overcomplicated the issue by attempting to combine breakage and reunion with the DNA synthesis that occurs during the actual replication process. This led him to a model for DNA replication which results in each polynucleotide in the daughter molecule being made up partly of parental DNA and partly of newly synthesized DNA (Figure 15.2A). This **dispersive** mode of replication contrasts with the **semiconservative** system proposed by Watson and Crick (Figure 15.2B). A third possibility is that replication is fully **conservative**, one of the daughter double helices being made entirely of newly synthesized DNA and the other comprising the two parental strands (Figure 15.2C). Models for conservative

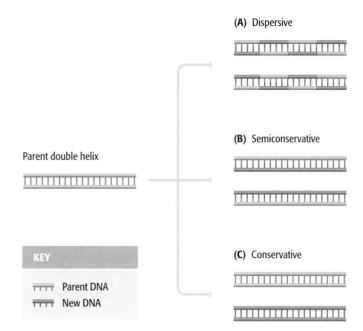

Figure 15.2 Three possible schemes for DNA replication. For the sake of clarity, the DNA molecules are drawn as ladders rather than helices.

replication are difficult to devise, but one can imagine that this type of replication might be accomplished without unwinding the parent helix.

The Meselson–Stahl experiment

Delbrück's breakage-and-reunion model was important because it stimulated experiments designed to test between the three modes of DNA replication illustrated in Figure 15.2. Radioactive isotopes had recently been introduced into molecular biology, so attempts were made to use DNA labeling (see Technical Note 2.1) to distinguish newly synthesized DNA from the parental polynucleotides. Each mode of replication predicts a different distribution of newly synthesized DNA, and hence of radioactive label, in the double helices resulting after two or more rounds of replication. Analysis of the radioactive contents of these molecules should therefore determine which replication scheme operates in living cells. Unfortunately, it proved impossible to obtain a clear-cut result, largely because of the difficulty in measuring the precise amount of radioactivity in the DNA molecules, the analysis being complicated by the rapid decay of the ^{32}P isotope that was used as the label.

The breakthrough was eventually made by Matthew Meselson and Franklin Stahl who, in 1958, carried out the required experiment not with a radioactive label but with ^{15}N, the nonradioactive "heavy" isotope of nitrogen. Now it was possible to analyze the replicated double helices by density gradient centrifugation (see Technical Note 7.1), because a DNA molecule labeled with ^{15}N has a higher buoyant density than an unlabeled molecule. Meselson and Stahl started with a culture of *E. coli* cells that had been grown with ^{15}NH$_4$Cl and whose DNA molecules therefore contained heavy nitrogen. The cells were transferred to normal medium, and samples taken after 20 minutes and 40 minutes, corresponding to one and two cell divisions, respectively. DNA was extracted from each sample and the molecules examined by density gradient centrifugation (Figure 15.3A). After one round of DNA replication, the daughter molecules synthesized in the presence of normal nitrogen formed a single band in the density gradient, indicating that each double helix was made up of equal amounts of newly synthesized and parental

DNA. This result immediately enabled the conservative mode of replication to be discounted, as this predicts that there would be two bands after one round of replication (Figure 15.3B), but did not provide a distinction between Delbrück's dispersive model and the semiconservative process favored by Watson and Crick. The distinction was, however, possible when the DNA molecules resulting from two rounds of replication were examined. Now the density gradient revealed two bands of DNA, the first corresponding to a hybrid

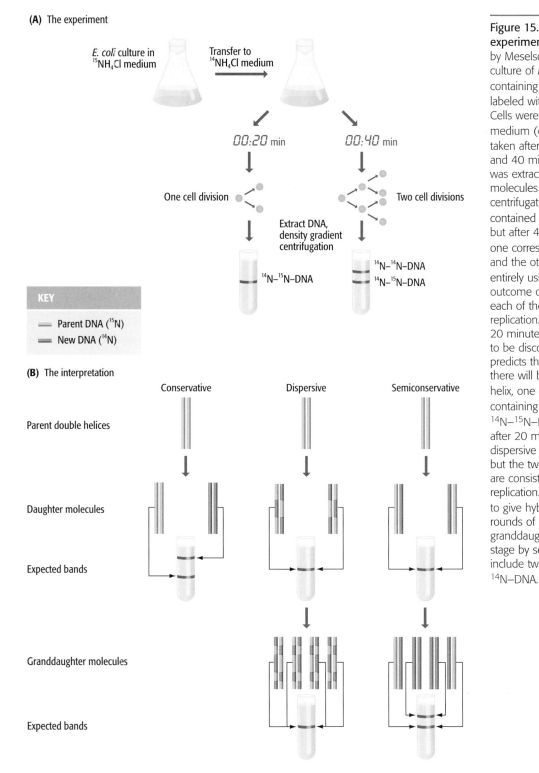

(A) The experiment

E. coli culture in $^{15}NH_4Cl$ medium

Transfer to $^{14}NH_4Cl$ medium

00:20 min

00:40 min

One cell division

Two cell divisions

Extract DNA, density gradient centrifugation

$^{14}N-^{15}N-DNA$

$^{14}N-^{14}N-DNA$
$^{14}N-^{15}N-DNA$

KEY

Parent DNA (^{15}N)
New DNA (^{14}N)

(B) The interpretation

Conservative   Dispersive   Semiconservative

Parent double helices

Daughter molecules

Expected bands

Granddaughter molecules

Expected bands

Figure 15.3 The Meselson–Stahl experiment. (A) The experiment carried out by Meselson and Stahl involved growing a culture of *Escherichia coli* in a medium containing $^{15}NH_4Cl$ (ammonium chloride labeled with the heavy isotope of nitrogen). Cells were then transferred to normal medium (containing $^{14}NH_4Cl$) and samples taken after 20 minutes (one cell division) and 40 minutes (two cell divisions). DNA was extracted from each sample and the molecules analyzed by density gradient centrifugation. After 20 minutes, all the DNA contained similar amounts of ^{14}N and ^{15}N, but after 40 minutes two bands were seen, one corresponding to hybrid $^{14}N-^{15}N-DNA$, and the other to DNA molecules made entirely using ^{14}N. (B) The predicted outcome of the experiment is shown for each of the three possible modes of DNA replication. The banding pattern seen after 20 minutes enables conservative replication to be discounted because this scheme predicts that after one round of replication there will be two different types of double helix, one containing just ^{15}N and the other containing just ^{14}N. The single $^{14}N-^{15}N-DNA$ band that was actually seen after 20 minutes is compatible with both dispersive and semiconservative replication, but the two bands seen after 40 minutes are consistent only with semiconservative replication. Dispersive replication continues to give hybrid $^{14}N-^{15}N$ molecules after two rounds of replication, whereas the granddaughter molecules produced at this stage by semiconservative replication include two that are made entirely of $^{14}N-DNA$.

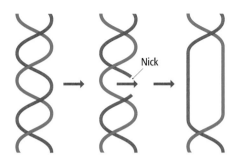

Figure 15.4 The mode of action of a Type I DNA topoisomerase. A Type I topoisomerase makes a nick in one strand of a DNA molecule, passes the intact strand through the nick, and reseals the gap.

composed of equal parts of newly synthesized and old DNA, and the second corresponding to molecules made up entirely of new DNA. This result agrees with the semiconservative scheme but is incompatible with dispersive replication, the latter predicting that after two rounds of replication all molecules would be hybrids.

15.1.2 DNA topoisomerases provide a solution to the topological problem

The Meselson–Stahl experiment proved that DNA replication in living cells follows the semiconservative scheme proposed by Watson and Crick, and hence indicated that the cell must have a solution to the topological problem. This solution was not understood by molecular biologists until some 25 years later, when the activities of the enzymes called DNA topoisomerases were characterized.

DNA topoisomerases are enzymes that carry out breakage-and-reunion reactions similar but not identical to that envisaged by Delbrück. Two types of DNA topoisomerase are recognized (Table 15.1):

- Type I topoisomerases introduce a break in one polynucleotide and pass the second polynucleotide through the gap that is formed (Figure 15.4). The two ends of the broken strand are then religated. This mode of action results in the linking number (the number of times one strand crosses the other in a circular molecule) being changed by one.

- Type II topoisomerases break both strands of the double helix, creating a "gate" through which a second segment of the helix is passed. This changes the linking number by two.

Breaking one or both DNA strands might appear to be a drastic solution to the topological problem, leading to the possibility that the topoisomerase might occasionally fail to rejoin a strand and hence inadvertently break a chromosome into two sections. This possibility is reduced by the mode of action of these enzymes. One end of each cut polynucleotide becomes covalently attached to a tyrosine amino acid at the active site of the enzyme, ensuring that this end of the polynucleotide is held tightly in place while the free end(s) is being manipulated. Type I and II topoisomerases are subdivided according to the precise chemical structure of the polynucleotide–tyrosine linkage: with IA and IIA enzymes the link involves a phosphate group attached to the free 5′ end of the cut polynucleotide, and with IB and IIB enzymes the linkage is via a 3′ phosphate group. The A and B topoisomerases probably evolved separately. Both types are present in eukaryotes but IB and IIB enzymes are very uncommon in prokaryotes.

Note that DNA topoisomerases do not themselves *unwind* the double helix. Instead they solve the topological problem by counteracting the overwinding

Table 15.1 DNA topoisomerases

Type	Substrate	Examples
Type I	Single-stranded DNA	*Escherichia coli* topoisomerases I and III; yeast and human topoisomerase III; archaeal reverse gyrase; eukaryotic topoisomerase I
Type II	Double-stranded DNA	*E. coli* topoisomerases II (DNA gyrase) and IV; eukaryotic topoisomerase II

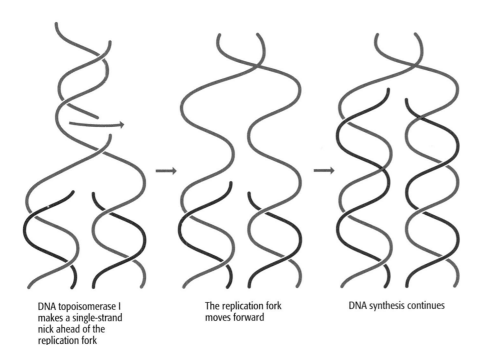

DNA topoisomerase I makes a single-strand nick ahead of the replication fork

The replication fork moves forward

DNA synthesis continues

Figure 15.5 Unzipping the double helix. During replication, the double helix is "unzipped" as a result of the action of DNA topoisomerases. The replication fork is therefore able to proceed along the molecule without the helix having to rotate.

that otherwise would be introduced into the molecule by the progression of the replication fork. The result is that the helix can be "unzipped," with the two strands pulled apart sideways without the molecule having to rotate (Figure 15.5). Replication is not the only activity that is complicated by the topology of the double helix, and it is becoming increasingly clear that DNA topoisomerases have equally important roles during transcription, recombination, and other processes that can result in over- or underwinding of DNA. In eukaryotes, topoisomerases form a major part of the nuclear matrix, the scaffold-like network that permeates the nucleus (Section 10.1.1), and are responsible for maintaining chromatin structure and unlinking DNA molecules during chromosome division. Most topoisomerases are only able to relax DNA, but some prokaryotic enzymes, such as the bacterial DNA gyrase and the archaeal reverse gyrase, can carry out the reverse reaction and introduce supercoils into DNA molecules.

15.1.3 Variations on the semiconservative theme

No exceptions to the semiconservative mode of DNA replication are known but there are several variations on this basic theme. DNA copying via a replication fork, as shown in Figure 15.1, is the predominant system, being used by chromosomal DNA molecules in eukaryotes and by the circular genomes of prokaryotes. Some smaller circular molecules, such as animal mitochondrial genomes (Section 8.3.2), use a slightly different process called **displacement replication**. In these molecules, the point at which replication begins is marked by a **D-loop**, a region of approximately 500 bp where the double helix is disrupted by the presence of an RNA molecule base-paired to one of the DNA strands (Figure 15.6). This RNA molecule acts as the starting point for synthesis of one of the daughter polynucleotides. This polynucleotide is synthesized by continuous copying of one strand of the helix, the second strand being displaced and subsequently copied after synthesis of the first daughter genome has been completed.

The advantage of displacement replication as performed by animal mitochondrial DNA is not clear. In contrast, the special type of displacement

Figure 15.6 Displacement replication.
The D-loop contains a short RNA molecule that primes DNA synthesis (see Section 15.2.2). After completion of the first strand synthesis a second RNA primer attaches to the displaced strand and initiates replication of this molecule. In this diagram, newly synthesized DNA is shown in red.

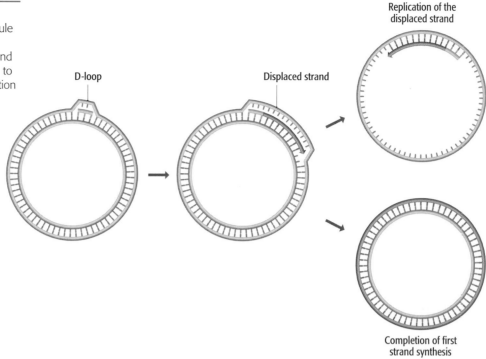

Figure 15.7 Rolling circle replication.

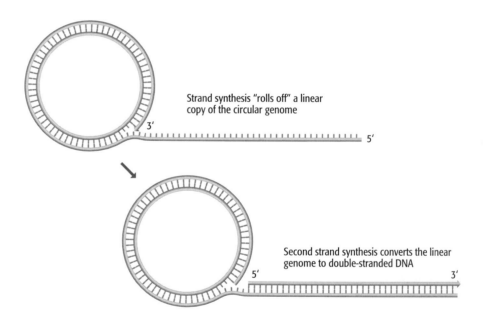

process called **rolling circle replication** is an efficient mechanism for the rapid synthesis of multiple copies of a circular genome. Rolling circle replication, which is used by λ and various other bacteriophages, initiates at a nick which is made in one of the parent polynucleotides. The free 3′ end that results is extended, displacing the 5′ end of the polynucleotide. Continued DNA synthesis "rolls off" a complete copy of the genome, and further synthesis eventually results in a series of genomes linked head to tail (Figure 15.7). These genomes are single stranded and linear, but can easily be converted to double-stranded circular molecules by complementary strand synthesis, followed by cleavage at the junction points between genomes, and circularization of the resulting segments.

15.2 The Replication Process

As with many processes in molecular biology, we conventionally look on genome replication as being made up of three phases—initiation, elongation, and termination:

- Initiation (Section 15.2.1) involves recognition of the position(s) on a DNA molecule where replication will begin.

- Elongation (Section 15.2.2) concerns the events occurring at the replication fork, where the parent polynucleotides are copied.

- Termination (Section 15.2.3), which in general is only vaguely understood, occurs when the parent molecule has been completely replicated.

As well as these three stages in replication, one additional topic demands attention. This relates to a limitation in the replication process that, if uncorrected, would lead to linear double-stranded DNA molecules getting shorter each time they are copied. The solution to this problem, which concerns the structure and synthesis of the telomeres at the ends of chromosomes (Section 7.1.2), is described in Section 15.2.4.

15.2.1 Initiation of genome replication

Initiation of replication is not a random process and always begins at the same position or positions on a DNA molecule, these points being called the **origins of replication**. Once initiated, two replication forks can emerge from the origin and progress in opposite directions along the DNA: replication is therefore bidirectional with most genomes (Figure 15.8). A circular bacterial genome has a single origin of replication, meaning that several thousand kilobases of DNA are copied by each replication fork. This situation differs from that seen with eukaryotic chromosomes, which have multiple origins and whose replication forks progress for shorter distances. The yeast *Saccharomyces cerevisiae*, for example, has about 332 origins, corresponding to 1 per 36 kb of DNA, and humans have some 20,000 origins, or 1 for every 150 kb of DNA.

Initiation at the E. coli *origin of replication*

We know substantially more about initiation of replication in bacteria than in eukaryotes. The *E. coli* origin of replication is referred to as *oriC*. By transferring segments of DNA from the *oriC* region into plasmids that lack their own origins, it has been estimated that the *E. coli* origin of replication spans approximately 245 bp of DNA. Sequence analysis of this segment shows that it contains two short repeat motifs, one of nine nucleotides and the other of 13 nucleotides (Figure 15.9A). The nine-nucleotide repeat, five copies of which are dispersed throughout *oriC*, is the binding site for a protein called DnaA. With five copies of the binding sequence, it might be imagined that five copies of DnaA attach to the origin, but in fact bound DnaA proteins cooperate with unbound molecules until some 30 copies are associated with the origin. Attachment occurs only when the DNA is negatively supercoiled, as is the normal situation for the *E. coli* chromosome (Section 8.1.1).

The result of DnaA binding is that the double helix opens up ("melts") within the tandem array of three AT-rich, 13-nucleotide repeats located at one end of the *oriC* sequence (Figure 15.9B). The exact mechanism is unknown but DnaA does not appear to possess the enzymatic activity needed to break base pairs, and it is therefore assumed that the helix is melted by torsional stresses

(A) Replication of a circular bacterial chromosome

← Direction of replication →

(B) Replication of a linear eukaryotic chromosome

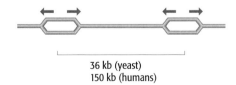

36 kb (yeast)
150 kb (humans)

Figure 15.8 Bidirectional DNA replication of (A) a circular bacterial chromosome and (B) a linear eukaryotic chromosome.

Figure 15.9 The *Escherichia coli* origin of replication. (A) The *E. coli* origin of replication is called *oriC* and is approximately 245 bp in length. It contains three copies of a 13-nucleotide repeat motif, consensus sequence 5′–GATCTNTTNTTTT–3′ where "N" is any nucleotide, and five copies of a nine-nucleotide repeat, consensus 5′–TT(A/T)T(A/C)CA(A/C)A–3′, where (A/T) is A or T, and (A/C) is A or C. The 13-nucleotide sequences form a tandem array of direct repeats at one end of *oriC*. The nine-nucleotide sequences are distributed through *oriC*, three units forming a series of direct repeats and two units in the inverted configuration, as indicated by the arrows. Three of the nine-nucleotide repeats—numbers 1, 3, and 5 when counted from the left-hand end of *oriC* as drawn here—are regarded as major sites for DnaA attachment; the other two repeats are minor sites. The overall structure of the origin is similar in all bacteria and the sequences of the repeats do not vary greatly. (B) Model for the attachment of DnaA proteins to *oriC*, resulting in melting of the helix within the AT-rich 13-nucleotide sequences.

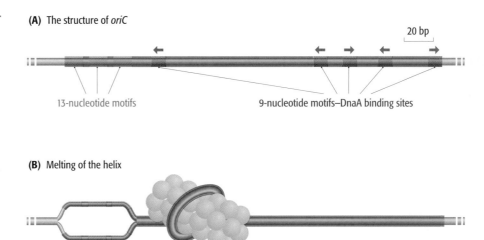

(A) The structure of *oriC*

13-nucleotide motifs   9-nucleotide motifs–DnaA binding sites

(B) Melting of the helix

Melted region   Barrel of DnaA proteins

introduced by attachment of the DnaA proteins. An attractive model imagines that the DnaA proteins form a barrel-like structure around which the helix is wound. Melting the helix is promoted by HU, the most abundant of the DNA packaging proteins of *E. coli* (Section 8.1.1).

Melting of the helix initiates a series of events that construct a nascent replication fork at either end of the open region. The first step is the attachment of a **prepriming complex** at each of these two positions. Each prepriming complex initially comprises 12 proteins, six copies of DnaB and six copies of DnaC, but DnaC has a transitory role and is released from the complex soon after it is formed, its function probably being simply to aid the attachment of DnaB. The latter is a **helicase**, an enzyme which can break base pairs (see Section 15.2.2). DnaB begins to increase the single-stranded region within the origin, enabling the enzymes involved in the elongation phase of genome replication to attach. This represents the end of the initiation phase of replication in *E. coli* as the replication forks now start to progress away from the origin and DNA copying begins.

Origins of replication in yeast have been clearly defined

The technique used to delineate the *E. coli oriC* sequence, involving transfer of DNA segments into a nonreplicating plasmid, has also proved valuable in identifying origins of replication in the yeast *Saccharomyces cerevisiae*. Origins identified in this way are called **autonomously replicating sequences**, or **ARSs**. A typical yeast origin is shorter than *E. coli oriC*, being usually less than 200 bp in length, but, like the *E. coli* origin, contains discrete segments with different functional roles, these "subdomains" having similar sequences in different origins (Figure 15.10A). Four subdomains are recognized. Two of these—subdomains A and B1—make up the **origin recognition sequence**, a stretch of some 40 bp in total that is the binding site for the **origin recognition complex** (**ORC**), a set of six proteins that attach to the origin (Figure 15.10B). ORCs have been described as yeast versions of the *E. coli* DnaA proteins, but this interpretation is probably not strictly correct because ORCs appear to remain attached to yeast origins throughout the cell cycle. Rather than being genuine initiator proteins, it is more likely that ORCs are involved in the regulation of genome replication, acting as mediators between replication origins and the regulatory signals that coordinate the initiation of DNA replication with the cell cycle (Section 15.3.1).

We must therefore look elsewhere in yeast origins for sequences with functions strictly equivalent to that of *oriC* of *E. coli*. This leads us to the two other conserved sequences in the typical yeast origin, subdomains B2 and B3 (see Figure 15.10A). Our current understanding suggests that these two subdomains function in a manner similar to the *E. coli* origin. Subdomain B2 appears to correspond to the 13-nucleotide repeat array of the *E. coli* origin, being the position at which the two strands of the helix are first separated. This melting is induced by torsional stress introduced by attachment of a DNA-binding protein, ARS binding factor 1 (ABF1), which attaches to subdomain B3 (see Figure 15.10B). As in *E. coli*, melting of the helix within a yeast replication origin is followed by attachment of the helicase and other replication enzymes to the DNA, completing the initiation process and enabling the replication forks to begin their progress along the DNA, as described in Section 15.2.2.

Replication origins in higher eukaryotes have been less easy to identify

Attempts to identify replication origins in humans and other higher eukaryotes have, until recently, been less successful. **Initiation regions** (parts of the chromosomal DNA where replication initiates) can be delineated by various biochemical methods, for example by allowing replication to initiate in the presence of labeled nucleotides, then arresting the process, purifying the newly synthesized DNA, and determining the positions of these nascent strands in the genome. These experiments have suggested that there are specific regions in mammalian chromosomes where replication begins, but some researchers have been doubtful whether these regions contain replication origins equivalent to those in yeast. One alternative hypothesis is that replication is initiated by protein structures that have specific positions in the nucleus, the chromosome initiation regions simply being those DNA segments located close to these protein structures in the three-dimensional organization of the nucleus.

Doubts about mammalian replication origins were increased by the failure of mammalian initiation regions to confer replicative ability on replication-deficient plasmids, although these experiments were not considered conclusive because it was recognized that a mammalian origin might be too long to be cloned in a plasmid or might function only when activated by distant sites in the chromosomal DNA. A key breakthrough has been the demonstration that an 8 kb segment of a human initiation region, when transferred to the monkey genome, still directs replication despite being removed from any hypothetical protein structure in the human nucleus. Analysis of this transferred initiation region has shown that there are primary sites within the region where initiation occurs at high frequency, surrounded by secondary sites, spanning the entire 8 kb region, at which replication initiates with lower frequency. The presence of discrete functional domains within the initiation region has also been demonstrated by examining the effects of deletions of parts of the region on the efficiency of replication initiation.

The demonstration that the human genome does in fact contain replication origins equivalent to those in yeast raises the question of whether mammals

(A) Structure of a yeast origin of replication

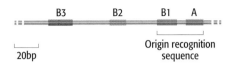

(B) Melting of the helix

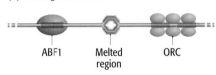

Figure 15.10 Structure of a yeast origin of replication. (A) Structure of ARS1, a typical autonomously replicating sequence (ARS) that acts as an origin of replication in *Saccharomyces cerevisiae*. The relative positions of the functional sequences A, B1, B2, and B3 are shown. (B) Melting of the helix occurs within subdomain B2, induced by attachment of the ARS binding protein 1 (ABF1) to subdomain B3. The proteins of the origin replication complex (ORC) are permanently attached to subdomains A and B1.

possess an equivalent of the yeast ORC. The answer appears to be yes, as several genes whose protein products have similar sequences to the yeast ORC proteins have been identified in higher eukaryotes, and some of these proteins have been shown to be able to replace the equivalent yeast protein in the yeast ORC. These results indicate that initiation of replication in yeast is a good model for events occurring in mammals, a conclusion that is very relevant to studies of the control of replication initiation, as we will see in Section 15.3.

15.2.2 The elongation phase of replication

Once replication has been initiated, the replication forks progress along the DNA and participate in the central activity of genome replication—the synthesis of new strands of DNA that are complementary to the parent polynucleotides. At the chemical level, the template-dependent synthesis of DNA (Figure 15.11) is very similar to the template-dependent synthesis of RNA that occurs during transcription (compare with Figure 12.1). However, this similarity should not mislead us into making an extensive analogy between

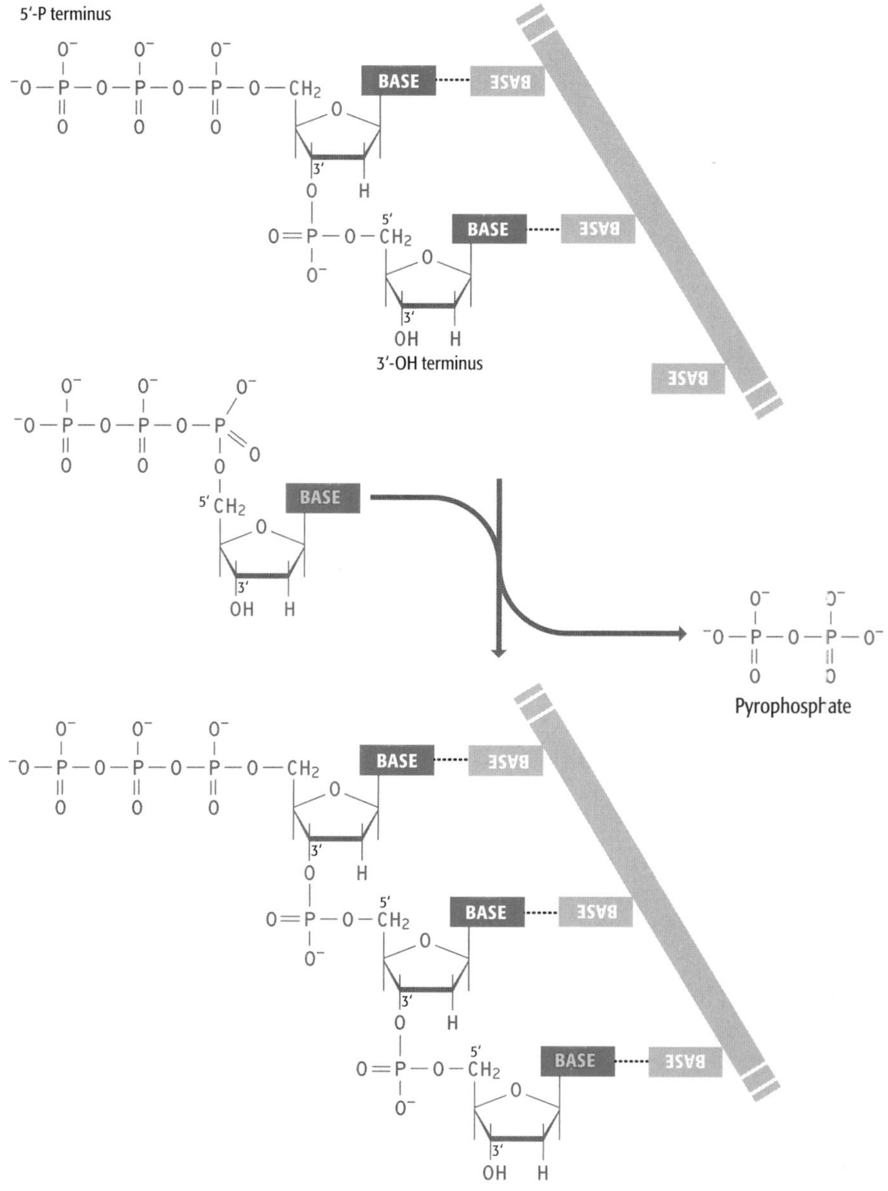

Figure 15.11 Template-dependent synthesis of DNA. Compare this reaction with template-dependent synthesis of RNA shown in Figure 12.1.

transcription and replication. The mechanics of the two processes are quite different, replication being complicated by two factors that do not apply to transcription:

- During DNA replication both strands of the double helix must be copied. This is an important complication because, as noted in Section 1.1.2, DNA polymerase enzymes are only able to synthesize DNA in the 5′→3′ direction. This means that one strand of the parent double helix, called the **leading strand**, can be copied in a continuous manner, but replication of the **lagging strand** has to be carried out in a discontinuous fashion, resulting in a series of short segments that must be ligated together to produce the intact daughter strand (Figure 15.12).

- The second complication arises because template-dependent DNA polymerases cannot initiate DNA synthesis on a molecule that is entirely single stranded: there must be a short double-stranded region to provide a 3′ end onto which the enzyme can add new nucleotides. This means that **primers** are needed, one to initiate complementary strand synthesis on the leading polynucleotide, and one for every segment of discontinuous DNA synthesized on the lagging strand (Figure 15.12).

Before dealing with these two complications we will first examine the DNA polymerase enzymes themselves.

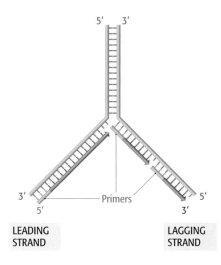

Figure 15.12 Complications with DNA replication. Two complications have to be solved when double-stranded DNA is replicated. First, only the leading strand can be continuously replicated by 5′→3′ DNA synthesis; replication of the lagging strand has to be carried out discontinuously. Second, initiation of DNA synthesis requires a primer. This is true both of cellular DNA synthesis, as shown here, and DNA synthesis reactions that are carried out in the test tube (Section 2.1.1).

The DNA polymerases of bacteria and eukaryotes

The principal chemical reaction catalyzed by a DNA polymerase is the 5′→3′ synthesis of a DNA polynucleotide, as shown in Figure 15.11. We learnt in Section 2.1.1 that some DNA polymerases combine this function with at least one exonuclease activity, which means that these enzymes can degrade polynucleotides as well as synthesize them (see Figure 2.7):

- A 3′→5′ exonuclease activity is possessed by many bacterial and eukaryotic template-dependent DNA polymerases (Table 15.2). This activity enables the enzyme to remove nucleotides from the 3′ end of the strand that it has just synthesized. It is looked on as a **proofreading** activity whose function is to correct the occasional base-pairing error that might occur during strand synthesis (see Section 16.1.1).

- A 5′→3′ exonuclease activity is less common but is possessed by some polymerases whose function in replication requires that they must be able to remove at least part of a polynucleotide that is already attached to the template strand that the polymerase is copying. This activity is utilized during the process that joins together the discontinuous DNA fragments synthesized on the lagging strand during bacterial DNA replication (see Figure 15.18).

The search for DNA polymerases began in the mid-1950s, as soon as it was realized that DNA synthesis was the key to replication of genes. It was thought that bacteria would probably have just a single DNA polymerase, and when the enzyme now called **DNA polymerase I** was isolated by Arthur Kornberg in 1957 there was a widespread assumption that this was the main replicating enzyme. The discovery that inactivation of the *E. coli polA* gene, which codes for DNA polymerase I, was not lethal (cells were still able to replicate their genomes) therefore came as something of a surprise, especially when a similar result was obtained with inactivation of *polB*, coding for a second enzyme, **DNA polymerase II**, which we now know is mainly

Table 15.2 DNA polymerases involved in replication of bacterial and eukaryotic genomes

Enzyme	Subunits	Exonuclease activities 3′→5′	5′→3′	Function
Bacterial DNA polymerases				
DNA polymerase I	1	Yes	Yes	DNA repair, replication
DNA polymerase III	At least 10	Yes	No	Main replicating enzyme
Eukaryotic DNA polymerases				
DNA polymerase α	4	No	No	Priming during replication
DNA polymerase γ	2	Yes	No	Mitochondrial DNA replication
DNA polymerase δ	2 or 3	Yes	No	Main replicative enzyme
DNA polymerase κ	1	?	?	Required for attachment of cohesin proteins which hold sister chromatids together until the anaphase stage of nuclear division (Section 15.2.3)

Bacteria and eukaryotes possess other DNA polymerases involved primarily in repair of damaged DNA. These enzymes include DNA polymerases II, IV, and V of *Escherichia coli* and the eukaryotic DNA polymerases β, ε, ζ, η, θ, and ι. DNA repair processes are described in Section 16.2.

involved in repair of damaged DNA rather than genome replication (Section 16.2). It was not until 1972 that the main replicating polymerase of *E. coli*, **DNA polymerase III**, was eventually isolated. Both DNA polymerases I and III are involved in genome replication, as we will see in the next section.

DNA polymerases I and II are single polypeptides but DNA polymerase III, befitting its role as the main replicating enzyme, is a multisubunit protein, with a molecular mass of approximately 900 kDa (Table 15.2). The three main subunits, which form the core enzyme, are called α, ε, and θ, with the polymerase activity specified by the α subunit and the 3′→5′ exonuclease activity by ε. The function of θ is not clear: it may have a purely structural role in bringing together the other two core subunits and in assembling the various accessory subunits. The latter include τ and γ, both coded by the same gene, with γ synthesized by translational frameshifting (Section 13.2.3), β, which acts as a "sliding clamp" and holds the polymerase complex tightly to the template, and the δ, δ′, χ, and ψ subunits.

Eukaryotes have at least nine DNA polymerases, which in mammals are distinguished by Greek suffixes (α, β, γ, δ, etc.), an unfortunate choice of nomenclature as it tempts confusion with the identically named subunits of *E. coli* DNA polymerase III. The main replicating enzyme is **DNA polymerase δ** (Table 15.2), which has two subunits (three according to some researchers) and works in conjunction with an accessory protein called the **proliferating cell nuclear antigen** (**PCNA**). PCNA is the functional equivalent of the β subunit of *E. coli* DNA polymerase III and holds the enzyme tightly to the template. **DNA polymerase α** also has an important function in DNA synthesis, being the enzyme that primes eukaryotic replication (see Figure 15.13B). **DNA polymerase γ**, although coded by a nuclear gene, is responsible for replicating the mitochondrial genome.

Discontinuous strand synthesis and the priming problem

The limitation that DNA polymerases can synthesize polynucleotides only in the 5'→3' direction means that the lagging strand of the parent molecule must be copied in a discontinuous fashion, as shown in Figure 15.12. The implication of this model—that the initial products of lagging-strand replication are short segments of polynucleotide—was confirmed in 1969 when **Okazaki fragments**, as these segments are now called, were first isolated from *E. coli*. In bacteria, Okazaki fragments are 1000–2000 nucleotides in length, but in eukaryotes the equivalent fragments appear to be much shorter, perhaps less than 200 nucleotides in length. This is an interesting observation that might indicate that each round of discontinuous synthesis replicates the DNA associated with a single nucleosome (between 140 and 150 bp wound around the core particle plus 50–70 bp of linker DNA; Section 7.1.1).

The second difficulty illustrated in Figure 15.12 is the need for a primer to initiate synthesis of each new polynucleotide. It is not known for certain why DNA polymerases cannot begin synthesis on an entirely single-stranded template, but it may relate to the proofreading activity of these enzymes, which is essential for the accuracy of replication. As described in Section 16.1.1, a nucleotide that has been inserted incorrectly at the extreme 3' end of a growing DNA strand, and hence is not base-paired to the template polynucleotide, can be removed by the 3'→5' exonuclease activity of a DNA polymerase. This means that the 3'→5' exonuclease activity must be more effective than the 5'→3' polymerase activity when the 3' nucleotide is not base-paired to the template. The implication is that the polymerase can extend a polynucleotide efficiently only if the 3' nucleotide is base-paired, which in turn could be the reason why an entirely single-stranded template, which by definition lacks a base-paired 3' nucleotide, cannot be used by a DNA polymerase.

Whatever the reason, priming is a necessity in DNA replication but does not present too much of a problem. Although DNA polymerases cannot deal with an entirely single-stranded template, RNA polymerases have no difficulty in this respect, so the primers for DNA replication are made of RNA. In bacteria, primers are synthesized by **primase**, a special RNA polymerase unrelated to the transcribing enzyme, with each primer being 4–15 nucleotides in length and most starting with the sequence 5'–AG–3'. Once the primer has been completed, strand synthesis is continued by DNA polymerase III (Figure 15.13A). In eukaryotes the situation is slightly more complex because the primase is tightly bound to DNA polymerase α, and cooperates with this enzyme in synthesis of the first few nucleotides of a new polynucleotide. This primase synthesizes an RNA primer of 8–12 nucleotides, and then hands over to DNA polymerase α, which extends the RNA primer by adding about 20 nucleotides of DNA. This DNA stretch often has a few ribonucleotides mixed in, but it is not clear if these are incorporated by DNA polymerase α or by intermittent activity of the primase. After completion of the RNA–DNA primer, DNA synthesis is continued by the main replicative enzyme, DNA polymerase δ (Figure 15.13B).

Priming needs to occur just once on the leading strand, within the replication origin, because once primed, the leading-strand copy is synthesized continuously until replication is completed. On the lagging strand, priming is a repeated process that must occur every time a new Okazaki fragment is initiated. In *E. coli*, which makes Okazaki fragments of 1000–2000 nucleotides in

Figure 15.13 **Priming of DNA synthesis in (A) bacteria and (B) eukaryotes.** In eukaryotes the primase forms a complex with DNA polymerase α, which is shown synthesizing the RNA primer followed by the first few nucleotides of DNA.

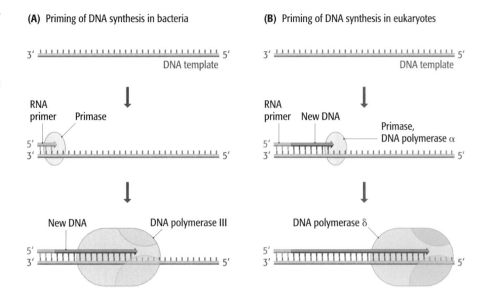

(A) Priming of DNA synthesis in bacteria

(B) Priming of DNA synthesis in eukaryotes

length, approximately 4000 priming events are needed every time the genome is replicated. In eukaryotes, the Okazaki fragments are much shorter and priming is a highly repetitive event.

Events at the bacterial replication fork

Now that we have considered the complications introduced by discontinuous strand synthesis and the priming problem, we can move on to study the combination of events occurring at the replication fork during the elongation phase of genome replication.

In Section 15.2.1 we identified attachment of the DnaB helicase, followed by extension of the melted region of the replication origin, as representing the end of the initiation phase of replication in *E. coli*. To a large extent, the division between initiation and elongation is artificial, the two processes running seamlessly one into the other. After the helicase has bound to the origin to form the prepriming complex, the primase is recruited, resulting in the **primosome**, which initiates replication of the leading strand. It does this by synthesizing the RNA primer that DNA polymerase III needs in order to begin copying the template.

DnaB is the main helicase involved in genome replication in *E. coli*, but it is by no means the only helicase that this bacterium possesses: in fact, there

Figure 15.14 **The role of the DnaB helicase during DNA replication in** *Escherichia coli*. DnaB is a 5'→3' helicase and so migrates along the lagging strand, breaking base pairs as it goes. It works in conjunction with a DNA topoisomerase to unwind the helix. To avoid confusion, the primase enzyme normally associated with the DnaB helicase is not shown in this drawing.

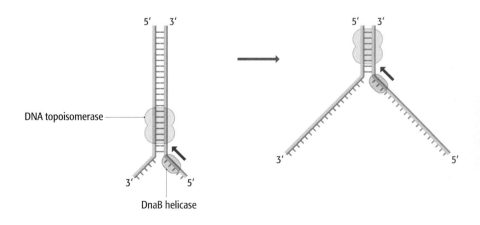

were eleven at the last count. The size of the collection reflects the fact that DNA unwinding is required not only during replication but also during diverse processes such as transcription, recombination, and DNA repair. The mode of action of a typical helicase has not been precisely defined, but it is thought that these enzymes bind to single-stranded rather than double-stranded DNA, and migrate along the polynucleotide in either the 5′→3′ or 3′→5′ direction, depending on the specificity of the helicase. Breakage of base pairs by the helicase requires energy, which is generated by hydrolysis of ATP. According to this model, a single DnaB helicase could migrate along the lagging strand (DnaB is a 5′→3′ helicase), unzipping the helix and generating the replication fork, the torsional stress generated by the unwinding activity being relieved by DNA topoisomerase action (Figure 15.14). This model is probably a good approximation of what actually happens, although it does not provide a function for the two other *E. coli* helicases thought to be involved in genome replication. Both of these, PriA and Rep, are 3′→5′ helicases and so could conceivably complement DnaB activity by migrating along the leading strand, but they may have lesser roles. The involvement of Rep in DNA replication might, in fact, be limited to participation in the rolling circle replication process used by λ and a few other *E. coli* bacteriophages (Section 15.1.3).

Single-stranded DNA is naturally "sticky" and the two separated polynucleotides produced by helicase action would immediately re-form base pairs after the enzyme has passed, if allowed to. The single strands are also highly susceptible to nuclease attack and are likely to be degraded if not protected in some way. To avoid these unwanted outcomes, **single-strand binding proteins (SSBs)** attach to the polynucleotides and prevent them from reassociating or being degraded (Figure 15.15A). The *E. coli* SSB is made up of four identical subunits and probably works in a similar way to the major eukaryotic SSB, called **replication protein A (RPA)**, by enclosing the polynucleotide in a channel formed by a series of SSBs attached side by side on the strand (Figure 15.15B). Detachment of the SSBs, which must occur when the replication complex arrives to copy the single strands, is brought about by a second set of proteins called **replication mediator proteins (RMPs)**. As with helicases, SSBs have diverse roles in different processes involving DNA unwinding.

After 1000–2000 nucleotides of the leading strand have been replicated, the first round of discontinuous strand synthesis on the lagging strand can begin. The primase, which is still associated with the DnaB helicase in the primosome, makes an RNA primer which is then extended by DNA polymerase III (Figure 15.16). This is the same DNA polymerase III complex that is synthesizing the leading-strand copy, the complex comprising, in effect, two copies of the polymerase, held together by a pair of τ subunits. It is not two complete copies of the polymerase because there is only a single γ **complex**, containing subunit γ in association with δ, δ′, χ, and ψ. The main function of the γ complex (sometimes called the "clamp loader") is to interact with the β subunit (the "sliding clamp") in each half of the complex, and hence control the attachment and removal of the enzyme from the template, a function that is required primarily during lagging-strand replication when the enzyme has to attach and detach repeatedly at the start and end of each Okazaki fragment. Some models of the DNA polymerase III complex place the two enzymes in opposite orientations to reflect the different directions in which DNA synthesis occurs, toward the replication fork on the leading strand and away from it

(A) SSBs attach to the unpaired polynucleotides

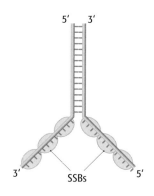

(B) Structure of RPA, a eukaryotic SSB

Figure 15.15 The role of single-strand binding proteins (SSBs) during DNA replication. (A) SSBs attach to the unpaired polynucleotides produced by helicase action and prevent the strands from base-pairing with one another or being degraded by single-strand-specific nucleases. (B) Structure of the eukaryotic SSB called RPA. The protein contains a β-sheet structure that forms a channel in which the DNA (shown in dark orange, viewed from the end) is bound. Reprinted with permission from Bochkarev *et al.* (1997), *Nature*, **385**, 176–181.

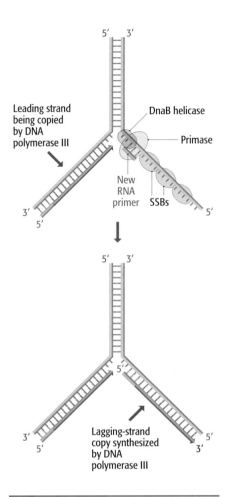

Figure 15.16 Priming and synthesis of the lagging-strand copy during DNA replication in *Escherichia coli*.

on the lagging strand. It is more likely, however, that the pair of enzymes face the same direction and the lagging strand forms a loop, so that DNA synthesis can proceed in parallel as the polymerase complex moves forward in pace with the progress of the replication fork (Figure 15.17).

The combination of the DNA polymerase III dimer and the primosome, migrating along the parent DNA and carrying out most of the replicative functions, is called the **replisome**. After its passage, the replication process must be completed by joining up the individual Okazaki fragments. This is not a trivial event because one member of each pair of adjacent Okazaki fragments still has its RNA primer attached at the point where ligation should take place (Figure 15.18). Table 15.2 shows us that this primer cannot be removed by DNA polymerase III, because this enzyme lacks the required 5′→3′ exonuclease activity. At this point, DNA polymerase III releases the lagging strand and its place is taken by DNA polymerase I, which does have a 5′→3′ exonuclease activity and so removes the primer, and usually the start of the DNA component of the Okazaki fragment as well, extending the 3′ end of the adjacent fragment into the region of the template that is exposed. The two Okazaki fragments now abut, with the terminal regions of both composed entirely of DNA. All that remains is for the missing phosphodiester bond to be put in place by a **DNA ligase**, linking the two fragments and completing replication of this region of the lagging strand.

The eukaryotic replication fork: variations on the bacterial theme

The elongation phase of genome replication is similar in bacteria and eukaryotes, although the details differ. The progress of the replication fork in eukaryotes is maintained by helicase activity, although which of the several eukaryotic helicases that have been identified are primarily responsible for DNA unwinding during replication has not been established. The separated polynucleotides are prevented from reattaching by single-strand binding proteins, the main one of these in eukaryotes being RPA.

We begin to encounter unique features of the eukaryotic replication process when we examine the method used to prime DNA synthesis. As described above, the eukaryotic DNA polymerase α cooperates with the primase enzyme to put in place the RNA–DNA primers at the start of the leading-strand copy and at the beginning of each Okazaki fragment. However, DNA polymerase α is not capable of lengthy DNA synthesis, presumably because it lacks the stabilizing effect of a sliding clamp equivalent to the β subunit of *E. coli* DNA polymerase III or the PCNA accessory protein that aids the eukaryotic DNA polymerase δ. This means that although DNA polymerase α can extend the initial RNA primer with about 20 nucleotides of DNA, it must then be replaced by the main replicative enzyme, DNA polymerase δ (see Figure 15.13B).

The DNA polymerase δ enzymes that copy the leading and lagging strands in eukaryotes do not associate into a dimeric complex equivalent to the one formed by DNA polymerase III during replication in *E. coli*. Instead, the two copies of the polymerase remain separate. The function performed by the γ complex of the *E. coli* polymerase—controlling attachment and detachment of the enzyme from the lagging strand—is carried out by a multisubunit accessory protein called **replication factor C** (**RFC**).

As in *E. coli*, completion of lagging-strand synthesis requires removal of the RNA primer from each Okazaki fragment. There appears to be no eukaryotic

DNA polymerase with the 5′→3′ exonuclease activity needed for this purpose and the process is therefore very different to that described for bacterial cells. The central player is the "flap endonuclease," **FEN1** (previously called MF1), which associates with the DNA polymerase δ complex at the 3′ end of an Okazaki fragment, in order to degrade the primer from the 5′ end of the adjacent fragment. Understanding exactly how this occurs is complicated by the inability of FEN1 to initiate primer degradation because it is unable to remove the ribonucleotide at the extreme 5′ end of the primer, because this ribonucleotide carries a 5′–triphosphate group which blocks FEN1 activity (Figure 15.19). One possibility is that most of the RNA component of the primer is removed by RNase H, which can degrade the RNA part of a base-paired RNA–DNA hybrid, but cannot cleave the phosphodiester bond between the last ribonucleotide and the first deoxyribonucleotide. However, this ribonucleotide will carry a 5′–monophosphate rather than 5′–triphosphate and so can be removed by FEN1 (Figure 15.20A). This model has the disadvantage that it assigns an essential role to RNase H, whereas experimental evidence suggests that cells lacking RNase H are still able to carry out lagging-strand replication. An alternative possibility is that a helicase breaks the base pairs holding the primer to the template strand, enabling the primer to be pushed aside by DNA polymerase δ as it extends the adjacent Okazaki fragment into the region thus exposed (Figure 15.20B). The flap that results can be cut off by FEN1, whose endonuclease activity can cleave the phosphodiester bond at the branch point where the displaced region attaches to the part of the fragment that is still base-paired. This scheme raises the possibility that both the RNA primer and all of the DNA originally synthesized by DNA polymerase α are removed. This is attractive because DNA polymerase α has no 3′→5′ proofreading activity (see Table 15.2) and therefore synthesizes DNA in a relatively error-prone manner. Removal of this region as part of the flap cleaved by FEN1, followed by resynthesis by DNA polymerase δ (which does have a proofreading activity and so makes a highly accurate copy of the template), would prevent these errors from becoming permanent features of the daughter double helix.

The final difference between replication in bacteria and in eukaryotes is that in eukaryotes there is no equivalent of the bacterial replisome. Instead, the enzymes and proteins involved in replication form sizeable structures within the nucleus, each structure containing hundreds or thousands of individual replication complexes. These structures are immobile because of attachments

Figure 15.17

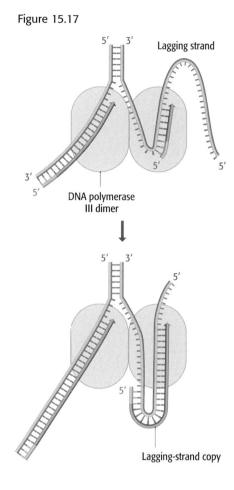

Figure 15.18

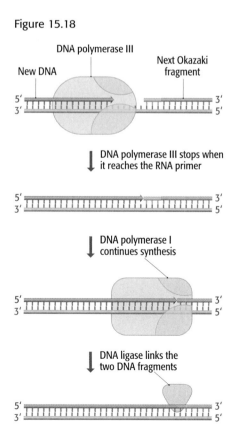

Figure 15.17 A model for parallel synthesis of the leading- and lagging-strand copies by a dimer of DNA polymerase III enzymes. It is thought that the lagging strand loops through its copy of the DNA polymerase III enzyme, in the manner shown, so that both the leading and lagging strands can be copied as the dimer moves along the molecule being replicated. The two components of the DNA polymerase III dimer are not identical because there is only one copy of the γ complex.

Figure 15.18 The series of events involved in joining up adjacent Okazaki fragments during DNA replication in *Escherichia coli*. DNA polymerase III lacks a 5′→3′ exonuclease activity and so stops making DNA when it reaches the RNA primer of the next Okazaki fragment. At this point, DNA synthesis is continued by DNA polymerase I, which does have a 5′→3′ exonuclease activity, and which works in conjunction with RNase H to remove the RNA primer and replace it with DNA. DNA polymerase I usually also replaces some of the DNA from the Okazaki fragment before detaching from the template. This leaves a single missing phosphodiester bond, which is synthesized by DNA ligase, completing this step in the replication process.

with the nuclear matrix, so DNA molecules are threaded through the complexes as they are replicated. The structures are referred to as **replication factories** and may in fact also be features of the replication process in at least some bacteria.

Genome replication in the archaea

We have little direct information about DNA replication in archaea, most of what we know having been deduced by searching archaeal genomes for genes and other sequences similar to the components of the replication apparatus

Figure 15.19 The "flap endonuclease" FEN1 cannot initiate primer degradation because its activity is blocked by the triphosphate group present at the 5′ end of the RNA primer.

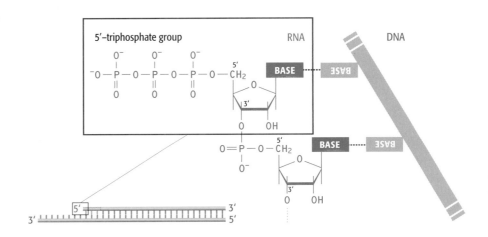

Figure 15.20 Two models for completion of lagging-strand replication in eukaryotes. The new DNA (red strand) is synthesized by DNA polymerase δ but this enzyme is not shown in order to increase the clarity of the diagrams.

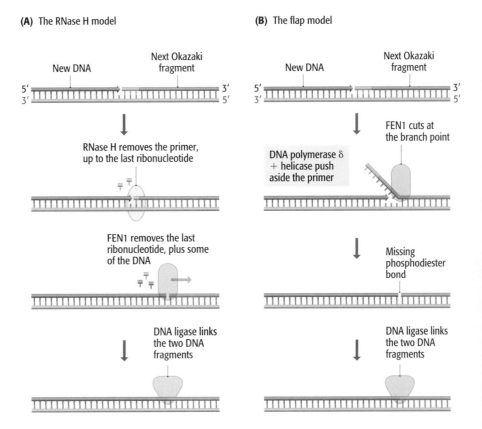

in bacteria and/or eukaryotes. Initial attempts to locate origins of replication in archaeal genomes by searching for sequence motifs found at bacterial or eukaryotic origins were unsuccessful. Subsequently, potential replication origins for a range of archaeal species were identified by statistical analysis of the frequencies of the four nucleotides in different parts of each archaeal genome, the rationale being that these frequencies might be significantly different to either side of an origin, as is the case in bacteria. For one species, *Pyrococcus abyssi*, a potential origin identified by nucleotide frequency analysis lies in the region of the genome that is replicated first, and so may be the true origin of replication.

The sequences of most of the proteins involved in the elongation phase of archaeal genome replication, as predicted from their genes, are similar to the equivalent eukaryotic versions. In particular, archaea have proteins that are homologs of the eukaryotic RFC and PCNA. The archaeal DNA polymerase is interesting because the subunit that specifies the DNA synthesis activity is similar to the equivalent subunit of the eukaryotic DNA polymerase δ, whereas the proofreading function is conferred by a protein that appears to be a homolog of subunit ε of *Escherichia coli* DNA polymerase III.

15.2.3 Termination of replication

Replication forks proceed along linear genomes, or around circular ones, generally unimpeded except when a region that is being transcribed is encountered. DNA synthesis occurs at approximately five times the rate of RNA synthesis, so the replication complex can easily overtake an RNA polymerase, but this probably does not happen: instead, it is thought that the replication fork pauses behind the RNA polymerase, proceeding only when the transcript has been completed.

Eventually the replication fork reaches the end of the molecule or meets a second replication fork moving in the opposite direction. What happens next is one of the least understood aspects of genome replication.

Replication of the E. coli *genome terminates within a defined region*
Bacterial genomes are replicated bidirectionally from a single point (see Figure 15.8), which means that the two replication forks should meet at a position diametrically opposite the origin of replication on the genome map. However, if one fork is delayed, possibly because it has to replicate extensive regions where transcription is occurring, then it might be possible for the other fork to overshoot the halfway point and continue replication on the "other side" of the genome (Figure 15.21). It is not immediately apparent why this should be undesirable, the daughter molecules presumably being unaffected, but it is not allowed to happen because of the presence of **terminator sequences**. Six of these have been identified in the *E. coli* genome (Figure 15.22A), each one acting as the recognition site for a sequence-specific DNA-binding protein called **Tus**.

The mode of action of Tus is quite unusual. When bound to a terminator sequence, a Tus protein allows a replication fork to pass if the fork is moving in one direction, but blocks progress if the fork is moving in the opposite direction around the genome. The directionality is set by the orientation of the Tus protein on the double helix. When approached from one direction, Tus blocks the passage of the DnaB helicase, which is responsible for progression

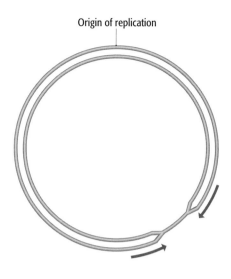

Origin of replication

Figure 15.21 A situation that is not allowed to occur during replication of the circular *Escherichia coli* genome. One of the replication forks has proceeded some distance past the halfway point. This does not happen during DNA replication in *E. coli* because of the action of the Tus proteins (see Figure 15.22B).

Figure 15.22 The role of terminator sequences during DNA replication in *Escherichia coli*. (A) The positions of the six terminator sequences on the *E. coli* genome are shown, with the arrowheads indicating the direction that each terminator sequence can be passed by a replication fork. (B) Bound Tus proteins allow a replication fork to pass when the fork approaches from one direction but not when it approaches from the other direction. The diagram shows a replication fork passing by the left-hand Tus, because the DnaB helicase that is moving the fork forward can disrupt the Tus when it approaches it from this direction. The fork is then blocked by the second Tus, because this one has its impenetrable wall of β-strands facing toward the fork.

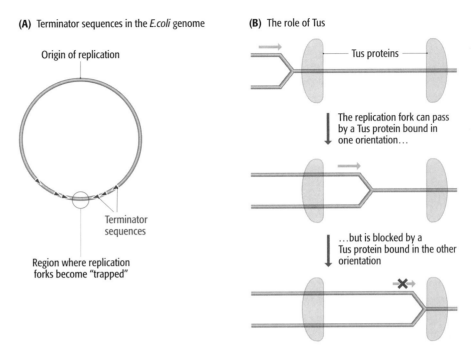

(A) Terminator sequences in the *E.coli* genome

Origin of replication

Terminator sequences

Region where replication forks become "trapped"

(B) The role of Tus

Tus proteins

The replication fork can pass by a Tus protein bound in one orientation…

…but is blocked by a Tus protein bound in the other orientation

of the replication fork, because the helicase is faced with a "wall" of β-strands which it is unable to penetrate. But when approaching from the other direction, DnaB is able to disrupt the structure of the Tus protein, probably because of the effect that unwinding of the double helix has on Tus, and so is able to pass by (Figure 15.22B).

The orientation of the termination sequences, and hence of the bound Tus proteins, in the *E. coli* genome is such that both replication forks become trapped within a relatively short region on the opposite side of the genome to the origin (see Figure 15.22A). This ensures that termination always occurs at or near the same position. Exactly what happens when the two replication forks meet is unknown, but the event is followed by disassembly of the replisomes, either spontaneously or in a controlled fashion. The result is two interlinked daughter molecules, which are separated by topoisomerase IV.

Little is known about termination of replication in eukaryotes

No sequences equivalent to bacterial terminators are known in eukaryotes, and proteins similar to Tus have not been identified. Quite possibly, replication forks meet at random positions and termination simply involves ligation of the ends of the new polynucleotides. We do know that the replication complexes do not break down, because these replication factories are permanent features of the nucleus.

Rather than concentrating on the molecular events occurring when replication forks meet, attention has been focused on the difficult question of how the daughter DNA molecules produced in a eukaryotic nucleus do not become impossibly tangled up. Although DNA topoisomerases have the ability to untangle DNA molecules, it is generally assumed that tangling is kept to a minimum so that extensive breakage-and-reunion reactions, as catalyzed by topoisomerases (see Figure 15.4), can be avoided. Various models have been proposed to solve this problem. One of these suggests that a eukaryotic chromosome is not randomly packed into its nuclear territory (Section

10.1.1), but is ordered around the replication factories, which appear to be present in only limited numbers. It is envisaged that each factory replicates a single region of the DNA, maintaining the daughter molecules in a specific arrangement that avoids their entanglement. Initially, the two daughter molecules are held together by **cohesin** proteins, which are attached immediately after passage of the replication fork by a process that appears to involve DNA polymerase κ, an enigmatic enzyme that is essential for replication but whose only known role does not obviously require a DNA polymerase activity. The cohesins maintain the alignment of the sister chromatids until the anaphase stage of nuclear division, when they are cleaved by cutting proteins, enabling the daughter chromosomes to separate (Figure 15.23).

15.2.4 Maintaining the ends of a linear DNA molecule

There is one final problem that we must consider before leaving the replication process. This concerns the steps that have to be taken to prevent the ends of a linear, double-stranded molecule from gradually getting shorter during successive rounds of chromosomal DNA replication. There are two ways in which this shortening might occur:

- The extreme 3′ end of the lagging strand might not be copied because the final Okazaki fragment cannot be primed, the natural position for the priming site being beyond the end of the template (Figure 15.24A). The absence of this Okazaki fragment means that the lagging-strand copy is shorter than it should be. If the copy remains this length then when it acts as a parent polynucleotide in the next round of replication the resulting daughter molecule will be shorter than its grandparent.

- If the primer for the last Okazaki fragment is placed at the extreme 3′ end of the lagging strand, then shortening will still occur, although to a lesser extent, because this terminal RNA primer cannot be converted into DNA by the standard processes for primer removal (Figure 15.24B). This is because the methods for primer removal (as illustrated in Figure 15.18 for bacteria and Figure 15.20 for eukaryotes) require extension of the 3′ end of an adjacent Okazaki fragment, which cannot occur at the very end of the molecule.

Once this problem had been recognized, attention was directed at the telomeres, the unusual DNA sequences at the ends of eukaryotic chromosomes. We noted in Section 7.1.2 that telomeric DNA is made up of a type of minisatellite sequence, being comprised of multiple copies of a short repeat motif, 5′–TTAGGG–3′ in most higher eukaryotes, a few hundred copies of this sequence occurring in tandem repeats at each end of every chromosome. The solution to the end-shortening problem lies with the way in which this telomeric DNA is synthesized.

Telomeric DNA is synthesized by the telomerase enzyme

Most of the telomeric DNA is copied in the normal fashion during DNA replication but this is not the only way in which it can be synthesized. To compensate for the limitations of the replication process, telomeres can be extended by an independent mechanism catalyzed by the enzyme **telomerase**. This is an unusual enzyme in that it consists of both protein and RNA. In the human enzyme, the RNA component is 450 nucleotides in length and contains near its 5′ end the sequence 5′–CUAACCCUAAC–3′, whose central region is the reverse complement of the human telomere repeat sequence 5′–TTAGGG–3′. This enables telomerase to extend the telomeric DNA at the 3′ end of a

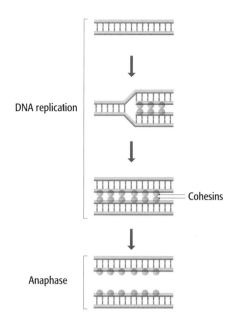

DNA replication

Anaphase

Cohesins

Figure 15.23 Cohesins. Cohesin proteins attach immediately after passage of the replication fork and hold the daughter molecules together until anaphase. During anaphase, the cohesins are cleaved, enabling the replicated chromosomes to separate prior to their distribution into daughter nuclei (see Figure 3.15).

polynucleotide by the copying mechanism shown in Figure 15.25, in which the telomerase RNA is used as a template for each extension step, the DNA synthesis being carried out by the protein component of the enzyme, which is a reverse transcriptase. The correctness of this model is indicated by comparisons between telomere repeat sequences and the telomerase RNAs of other species (Table 15.3): in all organisms that have been looked at, the telomerase RNA contains a sequence that enables it to make copies of the repeat motif present at the organism's telomeres. An interesting feature is

Figure 15.24 Two of the reasons why linear DNA molecules could become shorter after DNA replication. In both examples, the parent molecule is replicated in the normal way. A complete copy is made of its leading strand, but in (A) the lagging-strand copy is incomplete because the last Okazaki fragment is not made. This is because primers for Okazaki fragments are synthesized at positions approximately 200 bp apart on the lagging strand. If one Okazaki fragment begins at a position less than 200 bp from the 3′ end of the lagging strand then there will not be room for another priming site, and the remaining segment of the lagging strand is not copied. The resulting daughter molecule therefore has a 3′ overhang and, when replicated, gives rise to a granddaughter molecule that is shorter than the original parent. In (B) the final Okazaki fragment can be positioned at the extreme 3′ end of the lagging strand, but its RNA primer cannot be converted into DNA because this would require extension of another Okazaki fragment positioned beyond the end of the lagging strand. It is not clear if a terminal RNA primer can be retained throughout the cell cycle, nor is it clear if a retained RNA primer can be copied into DNA during a subsequent round of DNA replication. If the primer is not retained or is not copied into DNA, then one of the granddaughter molecules will be shorter than the original parent.

(A) The final Okazaki fragment cannot be primed

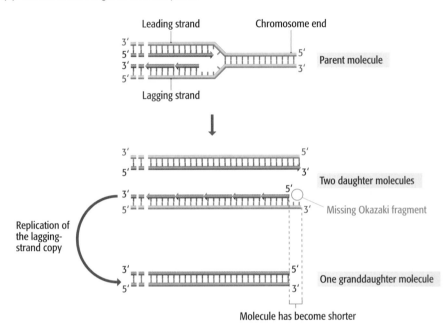

(B) The primer for the last Okazaki fragment is at the extreme 3′ end of the lagging strand

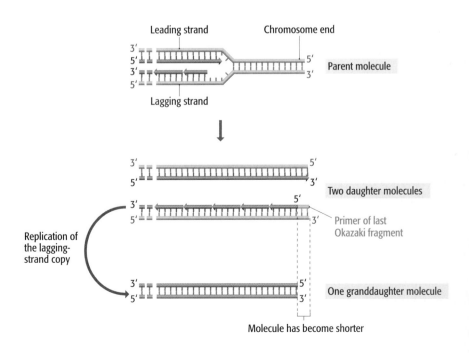

that, in all organisms, the strand synthesized by telomerase has a preponderance of G nucleotides, and is therefore referred to as the G-rich strand.

Telomerase can only synthesize this G-rich strand. It is not clear how the other polynucleotide—the C-rich strand—is extended, but it is presumed that when the G-rich strand is long enough, the primase–DNA polymerase α complex attaches at its end and initiates synthesis of complementary DNA in the normal way (Figure 15.26). This requires the use of a new RNA primer, so the C-rich strand will still be shorter than the G-rich one, but the important point is that the overall length of the chromosomal DNA has not been reduced.

The activity of telomerase must clearly be controlled very carefully to ensure that the appropriate length extension is made at each chromosome end. One part of this regulatory mechanism is provided by the TRF1 proteins that bind to the telomere repeat sequences (Section 7.1.2). TRF1 induces the formation of a folded chromatin structure that prevents the telomerase enzyme from accessing the end of the chromosome. As the telomere shortens, the number of bound TRF1 proteins decreases and the chromatin structure opens up, enabling telomerase to attach to the end of the chromosome and extend the telomere. As the telomere extends, TRF1 proteins reattach, inducing the chromatin to return to its folded structure so that the telomerase is once again excluded from the chromosome end. In effect, the TRF1 proteins mediate a negative feedback loop that regulates telomerase activity on a particular chromosome end. In mammalian cells, the closed chromatin structure may involve formation of a "t-loop," in which the free 3′ end of the telomere loops back, invades the double helix, and forms base pairs with its complementary sequence on the C-rich strand (Figure 15.27). This reaction is catalyzed by the second telomere-binding protein in humans, TRF2, and may provide additional stabilization of a chromosome end that does not require further extension.

Telomere length is implicated in cell senescence and cancer

Perhaps surprisingly, telomerase is not active in all mammalian cells. The enzyme is functional in the early embryo, but after birth is active only in the reproductive cells and **stem cells**. The latter are progenitor cells that divide continually throughout the lifetime of an organism, producing new cells to maintain organs and tissues in a functioning state. The best-studied examples are the hemopoietic stem cells of the bone marrow, which generate new blood cells.

Cells that lack telomerase activity undergo chromosome shortening every time they divide. Eventually, after many cell divisions, the chromosome ends

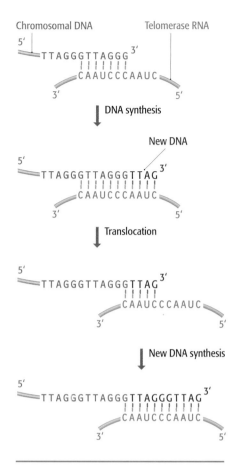

Figure 15.25 Extension of the end of a human chromosome by telomerase. The 3′ end of a human chromosomal DNA molecule is shown. The sequence comprises repeats of the human telomere motif 5′–TTAGGG–3′. The telomerase RNA base-pairs to the end of the DNA molecule which is extended a short distance, the length of this extension possibly determined by the presence of a stem-loop structure in the telomerase RNA. The telomerase RNA then translocates to a new base-pairing position slightly further along the DNA polynucleotide and the molecule is extended by a few more nucleotides. The process can be repeated until the chromosome end has been extended by a sufficient amount

Table 15.3 Sequences of telomere repeats and telomerase RNAs in various organisms

Species	Telomere repeat sequence	Telomerase RNA template sequence
Human	5′–TTAGGG–3′	5′–CUAACCCUAAC–3′
Oxytricha	5′–TTTTGGGG–3′	5′–CAAAACCCCAAAACC–3′
Tetrahymena	5′–TTGGGG–3′	5′–CAACCCCAA–3′

Oxytricha and *Tetrahymena* are protozoans which are particularly useful for telomere studies because at certain developmental stages their chromosomes break into small fragments, all of which have telomeres: they therefore have many telomeres per cell.

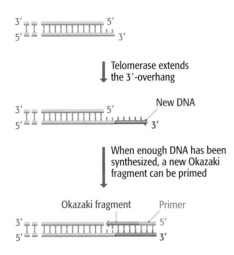

Figure 15.26 Completion of the extension process at the end of a chromosome. It is believed that after telomerase has extended the 3′ end by a sufficient amount, as shown in Figure 15.25, a new Okazaki fragment is primed and synthesized, converting the 3′ extension into a completely double-stranded end.

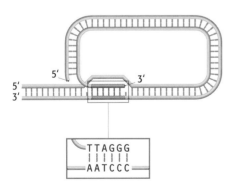

Figure 15.27 The "t-loop." The t-loop is formed when the free 3′ end of the telomere loops back and invades the double helix.

could become so truncated that essential genes are lost, but this is unlikely to be a major cause of the defects that can occur in cells lacking telomerase activity. Instead, the critical factor is the need to maintain a protein "cap" on each chromosome end, to protect these ends from the effects of the DNA repair enzymes that join together the uncapped ends that are produced by accidental breakage of a chromosome (Section 16.2.4). The proteins that form this protective cap, such as TRF2 in humans, recognize the telomere repeats as their binding sequences, and so have no attachment points after the telomeres have been deleted. If these proteins are absent then the repair enzymes can make inappropriate linkages between the ends of intact, although shortened, chromosomes; it is this that is probably the underlying cause of the disruption to the cell cycle that results from telomere shortening.

Telomere shortening will therefore lead to the termination of a cell lineage. For several years, biologists have attempted to link this process with **cell senescence**, a phenomenon originally observed in cell cultures. All normal cell cultures have a limited lifetime: after a certain number of divisions, the cells enter a senescent state in which they remain alive but cannot divide (Figure 15.28). With some mammalian cell lines, notably fibroblast cultures (connective tissue cells), senescence can be delayed by engineering the cells so that they synthesize active telomerase. These experiments suggest a clear relationship between telomere shortening and senescence, but the exactness of the link has been questioned, and any extrapolation from cell senescence to aging of the organism is fraught with difficulties.

Not all cell lines display senescence. Cancerous cells are able to divide continuously in culture, their immortality being looked upon as analogous to tumor growth in an intact organism. With several types of cancer, this absence of senescence is associated with activation of telomerase, sometimes to the extent that telomere length is maintained through multiple cell divisions, but often in such a way that the telomeres become longer than normal because the telomerase is overactive. It is not clear if telomerase activation is a *cause* or an *effect* of cancer, although the former seems more likely because at least one type of cancer, dyskeratosis congenita, appears to result from a mutation in the gene specifying the RNA component of human telomerase. The question is critical to understanding the etiology of a cancer but is less relevant to the therapeutic issue, which centers on whether telomerase could be a target for drugs designed to combat the cancer. Such a therapy could be successful even if telomerase activation is an effect of the cancer, because inactivation by drugs would induce senescence of the cancer cells and hence prevent their proliferation.

Telomeres in Drosophila

When the amino acid sequences of the protein subunits of telomerase enzymes are compared with those of other reverse transcriptases, the closest similarities are seen with the reverse transcriptases coded by the non-LTR retroelements called retroposons (Section 9.2.1). This is a fascinating observation when taken in conjunction with the unusual structure of the telomeres of *Drosophila*. These telomeres are not made up of the short repeated sequences seen in most other organisms, but instead consist of tandem arrays of much longer repeats, 6 or 10 kb in length. These repeats are full-length copies of two *Drosophila* retroposons, related to LINE-1 of humans, called *HeT-A* and *TART*. It is not known how these telomeres are maintained, but it is conceivable that the process is analogous to that carried out by telomerase, with a template RNA obtained by transcription of the telomeric retroposons being copied by the reverse transcriptase coded by the *TART* sequences (*HeT-A* does not have a reverse transcriptase gene).

The unusual structure of the *Drosophila* telomere could simply be a quirk of nature, but the attractive possibility that the telomeres of other organisms are degraded retroposons, as suggested by the similarities between telomerase and retroposon reverse transcriptases, cannot be discounted.

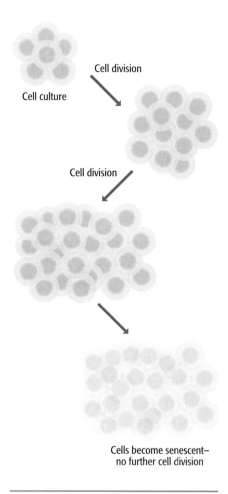

Figure 15.28 Cultured cells become senescent after multiple cell divisions.

15.3 Regulation of Eukaryotic Genome Replication

Genome replication in eukaryotic cells is regulated at two levels:

- Replication is coordinated with the cell cycle so that two copies of the genome are available when the cell divides.

- The replication process itself can be arrested under certain circumstances, for example if the DNA is damaged and must be repaired before copying can be completed.

We will end this chapter by looking at these regulatory mechanisms.

15.3.1 Coordination of genome replication and cell division

The concept of a **cell cycle** emerged from light microscopy studies carried out by the early cell biologists. Their observations showed that dividing cells pass through repeated cycles of mitosis—the period when nuclear and cell division occurs (see Figure 3.15)—and interphase, a less dramatic period when few dynamic changes can be detected with the light microscope. It was understood that chromosomes divide during interphase, so when DNA was identified as the genetic material, interphase took on a new importance as the period when genome replication must take place. This led to a reinterpretation of the cell cycle as a four-stage process (Figure 15.29), comprising:

- **Mitosis**, or **M phase**, the period when the nucleus and cell divide.

- Gap 1, or **G1 phase**, an interval when transcription, translation, and other general cellular activities occur.

- Synthesis, or **S phase**, when the genome is replicated.

- Gap 2, or **G2 phase**, a second interval period.

It is clearly important that the S phase and M phase are coordinated so that the genome is completely replicated, but replicated only once, before mitosis occurs. The periods immediately before entry into S phase and M phase are looked upon as key **cell cycle checkpoints**, and it is at one of these two points that the cycle becomes arrested if critical genes involved in cell cycle control are mutated, or if the cell undergoes trauma such as extensive DNA damage. Attempts to understand how genome replication is coordinated with mitosis have therefore concentrated on these two checkpoints, especially the pre-S checkpoint, the period immediately before replication.

Establishment of the prereplication complex enables genome replication to commence

Studies primarily with *Saccharomyces cerevisiae* have led to a model for controlling the timing of S phase, which postulates that genome replication requires construction of **prereplication complexes (pre-RCs)** at origins of

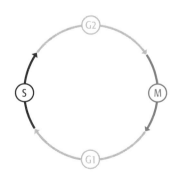

Figure 15.29 **The cell cycle.** The lengths of the individual phases vary in different cells. Abbreviations: G1 and G2, gap phases; M, mitosis; S, synthesis phase.

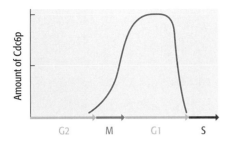

Figure 15.30 Graph showing the amount of Cdc6p in the nucleus at different stages of the cell cycle.

replication, these pre-RCs being converted to **post-RCs** as replication proceeds. A post-RC is unable to initiate replication and so cannot accidentally recopy a piece of the genome before mitosis has occurred. The ORC, the complex of six proteins that is assembled onto domains A and B1 of a yeast origin (see Figure 15.10B), was an early contender for the pre-RC but is probably not a central component because ORCs are present at origins of replication at all stages of the cell cycle. Instead, the ORC is looked on as the "landing pad" on which the pre-RC is constructed.

Various types of protein have been implicated as components of the pre-RC. The first is Cdc6p, which was originally identified in yeast and subsequently shown to have homologs in higher eukaryotes. Yeast Cdc6p is synthesized at the end of G2, as the cell enters mitosis, and becomes associated with chromatin in early G1 before disappearing at the end of G1, when replication begins (Figure 15.30). The involvement of Cdc6p in the pre-RC is suggested by experiments in which its gene is repressed, which results in an absence of pre-RCs, and other experiments in which Cdc6p is overproduced, which leads to multiple genome replications in the absence of mitosis. There is also biochemical evidence for a direct interaction between Cdc6p and yeast ORCs.

A second component of the pre-RC is thought to be the group of proteins called **replication licensing factors** (**RLFs**). As with Cdc6p, the first examples of these proteins were identified in yeast (the MCM family of proteins) with homologs in higher eukaryotes discovered at a later date. RLFs become bound to chromatin toward the end of M phase and remain in place until the start of S phase, after which they are gradually removed from the DNA as it is replicated. Their removal may be the key event that converts a pre-RC into a post-RC and so prevents reinitiation of replication at an origin that has already directed a round of replication.

Regulation of pre-RC assembly

Identification of the components of the pre-RC takes us some distance toward understanding how genome replication is initiated, but still leaves open the question of how replication is coordinated with other events in the cell cycle. Cell cycle control is a complex process, mediated largely by protein kinases which phosphorylate and activate enzymes and other proteins that have specific functions during the cell cycle. The same protein kinases are present in the nucleus throughout the cell cycle, so they must themselves be subject to control. This control is exerted partly by proteins called **cyclins** (which vary in abundance at different stages of the cell cycle), partly by other protein kinases that activate the cyclin-dependent protein kinases, and partly by inhibitory proteins. Even before we start looking for regulators of pre-RC assembly we can anticipate that the control system will be convoluted.

A number of cyclins have been linked with activation of genome replication and prevention of pre-RC reassembly after replication has been completed. These include the mitotic cyclins, whose main function was originally thought to be activation of mitosis but which also repress genome replication. When the effects of these cyclins are blocked by, for example, overproduction of proteins that inhibit their activity, the cell is not only incapable of entering M phase but also undergoes repeated genome replication. There are also more specific S-phase cyclins, such as Clb5p and Clb6p in *S. cerevisiae*, inactivation of which delays or prevents genome replication, and other

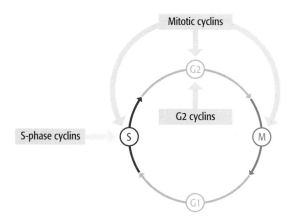

Figure 15.31 Cell-cycle control points for cyclins involved in regulation of genome replication.

cyclins that are active during G2 phase and prevent the assembly of pre-RCs in the period after genome replication and before cell division (Figure 15.31).

In addition to these cyclin-dependent control systems, genome replication is also regulated by a cyclin-independent protein kinase, Cdc7p-Dbf4p, found in organisms as diverse as yeasts and mammals. The proteins activated by this kinase have not been identified, separate lines of evidence suggesting that both RLFs and ORCs are targeted. Whatever the mechanism, Cdc7p-Dbf4p activity is a prerequisite for replication, the cyclin-dependent processes on their own being insufficient to push the cell into S phase.

15.3.2 Control within S phase

Regulation of the G1–S transition can be looked upon as the major control process affecting genome replication, but it is not the only one. The specific events occurring during S phase are also subject to regulation.

Early and late replication origins

Initiation of replication does not occur at the same time at all replication origins, nor is "origin firing" an entirely random process. Some parts of the genome are replicated early in S phase and some later, the pattern of replication being consistent from cell division to cell division. The general pattern is that actively transcribed genes and the centromeres are replicated early in S phase, and telomeres and nontranscribed regions of the genome are replicated later on. Early-firing origins are therefore tissue specific and reflect the pattern of genome expression occurring in a particular cell.

These features of genome replication were originally extrapolated from examination of just a few origins, but have recently been supplemented by studies utilizing microarray technology. Microarray analysis is based on hybridization probing, so to use a microarray to follow the pattern of genome replication it is necessary to devise a way of separating unreplicated DNA from replicated DNA, so that one or the other fraction can be used as a hybridization probe. If, for example, a sample of replicated DNA is obtained from cells that had just entered S phase, then this DNA could be used to probe the microarray in order to identify those genes that have already been replicated at this early stage. A second probe, prepared from replicated DNA from slightly later in S phase, would identify the next set of genes to be replicated, and so on. For *Saccharomyces cerevisiae*, these fractions are obtained by growing cells in medium with heavy nitrogen, transferring to normal

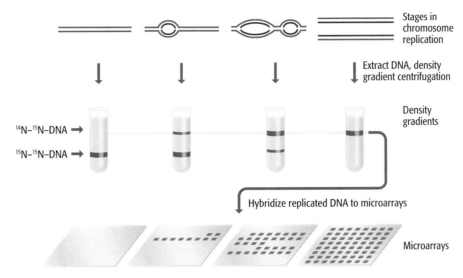

Stages in chromosome replication

Figure 15.32 Microarray analysis of origin firing in *S. cerevisiae*.

(A) Replication of chromosome VI

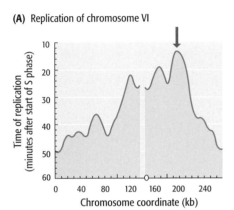

(B) Replication of chromosomes XII and XV

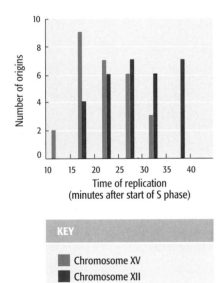

KEY

■ Chromosome XV

■ Chromosome XII

Figure 15.33 The dynamics of origin firing in *S. cerevisiae*. (A) Timing of origin firing along chromosome VI. The arrow indicates the point where replication of this chromosome begins, and the circle on the *x*-axis is the centromere. The blank area is a region of the chromosome that could not be examined. (B) Comparison between the replication dynamics for chromosomes XII and XV.

medium, and allowing the cells to enter S phase. The DNA is then extracted at appropriate times during S phase, treated with a restriction endonuclease, and fractionated by density gradient centrifugation (Figure 15.32). Two bands are seen, one made up of fragments of ^{15}N–^{15}N–DNA, derived from the unreplicated component of the genome, and the other made up of ^{14}N–^{15}N–DNA and hence derived from the regions that have undergone replication. The latter fraction is purified, labeled with a fluorescent marker, and applied to the microarray.

The analysis is simple but remarkably informative. Figure 15.33A charts the dynamics of origin firing along yeast chromosome VI and identifies a region midway along the shorter arm as the area where replication of this chromosome commences. As indicated by previous experiments, the centromere (indicated by the circle on the *x*-axis in Figure 15.33A) is replicated early in S phase, and the telomeres at the end of the replication cycle. Note that both telomeres are replicated at approximately the same time. This observation holds true for all yeast chromosomes, but not all chromosomes complete their replication cycle at the same time. Some, such as chromosomes XI and XV, are completely replicated early in S phase, and others, such as VIII and IX, are not fully replicated until much later. These time differences are not related to the length of the chromosome but instead indicate real variations in the kinetics of chromosome replication. Figure 15.33B emphasizes this point, showing that 15 minutes after the start of S phase, many of the origins in chromosome XV have been activated whereas replication of chromosome XII has only just begun. Microarray analysis also enables the migration rate of individual replication forks to be inferred, again showing variability. The mean speed is 2.9 kb per minute, but some forks move much more quickly, up to 11 kb per minute for the most active ones.

Understanding what determines the firing time of a replication origin is proving quite difficult. It is not simply the sequence of the origin, because transfer of a DNA segment from its normal position to another site in the same or a different chromosome can result in a change in the firing pattern of origins contained in that segment. This positional effect may be linked with chromatin organization and hence influenced by structures such as locus control regions (Section 10.1.2) that control DNA packaging. The position of the origin in the nucleus may also be important as origins that become active at similar periods within S phase appear to be clustered together, at least in mammals.

Checkpoints within S phase

The final aspect of the regulation of genome replication that we will consider is the function of the checkpoints that exist within S phase. These were first identified when it was shown that one of the responses of yeast cells to DNA damage is a slowing down and possibly a complete halting of the genome replication process. This is linked with the activation of genes whose products are involved in DNA repair (Section 16.2).

As with entry into S phase, cyclin-dependent kinases are implicated in the regulation of S-phase checkpoints. These kinases respond to signals from proteins associated with the replication fork. Replication-fork proteins, including the PCNA and the accessory protein RFC, were initially assigned roles in damage detection, but this role is probably not played by the versions of these proteins involved in DNA synthesis. The damage-sensing version of RFC has a different subunit composition to the standard RFC, and the sensing protein initially identified as a relative of PCNA appears to be a totally different protein (the 9-1-1 complex) that has a similar structure to PCNA and probably interacts with the double helix in a similar manner. The signals from the damage-sensing proteins are transduced by kinases such as ATM, ATR, Chk1, and Chk2 to effector proteins such as Cdc25 which interact with the cyclin-dependent kinases to elicit the appropriate cellular response (Figure 15.34). The replication process can be arrested by repressing the firing of origins of replication that are usually activated at later stages in S phase or by slowing the progression of existing replication forks. If the damage is not excessive then DNA repair processes are activated (Section 16.2); alternatively the cell may be shunted into the pathway of programmed cell death called **apoptosis**, the death of a single somatic cell as a result of DNA damage usually being less dangerous than allowing that cell to replicate its mutated DNA and possibly give rise to a tumor or other cancerous growth. In mammals, a central player in induction of cell cycle arrest and apoptosis is the protein called p53. This is classified as a tumor-suppressor protein, because when this protein is defective, cells with damaged genomes can avoid the S-phase checkpoints and possibly proliferate into a cancer. p53 is a sequence-specific DNA-binding protein that activates a number of genes thought to be directly responsible for arrest and apoptosis, and also represses expression of other genes that must be switched off to facilitate these processes.

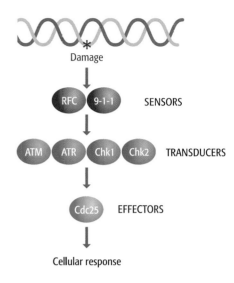

Figure 15.34 The cascade of events that initiate the appropriate cellular response to DNA damage.

Summary

In order to continue carrying out its function, the genome must replicate every time that a cell divides. Watson and Crick pointed out, when they first announced their discovery of the structure of DNA, that the specific base pairing that holds the two strands of the double helix together provides a means for accurate copying of each polynucleotide. They envisaged a semiconservative mode of replication in which each parent stand acts as a template for synthesis of a complementary daughter strand. The Meselson–Stahl experiment showed that this interpretation is correct, but there were still problems in understanding how the two strands of the helix were separated, especially in circular molecules which have little freedom to rotate. The discovery of DNA topoisomerases, which separate the strands of the double helix by repeated breakage and rejoining of one or both polynucleotides, solved this problem. No exceptions to the semiconservative mode of replication are known, though there are specialized versions, such as displacement replication and rolling circle replication. Initiation of genome replication occurs at discrete origins which have been well characterized in bacteria and in yeast, but which are less clearly understood in higher eukaryotes. Once replication has been initiated, a pair of replication forks travel in opposite directions along the DNA. DNA polymerase can only synthesize DNA in the 5′→3′ direction, which means that although one strand, called the leading strand, can be replicated in a continuous fashion, the second, lagging strand has to be copied in short segments. These are called Okazaki fragments. DNA synthesis must be primed by an RNA polymerase, and the helix must be unwound and the single strands stabilized by helicases and single-strand binding proteins. The replicating complex, called the replisome in bacteria, consists of the DNA polymerase enzyme along with ancillary proteins, such as the sliding clamp, which ensures that the connection between the polymerase and the DNA is secure but that the polymerase is still able to move along the DNA. Termination of replication occurs at specific regions within a bacterial chromosome but at less-well-defined areas in eukaryotic chromosomes. Eukaryotic chromosomes require special processes to maintain their ends, as replication results in a gradual shortening of the telomeres. These are elongated by the telomerase enzyme, which has an RNA subunit that acts as the template for synthesis of new telomere repeat units. Replication of the genome must be coordinated with the cell cycle. This is achieved by a combination of regulatory proteins, many of which are active only at specific periods of the cell cycle. Assembly of the prereplication complex at origins of replication is a critical step that is regulated to ensure that the genome is replicated just once per cell cycle. Once replication is underway, checkpoints during the synthesis phase respond to DNA damage in order to arrest or terminate genome replication.

Multiple Choice Questions

15.1.* The topological problem of DNA replication refers to which of the following?

  a. The blockage of DNA replication sites by nucleosomes.

  b. The difficulty of synthesizing DNA on the lagging strand.

  c. The unwinding of the double helix and the rotation of the DNA.

  d. The synchronization of DNA replication with cell division.

15.2. Which term describes the topological arrangement that prevents separation of the strands of the double helix without unwinding?

  a. Helinemic.

  b. Paranemic.

  c. Plectonemic.

  d. Toponemic.

15.3.* Which researcher(s) first proposed the "breakage-and-reunion" model for solving the topological problem of DNA replication?

  a. Delbrück.

  b. Kornberg.

  c. Meselson and Stahl.

  d. Watson and Crick.

15.4. Which of the following is a function of the *E. coli* DNA gyrase?

  a. To counteract the overwinding of the genome that occurs during DNA replication.

  b. To counteract the overwinding of the genome that occurs during transcription.

  c. To introduce supercoils into DNA molecules.

  d. All of the above.

15.5.* What types of DNA molecule are copied using the rolling circle replication process?

  a. Bacterial chromosomes.

  b. Some bacteriophage genomes (such as λ).

  c. Mitochondrial genomes.

  d. Yeast chromosomes.

15.6. The site at which DNA replication is initiated is called the:

  a. Enhancer.

  b. Initiator.

  c. Origin of replication.

  d. Promoter.

15.7.* What is the role of the primer in DNA synthesis?

  a. It provides a 5′–phosphate group for the addition of the next nucleotide.

  b. It provides 5′–phosphate groups that can be hydrolyzed to supply the energy required for DNA synthesis.

  c. It provides a 3′–hydroxyl group for the addition of the next nucleotide.

  d. It provides a source of nucleotides for the synthesis of the DNA strand.

15.8. Which of the following nuclease activities is utilized by DNA polymerases to provide the proofreading activity during DNA synthesis?

  a. 3′→5′ exonuclease.

  b. 5′→3′ exonuclease.

  c. Single-strand endonuclease.

  d. Double-strand endonuclease.

15.9.* What are Okazaki fragments?

  a. Short segments of polynucleotide synthesized on the leading strand of DNA.

  b. Short segments of polynucleotide synthesized on the lagging strand of DNA.

  c. The primers synthesized on the lagging strand that are required for DNA synthesis.

  d. The proteolytic fragments of DNA polymerase.

15.10. Which proteins prevent degradation or reassociation of single-stranded DNA at the replication fork?

  a. Helicases.

  b. Primases.

  c. Single-strand binding proteins.

  d. Topoisomerases.

15.11.* In bacteria, which of the following enzymes removes the RNA primers present at the start of each Okazaki fragment on the lagging strand?

  a. DNA polymerase I.

  b. DNA polymerase III.

  c. DNA ligase.

  d. RNase H.

15.12. What protein is involved in the separation of the two interlinked daughter chromosomes when DNA replication is terminated in *E. coli*?

  a. DnaB.

  b. DNA polymerase.

  c. Topoisomerase IV.

  d. Tus.

continued …

Multiple Choice Questions (continued) *Answers to odd-numbered questions can be found in the Appendix

15.13.* Which of the following statements about telomerase is TRUE?

    a. Telomerase is an RNA-dependent DNA polymerase.

    b. Telomerase is an RNA-dependent RNA polymerase.

    c. Telomerase is a DNA-dependent DNA polymerase.

    d. Telomerase is a DNA-dependent RNA polymerase.

15.14. The telomeres of *Drosophila* resemble which of the following?

    a. Centromeres.

    b. Microsatellites.

    c. Retroposons.

    d. DNA transposons.

15.15.* During which stage of the cell cycle does DNA replication occur?

    a. M.

    b. G1.

    c. S.

    d. G2.

15.16. What technique has been applied to studying the timing of replication initiation in different regions of the genome?

    a. Fluorescent *in situ* hybridization.

    b. Mass spectroscopy.

    c. Microarray analysis.

    d. PCR.

Short Answer Questions   *Answers to odd-numbered questions can be found in the Appendix

15.1.* Prior to the Meselson–Stahl experiment it was not known if DNA replication is dispersive, semiconservative, or conservative. Describe the differences in the DNA contents of the daughter molecules resulting from these different modes of replication.

15.2. Describe the mechanism for displacement replication of a DNA molecule.

15.3.* Describe the mechanism for the rolling circle replication process.

15.4. Where and how does the DnaA protein bind at the origin of replication in *E. coli*?

15.5.* How are the replication forks initiated at the origin of replication in *E. coli*?

15.6. What are the functions of the *E. coli* DNA polymerase III subunits α, β, and ε?

15.7.* Describe briefly the three enzymes that are involved in synthesizing the leading strand copy in eukaryotes.

15.8. What is known about the termination of genome replication in *E. coli*? What proteins and sequences are involved in this process?

15.9.* Why do the ends of linear chromosomes get shorter during successive rounds of DNA replication in eukaryotes?

15.10. How is telomerase activity regulated in eukaryotic cells?

15.11.* How do changes in the expression of the yeast Cdc6p protein affect the regulation of genome replication?

15.12. What general patterns have been observed regarding the timing of replication of different parts of the eukaryotic genome?

In-depth Problems

15.1.* Discuss why the semiconservative mode of DNA replication was favored even before the Meselson–Stahl experiment was carried out.

15.2. Evaluate the status of current research into mammalian replication origins.

15.3.* Write an extended report on "DNA helicases."

15.4. Our current knowledge of genome replication in eukaryotes is biased toward the events occurring at the replication fork. The next challenge is to convert this DNA-centered description of replication into a model that describes how replication is organized within the nucleus, addressing issues such as the role of replication factories and the processes used to avoid tangling of the daughter molecules. Devise a research plan to address one or more of these issues.

15.5.* Explore the links between telomeres, cell senescence, and cancer.

Figure Tests

15.1.* What type of DNA replication is shown in this figure? Discuss the process by which DNA is replicated using this system.

15.2. What two complications arise during DNA replication by a DNA polymerase?

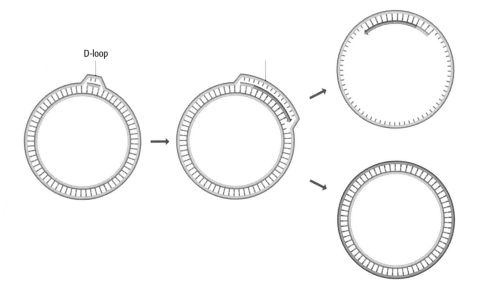

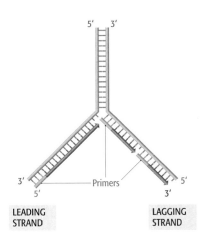

continued ...

Figure Tests (continued)

*Answers to odd-numbered questions can be found in the Appendix

15.3.* Discuss the mechanism by which the adjacent Okazaki fragments are joined during DNA replication of the lagging strand in *E. coli*.

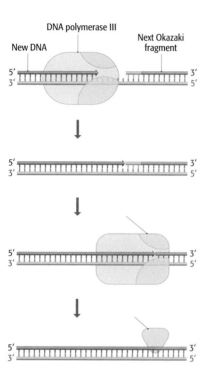

15.4. What is the mechanism by which telomerase is able to extend the 3' ends of chromosomes?

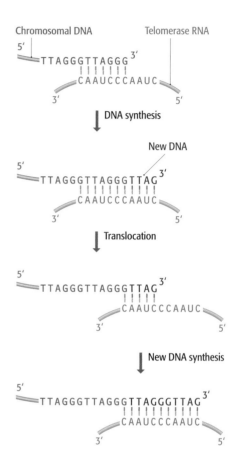

Further Reading

The history of research into genome replication

Crick, F.H.C., Wang, J.C. and Bauer, W.R. (1979) Is DNA really a double helix? *J. Mol. Biol.* **129:** 449–461. *Crick's response to suggestions that DNA has a side-by-side rather than helical conformation.*

Holmes, F.L. (1998) The DNA replication problem, 1953–1958. *Trends Biochem. Sci.* **23:** 117–120.

Kornberg, A. (1989) *For the Love of Enzymes: The Odyssey of a Biochemist.* Harvard University Press, Boston, Massachusetts. *A fascinating autobiography by the discoverer of DNA polymerase.*

Meselson, M. and Stahl, F. (1958) The replication of DNA in *Escherichia coli. Proc. Natl Acad. Sci. USA* **44:** 671–682. *The Meselson–Stahl experiment.*

Okazaki, T. and Okazaki, R. (1969) Mechanisms of DNA chain growth. *Proc. Natl Acad. Sci. USA* **64:** 1242–1248. *The discovery of Okazaki fragments.*

Rodley, G.A., Scobie, R.S., Bates, R.H.T. and Lewitt, R.M. (1976) A possible conformation for double-stranded polynucleotides. *Proc. Natl Acad. Sci. USA* **73:** 2959–2963. *A side-by-side model for DNA structure.*

Watson, J.D. and Crick, F.H.C. (1953) Genetical implications of the structure of deoxyribonucleic acid. *Nature* **171:** 964–967. *Describes possible processes for DNA replication, shortly after discovery of the double helix.*

DNA topoisomerases

Berger, J.M., Gamblin, S.J., Harrison, S.C. and Wang, J.C. (1996) Structure and mechanism of DNA topoisomerase II. *Nature* **379:** 225–232 and **380:** 179.

Champoux, J.J. (2001) DNA topoisomerases: structure, function, and mechanism. *Annu. Rev. Biochem.* **70:** 369–413.

Stewart, L., Redinbo, M.R., Qiu, X., Hol, W.G.J. and Champoux, J.J. (1998) A model for the mechanism of human topoisomerase I. *Science* **279:** 1534–1541.

Origins of replication

Aladjem, M.I., Rodewald, L.W., Kolman, J.L. and Wahl, G.M. (1998) Genetic dissection of a mammalian replicator in the human β-globin locus. *Science* **281:** 1005–1009.

Diffley, J.F.X. and Cocker, J.H. (1992) Protein–DNA interactions at a yeast replication origin. *Nature* **357:** 169–172.

Gilbert, D.M. (2001) Making sense of eukaryotic replication origins. *Science* **294:** 96–100.

DNA polymerases and events at the replication fork

Bochkarev, A., Pfuetzner, R.A., Edwards, A.M. and Frappier, L. (1997) Structure of the single-stranded-DNA-binding domain of replication protein A bound to DNA. *Nature* **385:** 176–181.

Hübscher, U., Nasheuer, H.-P. and Syväoja, J.E. (2000) Eukaryotic DNA polymerases: a growing family. *Trends Biochem. Sci.* **25:** 143–147.

Johnson, A. and O'Donnell, M. (2005) Cellular DNA replicases: components and dynamics at the replication fork. *Annu. Rev. Biochem.* **74:** 283–315. *Details of replication in bacteria and eukaryotes.*

Lemon, K.P. and Grossman, A.D. (1998) Localization of bacterial DNA polymerase: evidence for a factory model of replication. *Science* **282:** 1516–1519.

Liu, Y., Kao, H.I. and Bambara, R.A. (2004) Flap endonuclease I: a central component of DNA metabolism. *Annu. Rev. Biochem.* **73:** 589–615.

Myllykallio, H., Lopez, P., López-Garcia, P., Heilig, R., Saurin, W., Zivanovic, Y., Philippe, H. and Forterre, P. (2000) Bacterial mode of replication with eukaryotic-like machinery in a hyperthermophilic archaeon. *Science* **288:** 2212–2215.

Soultanas, P. and Wigley, D.B. (2001) Unwinding the 'Gordian knot' of helicase action. *Trends Biochem. Sci.* **26:** 47–54.

Trakselis, M.A. and Bell, S.D. (2004) The loader of the rings. *Nature* **429:** 708–709. *The sliding clamp and clamp loader.*

Telomeres

Blackburn, E.H. (2000) Telomere states and cell fates. *Nature* **408:** 53–56.

Cech, T.R. (2004) Beginning to understand the end of the chromosome. *Cell* **116:** 273–279. *Reviews all aspects of telomerase.*

McEachern, M.J., Krauskopf, A. and Blackburn, E.H. (2000) Telomeres and their control. *Annu. Rev. Genet.* **34:** 331–358. *Describes the processes involved in regulation of telomere length.*

Pardue, M.-L. and DeBaryshe, P.G. (2003) Retrotransposons provide an evolutionarily robust non-telomerase mechanism to maintain telomeres. *Annu. Rev. Genet.* **37:** 485–511.

Smogorzewska, A. and de Lange, T. (2004) Regulation of telomerase by telomeric proteins. *Annu. Rev. Biochem.* **73:** 177–208.

continued …

Control of genome replication

Kelly, T.J. and Brown, G.W. (2000) Regulation of chromosome replication. *Annu. Rev. Biochem.* **69:** 829–880.

Raghuraman, M.K., Winzeler, E.A., Collingwood, D., *et al.* (2001) Replication dynamics of the yeast genome. *Science* **294:** 115–121. *Microarray studies of origin firing.*

Sancar, A., Lindsey-Boltz, L.A., Ünsal-Kaçmaz, K. and Linn, S. (2004) Molecular mechanisms of mammalian DNA repair and the DNA damage checkpoints. *Annu. Rev. Biochem.* **73:** 39–85.

Stillman, B. (1996) Cell cycle control of DNA replication. *Science* **274:** 1659–1664.

Zhou, B.-B.S. and Elledge, S.J. (2000) The DNA damage response: putting checkpoints in perspective. *Nature* **408:** 433–439.

Mutations and DNA Repair

16

When you have read Chapter 16, you should be able to:

Define the term "mutation" as well as the various terms that are used to identify different types of mutation.

Describe, with specific examples, how mutations are caused by spontaneous errors in replication.

Give examples of chemical and physical mutagens and outline the alterations that they make to DNA molecules.

Recount, with specific examples, the effects of mutations on the coding and noncoding regions of genomes.

Describe the possible effects of mutations on multicellular organisms.

List and describe the various effects of mutations on microorganisms.

Discuss the biological significance of hypermutation and programmed mutations.

Distinguish between the various types of DNA repair mechanism.

Give detailed descriptions of the molecular events occurring during direct repair, base and nucleotide excision repair, and mismatch repair.

Describe how single- and double-stranded DNA breaks are repaired.

Outline how DNA damage can be bypassed during genome replication.

Summarize the link between DNA repair and human disease.

Genomes are dynamic entities that change over time as a result of the cumulative effects of small-scale sequence alterations caused by **mutation** (Section 16.1). A mutation is a change in the nucleotide sequence of a short region of a genome (Figure 16.1A). Many mutations are **point mutations** (also called simple mutations or single-site mutations) that replace one nucleotide with another. Point mutations are divided into two categories: **transitions**, which

(A) A mutation

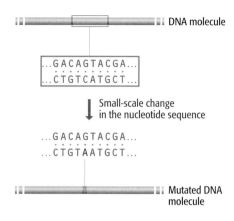

(B) DNA repair

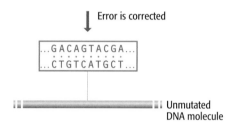

Figure 16.1 Mutation and DNA repair.
(A) A mutation is a small-scale change in the nucleotide sequence of a DNA molecule. A point mutation is shown but there are several other types of mutation, as described in the text. (B) DNA repair corrects mutations that arise as errors in replication and as a result of mutagenic activity.

are purine-to-purine or pyrimidine-to-pyrimidine changes (A→G, G→A, C→T, or T→C), and **transversions**, which are purine-to-pyrimidine or pyrimidine-to-purine changes (A→C, A→T, G→C, G→T, C→A, C→G, T→A, or T→G). Other mutations arise from **insertion** or **deletion** of one or a few nucleotides.

Mutations result either from errors in DNA replication or from the damaging effects of **mutagens**, such as chemicals and radiation, which react with DNA and change the structures of individual nucleotides. All cells possess **DNA repair** enzymes that attempt to minimize the number of mutations that occur (Section 16.2). These enzymes work in two ways. Some are prereplicative and search the DNA for nucleotides with unusual structures, these nucleotides being replaced before replication occurs; others are postreplicative and check newly synthesized DNA for errors, correcting any errors that they find (Figure 16.1B). A possible definition of mutation is therefore *a deficiency in DNA repair*.

Mutations can have dramatic effects on the cell in which they occur, a mutation in a key gene possibly resulting in a defective protein which could lead to death of the cell. Other mutations have a less-significant impact on the phenotype of the cell and many have none at all. As we will see in Chapter 18, all events that are not lethal have the potential to contribute to the evolution of the genome but for this to happen they must be inherited when the organism reproduces. With a single-celled organism such as a bacterium or yeast, all genome alterations that are not lethal or corrected are inherited by daughter cells and become permanent features of the lineage that descends from the original cell in which the alteration occurred. In a multicellular organism, only those events that occur in germ cells are relevant to genome evolution. Changes to the genomes of somatic cells are unimportant in an evolutionary sense, but they will have biological relevance if they result in a deleterious phenotype that affects the health of the organism.

16.1 Mutations

With mutations, we have to consider the following issues: how they arise; the effects they have on the genome and on the organism in which the genome resides; and whether it is possible for a cell to increase its mutation rate and induce programmed mutations under certain circumstances.

16.1.1 The causes of mutations

Mutations arise in two ways:

- Some mutations are **spontaneous** errors in replication that evade the proofreading function of the DNA polymerases that synthesize new polynucleotides at the replication fork (Section 15.2.2). These mutations are called **mismatches** because they are positions where the nucleotide that is inserted into the daughter polynucleotide does not match, by base pairing, the nucleotide at the corresponding position in the template DNA (Figure 16.2A). If the mismatch is not corrected in the daughter double helix then *one* of the granddaughter molecules produced during the next round of DNA replication will carry a permanent, double-stranded version of the mutation.

- Other mutations arise because a mutagen has reacted with the parent DNA, causing a structural change that affects the base-pairing capability of the altered nucleotide. Usually this alteration affects only one strand of the parent double helix, so only one of the daughter molecules carries the mutation, but two of the granddaughter molecules produced during the next round of replication will have it (Figure 16.2B).

(A) An error in replication

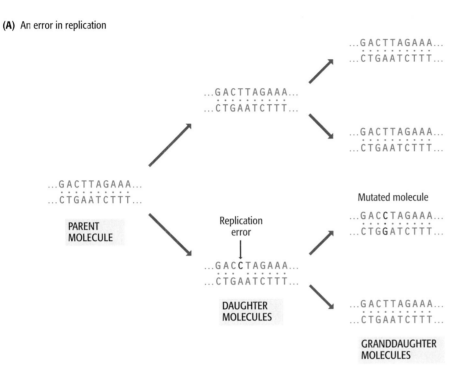

Figure 16.2 Examples of mutations. (A) An error in replication leads to a mismatch in one of the daughter double helices, in this case a T→C change because one of the As in the template DNA was miscopied. When the mismatched molecule is itself replicated, it gives one double helix with the correct sequence and one with a mutated sequence. (B) A mutagen has altered the structure of an A in the lower strand of the parent molecule, giving nucleotide X, which does not base-pair with the T in the other strand so, in effect, a mismatch has been created. When the parent molecule is replicated, X base-pairs with C, giving a mutated daughter molecule. When this daughter molecule is replicated, both granddaughters inherit the mutation.

(B) One possible effect of a mutagen

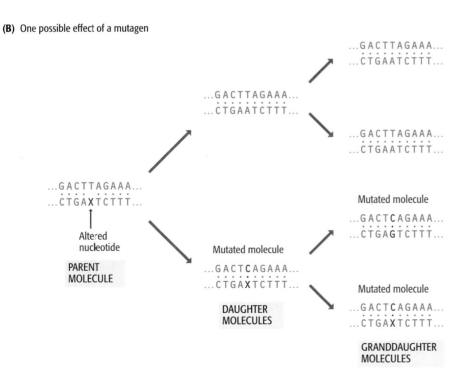

Technical Note 16.1 Mutation detection

Rapid procedures for detecting mutations in DNA molecules

Many genetic diseases are caused by point mutations that result in modification or inactivation of a gene product. Methods for detecting these mutations are important in two contexts. First, when the gene responsible for a genetic disease is initially identified it is usually necessary to examine many versions of that gene from different individuals in order to identify the mutation or mutations responsible for the disease state. Second, when a disease-causing mutation has been characterized, high-throughput detection methods are needed so that clinicians can screen many DNA samples in order to identify individuals who have the mutation and are at risk of developing the disease or passing it on to their children.

Any mutation can be identified by DNA sequencing, but sequencing is relatively slow and would be inappropriate for screening a large number of samples. DNA chip technology (Technical Note 3.1) could also be employed, but this is not yet a widely available option. For these reasons, a number of "low technology" methods have been devised. These can be divided into two categories: **mutation scanning** techniques, which require no prior information about the position of a mutation, and **mutation screening** techniques, which determine whether a specific mutation is present.

Most mutation scanning techniques involve analysis of the heteroduplex formed between a single strand of the DNA being examined and the complementary strand of a control DNA that has the unmutated sequence (Figure T16.1). If the test DNA contains a mutation then there will be a single mismatched position in the heteroduplex, where a base pair has not formed. Various techniques can be used to determine whether this mismatch is present or not:

- **Electrophoresis** or **high-performance liquid chromatography (HPLC)** can detect the mismatch by identifying the difference in the mobility of the mismatched

hybrid compared with the fully base-paired one in a polyacrylamide gel or HPLC column. This approach determines if a mismatch is present but does not provide information on where in the test DNA the mutation is located.

- **Cleavage** of the heteroduplex at the mismatch position followed by gel electrophoresis will locate the position of a mismatch. If the heteroduplex stays intact then no mismatch is present; if it is cleaved then it contains a mismatch, the position of the mutation in the test DNA being indicated by the sizes of the cleavage products. Cleavage is carried out by treatment with enzymes or chemicals that cut at single-stranded regions of mainly double-stranded DNA, or with a single-strand-specific ribonuclease such as S1 (see Figure 5.14) if the hybrid has been formed between the control DNA and an RNA version of the test DNA.

Most screening methods for detection of specific mutations make use of the ability of oligonucleotide hybridization to distinguish between target DNAs whose sequences differ at just one nucleotide position (see Figure 3.8A). In **allele-specific oligonucleotide (ASO) hybridization** the DNA samples are screened by probing with an oligonucleotide that hybridizes only to the mutant sequence (Figure T16.2). This is an efficient procedure but it is unnecessarily long-winded. The DNA samples are usually obtained by PCR of clinical isolates so a more rapid alternative is to use the diagnostic oligonucleotide as one of the PCR primers, so that the presence or absence of the mutation in the test DNA is indicated by the synthesis or otherwise of a PCR product.

Figure T16.1 Hybridization between complementary strands of DNA, one of which contains a mutation, will result in a double-stranded molecule with a mismatch.

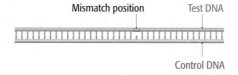

Figure T16.2 Allele-specific oligonucleotide (ASO) hybridization.

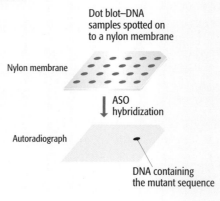

Errors in replication are a source of point mutations

When considered purely as a chemical reaction, complementary base pairing is not particularly accurate. Nobody has yet devised a way of carrying out the template-dependent synthesis of DNA without the aid of enzymes, but if the process could be carried out simply as a chemical reaction in a test tube then the resulting polynucleotide would probably have point mutations at 5–10 positions out of every hundred. This represents an error rate of 5%–10%, which would be completely unacceptable during genome replication. The template-dependent DNA polymerases that carry out DNA replication must therefore increase the accuracy of the process by several orders of magnitude. This improvement is brought about in two ways:

● A DNA polymerase operates a nucleotide-selection process that dramatically increases the accuracy of template-dependent DNA synthesis (Figure 16.3A). This selection process probably acts at three different stages during the polymerization reaction, discrimination against an incorrect nucleotide occurring when the nucleotide is first bound to the DNA polymerase, when it is shifted to the active site of the enzyme, and when it is attached to the 3′ end of the polynucleotide that is being synthesized.

● The accuracy of DNA synthesis is increased still further if the DNA polymerase possesses a 3′→5′ exonuclease activity and so is able to remove an incorrect nucleotide that evades the nucleotide-selection process and becomes attached to the 3′ end of the new polynucleotide (see Figure 2.7B). This is called **proofreading** (Section 15.2.2), but the name is a misnomer because the process is not a simple checking mechanism. Instead, each step in the synthesis of a polynucleotide should be viewed as a competition between the polymerase and exonuclease functions of the enzyme, the polymerase usually winning because it is more active than the exonuclease, at least when the 3′-terminal nucleotide is base-paired to the template. But the polymerase activity is less efficient if the terminal nucleotide is not base-paired, the resulting pause in polymerization allowing the exonuclease activity to predominate so that the incorrect nucleotide is removed (see Figure 16.3B).

Escherichia coli is able to synthesize DNA with an error rate of only 1 per 10^7 nucleotide additions. Interestingly, these errors are not evenly distributed between the two daughter molecules, the product of lagging-strand replication being prone to about 20 times as many errors as the leading-strand replicate. This asymmetry might indicate that DNA polymerase I, which is involved only in lagging-strand replication (Section 15.2.2), has a less-effective base selection and proofreading capability compared with DNA polymerase III, the main replicating enzyme.

Not all of the errors that occur during DNA synthesis can be blamed on the polymerase enzymes: sometimes an error occurs even though the enzyme adds the "correct" nucleotide, the one that base-pairs with the template. This is because each nucleotide base can occur as either of two alternative **tautomers**, structural isomers that are in dynamic equilibrium. For example, thymine exists as two tautomers, the *keto* and *enol* forms, with individual molecules occasionally undergoing a shift from one tautomer to the other. The equilibrium is biased very much toward the *keto* form but every now and then the *enol* version of thymine occurs in the template DNA at the precise time that the replication fork is moving past. This will lead to an "error," because *enol*-thymine base-pairs with G rather than A (Figure 16.4). The same problem can

(A) Nucleotide selection

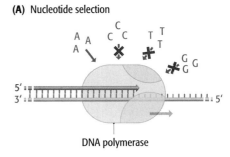

DNA polymerase

(B) "Proofreading"

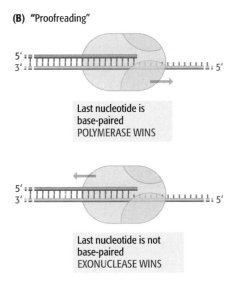

Last nucleotide is
base-paired
POLYMERASE WINS

Last nucleotide is not
base-paired
EXONUCLEASE WINS

Figure 16.3 Mechanisms for ensuring the accuracy of DNA replication. (A) The DNA polymerase actively selects the correct nucleotide to insert at each position. (B) Those errors that occur can be corrected by "proofreading" if the polymerase has a 3′→5′ exonuclease activity. If the last nucleotide that was inserted is base-paired to the template then the polymerase activity predominates, but if the last nucleotide is not base-paired then the exonuclease activity is favored.

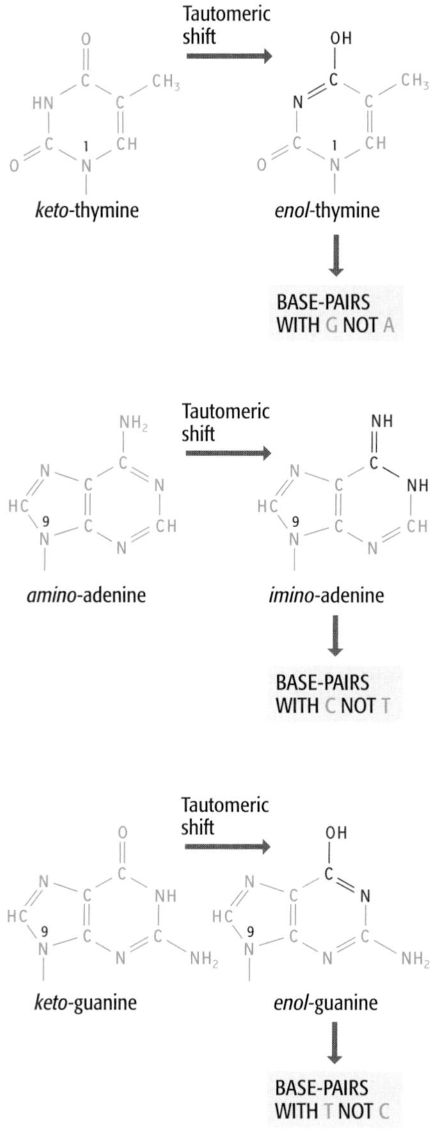

Figure 16.4 The effects of tautomerism on base pairing. In each of these three examples, the two tautomeric forms of the base have different pairing properties. Cytosine also has *amino* and *imino* tautomers but both pair with guanine.

occur with adenine, the rare *imino* tautomer of this base preferentially forming a pair with C, and with guanine, *enol*-guanine pairing with thymine. After replication, the rare tautomer will inevitably revert to its more common form, leading to a mismatch in the daughter double helix.

As stated above, the error rate for DNA synthesis in *E. coli* is 1 in 10^7. However, the overall error rate for replication of the *E. coli* genome is only 1 in 10^{10} to 1 in 10^{11}, the improvement compared with the polymerase error rate being the result of the mismatch repair system (Section 16.2.3) that scans newly replicated DNA for positions where the bases are unpaired and hence corrects the few mistakes that the replication enzymes make. This means that, on average, only one uncorrected replication error occurs every 2000 times that the *E. coli* genome is copied.

Replication errors can also lead to insertion and deletion mutations

Not all errors in replication are point mutations. Aberrant replication can also result in small numbers of extra nucleotides being inserted into the polynucleotide being synthesized, or some nucleotides in the template not being copied. An insertion or deletion that occurs within a coding region might result in a **frameshift** mutation, which changes the reading frame used for translation of the protein specified by the gene (see Figure 16.12). There is a tendency to use "frameshift" to describe all insertions and deletions, but this is inaccurate as inserting or deleting three nucleotides, or multiples of three, simply adds or removes codons or parts of adjacent codons without affecting the reading frame. Also, of course, many insertions/deletions occur outside of open reading frames, within the intergenic regions of a genome.

Insertion and deletion mutations can affect all parts of the genome but are particularly prevalent when the template DNA contains short repeated sequences, such as those found in microsatellites (Section 3.2.2). This is because repeated sequences can induce **replication slippage**, in which the template strand and its copy shift their relative positions so that part of the template is either copied twice or missed out. The result is that the new polynucleotide has a larger or smaller number, respectively, of the repeat units (Figure 16.5). This is the main reason why microsatellite sequences are so variable, replication slippage occasionally generating a new length variant, adding to the collection of alleles already present in the population.

Replication slippage is probably also responsible for the **trinucleotide repeat expansion diseases** that have been discovered in humans in recent years. Each of these neurodegenerative diseases is caused by a relatively short series of trinucleotide repeats becoming elongated to two or more times its normal length. For example, the human *HD* gene contains the sequence 5′–CAG–3′ repeated between 6 and 35 times in tandem, coding for a series of glutamines in the protein product. In Huntington's disease this repeat expands to a copy number of 36–121, increasing the length of the polyglutamine tract and resulting in a dysfunctional protein. Several other human diseases are also caused by expansions of polyglutamine codons (Table 16.1). Some diseases associated with mental retardation result from trinucleotide expansions in the leader region of a gene, giving a **fragile site**, a position where the chromosome is likely to break. Expansions involving intron and trailer regions are also known.

How triplet expansions are generated is not precisely understood. The size of the insertion is much greater than occurs with normal replication slippage, such as that seen with microsatellite sequences, and once the expansion reaches a certain length it appears to become susceptible to further expansion in subsequent rounds of replication, so that the disease becomes increasingly severe in succeeding generations. The possibility that expansion involves formation of hairpin loops in the DNA has been raised, based on the observation that only a limited number of trinucleotide sequences are known

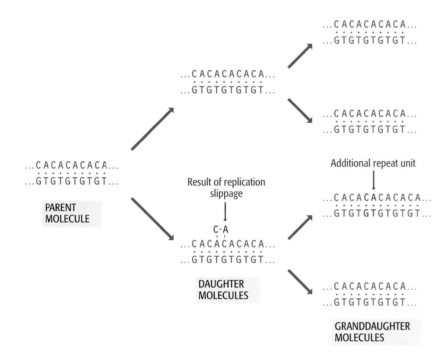

Figure 16.5 Replication slippage. The diagram shows replication of a five-unit CA repeat microsatellite. Slippage has occurred during replication of the parent molecule, inserting an additional repeat unit into the newly synthesized polynucleotide of one of the daughter molecules. When this daughter molecule replicates, it gives a granddaughter molecule whose microsatellite array is one unit longer than that of the original parent.

Table 16.1 Examples of human trinucleotide repeat expansions

| Locus | Repeat sequence | | Associated disease |
	Normal	Mutated	
Polyglutamine expansions (all in coding regions of genes)			
HD	$(CAG)_{6-35}$	$(CAG)_{36-121}$	Huntington's disease
AR	$(CAG)_{9-36}$	$(CAG)_{38-62}$	Spinal and bulbar muscular atrophy
DRPLA	$(CAG)_{6-35}$	$(CAG)_{49-88}$	Dentatorubral-pallidoluysian atrophy
SCA1	$(CAG)_{6-39}$	$(CAG)_{39-82}$	Spinocerebellar ataxia type 1
SCA3	$(CAG)_{12-40}$	$(CAG)_{55-84}$	Machado–Joseph disease
Fragile site expansions (both in the untranslated leader regions of genes)			
FRM1	$(CGG)_{6-53}$	$(CGG)_{60-over\ 230}$	Fragile X syndrome
FRM2	$(GCC)_{6-35}$	$(GCC)_{61-over\ 200}$	Fragile XE mental retardation
Other expansions (positions described below)			
DMPK	$(CTG)_{5-37}$	$(CTG)_{50-3000}$	Myotonic dystrophy
X25	$(GAA)_{7-34}$	$(GAA)_{34-over\ 200}$	Friedreich's ataxia

The *DMPK* and *X25* expansions are in the trailer and intron regions of their genes, respectively, and are thought to affect RNA processing. There are also a few disease-causing mutations that involve expansions of longer sequences, for example progressive myoclonus epilepsy caused by a $(CCCCGCCCCGCG)_{2-3}$ to $(CCCCGCCCCGCG)_{over\ 12}$ expansion in the promoter region of the *EPM1* locus.

to undergo expansion, and all of these sequences are GC–rich and so might form stable secondary structures. There is also evidence that at least one triplet expansion region—for Friedreich's ataxia—can form a triple helix structure. Studies of similar triplet expansions in yeast have shown that these are more prevalent when the *RAD27* gene is inactivated, an interesting observation as *RAD27* is the yeast version of the mammalian gene for FEN1, the protein involved in processing of Okazaki fragments (Section 15.2.2). This might indicate that a trinucleotide repeat expansion is caused by an aberration in lagging-strand synthesis.

Mutations are also caused by chemical and physical mutagens

Many chemicals that occur naturally in the environment have mutagenic properties and these have been supplemented in recent years with other chemical mutagens that result from human industrial activity. Physical agents such as radiation are also mutagenic. Most organisms are exposed to greater or lesser amounts of these various mutagens, their genomes suffering damage as a result.

The definition of the term "mutagen" is *a chemical or physical agent that causes mutations*. This definition is important because it distinguishes mutagens from other types of environmental agent that cause damage to cells in ways other than by causing mutations (Table 16.2). There are overlaps between these categories (for example, some mutagens are also carcinogens) but each type of agent has a distinct biological effect. The definition of "mutagen" also makes a distinction between true mutagens and other agents that damage DNA without causing mutations, for example by causing breaks in DNA molecules. This type of damage may block replication and cause the cell to die, but it is not a mutation in the strict sense of the term and the causative agents are therefore not mutagens.

Mutagens cause mutations in three different ways:

- Some act as **base analogs** and are mistakenly used as substrates when new DNA is synthesized at the replication fork.

- Some react directly with DNA, causing structural changes that lead to miscopying of the template strand when the DNA is replicated. These structural changes are diverse, as we will see when we look at individual mutagens.

- Some mutagens act indirectly on DNA. They do not themselves affect DNA structure, but instead cause the cell to synthesize chemicals such as peroxides that have a direct mutagenic effect.

Table 16.2 Categories of environmental agent that cause damage to living cells

Agent	Effect on living cells
Carcinogen	Causes cancer—the neoplastic transformation of eukaryotic cells
Clastogen	Causes fragmentation of chromosomes
Mutagen	Causes mutations
Oncogen	Induces tumor formation
Teratogen	Results in developmental abnormalities

The range of mutagens is so vast that it is difficult to devise an all-embracing classification. We will therefore restrict our study to the most common types. For chemical mutagens these are as follows:

- **Base analogs** are purine and pyrimidine bases that are similar enough to the standard bases of DNA to be incorporated into nucleotides when these are synthesized by the cell. The resulting unusual nucleotides can then be used as substrates for DNA synthesis during genome replication. For example, **5-bromouracil** (**5-bU**; Figure 16.6A) has the same base-pairing properties as thymine, and nucleotides containing this base can be added to the daughter polynucleotide at positions opposite As in the template. The mutagenic effect arises because the equilibrium between the two tautomers of 5-bU is shifted more toward the rarer *enol* form than is the case with thymine. This means that during the next round of replication there is a relatively high chance of the polymerase encountering *enol*-5bU, which (like *enol*-thymine) pairs with G rather than A (Figure 16.6B). This results in a point mutation (Figure 16.6C). **2-Aminopurine** acts in a similar way: it is an analog of adenine with an *amino*-tautomer that pairs with thymine, and an *imino*-tautomer that pairs with cytosine, the *imino* form being more common than *imino*-adenine and hence inducing T-to-C transitions during DNA replication.

(A) 5-Bromouracil

Figure 16.6 5-Bromouracil and its mutagenic effect.

(B) Base pairing with 5-bromouracil

5-Bromouracil *keto* form   Adenine   5-Bromouracil *enol* form   Guanine

(C) The mutagenic effect of 5-bromouracil

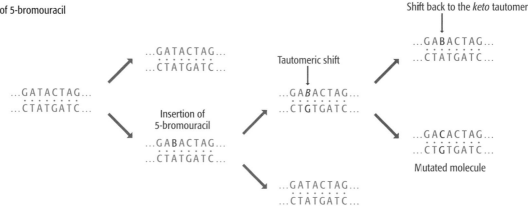

Figure 16.7 Hypoxanthine is a deaminated version of adenine. The nucleoside that contains hypoxanthine is called inosine (see Figure 12.18).

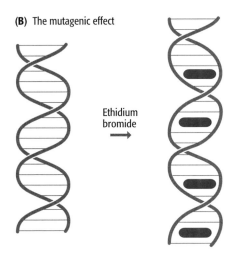

Figure 16.8 The mutagenic effect of ethidium bromide. (A) Ethidium bromide is a flat, plate-like molecule that is able to slot in between the base pairs of the double helix. (B) Ethidium bromide molecules are shown intercalated into the helix: the molecules are viewed sideways on. Note that intercalation increases the distance between adjacent base pairs.

- **Deaminating agents** also cause point mutations. A certain amount of base deamination (removal of an amino group) occurs spontaneously in genomic DNA molecules, with the rate being increased by chemicals such as nitrous acid, which deaminates adenine, cytosine, and guanine (thymine has no amino group and so cannot be deaminated), and sodium bisulfite, which acts only on cytosine. Deamination of guanine is not mutagenic because the resulting base, xanthine, blocks replication when it appears in the template polynucleotide. Deamination of adenine gives hypoxanthine (Figure 16.7), which pairs with C rather than T, and deamination of cytosine gives uracil, which pairs with A rather than G. Deamination of these two bases therefore results in point mutations when the template strand is copied.

- **Alkylating agents** are a third type of mutagen that can give rise to point mutations. Chemicals such as **ethylmethane sulfonate** (**EMS**) and dimethylnitrosamine add alkyl groups to nucleotides in DNA molecules, as do methylating agents such as methyl halides, which are present in the atmosphere, and the products of nitrite metabolism. The effect of alkylation depends on the position at which the nucleotide is modified and the type of alkyl group that is added. Methylations, for example, often result in modified nucleotides with altered base-pairing properties and so lead to point mutations. Other alkylations block replication by forming crosslinks between the two strands of a DNA molecule, or by adding large alkyl groups that prevent progress of the replication complex.

- **Intercalating agents** are usually associated with insertion mutations. The best-known mutagen of this type is **ethidium bromide**, which fluoresces when exposed to ultraviolet (UV) radiation and so is used to reveal the positions of DNA bands after agarose gel electrophoresis (see Technical Note 2.2). Ethidium bromide and other intercalating agents are flat molecules that can slip between base pairs in the double helix, slightly unwinding the helix and hence increasing the distance between adjacent base pairs (Figure 16.8).

The most important types of physical mutagen are as follows:

- Ultraviolet radiation of wavelength 260 nm induces dimerization of adjacent pyrimidine bases, especially if these are both thymines (Figure 16.9A), resulting in a **cyclobutyl dimer**. Other pyrimidine combinations also form dimers, the order of frequency with which this occurs being 5′–CT–3′ greater than 5′–TC–3′ greater than 5′–CC–3′. Purine dimers are much less common. UV-induced dimerization usually results in a deletion mutation when the modified strand is copied. Another type of UV-induced **photoproduct** is the **(6–4) lesion**, in which carbons number 4 and 6 of adjacent pyrimidines become covalently linked (Figure 16.9B).

- Ionizing radiation has various effects on DNA depending on the type of radiation and its intensity. Point, insertion, and/or deletion mutations might arise, as well as more severe forms of DNA damage that prevent subsequent replication of the genome. Some types of ionizing radiation act directly on DNA, others act indirectly by stimulating the formation of reactive molecules such as peroxides in the cell.

- Heat stimulates the water-induced cleavage of the β-*N*-glycosidic bond that attaches the base to the sugar component of the nucleotide (Figure 16.10A). This occurs more frequently with purines than with pyrimidines and results in an **AP** (**apurinic/apyrimidinic**) **site**, or **baseless site**. The

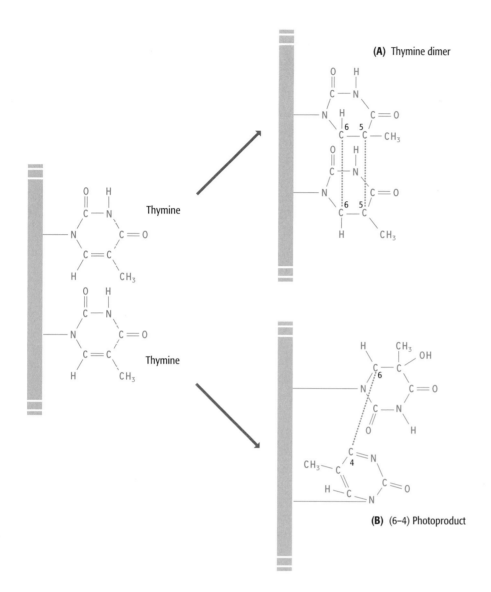

(A) Thymine dimer

Thymine

Thymine

(B) (6–4) Photoproduct

Figure 16.9 Photoproducts induced by UV irradiation. A segment of a polynucleotide containing two adjacent thymine bases is shown. (A) A thymine dimer contains two UV-induced covalent bonds, one linking the carbons at position 6 and the other linking the carbons at position 5. (B) The (6–4) lesion involves formation of a covalent bond between carbons 4 and 6 of the adjacent nucleotides.

sugar–phosphate that is left is unstable and rapidly degrades, leaving a gap if the DNA molecule is double stranded (Figure 16.10B). This reaction is not normally mutagenic because cells have effective systems for repairing gaps (Section 16.2.2), which is reassuring when one considers that 10,000 AP sites are generated in each human cell per day. Gaps do, however, lead to mutations under certain circumstances, for example in *E. coli* when the SOS response is activated, when gaps are filled with As regardless of the identity of the nucleotide in the other strand (Section 16.2.5).

16.1.2 The effects of mutations

When considering the effects of mutations we must make a distinction between the *direct* effect that a mutation has on the functioning of a genome and its *indirect* effect on the phenotype of the organism in which it occurs. The direct effect is relatively easy to assess because we can use our understanding of gene structure and expression to predict the impact that a mutation will have on genome function. The indirect effects are more complex because these relate to the phenotype of the mutated organism which, as described in Section 5.2.2, is often difficult to correlate with the activities of individual genes.

Figure 16.10 The mutagenic effect of heat. (A) Heat induces hydrolysis of a β-*N*-glycosidic bond, resulting in a baseless site in a polynucleotide. (B) Schematic representation of the effect of heat-induced hydrolysis on a double-stranded DNA molecule. The baseless site is unstable and degrades, leaving a gap in one strand.

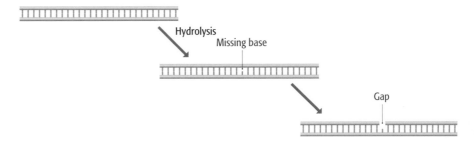

(A) Heat-induced hydrolysis of a β-*N*-glycosidic bond

Hydrolysis

Cleaved base

(B) The effect of hydrolysis on double-stranded DNA

Hydrolysis

Missing base

Gap

The effects of mutations on genomes

Many mutations result in nucleotide sequence changes that have no effect on the functioning of the genome. These **silent mutations** include virtually all of those that occur in intergenic DNA and in the noncoding components of genes and gene-related sequences. In other words, some 98.5% of the human genome can be mutated without significant effect.

Mutations in the coding regions of genes are much more important. First, we will look at point mutations that change the sequence of a triplet codon. A mutation of this type will have one of four effects (Figure 16.11):

- It may result in a **synonymous** change, the new codon specifying the same amino acid as the unmutated codon. A synonymous change is therefore a silent mutation because it has no effect on the coding function of the genome: the mutated gene codes for exactly the same protein as the unmutated gene.

- It may result in a **nonsynonymous** change, the mutation altering the codon so that it specifies a different amino acid. The protein coded by the mutated gene therefore has a single amino acid change. This often has no significant effect on the biological activity of the protein because most proteins can tolerate at least a few amino acid changes without noticeable effect on their ability to function in the cell, but changes to some amino acids, such as those at the active site of an enzyme, have a greater impact. A nonsynonymous change is also called a **missense** mutation.

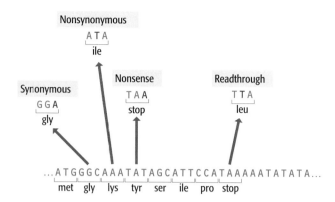

Figure 16.11 Effects of point mutations on the coding region of a gene. Four different effects of point mutations are shown. The readthrough mutation results in the gene being extended beyond the end of the sequence shown here, the leucine codon created by the mutation being followed by AAA = lys, TAT = tyr, and ATA = ile. See Figure 1.20 for the genetic code.

- The mutation may convert a codon that specifies an amino acid into a termination codon. This is a **nonsense** mutation and it results in a shortened protein because translation of the mRNA stops at this new termination codon rather than proceeding to the correct termination codon further downstream. The effect of this on protein activity depends on how much of the polypeptide is lost: usually the effect is drastic and the protein is nonfunctional.

- The mutation could convert a termination codon into one specifying an amino acid, resulting in **readthrough** of the stop signal so the protein is extended by an additional series of amino acids at its C terminus. Most proteins can tolerate short extensions without an effect on function, but longer extensions might interfere with folding of the protein and so result in reduced activity.

Deletion and insertion mutations also have distinct effects on the coding capabilities of genes (Figure 16.12). If the number of deleted or inserted nucleotides is three or a multiple of three, then one or more codons are removed or added, the resulting loss or gain of amino acids having varying effects on the function of the encoded protein. Deletions or insertions of this type are often inconsequential but will have an impact if, for example, amino acids involved in an enzyme's active site are lost, or if an insertion disrupts an important secondary structure in the protein. On the other hand, if the number of deleted or inserted nucleotides is not three or a multiple of three, then a frameshift results, all of the codons downstream of the mutation being taken from a different reading frame from that used in the unmutated gene.

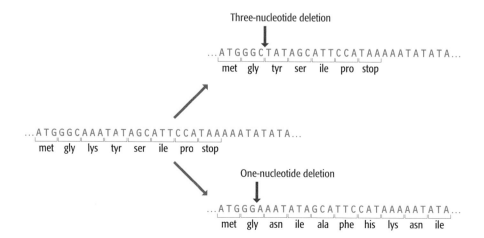

Figure 16.12 Deletion mutations. In the top sequence, three nucleotides comprising a single codon are deleted. This shortens the resulting protein product by one amino acid but does not affect the rest of its sequence. In the lower section, a single nucleotide is deleted. This results in a frameshift so that all the codons downstream of the deletion are changed, including the termination codon which is now read through. Note that if a three-nucleotide deletion removes parts of adjacent codons then the result is more complicated than shown here. Consider, for example, deletion of the trinucleotide GCA from the sequence ...ATGGGCAAATAT... coding for met–gly–lys–tyr. The new sequence is ...ATGGAATAT... coding for met–glu–tyr. Two amino acids have been replaced by a single, different one.

Figure 16.13 Two possible effects of deletion mutations in the region upstream of a gene.

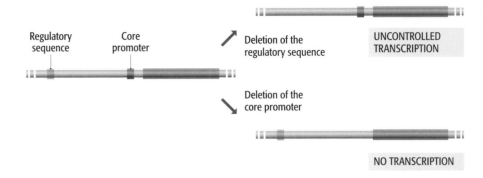

This usually has a significant effect on the protein function, because a greater or lesser part of the mutated polypeptide has a completely different sequence to the normal polypeptide.

It is less easy to make generalizations about the effects of mutations that occur outside of the coding regions of the genome. Any protein-binding site is susceptible to point, insertion, or deletion mutations that change the identity or relative positioning of nucleotides involved in the DNA–protein interaction. These mutations therefore have the potential to inactivate promoters or regulatory sequences, with predictable consequences for gene expression (Figure 16.13; Sections 11.2 and 11.3). Origins of replication could conceivably be made nonfunctional by mutations that change, delete, or disrupt sequences recognized by the relevant binding proteins (Section 15.2.1) but these possibilities are not well documented. There is also little information about the potential impact on gene expression of mutations that affect nucleosome positioning (Section 10.2.2).

One area that has been better researched concerns mutations that occur in introns or at exon–intron boundaries. In these regions, individual point mutations will be important if they change nucleotides involved in the RNA–protein and RNA–RNA interactions that occur during splicing of different types of intron (Sections 12.2.2 and 12.2.4). For example, mutation of either the G or T in the DNA copy of the 5′ splice site of a GU–AG intron, or of the A or G at the 3′ splice site, will disrupt splicing because the correct exon–intron boundary will no longer be recognized. This may mean that the intron is not removed from the pre-mRNA, but it is more likely that a cryptic splice site (see Figure 12.28B) will be used as an alternative. It is also possible for a mutation within an intron or an exon to create a new cryptic site that is preferred over a genuine splice site that is not itself mutated. Both types of event have the same result: relocation of the active splice site, leading to aberrant splicing. This might delete part of the resulting protein, add a new stretch of amino acids, or lead to a frameshift. Several versions of the blood disease β-thalassemia are caused by mutations that lead to cryptic splice site selection during processing of β-globin transcripts.

The effects of mutations on multicellular organisms

Now we turn to the indirect effects that mutations have on organisms, beginning with multicellular diploid eukaryotes such as humans. The first issue to consider is the relative importance of the same mutation in a somatic cell compared with a germ cell. Because somatic cells do not pass copies of their genomes to the next generation, a somatic cell mutation is important only for the organism in which it occurs: it has no potential evolutionary impact. In fact, most somatic cell mutations have no significant effect, even if they result in cell death, because there are many other identical cells in the same tissue and the loss of one cell is immaterial. An exception is when a mutation causes a somatic cell to malfunction in a way that is harmful to the organism, for instance by inducing tumor formation or other cancerous activity.

Mutations in germ cells are more important because they can be transmitted to members of the next generation and will then be present in all the cells of any individual who inherits the mutation. Most mutations, including all silent ones and many in coding regions, will still not change the phenotype of the organism in any significant way. Those that do have an effect can be divided into two categories:

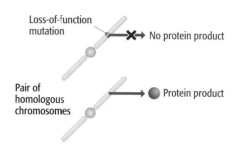

Figure 16.14 A loss-of-function mutation is usually recessive because a functional version of the gene is present on the second chromosome copy.

- **Loss-of-function** is the normal result of a mutation that reduces or abolishes a protein activity. Most loss-of-function mutations are recessive, because in a heterozygote the second chromosome copy carries an unmutated version of the gene coding for a fully functional protein whose presence compensates for the effect of the mutation (Figure 16.14). There are some exceptions where a loss-of-function mutation is dominant, one example being **haploinsufficiency**, where the organism is unable to tolerate the approximately 50% reduction in protein activity suffered by the heterozygote. This is the explanation for a few genetic diseases in humans, including Marfan syndrome which results from a mutation in the gene for the connective tissue protein called fibrillin.

- **Gain-of-function** mutations are much less common. The mutation must be one that confers an abnormal activity on a protein. Many gain-of-function mutations are in regulatory sequences rather than in coding regions, and can therefore have a number of consequences. For example, a mutation might lead to one or more genes being expressed in the wrong tissues, these tissues gaining functions that they normally lack. Alternatively the mutation could lead to overexpression of one or more genes involved in control of the cell cycle, thus leading to uncontrolled cell division and hence to cancer. Because of their nature, gain-of-function mutations are usually dominant.

Assessing the effects of mutations on the phenotypes of multicellular organisms can be difficult. Not all mutations have an immediate impact: some are **delayed onset** and only confer an altered phenotype later in the individual's life. Other mutations display **nonpenetrance** in some individuals, never being expressed even though the individual has a dominant mutation or is a homozygous recessive. With humans, these factors complicate attempts to map disease-causing mutations by pedigree analysis (Section 3.2.4) because they introduce uncertainty about which members of a family carry a mutant allele.

The effects of mutations on microorganisms

Mutations in microbes such as bacteria and yeast can also be described as loss-of-function or gain-of-function, but with microorganisms this is neither the normal nor the most useful classification scheme. Instead, a more detailed description of the phenotype is usually attempted on the basis of the growth properties of mutated cells in various culture media. This enables most mutations to be assigned to one of four categories:

(A) A tryptophan auxotroph

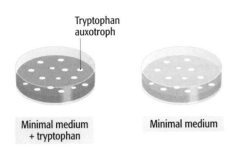

(B) Replica plating

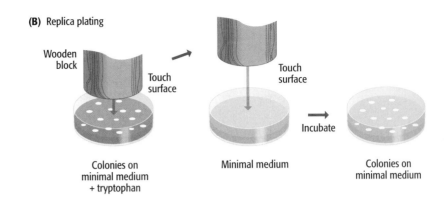

Figure 16.15 A tryptophan auxotrophic mutant. (A) Two Petri-dish cultures are shown. Both contain minimal medium, which provides just the basic nutritional requirements for bacterial growth (nitrogen, carbon, and energy sources, plus some salts). The medium on the left is supplemented with tryptophan but the medium on the right is not. Unmutated bacteria, plus tryptophan auxotrophs, can grow on the plate on the left, the auxotrophs growing because the medium supplies the tryptophan that they cannot make themselves. Tryptophan auxotrophs cannot grow on the plate on the right because this does not contain tryptophan. (B) To identify a tryptophan auxotroph, colonies are first grown on the minimal medium + tryptophan plate and then transferred to the minimal medium plate by replica plating. After incubation, colonies appear on the minimal medium plate in the same relative positions as on the plate containing tryptophan, except for the tryptophan auxotrophs which do not grow. These colonies can therefore be identified and samples of the tryptophan auxotrophic bacteria recovered from the minimal medium + tryptophan plate.

- **Auxotrophs** are cells that will only grow when provided with a nutrient not required by the unmutated organism. For example, *E. coli* normally makes its own tryptophan, courtesy of the enzymes coded by the five genes in the tryptophan operon (Figure 8.8B). If one of these genes is mutated in such a way that its protein product is inactivated, then the cell is no longer able to make tryptophan and so becomes a tryptophan auxotroph. It cannot survive on a medium that lacks tryptophan, and can grow only when this amino acid is provided as a nutrient (Figure 16.15). Unmutated bacteria, which do not require extra supplements in their growth media, are called **prototrophs**.

- **Conditional-lethal mutants** are unable to withstand certain growth conditions: under **permissive conditions** they appear to be entirely normal but when transferred to **restrictive conditions** the mutant phenotype is seen. **Temperature-sensitive mutants** are typical examples of conditional-lethal mutants. Temperature-sensitive mutants behave like wild-type cells at low temperatures but exhibit their mutant phenotype when the temperature is raised above a certain threshold, which is different for each mutant. Usually this is because the mutation reduces the stability of a protein, so the protein becomes unfolded and hence inactive when the temperature is raised.

- **Inhibitor-resistant mutants** are able to resist the toxic effects of an antibiotic or another type of inhibitor. There are various molecular explanations for this type of mutant. In some cases the mutation changes the structure of the protein that is targeted by the inhibitor, so the latter can no longer bind to the protein and interfere with its function. This is the basis of streptomycin resistance in *E. coli*, which results from a change in the structure of ribosomal protein S12. Another possibility is that the mutation changes the properties of a protein responsible for transporting the inhibitor into the cell, this often being the way in which resistance to toxic metals is acquired.

- **Regulatory mutants** have defects in promoters and other regulatory sequences. This category includes **constitutive mutants**, which continuously express genes that are normally switched on and off under different conditions. For example, a mutation in the operator sequence of the lactose operon (Section 11.3.1) can prevent the repressor from binding and so result in the lactose operon being expressed all the time, even when lactose is absent and the genes should be switched off (Figure 16.16).

In addition to these four categories, many mutations are lethal and so result in death of the mutant cell, whereas others have no effect. The latter are less common in microorganisms than in higher eukaryotes, because most microbial genomes are relatively compact, with little noncoding DNA. Mutations can also be **leaky**, meaning that a less extreme form of the mutant phenotype is expressed. For example, a leaky version of the tryptophan auxotroph illustrated in Figure 16.15 would grow slowly on minimal medium, rather than not growing at all.

16.1.3 Hypermutation and the possibility of programmed mutations

Is it possible for cells to utilize mutations in a positive fashion, either by increasing the rate at which mutations appear in their genomes, or by directing mutations toward specific genes? Both types of event might appear, at first glance, to go against the accepted wisdom that mutations occur randomly. The randomness of mutations is an important concept in biology because it is a requirement of the Darwinian view of evolution, which holds that changes in the characteristics of an organism occur by chance and are not influenced by the environment in which the organism is placed. In contrast, the Lamarckian theory of evolution, which biologists rejected well over a century ago, states that organisms acquire changes that enable them to adapt to their environment. The Darwinian view requires that mutations occur at random, whereas Lamarckian evolution demands that mutations occur in response to the environment.

Two phenomena that, at first glance, appear to contravene the notion that mutations occur at random are **hypermutation** and **programmed mutations**.

Hypermutation results from abnormal DNA repair processes

Hypermutation occurs when a cell allows the rate at which mutations occur in its genome to increase. Several examples of hypermutation are known, one of these forming part of the mechanism used by some vertebrates, including humans, to generate a diverse array of immunoglobulin proteins. We have already touched on this phenomenon in Section 14.2.1 when we examined the genome rearrangements that result in joining of the V, D, J, and C segments of the immunoglobulin heavy- and light-chain genes (see Figure 14.19). Additional diversity is produced by hypermutation of the V gene segments after assembly of the intact immunoglobulin gene (Figure 16.17), the mutation rate for these segments being 6–7 orders of magnitude greater than the background mutation rate experienced by the rest of the genome.

The precise mechanism for V gene segment hypermutation is unknown and several models have been proposed based on the available experimental

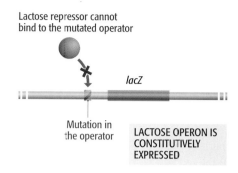

Figure 16.16 The effect of a constitutive mutation in the lactose operator. The operator sequence has been altered by a mutation, and the lactose repressor can no longer bind to it. The result is that the lactose operon is transcribed all the time, even when lactose is absent from the medium. This is not the only way in which a constitutive *lac* mutant can arise. For example, the mutation could be in the gene coding for the lactose repressor, changing the tertiary structure of the repressor protein so that its DNA-binding motif is disrupted and it can no longer recognize the operator sequence, even when the latter is unmutated. See Figure 11.24 for more details about the lactose repressor and its regulatory effect on expression of the lactose operon.

Figure 16.17 Hypermutation of the V gene segment of an intact immunoglobulin gene. See Figure 14.19 for a description of the events leading to assembly of an immunoglobulin gene.

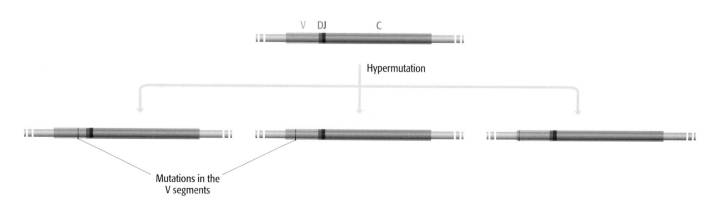

(A) Hypermutation by incorrect repair of mismatches

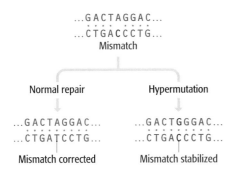

(B) Hypermutation by conversion of cytosine to uracil

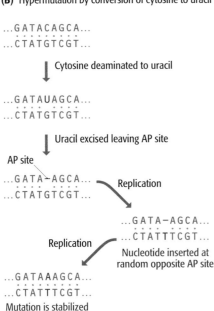

Figure 16.18 Two alternative schemes for hypermutation of immunoglobulin V gene segments.

evidence. At first, the enhanced mutation rate was thought to result from unusual behavior of the mismatch repair system which normally corrects replication errors (see Section 16.2.3). At all other positions within the genome, mismatch repair corrects errors of replication by searching for mismatches and replacing the nucleotide in the daughter strand, this being the strand that has just been synthesized and so contains the error. It was thought that at V gene segments, the repair system changes the nucleotide in the parent strand, and so stabilizes the mutation rather than correcting it (Figure 16.18A). More recently, it has been shown that two enzymes—a cytosine deaminase and uracil-DNA glycosylase—are required for V segment hypermutation. This has led to the suggestion that hypermutation is brought about by conversion of some cytosine bases into uracils (by the deaminase) followed by excision of the uracils (by the glycosylase) to produce AP sites (Figure 16.18B). This is similar to the early steps in the base excision repair process (Section 16.2.2), in which a uracil resulting from the action of a deaminating mutagen is excised by uracil-DNA glycosylase. In base excision repair, the AP site would then be filled in by DNA polymerase β, to restore the original sequence, but in hypermutation the AP sites are not repaired. This means that during the next round of replication, any of the four nucleotides could be placed in the daughter strand opposite each AP site—a T in the example shown in Figure 16.18B. A further round of replication then stabilizes the mutation.

Programmed mutations appear to support the Lamarckian theory of evolution

An apparent increase in mutation rate arising from modifications to the normal DNA repair process does not contradict the dogma regarding the randomness of mutations. On the other hand, reports dating back to 1988, suggesting that *E. coli* is able to direct mutations toward genes whose mutation would be advantageous under the environmental conditions that the bacterium is encountering, are much more difficult to explain.

The randomness of mutations in bacteria was first demonstrated by Luria and Delbrück in 1943. They grew a series of *E. coli* cultures in different flasks and then added T1 bacteriophages to each one. Most of the bacteria were killed by the phages, but a few T1-resistant mutants were able to survive. These were identified by plating samples from each culture, soon after T1 infection, onto an agar medium. If mutations leading to T1 resistance occurred randomly in the cultures before the bacteriophages were added, each culture would contain a different number of resistant mutants, the number depending on how early during the growth period the first mutant cells arose (Figure 16.19). Those that arose early would divide many times to give rise to a large number of resistant progeny in the culture at the end of the growth period, whereas those that arose later would give rise to just a few progeny. Some cultures would therefore contain many T1-resistant cells and others would contain just a few. Alternatively, if resistant bacteria arose by programmed mutation only when the T1 phage was added, then all cultures would have similar numbers of mutants. Luria and Delbrück found that each of their cultures contained a different number of T1-resistant bacteria; thus, they concluded that mutations occur randomly and not in response to T1 phage.

The possibility that Luria and Delbrück's conclusion might not be universally true for *E. coli* mutations was first suggested by studies of an *E. coli* strain

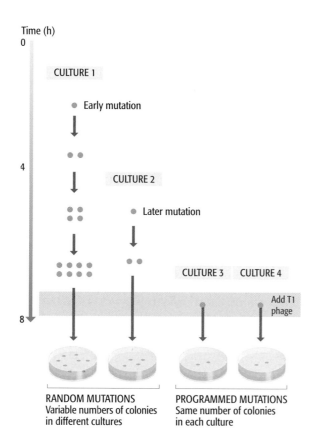

Time (h)
0

CULTURE 1

• Early mutation

4

CULTURE 2

• Later mutation

CULTURE 3   CULTURE 4

Add T1
phage

8

RANDOM MUTATIONS
Variable numbers of colonies
in different cultures

PROGRAMMED MUTATIONS
Same number of colonies
in each culture

Figure 16.19 Random and programmed mutations. The results obtained by Luria and Delbrück are shown on the left. During growth of the *E. coli* cultures, mutations that give resistance to T1 bacteriophage occur randomly at various times, meaning that each culture has a different number of resistant cells when the phages are added. Variable numbers of colonies are therefore seen when the bacteria are plated onto solid medium. The result that would have been obtained if programmed mutations had occurred is shown on the right. Now the T1-resistant bacteria acquire their mutations in response to exposure to the phages, and each plate therefore has a similar number of colonies.

that carries a nonsense mutation in its *lacZ* gene. The presence of the termination codon in *lacZ* means that these cells are unable to synthesize functional β-galactosidase enzymes and so cannot use lactose as a carbon and energy source—they are therefore lactose auxotrophs. This is not necessarily a permanent situation because a cell could undergo a mutation that converts the termination codon back into one specifying an amino acid. These new mutants would be able to make β-galactosidase and use any lactose that is available. According to Luria and Delbrück's results, such mutations should occur at random and should not be influenced by the presence of lactose in the medium. But studies have shown that when the lactose auxotrophs are plated onto a minimal medium containing lactose as the only sugar—circumstances that require that the bacteria must mutate into lactose prototrophs in order to survive—then the number of lactose prototrophs that arise is significantly higher than that expected if mutations occurred randomly. In other words, some cells undergo programmed mutation and acquire the specific change in DNA sequence needed to withstand the selective pressure.

These experiments suggest that bacteria can program mutations according to the selective pressures to which they are exposed. In other words, the environment can directly affect the phenotype of the organism, as suggested by Lamarck, rather than operating through the random processes postulated by Darwin. With such radical implications, it is not surprising that the experiments have been debated at length, with numerous attempts to discover flaws in their design or alternative explanations for the results. Variations of the *lacZ* experimental system have suggested that the original results are authentic, and similar events in other bacteria have been described. Models based on gene amplification rather than selective mutation are being tested, and attention has also been directed at the possible roles of recombination events, such as transposition of insertion elements, in the generation of programmed mutations.

16.2 DNA Repair

In view of the thousands of damage events that genomes suffer every day, coupled with the errors that occur when the genome replicates, it is essential that cells possess efficient repair systems. Without these repair systems a genome would not be able to maintain its essential cellular functions for more than a few hours before key genes became inactivated by DNA damage. Similarly, cell lineages would accumulate replication errors at such a rate that their genomes would become dysfunctional after a few cell divisions.

Most cells possess four different categories of DNA repair system (Figure 16.20):

- **Direct repair** systems (Section 16.2.1), as the name suggests, act directly on damaged nucleotides, converting each one back to its original structure.

- **Excision repair** (Section 16.2.2) involves excision of a segment of the polynucleotide containing a damaged site, followed by resynthesis of the correct nucleotide sequence by a DNA polymerase.

- **Mismatch repair** (Section 16.2.3) corrects errors of replication, again by excising a stretch of single-stranded DNA containing the offending nucleotide and then repairing the resulting gap.

- **Nonhomologous end-joining** (Section 16.2.4) is used to mend double-strand breaks.

Most, if not all, organisms also possess systems that enable them to replicate damaged regions of their genome without prior repair. We will examine these systems in Section 16.2.5, and in Section 16.2.6 we will survey the human diseases that result from defects in DNA repair processes.

Figure 16.20 Four categories of DNA repair system.

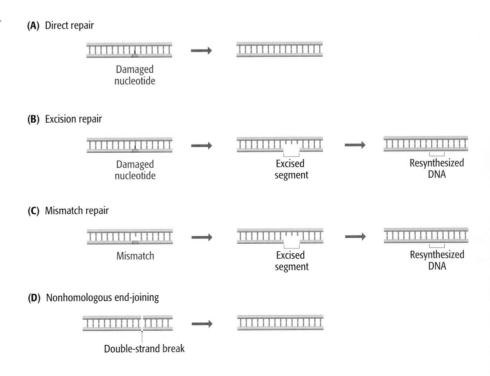

(A) Direct repair

Damaged nucleotide

(B) Excision repair

Damaged nucleotide

Excised segment

Resynthesized DNA

(C) Mismatch repair

Mismatch

Excised segment

Resynthesized DNA

(D) Nonhomologous end-joining

Double-strand break

16.2.1 Direct repair systems fill in nicks and correct some types of nucleotide modification

Most of the types of DNA damage that are caused by chemical or physical mutagens (Section 16.1.1) can only be repaired by excision of the damaged nucleotide followed by resynthesis of a new stretch of DNA, as shown in Figure 16.20B. Only a few types of damaged nucleotide can be repaired directly:

- **Nicks** can be repaired by a DNA ligase if all that has happened is that a phosphodiester bond has been broken, without damage to the 5′–phosphate and 3′–hydroxyl groups of the nucleotides either side of the nick (Figure 16.21). This is often the case with nicks resulting from the effects of ionizing radiation.

- Some forms of **alkylation** damage are directly reversible by enzymes that transfer the alkyl group from the nucleotide to their own polypeptide chains. Enzymes capable of doing this are known in many different organisms and include the **Ada enzyme** of *E. coli*, which is involved in an adaptive process that this bacterium is able to activate in response to DNA damage. Ada removes alkyl groups attached to the oxygen groups at positions 4 and 6 of thymine and guanine, respectively, and can also repair phosphodiester bonds that have become methylated. Other alkylation repair enzymes have more restricted specificities, an example being human **MGMT** (O^6-methylguanine–DNA methyltransferase) which, as its name suggests, only removes alkyl groups from position 6 of guanine.

- **Cyclobutyl dimers** are repaired by a light-dependent direct system called **photoreactivation**. In *E. coli*, the process involves the enzyme called **DNA photolyase** (more correctly named deoxyribodipyrimidine photolyase). When stimulated by light with a wavelength between 300 nm and 500 nm the enzyme binds to cyclobutyl dimers and converts them back to the original monomeric nucleotides. Photoreactivation is a widespread but not universal type of repair: it is known in many but not all bacteria and also in quite a few eukaryotes, including some vertebrates, but is absent in humans and other placental mammals. A similar type of photoreactivation involves the **(6–4) photoproduct photolyase** and results in repair of (6–4) lesions. Neither *E. coli* nor humans have this enzyme but it is possessed by a variety of other organisms.

16.2.2 Excision repair

The direct types of damage reversal described above are important, but they form a very minor component of the DNA repair mechanisms of most organisms. This point is illustrated by the human genome sequence, which appears to contain just a single gene coding for a protein involved in direct repair (the *MGMT* gene), but which has at least 40 genes for components of the excision repair pathways. These pathways fall into two categories:

- **Base excision repair** involves removal of a damaged nucleotide base, excision of a short piece of the polynucleotide around the AP site thus created, and resynthesis with a DNA polymerase.

- **Nucleotide excision repair** is similar to base excision repair but is not preceded by removal of a damaged base and can act on more substantially damaged areas of DNA.

We will examine each of these pathways in turn.

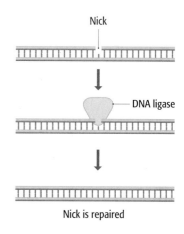

Figure 16.21 Repair of a nick by DNA ligase.

Figure 16.22 Base excision repair.
(A) Excision of a damaged nucleotide by a DNA glycosylase. (B) Schematic representation of the base excision repair pathway. Alternative versions of the pathway are described in the text.

(A) Removal of a damaged base by DNA glycosylase

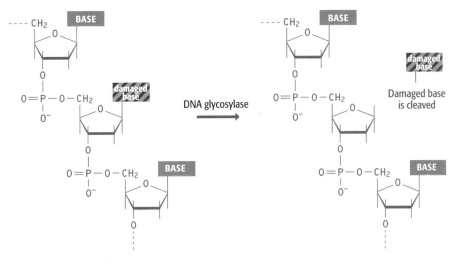

(B) Outline of the pathway

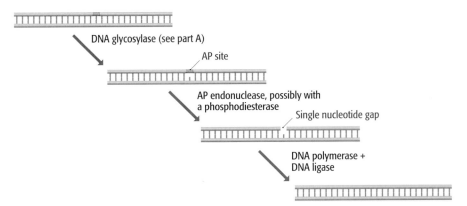

Base excision repairs many types of damaged nucleotide

Base excision is the least complex of the various repair systems that involve removal of one or more nucleotides followed by resynthesis of DNA to span the resulting gap. It is used to repair many modified nucleotides whose bases have suffered relatively minor damage resulting from, for example, exposure to alkylating agents or ionizing radiation (Section 16.1.1). The process is initiated by a **DNA glycosylase** which cleaves the β-*N*-glycosidic bond between a damaged base and the sugar component of the nucleotide (Figure 16.22A). Each DNA glycosylase has a limited specificity (Table 16.3), the specificities of the glycosylases possessed by a cell determining the range of damaged nucleotides that can be repaired by the base excision pathway. Most organisms are able to deal with deaminated bases such as uracil (deaminated cytosine) and hypoxanthine (deaminated adenine), oxidation products such as 5-hydroxycytosine and thymine glycol, and methylated bases such as 3-methyladenine, 7-methylguanine, and 2-methylcytosine. Other DNA glycosylases remove normal bases as part of the mismatch repair system (Section 16.2.3). Most of the DNA glycosylases involved in base excision repair are thought to diffuse along the minor groove of the DNA double helix in search of damaged nucleotides, but some may be associated with the replication enzymes.

A DNA glycosylase removes a damaged base by "flipping" the structure to a position outside the helix and then detaching it from the polynucleotide. This

Table 16.3 Examples of human DNA glycosylases

DNA glycosylase	Specific for
MBD4	Uracil
MPG	Ethenoadenine, hypoxanthine, 3-methyladenine
NTH1	Cytosine glycol, dihydrouracil, formamidopyrimidine, thymine glycol
OGG1	Formamidopyrimidine, 8-oxoguanine
SMUG1	Uracil
TDG	Ethenocytosine, uracil
UNG	Uracil, 5-hydroxyuracil

creates an AP site, or baseless site, which is converted into a single nucleotide gap in the second step of the repair pathway (Figure 16.22B). This step can be carried out in a variety of ways. The standard method makes use of an **AP endonuclease**, such as exonuclease III or endonuclease IV of *E. coli* or APE1 of humans, which cuts the phosphodiester bond on the 5′ side of the AP site. Some AP endonucleases can also remove the sugar from the AP site, this being all that remains of the damaged nucleotide, but others lack this ability and so work in conjunction with a separate **phosphodiesterase**. An alternative pathway for converting the AP site into a gap utilizes the endonuclease activity possessed by some DNA glycosylases, which can make a cut at the 3′ side of the AP site, probably at the same time that the damaged base is removed, followed again by removal of the sugar by a phosphodiesterase.

The single nucleotide gap is filled in by a DNA polymerase, using base pairing with the undamaged base in the other strand of the DNA molecule to ensure that the correct nucleotide is inserted. In *E. coli* the gap is filled in by DNA polymerase I, and in mammals by DNA polymerase β (see Table 15.2). Yeast seems to be unusual in that it uses its main DNA replicating enzyme, DNA polymerase δ, for this purpose. After gap filling, the final phosphodiester bond is put in place by a DNA ligase.

Nucleotide excision repair is used to correct more extensive types of damage

Nucleotide excision repair has a much broader specificity than the base excision system and is able to deal with more extreme forms of damage such as intrastrand cross-links and bases that have become modified by attachment of large chemical groups. It is also able to correct cyclobutyl dimers by a **dark repair** process, providing those organisms that do not have the photoreactivation system (such as humans) with a means of repairing this type of damage.

In nucleotide excision repair, a segment of single-stranded DNA containing the damaged nucleotide(s) is excised and replaced with new DNA. The process is therefore similar to base excision repair except that it is not preceded by selective base removal, and a longer stretch of polynucleotide is cut out. The best-studied example of nucleotide excision repair is the **short patch** process of *E. coli*, so-called because the region of polynucleotide that is excised and subsequently "patched" is relatively short, usually 12 nucleotides in length.

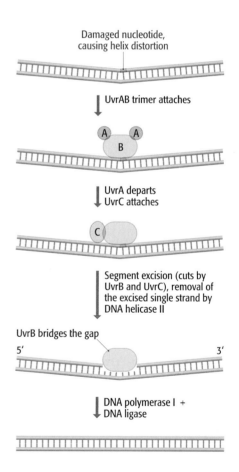

Figure 16.23 Short patch nucleotide excision repair in *Escherichia coli*. The damaged nucleotide is shown distorting the helix because this is thought to be one of the recognition signals for the UvrAB trimer that initiates the short patch process. See the text for details of the events occurring during the repair pathway. Abbreviations: A, UvrA; B, UvrB; C, UvrC.

Short patch repair is initiated by a multienzyme complex called the **UvrABC endonuclease**, sometimes also referred to as the "excinuclease." In the first stage of the process a trimer comprising two UvrA proteins and one copy of UvrB attaches to the DNA at the damaged site. How the site is recognized is not known but the broad specificity of the process indicates that individual types of damage are not directly detected and that the complex must search for a more general attribute of DNA damage such as distortion of the double helix. UvrA may be the part of the complex most involved in damage location because it dissociates once the site has been found and plays no further part in the repair process. Departure of UvrA allows UvrC to bind (Figure 16.23), forming a UvrBC dimer that cuts the polynucleotide either side of the damaged site. The first cut is made by UvrB at the fifth phosphodiester bond downstream of the damaged nucleotide, and the second cut is made by UvrC at the eighth phosphodiester bond upstream, resulting in the 12-nucleotide excision, although there is some variability, especially in the position of the UvrB cut site. The excised segment is then removed, usually as an intact oligonucleotide, by DNA helicase II, which presumably detaches the segment by breaking the base pairs holding it to the second strand. UvrC also detaches at this stage, but UvrB remains in place and bridges the gap produced by the excision. The bound UvrB is thought to prevent the single-stranded region that has been exposed from base-pairing with itself, but the role of UvrB could be to prevent this strand from becoming damaged, or possibly to direct the DNA polymerase to the site that needs to be repaired. As in base excision repair, the gap is filled in by DNA polymerase I and the last phosphodiester bond is synthesized by DNA ligase.

E. coli also has a **long patch** nucleotide excision repair system that involves Uvr proteins but differs in that the piece of DNA that is excised can be anything up to 2 kb in length. Long patch repair has been less well studied and the process is not understood in detail, but it is presumed to work on more extensive forms of damage, possibly regions where groups of nucleotides, rather than just individual ones, have become modified. The eukaryotic nucleotide excision repair process is also called "long patch" but results in replacement of only 24–29 nucleotides of DNA. In fact, there is no "short patch" system in eukaryotes and the name "long patch" is used to distinguish the process from base excision repair. The system is more complex than in *E. coli* and the relevant enzymes do not seem to be homologs of the Uvr proteins. In humans, at least 16 proteins are involved, with the downstream cut being made at the same position as in *E. coli*—the fifth phosphodiester bond—but with a more distant upstream cut, resulting in the longer excision. Both cuts are made by endonucleases that attack single-stranded DNA specifically at its junction with a double-stranded region, indicating that before the cuts are made the DNA around the damage site has been melted, presumably by a helicase (Figure 16.24). This activity is provided at least in part by TFIIH, one of the components of the RNA polymerase II initiation complex (see Table 11.5). At first it was assumed that TFIIH simply had a dual role in the cell, functioning separately in both transcription and repair, but now it is thought that there is a more direct link between the two processes. This view is supported by the discovery of **transcription-coupled repair**, which repairs some forms of damage in the template strands of genes that are being actively transcribed. The first type of transcription-coupled repair to be discovered was a modified version of nucleotide excision, but it is now known that base excision repair is also coupled with transcription. These discoveries

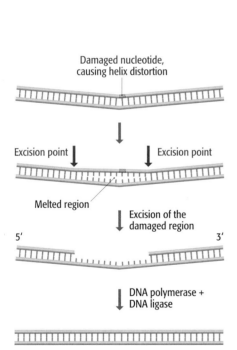

Damaged nucleotide, causing helix distortion

Excision point   Excision point

Melted region

Excision of the damaged region

5′   3′

DNA polymerase + DNA ligase

Figure 16.24 Outline of the events involved during nucleotide excision repair in eukaryotes. The endonucleases that remove the damaged region make cuts specifically at the junction between single-stranded and double-stranded regions of a DNA molecule. The DNA is therefore thought to melt either side of the damaged nucleotide, as shown in the diagram, possibly as a result of the helicase activity of TFIIH.

do not imply that nontranscribed regions of the genome are not repaired. The excision repair processes protect the entire genome from damage, but it is entirely logical that special mechanisms should exist for directing the processes at genes that are being transcribed. The template strands of these genes contain the genome's biological information and maintaining their integrity should be the highest priority for the repair systems.

16.2.3 Mismatch repair: correcting errors of replication

Each of the repair systems that we have looked at so far—direct, base excision, and nucleotide excision repair—recognize and act upon DNA damage caused by mutagens. This means that they search for abnormal chemical structures such as modified nucleotides, cyclobutyl dimers, and intrastrand cross-links. They cannot, however, correct mismatches resulting from errors in replication because the mismatched nucleotide is not abnormal in any way, it is simply an A, C, G, or T that has been inserted at the wrong position. As these nucleotides look exactly like any other nucleotide, the mismatch repair system that corrects replication errors has to detect not the mismatched nucleotide itself but the absence of base pairing between the parent and daughter strands. Once it has found a mismatch, the repair system excises part of the daughter polynucleotide and fills in the gap, in a manner similar to base and nucleotide excision repair.

The scheme described above leaves one important question unanswered. The repair must be made in the daughter polynucleotide because it is in this newly synthesized strand that the error has occurred: the parent polynucleotide has the correct sequence. How does the repair process know which strand is which? In *E. coli* the answer is that the daughter strand is, at this stage, undermethylated and can therefore be distinguished from the parent polynucleotide, which has a full complement of methyl groups. *E. coli* DNA is methylated because of the activities of the **DNA adenine methylase** (**Dam**), which converts adenines to 6-methyladenines in the sequence 5′–GATC–3′, and the **DNA cytosine methylase** (**Dcm**), which converts cytosines to 5-methylcytosines in 5′–CCAGG–3′ and 5′–CCTGG–3′. These methylations are not mutagenic, the modified nucleotides having the same base-pairing properties as the unmodified versions. There is a delay between DNA replication and methylation of the daughter strand, and it is during this window of opportunity that the repair system scans the DNA for mismatches and makes the required corrections in the undermethylated daughter strand (Figure 16.25).

E. coli has at least three mismatch repair systems, called "long patch," "short patch," and "very short patch," the names indicating the relative lengths of the excised and resynthesized segments of DNA. The long patch system replaces up to a kilobase or more of DNA and requires the MutH, MutL, and MutS proteins, as well as the DNA helicase II that we met during nucleotide excision repair. MutS recognizes the mismatch and MutH distinguishes the two strands by binding to unmethylated 5′–GATC–3′ sequences (Figure 16.26). The role of MutL is unclear but it might coordinate the activities of the other two proteins so that MutH binds to unmethylated 5′–GATC–3′ sequences only in the vicinity of mismatch sites recognized by MutS. After binding, MutH cuts the phosphodiester bond immediately upstream of the G in the methylation sequence and DNA helicase II detaches the single strand. There does not appear to be an enzyme that cuts the strand downstream of the mismatch; instead the detached single-stranded region is degraded by an

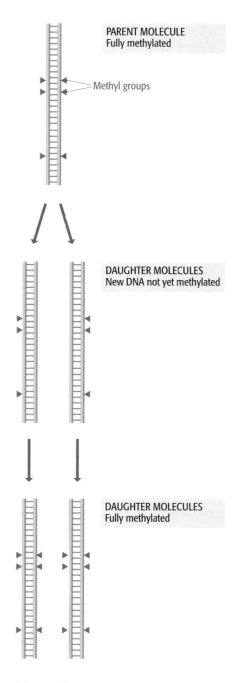

Figure 16.25 Methylation of newly synthesized DNA in *Escherichia coli* does not occur immediately after replication, providing a window of opportunity for the mismatch repair proteins to recognize the daughter strands and correct replication errors.

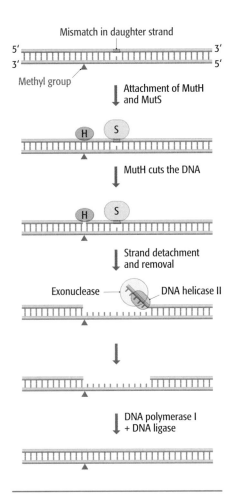

Figure 16.26 **Long patch mismatch repair** *in Escherichia coli.* Abbreviations: H, MutH; S, MutS.

exonuclease that follows the helicase and continues beyond the mismatch site. The gap is then filled in by DNA polymerase I and DNA ligase. Similar events are thought to occur during short patch and very short patch mismatch repair, the difference being the specificities of the proteins that recognize the mismatch. The short patch system, which results in excision of a segment less than 10 nucleotides in length, begins when MutY recognizes an A–G or A–C mismatch, and the very short repair system corrects G–T mismatches which are recognized by the Vsr endonuclease.

Eukaryotes have homologs of the *E. coli* MutS and MutL proteins and their mismatch repair processes probably work in a similar way. The one difference is the absence of a homolog of MutH, which suggests that methylation might not be the method used to distinguish between the parent and daughter polynucleotides. Methylation has been implicated in mismatch repair in mammalian cells, but the DNA of some eukaryotes, including fruit flies and yeast, is not extensively methylated. It is therefore thought that in these organisms a different method must be used to identify the daughter strand. Possibilities include an association between the repair enzymes and the replication complex, so that repair is coupled with DNA synthesis, or use of single-strand binding proteins that mark the parent strand.

16.2.4 Repair of DNA breaks

A single-strand break in a double-stranded DNA molecule, such as is produced by some types of oxidative damage, does not present the cell with a critical problem as the double helix retains its overall intactness. The exposed single strand is coated with PARP1 proteins, which protect this intact strand from breaking and prevent it from participating in unwanted recombination events. The break is then filled in by the enzymes involved in the excision repair pathways (Figure 16.27).

A double-strand break is more serious because this converts the original double helix into two separate fragments which have to be brought back together again in order for the break to be repaired. The two broken ends must be protected from further degradation which could result in a deletion mutation appearing at the repaired break point. The repair processes must also ensure that the correct ends are joined: if there are two broken chromosomes in the nucleus, then the correct pairs must be brought together so that the original structures are restored. Experimental studies of mouse cells indicate that achieving this outcome is difficult and if two chromosomes are broken then misrepair resulting in hybrid structures occurs relatively frequently. Even if only one chromosome is broken, there is still a possibility that a natural chromosome end could be confused as a break and an incorrect repair made. This type of error is not unknown, despite the presence of special telomere-binding proteins that mark the natural ends of chromosomes (Section 7.1.2).

Double-strand breaks are generated by exposure to ionizing radiation and some chemical mutagens, and breakages can also occur during DNA replication. Most, if not all, organisms have two distinct pathways for repair of double-strand breaks. The first involves homologous recombination and we will therefore postpone our examination of it until after we have dealt with the basic pathway for homologous recombination in Chapter 17. The second system is called **nonhomologous end-joining** (**NHEJ**). Progress in understanding NHEJ has been stimulated by studies of mutant human cell lines, which have resulted in the identification of various sets of genes involved in the

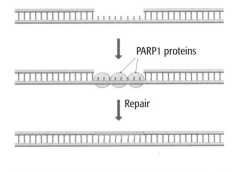

Figure 16.27 **Single-strand break repair.**

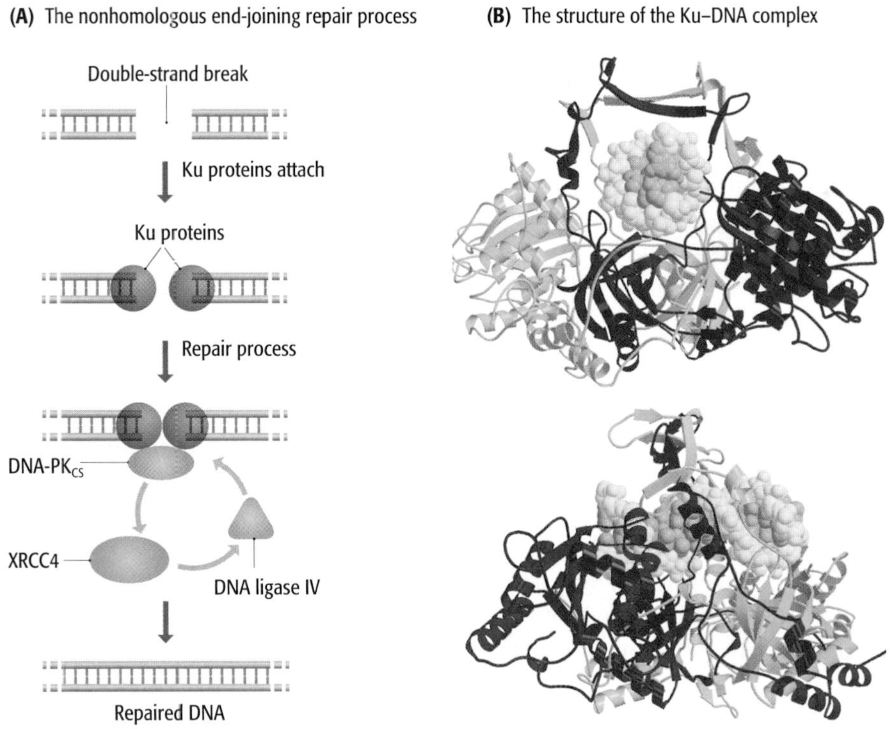

(A) The nonhomologous end-joining repair process

(B) The structure of the Ku–DNA complex

Figure 16.28 **Nonhomologous end-joining (NHEJ) in humans.** (A) The repair process. Additional proteins not shown in the diagram are also involved in NHEJ. These include the protein kinases ATM and ATR (Section 15.3.2), whose main role may be to signal to the cell the fact that a double-strand break has occurred and the cell cycle should be arrested until the break is repaired. If the cell enters mitosis with a broken chromosome then part of that chromosome will be lost, because only one of the fragments will contain a centromere, and centromeres are essential for distribution of chromosomes to the daughter nuclei during anaphase (see Figure 3.15). (B) Structure of a Ku–DNA complex. On the left is the view looking down onto the broken end of the DNA double helix, and on the right is the side view, with the broken end of the DNA molecule on the left. Ku is a heterodimer made up of the Ku70 and Ku80 subunits, the numbers indicating the molecular masses in kDa. Ku70 is colored red and Ku80 yellow. The DNA molecule is shown in gray. Reprinted with permission from Walker *et al.* (2001) *Nature,* **412,** 607–614.

process. These genes specify a multicomponent protein complex that directs a DNA ligase to the break (Figure 16.28A). The complex includes two copies of the Ku protein, one copy attaching to each broken DNA end. Ku proteins can only bind to cut ends, not to the internal regions of a DNA molecule, because the DNA molecule must fit into a loop formed by the association between the two subunits that make up each Ku protein (Figure 16.28B). Individual Ku proteins have an affinity for one another, which means that the two broken ends of the DNA molecule are brought into proximity. Ku binds to the DNA in association with the DNA-PK$_{CS}$ protein kinase, which activates a third protein, XRCC4, which interacts with the mammalian DNA ligase IV, directing this repair protein to the double-strand break.

NHEJ was originally thought to be restricted to eukaryotes, but searches of the protein databases have uncovered bacterial homologs of the mammalian Ku proteins, and experimental studies have indicated that these act in conjunction with bacterial ligases in a simplified version of the double-strand break repair process.

16.2.5 Bypassing DNA damage during genome replication

If a region of the genome has suffered extensive damage then it is conceivable that the repair processes will be overwhelmed. The cell then faces a stark choice between dying or attempting to replicate the damaged region even though this replication may be error-prone and result in mutated daughter molecules. When faced with this choice *E. coli* cells invariably take the second option, by inducing one of several emergency procedures for bypassing sites of major damage.

The SOS response is an emergency measure for coping with a damaged genome

The best studied of these bypass processes occurs as part of the **SOS response**, which enables an *E. coli* cell to replicate its DNA even though the template

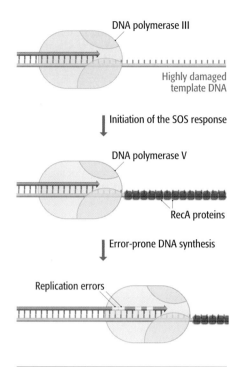

DNA polymerase III

Highly damaged
template DNA

Initiation of the SOS response

DNA polymerase V

RecA proteins

Error-prone DNA synthesis

Replication errors

Figure 16.29 The SOS response of *Escherichia coli.*

polynucleotides contain AP sites and/or cyclobutyl dimers and other photo-products resulting from exposure to chemical mutagens or UV radiation that would normally block, or at least delay, the replication complex. Bypass of these sites requires construction of a **mutasome**, comprising the UmuD$'_2$C complex (also called DNA polymerase V, a trimer made up of two UmuD$'$ proteins and one copy of UmuC) and several copies of the **RecA protein**. The latter is primarily a single-strand binding protein that plays many roles in DNA repair and recombination. In this bypass system, RecA coats the damaged polynucleotide strands, enabling the UmuD$'_2$C complex to displace DNA polymerase III and carry out error-prone DNA synthesis until the damaged region has been passed and DNA polymerase III can take over once again (Figure 16.29).

As well as acting as a single-strand binding protein that facilitates the mutasome bypass process, RecA also has a second function as an activator of the overall SOS response. The protein is stimulated by chemical signals (not yet identified) that indicate the presence of extensive DNA damage. In response, RecA cleaves a number of target proteins, directly or indirectly, including UmuD, the cleavage converting this protein into its active form UmuD$'$, and initiating the mutasome repair process. RecA also cleaves a repressor protein called LexA, switching on or increasing the expression of a number of genes normally repressed by LexA, these including both the *recA* gene itself (leading to a 50-fold increase in RecA synthesis) and several other genes whose products are involved in DNA repair pathways. RecA also cleaves the *c*I repressor of λ bacteriophage, so if an integrated λ prophage is present in the genome it can excise and leave the sinking ship (Section 14.3.1).

The SOS response is primarily looked on as the last best chance that the bacterium has to replicate its DNA and hence survive under adverse conditions. However, the price of survival is an increased mutation rate because the mutasome does not repair damage, it simply allows a damaged region of a polynucleotide to be replicated. When it encounters a damaged position in the template DNA, the polymerase selects a nucleotide more or less at random, although with some preference for placing an A opposite an AP site: in effect the error rate of the replication process increases. It has been suggested that this increased mutation rate is the purpose of the SOS response, mutation being in some way an advantageous response to DNA damage, but this idea remains controversial.

For some time, the SOS response was thought to be the only damage-bypass process in bacteria, but we now appreciate that at least two other *E. coli* polymerases act in a similar way, although with different types of damage. These are DNA polymerase II, which can bypass nucleotides bound to mutagenic chemicals such as *N*-2-acetylaminofluorene, and DNA polymerase IV (also called DinB), which can replicate through a region of template DNA in which the two parent polynucleotides have become misaligned. Bypass polymerases have also been discovered in eukaryotic cells. These include DNA polymerases ε and η, which can bypass cyclobutyl dimers, and DNA polymerases ι and ζ, which work together to replicate through photoproducts and AP sites.

16.2.6 Defects in DNA repair underlie human diseases, including cancers

The importance of DNA repair is emphasized by the number and severity of inherited human diseases that have been linked with defects in one of the repair processes. One of the best characterized of these is xeroderma pigmentosum, which results from a mutation in any one of several genes for proteins involved in nucleotide excision repair. Nucleotide excision is the only way in which human cells can repair cyclobutyl dimers and other photoproducts, so it is no surprise that the symptoms of xeroderma pigmentosum include hypersensitivity to UV radiation, patients suffering more mutations than normal on exposure to sunlight, and often suffering from skin cancer. Trichothiodystrophy is also caused by defects in nucleotide excision repair, but this is a more complex disorder which, although not involving cancer, usually includes problems with both the skin and nervous system.

A few diseases have been linked with defects in the transcription-coupled component of nucleotide excision repair. These include breast and ovarian cancers, the *BRCA1* gene that confers susceptibility to these cancers coding for a protein that has been implicated, at least indirectly, with transcription-coupled repair, and Cockayne syndrome, a complex disease manifested by growth and neurologic disorders. A deficiency in transcription-coupled repair has also been identified in humans suffering from the cancer-susceptibility syndrome called HNPCC (hereditary nonpolyposis colorectal cancer), although this disease was originally identified as a defect in mismatch repair. Ataxia telangiectasia, the symptoms of which include sensitivity to ionizing radiation, results from defects in the *ATX* gene, which is involved in the damage-detection process. Other diseases that are associated with a breakdown in DNA repair are Bloom's and Werner's syndromes, which are caused by inactivation of a DNA helicase that may have a role in NHEJ, Fanconi's anemia, which confers sensitivity to chemicals that cause cross-links in DNA but whose biochemical basis is not yet known, and spinocerebellar ataxia, which results from defects in the pathway used to repair single-strand breaks.

Summary

A mutation is a change in the nucleotide sequence of a DNA molecule, either a point mutation which affects just a single nucleotide, or an insertion or deletion of one or more adjacent nucleotides. Mutations can result from errors made during DNA replication, even though DNA polymerases have nucleotide-selection processes and proofreading activities that maintain a high degree of accuracy. These checking mechanisms can, however, be evaded if an unusual tautomeric version of a nucleotide is present in the template. A second type of replication error, called slippage, can lead to insertion or deletion mutations. There are also many types of chemical or physical agent that can cause mutations. Some compounds act as base analogs and cause mutations after being mistaken for genuine nucleotides by the replication machinery. Deaminating and alkylating agents directly attack DNA molecules, and intercalating agents such as ethidium bromide slide between base pairs, causing insertions and deletions when the helix is replicated. UV radiation causes adjacent nucleotides to link together into dimers, and ionizing radiation and heat cause various types of damage. Within a gene, a point mutation might have no effect on the coding properties because of the degeneracy of the genetic code, but some mutations will change the meaning of a codon, so that the codon specifies a different amino acid or perhaps becomes a termination codon. Insertions and deletions might cause frameshifts that lead to premature termination or to readthrough of the correct termination codon. Any of these mutations could lead to a loss-of-function mutation, and a few might cause a gain-of-function mutation, possibly leading to cancer. In bacteria, a mutation might give rise to an auxotroph, a cell that requires a growth supplement not needed by the wild type, or to a mutant that displays resistance to an antibiotic or other inhibitor. The possibility that cells can increase their mutation rates under some circumstances, or programme mutations in response to environmental changes, is being hotly debated. All cells possess DNA repair processes that enable many mutations to be corrected. Direct repair systems are uncommon but the few that are known correct some types of base damage including removal of UV-induced nucleotide dimers. Excision repair processes involve excision of a segment of a polynucleotide containing a damaged site, followed by resynthesis of the correct nucleotide sequence by a DNA polymerase. Mismatch repair corrects errors of replication, again by excising a stretch of single-stranded DNA containing the mutation and repairing the resulting gap. Nonhomologous end-joining is used to mend double-strand breaks. There are also processes for bypassing sites of DNA damage during replication, many of these acting as an emergency system for rescuing a genome that has become heavily mutated.

Multiple Choice Questions

16.1.* Which of the following statements is INCORRECT?

    a. A mutation is a change in the nucleotide sequence of a short region of the genome.

    b. All mutations are caused by environmental agents.

    c. Many mutations can be repaired.

    d. Some mutations are caused by errors in replication.

16.2. Which of the following results in replacement of one nucleotide by another?

    a. Deletion mutation.

    b. Insertion mutation.

    c. Point mutation.

    d. Translocation.

16.3.* Spontaneous mutations arise from which of the following?

    a. Chemical mutagens.

    b. Errors in DNA replication.

    c. Heat.

    d. Radiation.

16.4. How does proofreading increase the accuracy of genome replication?

    a. The DNA polymerase discriminates against incorrect nucleotides when these first become bound to the enzyme.

    b. The 5'→3' exonuclease activity of the DNA polymerase removes a nucleotide that has been incorrectly added to the end of the polynucleotide being synthesized.

    c. When the 3'-terminal nucleotide is not base-paired to the template DNA, it is removed by the exonuclease activity of the DNA polymerase.

    d. All of the above.

16.5.* Which of the following is a common cause of errors in genome replication?

    a. Formation of G-U base pairs at the replication fork.

    b. Replication of regions of the genome that are being transcribed.

    c. A tautomeric shift within a nucleotide in the template DNA.

    d. The presence of nucleosomes attached to the DNA being replicated.

16.6. Variations in which types of repeat sequence commonly arise by replication slippage?

    a. Microsatellites.

    b. Minisatellites.

    c. Retroposons.

    d. DNA transposons.

16.7.* Which types of chemical mutagen are incorporated into the genome by the DNA polymerase during genome replication?

    a. Alkylating agents.

    b. Base analogs.

    c. Deaminating agents.

    d. Intercalating agents.

16.8. Ultraviolet radiation causes which type of DNA damage?

    a. Cyclobutyl dimers.

    b. AP (apurinic/apyrimidinic) sites.

    c. Base deamination.

    d. Base tautomerization.

16.9.* Which type of mutation converts a codon specifying an amino acid into a termination codon?

    a. Nonsense.

    b. Nonsynonymous.

    c. Readthrough.

    d. Synonymous.

16.10. What is an auxotrophic mutant?

    a. A mutant that can grow on minimal medium.

    b. A mutant that requires an antibiotic for growth.

    c. A mutant that requires nutrients that are not required by the wild-type organism.

    d. A mutant that can grow at restrictive temperatures.

16.11.* Which of the following types of DNA damage CANNOT be repaired by *E. coli* using a direct repair system?

    a. Alkylated bases.

    b. AP sites.

    c. Cyclobutyl dimers.

    d. Missing phosphodiester bonds.

16.12. Which of the following describes nucleotide excision repair?

    a. A region of double-stranded DNA containing damaged nucleotides is removed and replaced with new DNA.

    b. A single damaged nucleotide is removed and replaced with a new nucleotide.

    c. A single damaged base is removed and replaced with a new base.

    d. A region of single-stranded DNA containing damaged nucleotides is removed and replaced with new DNA.

continued …

Multiple Choice Questions (continued) *Answers to odd-numbered questions can be found in the Appendix

16.13.* Which of the following is a feature of mismatch repair?

   a. Modified nucleotides are recognized.

   b. Cyclobutyl dimers are removed.

   c. The parent and daughter strands of newly replicated DNA are distinguished.

   d. The correct reading frame is identified.

16.14. How are the parent and daughter strands of newly replicated DNA distinguished in *E. coli*?

   a. The daughter strands are methylated as soon as they are synthesized.

   b. The daughter strands are not immediately methylated.

   c. The daughter strands are not immediately attached to histone proteins.

   d. The daughter strands contain ribonucleotides from the RNA primers used to initiate DNA synthesis.

16.15.* How does an *E. coli* cell attempt to replicate damaged DNA during the SOS response?

   a. Regions of damaged DNA are deleted from the genome.

   b. Nucleotides are incorporated at random at damaged sites.

   c. All DNA synthesis is stopped until the damage can be repaired.

   d. Messenger RNA is converted into DNA which is inserted at the damaged sites by recombination.

Short Answer Questions   *Answers to odd-numbered questions can be found in the Appendix

16.1.* What are the mechanisms by which mutations arise in a genome?

16.2. What are the differences between prereplicative and postreplicative DNA repair processes?

16.3.* How does a DNA polymerase select the correct nucleotide during DNA synthesis?

16.4. How does the base analog 2-aminopurine produce mutations in DNA?

16.5.* How does heat affect the structure of DNA? How common is heat-induced damage to DNA and what are the effects of this damage?

16.6. How can mutations in noncoding DNA sequences affect the expression of a genome?

16.7.* In humans, familial hypercholesterolemia occurs from a loss-of-function mutation in a gene encoding the LDL receptor-binding domain of apolipoprotein B-100 and is inherited as a dominant trait. Why is this?

16.8. How could conditional-lethal mutations be used to identify or characterize essential gene products in a microorganism?

16.9.* What is the basis to hypermutation of the V segments of the human immunoglobulin genes?

16.10. What are the steps that occur during base excision repair?

16.11.* What is the process by which double-strand breaks in DNA are repaired by the nonhomologous end-joining (NHEJ) system?

16.12. What is the role of the LexA protein in the SOS response of *E. coli*?

In-depth Problems

*Guidance to odd-numbered questions can be found in the Appendix

16.1.* Explain why a purine-to-purine or pyrimidine-to-pyrimidine point mutation is called a transition, whereas a purine-to-pyrimidine (or vice versa) change is called a transversion.

16.2. What would be the anticipated ratio of transitions to transversions in a large number of mutations?

16.3.* Explore the current knowledge concerning trinucleotide repeat expansion diseases, including hypotheses that attempt to explain why triplet expansion in these genes leads to a disease.

16.4. Evaluate the evidence for programmed mutations.

16.5.* The bacterium *Deinococcus radiodurans* is highly resistant to radiation and to other physical and chemical mutagens. Discuss how these special properties of *D. radiodurans* are reflected in its genome sequence.

Figure Tests

*Answers to odd-numbered questions can be found in the Appendix

16.1.* What type of mutation (see the red-colored nucleotides) is depicted in this figure? What is the cause of this mutation?

16.2. Discuss the potential impact of the different types of mutations depicted in the figure.

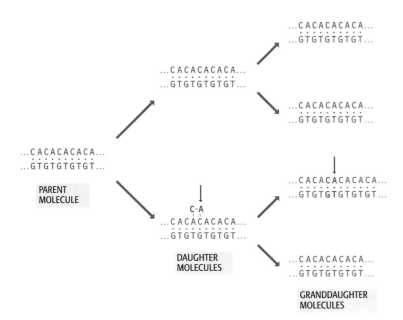

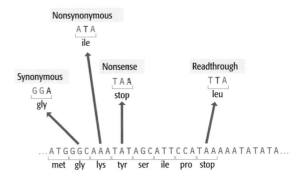

continued ...

Figure Tests (continued)

*Answers to odd-numbered questions can be found in the Appendix

16.3.* What is the name of the enzyme that removes the damaged base from a DNA molecule? Which pathway is used to repair the DNA after the base has been removed?

16.4. Which mode of DNA replication is depicted in this figure? When does this occur?

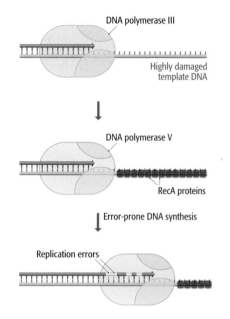

Further Reading

Causes of mutations

Kunkel, T.A. and Bebenek, K. (2000) DNA replication fidelity. *Annu. Rev. Biochem.* **69:** 497–529. *Covers the processes that ensure that the minimum number of errors are made during DNA replication.*

Trinucleotide expansion diseases

Ashley, C.T. and Warren, S.T. (1995) Trinucleotide repeat expansion and human disease. *Annu. Rev. Genet.* **29:** 703–728.

Perutz, M.F. (1999) Glutamine repeats and neurodegenerative diseases: molecular aspects. *Trends Biochem. Sci.* **24:** 58–63.

Sutherland, G.R., Baker, E. and Richards, R.I. (1998) Fragile sites still breaking. *Trends Genet.* **14:** 501–506.

Hypermutation and programmed mutations

Andersson, D.I., Slechta, E.S. and Roth, J.R. (1998) Evidence that gene amplification underlies adaptive mutability of the bacterial *lac* operon. *Science* **282:** 1133–1135.

Cairns, J., Overbaugh, J. and Miller, S. (1988) The origin of mutants. *Nature* **335:** 142–145. *The original experiments suggesting that bacteria can program mutations.*

Chicurel, M. (2001) Can organisms speed their own mutation? *Science* **292:** 1824–1827.

Nola, J.D. and Neuberger, M.S. (2002) Altering the pathway of immunoglobulin hypermutation by inhibiting uracil-DNA glycosylase. *Nature* **419:** 43–48.

Excision repair

Lehmann, A.R. (1995) Nucleotide excision repair and the link with transcription. *Trends Biochem. Sci.* **20:** 402–405.

Seeberg, E., Eide, L. and Bjørås, M. (1995) The base excision repair pathway. *Trends Biochem. Sci.* **20:** 391–397.

Mismatch repair

Kolodner, R.D. (1995) Mismatch repair: mechanisms and relationship to cancer susceptibility. *Trends Biochem. Sci.* **20:** 397–401.

Kunkel, T.A. and Erie, D.A. (2005) DNA mismatch repair. *Annu. Rev. Biochem.* **74:** 681–710.

Shannon, M. and Weigert, M. (1998) Fixing mismatches. *Science* **279:** 1159–1160.

Repair of DNA breaks

Critchlow, S.E. and Jackson, S.P. (1998) DNA end-joining: from yeast to man. *Trends Biochem. Sci.* **23:** 394–398.

Walker, J.R., Corpina, R.A. and Goldberg, J. (2001) Structure of the Ku heterodimer bound to DNA and its implications for double-strand break repair. *Nature* **412:** 607–614.

Wilson, T.E., Topper, L.M. and Palmbos, P.L. (2003) Non-homologous end-joining: bacteria join the chromosome breakdance. *Trends Biochem. Sci.* **28:** 62–66. *Evidence for NHEJ in bacteria.*

Bypassing DNA damage

Hanaoka, F. (2001) SOS polymerases. *Nature* **409:** 33–34.

Johnson, R.E., Prakash, S. and Prakash, L. (1999) Efficient bypass of a thymine-thymine dimer by yeast DNA polymerase, Polε. *Science* **283:** 1001–1004.

Johnson, R.E., Washington, M.T., Haracska, L., Prakash, S. and Prakash, L. (2000) Eukaryotic polymerases ι and ζ act sequentially to bypass DNA lesions. *Nature* **406:** 1015–1019.

Sutton, M.D., Smith, B.T., Godoy, V.G. and Walker, G.C. (2000) The SOS response: recent insights into *umuDC*-dependent mutagenesis and DNA damage tolerance. *Annu. Rev. Genet.* **34:** 479–497.

Repair and disease

Gowen, L.C., Avrutskaya, A.V., Latour, A.M., Koller, B.H. and Leadon, S.A. (1998) BRCA1 required for transcription-coupled repair of oxidative DNA damage. *Science* **281:** 1009–1012.

Hanawalt, P.C. (2000) The bases for Cockayne syndrome. *Nature* **405:** 415–416.

Recombination

When you have read Chapter 17, you should be able to:

Describe and distinguish between the various models for homologous recombination.

Give a detailed account of the RecBCD pathway for homologous recombination in *Escherichia coli*.

Summarize the key features of the RecE and RecF pathways of *E. coli*, and give the names and functions of those eukaryotic proteins thought to be involved in homologous recombination.

Describe how homologous recombination is used to repair DNA breaks.

Give a detailed account of the role of site-specific recombination in the lysogenic infection cycle of bacteriophage λ, and explain why site-specific recombination is important in construction of genetically modified crops.

Illustrate the Shapiro models for conservative and replicative transposition of a DNA transposon.

Describe the transposition pathway for an LTR retroelement.

Briefly discuss the possible ways by which cells minimize the harmful effects of transposition.

Recombination is the term originally used by geneticists to describe the outcome of crossing-over between pairs of homologous chromosomes during meiosis. Crossing-over results in daughter chromosomes that have different combinations of alleles compared with their parent chromosomes (Section 3.2.3). In the 1960s, models were proposed for the molecular events that underlie crossing-over, and it was realized that a key part of molecular recombination is the breakage and subsequent rejoining of DNA molecules. Biologists now use "recombination" to refer to a variety of processes that involve the breakage and reunion of polynucleotides. These include:

- **Homologous recombination**, also called **general** (or generalized) **recombination**, which occurs between segments of DNA molecules that share extensive sequence homology. These segments might be present on

Figure 17.1 Three different types of recombination event.

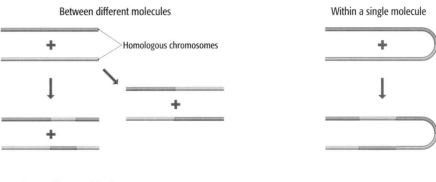

(A) Homologous recombination

Between different molecules   Within a single molecule

+   Homologous chromosomes   +

+   +

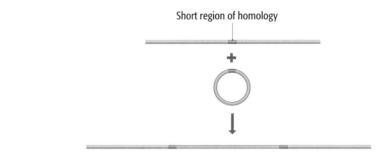

(B) Site-specific recombination

Short region of homology

+

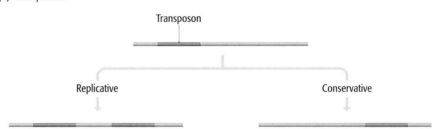

(C) Transposition

Transposon

Replicative   Conservative

different chromosomes, or might be two parts of a single chromosome (Figure 17.1A). Homologous recombination is responsible for crossing-over during meiosis, and was initially studied in this context, but we now believe that its primary cellular role is in DNA repair.

- **Site-specific recombination**, which occurs between DNA molecules that have only short regions of sequence similarity, possibly just a few base pairs (Figure 17.1B). Site-specific recombination is responsible for the insertion of phage genomes, such as that of λ, into bacterial chromosomes.

- **Transposition**, which results in the transfer of a segment of DNA from one position in the genome to another (Figure 17.1C).

Various other events that we have studied, including mating-type switching in yeast (see Figure 14.16) and construction of immunoglobulin genes (see Figure 14.19), are also the results of recombination.

Without recombination, genomes would be relatively static structures, undergoing very little change. The gradual accumulation of mutations over a long period of time would result in small-scale alterations in the nucleotide sequence of the genome, but more extensive restructuring, which is the role of recombination, would not occur, and the evolutionary potential of the genome would be severely restricted.

17.1 Homologous Recombination

The study of homologous recombination has presented two significant challenges for molecular biologists, neither of which has yet been fully met. The first challenge has been to describe the series of interactions, involving breakage and reunion of polynucleotides, that occur during recombination. The models for homologous recombination that have resulted from this work are described in Section 17.1.1. The second challenge relates to the fact that recombination is a cellular process which, like other cellular processes involving DNA (e.g., transcription and replication), is carried out and regulated by enzymes and other proteins. Biochemical studies have revealed a series of related recombination pathways, which are outlined in Section 17.1.2. Work on these pathways led to the realization that homologous recombination underlies several important types of DNA repair, this repair function probably being more important to the cell (especially bacterial cells) than the capacity that homologous recombination provides for crossing-over between chromosomes. We will examine these repair processes in Section 17.1.3.

17.1.1 Models for homologous recombination

Many of the breakthroughs in understanding homologous recombination were made by Robin Holliday, Matthew Meselson, and their colleagues in the 1960s and 1970s. This work resulted in a series of models that showed how breakage and reunion of DNA molecules could lead to the exchange of chromosome segments known to occur during crossing-over. We will therefore begin our study of homologous recombination by examining these models.

The Holliday and Meselson–Radding models for homologous recombination

These models describe recombination between two homologous, double-stranded molecules, ones with identical or nearly identical sequences. The central feature of these models is formation of a **heteroduplex** resulting from the exchange of polynucleotide segments between the two homologous molecules (Figure 17.2). The heteroduplex is initially stabilized by base-pairing between each transferred strand and the intact polynucleotides of the recipient molecules, this base pairing being possible because of the sequence similarity between the two molecules. Subsequently the gaps are sealed by DNA ligase, giving a **Holliday structure**. This structure is dynamic, **branch migration** resulting in exchange of longer segments of DNA if the two helices rotate in the same direction.

Separation, or **resolution**, of the Holliday structure back into individual, double-stranded molecules occurs by cleavage across the branch point. This is the key to the entire process because the cut can be made in either of two orientations, as becomes apparent when the three-dimensional configuration, or **chi form**, of the Holliday structure is examined (see Figure 17.2). These two cuts have very different results. If the cut is made left–right across the chi form as drawn in Figure 17.2, then all that happens is that a short segment of polynucleotide, corresponding to the distance migrated by the branch of the Holliday structure, is transferred between the two molecules. On the other

Figure 17.2 The Holliday model for homologous recombination.

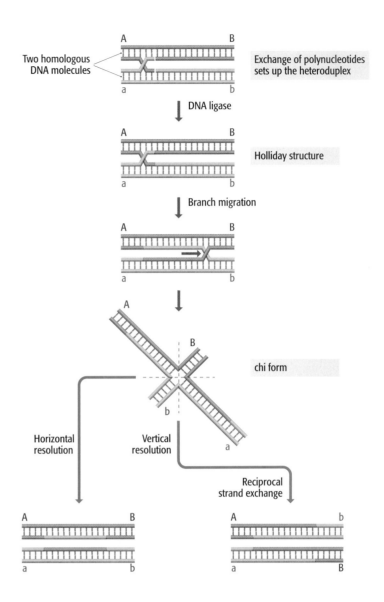

hand, an up–down cut results in **reciprocal strand exchange**, double-stranded DNA being transferred between the two molecules so that the end of one molecule is exchanged for the end of the other molecule. This is the DNA transfer seen in crossing-over.

So far we have ignored one aspect of this model. This is the way in which the two double-stranded molecules interact at the beginning of the process to produce the heteroduplex. In Holliday's original scheme, the two molecules lined up with one another and single-strand nicks appeared at equivalent positions in each helix. This produced free, single-stranded ends that could be exchanged, resulting in the heteroduplex (Figure 17.3A). This feature of the model was criticized because no mechanism could be proposed for ensuring that the nicks occurred at precisely the same position on each molecule. The Meselson–Radding modification proposes a more satisfactory scheme whereby a single-strand nick occurs in just one of the double helices, the free end that is produced "invading" the unbroken double helix at the homologous position and displacing one of its strands, forming a **D-loop** (Figure 17.3B). Subsequent cleavage of the displaced strand at the junction between its single-stranded and base-paired regions produces the heteroduplex.

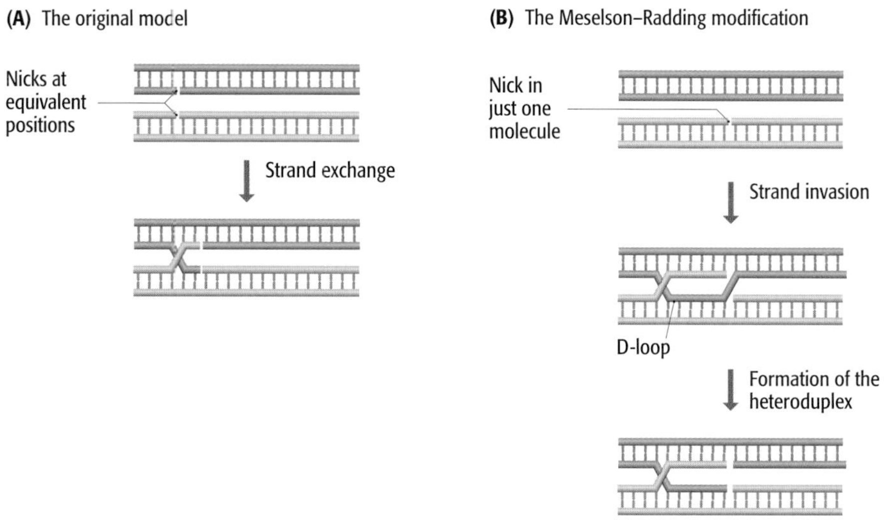

(A) The original model

Nicks at equivalent positions

↓ Strand exchange

(B) The Meselson–Radding modification

Nick in just one molecule

↓ Strand invasion

D-loop

↓ Formation of the heteroduplex

Figure 17.3 Two schemes for initiation of homologous recombination. (A) Initiation as described by the original model for homologous recombination. (B) The Meselson–Radding modification, which proposes a more plausible series of events for formation of the heteroduplex.

The double-strand break model for homologous recombination

Although the Holliday model for homologous recombination, either in its original form or as modified by Meselson and Radding, explained how crossing-over could occur during meiosis, it had inadequacies which prompted the development of alternative schemes. In particular, it was thought that the Holliday model could not explain **gene conversion**, a phenomenon first described in yeast and fungi but now known to occur with many eukaryotes. In yeast, fusion of a pair of gametes results in a zygote that gives rise to an ascus containing four haploid spores whose genotypes can be individually determined (see Figure 14.15). If the gametes have different alleles at a particular locus then under normal circumstances two of the spores will display one genotype and two will display the other genotype, but sometimes this expected 2:2 segregation pattern is replaced by an unexpected 3:1 ratio (Figure 17.4). This is called gene conversion because the ratio can only be explained by one of the alleles "converting" from one type to the other, presumably by recombination during the meiosis that occurs after the gametes have fused.

The double-strand break (DSB) model provides an opportunity for gene conversion to take place during recombination. According to this model, homologous recombination initiates not with a single-strand nick, as in the Meselson–Radding scheme, but with a double-strand cut that breaks one of the partners in the recombination into two pieces (Figure 17.5). After the double-strand cut, one strand in each half of the molecule is shortened, so each end now has a 3′ overhang. One of these overhangs invades the homologous DNA molecule in a manner similar to that envisaged by the Meselson–Radding scheme, setting up a Holliday junction that can migrate along the heteroduplex if the invading strand is extended by a DNA polymerase. To complete the heteroduplex, the other broken strand (the one not involved in the Holliday junction) is also extended. Note that both DNA syntheses involve extension of strands from the partner that suffered the double-strand cut, using as templates the equivalent regions of the uncut partner. This is the basis of the gene conversion because it means that the polynucleotide segments removed from the cut partner have been replaced with copies of the DNA from the uncut partner. After ligation, the resulting heteroduplex has a pair of Holliday structures that can be resolved in a number

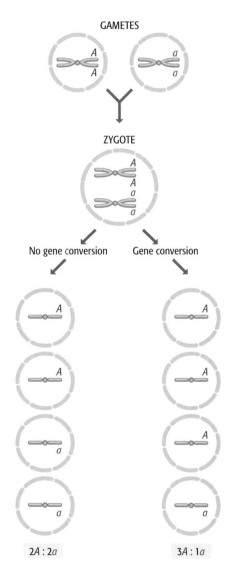

Figure 17.4 Gene conversion. One gamete contains allele *A* and the other contains allele *a*. These fuse to produce a zygote that gives rise to four haploid spores, all contained in a single ascus. Normally, two of the spores will have allele *A* and two will have allele *a*, but if gene conversion occurs the ratio will be changed, possibly to 3*A*:1*a* as shown here.

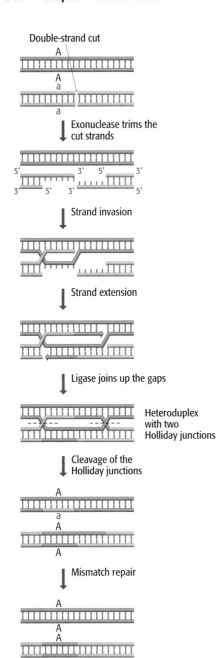

Heteroduplex with two Holliday junctions

Figure 17.5 **The double-strand break model for homologous recombination.** This model explains how gene conversion can occur.

of ways, some resulting in gene conversion and others giving a standard reciprocal strand exchange. An example leading to gene conversion is shown in Figure 17.5.

Although initially proposed as a mechanism for explaining gene conversion in yeast, the DSB model is now looked on as at least a close approximation to the way in which homologous recombination operates in all organisms. Acceptance of this model has come about for two reasons. First, in 1989 it was discovered that during meiosis chromosomes undergo double-strand breakages at 100–1000 times the rate seen in vegetative cells. The implication that formation of double-strand breaks is an inherent part of meiosis clearly favors the DSB model at the expense of schemes in which recombination is initiated by one or more single-strand nicks. The second factor leading to acceptance of the DSB model was the realization that homologous recombination is involved in DNA repair, and specifically is responsible for repairing double-strand breaks that occur as aberrations in the replication process. The Holliday and Meselson–Radding models do not explain this aspect of homologous recombination, whereas double-strand break repair is implicit in the DSB model. We return to the role of homologous recombination in DNA repair in Section 17.1.3.

17.1.2 The biochemistry of homologous recombination

Homologous recombination occurs in all organisms but, as with many aspects of molecular biology, the initial progress in understanding how the process is carried out in the cell was made with *E. coli*. Mutation studies have identified a number of *E. coli* genes that, when inactivated, give rise to defects in homologous recombination, indicating that their protein products are involved in the process in some way. Three distinct recombination systems have been described, these being the RecBCD, RecE, and RecF pathways, with RecBCD apparently being the most important in the bacterium.

The RecBCD pathway of Escherichia coli

In the RecBCD pathway, recombination is mediated by the **RecBCD enzyme** which, as its name implies, is made up of three different proteins. Two of these—RecB and RecD—are helicases. To initiate homologous recombination, one copy of the RecBCD enzyme attaches to the free ends of a chromosome at a double-strand break. The DNA is unwound from each end, through the action of RecB, which travels along one strand in the 3′→5′ direction, and RecD, traveling 5′→3′ along the other strand. The RecB protein, as well as being a helicase, also has 3′→5′ exonuclease activity and so progressively degrades the strand it is traveling along—the one with the free 3′ end (Figure 17.6).

RecBCD progresses along the DNA molecule at a rate of approximately 1 kb per second until it reaches the first copy of the eight-nucleotide consensus sequence 5′–GCTGGTGG–3′ called the **chi site**, which occurs on average once every 6 kb in *E. coli* DNA. At the chi site, the conformation of RecBCD changes so that the RecD helicase becomes uncoupled and the progress of the RecBCD complex slows to about half its initial rate. Exactly what happens next is unclear, but according to the most recent model, the change in conformation of the enzyme also reduces, or completely abolishes, the 3′→5′ exonuclease activity of RecB, this protein now making a single endonucleolytic cut in the other strand of the DNA molecule, at a position close to the chi site (see Figure 17.6). Whatever the precise mechanism, the result is that the RecBCD enzyme produces a double-stranded molecule with a 3′ overhang, exactly as envisaged by the DSB model (see second panel of Figure 17.5).

The next step is establishment of the heteroduplex. This stage is mediated by the RecA protein, which forms a protein-coated DNA filament that is able to invade the intact double helix and set up the D-loop (Figure 17.7). An intermediate in formation of the D-loop is a **triplex** structure (see third panel of Figure 17.5), a three-stranded DNA helix in which the invading polynucleotide lies within the major groove of the intact helix and forms hydrogen bonds with the base pairs it encounters.

Branch migration is catalyzed by the RuvA and RuvB proteins, both of which attach to the branch point of the heteroduplex formed by invasion of the 3′ overhang into the partner molecule. X-ray crystallography studies suggest that four copies of RuvA bind directly to the branch, forming a core to which two RuvB rings (each consisting of eight proteins) attach, one to either side (Figure 17.8). The resulting structure might act as a "molecular motor," rotating the helices in the required manner so that the branch point moves. Branch migration does not appear to be a random process, but instead stops preferentially at the sequence 5′–(A/T)TT(G/C)–3′ (where (A/T) etc. denotes that either of two nucleotides can be present at the position indicated). This sequence occurs frequently in the *E. coli* genome, so presumably migration does not always halt at the first instance of the motif that is reached. When branch migration has ended, the RuvAB complex detaches and is replaced by two RuvC proteins (see Figure 17.8) which carry out the cleavage that resolves the Holliday structure. The cuts are made between the second T and the (G/C) components of the recognition sequence.

Note that the above description provides no precise role for the RecC protein in homologous recombination by the RecBCD pathway. RecC is, in fact, rather enigmatic as its amino acid sequence is dissimilar from that of any other protein synthesized by *E. coli*. Recent X-ray crystallography studies have shown that the tertiary structure of RecC resembles that of the SF1 family of helicases. The amino acids that are central to the helicase activity are absent from RecC, although the equivalent region of the RecC protein still makes contact with the DNA molecule. It seems unlikely that RecC possesses any residual helicase activity, and it therefore does not aid RecB and RecD in unwinding the double helix, but one possibility is that RecC has acquired a scanning function and is responsible for identifying the chi site and hence initiating the conformational change in RecBCD that leads to formation of the heteroduplex.

Other homologous recombination pathways in E. coli

Mutants of *E. coli* that lack components of the RecBCD system are still able to carry out homologous recombination, albeit with lowered efficiency. This is because the bacterium possesses at least two other homologous recombination pathways, called RecE and RecF. In normal *E. coli* cells most homologous recombination takes place via RecBCD, but if this pathway is inactivated by mutation the RecE system is able to take over, to be replaced by RecF when RecE is inactivated.

The details of the RecF pathway are beginning to emerge, and the general mechanism appears to be similar to that described for RecBCD. The helicase

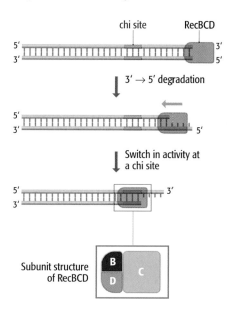

Progression of RecBCD along a DNA molecule

Figure 17.6 The RecBCD pathway for homologous recombination in E. coli. These events are responsible for the first step in the double-strand break model for homologous recombination shown in Figure 17.5.

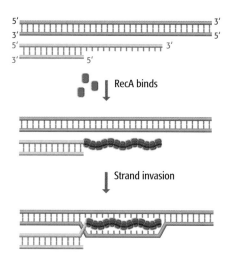

Figure 17.7 The role of the RecA protein in formation of the D-loop during homologous recombination in E. coli.

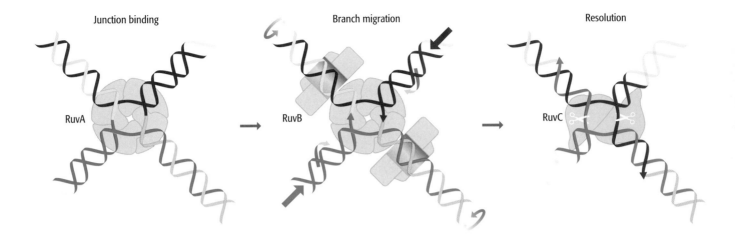

Junction binding   Branch migration   Resolution

RuvA   RuvB   RuvC

Figure 17.8 The role of the Ruv proteins in homologous recombination in *Escherichia coli*. Branch migration is induced by a structure comprising four copies of RuvA bound to the Holliday junction with an RuvB ring on either side. After RuvAB has detached, two RuvC proteins bind to the junction, the orientation of their attachment determining the direction of the cuts that resolve the structure.

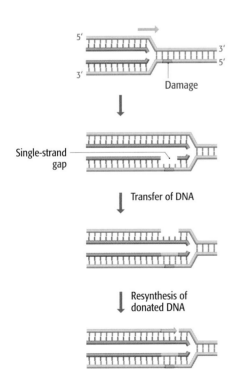

Damage

Single-strand gap

Transfer of DNA

Resynthesis of donated DNA

Figure 17.9 Single-strand gap repair by the RecF pathway of *E. coli*.

activity for the RecF pathway is provided by RecQ, and the 5′ end of the strand is removed by RecJ, leaving a 3′ overhang which becomes coated in RecA proteins through the combined actions of RecF, RecJ, RecO, RecQ, and RecR. There is considerable interchangeability between the components of the RecBCD and RecF pathways, and it is thought that hybrid systems operate in some mutants that lack one or other components of the standard processes. There are differences, however, as only the RecBCD pathway initiates recombination at the chi sites scattered around the *E. coli* genome, and only RecF is able to induce recombination between a pair of plasmids. RecF is also the primary pathway responsible for recombination repair of single-strand gaps resulting from replication of heavily damaged DNA (Section 17.1.3). The RecE pathway is much less well studied but the indications are that it has substantial overlap with the RecF system, with RecJ, RecO, RecQ, and RecF itself appearing to be common to both pathways.

As well as the RecBCD, RecE, and RecF variants, whose functions are to set up the heteroduplex structure, *E. coli* also has alternative means for carrying out the branch migration step. Mutants that lack RuvA or RuvB are still able to carry out homologous recombination because the function of RuvAB can also be provided by a helicase called RecG. It is not yet clear whether RuvAB and RecG are simply interchangeable or if they are specific for different recombination scenarios. RuvC mutants are also able to carry out homologous recombination, suggesting that *E. coli* possesses other proteins that are able to resolve Holliday structures, but the identity of these protein(s) is unknown.

Homologous recombination pathways in eukaryotes

The double-strand break model for homologous recombination is thought to hold for all organisms, not just *E. coli*—recall that it was initially devised to explain gene conversion in *Saccharomyces cerevisiae* (Section 17.1.1). The biochemical events underlying the process appear to be similar in all organisms, and a number of yeast proteins have been identified that carry out functions equivalent to those occurring during the RecBCD pathway in *E. coli*. In particular, two yeast proteins called RAD51 and DMC1 are the homologs of RecA of *E. coli*. Although specific roles for RAD51 and DMC1 are suspected, they are thought to work together in many homologous recombination events. This conclusion arises because mutants that lack one or other protein have similar phenotypes, and the two proteins are found together at the same locations within nuclei that are undergoing meiosis. Interestingly, there are about 100 recombination hotspots in the *S. cerevisiae* genome (Section 3.2.3), suggesting that sequences equivalent to *E. coli* chi sites might be present, albeit at a lower frequency—one per 40 kb in the yeast genome compared with one per 6 kb for *E. coli*. Proteins homologous to RAD51 and DMC1 are also known in several other eukaryotes including humans.

One puzzling aspect of homologous recombination in eukaryotes has been the mechanism by which Holliday structures are resolved, because for many years proteins homologous to *E. coli* RuvC have been sought but not found. In fact, RuvC is not universal in all bacteria, some species apparently using a totally different type of nuclease to resolve Holliday structures. In the archaea, the equivalent function is thought to be performed by the Hjc protein, which has no sequence homology with RuvC but has been shown, by biochemical tests, to bind to Holliday structures. Several eukaryotic proteins have been identified which have structural similarity with Hjc, including RAD51C of *S. cerevisiae*, which is a member of the RAD51 multigene family, and Mus81 of humans. Models have also been proposed for resolving Holliday structures by a combination of helicase and topoisomerase activity, these models suggesting that direct RuvC equivalents might not be needed in all eukaryotes.

17.1.3 Homologous recombination and DNA repair

The attention paid by geneticists to crossing-over as a central feature of sexual reproduction inevitably biased the initial studies of homologous recombination toward events occurring during meiosis. An alternative role for homologous recombination became apparent when *E. coli* mutants defective for components of the RecBCD and other recombination pathways were first examined and shown to have deficiencies in DNA repair. Today we believe that the principal function of homologous recombination is in **postreplicative repair**, its role in crossing-over being of secondary importance in most cells.

Postreplicative repair occurs when breaks arise in daughter DNA molecules as a result of aberrations in the replication process. One such aberration can occur when the replication machinery is attempting to copy a segment of the genome that is heavily damaged, in particular regions in which cyclobutyl dimers are prevalent. When a cyclobutyl dimer is encountered, the template strand cannot be copied and the DNA polymerase simply jumps ahead to the nearest undamaged region, where it restarts the replication process. The result is that one of the daughter polynucleotides has a gap (Figure 17.9). One way in which this gap could be repaired is by a recombination event that transfers the equivalent segment of DNA from the parent polynucleotide present in the second daughter double helix. The gap that is now present in the second double helix is refilled by a DNA polymerase, using the undamaged daughter polynucleotide within this helix as the template. In *E. coli*, this type of single-strand gap repair utilizes the RecF recombination pathway.

If the damaged site cannot be bypassed then the daughter polynucleotide, rather than having a gap, will terminate (Figure 17.10). There are several ways in which this gap could be repaired. One possibility is that the replication fork stalls and reverses a short distance, so that a duplex is formed between the daughter polynucleotides. The incomplete polynucleotide is then extended by a DNA polymerase, using the undamaged daughter strand as a template.

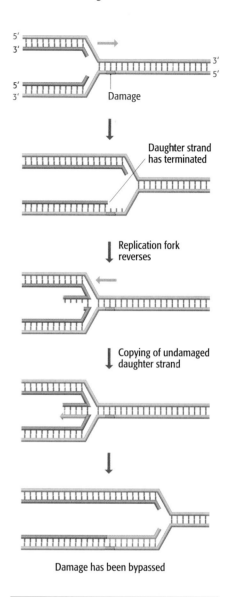

Figure 17.10 A terminated daughter polynucleotide can be rescued by reversal of the replication fork.

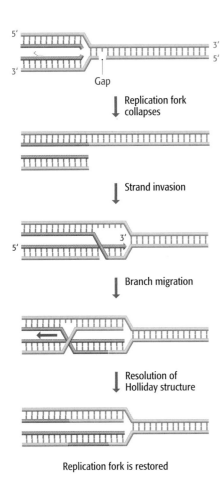

Gap

Replication fork
collapses

Strand invasion

Branch migration

Resolution of
Holliday structure

Replication fork is restored

Figure 17.11 One mechanism for
recovery of a collapsed replication fork by
homologous recombination.

Figure 17.12 The core sequence of the
***att* sites present in bacteriophage λ and**
in the *E. coli* chromosome. The red line
indicates the staggered cut made in each
att site during integration and excision of
the phage genome.

The replication fork then moves forward again, by a process equivalent to the
branch migration step of homologous recombination. As a result the damaged site is bypassed and replication can continue.

A more serious aberration occurs if one of the parent polynucleotides being
replicated contains a single-strand nick. Now the replication process leads to
a double-strand break in one of the daughter double helices, and the replication fork is lost (Figure 17.11). The break can be repaired by a form of
homologous recombination between the broken end and the second,
undamaged molecule. In the scheme shown in Figure 17.11, the daughter
polynucleotide at the double-strand break is extended via a strand exchange
reaction which enables it to use the other parent strand as a template. Branch
migration followed by resolution of the Holliday structure then restores the
replication fork.

17.2 Site-Specific Recombination

A region of extensive homology is not a prerequisite for recombination: the
process can also be initiated between two DNA molecules that have only very
short sequences in common. This is called site-specific recombination and it
has been extensively studied because of the part that it plays during the infection cycle of bacteriophage λ.

17.2.1 Integration of λ DNA into the *E. coli* genome

After injecting its DNA into an *E. coli* cell, bacteriophage λ can follow either
of two infection pathways (Section 14.3.1). One of these, the lytic pathway,
results in the rapid synthesis of λ coat proteins, combined with replication of
the λ genome, leading to death of the bacterium and release of new phages
within about 45 minutes of the initial infection. In contrast, if the phage follows the lysogenic pathway, new phages do not immediately appear. The bacterium divides as normal, possibly for many cell divisions, with the phage in
a quiescent form called the prophage. Eventually, possibly as the result of
DNA damage or some other stimulus, the phage becomes active again.

During the lysogenic phase the λ genome becomes integrated into the *E. coli*
chromosome. It is therefore replicated whenever the *E. coli* DNA is copied,
and is passed on to daughter cells just as though it was a standard part of the
bacterium's genome. Integration occurs by site-specific recombination
between the attachment or *att* sites, *attP* on the λ genome and *attB* on the *E.
coli* chromosome. Each of these attachment sites has at its center an identical 15 bp core sequence referred to as O (Figure 17.12), flanked by variable
sequences called B and B′ in the bacterial genome and P and P′ in the phage
DNA. B and B′ are quite short, just 4 bp each, meaning that *attB* covers just
23 bp of DNA, but P and P′ are much longer with the entire *attP* sequence
spanning over 250 bp. Mutations in the core sequence inevitably lead to
inactivation of the *att* site so that it can no longer participate in recombination, but mutations in the flanking sequences have a less severe consequence
and only decrease the efficiency of recombination. If *attB*, the attachment site
in the *E. coli* genome, is inactivated then insertion of the λ DNA can occur at
secondary sites which share some sequence similarity with the genuine *attB*
locus. If a secondary site is being used then the frequency of lysogeny is
greatly reduced, integration possibly occurring at less than 0.01% of the frequency observed with unmutated *E. coli* cells.

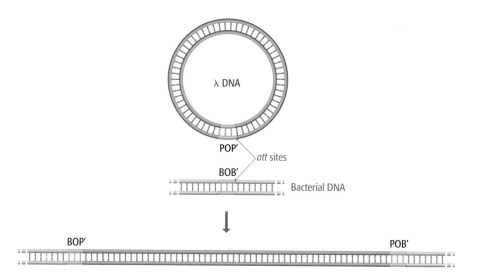

Figure 17.13 Integration of the bacteriophage λ genome into *E. coli* chromosomal DNA. Both λ and *E. coli* DNA have a copy of the *att* site, each one comprising an identical central sequence called "O" and flanking sequences P and P' (for the phage *att* site) or B and B' (bacterial *att* site). Recombination between the O regions integrates the λ genome into the bacterial DNA.

Because this is recombination between two circular molecules, the result is that one bigger circle is formed: in other words, the λ DNA becomes integrated into the bacterial genome (Figure 17.13). The recombination event is catalyzed by a specialized Type I topoisomerase (Section 15.1.2) called **integrase**, a member of a diverse family of **recombinases** present in bacteria, archaea, and yeast. There are at least four binding sites for integrase within *attP*, as well as at least three sites for a second protein, the integration host factor, or IHF. Together these proteins coat the phage attachment site. The integrase then makes a staggered, double-strand cut at equivalent positions in the λ and bacterial *att* sites. The two short, single-strand overhangs are then exchanged between the DNA molecules, producing a Holliday junction which migrates a few base pairs along the heteroduplex before being cleaved. This cleavage, providing that it is made in the appropriate orientation, resolves the Holliday structure in such a way that the λ DNA becomes inserted into the *E. coli* genome.

Integration creates hybrid versions of the attachment sites, which are now called *attR* (which has the structure BOP') and *attL* (whose structure is POB'). A second site-specific recombination between the two *att* sites, now both contained in the same molecule, reverses the original process and releases the λ DNA. This recombination is also catalyzed by the integrase, but in conjunction with a protein called "excisionase," coded by the λ *xis* gene, rather than IHF. The functions of Xis and IHF in, respectively, excision and integration are probably quite different, and the two proteins should not be looked on as playing equivalent roles in the two processes. The key point is that the combination of integrase and excisionase is able to draw the *attR* and *attL* sites together in order to initiate the *intra*molecular recombination that excises the λ genome. After excision, the λ genome returns to the lytic mode of infection and directs synthesis of new phages.

17.2.2 Site-specific recombination is an aid in genetic engineering

The processes responsible for integration and excision of the λ genome are fairly typical of the strategies used by phages to establish lysogeny, though with some phages the molecular events are less complex than those seen with λ. Integration and excision of the bacteriophage P1 genome, for example,

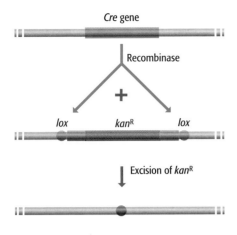

Figure 17.14 The use of the Cre recombinase in plant genetic engineering. Expression of the *Cre* gene results in excision of the *kan^R* gene from the plant DNA.

requires just a single enzyme, the Cre recombinase, which recognizes 34 bp target sites, called *loxB* and *loxP*, which are identical to one another and have no flanking sequences equivalent to B, B′, and so forth.

The simplicity of the P1 system has led to its utilization in genetic engineering projects in which site-specific recombination is a requirement. An important application has arisen in the technology used for the generation of genetically modified crops. One of the main areas of concern to emerge from the debate over genetically modified plants is the possible harmful effects of the marker genes used with plant cloning vectors. Most plant vectors carry a copy of a gene for kanamycin resistance (see Figure 2.27), enabling transformed plants to be identified during the cloning process. The *kan^R* gene is bacterial in origin and codes for the enzyme neomycin phosphotransferase II. This gene and its enzyme product are present in all cells of an engineered plant. The fear that neomycin phosphotransferase might be toxic to humans has been allayed by tests with animal models, but there are still concerns that the *kan^R* gene contained in a genetically modified foodstuff could be passed to bacteria in the human gut, making these resistant to kanamycin and related antibiotics, or that the *kan^R* gene could be passed to other organisms in the environment, possibly resulting in damage to the ecosystem.

The fears surrounding the use of *kan^R* and other marker genes have prompted biotechnologists to devise ways of removing these genes from plant DNA after the transformation event has been verified. One of the strategies makes use of Cre recombinase. To use this system the plant is transformed with two cloning vectors, the first carrying the gene being added to the plant along with its *kan^R* selectable marker gene surrounded by the *lox* target sequences, and the second carrying the Cre recombinase gene. After transformation, expression of the *Cre* gene therefore results in excision of the *kan^R* gene from the plant DNA (Figure 17.14).

17.3 Transposition

Transposition is not a type of recombination but a process that often utilizes recombination, the end result being the transfer of a segment of DNA from one position in the genome to another. A characteristic feature of transposition is that the transferred segment is flanked by a pair of short direct repeats (Figure 17.15) which are formed during the transposition process.

In Section 9.2 we examined the various types of transposable element known in eukaryotes and prokaryotes and discovered that these could be broadly divided into three categories on the basis of their transposition mechanism:

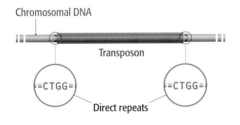

Figure 17.15 Integrated transposable elements are flanked by short direct repeat sequences. This particular transposon is flanked by the tetranucleotide repeat 5′–CTGG–3′. Other transposons have different direct repeat sequences.

- DNA transposons that transpose replicatively, the original transposon remaining in place and a new copy appearing elsewhere in the genome (Figure 17.16).
- DNA transposons that transpose conservatively, the original transposon moving to a new site by a cut-and-paste process.
- Retroelements, all of which transpose replicatively via an RNA intermediate.

We will now examine the recombination events that are responsible for each of these three types of transposition.

Figure 17.16 **Replicative and conservative transposition.** DNA transposons use either the replicative or conservative pathway (some can use both). Retroelements transpose replicatively via an RNA intermediate.

17.3.1 Replicative and conservative transposition of DNA transposons

A number of models for replicative and conservative transposition of DNA transposons have been proposed over the years but most are modifications of a scheme originally outlined by Shapiro in 1979. According to this model, the replicative transposition of a bacterial element such as a Tn3-type transposon or a transposable phage (Section 9.2.2) is initiated by one or more endonucleases that make single-strand cuts either side of the transposon and in the target site where the new copy of the element will be inserted (Figure 17.17). At the target site the two cuts are separated by a few base pairs, so that the cleaved double-stranded molecule has short 5′ overhangs.

Ligation of these 5′ overhangs to the free 3′ ends either side of the transposon produces a hybrid molecule in which the original two DNAs—the one containing the transposon and the one containing the target site—are linked together by the transposable element flanked by a pair of structures resembling replication forks. DNA synthesis at these replication forks copies the transposable element and converts the initial hybrid into a **cointegrate**, in which the two original DNAs are still linked. Homologous recombination between the two copies of the transposon uncouples the cointegrate, separating the original DNA molecule (with its copy of the transposon still in place) from the target molecule, which now contains a copy of the transposon. Replicative transposition has therefore occurred.

A modification of the process just described changes the mode of transposition from replicative to conservative (see Figure 17.17). Rather than carrying out DNA synthesis, the hybrid structure is converted back into two separate DNA molecules simply by making additional single-strand nicks either side of the transposon. This cuts the transposon out of its original molecule, leaving it "pasted" into the target DNA.

17.3.2 Transposition of retroelements

From the human perspective, the most important retroelements are the retroviruses, which include the human immunodeficiency viruses that cause AIDS and various other virulent types. Most of what we know about retrotransposition refers specifically to retroviruses, although it is believed that other retroelements, such as retrotransposons of the *Ty1/copia* and *Ty3/gypsy* families, transpose by similar mechanisms. These mechanisms do not involve recombination but, for convenience, we will consider them here.

The first step in retrotransposition is synthesis of an RNA copy of the inserted retroelement (Figure 17.18). The long terminal repeat (LTR) at the 5′ end of the element contains a TATA sequence which acts as a promoter for transcription by RNA polymerase II (Section 11.2.2). Some retroelements also have enhancer sequences (Section 11.3) that are thought to regulate the amount of transcription that occurs. Transcription continues through the entire length of the element, up to a polyadenylation sequence (Section 12.2.1) in the 3′ LTR.

The transcript now acts as the template for RNA-dependent DNA synthesis, catalyzed by a reverse transcriptase enzyme coded by part of the *pol* gene of the retroelement (see Figure 9.14). Because this is synthesis of DNA, a primer is required (Section 15.2.2) and, as during genome replication, the primer is made of RNA rather than DNA. During genome replication, the primer is synthesized *de novo* by a polymerase enzyme (see Figure 15.13), but retroelements do not code for RNA polymerases and so cannot make primers in this

Figure 17.17 A model for the process resulting in replicative and conservative transposition.

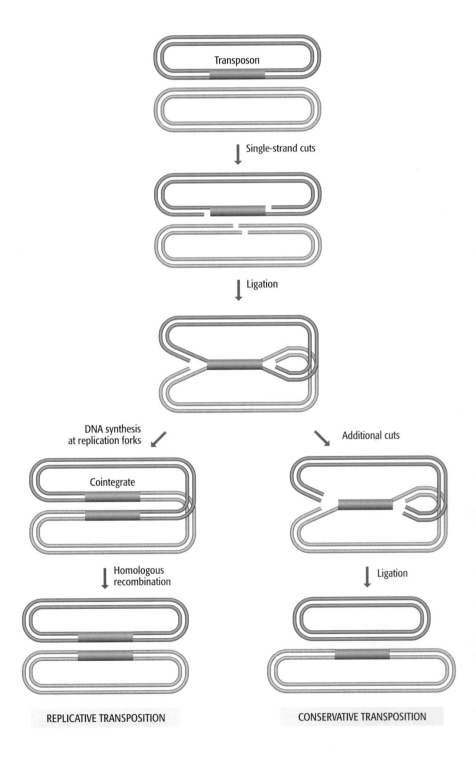

way. Instead they use one of the cell's tRNA molecules as a primer, which one depending on the retroelement: the *Ty1/copia* family of elements always use tRNA^Met but other retroelements use different tRNAs.

The tRNA primer anneals to a site within the 5′ LTR (see Figure 17.18). At first glance this appears to be a strange location for the priming site because it means that DNA synthesis is directed away from the central region of the

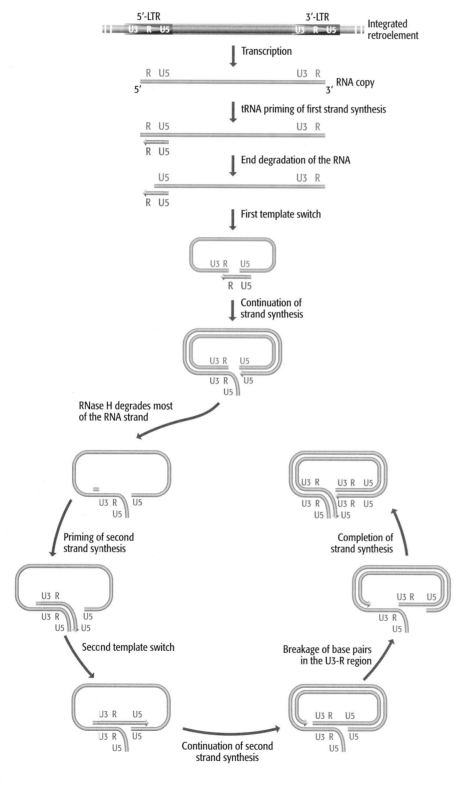

Figure 17.18 RNA and DNA replication during transposition of a retroelement. This diagram shows how an integrated retroelement is copied into a free double-stranded DNA version. The first step is synthesis of an RNA copy, which is then converted to double-stranded DNA by a series of events that involves two template switches, as described in the text.

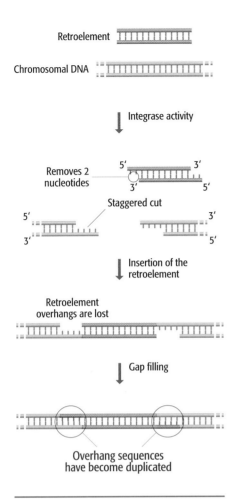

Figure 17.19 Integration of the double-stranded DNA version of a retroelement into the host genome. Integration of the retroelement results in a four-nucleotide direct repeat either side of the inserted sequence. With retroviruses, this stage in the transposition pathway requires both integrase and the DNA-PK$_{CS}$ protein kinase and Ku proteins that are also involved in NHEJ (Section 16.2.4).

retroelement and so results in only a short copy of part of the 5′ LTR. In fact, when the DNA copy has been extended to the end of the LTR, a part of the RNA template is degraded and the DNA overhang that is produced reanneals to the 3′ LTR of the retroelement which, being a long terminal *repeat*, has the same sequence as the 5′ LTR and so can base-pair with the DNA copy. DNA synthesis now continues along the RNA template, eventually displacing the tRNA primer. Note that the result is a DNA copy of the entire template, including the priming site: the template switching is, in effect, the strategy that the retroelement uses to solve the "end-shortening" problem, the same problem that chromosomal DNAs address through telomere synthesis (Section 15.2.4).

Completion of synthesis of the first DNA strand results in a DNA–RNA hybrid. The RNA is partially degraded by an RNase H enzyme, coded by another part of the *pol* gene. The RNA that is not degraded, usually just a single fragment attached to a short polypurine sequence adjacent to the 3′ LTR, primes synthesis of the second DNA strand, again by reverse transcriptase, which is able to act as both an RNA- and DNA-dependent DNA polymerase. As with the first round of DNA synthesis, second-strand synthesis initially results in a DNA copy of just the LTR, but a second template switch, to the other end of the molecule, enables the DNA copy to be extended until it is full length. This creates a template for further extension of the first DNA strand, so that the resulting double-stranded DNA is a complete copy of the internal region of the retroelement plus the two LTRs.

All that remains is to insert the new copy of the retroelement into the genome. It was originally thought that insertion occurred randomly, but it now appears that although no particular sequence is used as a target site, integration occurs preferentially at certain positions. Insertion involves removal of two nucleotides from the 3′ ends of the double-stranded retroelement by the integrase enzyme (coded by yet another part of *pol*). The integrase also makes a staggered cut in the genomic DNA so that both the retroelement and the integration site now have 5′ overhangs (Figure 17.19). These overhangs might not have complementary sequences but they still appear to interact in some way so that the retroelement becomes inserted into the genomic DNA. The interaction results in loss of the retroelement overhangs and filling in of the gaps that are left, which means that the integration site becomes duplicated into a pair of direct repeats, one at either end of the inserted retroelement.

17.3.3 How do cells minimize the harmful effect of transposition?

Transposition can have deleterious effects on a genome. These effects go beyond the obvious disruption of gene activity that will occur if a transposable element takes up a new position that lies within the coding region of a gene. Some elements, notably retrotransposons, contain promoter and enhancer sequences that can modify the expression patterns of adjacent genes, and transposition often involves the creation of double-strand breaks which, as we saw in Section 16.2.4, can have seriously harmful effects on the integrity of a genome. Interspersed repeats, at least some of which are mobile, make up almost half of each mammalian genome and a substantially larger fraction of some plant genomes. It might therefore be expected that these and other genomes have evolved mechanisms for limiting the movement of their transposable elements.

One way in which transposition of both DNA transposons and retroelements could be prevented is by methylation of their DNA sequences, methylation being a common means of silencing genomic regions (Section 10.3.1). Many transposable element sequences are indeed hypermethylated—90% of the methylated cytosines in the human genome are located in the interspersed repeat sequences—but experimental evidence linking this methylation with suppression of transposition has been difficult to obtain. Recently it has been shown that mutants of the plant *Arabidopsis thaliana* that have deficient methylation systems suffer from a higher than normal amount of transposition, but this heightened transposition might not be displayed by all types of transposon in the plant genome. Further work is underway to explore the role of methylation in *Arabidopsis* and to extend the studies to other types of organism.

Summary

Recombination was originally used to describe the outcome of crossing-over between pairs of homologous chromosomes during meiosis. The term is now also used to refer to the molecular events which underlie this process. Homologous recombination occurs between segments of DNA molecules that share extensive sequence homology. The initial models for homologous recombination envisaged the recombination process being initiated by nicks that occur in one or both of the double-stranded molecules, but it is now thought that the start point is a double-strand break in one of these molecules. Strand exchange leads to a heteroduplex structure which is resolved by cleavage, possibly leading to exchange of DNA segments or to gene conversion. *Escherichia coli* possesses at least three molecular pathways for homologous recombination. The one that has been studied in greatest detail, the RecBCD pathway, involves the unwinding of one partner in the recombination by a pair of helicases, which attach to the double-strand break and progress along the molecule. At a recognition sequence called a chi site the RecBCD complex initiates the strand exchange, with the RecA protein playing a central role in transfer of the invading strand into the intact double helix. Branch migration within the heteroduplex, and resolution of the structure, is catalyzed by the Ruv proteins. The RecE and RecF pathways operate in similar ways, with a number of proteins shared between the three recombination systems. Equivalent proteins are known in eukaryotes. Homologous recombination is responsible for postreplicative repair of DNA breaks. Site-specific recombination does not require regions of extensive homology between the partner molecules. This type of recombination is responsible for insertion of bacteriophage genomes, such as the λ genome, into the host bacterial chromosome. Integration of the λ genome into *E. coli* DNA occurs by recombination between a pair of 15 bp sequences contained within longer attachment sites. Integration requires the integrase coded by the λ genome and the *E. coli* integration host factor. Excision requires the integrase and an excisionase protein. Transposition of DNA transposons occurs by recombination. The process can be either replicative or conservative, both occurring by a series of events initially described by Shapiro in 1979. Retroelements transpose via an RNA intermediate which is transcribed from the parent copy of the transposon. After copying into double-stranded DNA, the retroelement is reinserted into the host chromosome.

Multiple Choice Questions

17.1.* Which of the following is an example of site-specific recombination?

a. Crossing-over during meiosis.

b. Gene conversion.

c. Integration of the bacteriophage λ genome into the E. coli chromosome.

d. Insertion of a transposon into a new site in a genome.

17.2. What type of DNA exchange occurs during crossing-over in meiosis?

a. Single strand exchange.

b. Reciprocal strand exchange.

c. Integrative strand exchange.

d. Replicative strand exchange.

17.3.* How does the Meselson–Radding model for recombination explain how the two DNA molecules interact at the beginning of homologous recombination?

a. Single-strand nicks appear at equivalent positions in each molecule.

b. A specialized topoisomerase produces single-strand breaks in both DNA molecules.

c. The two DNA molecules initiate recombination without breakage.

d. A single-strand nick occurs in one molecule creating a free end that invades the other molecule to displace one of its strands.

17.4. Gene conversion is observed when:

a. During meiosis, one allele is replaced by another allele.

b. During meiosis, the expected 2:2 ratio of alleles is replaced with a 4:0 ratio.

c. During meiosis, the expected 2:2 ratio of alleles is replaced with a 3:1 ratio.

d. All of the above.

17.5.* What occurs at chi sites during homologous recombination mediated by the RecBCD enzyme in E. coli?

a. This is where the RecBCD proteins bind to DNA to initiate recombination.

b. This is the site that causes the enzyme to produce a double-strand break in the DNA.

c. This is the site where the helicase activity of the enzyme begins to degrade the DNA.

d. This is the site where branch migration is terminated.

17.6. Which proteins catalyze branch migration during homologous recombination in E. coli?

a. RecA.

b. RecBCD.

c. RuvA and RuvB.

d. Topoisomerase IIB.

17.7.* What is thought to be the primary function of homologous recombination?

a. Crossing-over in meiosis.

b. Gene conversion.

c. Integration of lysogenic phage genomes.

d. Postreplicative DNA repair.

17.8. The Cre system, utilized by plant genetic engineers, is an example of which type of recombination?

a. Homologous recombination.

b. Retrotransposition.

c. Site-specific recombination.

d. Transposition.

17.9.* What happens if inactivating mutations occur in the E. coli attB sequences?

a. The λ phage DNA forms a partial Holliday structure with the E. coli genome and is then degraded.

b. The λ phage can only follow the lytic cycle.

c. The λ phage DNA can be inserted into the E. coli genome at secondary sites at a much lower frequency.

d. The λ phage DNA is inserted into the attB site but it cannot be excised.

17.10. The excision of the λ prophage from the E. coli genome is:

a. An intramolecular recombination event.

b. An intermolecular recombination event.

c. A nuclease release event.

d. A transposon-mediated event.

17.11.* How does homologous recombination play a role in replicative transposition?

a. Replicative transposition can only occur between homologous sequences.

b. The proteins involved in homologous recombination are required for the initiation of replicative transposition.

c. After the transposon is replicated, the free copy of the sequence is integrated into the genome at the new site via homologous recombination.

Multiple Choice Questions (continued) *Answers to odd-numbered questions can be found in the Appendix

    d. Replication of the transposon sequence converts the hybrid DNA molecule into a cointegrate, which is uncoupled via homologous recombination.

17.12. How do cells minimize the potentially harmful effects of transposition?

    a. Immunoglobulins bind to the transposon-encoded proteins.

    b. Transposon sequences are condensed into tightly packed chromatin.

    c. Transposon sequences are methylated.

    d. Transposon proteins are targeted for degradation within proteasomes by ubiquitin.

Short Answer Questions *Answers to odd-numbered questions can be found in the Appendix

17.1.* What is the role of recombination in genome evolution?

17.2. How can the resolution of a Holliday structure yield two different results?

17.3.* Describe how the double-strand break model explains how gene conversion occurs.

17.4. Some *E. coli* strains that are used for propagating recombinant plasmids contain *recA* mutations. Why might *recA* defects be useful for researchers working with recombinant plasmids?

17.5.* As *E. coli* is a haploid organism, when might it have two homologous DNA molecules that can participate in recombination events?

17.6. Which of the *E. coli* proteins involved in homologous recombination have homologs in yeast? Which *E. coli* protein involved in this process does not appear to have a homolog in yeast?

17.7.* What are the properties of the *attP* and *attB* sites that mediate integration of λ DNA into the *E. coli* genome?

17.8. How is the new copy of a retroelement inserted into a genome?

17.9.* What is the role of tRNA molecules in the replication of retroelements?

17.10. Give examples of the harmful effects that transposons can have on a genome.

In-depth Problems *Guidance to odd-numbered questions can be found in the Appendix

17.1.* Write a detailed essay on the various roles of the RecA protein in biology.

17.2. Discuss the importance of homologous recombination in biology.

17.3.* Determination of the structure of the RecBCD complex was looked on as a key step in understanding the molecular basis of homologous recombination. Explain why knowing the structure of this complex was so important.

17.4. The Cre recombination system underlies one of the

more controversial aspects of plant genetic engineering, the so-called terminator technology. This is one of the processes by which the companies who market genetically modified crops attempt to protect their financial investment by ensuring that farmers must buy new seed every year, rather than simply collecting seed from the crop and sowing this second generation seed the following year. The terminator technology centers on the gene for ribosome inactivating protein (RIP). The RIP protein blocks protein synthesis by cutting one of the ribosomal RNA molecules into two segments, which means that any

continued …

cell in which the RIP protein is active will quickly die. Use this information to deduce exactly how the terminator technology works.

17.5.* Give a detailed answer to the question "how do cells minimize the harmful effect of transposition?"

Figure Tests *Answers to odd-numbered questions can be found in the Appendix

17.1.* Does the figure depict an example of homologous recombination, site-specific recombination, or transposition?

17.2. What is the event that is depicted in this figure, and what steps in this process are shown?

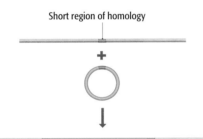

Short region of homology

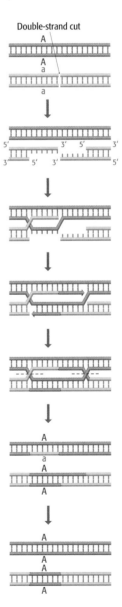

Double-strand cut

17.3.* This figure depicts the activity of a protein in the RecBCD pathway in *E. coli*. What is the protein that binds to the single-stranded DNA and mediates the formation of the D-loop?

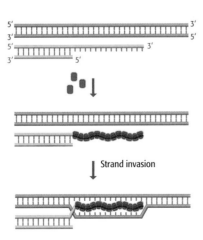

Strand invasion

Figure Tests (continued)

17.4. The figure shows part of the transposition pathway of a retroelement. Discuss the steps of the pathway shown in the figure.

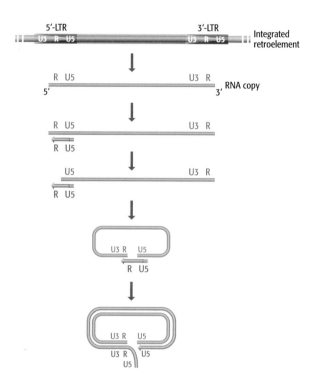

Further Reading

Models for homologous recombination

Eggleston, A.K. and West, S.C. (1996) Exchanging partners in *E. coli. Trends Genet.* **12:** 20–26.

Heyer, W.D., Ehmsen, K.T. and Solinger, J.A. (2003) Holliday junctions in the eukaryotic nucleus: resolution in sight? *Trends Biochem. Sci.* **28:** 548–557.

Holliday, R. (1964) A mechanism for gene conversion in fungi. *Genet. Res.* **5:** 282–304.

Kowalczykowski, S.C. (2000) Initiation of genetic recombination and recombination-dependent replication. *Trends Biochem. Sci.* **25:** 156–165.

Meselson, M. and Radding, C.M. (1975) A general model for genetic recombination. *Proc. Natl Acad. Sci. USA* **72:** 358–361.

Shinagawa, H. and Iwasaki, H. (1996) Processing the Holliday junction in homologous recombination. *Trends Biochem. Sci.* **21:** 107–111.

Molecules for homologous recombination

Amundsen, S.K. and Smith, G.R. (2003) Interchangeable parts of the *Escherichia coli* recombination machinery. *Cell* **112:** 741–744. *Hybrid pathways involving parts of the RecBCD and RecF systems.*

Baumann, P. and West, S.C. (1998) Role of the human RAD51 protein in homologous recombination and double-stranded-break repair. *Trends Biochem. Sci.* **23:** 247–251.

Masson, J.-Y. and West, S.C. (2001) The Rad51 and Dmc1 recombinases: a non-identical twin relationship. *Trends Biochem. Sci.* **26:** 131–136.

Pyle, A.M. (2004) Big engine finds small breaks. *Nature* **432:** 157–158. *The structure of the RecBCD complex.*

Rafferty, J.B., Sedelnikova, S.E., Hargreaves, D., Artymiuk, P.J., Baker, P.J., Sharples, G.J., Mahdi, A.A., Lloyd, R.G. and Rice, D.W. (1996) Crystal structure of DNA recombination protein RuvA and a model for its binding to the Holliday junction. *Science* **274:** 415–421.

Symington, L.S. and Holloman, W.K. (2004) Resolving resolvases. *Science* **303:** 184–185. *Proteins for resolution of Holliday structures in eukaryotes.*

West, S.C. (1997) Processing of recombination intermediates by the RuvABC proteins. *Annu. Rev. Genet.* **31:** 213–244.

Site-specific recombination

Kwon, H.J., Tirumalai, R., Landy, A. and Ellenberger, T. (1997) Flexibility in DNA recombination: structure of the lambda integrase catalytic core. *Science* **276:** 126–131.

Transposition

Bushman, F.D. (2003) Targeting survival: integration site selection by retroviruses and LTR-retrotransposons. *Cell* **115:** 135–138.

Shapiro, J.A. (1979) Molecular model for the transposition and replication of bacteriophage Mu and other transposable elements. *Proc. Natl Acad. Sci. USA* **76:** 1933–1937.

How Genomes Evolve

18

When you have read Chapter 18, you should be able to:

Explain why biologists believe that the first genomes were made of RNA.

Describe the events thought to have led to adoption of protein catalysts and DNA genomes by the earliest cells.

Distinguish between the various ways in which genomes can acquire new genes.

Give examples to illustrate the role of gene duplication in the evolution of multigene families.

Discuss the evidence for whole genome duplications in the evolutionary histories of *Saccharomyces cerevisiae*, *Arabidopsis thaliana*, and humans.

Describe the impact that segment duplications have had on the recent evolution of the human genome.

Explain how new genes can arise by domain duplication and domain shuffling.

Assess the likely impact of lateral gene transfer on genome evolution in bacteria and in eukaryotes.

Outline how transposable elements may have influenced genome evolution.

Define and evaluate the "introns early" and "introns late" hypotheses.

List the differences between the human and chimpanzee genomes and discuss how such similar genomes can give rise to such different biological attributes.

Mutation and recombination provide the genome with the means to evolve, but we learn very little about the evolutionary histories of genomes simply by studying these events in living cells. Instead we must combine our understanding of mutation and recombination with comparisons between the genomes of different organisms in order to infer the patterns of genome evolution that have occurred. Clearly, this approach is imprecise and uncertain but, as we will see, it is based on a surprisingly large amount of hard data and we can be reasonably confident that, at least in outline, the picture that emerges is not too far from the truth.

In this chapter we will explore the evolution of genomes from the very origins of biochemical systems through to the present day. We will look at ideas regarding the **RNA world**, prior to the appearance of the first DNA molecules, and then examine how DNA genomes have gradually become more complex. Finally, in Section 18.4 we will compare the human genome with the genome of the chimpanzee in order to identify the evolutionary changes that have occurred during the last five million years and which must, somehow, make us what we are.

18.1 Genomes: the First Ten Billion Years

Cosmologists believe that the universe began some 14 billion years ago with the gigantic "primordial fireball" called the Big Bang. Mathematical models suggest that after about 4 billion years galaxies began to fragment from the clouds of gas emitted by the Big Bang, and that within our own galaxy the solar nebula condensed to form the Sun and its planets about 4.6 billion years ago (Figure 18.1). The early Earth was covered with water and it was in this vast planetary ocean that the first biochemical systems appeared, cellular life being well established by the time land masses began to appear, some 3.5 billion years ago. But cellular life was a relatively late stage in biochemical evolution, being preceded by self-replicating polynucleotides that were the progenitors of the first genomes. We must begin our study of genome evolution with these precellular systems.

Figure 18.1 The origins of the universe, galaxies, solar system, and cellular life.

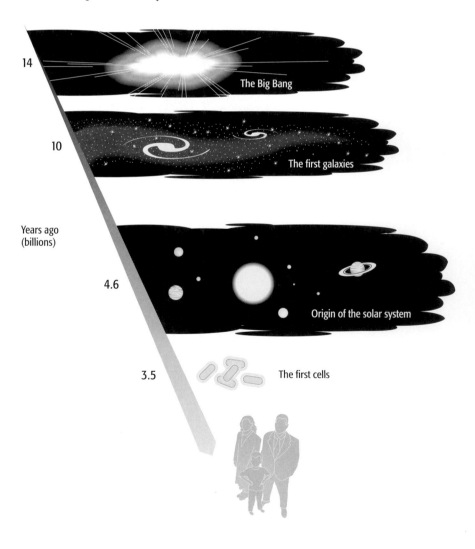

Years ago (billions)

14 — The Big Bang

10 — The first galaxies

4.6 — Origin of the solar system

3.5 — The first cells

18.1.1 The origins of genomes

The first oceans are thought to have had a similar salt composition to those of today but the Earth's atmosphere, and hence the dissolved gases in the oceans, was very different. The oxygen content of the atmosphere remained very low until photosynthesis evolved, and to begin with the most abundant gases were probably methane and ammonia. Experiments attempting to recreate the conditions in the ancient atmosphere have shown that electrical discharges in a methane–ammonia mixture result in chemical synthesis of a range of amino acids, including alanine, glycine, valine, and several of the others found in proteins. Hydrogen cyanide and formaldehyde are also formed, these participating in additional reactions to give other amino acids, as well as purines, pyrimidines, and, in less abundance, sugars. At least some of the building blocks of biomolecules could therefore have accumulated in the ancient chemosphere.

The first biochemical systems were centered on RNA

Polymerization of the building blocks into biomolecules might have occurred in the oceans or could have been promoted by the repeated condensation and drying of droplets of water in clouds. Alternatively, polymerization might have taken place on solid surfaces, perhaps making use of monomers immobilized on clay particles, or in hydrothermal vents. The precise mechanism need not concern us: what is important is that it is possible to envisage purely geochemical processes that could lead to synthesis of polymeric biomolecules similar to the ones found in living systems. It is the next steps that we must worry about. We have to go from a random collection of biomolecules to an ordered assemblage that displays at least some of the biochemical properties that we associate with life. These steps have never been reproduced experimentally and our ideas are therefore based mainly on speculation tempered by a certain amount of computer simulation. One problem is that the speculations are unconstrained because the global ocean could have contained as many as 10^{10} biomolecules per liter and we can allow a billion years for the necessary events to take place. This means that even the most improbable scenarios cannot be dismissed out of hand.

Progress in understanding the origins of life was initially stalled by the apparent requirement that polynucleotides and polypeptides must work in harness in order to produce a self-reproducing biochemical system. This is because proteins are required to catalyze biochemical reactions but cannot carry out their own self-replication. Polynucleotides can specify the synthesis of proteins and self-replicate, but it was thought that they could do neither without the aid of proteins. It appeared that the biochemical system would have to spring fully formed from the random collection of biomolecules because any intermediate stage could not be perpetuated. The major breakthrough came in the mid-1980s when it was discovered that RNA can have catalytic activity.

Those ribozymes that are known today carry out three types of biochemical reaction:

- Self-cleavage, as displayed by the self-splicing Group I, II, and III introns and by some virus genomes (Table 12.4 and Section 12.2.4).

- Cleavage of other RNAs, as carried out by, for example, RNase P (Table 12.4 and Section 12.1.3).

- Synthesis of peptide bonds, carried out by the rRNA component of the ribosome (Section 13.2.3).

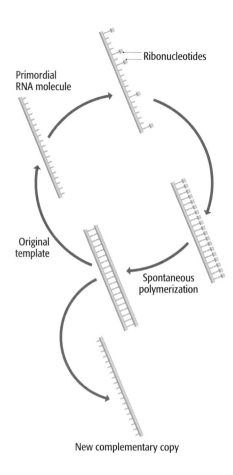

Primordial RNA molecule

Ribonucleotides

Original template

Spontaneous polymerization

New complementary copy

Figure 18.2 Copying of RNA molecules in the early RNA world. Before the evolution of RNA polymerases, ribonucleotides that became associated with an RNA template would have had to polymerize spontaneously. This process would have been inaccurate and many RNA sequences would have been generated

In the test tube, synthetic RNA molecules have been shown to carry out other biologically relevant reactions such as synthesis of ribonucleotides, synthesis and copying of RNA molecules, and transfer of an RNA-bound amino acid to a second amino acid forming a dipeptide, in a manner analogous to the role of tRNA in protein synthesis. The discovery of these catalytic properties solved the polynucleotide–polypeptide dilemma by showing that the first biochemical systems could have been centered entirely on RNA.

Ideas about the RNA world have taken shape in recent years. We now envisage that RNA molecules initially replicated in a slow and haphazard fashion simply by acting as templates for binding of complementary nucleotides which polymerized spontaneously (Figure 18.2). This replication process would have been very inaccurate so a variety of RNA sequences would have been generated, eventually leading to one or more with nascent ribozyme properties that were able to direct their own, more accurate self-replication. It is possible that a form of natural selection operated so that the most efficient replicating systems began to predominate, as has been shown to occur in experimental systems. A greater accuracy in replication would have enabled RNAs to increase in length without losing their sequence specificity, providing the potential for more sophisticated catalytic properties, possibly culminating in structures as complex as present-day Group I introns (see Figure 12.39) and ribosomal RNAs (see Figure 13.11).

To call the early RNAs "genomes" is a little fanciful, but the term **protogenome** has attractions as a descriptor for molecules that were self-replicating and able to direct simple biochemical reactions. These reactions might have included energy metabolism based, as today, on the release of free energy by hydrolysis of the phosphate–phosphate bonds in the ribonucleotides ATP and GTP, and the reactions might have become compartmentalized within lipid membranes, forming the first cell-like structures. There are difficulties in envisaging how long-chain unbranched lipids could form by chemical- or ribozyme-catalyzed reactions, but once present in sufficient quantities they would have assembled spontaneously into membranes, possibly encapsulating one or more protogenomes and providing the RNAs with an enclosed environment in which more controlled biochemical reactions could be carried out.

The first DNA genomes

How did the RNA world develop into the DNA world? The first major change was probably the development of protein enzymes, which supplemented, and eventually replaced, most of the catalytic activities of ribozymes. There are several unanswered questions relating to this stage of biochemical evolution, including the reason why the transition from RNA to protein occurred in the first place. Originally, it was assumed that the 20 amino acids in polypeptides provided proteins with greater chemical variability than the four ribonucleotides in RNA, enabling protein enzymes to catalyze a broader range of biochemical reactions, but this explanation has become less attractive as more and more ribozyme-catalyzed reactions have been demonstrated in the test tube. A more recent suggestion is that protein-mediated catalysis is more efficient because of the inherent flexibility of folded polypeptides compared with the greater rigidity of base-paired RNAs. Alternatively, enclosure of RNA protogenomes within membrane vesicles

(A) A ribozyme that is also a coding molecule

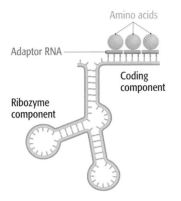

(B) A ribozyme that synthesizes coding molecules

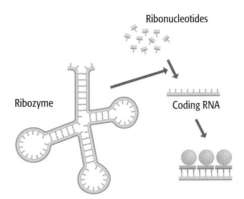

Figure 18.3 Two scenarios for the evolution of the first coding RNA. A ribozyme could have evolved to have a dual catalytic and coding function (A), or a ribozyme could have synthesized a coding molecule (B). In both examples, the amino acids are shown attaching to the coding molecule via small adaptor RNAs, the presumed progenitors of today's tRNAs.

could have prompted the evolution of the first proteins, because RNA molecules are hydrophilic and must be given a hydrophobic coat, for instance by attachment to peptide molecules, before being able to pass through or become integrated into a membrane.

The transition to protein-mediated catalysis demanded a radical shift in the function of the RNA protogenomes. Rather than being directly responsible for the biochemical reactions occurring in the early cell-like structures, the protogenomes became coding molecules whose main function was to specify the construction of the catalytic proteins. Whether the ribozymes themselves became coding molecules, or coding molecules were synthesized by the ribozymes is not known, although the most persuasive theories about the origins of protein synthesis and the genetic code suggest that the latter alternative is more likely to be correct (Figure 18.3). Whatever the mechanism, the result was the paradoxical situation whereby the RNA protogenomes had abandoned their roles as enzymes, which they were good at, and taken on a coding function for which they were less well suited because of the relative instability of the RNA phosphodiester bond (Section 1.2.1). A transfer of the coding function to the more stable DNA seems almost inevitable and would not have been difficult to achieve, reduction of ribonucleotides giving deoxyribonucleotides which could then be polymerized into copies of the RNA protogenomes by a reverse-transcriptase-catalyzed reaction (Figure 18.4). The replacement of uracil with its methylated derivative thymine probably conferred even more stability on the DNA polynucleotide, and the adoption of double-stranded DNA as the coding molecule was almost certainly prompted by the possibility of repairing DNA damage by copying the partner strand (Sections 16.2.2 and 16.2.3).

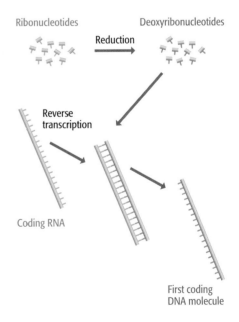

Figure 18.4 Conversion of a coding RNA molecule into the progenitor of the first DNA genome.

According to this scenario, the first DNA genomes comprised many separate molecules, each specifying a single protein and each therefore equivalent to a single gene. The linking together of these genes into the first chromosomes, which could have begun before the transition to DNA, would have improved the efficiency of gene distribution during cell division, as it is easier to organize the equal distribution of a few large chromosomes than many separate genes. As with most stages in early genome evolution, several different mechanisms by which genes might have become linked have been proposed.

How unique is life?

If the experimental simulations and computer models are correct then it is likely that the initial stages in biochemical evolution occurred many times in parallel in the oceans or atmosphere of the early Earth. It is therefore quite possible that "life" arose on more than one occasion, even though all present-day organisms appear to derive from a single origin. This single origin is indicated by the remarkable similarity between the basic molecular biological and biochemical mechanisms in bacterial, archaeal, and eukaryotic cells. To take just one example, there appears to be no obvious biological or chemical reason why any particular triplet of nucleotides should code for any particular amino acid, but the genetic code, although not universal, is virtually the same in all organisms that have been studied. If these organisms derived from more than one origin then we would anticipate two or more very different codes.

If multiple origins are possible, but modern life is derived from just one origin, then at what stage did a single biochemical system begin to predominate? The question cannot be answered precisely, but the most likely scenario is that the predominant system was the first to develop the means to synthesize protein enzymes and therefore probably also the first to adopt a DNA genome. The greater catalytic potential and more accurate replication conferred by protein enzymes and DNA genomes would have given these cells a significant advantage compared with those still containing RNA protogenomes. The DNA–RNA–protein cells would have multiplied more rapidly, enabling them to out-compete the RNA cells for nutrients which, before long, would have included the RNA cells themselves.

Are life forms based on informational molecules other than DNA and RNA possible? It is not impossible that RNA was preceded by some other informational molecule at the very earliest period of biochemical evolution. In particular, a pyranosyl version of RNA, in which the sugar takes on a slightly different structure, might be a better choice than normal RNA for an early protogenome because the base-paired molecules that it forms are more stable. The same is true of **peptide nucleic acid** (**PNA**), a polynucleotide analog in which the sugar–phosphate backbone is replaced by amide bonds (Figure 18.5). PNAs have been synthesized in the test tube and have been shown to form base pairs with normal polynucleotides. However, there are no indications that either pyranosyl RNA or PNA were more likely than RNA to have evolved in the prebiotic soup.

18.2 Acquisition of New Genes

Although the very old fossil record is difficult to interpret, there is reasonably convincing evidence that by 3.5 billion years ago biochemical systems had evolved into cells similar in appearance to modern bacteria. We cannot tell from the fossils what kinds of genomes these first real cells had, but from the discussion in the preceding section we can infer that they were made of double-stranded DNA and consisted of a small number of chromosomes, possibly just one, each containing many linked genes.

If we follow the fossil record forward in time we see the first evidence for eukaryotic cells—structures resembling single-celled algae—about 1.4 billion years ago (Figure 18.6) and for multicellular algae about 0.9 billion years ago. Multicellular animals appeared around 640 million years ago, although

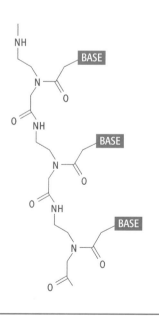

Figure 18.5 A short stretch of peptide nucleic acid. A peptide nucleic acid has an amide backbone instead of the sugar–phosphate structure found in a standard nucleic acid.

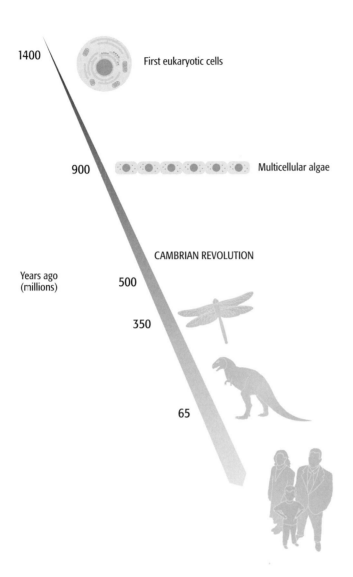

Figure 18.6 The evolution of life.

there are enigmatic burrows suggesting that animals lived earlier than this. The Cambrian Revolution, when invertebrate life proliferated into many novel forms, occurred 530 million years ago and ended with the disappearance of many of the novel forms in a mass extinction 500 million years ago. Since then, evolution has continued apace and with increasing diversification: the first terrestrial insects, animals, and plants were established by 350 million years ago, the dinosaurs had been and gone by the end of the Cretaceous, 65 million years ago, and the first hominoids appeared a mere 4.5 million years ago.

Morphological evolution was accompanied by genome evolution. It is dangerous to equate evolution with "progress" but it is undeniable that as we move up the evolutionary tree we see increasingly complex genomes. One indication of this complexity is gene number, which varies from less than 1000 in some bacteria to 30,000–40,000 in vertebrates such as humans. Within individual lineages—within the bacteria, for example—changes in gene number have probably been gradual, with the acquisition of new genes balanced at least in part by the loss of existing ones. We should also remember that the evolutionary pathways of some organisms have involved a *decrease* rather than increase in gene number, leading to the minimal genomes of

Mycoplasma and other parasitic species (Section 8.2.2). The gradual changes within lineages were, however, punctuated by at least two transition periods when new organisms with greatly increased gene numbers appeared. One of these transitions accompanied the arrival of the first eukaryotes about 1.4 billion years ago, these cells probably containing at least 10,000 genes (the minimum in modern eukaryotes) compared with the 5000 or fewer typical of prokaryotes. The second transition was associated with the arrival of the first vertebrates soon after the end of the Cambrian, these, like modern vertebrates, probably having at least 30,000 genes.

There are two fundamentally different ways in which new genes could be acquired by a genome:

- By duplicating some or all of the existing genes in the genome.
- By acquiring genes from other species.

Both events have been important in genome evolution, as we will see in the next two sections.

18.2.1 Acquisition of new genes by duplication events

A central role for gene duplication in genome evolution was first proposed in 1970. The initial result of gene duplication will be two identical genes. Selective constraints will ensure that one of these genes retains its original nucleotide sequence, or something very similar to it, so that it can continue to provide the protein function that was originally supplied by the single gene copy before the duplication took place. It is possible that the same selective constraints will apply to the second gene, especially if the increase in the rate of synthesis of the gene product, made possible by the duplication, confers a benefit on the organism (Figure 18.7). More frequently, however, the second copy will confer no benefit and hence will not be subject to the same selective pressures and so will accumulate mutations at random. Evidence shows that the majority of new genes that arise by duplication acquire deleterious mutations that inactivate them so that they become pseudogenes, examination of existing pseudogenes suggesting that the commonest inactivating mutations are frameshifts and nonsense mutations that occur within the coding region of the gene. Occasionally, however, mutations might not lead to inactivation of the gene but instead result in a new gene function that is useful to the organism (see Figure 18.7).

Figure 18.7 Three scenarios for the outcome of a gene duplication.

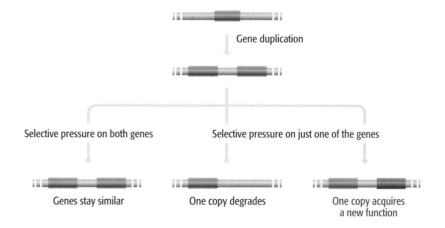

Gene duplication

Selective pressure on both genes   Selective pressure on just one of the genes

Genes stay similar   One copy degrades   One copy acquires a new function

We will first consider the evidence for past gene duplications that is contained in present-day genome sequences, and then look at the mechanisms by which gene duplication can occur.

Genome sequences provide extensive evidence of past gene duplications

The most cursory examination of a genome sequence provides ample evidence that many genes have arisen by duplication events. The importance of the first scenario illustrated in Figure 18.7, where the increased amount of gene product resulting from a gene duplication is beneficial and stabilizes the duplication, is supported by the many examples of multigene families made up of genes with identical or near-identical sequences. The prime examples are the rRNA genes, whose copy numbers range from two in *Mycoplasma genitalium* to 500 or more in *Xenopus laevis*, with all of the copies having virtually the same sequence. These multiple copies of identical genes presumably reflect the need for rapid synthesis of rRNAs at certain stages of the cell cycle. Note that the existence of these multigene families indicates not only that gene duplications have occurred in the past but also that there must be a molecular mechanism that ensures that the family members retain their identity over evolutionary time. This is called **concerted evolution**. If one copy of the family acquires an advantageous mutation then it is possible for that mutation to spread throughout the family until all members possess it. The most likely way in which this can be achieved is by gene conversion which, as described in Section 17.1.1, can result in the sequence of one copy of a gene being replaced with all or part of the sequence of a second copy. Multiple gene conversion events could therefore maintain identity among the sequences of the individual members of a multigene family, especially if those members are arranged in tandem arrays.

Figure 18.8 Evolution of the globin gene superfamily of humans. The members of the superfamily are now on different chromosomes. The neuroglobin gene is on chromosome 14, the cytoglobin gene is on chromosome 17, and the myoglobin gene is on chromosome 22. The α-globin cluster is on chromosome 16 and the β-globin cluster is on chromosome 11. Abbreviation: MYr, millions of years ago.

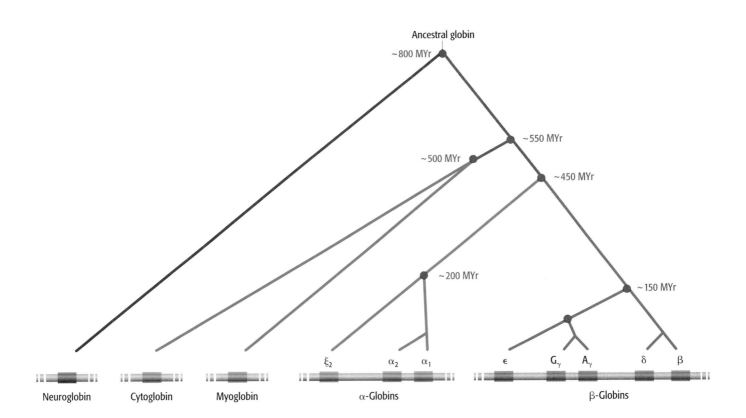

The third scenario in Figure 18.7 results in the duplicated gene accumulating mutations that give it a new, useful function. Again, multigene families provide many indications that such events have occurred frequently in the past. We have already seen that gene duplications in the globin gene families led to the evolution of new globin proteins that are used by the organism at different stages in its development (see Figure 7.19). We noted that all the globin genes, both the α- and β-types, have related nucleotide sequences and hence form part of a superfamily that includes genes specifying various other proteins which, like the blood globins, have the capacity to bind oxygen molecules. From the degrees of similarity displayed by pairs of genes in the superfamily, it is possible to deduce the pattern of gene duplications that gave rise to the genes we see today, and by applying the **molecular clock** (Section 19.2.2) to the data we can estimate how many millions of years ago each duplication took place. These analyses tell us that a duplication some 800 million years ago resulted in a pair of ancestral genes, one of which evolved into the modern gene for the brain protein neuroglobin, and the other of which gave rise to all the other members of the superfamily (Figure 18.8). Some 250 million years later there was a second duplication on the path leading to the blood globins, one of the products of this duplication being a gene that, via another duplication, gave rise to myoglobin, which is active in muscle, and cytoglobin, which is present in many tissues but whose function is not yet understood. The proto-α and proto-β lineages split by a duplication that occurred 450 million years ago and the duplications within the α- and β-globin gene families took

Figure 18.9 The evolution of the mammalian β-globin genes. Reprinted from *Genomics*, Vol. 13, Tagle *et al.*, 'The β-globin gene cluster …,' 741–760, 1992, wither permission from Elsevier.

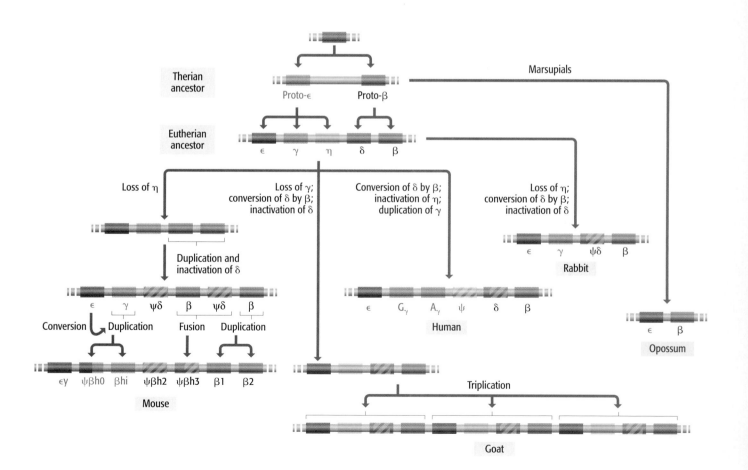

place during the last 200 million years. Within this more recent time frame it is possible to deduce not only the pattern of gene duplication but also some of the more detailed changes that have occurred within individual genes. Hence the events leading to the various groups of β-globin genes present in different mammals have been inferred (Figure 18.9).

We observe similar patterns of evolution when we compare the sequences of other genes. The trypsin and chymotrypsin genes, for example, are related by a common ancestral gene that duplicated approximately 1500 million years ago. Both now code for proteases involved in protein breakdown in the vertebrate digestive tract, trypsin cutting other proteins at arginine and lysine amino acids, and chymotrypsin cutting at phenylalanines, tryptophans, and tyrosines. Genome evolution has therefore produced two complementary protein functions where originally there was just one.

Another striking example of gene evolution by duplication is provided by the homeotic selector genes, the key developmental genes responsible for specification of the body plans of animals. As described in Section 14.3.4, *Drosophila* has a single cluster of homeotic selector genes, called HOM-C, which consists of eight genes each containing a homeodomain sequence coding for a DNA-binding motif in the protein product (see Figure 14.37). These eight genes, as well as other homeodomain genes in *Drosophila*, are believed to have arisen by a series of gene duplications that began with an ancestral gene that existed about 1000 million years ago. The functions of the modern genes, each specifying the identity of a different segment of the fruit fly, gives us a tantalizing glimpse of how gene duplication and sequence divergence could, in this case, have been the underlying processes responsible for increasing the morphological complexity of the series of organisms in the *Drosophila* evolutionary tree. If we then move further up the evolutionary tree we see that vertebrates have four Hox gene clusters (see Figure 14.37), each a recognizable copy of the *Drosophila* cluster, with sequence similarities between genes in equivalent positions. The implication is that in the vertebrate lineage there were two duplications, not of individual Hox genes but of the entire cluster (Figure 18.10). Not all of the vertebrate Hox genes have been ascribed functions, but we believe that the additional versions possessed by vertebrates relate to the added complexity of the vertebrate body plan. Two observations support this conclusion. The amphioxus, an invertebrate that displays some primitive vertebrate features, has two Hox clusters, which is what we might expect for a primitive "protovertebrate." Ray-finned fishes, probably the most diverse group of vertebrates with a vast range of different variations of the basic body plan, have seven Hox clusters.

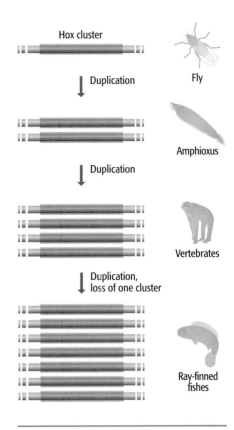

Figure 18.10 Evolution of the Hox gene cluster from flies to ray-finned fishes.

A variety of processes could result in gene duplication

There are several ways in which a short segment of a DNA molecule, possibly containing one or a small group of genes, could be duplicated. These include:

- **Unequal crossing-over**, which is a recombination event that is initiated by similar nucleotide sequences that are not at identical places in a pair of homologous chromosomes. As shown in Figure 18.11A, the result of unequal crossing-over can be duplication of a segment of DNA in one of the recombination products.

- **Unequal sister chromatid exchange**, which occurs by the same mechanism as unequal crossing-over, but involves a pair of chromatids from a single chromosome (Figure 18.11B).

(A) Unequal crossing-over

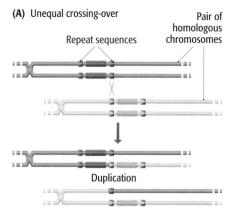

Repeat sequences

Pair of homologous chromosomes

Duplication

(B) Unequal sister chromatid exchange

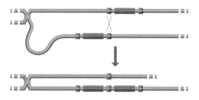

(C) During DNA replication

Replication fork   Replication fork

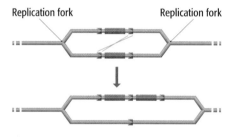

Figure 18.11 Models for gene duplication by (A) unequal crossing-over between homologous chromosomes, (B) unequal sister chromatid exchange, and (C) during replication of a bacterial genome. In each case, recombination occurs between two different copies of a short repeat sequence, leading to duplication of the sequence between the repeats. Unequal crossing-over and unequal sister chromatid exchange are essentially the same except that the first involves chromatids from a pair of homologous chromosomes and the second involves chromatids from a single chromosome. In (C), recombination occurs between two daughter double helices that have just been synthesized by DNA replication.

- DNA amplification, which is sometimes used in this context to describe duplication of segments of DNA in bacteria and other haploid organisms, the duplications arising by unequal recombination between the two daughter DNA molecules in a replication bubble (Figure 18.11C).

- Replication slippage (see Figure 16.5), which can result in duplication of short segments such as microsatellite sequences. This process could conceivably duplicate a region large enough to contain an entire gene, but in practice this is unlikely.

Each of the four processes listed above lead to tandem duplications—ones in which the two duplicated segments lie adjacent to one another in the genome. This is the pattern seen with many multigene families, such as the α-globin gene family on human chromosome 16 and the β-globin family on chromosome 11, but it is not the only possibility. Family members are not always colocated: for example, in the human genome there are three functional genes for the metabolic enzyme aldolase, each on a different chromosome. These copies might have once been present as a tandem array, and become dispersed as part of large-scale genome reorganizations, but it is also possible that the distant locations are a consequence of the duplication process. This would be the case if the duplication occurred by retrotransposition in a manner similar to that thought to lead to formation of processed pseudogenes (see Figure 7.20). A processed pseudogene arises when the mRNA copy of a gene is converted into cDNA and reinserted into the genome. The resulting structure is a pseudogene because it lacks a promoter sequence, this being absent from the mRNA. The pseudogene could conceivably be inserted adjacent to the promoter of an existing gene and hence become active by subverting this promoter for its own use. Gene duplicates that arise in this way are called **retrogenes** and, again, many examples are known in different genomes (in the human genome the testis-specific version of the pyruvate dehydrogenase gene is a retrogene). A distinctive feature of a retrogene is that it lacks any introns present in the parent copy of the gene, as these are not present in the mRNA. Recently it has been realized that complete copies of genes, including not only the introns but also some or all of the promoter sequences, can also be made by reverse transcription. This can occur when the RNA that is reverse-transcribed is not an mRNA, but an antisense copy of the gene made by transcription of the "wrong" polynucleotide (Figure 18.12). We are beginning to realize that antisense RNAs are not uncommon and may indeed have a role in gene regulation, similar perhaps to the mode of action of microRNAs (Section 12.2.6). If converted into cDNA, then an antisense RNA could provide a full-length, functional gene duplicate for insertion into the genome, at a position that could be distant from the original copy.

Whole genome duplication is also possible

The processes described above give rise to relatively short DNA duplications, perhaps a few tens of kilobases in length. Are larger duplications possible? It seems unlikely that duplication of entire chromosomes has played any major role in genome evolution, because we know that duplication of individual human chromosomes, resulting in a cell that contains three copies of one chromosome and two copies of all the others (the condition called **trisomy**), is either lethal or results in a genetic disease, such as Down syndrome, and similar harmful effects have been observed in artificially generated trisomic mutants of *Drosophila*. Probably, the resulting increase in copy numbers for

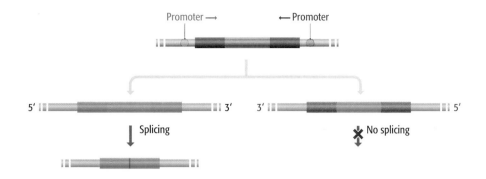

Figure 18.12 **An antisense RNA copy of a gene will retain the gene's introns.** On the left, the gene is transcribed from its standard promoter to give a pre-mRNA whose intron is removed by splicing. On the right, an antisense RNA is synthesized from a downstream promoter. This RNA is not spliced because the sequences at its exon–intron junctions are not the standard ones recognized by the spliceosome (Section 12.2.2).

some genes but not others leads to an imbalance of the gene products and disruption of the cellular biochemistry.

The harmful effects of trisomy do not mean that duplication of the entire set of chromosomes in a nucleus must be discounted. Genome duplication can occur if an error during meiosis leads to the production of gametes that are diploid rather than haploid (Figure 18.13). If two diploid gametes fuse then the result will be a type of **autopolyploid**, in this case a tetraploid cell whose nucleus contains four copies of each chromosome. Autopolyploidy, as with other types of polyploidy (Section 18.2.2), is not uncommon, especially among plants. Autopolyploids are often viable because each chromosome still has a homologous partner and so can form a bivalent during meiosis. This allows an autopolyploid to reproduce successfully, but generally prevents interbreeding with the original organism from which it was derived. This is because a cross between, for example, a tetraploid and diploid would

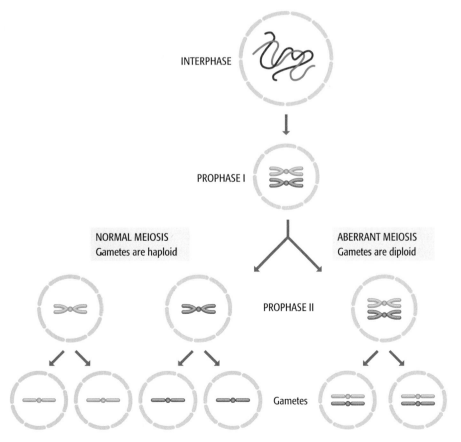

Figure 18.13 **The basis of autopolyploidization.** The normal events occurring during meiosis are shown, in abbreviated form, on the left (compare with Figure 3.16). On the right, an aberration has occurred between prophase I and prophase II and the pairs of homologous chromosomes have not separated into different nuclei. The resulting gametes will be diploid rather than haploid.

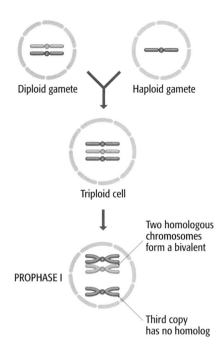

Diploid gamete   Haploid gamete

Triploid cell

PROPHASE I

Two homologous chromosomes form a bivalent

Third copy has no homolog

Figure 18.14 Autopolyploids cannot interbreed successfully with their parents. Fusion of a diploid gamete produced by the aberrant meiosis shown in Figure 18.13 with a haploid gamete produced by normal meiosis leads to a triploid nucleus, one that has three copies of each homologous chromosome. During prophase I of the next meiosis, two of these homologous chromosomes will form a bivalent but the third will have no partner. This has a disruptive effect on the segregation of chromosomes during anaphase (see Figure 3.16) and usually prevents meiosis from reaching a successful conclusion. This means that gametes are not produced and the triploid organism is sterile. Note that the bivalent could have formed between any two of the three homologous chromosomes, not just between the pair shown in the diagram.

give a triploid offspring which would not itself be able to reproduce because one full set of its chromosomes would lack homologous partners (Figure 18.14). Autopolyploidy is therefore a mechanism by which speciation can occur, a pair of species usually being defined as two organisms that are unable to interbreed. The generation of new plant species by autopolyploidy has in fact been observed, notably by Hugo de Vries, one of the rediscoverers of Mendel's experiments. During his work with evening primrose, *Oenothera lamarckiana*, de Vries isolated a tetraploid version of this normally diploid plant, which he named *Oenothera gigas*. Autopolyploidy among animals is less common, especially in those with two distinct sexes, possibly because of problems that arise if a nucleus possesses more than one pair of sex chromosomes. It is not impossible though, and at least one mammal, the red viscacha rat of Argentina, has a tetraploid genome.

Analysis of modern genomes provides evidence for past genome duplications

Autopolyploidy does not lead directly to an increase in gene number because the initial product is an organism that simply has extra copies of every gene, rather than any new genes. It does, however, provide the *potential* for an increase because the extra genes are not essential to the functioning of the cell and so can undergo mutational change without harming the viability of the organism. With this in mind, is there any evidence that whole genome duplication has been important in the large-scale acquisition of new genes during the evolutionary histories of present-day genomes?

From what we understand about the way in which genomes change over time, we might anticipate that evidence for whole genome duplication would be quite difficult to obtain. Many of the extra gene copies resulting from genome duplication would decay to the extent that they are no longer visible in the DNA sequence. Those genes that are retained, because their duplicated function is useful to the organism or because they have evolved new functions, should be identifiable, but it would be difficult to distinguish if they have arisen by duplication of the entire genome or by duplication of much smaller segments. For a genome duplication to be signaled it would be necessary to find large, duplicated *sets* of genes, with the same order of genes in both sets. To what extent these duplicated sets are still visible in the genome will depend on how frequently past recombination events have moved genes to new positions.

To search the *Saccharomyces cerevisiae* genome for evidence of past duplications, homology analysis (Section 5.2.1) was carried out with every yeast gene tested against every other yeast gene. To be considered descendants of a duplication event, two genes had to display at least 25% identity when the predicted amino acid sequences of their protein products were compared. About 800 gene pairs were identified in this way, 376 of which could be placed in 55 duplicate sets, each of these sets containing at least three genes in the same order (Figure 18.15), possibly with other genes interspersed between them, the sets altogether covering half the genome. These sets could have arisen by duplication of segments rather than the entire genome, but if this was the case then it might be anticipated that some of the genes would have been duplicated more than once. The fact that there were just two copies of each gene, and never three or four, therefore supported the notion that the copies arose by whole genome duplication. This possibility became more certain when the complete genome sequences of other yeast species were obtained. Comparisons between the genomes of *S. cerevisiae, Kluyveromyces lactis,* and *Ashbya gossypii* have been particularly informative. These three species shared a common ancestor that lived over 100 million years ago, previous to the time of the genome duplication event inferred from the homology analysis. If that duplication had indeed occurred in the lineage leading to *S. cerevisiae* then it would be anticipated that this species would have duplicated copies of many genes present as singletons in the *K. lactis* and *A. gossypii* genomes. This turns out to be the case, this new analysis suggesting that some 10% of the genes in the modern *S. cerevisiae* genome derive from a whole genome duplication that occurred just under 100 million years ago.

Equivalent work has been carried out with other genomes and the picture that is emerging is that whole genome duplication has been a relatively frequent event in the evolution of many groups of organisms. For example, comparisons between the *Arabidopsis thaliana* genome sequence and segments of other plant genomes suggest that the ancestor of the *A. thaliana* genome underwent four rounds of genome duplication between 100 million and 200 million years ago. The human genome and other mammalian genomes also contain so many gene duplicates that at least one whole genome duplication event is thought to have occurred in this lineage between 350 million and 600 million years ago.

Smaller duplications can also be identified in the human genome and other genomes

Although the most recent whole genome duplication in the human lineage occurred a considerable time ago, the human genome has not been quiescent in the intervening period. Indeed, the opposite is true, one surprise provided by the human genome sequence being the realization that there has been extensive and frequent duplication of short segments of the genome in the relatively recent past. This is illustrated by Figure 18.16, which depicts the duplication events that are inferred for the last 35 million years of evolution of the long arm of human chromosome 22. As in the yeast study described above, these duplications have been identified by comparing different parts of the genome, in this case to identify regions of more than 1 kb in length that display 90% or higher nucleotide sequence similarity. This analysis ignores interspersed repeat sequences and so locates regions that

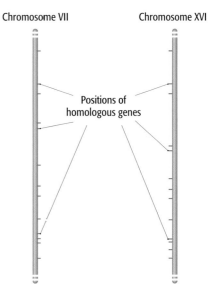

Figure 18.15 An example of a duplicated set of genes in the *S. cerevisiae* genome. Each of the three pairs of genes that are indicated have high sequence homology. This and other duplicated sets provide evidence for a genome duplication just under 100 million years ago in the *S. cerevisiae* lineage.

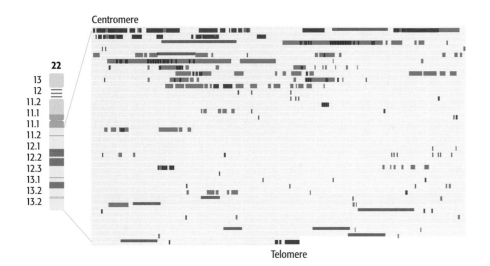

Figure 18.16 Segment duplications in the long arm of human chromosome 22. The diagram depicts the 34 Mb of the long arm of chromosome 22 as a series of thin horizontal lines, each one representing 1 Mb of DNA sequence, running from the centromere at the top to the telomere at the bottom. This analysis was based on the draft human genome sequence which contained eleven gaps in this region, shown as black bars. The pink boxes, which make up 3.9% of the 34 Mb, are sequences that are duplicated within this chromosome arm, and the blue boxes (6.4% of the total) are duplications of regions in other chromosomes.

are probably the products of relatively recent duplication events. Almost 200 segments, making up over 10% of this 34 Mb region of the human genome, appear to have arisen by duplications that occurred within the last 35 million years. Over 100 of these segments are duplicates of sequences present on other chromosomes, and the remainder are duplicates with both copies located within this arm.

The pattern of duplications in the long arm of chromosome 22 is fairly typical of the human genome as a whole. Individual duplications range from 1 to 400 kb in length and there is a distinct bias toward the areas adjacent to the centromeres, with relatively few duplications in the more distal regions of each chromosome arm. When considering the size of these segment duplications, we must bear in mind that the average human gene is 20–25 kb in length and that the genes are scattered sparsely throughout the genome. When the sequences of the segment duplications are examined it emerges that few of these duplication events result in duplication of entire genes, but that several involve parts of genes, and that some of these duplications lead to the upstream exons of one gene being placed alongside downstream exons of a second gene. Some of these new combinations are transcribed, but it is not clear whether the transcripts are functional. The evolutionary potential of segment duplications is therefore uncertain. What is clear, however, is that recombination between a pair of intrachromosomal duplications can result in deletion of the region between the duplications, and that deletions generated in this way can give rise to genetic disease. An example is Charcot–Marie–Tooth syndrome, which develops 5–15 years after birth and is characterized by degeneration of the peripheral nervous system, leading to weakness and difficulty in walking. The syndrome is caused by recombination between a pair of 24 kb segment duplications on human chromosome 17, deleting a 1.5 Mb segment of the genome. The loss of the genes contained in this segment gives rise to the disease.

Genome evolution also involves rearrangement of existing genes

The range of sizes observed for segment duplications in the human genome raises the possibility that the duplication events illustrated in Figure 18.11 could, as well as giving rise to new copies of genes, cause alterations *within* existing genes. This would be an alternative way in which novel protein functions could evolve. This is possible because most proteins are made up of structural domains, each comprising a segment of the polypeptide chain and hence encoded by a contiguous series of nucleotides (Figure 18.17). Rearrangement of domain-encoding gene segments could result in novel protein functions:

- **Domain duplication** would occur when the gene segment coding for a structural domain is duplicated by unequal crossing-over, replication slippage, or one of the other methods that we have considered for duplication of DNA sequences (Figure 18.18A). Duplication could result in the structural domain being repeated in the protein, which might itself be

Figure 18.17 Each structural domain is an individual unit in a polypeptide chain and is coded by a contiguous series of nucleotides. In this simplified example, each secondary structure in the polypeptide is looked upon as an individual structural domain. In reality, most structural domains comprise two or more secondary structural units.

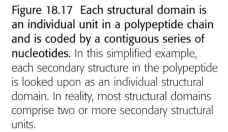
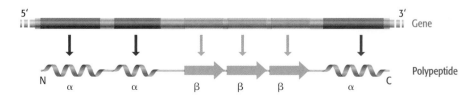

(A) Domain duplication

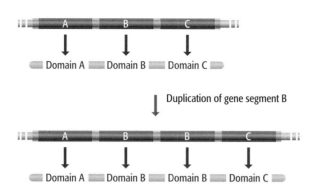

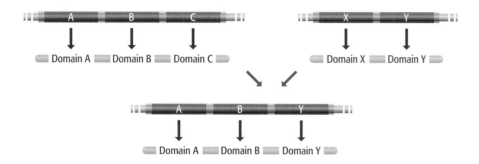

advantageous, for example by making the protein more stable. The duplicated domain might also change over time as its coding sequence becomes mutated, leading to a modified structure that might provide the protein with a new activity. Note that domain duplication causes the gene to become longer. Gene elongation appears to be a general consequence of genome evolution, the genes of higher eukaryotes being longer, on average, than those of lower organisms.

● **Domain shuffling** would occur when segments coding for structural domains from completely different genes are joined together to form a new coding sequence that specifies a hybrid or mosaic protein, one that would have a novel combination of structural features and might provide the cell with an entirely new biochemical function (Figure 18.18B).

Implicit in these models of domain duplication and shuffling is the need for the relevant gene segments to be separated so that they can themselves be rearranged and shuffled. This requirement has led to the attractive suggestion that exons might code for structural domains. With some proteins, duplication or shuffling of exons does seem to have resulted in the structures seen today. An example is provided by the α2 Type I collagen gene of vertebrates, which codes for one of the three polypeptide chains of collagen. Each of the three collagen polypeptides has a highly repetitive sequence made up of repeats of the tripeptide glycine–X–Y, where X is usually proline and Y is usually hydroxyproline (Figure 18.19). The chicken α2 Type I gene is split into 52 exons, 42 of which cover the part of the gene coding for the glycine–X–Y repeats. Within

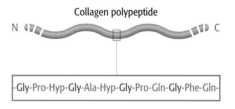

Figure 18.19 The α2 Type I collagen polypeptide has a repetitive sequence described as Gly–X–Y. Every third amino acid is glycine, X is often proline, and Y is often hydroxyproline (Hyp). Hydroxyproline is synthesized by posttranslational modification of proline (Section 13.3.3). The collagen polypeptide has a helical conformation, but one that is more extended than the standard α-helix.

Figure 18.20 The modular structure of the tissue plasminogen activator gene.

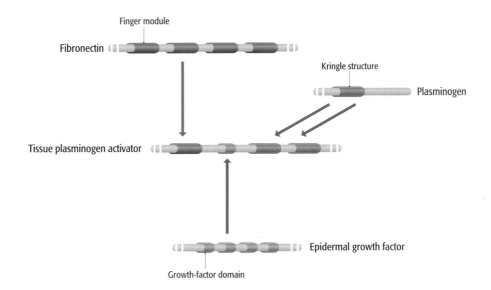

Figure 18.20 The modular structure of the tissue plasminogen activator gene.

this region, each exon encodes a set of complete tripeptide repeats. The number of repeats per exon varies but is 5 (in 5 exons), 6 (23 exons), 11 (5 exons), 12 (8 exons), or 18 (1 exon). Clearly this gene could have evolved by duplication of exons leading to repetition of the structural domains.

Domain shuffling is illustrated by tissue plasminogen activator (TPA), a protein found in the blood of vertebrates and which is involved in the blood clotting response. The TPA gene has four exons, each coding for a different structural domain (Figure 18.20). The upstream exon codes for a "finger" module that enables the TPA protein to bind to fibrin, a fibrous protein found in blood clots and which activates TPA. This exon appears to be derived from a second fibrin-binding protein, fibronectin, and is absent from the gene for a related protein, urokinase, which is not activated by fibrin. The second TPA exon specifies a growth factor domain which has apparently been obtained from the gene for epidermal growth factor and which may enable TPA to stimulate cell proliferation. The last two exons code for "kringle" structures which TPA uses to bind to fibrin clots; these kringle exons come from the plasminogen gene.

Type I collagen and TPA provide elegant examples of gene evolution but, unfortunately, the clear links that they display between structural domains and exons are exceptional and are rarely seen with other genes. Many other genes appear to have evolved by duplication and shuffling of segments, but in these the structural domains are coded by segments of genes that do not coincide with individual exons or even groups of exons. Domain duplication and shuffling still occur, but presumably in a less precise manner and with many of the rearranged genes having no useful function. Despite being haphazard, the process clearly works, as indicated by, among other examples, the number of proteins that share the same DNA-binding motifs (Section 11.1.1). Several of these motifs probably evolved *de novo* on more than one occasion, but it is clear that in many cases the nucleotide sequence coding for the motif has been transferred to a variety of different genes.

One possible mechanism for moving gene segments around a genome is in association with transposable elements. The transposition of a LINE-1 element (Section 9.2.1) can occasionally result in a short piece of the adjacent DNA

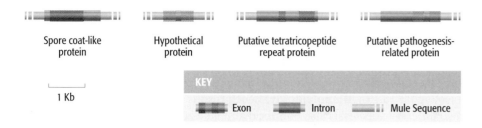

Spore coat-like protein Hypothetical protein Putative tetratricopeptide repeat protein Putative pathogenesis-related protein

1 Kb

KEY

Exon Intron Mule Sequence

Figure 18.21 MULEs often contain segments of genes captured from the host chromosomes. The diagram shows five adjacent MULEs from chromosome 1 of rice. Colored segments are gene exons and gray segments between and adjacent to exons are introns. The thin gray bars at the ends of the structures are the terminal inverted repeats of the MULE transposons. The identities of the gene segments are given when known.

being transferred along with the transposon, a process called **3′ transduction**, the transferred segment being located at the 3′ end of the element. LINE-1 elements are sometimes found in introns so 3′ transduction could conceivably move downstream exons to new sites in a genome. Movement of exons and other gene segments might also be brought about by DNA transposons called *Mutator*-like transposable elements (MULEs), which are found in many eukaryotes but are especially common in plants. MULEs often contain within their DNA sequence segments of genes captured from the host genome (Figure 18.21). Transposition of a MULE would therefore move the captured segments to a new location. MULEs can collect segments of different genes as they travel around a genome, assembling new hybrid genes as they go. MULEs therefore provide an attractive way of driving gene evolution, but there are still several unanswered questions about their impact. In particular, it is not yet clear how frequently gene segments are able to escape from MULEs.

18.2.2 Acquisition of new genes from other species

The second possible way in which a genome can acquire new genes is to obtain them from another species. Comparisons of bacterial and archaeal genome sequences suggest that **lateral gene transfer** has been a major event in the evolution of prokaryotic genomes (Section 8.2.3). The genomes of most bacteria and archaea contain at least a few hundred kilobases of DNA, representing tens of genes, that appears to have been acquired from a second prokaryote.

There are several mechanisms by which genes can be transferred between prokaryotes but it is difficult to be sure how important these various processes have been in shaping the genomes of these organisms. Conjugation (Section 3.2.4), for example, enables plasmids to move between bacteria and frequently results in the acquisition of new gene functions by the recipients. On a day-to-day basis, plasmid transfer is important because it is the means by which genes for resistance to antibiotics such as chloramphenicol, kanamycin, and streptomycin spread through bacterial populations and across species barriers, but its evolutionary relevance is questionable. It is true that the genes transferred by conjugation can become integrated into the recipient bacterium's genome, but usually the genes are carried by composite transposons (see Figure 9.17B), which means that the integration is reversible and so might not result in a permanent change to the genome. A second process for DNA transfer between prokaryotes, transformation (Section 3.2.4), is more likely to have had an influence on genome evolution. Only a few bacteria, notably members of the *Bacillus*, *Pseudomonas*, and *Streptococcus* genera, have efficient mechanisms for the uptake of DNA from the surrounding environment, but efficiency of DNA uptake is probably not relevant when we are dealing with an evolutionary timescale. More important is the fact that gene flow by transformation can occur between any pair of prokaryotes, not just closely related ones (as is the case with conjugation), and so could account for the transfers that appear to have occurred between bacterial and archaeal genomes (Section 8.2.3).

In plants, new genes can be acquired by polyploidization. We have already seen how autopolyploidization can result in genome duplication in plants (see Figure 18.13). **Allopolyploidy**, which results from interbreeding between two different species, is also common and, like autopolyploidy, can result in a viable hybrid. Usually, the two species that form the allopolyploid are closely related and have many genes in common, but each parent will possess a few novel genes or at least distinctive alleles of shared genes. For example, the bread wheat, *Triticum aestivum*, is a hexaploid that arose by allopolyploidization between cultivated emmer wheat, *T. turgidum*, which is a tetraploid, and a diploid wild grass, *Aegilops squarrosa*. The wild grass nucleus contained novel alleles for the high-molecular-weight glutenin genes which, when combined with the glutenin alleles already present in emmer wheat, resulted in the superior properties for breadmaking displayed by the hexaploid wheats. Allopolyploidization can therefore be looked upon as a combination of genome duplication and interspecies gene transfer.

Among animals, the species barriers are less easy to cross and it is difficult to find clear evidence for lateral gene transfer of any kind. Several eukaryotic genes have features associated with archaeal or bacterial sequences, but rather than being the result of lateral gene transfer, these similarities are thought to result from conservation during millions of years of parallel evolution. Most proposals for gene transfer between animal species center on retroviruses and transposable elements. Transfer of retroviruses between animal species is well documented, as is their ability to carry animal genes between individuals of the same species, suggesting that they might be possible mediators of lateral gene transfer. The same could be true of transposable elements such as P elements, which are known to spread from one *Drosophila* species to another, and *mariner*, which has also been shown to transfer between *Drosophila* species and which may have crossed from other species into humans.

18.3 Noncoding DNA and Genome Evolution

So far we have concentrated our attention on the evolution of the coding component of the genome. As coding DNA makes up only 1.5% of the human genome, our view of genome evolution would be very incomplete if we did not devote some time to considering noncoding DNA. The presence of extensive amounts of noncoding DNA in eukaryotic genomes is a puzzle for molecular evolutionists. Why is this apparently superfluous DNA tolerated? One possibility is that the noncoding DNA has a function that has not yet been identified and, as such, must be maintained because without it the cell would be nonviable. Plausible functions are not as difficult to identify as might be imagined. In several places in the preceding chapters the importance of chromatin structure has been stressed, including the attachment of chromatin to sites within the nucleus. Possibly, some of the noncoding component of a genome is involved in these aspects of genome organization. Alternatively, noncoding DNA might have a broad-ranging control function that so far has eluded discovery by molecular biologists. A final possibility is that noncoding DNA has no function but is tolerated by a genome because there is no selective pressure to get rid of it. If this view is correct then possession of noncoding DNA is neither an advantage nor a disadvantage and so the noncoding DNA is simply propagated along with the coding DNA. According to this hypothesis, the noncoding DNA could simply be "junk" or might be parasitic "selfish DNA."

There is little that can be said about the evolution of much of the noncoding component of a genome. We envisage that duplications and other rearrangements have occurred through recombination and replication slippage, and that sequences have diverged through accumulation of mutations unfettered by the restraining selective forces acting on functional regions of the genome. We recognize that some parts of the noncoding DNA, for example the regulatory regions upstream of genes, have important functions, but for many other parts of the noncoding DNA, all we can say is that it evolves in an apparently random fashion. But this randomness does not apply to all components of the noncoding DNA. In particular, transposable elements and introns have interesting evolutionary histories and are of general importance in genome evolution.

18.3.1 Transposable elements and genome evolution

Transposable elements have a number of effects on evolution of the genome as a whole. The most significant of these is the ability of transposons to initiate recombination events that lead to genome rearrangements. This has nothing to do with the transposable activity of these elements, it simply relates to the fact that different copies of the same element have similar sequences and can therefore initiate recombination between two parts of the same chromosome or between different chromosomes (Figure 18.22). In many cases, the resulting rearrangement will be harmful because important genes will be deleted, but some instances where the result has been beneficial have been documented. Recombination between a pair of LINE-1 elements (Section 9.2.1) approximately 35 million years ago is thought to be the cause of the β-globin gene duplication that resulted in the Gγ and Aγ members of this gene family (see Figure 18.9).

Movement of transposons from one site to another can also have an impact on genome evolution. The possible relocation of gene segments during transposition of LINE elements and MULEs was mentioned in Section 18.2.1. Transposition has also been associated with altered patterns of gene expression. For example, the efficiency with which DNA-binding proteins that are attached to upstream regulatory sequences can activate transcription of a gene might be affected if a transposon moves into a new site immediately upstream of the gene (Figure 18.23). The presence of promoters and/or enhancers within the transposon might also affect transcription by subjecting the adjacent gene to an entirely new regulatory regime. An interesting example of transposon-directed gene expression occurs with the mouse gene *Slp*, which codes for a protein involved in the immune response. The tissue specificity of *Slp* is conferred by an enhancer located within an adjacent retrotransposon. There are also examples where insertion of a transposon into a gene has resulted in an altered splicing pattern.

18.3.2 The origins of introns

Ever since introns were first discovered in the 1970s their origins have been debated. There are few controversies surrounding the Group I, II, and III types (see Table 12.2) as it is generally accepted that all these self-splicing

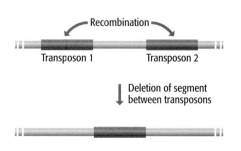

Figure 18.22 Recombination between pairs of repeated sequences, such as transposons, can result in deletion of segments of the genome.

Figure 18.23 Insertion of a transposon into the region upstream of a gene could affect the ability of DNA-binding proteins to activate transcription.

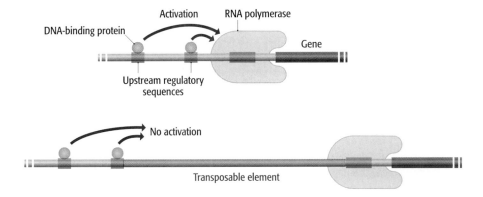

introns evolved in the RNA world and have survived ever since without undergoing a great deal of change. The problems surround the origin of the GU–AG introns, the ones that are found in large numbers in eukaryotic nuclear genomes.

"Introns early" and "introns late": two competing hypotheses

A number of proposals for the origin of GU–AG introns have been put forward but the debate is generally considered to be between two opposing hypotheses:

- **"Introns early"** states that introns are very ancient and are gradually being lost from eukaryotic genomes.

- **"Introns late"** states that introns evolved relatively recently and are gradually accumulating in eukaryotic genomes.

There are several different models for each hypothesis. For "introns early" the most persuasive model is the one also called the "exon theory of genes," which holds that introns were formed when the first DNA genomes were constructed, soon after the end of the RNA world. These genomes would have contained many short genes, each derived from a single coding RNA molecule and each specifying a very small polypeptide, perhaps just a single structural domain. These polypeptides would probably have had to associate together into larger multidomain proteins in order to produce enzymes with specific and efficient catalytic mechanisms (Figure 18.24). To aid the synthesis of a multidomain enzyme it would have been beneficial for the enzyme's individual polypeptides to become linked into a single protein, such as we see today. It is envisaged that this was achieved by splicing together the transcripts of the relevant minigenes, a process that was aided by rearranging the genome so that groups of minigenes specifying the different parts of individual multidomain proteins were positioned next to each other. In other words, the minigenes became exons and the DNA sequences between them became introns.

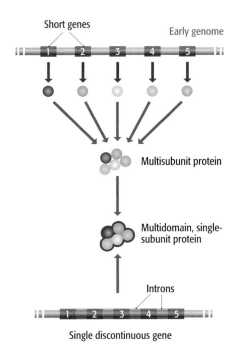

Figure 18.24 The "exon theory of genes." The short genes of the first genomes probably coded for single-domain polypeptides that would have had to associate together to form a multisubunit protein to produce an effective enzyme. Later the synthesis of this enzyme could have been made more efficient by linking the short genes together into one discontinuous gene coding for a multidomain single-subunit protein.

According to the exon theory of genes and other "introns early" hypotheses, all genomes originally possessed introns. But we know that bacterial genomes do not have GU–AG introns, so if these hypotheses are correct then we must assume that for some reason introns became lost from the ancestral bacterial genome at an early stage in its evolution. This is a stumbling block because it is difficult to envisage how a large number of introns could be lost from a genome without risking the disruption of many gene functions. If an intron is removed from a gene with any imprecision then a part of the coding region will be lost or a frameshift mutation will occur, both of which would be expected to inactivate the gene. The "introns late" hypothesis avoids this

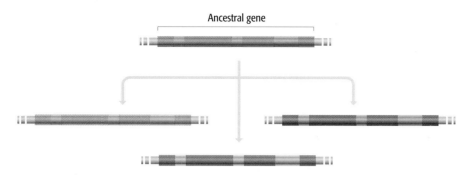

Ancestral gene

Modern genes–in unrelated organisms

Figure 18.25 One prediction of the "introns early" hypothesis is that the positions of introns in homologous genes should be similar in unrelated organisms, because all these genes are descended from an ancestral intron-containing gene.

problem by proposing that, to begin with, no genes had introns, these structures invading the early eukaryotic nuclear genome and subsequently proliferating into the numbers seen today. The similarities between the splicing pathways for GU–AG and Group II introns (Section 12.2.4) suggest that the invaders that gave rise to GU–AG introns might well have been Group II sequences that escaped from organelle genomes. However, the similarity between GU–AG and Group II introns does not prove the "introns late" view, because it is equally possible to devise an "introns early" model, different to the exon theory of genes, in which Group II sequences gave rise to GU–AG introns, but at a very early stage in genome evolution.

The current evidence disproves neither hypothesis

One of the reasons why the debate regarding the origin of GU–AG introns has continued for over 25 years is because evidence in support of either hypothesis has been difficult to obtain and is often ambiguous. One prediction of "introns early" is that there should be a close similarity between the positions of introns in homologous genes from unrelated organisms, because all these genes are descended from an ancestral intron-containing gene (Figure 18.25). Early support for "introns early" came when this was shown to be the case for four introns in animal and plant genes for triosephosphate isomerase. However, when a larger number of species was examined, the positions of the introns in this gene became less easy to interpret: it appeared that introns had been lost in some lineages but gained in others. This scenario fits both "introns early" and "introns late," as both hypotheses allow for the loss, gain, or repositioning of introns by recombination events occurring in individual lineages. When many genes in many organisms are examined, the general picture that emerges is that intron numbers have gradually increased during the evolution of animal genomes, this being put forward as evidence for "introns late," despite the fact that animal mitochondrial genomes do not contain Group II introns that could supplement the existing nuclear introns by repeated invasions. Intron numbers must therefore have increased by recombination events.

An alternative approach has been to try to correlate exons with protein structural domains, as the "introns early" hypothesis predicts that such a link should be evident, even allowing for the fuzzying effects of evolution since the primitive minigenes were assembled into the first real genes. Again, the first evidence to be obtained supported "introns early." A study of vertebrate globin proteins concluded that each of these comprises four structural domains, the first corresponding to exon 1 of the globin gene, the second and third to exon 2, and the fourth to exon 3 (Figure 18.26). The prediction that there should be globin genes with another intron that splits the second and

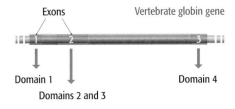

Figure 18.26 A vertebrate globin gene showing the relationship between the three exons and the four domains of the globin protein.

third domains was found to be correct when the leghemoglobin gene of soybean was shown to have an intron at exactly the expected position. Unfortunately, as more globin genes were sequenced, more introns were discovered—more than ten in all. The positions of the majority of these do not correspond to junctions between domains.

The globin genes therefore conform with the general principle that emerged from our discussion of domain shuffling (Section 18.2.1): that in most cases there are no clear links between gene exons and protein structural domains. But is our definition of "structural domain" correct? A structural domain within a protein may not simply correspond with a group of secondary structures such as α-helices and β-sheets. A more subtle interpretation might be that a structural domain is a polypeptide segment whose amino acids are less than a certain distance apart in the protein's tertiary structure. It has been suggested that when this definition is adopted there is a better correlation between structural domain and exons.

18.4 The Human Genome: the Last Five Million Years

Although the evolutionary history of humans is controversial, it is generally accepted that our closest relative among the primates is the chimpanzee and that the most recent ancestor that we share with the chimpanzee lived 4.6–5.0 million years ago. Since the split, the human lineage has embraced two genera—*Australopithecus* and *Homo*—and a number of species, not all of which were on the direct line of descent to *Homo sapiens* (Figure 18.27). The result is us, a novel species in possession of what are, at least to our eyes, important biological attributes that make us very different from all other animals. So how different are we from the chimpanzees?

As far as our genomes are concerned the answer is 1.73%, this being the extent of the nucleotide sequence dissimilarity between humans and chimpanzees. Indeed, when the human and chimpanzee genomes are compared it is much easier to find similarities than differences. The degree of nucleotide sequence identity within the coding DNA is greater than 98.5%, with 29% of

Figure 18.27 One possible scheme for the evolution of modern humans from australopithecine ancestors. There are many controversies in this area of research and several different hypotheses have been proposed for the evolutionary relationships between different fossils. Abbreviation: MYr, million years.

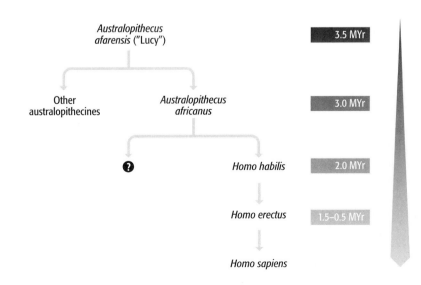

the genes in the human genome coding for proteins whose amino acid sequences are identical to the sequences of their counterparts in chimpanzees. Even in the noncoding regions of the human and chimpanzee genomes the nucleotide identity is rarely less than 97%. Gene order is almost the same in the two genomes, and the chromosomes have very similar appearances. At this level, the most dramatic difference is that human chromosome 2 is two separate chromosomes in chimpanzees (Figure 18.28), so chimpanzees, as well as other apes, have 24 pairs of chromosomes whereas humans have just 23 pairs. The alphoid DNA sequences present at human centromeres (Section 7.1.2) are quite different from the equivalent sequences in chimpanzee and gorilla chromosomes, and Alu elements (Section 9.2.1) are more prevalent in the human genome, but these features probably tell us more about the evolution of repetitive DNA than about the differences between humans and chimpanzees.

Comparisons of the human and chimpanzee genomes have also failed to reveal changes to individual genes which might somehow be key to the special attributes of humans. Analyses designed to reveal genes that have been under positive selection in the human lineage have identified several associated with amino acid breakdown, in line with the greater proportion of meat eaten by humans compared with chimpanzees. Genes providing protection against human diseases such as tuberculosis and malaria have also been subject to positive selection. But no genes with clear roles in brain or neuronal development have been uncovered by this type of analysis. The only substantial difference in gene structure is that humans lack a 92 bp segment of the gene for N-glycolyl-neuraminic acid hydroxylase and so cannot synthesize the hydroxylated form of N-glycolyl-neuraminic acid, which is present on the surfaces of some chimpanzee cells. This may have an effect on the ability of certain pathogens to enter human cells, and could possibly influence some types of cell–cell interaction, but the difference is not thought to be particularly significant. Two other genes that are functional in chimpanzees appear to have been inactivated by point mutations in the human genome, one of these coding for a T-cell receptor and the other for a hair keratin protein, but neither of these changes is likely to have had any substantial impact on human evolution. In any case, it seems impossible that the special features of humans could have arisen as a result of a gene knockout. Instead we need to find genes whose activities have changed rather than been lost. In this regard, considerable interest has centered on the gene for the FOXP2 transcription factor, as defects in this protein result in the human disability called dysarthria, characterized by a difficulty in articulating speech. This gene might therefore underlie the human ability for language. There are indeed two amino acid differences between the FOXP2 genes in humans and chimpanzees, suggesting that this gene has undergone relatively recent change, but a direct link between these amino acid differences and human linguistic ability is elusive.

Human

Chimpanzee

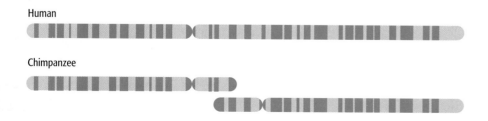

Figure 18.28 Human chromosome 2 is a fusion of two chromosomes that are separate in chimpanzees.

It is now becoming clear that many, if not all, of the key differences between humans and chimpanzees are likely to lie not with the genomes themselves but with the way the genomes are expressed. Attention is therefore moving from the genomes to the transcriptomes and proteomes, and these studies are beginning to suggest that the pattern of genome expression in the brain has undergone significant change in the human lineage since the divergence of humans and chimpanzees. Of course, this is precisely what we might expect, as it is clearly our brains that distinguish us from chimpanzees and other animals. The key question, which has not yet been answered, is whether the identities of the genes that are up- or downregulated in the human brain are informative in any way.

Summary

It is thought that the first polynucleotides to evolve, several billion years ago, were made of RNA rather than DNA. These RNA molecules probably combined a self-replicating ability with some enzymatic activity, and it is possible that enclosure of these in simple, lipid envelopes gave rise to the progenitors of the first cells. Experiments have suggested that short peptides might be constructed by catalytic RNAs, these peptides taking over some of the enzymatic functions of the ribozymes. DNA probably evolved as a more stable version of the RNA protogenomes. During evolution, there have been at least two periods when more complex genomes have emerged, in each case ones with increased gene numbers. Gene duplication is known to be an important event that can result in a genome acquiring new genes. The globin gene superfamily arose by a series of gene duplications whose pattern and timing can be inferred by making comparisons between the sequences of the globin genes in existence today. Duplication events have also played an important role during evolution of the homeotic selector genes that specify the body plan of eukaryotes. Duplication of entire genomes is also possible and is thought to have occurred in the lineages leading to *Saccharomyces cerevisiae*, to *Arabidopsis thaliana*, and also to vertebrates. Smaller duplications, of a few tens of kilobases, have occurred regularly in the recent evolution of the human genome. Some of these have resulted in new combinations of exons, and some genes have clearly arisen by duplication or shuffling of protein domains coded by individual exons. Exons can also be transported around a genome by attachment to transposable elements. Lateral gene transfer results in the acquisition of genes from other species. This has been a regular event in the evolution of prokaryotic genomes but is much less common in eukaryotes, except possibly in plants which can form new polyploids by fusion of gametes from related species. The origin of introns is unclear, with evidence supporting both the "introns early" and "introns late" hypotheses. The former suggests that the first genomes had introns which are gradually being lost, and the latter proposes there has been a gradual increase in intron numbers. Five million years ago the human and chimpanzee lineages diverged. The genomes of humans and chimpanzees still display 98.3% nucleotide sequence identity, with many genes giving rise to identical protein products. Identifying specific features of the human genome that make us human is proving difficult, and it is now thought that the most important differences between humans and chimpanzees might not be the genomes themselves but the way in which the genomes are expressed.

Multiple Choice Questions

18.1.* The first biochemical systems on Earth were probably centered on which types of biomolecule?

  a. Carbohydrates.

  b. DNA.

  c. Proteins.

  d. RNA.

18.2. The term protogenome describes:

  a. The first DNA genomes.

  b. The first cellular RNA genomes.

  c. Early RNA molecules that could self-replicate and direct biochemical reactions.

  d. The first polymeric RNA molecules.

18.3.* Which of the following statements about the transition from RNA to DNA genomes is FALSE?

  a. The phosphodiester bonds in DNA are more stable than those in RNA.

  b. RNA was readily oxidized by the oxygen in the Earth's atmosphere to form DNA.

  c. The replacement of uracil with thymine provided more stability to DNA.

  d. Double-stranded DNA provided a mechanism for repairing genetic material.

18.4. What is concerted evolution?

  a. A process by which two gene products evolve to interact with each other.

  b. A process by which genes are duplicated to provide additional gene products.

  c. A process by which genes are mutated so that they can be recruited to new gene families.

  d. A process by which the members of a gene family retain the same or similar nucleotide sequences.

18.5.* Which of the following processes is thought to underlie concerted evolution?

  a. Gene conversion.

  b. Lateral gene transfer.

  c. Programmed mutation.

  d. Transposition.

18.6. How can the pattern of gene duplications within a multigene family be determined?

  a. By comparing the nucleotide sequences of the genes.

  b. By comparing the physiological functions of the gene products.

  c. By comparing the structures of the gene products.

  d. By comparing the locations of the genes in the genome.

18.7.* Which of the following processes, resulting in gene duplication, occurs when DNA is exchanged between a pair of chromatids within a single chromosome?

  a. DNA amplification.

  b. Replication slippage.

  c. Unequal crossing-over.

  d. Unequal sister chromatid exchange.

18.8. What is the result of autopolyploidy?

  a. A nucleus derived from the fusion of gametes from two different species.

  b. A nucleus containing extra copies of a single chromosome.

  c. A nucleus derived from the fusion of two diploid gametes from the same species.

  d. A nucleus that contains extra copies of the sex chromosomes.

18.9.* Which of the following events could lead to the evolution of a new gene that contains exons from two or more other genes?

  a. Domain duplication.

  b. Domain shuffling.

  c. Gene conversion.

  d. Gene duplication.

18.10. Which process describes the transposition of a DNA segment along with an adjacent transposon?

  a. 3' transduction.

  b. Gene conversion.

  c. Retrotransposition.

  d. Transformation.

18.11.* Within prokaryotes, lateral gene transfer is most likely to occur as a result of which of the following processes?

  a. Conjugation.

  b. Transduction.

  c. Transformation.

  d. Transposition.

18.12. The splicing pathway for CU–AG introns is most similar to the self-splicing pathway of which group of introns?

  a. Group I.

  b. Group II.

continued …

Multiple Choice Questions (continued) *Answers to odd-numbered questions can be found in the Appendix

c. Group III.

d. None of the above.

18.13.* The differences between humans and chimpanzees are most likely due to:

a. The presence of extra genes in the human genome.

b. The deletion of chimpanzee genes from the human genome.

c. Changes in the amino acid sequences of proteins involved in speech.

d. Differences between the expression patterns of the two genomes.

Short Answer Questions *Answers to odd-numbered questions can be found in the Appendix

18.1.* How does the atmosphere of Earth today differ from the atmosphere present when life first evolved?

18.2. How might some of the amino acids, nucleotide bases, and sugars have been synthesized before life evolved?

18.3.* If all present-day living organisms evolved from a single origin, is it possible or likely that there were other origins of biological systems on the ancient Earth? Explain your answer.

18.4. Provide a time line for the evolution of living organisms from the formation of the Earth to the appearance of the first hominoids.

18.5.* Which evolutionary periods are associated with increases in gene numbers in living organisms?

18.6. Discuss how the homeotic selector genes provide a good example of genome evolution by gene duplication.

18.7.* What are the characteristics of retrogenes and how do these arise in the genome?

18.8. What is the evidence for whole genome duplication in the evolutionary lineage leading to *Saccharomyces cerevisiae*?

18.9.* How can segment duplications give rise to new genes?

18.10. In which organisms is allopolyploidy likely to occur?

18.11.* What is the "exon theory of genes"?

18.12. Why do humans and chimpanzees have different numbers of chromosomes?

In-depth Problems *Guidance to odd-numbered questions can be found in the Appendix

18.1.* Give the evidence for the conclusion that the *Arabidopsis thaliana* genome underwent four rounds of genome duplication between 100 million and 200 million years ago.

18.2. Are the examples of domain duplication and domain shuffling given in Section 18.2.1 special cases or are they representative of genome evolution in general?

18.3.* One of the initial publications of the draft human genome sequence (IHGSC [International Human Genome Sequencing Consortium] [2001] Initial sequencing and analysis of the human genome. *Nature* **409**: 860–921) suggested that between 113

and 223 human genes might have been acquired from bacteria by lateral gene transfer. Subsequently it was concluded that this interpretation is incorrect and these genes are not bacterial in origin. What was the evidence that supported lateral transfer of these genes and why was this evidence subsequently discounted?

18.4. Evaluate the "introns early" and "introns late" hypotheses.

18.5.* To what extent do you believe it will be possible to determine the genetic basis to the special attributes of humans from comparisons between the genome sequences of humans and other primates?

Figure Tests

18.1.* Discuss how new genes can arise from gene duplication, based on the pathways shown in the figure. Which of the pathways would give rise to a pseudogene?

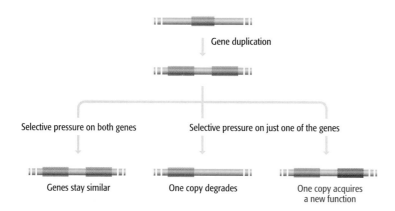

Gene duplication

Selective pressure on both genes   Selective pressure on just one of the genes

Genes stay similar   One copy degrades   One copy acquires a new function

18.3.* Discuss how the processes shown in the figures result in evolution of new genes.

(A)

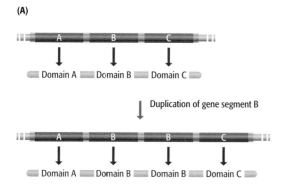

Domain A   Domain B   Domain C

Duplication of gene segment B

Domain A   Domain B   Domain B   Domain C

(B)

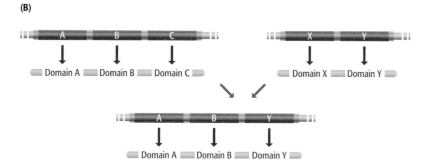

Domain A   Domain B   Domain C   Domain X   Domain Y

Domain A   Domain B   Domain Y

18.2. Which of the following models, (A), (B), or (C), represents gene duplication arising by unequal sister chromatid exchange?

(A)

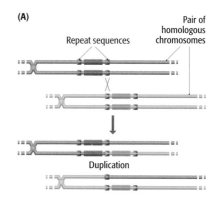

Repeat sequences   Pair of homologous chromosomes

Duplication

(B)

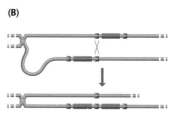

(C)

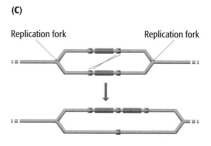

Replication fork   Replication fork

continued ...

Figure Tests (continued)

18.4. Are the evolutionary events depicted in this figure supportive of the "introns early" or "introns late" hypothesis? What is represented by this figure?

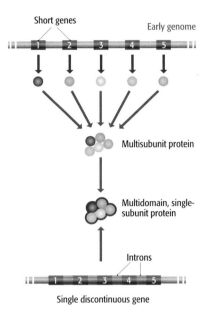

Further Reading

Landmark books

Maynard Smith, J. and Szathmáry, E. (1995) *The Major Transitions in Evolution.* WH Freeman, Oxford. *Begins with the origin of life and ends with the evolution of human language.*

Ohno, S. (1970) *Evolution by Gene Duplication.* George Allen and Unwin, London.

The RNA world and the origins of genomes

Bartel, D.P. and Unrau, P.J. (1999) Constructing an RNA world. *Trends Genet.* **12**: M9–M13.

Freeland, S.J., Knight, R.D. and Landweber, L.F. (1999) Do proteins predate DNA? *Science* **286**: 690–692.

Lohse, P.A. and Szostak, J.W. (1996) Ribozyme-catalysed amino-acid transfer reactions. *Nature* **381**: 442–444.

Miller, S.L. (1953) A production of amino acids under possible primitive Earth conditions. *Science* **117**: 528–529.

Orgel, L.E. (2000) A simpler nucleic acid. *Science* **290**: 1306–1307. *Pyranosyl RNA.*

Robertson, M.P. and Ellington, A.D. (1998) How to make a nucleotide. *Nature* **395**: 223–225.

Unrau, P.J. and Bartel, D.P. (1998) RNA-catalysed nucleotide synthesis. *Nature* **395**: 260–263.

Gene duplications

Amores, A., Force, A., Yan, Y.-L., *et al.* (1998) Zebrafish *hox* clusters and vertebrate genome evolution. *Science* **282**: 1711–1714.

Wagner, A. (2001) Birth and death of duplicated genes in completely sequenced eukaryotes. *Trends Genet.* **17**: 237–239.

Genome and segment duplications

Eichler, E.E. (2001) Recent duplication, domain accretion and dynamic mutation of the human genome. *Trends Genet.* **17**: 661–669.

Goffeau, A. (2004) Seeing double. *Nature* **430**: 25–26. *Comparisons between different yeast species indicate a genome duplication in the Saccharomyces cerevisiae lineage.*

Vision, T.J., Brown, D.G. and Tanksley, S.D. (2000) The origins of genomic duplications in *Arabidopsis. Science* **290**: 2114–2117.

Wolfe, K.H. and Shields, D.C. (1997) Molecular evidence for an ancient duplication of the entire yeast genome. *Nature* **387**: 708–713.

Transposable elements and genome evolution

Jiang, N., Bao, Z., Zhang, X., Eddy, S.R. and Wessler, S.R. (2004) Pack-MULE transposable elements mediate gene evolution in plants. *Nature* **431**: 569–573.

Kazazian, H.H. (2000) L1 retrotransposons shape the mammalian genome. *Science* **289**: 1152–1153.

Origins of introns

de Souza, S.J., Long, M., Schoenbach, L., Roy, S.W. and Gilbert, W. (1996) Intron positions correlate with module boundaries in ancient proteins. *Proc. Natl Acad. Sci. USA* **93**: 14632–14636.

Gilbert, W. (1987) The exon theory of genes. *Cold Spring Harb. Symp. Quant. Biol.* **52**: 901–905.

Gilbert, W., Marchionni, M. and McKnight, G. (1986) On the antiquity of introns. *Cell* **46**: 151–153.

Palmer, J.D. and Logsdon, J.M. (1991) The recent origin of introns. *Curr. Opin. Genet. Dev.* **1**: 470–477.

Humans and chimpanzees

Balter, M. (2005) Are human brains still evolving? Brain genes show signs of selection. *Science* **309**: 1662–1663.

Chou, H.H., Takematsu, H., Diaz, S., *et al.* (1998) A mutation in human CMP-sialic acid hydroxylase occurred after the *Homo–Pan* divergence. *Proc. Natl Acad. Sci. USA* **95**: 11751–11756.

Khaitovich, P., Hellmann, I., Enard, W., Nowick, K., Leinweber, M., Franz, H., Weiss, G., Lachmann, M. and Paabo, S. (2005) Parallel patterns of evolution in the genomes and transcriptomes of humans and chimpanzees. *Science* **309**: 1850–1854.

Li, W.H. and Saunders, M.A. (2005) The chimpanzee and us. *Nature* **437**: 50–51. *Describes the key differences between the human and chimpanzee genomes.*

Muchmore, E.A., Diaz, S. and Varki, A. (1998) A structural difference between the cell surfaces of humans and the great apes. *Am. J. Phys. Anthropol.* **107**: 187–198.

Molecular Phylogenetics

19

When you have read Chapter 19, you should be able to:

Recount how taxonomy led to phylogeny and discuss the reasons why molecular markers are important in phylogenetics.

Describe the key features of a phylogenetic tree and distinguish between inferred trees, true trees, gene trees, and species trees.

Explain how data from which a tree can be reconstructed are obtained from a multiple alignment of DNA sequences.

Outline the basis to the neighbor-joining and maximum parsimony methods for tree reconstruction.

Describe how the accuracy of a tree is assessed.

Discuss, with examples, the applications and limitations of molecular clocks.

Explain why the evolutionary relationships of some DNA sequences cannot be described by a conventional tree, and describe how networks are used to overcome these problems.

Give examples of the use of phylogenetic trees in studies of human evolution and the evolution of the human and simian immunodeficiency viruses.

Outline how genes are studied in populations.

Describe how molecular phylogenetics is being used to study the origins of modern humans, and the migrations of modern humans into Europe and the New World.

If genomes evolve by the gradual accumulation of mutations, then the amount of difference in nucleotide sequence between a pair of genomes should indicate how recently those two genomes shared a common ancestor. Two genomes that diverged in the recent past would be expected to have fewer differences than a pair of genomes whose common ancestor is more ancient. This means that by comparing three or more genomes it should be possible to work out the evolutionary relationships between them. These are the objectives of **molecular phylogenetics**.

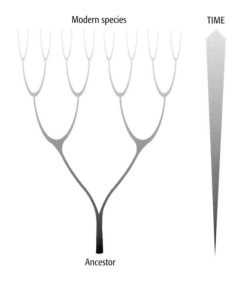

Modern species   TIME

Ancestor

Figure 19.1 The tree of life. An ancestral species is at the bottom of the "trunk" of the tree. As time passes, new species evolve from earlier ones so the tree repeatedly branches until we reach the present time, when there are many species descended from the ancestor.

19.1 From Classification to Molecular Phylogenetics

Molecular phylogenetics predates DNA sequencing by several decades. It is derived from the traditional method for classifying organisms according to their similarities and differences, as first practiced in a comprehensive fashion by Linnaeus in the eighteenth century. Linnaeus was a systematicist not an evolutionist, his objective being to place all known organisms into a logical classification which he believed would reveal the great plan used by the Creator—the *Systema Naturae*. However, he unwittingly laid the framework for later evolutionary schemes by dividing organisms into a hierarchic series of taxonomic categories, starting with kingdom and progressing down through phylum, class, order, family, and genus to species. The naturalists of the eighteenth and early nineteenth centuries likened this hierarchy to a "tree of life" (Figure 19.1), an analogy that was adopted by Darwin in *The Origin of Species* as a means of describing the interconnected evolutionary histories of living organisms. The classificatory scheme devised by Linnaeus therefore became reinterpreted as a **phylogeny** indicating not just the similarities between species but also their evolutionary relationships.

19.1.1 The origins of molecular phylogenetics

Whether the objective is to construct a classification or to infer a phylogeny, the relevant data are obtained by examining variable characters in the organisms being compared. Originally, these characters were morphological features, but molecular data were introduced at a surprisingly early stage. In 1904, Nuttall used immunological tests to deduce relationships between a variety of animals, one of his objectives being to place humans in their correct evolutionary position relative to other primates, an issue that we will return to in Section 19.3.1. Nuttall's work showed that molecular data can be used in phylogenetics, but the approach was not widely adopted until the late 1950s, the delay being due largely to technical limitations, but also partly because classification and phylogenetics had to undergo their own evolutionary changes before the value of molecular data could be fully appreciated.

Phenetics and cladistics require large datasets

These changes began in 1957 with the introduction of **phenetics**, a novel phylogenetic method which challenged the prevailing view that classifications should be based on comparisons between a limited number of characters that taxonomists believed to be important for one reason or another. Pheneticists argued that classifications should encompass as many variable characters as possible, these characters being scored numerically and analyzed by rigorous mathematical methods.

The introduction of phenetics was followed ten years later by another new phylogenetic approach called **cladistics**, which also emphasizes the need for large datasets but differs from phenetics in that it does not give equal weight to all characters. The argument is that in order to infer the branching order in a phylogeny it is necessary to distinguish those characters that provide a good indication of evolutionary relationships from other characters that might be misleading. This might appear to take us back to the prephenetic approach but cladistics is much less subjective: rather than making assumptions about which characters are "important," cladistics demands that the evolutionary relevance of individual characters be defined. In particular, errors in the branching pattern within a phylogeny are minimized by recognizing two types of anomalous data:

- **Convergent evolution**, or **homoplasy**, occurs when the same character state evolves in two separate lineages. For example, both birds and bats possess wings, but bats are more closely related to wingless mammals than they are to birds (Figure 19.2A). The character state "possession of wings" is therefore misleading in the context of vertebrate phylogeny.

- **Ancestral character states** must be distinguished from **derived character states**. An ancestral (or **plesiomorphic**) character state is one possessed by a remote common ancestor of a group of organisms, an example being five toes in vertebrates. A derived (or **apomorphic**) character state is one that evolved from the ancestral state in a more recent common ancestor, and so is seen in only a subset of the species in the group being studied. Among vertebrates, the possession of a single toe, as displayed by modern horses, is a derived character state (Figure 19.2B). If we did not realize this then we might conclude that humans are more closely related to lizards, which like us have five toes, rather than to horses.

Large datasets can be obtained by studying molecular characters

Phenetics and cladistics have had an uneasy relationship over the last 40 years. Most of today's evolutionary biologists favor cladistics, even though a strictly cladistic approach throws up some apparently counterintuitive results, a notable example being the conclusion that the birds should not have their own class (Aves) but be included among the reptiles. But both methodologies share a need for large mathematical datasets as the basic material from which phylogenies are reconstructed. The difficulty in obtaining datasets of this type when morphological characters are used was one of the main driving forces behind the gradual shift toward use of molecular data, which have three advantages compared with other types of phylogenetic information:

- When molecular data are used, a single experiment can provide information on many different characters: in a DNA sequence, for example, every nucleotide position is a character with four **character states**—A, C, G, and T. Large molecular datasets can therefore be generated relatively quickly.

- Molecular character states are unambiguous: A, C, G, and T are easily recognizable and one nucleotide cannot be confused with another. Some

Figure 19.2 Anomalies in phylogenetics. (A) Possession of wings by birds and bats is an example of convergent evolution. (B) The single toe of horses is a derived character state.

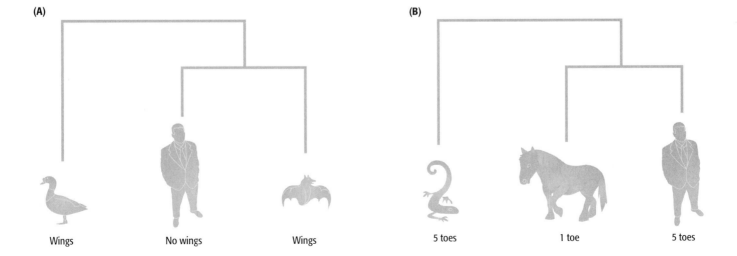

(A)

Wings   No wings   Wings

(B)

5 toes   1 toe   5 toes

morphological characters, such as those based on the shape of a structure, can be less easy to distinguish because of overlaps between different character states.

- Molecular data are easily converted to numerical form and hence are amenable to mathematical and statistical analysis.

The sequences of protein and DNA molecules provide the most detailed and unambiguous data for molecular phylogenetics, but techniques for protein sequencing did not become routine until the late 1960s, and rapid DNA sequencing was not developed until 10 years after that. Early studies therefore depended largely on indirect assessments of DNA or protein variations, using one of three methods:

- Immunological data, such as those obtained by Nuttall, involve measurements of the amount of cross-reactivity seen when an antibody specific for a protein from one organism is mixed with the same protein from a different organism. Remember that in Section 14.2.1 we learned that antibodies are immunoglobulin proteins that help to protect the body against invasion by bacteria, viruses, and other unwanted substances by binding to these "antigens." Proteins also act as antigens, so if human β-globin, for example, is injected into a rabbit then the rabbit makes an antibody that binds specifically to that protein. The antibody will also cross-react with β-globins from other vertebrates, because these β-globins have similar structures to the human version. The degree of cross-reactivity depends on how similar the β-globin being tested is to the human protein, providing the similarity data used in the phylogenetic analysis.

- Protein electrophoresis is used to compare the electrophoretic properties, and hence degree of similarity, of proteins from different organisms. This technique has proved useful for comparing closely related species and variations between members of a single species.

- DNA–DNA hybridization data are obtained by hybridizing DNA samples from the two organisms being compared. The DNA samples are denatured and mixed together so that hybrid molecules form. The stability of these hybrid molecules depends on the degree of similarity between the nucleotide sequences of the two DNAs, and is measured by determining the melting temperature (see Figure 3.8), a stable hybrid having a higher melting temperature than a less stable one. The melting temperatures obtained with DNAs from different pairs of organisms provide the data used in the phylogenetic analysis.

By the end of the 1960s these indirect methods had been supplemented with an increasing number of protein sequence studies, and during the 1980s DNA-based phylogenetics began to be carried out on a large scale. Protein sequences are still used today in some contexts, but DNA has now become by far the predominant molecule studied. This is mainly because DNA yields more phylogenetic information than protein, the nucleotide sequences of a pair of homologous genes having a higher information content than the amino acid sequences of the corresponding proteins, because mutations that result in synonymous changes alter the DNA sequence but do not affect the amino acid sequence (Figure 19.3). Entirely novel information can also be obtained by DNA sequence analysis because variability in both the coding and noncoding regions of the genome can be examined. The ease with which DNA samples for sequence analysis can be prepared by PCR (Section 2.3) is

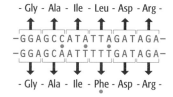

Figure 19.3 DNA yields more phylogenetic information than protein. The two DNA sequences differ at three positions but the amino acid sequences differ at only one position. These positions are indicated by blue dots. Two of the nucleotide substitutions are therefore synonymous and one is nonsynonymous (see Figure 16.11).

another key reason behind the predominance of DNA analysis in modern molecular phylogenetics.

As well as DNA sequences, molecular phylogenetics also makes use of DNA markers such as RFLPs, SSLPs, and SNPs (Section 3.2.2), particularly for intraspecific studies such as those aimed at understanding migrations of prehistoric human populations (Section 19.3.2). Later in this chapter we will consider various examples of the use of both DNA sequences and DNA markers in molecular phylogenetics, but first we must make a more detailed study of the methodology used in this area of genome research.

19.2 The Reconstruction of DNA-based Phylogenetic Trees

The objective of most phylogenetic studies is to reconstruct the tree-like pattern that describes the evolutionary relationships between the organisms being studied. Before examining the methodology for doing this, we must first take a closer look at a typical tree in order to familiarize ourselves with the basic terminology used in phylogenetic analysis.

19.2.1 The key features of DNA-based phylogenetic trees

A typical phylogenetic tree is shown in Figure 19.4A. This tree could have been reconstructed from any type of comparative data, but as we are interested in DNA sequences we will assume that the tree shows the relationships between four homologous genes, called *A*, *B*, *C*, and *D*. The **topology** of this tree comprises four **external nodes**, each representing one of the four genes that we have compared, and two **internal nodes** representing ancestral genes. The lengths of the **branches** indicate the degree of difference between the genes represented by the nodes. The degree of difference is calculated when the sequences are compared, as described in Section 19.2.2.

The tree in Figure 19.4A is **unrooted**, which means that it is only an illustration of the relationships between *A*, *B*, *C*, and *D* and does not tell us anything about the series of evolutionary events that led to these genes. Five different evolutionary pathways are possible, each depicted by a different **rooted** tree, as shown in Figure 19.4B. To distinguish between them the phylogenetic analysis must include at least one **outgroup**, this being a homologous gene that we know is less closely related to *A*, *B*, *C*, and *D* than these four genes are to each other. The outgroup enables the root of the tree to be located and the correct evolutionary pathway to be identified. The criteria used when choosing an outgroup depend very much on the type of analysis that is being carried out. As an example, let us say that the four homologous genes in our tree come from human, chimpanzee, gorilla, and orangutan. We could then use as an outgroup the homologous gene from another primate, such as the baboon, which we know from paleontological evidence branched away from the lineage leading to human, chimpanzee, gorilla, and orangutan before the time of the common ancestor of those four species (Figure 19.5).

The phylogenetic tree constructed from primate genes displays a relatively simple pattern of relationships. Most trees are more complex and we need terms to distinguish between the different types of relationship that are seen (Figure 19.6). Two or more sequences are said to be **monophyletic** if they are

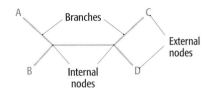

(A) An unrooted tree

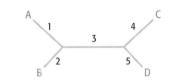

(B) Rooted trees

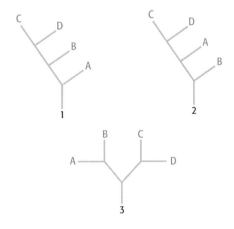

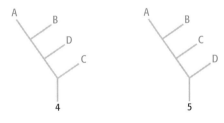

Figure 19.4 Phylogenetic trees. (A) An unrooted tree with four external nodes. (B) The five rooted trees that can be drawn from the unrooted tree shown in part A. The positions of the roots are indicated by the numbers on the outline of the unrooted tree.

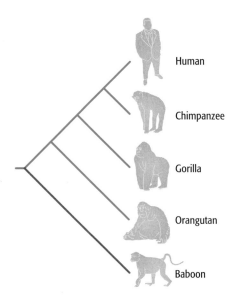

Figure 19.5 The use of an outgroup to root a phylogenetic tree. The tree of human, chimpanzee, gorilla, and orangutan genes is rooted with a baboon gene because we know from the fossil record that baboons split away from the primate lineage before the time of the common ancestor of the other four species. For more information on phylogenetic analysis of humans and other promates see Section 19.3.1.

derived from a single common ancestral DNA sequence. A group of monophyletic sequences is called a **clade** if it comprises all of the sequences included in the analysis that are descended from the ancestral sequence at the root of the clade. If the group excludes some members of the clade then that group is **paraphyletic**. Alternatively, if two or more DNA sequences are derived from different ancestral sequences then that group is said to be **polyphyletic**.

We refer to the rooted tree that we obtain by phylogenetic analysis as an **inferred tree**. This is to emphasize that it depicts the series of evolutionary events that are inferred from the data that were analyzed, and may not be the same as the **true tree**, the one that depicts the actual series of events that occurred. Sometimes we can be fairly confident that the inferred tree is the true tree, but most phylogenetic data analyses are prone to uncertainties which are likely to result in the inferred tree differing in some respects from the true tree. In Section 19.2.2 we will look at the various methods used to assign degrees of confidence to the branching pattern in an inferred tree, and later in the chapter we will examine some of the controversies that have arisen as a result of the imprecise nature of phylogenetic analysis.

Gene trees are not the same as species trees

The tree shown in Figure 19.5 illustrates a common type of molecular phylogenetics project, where the objective is to use a **gene tree**, reconstructed from comparisons between the sequences of **orthologous** genes (those derived from the same ancestral sequence), to make inferences about the evolutionary history of the species from which the genes are obtained. The assumption is that the gene tree, based on molecular data with all its advantages, will be a more accurate and less ambiguous representation of the **species tree** than that obtainable by morphological comparisons. This assumption is often correct, but it does not mean that the gene tree is the *same* as the species tree.

Figure 19.6 The mammalian phylogeny, illustrating the key terms used to describe different types of relationship seen in phylogenetic trees. The kangaroo is used as the outgroup. The numbers are the estimated divergence times for the indicated lineages, in millions of years. Humans and chimpanzees are monophyletic because they share a common ancestor that lived 4.6–5.0 million years ago. Humans, chimpanzees, gorillas and orangutans form a clade as they all derive from a single common ancestor that lived 13 million years ago. The indicated group is paraphyletic as it excludes sheep and goats, which share the same common ancestor as the other members of this group. Pigs and horses are polyphyletic as they have different common ancestors. For pigs, this ancestor lived 65 million years ago and also gave rise to sheep, goats and cattle, and for horses the ancestor is from 75 million years ago and also gave rise to cats and dogs.

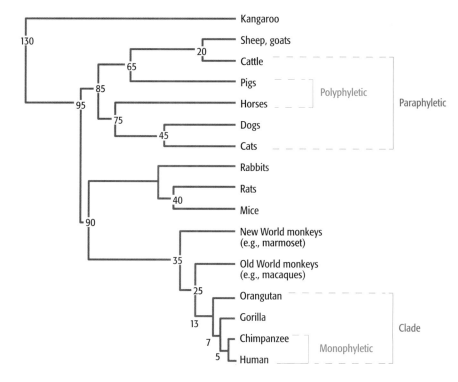

(A) A gene tree

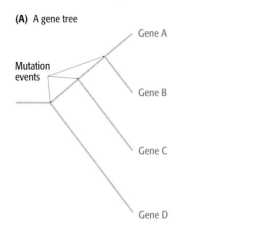

(B) A species tree

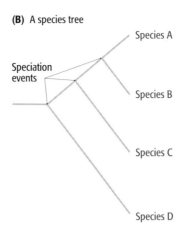

Figure 19.7 The difference between a gene tree and a species tree.

For that to be the case, the internal nodes in the gene and species trees would have to be precisely equivalent. Usually they are not equivalent, because:

- An internal node in a gene tree represents the divergence of an ancestral gene into two alleles with different DNA sequences: this occurs by mutation (Figure 19.7A).

- An internal node in a species tree represents a speciation event (Figure 19.7B): this occurs by the population of the ancestral species splitting into two groups that are unable to interbreed, often because they become geographically isolated.

The important point is that these two events—mutation and speciation—are not expected to occur at the same time. For example, the mutation event could precede the speciation. This would mean that, to begin with, both alleles of the gene are present in the unsplit population of the ancestral species (Figure 19.8). When the population split occurs, it is likely that both alleles will still be present in each of the two resulting groups. After the split, the new populations evolve independently. One possibility is that the results of **random genetic drift** (Section 19.3.2) lead to one allele being lost from one population and the other being lost from the other population. This establishes the two separate genetic lineages that we infer from phylogenetic analysis of the gene sequences present in the modern species that result from the continued evolution of the two populations.

How do these considerations affect the equivalence of the gene and species trees? There are various implications, two of which are as follows:

- If a **molecular clock** (Section 19.2.2) is used to date the time at which the gene divergence took place, then it cannot be assumed that this is also the time of the speciation event. If the node being dated is ancient, say 50 million or more years ago, then the error may not be noticeable. But if the speciation event is relatively recent, as when primates are being compared, then the date for the gene divergence might be significantly different to that for the speciation event.

- If the first speciation event is quickly followed by a second speciation event in one of the two resulting populations, then the branching order of the gene tree might be different from that of the species tree. This can occur if the genes in the modern species are derived from alleles that had already appeared before the first of the two speciation events, as illustrated in Figure 19.9.

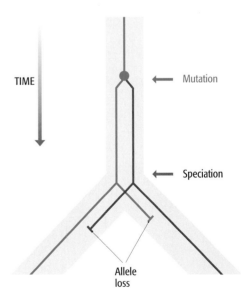

Figure 19.8 Mutation might precede speciation, giving an incorrect time for a speciation event if a molecular clock is used.

Figure 19.9 A gene tree can have a different branching order from a species tree. In this example, the gene has undergone two mutations in the ancestral species, the first mutation giving rise to the "red" allele and the second to the "blue" allele. Random genetic drift in association with the two subsequent speciation events results in the green allele lineage appearing in species A, the blue allele lineage in species B, and the red allele lineage in species C. Molecular phylogenetics based on the gene sequences will reveal that the green–red split occurred before the red–blue split, giving the gene tree shown on the right. However, the actual species tree is different, as shown on the left.

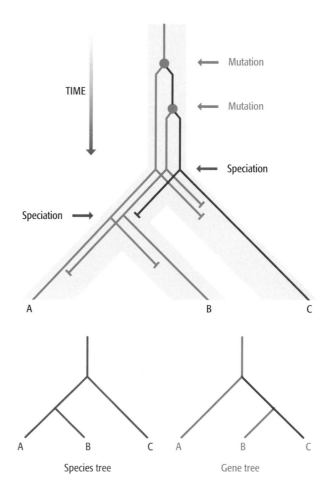

19.2.2 Tree reconstruction

In this section we will look at how tree reconstruction is carried out with DNA sequences, concentrating on the four main steps in the procedure:

- Aligning the DNA sequences and obtaining the comparative data that will be used to reconstruct the tree.

- Converting the comparative data into a reconstructed tree.

- Assessing the accuracy of the reconstructed tree.

- Using a molecular clock to assign dates to branch points within the tree.

Sequence alignment is the essential preliminary to tree reconstruction

The data used in reconstruction of a DNA-based phylogenetic tree are obtained by comparing nucleotide sequences. These comparisons are made by aligning the sequences so that nucleotide differences can be scored. This is the critical part of the entire enterprise because if the alignment is incorrect then the resulting tree will definitely not be the true tree.

The first issue to consider is whether the sequences being aligned are homologous. If they are homologous then they must, by definition, be derived from a common ancestral sequence (Section 5.2.1) and so there is a sound basis for the phylogenetic study. If they are not homologous then they do not share a common ancestor. The phylogenetic analysis will find a common ancestor because the methods used for tree reconstruction always produce a tree of some description, even if the data are completely erroneous, but the resulting tree will have no biological relevance. With some DNA sequences—for example, the β-globin genes of different vertebrates—there is little difficulty in being sure that the sequences being compared are homologous, but this is not always the case, and one of the commonest errors that arises during phylogenetic analysis is the inadvertent inclusion of a nonhomologous sequence.

Once it has been established that two DNA sequences are indeed homologous, the next step is to align the sequences so that homologous nucleotides can be compared. With some pairs of sequences this is a trivial exercise (Figure 19.10A), but it is not so easy if the sequences are relatively dissimilar and/or have diverged by the accumulation of insertions and deletions as well as point mutations. Insertions and deletions cannot be distinguished from one another when pairs of sequences are compared so we refer to them as **indels**. Placing indels at their correct positions is often the most difficult part of sequence alignment (Figure 19.10B).

Some pairs of sequences can be aligned reliably by eye. For more complex pairs, alignment might be possible by the **dot matrix** method (Figure 19.11). The two sequences are written out on the *x*- and *y*-axes of a graph, and dots placed in the squares of the graph paper at positions corresponding to identical nucleotides in the two sequences. The alignment is indicated by a diagonal series of dots, broken by empty squares where the sequences have nucleotide differences, and shifting from one column to another at places where indels occur.

More rigorous mathematical approaches to sequence alignment have also been devised. The first of these is the **similarity approach**, which aims to maximize the number of matched nucleotides—those that are identical in the two sequences. The complementary approach is the **distance method**, in which the objective is to minimize the number of mismatches. Often the two procedures will identify the same alignment as being the best one. Usually the comparison involves more than just two sequences, meaning that a **multiple alignment** is required. This can rarely be done effectively with pen and paper so, as in all steps in a phylogenetic analysis, a computer program is used. For multiple alignments, Clustal is often the most popular choice. Clustal and other software packages for phylogenetic analysis are described in Technical Note 19.1.

Converting alignment data into a phylogenetic tree

Once the sequences have been aligned accurately, an attempt can be made to reconstruct the phylogenetic tree. To date, nobody has devised a perfect method for tree reconstruction, and several different procedures are used routinely. Comparative tests have been run with artificial data, for which the true tree is known, but these have failed to identify any particular method as being better than any of the others.

The main distinction between the different tree-building methods is the way in which the multiple sequence alignment is converted into numerical data that can be analyzed mathematically in order to reconstruct the tree. The simplest approach is to convert the sequence information into a **distance matrix**, which is simply a table showing the evolutionary distances between

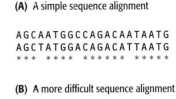

(A) A simple sequence alignment

```
AGCAATGGCCAGACAATAATG
AGCTATGGACAGACATTAATG
*** **** ****** *****
```

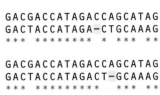

(B) A more difficult sequence alignment

```
GACGACCATAGACCAGCATAG
GACTACCATAGA-CTGCAAAG
*** ******** * *** **
```

```
GACGACCATAGACCAGCATAG
GACTACCATAGACT-GCAAAG
*** ******** *** **
```

— Possible positions for the indel

Figure 19.10 Sequence alignment. (A) Two sequences that have not diverged to any great extent can be aligned easily by eye. (B) A more complicated alignment in which it is not possible to determine the correct position for an indel. If errors in indel placement are made in a multiple alignment then the tree reconstructed by phylogenetic analysis is unlikely to be correct. In this diagram, the red asterisks indicate nucleotides that are the same in both sequences.

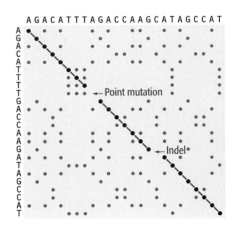

Figure 19.11 The dot matrix technique for sequence alignment. The correct alignment stands out because it forms a diagonal of continuous dots, broken at point mutations and shifting to a different diagonal at indels.

Technical Note 19.1 Phylogenetic analysis

Software packages for construction of phylogenetic trees

Few sets of DNA sequences are simple enough to be converted into phylogenetic trees entirely by hand. Virtually all research in this area is carried out using computers with the aid of any one of a variety of software packages designed specifically for one or other of the steps in tree reconstruction.

One of the easiest to use and most popular packages is Clustal, which was originally written in 1988 and has undergone several upgrades in the intervening years. Clustal is primarily a program for carrying out multiple alignments of protein or DNA sequences, which it is able to do very effectively provided that the sequences being compared do not contain extensive internal repeat motifs. Clustal is usually used in conjunction with NJplot, a simple program for tree reconstruction by the neighbor-joining method. One important

advantage of Clustal and NJplot is that they do not require huge amounts of computer memory and so can be run on small PCs or Macintosh computers.

More comprehensive software packages enable the researcher to choose between a variety of different methods for tree reconstruction and to carry out more sophisticated types of phylogenetic analysis. The most widely used of these packages are PAUP and PHYLIP. The tree-building programs in PAUP are often looked upon as the most accurate ones currently available and are able to handle relatively large datasets. PHYLIP has the advantage of including a number of software tools not readily available from other sources. Other popular packages include PAML, MacClade, and HENNIG86.

all pairs of sequences in the dataset (Figure 19.12). The evolutionary distance is calculated from the number of nucleotide differences between a pair of sequences and is used to establish the lengths of the branches connecting these two sequences in the reconstructed tree. In the example shown in Figure 19.12 the evolutionary distance is expressed as the number of nucleotide differences per nucleotide site for each sequence pair. For example, sequences 1 and 2 are 20 nucleotides in length and have four differences, corresponding to an evolutionary distance of 4/20 = 0.2. Note that this analysis assumes that there are no **multiple substitutions** (also called **multiple hits**). Multiple substitution occurs when a single site undergoes two or more changes (e.g., the ancestral sequence ...ATGT... gives rise to two modern sequences: ...AGGT... and ...ACGT...). There is only one nucleotide difference between the two modern sequences, but there have been two nucleotide substitutions. If this multiple hit is not recognized then the evolutionary distance between the two modern sequences will be significantly underestimated. To avoid this problem, distance matrices for phylogenetic analysis are usually constructed using mathematical methods that include statistical devices for estimating the amount of multiple substitution that has occurred.

Figure 19.12 A simple distance matrix. The matrix shows the evolutionary distance between each pair of sequences in the alignment.

Multiple alignment

1 A G G C C A A G C C A T A G C T G T C C
2 A G G C A A A G A C A T A C C T G A C C
3 A G G C C A A G A C A T A G C T G T C C
4 A G G C A A A G A C A T A C C T G T C C

Distance matrix

	1	2	3	4
1	–	0.20	0.05	0.15
2		–	0.15	0.05
3			–	0.10
4				–

The **neighbor-joining method** is a popular tree-building procedure that uses the distance matrix approach. To begin the reconstruction, it is initially assumed that there is just one internal node from which branches leading to all the DNA sequences radiate in a star-like pattern (Figure 19.13). This is virtually impossible in evolutionary terms but the pattern is just a starting point. Next, a pair of sequences is chosen at random, removed from the star, and attached to a second internal node, connected by a branch to the center of the star. The distance matrix is then used to calculate the total branch length in this new "tree." The sequences are then returned to their original positions and another pair attached to the second internal node, and again the total branch length is calculated. This operation is repeated until all the possible pairs have been examined, enabling the combination that gives the tree with the shortest total branch length to be identified. This pair of sequences will be neighbors in the final tree; in the interim, they are combined into a single unit, creating a new star with one branch fewer than the original one. The whole process of pair selection and tree-length calculation is now repeated so that a second pair of neighboring sequences is identified, and then repeated again so that a third pair is located, and so on. The result is a complete reconstructed tree.

The advantage of the neighbor-joining method is that the data handling is relatively easy to carry out, largely because the information content of the multiple alignment has been reduced to its simplest form. The disadvantage is that some of the information is lost, in particular that pertaining to the identities of the ancestral and derived nucleotides (equivalent to ancestral and derived character states, defined in Section 19.1.1) at each position in the multiple alignment. The **maximum parsimony** method takes account of this information, utilizing it to recreate the series of nucleotide changes that are most likely to have resulted in the pattern of variation revealed by the multiple alignment. **Parsimony** was originally a philosophical term stating that, when deciding between competing hypotheses, preference should be given to the one that involves the fewest unconnected assumptions. In molecular phylogenetics, parsimony is an approach that decides between different tree topologies by identifying the one that involves the shortest evolutionary pathway, this being the pathway that requires the smallest number of nucleotide changes to go from the ancestral sequence, at the root of the tree, to all the present-day sequences that have been compared. Trees are therefore constructed at random and the number of nucleotide changes that they involve calculated until all possible topologies have been examined and the one requiring the smallest number of steps identified. This is presented as the most likely inferred tree.

The maximum parsimony method is more rigorous in its approach compared with the neighbor-joining method, but this increase in rigor inevitably extends the amount of data handling that is involved. This is a significant problem because the number of possible trees that must be scrutinized increases rapidly as more sequences are added to the dataset. With just five sequences there are only 15 possible unrooted trees, but for ten sequences there are 2,027,025 unrooted trees and for 50 sequences the number exceeds the number of atoms in the universe. Even with a high-speed computer it is not possible to check every one of these trees in a reasonable time, if at all, so often the maximum parsimony method is unable to carry out a comprehensive analysis. The same is true with many of the other more sophisticated methods for tree reconstruction.

(A) The starting point for the neighbor-joining method

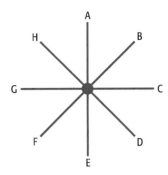

(B) Removal of two sequences from the star

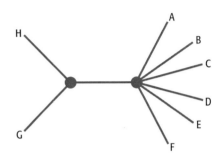

Figure 19.13 Manipulations carried out when using the neighbor-joining method for tree reconstruction.

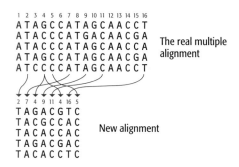

Figure 19.14 Constructing a new multiple alignment in order to bootstrap a phylogenetic tree. The new alignment is built up by taking columns at random from the real alignment. Note that the same column can be sampled more than once.

Assessing the accuracy of a reconstructed tree

The limitations to the methods used in phylogenetic reconstruction lead inevitably to questions about the veracity of the resulting trees. Statistical tests of the accuracy of a reconstructed tree have been devised, but these are necessarily complex because a tree is geometric rather than numeric, and the accuracy of one part of the topology may be more or less than the accuracy of the other parts.

The routine method for assigning confidence limits to different branch points within a tree is to carry out a **bootstrap analysis**. To do this we need a second multiple alignment that is different from, but equivalent to, the real alignment. This new alignment is built up by taking columns, at random, from the real alignment, as illustrated in Figure 19.14. The new alignment therefore comprises sequences that are different from the original, but it has a similar pattern of variability. This means that when we use the new alignment in tree reconstruction we do not simply reproduce the original analysis, but we should obtain the same tree.

In practice, 1000 new alignments are created so 1000 replicate trees are reconstructed. A **bootstrap value** can then be assigned to each internal node in the original tree, this value being the number of times that the branch pattern seen at that node was reproduced in the replicate trees. If the bootstrap value is greater than 700/1000 then we can assign a reasonable degree of confidence to the topology at that particular internal node.

Molecular clocks enable the time of divergence of ancestral sequences to be estimated

When we carry out a phylogenetic analysis our primary objective is to infer the pattern of the evolutionary relationships between the DNA sequences that are being compared. These relationships are revealed by the topology of the tree that is reconstructed. Often we also have a secondary objective: to discover when the ancestral sequences diverged to give the modern sequences. This information is interesting in the context of genome evolution, as we discovered when we looked at the evolutionary history of the human globin genes (see Figure 18.8). The information is even more interesting on occasions when we are able to equate a gene tree with a species tree, because now the times at which the ancestral sequences diverged approximate to the dates of speciation events.

To assign dates to branch points in a phylogenetic tree we must make use of a molecular clock. The molecular clock hypothesis, first proposed in the early 1960s, states that nucleotide substitutions (or amino acid substitutions if protein sequences are being compared) occur at a constant rate. This means that the degree of difference between two sequences can be used to assign a date to the time at which their ancestral sequence diverged. However, to be able to do this, the molecular clock must be calibrated so that we know how many nucleotide substitutions to expect per million years. Calibration is usually achieved by reference to the fossil record. For example, fossils suggest that the most recent common ancestor of humans and orangutans lived 13 million years ago. To calibrate the human molecular clock we therefore compare human and orangutan DNA sequences to determine the amount of nucleotide substitution that has occurred, and then divide this figure by 13, followed by 2, to obtain a rate of substitution per million years (Figure 19.15).

At one time it was thought that there might be a universal molecular clock that applied to all genes in all organisms. Now we realize that molecular clocks are different in different organisms and are variable even within a single organism. The differences between organisms might be the result of generation times, because a species with a short generation time is likely to accumulate DNA replication errors at a faster rate than a species with a longer generation time. This probably explains the observation that rodents have a faster molecular clock than primates. Within a single genome the variations are as follows:

- Nonsynonymous substitutions occur at a slower rate than synonymous ones. This is because a mutation that results in a change in the amino acid sequence of a protein might be deleterious to the organism, so the accumulation of nonsynonymous mutations in the population is reduced by the processes of natural selection. This means that when gene sequences in two species are compared, there are usually fewer nonsynonymous than synonymous substitutions.

- The molecular clock for mitochondrial genes is faster than that for genes in the nuclear genome. This is probably because mitochondria lack many of the DNA repair systems that operate on nuclear genes.

- Molecular clocks appear to have increased in speed over the last 1–2 million years. This is thought to be an artifact, due perhaps to the persistence during this period of mutations that are only slightly deleterious and which are therefore removed relatively slowly by natural selection. The apparent acceleration means that inaccuracies might arise if a molecular clock is calibrated against relatively ancient events in the fossil record and then used to assign dates to more recent events.

Despite these complications, molecular clocks have become an immensely valuable adjunct to tree reconstruction, as we will see in Section 19.3 when we look at some typical molecular phylogenetics projects.

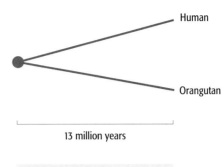

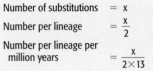

Figure 19.15 Calibrating the human molecular clock. The number of substitutions is determined for a pair of homologous genes from human and orangutan: call this number "x." The number of substitutions per lineage is therefore x/2, and the number per million years is x/(2 × 13).

Standard tree reconstruction is not appropriate for all DNA sequence datasets

Conventional tree building assumes that the relationships between a group of DNA sequences can be described by a simple branching pattern. This assumption is not always correct. For example, if a group of sequences are able to recombine then new sequences might be generated that share features of, and indeed are descendants of, pairs of ancestral sequences that occupy distant positions in a conventional phylogenetic tree (Figure 19.16A). If the products of recombination are included in the tree then the dichotomous branching

(A) Problems with recombination

(B) Problems when ancestral sequences still exist

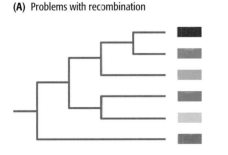

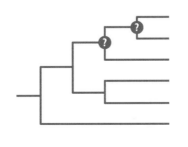

Figure 19.16 Two problems that complicate conventional tree building.

pattern begins to break down and the relationships become difficult to distinguish. This is a particular problem when attempts are made to determine the evolutionary relationships between the members of a multigene family, as these sequences reside in a single genome, possibly as a tandem array, and it is therefore possible that recombination between them has occurred in the past. Conventional tree building is also inappropriate if ancestral sequences, represented by internal nodes, still exist (Figure 19.16B). This is rarely a problem when the genes being compared come from different species, as the internal nodes now represent sequences that were present in ancestral species that are now extinct, but it is very relevant when the sequences being compared are taken from a single population with a short evolutionary history. This problem often arises in studies of recent human evolution.

One solution to these problems is to use a network rather than a tree to illustrate the evolutionary relationships between the sequences being examined (Figure 19.17). A network enables ancestral sequences that still exist to be identified (such as Sequence 1 in Figure 19.17) and clearly illustrates their relationship with the sequences descended from them. Sequences that are extinct (or which exist in the population but are not present in the dataset) can also be identified. Reticulations in the network, such as that linking Sequences 1, 3, 7, and 8 in Figure 19.17, are particularly interesting as these indicate sequences that have arisen either by recombination or by parallel mutations (homoplasy) in different lineages.

Simple networks can be constructed manually, but their topologies quickly become complex as more sequences are added (Figure 19.18), especially if all

Figure 19.17 Phylogenetic analysis by network construction. Panel (A) shows a multiple alignment from which the network is constructed, with dots indicating identities with Sequence 1. (B) The network is constructed as follows. (i) Sequence 1 is used as the starting point. (ii) Sequence 2 differs from Sequence 1 at four positions and is linked to Sequence 1 by a line in the network. (iii) Sequence 3 differs from Sequence 1 at one position, this position different from any of those in Sequence 2. Sequence 3 is therefore linked directly to Sequence 1. (iv) Sequence 4 shares with Sequence 3 the C→T substitution at position 11, but has an additional substitution at position 2. The line leading to Sequence 3 is therefore extended and Sequence 4 placed at its terminus. (v) Sequence 5 has one substitution compared with Sequence 1, this being a unique substitution not seen so far. Sequence 5 is therefore linked directly to Sequence 1. (vi) Similarly, Sequence 6 has three unique substitutions and is linked directly to Sequence 1. (vii) Sequence 7 has two differences from Sequence 1, one of these being the C→T at position 11 seen in Sequences 3 and 4. Sequence 7 therefore connects directly with Sequence 3. In this part of the network, the line between Sequences 1 and 3 represents the substitution at position 11, that between Sequences 3 and 4 is the substitution at position 2, and that between Sequences 3 and 7 is the substitution at position 24. (viii) Sequence 8 shares with Sequence 7 the A→G substitution at position 24, and has a unique substitution, not seen in any previous sequence, at position 9. Sequence 8 must therefore form a branch off a line between Sequences 1 and 7. This branch creates an "empty node", shown as a small dot, which represents a sequence not present in the dataset, possibly because it is extinct.

(A)

```
           1    5   10    15   20    25   30    35   40
  1  GCATTACTTTCGGTAGCGCGGAAAGGCGACGGGACTTATA
  2  .............C...C.......C.....A........
  3  ..........T............................
  4  .T........T............................
  5  .........................T.............
  6  ......T.....C.........................G......
  7  ..........T...........G................
  8  ........C.............G................
```

(B)

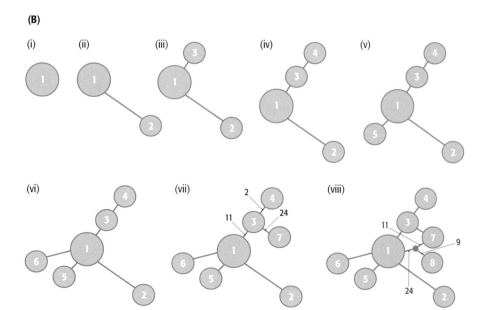

possible relationships between sequences are included. Algorithms for reducing the complexity of a network, without losing phylogenetic information, have therefore been devised. These include pruning methods which assign weights to the various pathways between pairs of sequences and remove those links which are less likely to represent true relationships.

19.3 The Applications of Molecular Phylogenetics

Molecular phylogenetics has grown in stature since the start of the 1990s, largely because of the development of more rigorous methods for tree building, combined with the explosion of DNA sequence information obtained initially by PCR analysis and more recently by genome projects. The importance of molecular phylogenetics has also been enhanced by the successful application of tree reconstruction and other phylogenetic techniques to some of the more perplexing issues in biology. In this final section we will survey some of these successes.

19.3.1 Examples of the use of phylogenetic trees

First, we will consider two projects that illustrate the various ways in which conventional tree reconstruction is being used in modern molecular biology.

DNA phylogenetics has clarified the evolutionary relationships between humans and other primates

Darwin was the first biologist to speculate on the evolutionary relationships between humans and other primates. His view—that humans are closely related to the chimpanzee, gorilla, and orangutan—was controversial when it was first proposed and fell out of favor, even among evolutionists, in the following decades. Indeed, biologists were among the most ardent advocates of an anthropocentric view of our place in the animal world.

From studies of fossils, paleontologists had concluded prior to 1960 that chimpanzees and gorillas are our closest relatives but that the relationship was distant, the split, leading to humans on the one hand and chimpanzees and gorillas on the other, having occurred some 15 million years ago. The first detailed molecular data, obtained by immunological studies in the 1960s, confirmed that humans, chimpanzees, and gorillas do indeed form a single clade, but suggested that the relationship is much closer, a molecular clock indicating that this split occurred only 5 million years ago. This was one of the first attempts to apply a molecular clock to phylogenetic data and the result was, quite naturally, treated with some suspicion. In fact, an acrimonious debate opened up between paleontologists, who believed in the ancient split indicated by the fossil evidence, and biologists, who had more confidence in the recent date suggested by the molecular data. This debate was eventually "won" by the molecular biologists, whose view that the split occurred about 5 million years ago became generally accepted.

As more and more molecular data were obtained, the difficulties in establishing the exact pattern of the evolutionary events that led to humans, chimpanzees, and gorillas became apparent. Comparisons of the mitochondrial genomes of the three species by restriction mapping (Section 3.3.1) and DNA sequencing suggested that the chimpanzee and gorilla are more closely related to each other than either is to humans (Figure 19.19A), whereas DNA–DNA hybridization data supported a closer relationship between humans and

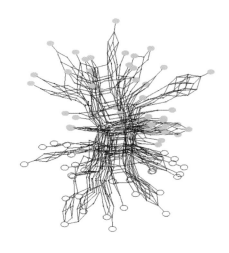

Figure 19.18 A complex network. This network has been constructed from 5S rRNA sequences from sea beets (*Beta vulgaris* ssp. *maritima*). The grey circles are sequences from plants growing around Poole Harbour, UK, and the open circles are plants from cliff tops a few kilometers to the south. The network was constructed from an alignment of 110 × 289 bp sequences. The network originally contained 6808 links, but these were reduced to the 655 shown here by pruning, as described in the text. Network courtesy of Sarah Dyer.

(A) Mitochondrial DNA data

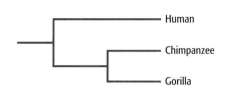

(B) DNA–DNA hybridization data

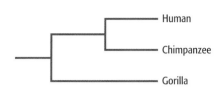

(C) Combined molecular datasets

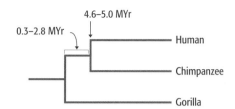

Figure 19.19 Different interpretations of the evolutionary relationships between humans, chimpanzees, and gorillas. Abbreviation: MYr, million years.

chimpanzees (Figure 19.19B). The reason for these conflicting results is the close similarity between DNA sequences in the three species, the differences being less than 3% for even the most divergent regions of the genomes (Section 18.4). This makes it difficult to establish relationships unambiguously.

The solution to the problem has been to make comparisons between as many different genes as possible and to target those loci that are expected to show the greatest amount of dissimilarity. By 1997, 14 different molecular datasets had been obtained, including sequences of variable loci such as pseudogenes and noncoding sequences. Analysis of these datasets confirmed that the chimpanzee is the closest relative to humans, with our lineages diverging 4.6–5.0 million years ago. The gorilla is a slightly more distant cousin, its lineage having diverged from the human–chimpanzee one between 0.3 million and 2.8 million years earlier (Figure 19.19C).

The origins of AIDS

The global epidemic of acquired immune deficiency syndrome (AIDS) has touched everyone's lives. AIDS is caused by human immunodeficiency virus 1 (HIV-1), a retrovirus (Section 9.1.2) that infects cells involved in the immune response. The demonstration in the early 1980s that HIV-1 is responsible for AIDS was quickly followed by speculation about the origin of the disease. Speculation centered around the discovery that similar immunodeficiency viruses are present in primates such as the chimpanzee, sooty mangabey, mandrill, and various monkeys. These simian immunodeficiency viruses (SIVs) are not pathogenic in their normal hosts but it was thought that if one had become transferred to humans then within this new species the virus might have acquired new properties, such as the ability to cause disease and to spread rapidly through the population.

Retrovirus genomes accumulate mutations relatively quickly because reverse transcriptase, the enzyme that copies the RNA genome contained in the virus particle into the DNA version that integrates into the host genome (see Section 9.1.2), lacks an efficient proofreading activity (Section 15.2.2) and so tends to make errors when it carries out RNA-dependent DNA synthesis. This means that the molecular clock runs rapidly in retroviruses, and genomes that diverged quite recently display sufficient nucleotide dissimilarity for a phylogenetic analysis to be carried out. Even though the evolutionary period we are interested in is less than 100 years, HIV and SIV genomes contain sufficient data for their relationships to be inferred by phylogenetic analysis.

The starting point for this phylogenetic analysis is RNA extracted from virus particles. RT-PCR (see Technical Note 5.1) is therefore used to convert the RNA into a DNA copy and then to amplify the DNA so that sufficient amounts for nucleotide sequencing are obtained. Comparison between virus DNA sequences has resulted in the reconstructed tree shown in Figure 19.20. This tree has a number of interesting features. First, it shows that different samples of HIV-1 have slightly different sequences, the samples as a whole forming a tight cluster, almost a star-like pattern, that radiates from one end of the unrooted tree. This star-like topology implies that the global AIDS epidemic began with a very small number of viruses, perhaps just one, which have spread and diversified since entering the human population. The closest relative to HIV-1 among primates is the SIV of chimpanzees, the implication being that this virus jumped across the species barrier between chimpanzees and humans and initiated the AIDS epidemic. However, this epidemic did not begin immediately: a relatively long, uninterrupted branch links the center of the HIV-1 radiation with the internal node leading to the relevant SIV sequence, suggesting that after transmission to humans, HIV-1 underwent a latent period when it remained restricted to a small part of the global human population, presumably in Africa, before beginning its rapid spread to other parts of the world. Other primate SIVs are less closely related to HIV-1, but one, the SIV from sooty mangabey, clusters in the tree with the second human immunodeficiency virus, HIV-2. It appears that HIV-2 was transferred to the human population independently of HIV-1, and from a different simian host. HIV-2 is also able to cause AIDS, but has not, as yet, become globally epidemic.

An intriguing addition to the HIV/SIV tree was made in 1998 when the RNA of an HIV-1 isolate from a blood sample taken in 1959 from an African male was sequenced. The RNA was highly fragmented and only a short DNA sequence could be obtained, but this was sufficient for the sequence to be placed on the phylogenetic tree (see Figure 19.20). This sequence, called ZR59, attaches to the tree by a short branch that emerges from near the center of the HIV-1 radiation. The positioning indicates that the ZR59 sequence represents one of the earliest versions of HIV-1 and shows that the global spread of HIV-1 was already underway by 1959. A later and more comprehensive analysis of HIV-1 sequences has suggested that the spread began in the period between 1915 and 1941, with a best estimate of 1931. Pinning down the date in this way has enabled epidemiologists to begin an investigation of the historic and social conditions that might have been responsible for the start of the AIDS epidemic.

19.3.2 Molecular phylogenetics as a tool in the study of human prehistory

Now we will turn our attention to the use of molecular phylogenetics in intraspecific studies: the study of the evolutionary history of members of the same species. We could choose any one of several different organisms to illustrate the approaches and applications of intraspecific studies, but many people look on *Homo sapiens* as the most interesting organism so we will investigate how molecular phylogenetics is being used to deduce the origins of modern humans and the geographic patterns of their recent migrations in the Old and New Worlds.

Studying genes in populations

In any application of molecular phylogenetics, the genes chosen for analysis must display variability in the organisms being studied. If there is no variability then there is no phylogenetic information. This presents a problem in intraspecific studies because the organisms being compared are all members of the same species and so share a great deal of genetic similarity, even if the species has split into populations that interbreed only intermittently. This

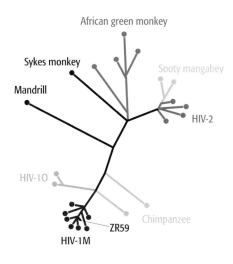

Figure 19.20 The phylogenetic tree reconstructed from HIV and SIV genome sequences. The AIDS epidemic is due to the HIV-1M type of immunodeficiency virus. ZR59 is positioned near the root of the star-like pattern formed by genomes of this type.

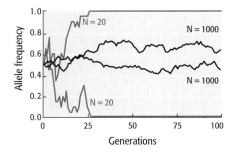

Figure 19.21 The effects of genetic drift on allele frequency. The graph shows the results of simulations of the frequencies of a pair of alleles over 100 generations, for populations containing 20 or 1000 individuals. In both populations, one allele eventually becomes fixed and the other lost, but the time taken for this to occur is greatly influenced by the population size. Adapted from M. Jobling *et al.*, *Human Evolutionary Genetics*.

means that the DNA sequences that are used in the phylogenetic analysis must be the most variable ones that are available. In humans there are three main possibilities.

- Multiallelic genes, such as members of the HLA family (Section 3.2.1), which exist in many different sequence forms.

- Microsatellites, which evolve not through mutation but by replication slippage (Section 16.1.1). Cells do not appear to have any repair mechanism for reversing the effects of replication slippage, so new microsatellite alleles are generated relatively frequently.

- Mitochondrial DNA, which as mentioned in Section 19.2.2, accumulates nucleotide substitutions relatively rapidly because mitochondria lack many of the repair systems that slow down the molecular clock in the human nucleus. The mitochondrial DNA variants present in a single species are called **haplogroups**.

It is important to note that it is not the potential for change that is critical to the application of these loci in phylogenetic analysis, it is the fact that different alleles or haplogroups of each locus coexist in the population as a whole. The loci are therefore **polymorphic** and information pertaining to the relationships between different individuals can be obtained by comparing the combinations of alleles and/or haplogroups that those individuals possess. New alleles and haplogroups appear in a population because of mutations that occur in the reproductive cells of individual organisms. Each allele has its own **allele frequency**, which changes over time as a result of **natural selection** and **random genetic drift**. Natural selection occurs because of differences in **fitness** (the ability of an organism to survive and reproduce) that, according to Darwin, result in the "preservation of favourable variations and the rejection of injurious variations." Natural selection therefore decreases the frequencies of alleles that reduce the fitness of an organism, and increases the frequency of alleles that improve fitness. In reality, few new alleles that arise in a population have a significant impact on fitness so most are not affected by natural selection, but their frequencies still change because of random genetic drift caused by the arbitrary nature of birth, death, and reproduction (Figure 19.21).

Either because of natural selection or random genetic drift, one allele can begin to predominate in a population and eventually achieve a frequency of 100%. This allele has now become **fixed**. Mathematical models predict that over time different alleles become fixed in a population, resulting in a series of **gene substitutions**. If a species splits into two populations that do not interbreed extensively, then the allele frequencies in the two populations will change differently so that after a few tens of generations the two populations will have distinctive genetic features. Eventually, different gene substitutions occur in the two populations, but even before this happens we can distinguish between them because of the differences in allele frequencies. These differences can be used to date the time when the population split occurred and to determine if one or both populations experienced a **bottleneck**, a period when the population size became substantially reduced.

The origins of modern humans—out of Africa or not?

It seems reasonably certain that the origin of humans lies in Africa because it is here that all the oldest prehuman fossils have been found. The paleontological evidence reveals that hominids first moved outside of Africa over 1 million years ago, but these were not modern humans, they were an earlier species called *Homo erectus* (Figure 19.22). These were the first hominids to become geographically dispersed, eventually spreading to all parts of the Old World.

The events that followed the dispersal of *Homo erectus* are controversial. From comparisons using fossil skulls and bones, paleontologists have concluded that the *Homo erectus* populations that became located in different parts of the Old World gave rise to the modern human populations of those areas by a process called **multiregional evolution** (Figure 19.23A). There may have been a certain amount of interbreeding between humans from different geographic regions but, to a large extent, these various populations remained separate throughout their evolutionary history.

Doubts about the multiregional hypothesis were first raised by reinterpretations of the fossil evidence and were subsequently brought to a head by publication in 1987 of a phylogenetic tree reconstructed from mitochondrial RFLP data obtained from 147 humans representing populations from all parts of the world. The tree (Figure 19.24) confirmed that the ancestors of modern humans lived in Africa but suggested that they were still there about 200,000 years ago. This inference was made by applying the mitochondrial molecular clock to the tree, which showed that the ancestral mitochondrial DNA, the one from which all modern mitochondrial DNAs are descended, existed between 140,000 and 290,000 years ago. The tree showed that this mitochondrial genome was located in Africa, so the person who possessed it, the so-called mitochondrial Eve (she had to be female because mitochondrial DNA is only inherited through the female line), must have been African.

The discovery of mitochondrial Eve prompted a new scenario for the origin of modern humans. Rather than evolving in parallel throughout the world, as suggested by the multiregional hypothesis, **Out of Africa** states that *Homo sapiens* originated in Africa, members of this species then moving into the rest of the Old World between 100,000 and 50,000 years ago, displacing the descendants of *Homo erectus* that they encountered (see Figure 19.23B).

Such a radical change in thinking inevitably did not go unchallenged. When the RFLP data were examined by other molecular phylogeneticists it became clear that the original computer analysis had been flawed, and that several quite different trees could be reconstructed from the data, some of which did not have a root in Africa. These criticisms were countered by more detailed mitochondrial DNA sequence datasets, most of which are compatible with a relatively recent African origin and so support the Out of Africa hypothesis rather than multiregional evolution. An interesting complement to "mitochondrial Eve" has been provided by studies of the Y chromosome, which suggest that "Y chromosome Adam" also lived in Africa some 200,000 years ago. Of course, this Eve and Adam were not equivalent to the biblical characters and were by no means the only people alive at that time: they were simply the individuals who carried the ancestral mitochondrial DNA and Y chromosomes that gave rise to all the mitochondrial DNAs and Y chromosomes in existence today. The important point is that these ancestral DNAs were still in Africa well after the spread of *Homo erectus* into Eurasia.

The mitochondrial DNA and Y chromosome studies appear to provide strong evidence in support of the Out of Africa theory. But complications have arisen from examination of nuclear genes other than those on the Y chromosome.

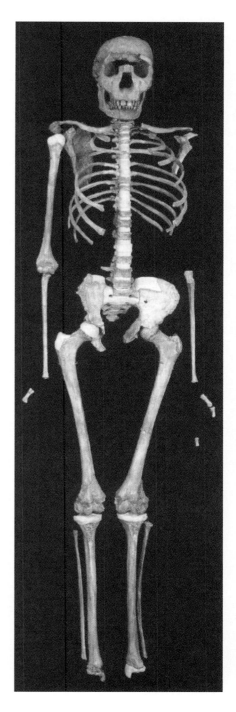

Figure 19.22 Skeleton of an adolescent male *Homo erectus*. This specimen, also called the "Nariokotome boy," is about 1.6 million years in age and was found near Lake Turkana, Kenya. Image courtesy of the Library, American Museum of Natural History.

Figure 19.23

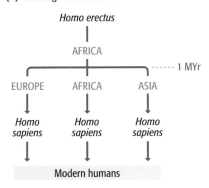

(A) Multiregional evolution

Figure 19.23

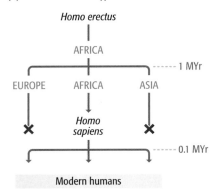

(B) The Out of Africa hypothesis

Figure 19.24

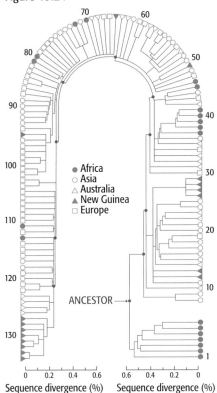

For example, β-globin sequences give a much earlier date, 800,000 years ago, for the common ancestor, and studies of an X-chromosome gene, *PDHA1*, place the ancestral sequence at 1,900,000 years ago. Molecular anthropologists are currently debating the significance of these results. More datasets, and hopefully some sort of Grand Synthesis, are eagerly awaited.

Neandertals are not the ancestors of modern Europeans

Neandertals are extinct hominids who lived in Europe between 300,000 and 30,000 years ago (Figure 19.25). They were descended from the *Homo erectus* populations who left Africa about 1 million years ago and, according to the Out of Africa hypothesis, were displaced when modern humans reached Europe about 50,000 years ago. Therefore, one prediction of the Out of Africa hypothesis is that there is no genetic continuity between Neandertals and the modern humans that live in Europe today. Bearing in mind that the last Neandertal died out 30,000 years ago, is there any way that we can test this hypothesis?

Ancient DNA has provided an answer. It has been known for some years that DNA molecules can survive the death of the organism in which they are contained, being recoverable centuries and possibly millennia later as short, degraded fragments preserved in bones and other biological remains. There is never very much ancient DNA in a specimen, possibly no more than a few hundred genomes in a gram of bone, but that need not concern us because we can always use PCR to amplify these tiny amounts into larger quantities from which we can obtain DNA sequences.

The study of ancient DNA has been plagued with controversies over the last 15 years. In the early 1990s there were many reports of ancient human DNA being detected in bones and other archaeological specimens, but often it turned out that what had been amplified by PCR was not ancient DNA at all, but contaminating modern DNA left on the specimen by the archaeologist who dug it up or by the molecular biologist who carried out the DNA extraction. The worldwide success of the film *Jurassic Park* led to reports of DNA in insects preserved in amber and even in dinosaur bones, but all these claims are now doubted. Many biologists started to wonder if ancient DNA existed at all, but gradually it became clear that if the work is carried out with extreme care, it is sometimes possible to extract authentic ancient DNA from specimens up to about 50,000 years old. This is just old enough to include a few Neandertal bones.

The first Neandertal specimen selected for study was believed to be between 30,000 and 100,000 years old. DNA extraction was carried out with a fragment

---

Figure 19.23 Two competing hypotheses for the origins of modern humans.

Figure 19.24 A phylogenetic tree reconstructed from mitochondrial RFLP data obtained from 147 modern humans. The ancestral mitochondrial DNA is inferred to have existed in Africa because of the split in the tree between the seven modern African mitochondrial genomes placed below the ancestral sequence and all the other genomes above it. Because this lower branch is purely African it is deduced that the ancestor was also African. The scale bars at the bottom indicate sequence divergence from which, using the mitochondrial molecular clock, it is possible to assign dates to the branch points in the tree. The clock suggests that the ancestral sequence existed between 140,000 and 290,000 years ago. Reprinted with permission from Cann *et al.* (1997), *Nature*, **325**, 31–36.

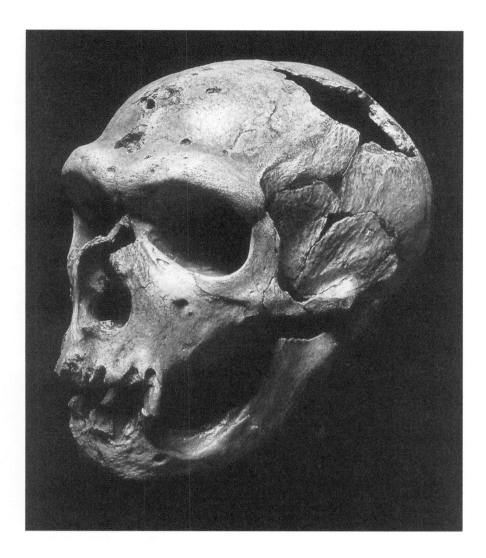

Figure 19.25 **Skull of a 40- to 50-year-old male Neandertal.** The specimen is about 50,000 years old and was found at La Chapelle-aux-Saints, France. Image courtesy of John Reader/Science Photo Library.

of bone weighing about 400 mg and a technique called **quantitative PCR** was used to determine if human DNA molecules were present and, if so, how many. The results indicated that the bone fragment contained about 1300 copies of the Neandertal mitochondrial genome, enough to suggest that sequence analysis was worth attempting. PCRs were therefore directed at what was expected to be the most variable part of the Neandertal mitochondrial genome. Because it was anticipated that the DNA would be broken into very short pieces, a sequence was built up in sections by carrying out nine overlapping PCRs, none amplifying more than 170 bp of DNA but together giving a total length of 377 bp.

A phylogenetic tree was then constructed to compare the sequence obtained from the Neandertal bone with the sequences of six mitochondrial DNA haplogroups from modern humans (Figure 19.26). The Neandertal sequence was positioned on a branch of its own, connected to the root of the tree but not linked directly to any of the modern human sequences. Next, a huge multiple alignment was made in order to compare the Neandertal sequence with 994 sequences from modern humans. The differences were striking. The Neandertal sequence differed from the modern sequences at an average of 27.2 ± 2.2 nucleotide positions whereas the modern sequences, which came from all over the world, not just Europe, differed from each other at only

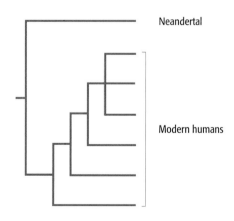

Figure 19.26 **Relationships between Neandertals and modern humans,** as deduced from the ancient DNA sequence obtained from a Neandertal bone.

8.0 ± 3.1 positions. Similar results were obtained when mitochondrial DNA from a second Neandertal skeleton was examined. The degree of difference between Neandertal and modern European DNA is incompatible with the notion that modern Europeans are descended from Neandertals and strongly supports the Out of Africa hypothesis. Supporters of multiregional evolution are not convinced, however, and the debate over modern human origins continues.

The patterns of more recent migrations into Europe are also controversial

By whatever evolutionary pathway, modern humans were present throughout most of Europe by 40,000 years ago. This is clear from the fossil and archaeological records. The next controversial issue in human prehistory concerns whether these populations were displaced about 30,000 years later by other humans migrating into Europe from the Middle East.

The question centers on the process by which agriculture spread into Europe. The transition from hunting and gathering to farming occurred in the Middle East some 9000–10,000 years ago, when early Neolithic villagers began to cultivate crops such as wheat and barley. After becoming established in the Middle East, farming spread into Asia, Europe, and North Africa. By searching for evidence of agriculture at archaeological sites, for example by looking for the remains of cultivated plants or for implements used in farming, it has been possible to trace the expansion of farming along two routes through Europe, one around the coast to Italy and Spain and the second through the Danube and Rhine valleys to northern Europe (Figure 19.27).

How did farming spread? The simplest explanation is that farmers migrated from one part of Europe to another, taking with them their implements, animals, and crops, and displacing the indigenous preagricultural communities that were present in Europe at that time. This **wave of advance** model was initially favored by geneticists because of the results of a large-scale analysis of

Figure 19.27 The spread of agriculture from the Middle East to Europe. The dark-green area is the "Fertile Crescent," the area of the Middle East where many of today's crops—wheat, barley, and so on—grow wild and where these plants are thought to have first been taken into cultivation.

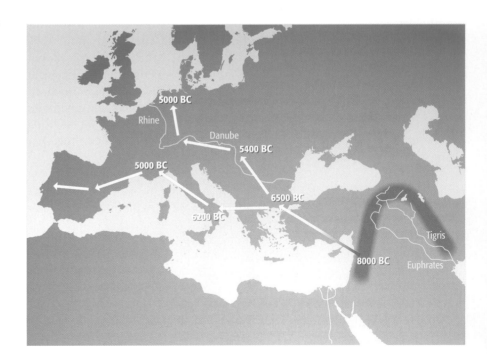

the allele frequencies for 95 nuclear genes in populations from across Europe. Such a large and complex dataset cannot be analyzed in any meaningful way by conventional tree building but instead has to be examined by more advanced statistical methods, ones based more in population biology than phylogenetics. One such procedure is **principal component analysis**, which attempts to identify in a dataset patterns that correspond to an uneven geographic distribution of alleles, an uneven distribution possibly being indicative of a past population movement. The most striking pattern within the European dataset, accounting for about 28% of the total genetic variation, is a gradation of allele frequencies across Europe (Figure 19.28). This pattern implies that a migration of people occurred either from the Middle East to northeast Europe, or in the opposite direction. Because the former coincides with the expansion of farming, as revealed by the archaeological record, this first principal component was looked upon as providing strong support for the wave of advance model.

Figure 19.28 A genetic gradation across modern Europe.

The analysis looked convincing but two criticisms were raised. The first was that the data provided no indication of when the inferred migration took place, so the link between the first principal component and the spread of agriculture was based solely on the pattern of the allele gradation, not on any complementary evidence relating to the period when this gradation was set up. The second criticism arose because of the results of a second study of European human populations, one that did include a time dimension. This study looked at mitochondrial DNA haplogroups in 821 individuals from various populations across Europe. It failed to confirm the gradation of allele frequencies detected in the nuclear DNA dataset, and instead suggested that European populations have remained relatively static over the last 20,000 years. A refinement of this work led to the discovery that eleven mitochondrial DNA haplogroups predominate in the modern European population. These eleven haplogroups are a subset of the 100 or so known throughout the world (Figure 19.29). Most of these haplogroups are associated with particular geographic regions and detailed statistical analysis of their distributions and relationships can be used to infer their times of origin, which for the European haplogroups are thought to indicate the date at which each one entered Europe (Figure 19.30). The most ancient haplogroup, called U, first appeared in Europe approximately 50,000 years ago, coinciding with the period when, according to the archaeological record, the first modern humans moved into the continent as the ice sheets withdrew to the north at the end of the last major glaciation. The youngest haplogroups, J and T1, which at 9000 years in age could correspond to the origins of agriculture, are possessed by just 8.3% of the modern European population, suggesting that the spread of farming into Europe was not the huge wave of advance indicated by the principal component study. Instead, it is now thought that farming was brought into Europe by a smaller group of "pioneers" who interbred with the existing prefarming communities rather than displacing them.

Prehistoric human migrations into the New World

Finally we will examine the completely different set of controversies surrounding the hypotheses regarding the patterns of human migration that led to the first entry of people into the New World. There is no evidence for the spread of *Homo erectus* into the Americas, so it is presumed that humans did not enter the New World until after modern *Homo sapiens* had evolved in, or migrated into, Asia. The Bering Strait between Asia and North America is quite

Figure 19.29 **Human mitochondrial DNA haplogroups.** The network shows the relationships between the major mitochondrial haplogroups in the present-day human population. The color coding indicates the geographical region within which each haplogroup is most common.

shallow and if the sea level dropped by 50 meters it would be possible to walk across from one continent to the other. It is believed that this Beringian land bridge was the route taken by the first humans to venture into the New World.

The sea was 50 meters or more below its current level for most of the last Ice Age, between about 60,000 and 11,000 years ago, but for most of this time the route would have been impassable because of the build-up of ice, not on the land bridge itself but in the areas that are now Alaska and northwest Canada.

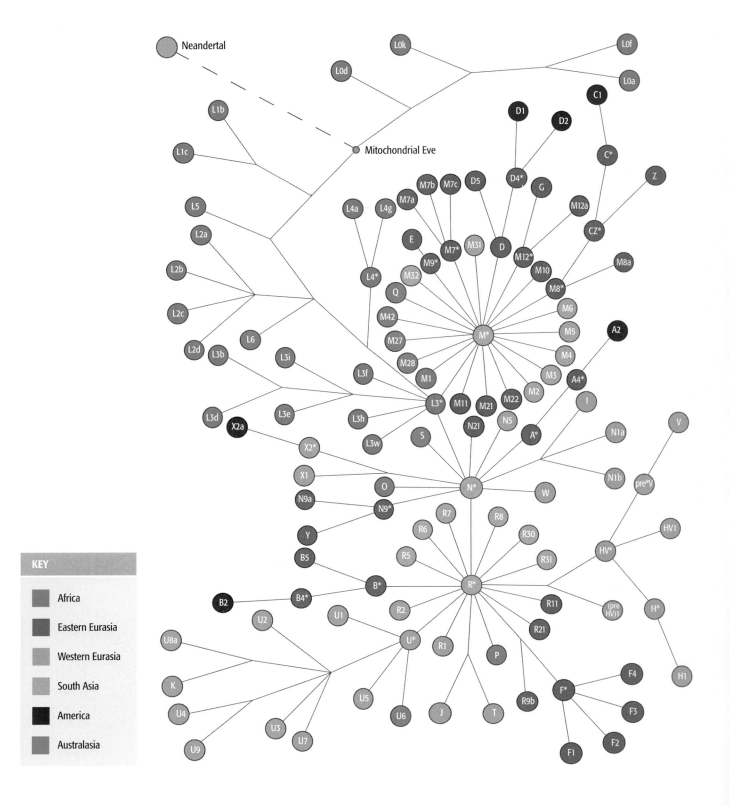

KEY

- Africa
- Eastern Eurasia
- Western Eurasia
- South Asia
- America
- Australasia

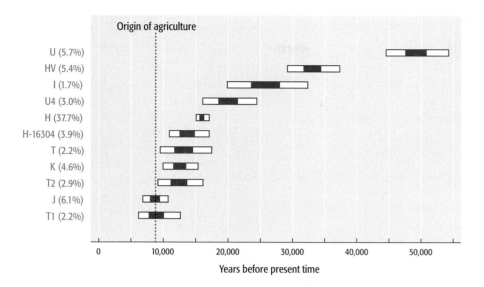

Figure 19.30 The eleven major European mitochondrial haplotypes. The calculated time of origin for each haplotype is shown, the closed and open parts of each bar indicating different degrees of confidence. The percentages refer to the proportions of the modern European population with each haplotype. All the haplotypes except J and T1 had entered Europe before the origin of agriculture 9000–10,000 years ago.

Also, the glacier-free parts of northern America would have been arctic during much of this period, providing few game animals for the migrants to hunt and very little wood with which they could make fires. But for a brief period around 12,000 years ago the Beringian land bridge was open at a time when the climate was warming and the glaciers were receding, such that there was an ice-free corridor leading from Beringia to central North America (Figure 19.31). These considerations, together with the absence of archaeological evidence of humans in North America before 11,500 years ago, led to the adoption of "about 12,000 years ago" as the date for the first entry of humans into the New World. Recent evidence for human occupation before 11,500 years ago has prompted some rethinking, but it is still generally assumed that a substantial population migration into North America, from which all modern Native Americans are descended, occurred about 12,000 years ago.

What information does molecular phylogenetics provide? The first relevant studies were carried out in the late 1980s using mitochondrial RFLP data. These indicated that Native Americans are descended from Asian ancestors and identified four distinct mitochondrial haplogroups, called A, B, C, and D, among the population as a whole. Linguistic studies had already shown that American languages can be divided into three different groupings, suggesting that modern Native Americans are descended from three sets of people, each speaking a different language. The inference from the molecular data that there may in fact have been four ancestral populations was not too disquieting. The first significant dataset of mitochondrial DNA sequences was obtained in 1991, enabling the rigorous application of a molecular clock. This indicated that the migrations into North America occurred between 15,000 and 8000 years ago, which is consistent with the archaeological evidence that humans were absent from the continent before 11,500 years ago.

These early phylogenetic analyses confirmed, or at least were not too discordant with, the complementary evidence provided by archaeological and linguistic studies. However, the additional molecular data that have been acquired since 1992 have tended to confuse rather than clarify the issue. For example, different datasets have provided a variety of estimates for the number of migrations into North America. One particularly comprehensive analysis, based on mitochondrial DNA, put the figure at just one migration, and

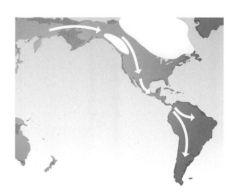

Figure 19.31 A possible route for migration of humans into the New World. For a short period some 12,000 years ago, the Beringian land bridge was open at the same time as the ice-free corridor into central North America. The dark green areas show regions that were exposed land at that time but are currently below sea level.

suggested that it occurred between 25,000 and 20,000 years ago, much earlier than the traditional date for the migration period. The first studies of Y chromosomes assigned a date of approximately 22,500 years ago to the "Native American Adam," the carrier of the Y chromosome that is ancestral to most, if not all, of the Y chromosomes in modern Native Americans. These results are still being hotly debated. Some molecular anthropologists accept the implication that humans became established in North America about 20,000 years ago, much earlier than indicated by the archaeological and previous genetic evidence. Others believe that the genetic data as a whole are consistent with two migrations, one 20,000 to 15,000 years ago, comprising all four haplogroups, followed by a second substantial migration at a more recent period, involving the same haplogroups and resulting in a modification of the geographic distributions of these haplogroups in North America.

Summary

Molecular phylogenetics uses molecular information to infer the evolutionary relationships between genes and between organisms. Molecular information, such as immunological data, protein electrophoresis patterns, and DNA–DNA hybridization data, has been used in phylogenetics for many years, but today most comparisons are made between DNA sequences. Groups of sequences are first aligned and polymorphisms identified. This information is converted into numerical data, such as a distance matrix, which can be analyzed mathematically to reconstruct a tree showing the evolutionary relationships between the sequences. There are several methods for tree reconstruction, such as neighbor-joining and maximum parsimony, which use different approaches to identifying the most likely tree for a particular set of data. Some of these methods are computer-intensive with a major limitation being the vast amount of computer memory needed to make a rigorous examination of an alignment. Sometimes it is possible to apply a molecular clock to a tree in order to assign a date to a branch point, but the rate of the molecular clock is not the same for every organism nor even for every part of a genome, hence care is needed to avoid errors. Some sets of phylogenetic data, such as the sequences of members of a multigene family which can recombine with one another, cannot be represented by a conventional tree but can be depicted as a network. Molecular phylogenetics has resolved long-standing issues regarding the evolutionary relationships between the primates and is providing valuable information on the evolution of HIV and, by inference, the history of the AIDS epidemic. Molecular phylogenetics can also be applied to studies of human prehistory. By comparing mitochondrial and Y chromosome sequences it has been deduced that all modern humans are descended from a population that lived in Africa some 200,000 years ago, and which migrated out of Africa 100,000–50,000 years ago, displacing the descendants of *Homo erectus* that were living throughout the Old World at that time. Studies of preserved DNA from fossils have shown that modern Europeans are not directly descended from Neandertals. The introduction of farming into Europe coincides with a relatively small-scale movement of humans from the Middle East, not a large-scale migration followed by displacement of the indigenous hunter-gatherers as first thought. Humans entered the New World from Asia during the last Ice Age, across a land bridge in what is now the Bering Strait.

Multiple Choice Questions

*Answers to odd-numbered questions can be found in the Appendix

19.1.* Which of the following is NOT a feature of molecular phylogenetics?

   a. The use of molecular data to reconstruct a phylogenetic tree.

   b. The use of molecular data to understand the genetic basis to variable phenotypes.

   c. The use of molecular data to infer the evolutionary relationships between genomes.

   d. The application of rigorous mathematical methods to analysis of variable characters.

19.2. What was the first molecular method used to infer relationships between organisms?

   a. DNA sequencing.

   b. Protein electrophoresis.

   c. Immunological tests.

   d. DNA–DNA hybridization.

19.3.* Which of the following is an example of convergent evolution (homoplasy)?

   a. The wings of birds and bats.

   b. The hemoglobin gene family.

   c. The ribosomal RNA genes.

   d. The number of toes on horses and humans.

19.4. Early molecular approaches to phylogenetics in the 1950s–1960s utilized all of the following EXCEPT:

   a. DNA–DNA hybridization data.

   b. Immunological data.

   c. Protein electrophoresis.

   d. Protein sequencing.

19.5.* The lengths of the branches of a phylogenetic tree constructed from DNA sequence data indicate:

   a. The length of time since the organisms diverged.

   b. The number of synonymous changes between the genes.

   c. The degree of difference between the genes represented by the nodes.

   d. None of the above.

19.6. If two or more DNA sequences are derived from different ancestral sequences, they are said to be:

   a. Monophyletic.

   b. Orthophyletic.

   c. Paraphyletic.

   d. Polyphyletic.

19.7.* Orthologous genes are which of the following:

   a. Genes that do not share a common ancestral origin.

   b. Homologous genes that are present in the genomes of different organisms.

   c. Homologous genes that are present in the same genome.

   d. Nonhomologous genes that arose from convergent evolution.

19.8. Which method of tree reconstruction identifies the topology that involves the shortest evolutionary path?

   a. Distance matrix.

   b. Maximum parsimony.

   c. Neighbor-joining.

   d. Principal component analysis.

19.9.* Which of the following is NOT a complication in the use of molecular clocks?

   a. Nonsynonymous mutations occur at a slower rate than synonymous mutations.

   b. The molecular clocks for eukaryotic genes are faster than those for prokaryotic genes.

   c. The molecular clocks for mitochondrial genes are faster than those for nuclear genes.

   d. Molecular clocks appear to have increased in speed over the past 1–2 million years.

19.10. Which of the following events complicates conventional tree building but is not a problem when networks are constructed?

   a. The presence of ancestral sequences in the dataset being examined.

   b. The presence of sequences that have arisen by recombination.

   c. Both a and b.

   d. Neither a nor b.

19.11.* Phylogenetic studies of HIV genome sequences have revealed all of the following EXCEPT:

   a. HIV-1 is most closely related to the SIV of chimpanzees.

   b. HIV-1 and HIV-2 were transferred to humans independently.

   c. HIV-1 and HIV-2 were transferred to humans from the same primate species.

   d. The global spread of HIV-1 was underway by 1959.

19.12. Which of the following is NOT useful for phylogenetic analysis of human populations?

   a. Ribosomal RNA genes.

   b. Microsatellites.

   c. Mitochondrial DNA.

   d. Multiallelic genes.

continued …

Multiple Choice Questions (continued) *Answers to odd-numbered questions can be found in the Appendix

19.13. *Which of the following are only inherited maternally?
  a. Globin genes.
  b. Mitochondrial DNA.
  c. X chromosomes.
  d. Y chromosomes.

19.14. Which of the following methods can be used to identify an uneven geographic distribution of alleles?
  a. Distance matrix.
  b. Maximum parsimony.
  c. Neighbor-joining.
  d. Principal component analysis.

Short Answer Questions   *Answers to odd-numbered questions can be found in the Appendix

19.1.* How does phenetics differ from the traditional classification methods used before 1957?

19.2. How does cladistics differ from phenetics?

19.3.* What is the difference between an ancestral and a derived character state?

19.4. Why are DNA sequences favored over protein sequences for phylogenetic studies?

19.5.* How do the internal nodes in gene and species trees differ?

19.6. What are the differences between the similarity approach and distance method for aligning sequences?

19.7.* What factors affect the frequency of alleles in a population?

19.8. What are the differences between the multiregional and Out of Africa hypotheses for the evolution of modern humans?

19.9.* How was phylogenetic analysis of Neandertal DNA sequences first performed and what conclusion was reached about the relationship between Neandertals and modern humans?

19.10. Discuss how principal component analysis has been used to study the prehistoric migration of farmers into Europe. What were the conclusions of these studies?

19.11.* What information has molecular phylogenetics provided about the prehistoric migration of human populations into North America?

In-depth Problems   *Guidance to odd-numbered questions can be found in the Appendix

19.1.* Can a gene tree ever be equivalent to a species tree?

19.2. How reliable are molecular clocks?

19.3.* Phylogenetic studies of mitochondrial DNA assume that this genome is inherited through the maternal line and that there is no recombination between maternal and paternal genomes. Assess the validity of this assumption and describe how the hypotheses regarding the origins and migrations of modern

humans would be affected if recombination between maternal and paternal genomes was shown to occur.

19.4. What is the potential of ancient DNA in studies of human evolution?

19.5.* Explore the ways in which molecular phylogenetics has been used to study the mitochondrial DNA haplotypes present in modern European populations.

Figure Tests

19.1.* Which figure, (A) or (B), gives an example of a derived character state?

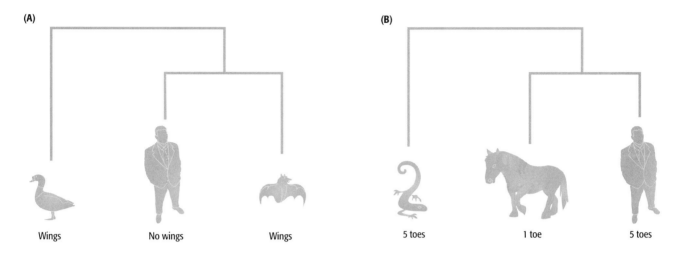

(A)

Wings   No wings   Wings

(B)

5 toes   1 toe   5 toes

19.2. Which is the outgroup in this phylogenetic tree? Using this tree as an example, discuss the terms monophyletic, polyphyletic, and paraphyletic.

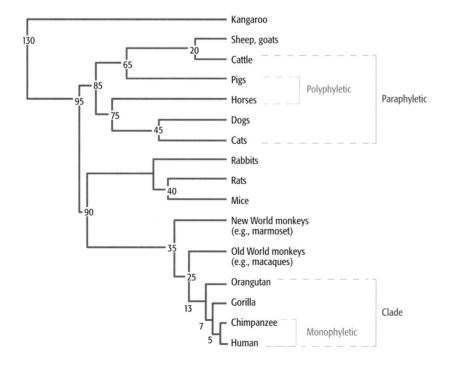

Figure Tests (continued)

19.3.* How is it possible for a species tree and a gene tree to give different branching patterns?

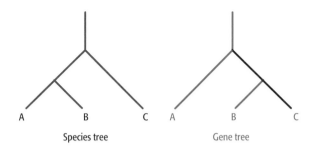

19.4. What type of phylogenetic analysis is shown in the figure? What is the purpose of this type of analysis?

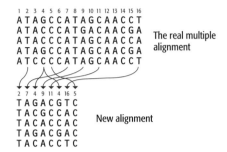

Further Reading

Key textbooks and reviews

Avise, J.C. (2004) *Molecular Markers, Natural History and Evolution*, 2nd Ed. Chapman and Hall, New York. *A detailed description of the use of molecular data in studies of evolution.*

Futuyama, D.J. (1998) *Evolutionary Biology*, 3rd Ed. Sinauer, Sunderland, Massachusetts.

Hall, B.G. (2004) *Phylogenetic Trees Made Easy: A How-To Manual for Molecular Biologists*, 2nd Ed. Sinauer, Sunderland, Massachusetts.

Nei, M. (1996) Phylogenetic analysis in molecular evolutionary genetics. *Annu. Rev. Genet.* **30:** 371–403. *Brief review of tree-building techniques.*

Tree reconstruction

Doolittle, W.F. (1999) Phylogenetic classification and the universal tree. *Science* **284:** 2124–2128. *Discusses the strengths and weaknesses of molecular phylogenetics as a means of inferring species trees.*

Felsenstein, J. (1989) PHYLIP – Phylogeny Inference Package (Version 3.20). *Cladistics* **5:** 164–166.

Jeanmougin, F., Thompson, J.D., Gouy, M., Higgins, D.G. and Gibson, T.J. (1998) Multiple sequence alignment with Clustal X. *Trends Biochem. Sci.* **23:** 403–405.

Saitou, N. and Nei, M. (1987) The neighbor-joining method: a new method for reconstructing phylogenetic trees. *Mol. Biol. Evol.* **4:** 406–425.

Swofford, D.L. (1993) *PAUP: Phylogenetic Analysis Using Parsimony*. Illinois Natural History Survey, Champaign, Illinois.

Whelan, S., Liò, P. and Goldman, N. (2001) Molecular phylogenetics: state-of-the-art methods for looking into the past. *Trends Genet.* **17:** 262–272.

Yang, Z. (1997) PAML: a program package for phylogenetic analysis by maximum likelihood. *CABIOS* **13:** 555–556.

The molecular clock

Gu, X. and Li, W.-H. (1992) Higher rates of amino acid substitution in rodents than in humans. *Mol. Phylogenet. Evol.* **1:** 211–214.

Penny, D. (2005) Relativity for molecular clocks. *Nature* **436:** 183–184. *The apparent increase in the rates of molecular clocks over the last few million years.*

Strauss, E. (1999) Can mitochondrial clocks keep time? *Science* **283:** 1435–1438.

Relationships among primates

Ruvolo, M. (1997) Molecular phylogeny of the hominoids: inferences from multiple independent DNA sequence data sets. *Mol. Biol. Evol.* **14:** 248–265.

Sarich, V.M. and Wilson, A.C. (1967) Immunological time scale for hominid evolution. *Science* **158:** 1200–1203.

Origins of HIV

Korber, B., Muldoon, M., Theiler, J., Gao, F., Gupta, R., Lapedes, A., Hahn, B.H., Wolinsky, S. and Bhattacharya, T. (2000) Timing the ancestor of the HIV-1 pandemic strains. *Science* **288:** 1789–1796.

Leitner, T., Escanilla, D., Franzen, C., Uhlen, M. and Albert, J. (1996) Accurate reconstruction of a known HIV-1 transmission history by phylogenetic tree analysis. *Proc. Natl Acad. Sci. USA* **93:** 10864–10869.

Zhu, T., Korber, B.T., Nahmias, A.J., Hooper, E., Sharp, P.M. and Ho, D.D. (1998) An African HIV-1 sequence from 1959 and implications for the origin of the epidemic. *Nature* **391:** 594–597.

Origins of modern humans

Cann, R.L., Stoneking, M. and Wilson, A.C. (1987) Mitochondrial DNA and human evolution. *Nature* **325:** 31–36. *The first discovery of mitochondrial Eve.*

Harding, R.M., Fullerton, S.M., Griffiths, R.C., Bond, J., Cox, M.J., Schneider, J.A., Moulin, D.S. and Clegg, J.B. (1997) Archaic African and Asian lineages in the genetic ancestry of modern humans. *Am. J. Hum. Genet.* **60:** 772–789. *Studies of nuclear genes.*

Ingman, M., Kaessmann, H., Pääbo, S. and Gyllensten, U. (2000) Mitochondrial genome variation and the origin of modern humans. *Nature* **408:** 708–713.

Krings, M., Stone, A., Schmitz, R.W., Krainitzki, H., Stoneking, M. and Pääbo, S. (1997) Neandertal DNA sequences and the origin of modern humans. *Cell* **90:** 19–30.

Human migrations

Cavalli-Sforza, L.L. (1998) The DNA revolution in population genetics. *Trends Genet.* **14:** 60–65. *Principal component analysis of human nuclear genes.*

Chikhi, L., Destro-Bisol, G., Bertorelle, G., Pascali, V. and Barbujana, G. (2002) Y genetic data support the Neolithic demic diffusion model. *Proc. Natl Acad. Sci. USA* **99:** 11008–11013. *Migrations in Europe.*

Forster, P., Harding, R., Torroni, A. and Bandelt, H.J. (1996) Origin and evolution of native American mtDNA variation: a reappraisal. *Am. J. Hum. Genet.* **59:** 935–945.

Richards, M. (2003) The Neolithic invasion of Europe. *Annu. Rev. Anthropol.* **32:** 135–162.

Semino, O., Passarino, G., Oefner, P.J., *et al.* (2000) The genetic legacy of paleolithic *Homo sapiens sapiens* in extant Europeans: a Y chromosome perspective. *Science* **290:** 1155–1159.

Silva, W.A., Bonatto, S.L., Holanda, A.J., *et al.* (2002) Mitochondrial genome diversity of Native Americans supports a single early entry of founder populations into America. *Am. J. Hum. Genet.* **71:** 187–192.

Appendix

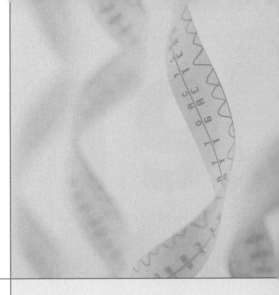

Answers

Chapter 1: Genomes, Transcriptomes, and Proteomes

Multiple Choice Questions
1.1-c; 1.3-a; 1.5-c; 1.7-c; 1.9-a; 1.11-c; 1.13-a; 1.15-b

Short Answer Questions
1.1. DNA was first discovered in 1869 (Miescher) and shown to contain genetic information in the 1940s (Avery, MacLeod, and McCarty, and Hershey and Chase). The structure of the double helix was determined in 1953 (Watson and Crick) and the first complete genome sequence for a cellular organism was completed in 1995.

1.3. The limitation that A can only base-pair with T, and G can only base-pair with C, means that DNA replication can result in perfect copies of a parent molecule through the simple expedient of using the sequences of the pre-existing strands to dictate the sequences of the new strands. Base pairing therefore enables DNA molecules to be replicated into perfect copies.

1.5. Bacterial mRNAs have half-lives of no more than a few minutes and in eukaryotes most mRNAs are degraded a few hours after synthesis. This rapid turnover means that the composition of the transcriptome is not fixed and can quickly be restructured by changing the rate of synthesis of individual mRNAs. The composition of the transcriptome can therefore be rapidly adjusted in accordance with the needs of the cell.

1.7. Every cell receives part of its parent's transcriptome when it is first brought into existence by cell division, and maintains a transcriptome throughout its lifetime. Transcription of individual protein-coding genes does not therefore result in synthesis of the transcriptome but instead maintains the transcriptome by replacing mRNAs that have been degraded, and brings about changes to the composition of the transcriptome via the switching on and off of different sets of genes.

1.9. Proteins are structurally and functionally diverse because the amino acids from which they are made are chemically diverse. Different sequences of amino acids therefore result in different combinations of chemical reactivities, these combinations dictating not only the overall structure of the resulting protein but also the positioning on the surface of the structure of reactive groups that determine the chemical properties of the protein.

1.11. The location of the codon is important in determining whether it functions as a stop codon or specifies selenocysteine. There is a hairpin loop structure just downstream of the selenocysteine codon that allows this amino acid to be incorporated into the growing polypeptide chain.

In-depth Problems

1.1. The justification for this statement can be found in *The Double Helix* by James Watson or in most of the various books written about the history of DNA, such as *The Eighth Day of Creation* by Horace Freeland Judson (see Further Reading). It was during the evening of Saturday 7 March 1953 that Watson and Crick completed their model of the double helix structure, made out of thin pieces of galvanized metal and brass rods, with a scale of 50 cm per nm, a complete turn of the model helix being almost 2 m. It could be argued, however, that the actual *discovery* was made one week earlier when Watson and Crick realized that the hydrogen-bonded structures formed by A with T and G with C have the same outline shapes (see Figure 1.8B), enabling these base pairs to be stacked to form a regular helix of unvarying width. Crick recalls that this discovery came when they realized the significance of Chargaff's base ratios, but Watson maintains that this significance was appreciated only after they had built their first models of the nucleotide pairs.

1.3. The elucidation of the genetic code was the most important breakthrough in biology in the 1960s, and although this work was carried out almost half a century ago it is still an excellent example of the scientific method—how to plan a research strategy toward a defined end, and how to modify that strategy to utilize new techniques that are developed during the course of a project. Books such as *The Eighth Day of Creation* (see Further Reading) give a detailed account of how the code was cracked, but the best starting points for a tutorial discussion of this subject are the three reviews in *Scientific American* that were written during and at the end of the project. These are Crick, F.H.C. (1962) The genetic code. *Sci. Am.* **207(4):** 66–74; Nirenberg, M.W. (1963) The genetic code II. *Sci. Am.* **208(3):** 80–94; and Crick, F.H.C. (1966) The genetic code III. *Sci Am.* **215(4):** 55–62.

Figure Tests

1.1. Part A shows that treatment of the transforming principle with protease or ribonuclease has no effect, but that the transforming principle is inactivated by deoxyribonuclease. The transforming principle, which contains the genetic information needed to convert the harmless bacteria into the virulent form, must therefore be made of DNA. Part B shows that when bacteriophages are labeled with ^{32}P and ^{35}S, most of the ^{32}P-labeled material (the DNA) but only 20% of the ^{35}S-labeled material (the phage protein) enters the cells during infection. As the bacteriophage genes must enter the bacteria to direct synthesis of new bacteriophages, these genes must be made of DNA.

1.3. The model shows that the sugar–phosphate backbone is on the outside of the molecule and that the bases are present on the inside. The bases are clearly exposed in the major and minor grooves where they can be recognized by DNA-binding proteins.

Chapter 2: Studying DNA

Multiple Choice Questions

2.1-b; 2.3-d; 2.5-a; 2.7-a; 2.9-b; 2.11-d; 2.13-d; 2.15-a

Short Answer Questions

2.1. A gene is cloned when the DNA fragment containing it is inserted into a vector DNA molecule (such as a plasmid or bacteriophage) and then replicated in a host cell.

2.3. One can attach linker or adapter molecules to the ends of the blunt-ended molecules to generate sticky ends that facilitate ligation.

2.5. This allows for easy selection of bacterial cells that have been transformed with the plasmid.

2.7. The bacteriophage genome contains genes for the lysogenic infection of *E. coli* that are not essential and can be replaced with new DNA. The bacteriophage can be used to close DNA molecules up to 18 kb in length.

2.9. These require the presence of a centromere, telomeres, and at least one origin of replication.

2.11. The primers hybridize to specific sequences in the template DNA and define the regions to be amplified.

In-depth Problems

2.1. The moratorium arose from the Asilomar Conference in 1975, and was proposed by Paul Berg and others in Berg, P., Baltimore, D., Brenner, S., Roblin, R.O. and Singer, M.F. (1975) Summary statement of the Asilomar Conference on recombinant DNA molecules. *Proc. Natl Acad. Sci. USA* **72:** 1981–1984. A less specialist account of the background to the moratorium can be found in Cherfas, J. (1982) *Man Made Life*. Blackwell Scientific Publishers, Oxford. It is tempting to conclude that because the fears of these scientists have never become reality then those fears were unjustified, but a full debate of this issue must also take account of the outcomes of the moratorium (e.g. the development of strategies for preventing the survival of genetically engineered bacteria in the natural environment) which have been greatly instrumental in ensuring that the dangers that prompted the moratorium have been avoided.

2.3. This question looks ahead to the discussion of restriction mapping in Section 3.3.1. Attempting to "invent" restriction mapping in a class discussion is an excellent way of ensuring that the principles of restriction, gel electrophoresis, and such like have been grasped.

2.5. Assuming the target sequence for the PCR is single copy in the genome being studied, then the important issue is the length of the primers. If the primers are too short they might hybridize to nontarget sites and give undesired amplification products. The best way to illustrate this point is to imagine that total human DNA is used in a PCR experiment with a pair of primers eight nucleotides in length. The likely result is that a number of different fragments will be amplified. This is because attachment sites for these primers are expected to occur, on average, once every $4^8 = 65,536$ bp, giving approximately 49,000 possible sites in the 3,200,000 kb of nucleotide sequence that makes up the human genome. This means that it would be very unlikely that a pair of eight-nucleotide primers would give a single, specific amplification product with human DNA as there would be several positions where annealing sites would, by chance, lie close enough together to give an amplification product. In contrast, the expected frequency of a 17-nucleotide sequence is once every $4^{17} = 17,179,869,184$ bp. This figure is over five times greater than the length of the human genome, so a pair of 17-nucleotide primers should therefore give a single, specific amplification product. The ideal annealing temperature must be low enough to enable hybridization between primer and template, but high enough to prevent mismatched hybrids from forming. The information needed to understand how to determine the appropriate temperature for a pair of primers is given in the legend to Figure 3.8. For a more detailed discussion, see Brown, T.A. (2006) *Gene Cloning and DNA Analysis: An Introduction*, 5th Ed. Blackwell Scientific Publishers, Oxford.

Figure Tests

2.1. The primer initiates DNA synthesis, by providing the 3′–OH group needed for the nucleotide addition. The primer can also be used to specify the location of DNA synthesis on the template molecules (as in DNA sequencing and PCR).
2.3. This is a cosmid that can carry insert molecules up to 44 kb in length.

Chapter 3: Mapping Genomes

Multiple Choice Questions
3.1-d; 3.3-a; 3.5-d; 3.7-c; 3.9-b; 3.11-c; 3.13-a; 3.15-d

Short Answer Questions

3.1. A genome map provides a guide for the sequencing experiments by showing the positions of genes and other distinctive features. If a map is unavailable then it is likely that errors will be made in assembling the genome sequence, especially in regions that contain repetitive DNA.
3.3. The primers for the PCR are designed so that they anneal either side of the polymorphic site, and the RFLP is typed by treating the amplified fragment with the restriction enzyme and then running a sample in an agarose gel. Before the invention of PCR, RFLPs were typed by Southern hybridization, which is time-consuming.
3.5. If a pair of genes display linkage then they must be on the same chromosome. If crossing-over is a random event then the recombination frequency between a pair of linked genes is a measure of their distance apart on the chromosome. The recombination frequencies for different pairs of genes can be used to construct a map of their relative positions on the chromosome.
3.7. The double homozygote will produce gametes that are all the same genetically and if they are recessive then this parent will not contribute to the phenotype of the offspring.
3.9. FISH uses a fluorescently labeled DNA fragment as a probe to bind to an intact chromosome. The binding position can be determined and this information used to create a physical map of the chromosome.
3.11. Individual chromosomes can be separated by flow cytometry. Dividing cells are carefully broken open so that a mixture of intact chromosomes is obtained. The chromosomes are then stained with a fluorescent dye. The amount of dye that a chromosome binds depends on its size, so larger chromosomes bind more dye and fluoresce more brightly than smaller ones. The chromosome preparation is diluted and passed through a fine aperture, producing a stream of droplets, each one containing a single chromosome. The droplets pass through a detector that measures the amount of fluorescence, and hence identifies which droplets contain the particular chromosome being sought. An electric charge is applied to these drops, and no others, enabling the droplets containing the desired chromosome to be deflected and separated from the rest.

In-depth Problems
3.1. The text indicates that the ideal features include high frequency in the genome being studied, ease of typing, and the presence of multiple alleles. This implies that SSLPs should be the "ideal" markers, but in reality SNPs are more popular. A discussion of this apparent paradox demands consideration of the relative importance of each of the three criteria, and in particular a realization that the critical feature of an "ideal" marker is high density.

3.3. Many teachers will remember tackling this question during their own student days, and the answer has not changed: short generation time, large number of offspring, easily scored phenotypes, and such like. It is instructive to consider to what extent genomics has added new criteria to this list: is a complete genome sequence a useful feature of an organism to be used in studies of heredity?

3.5. This is a very open-ended question that is designed to prompt discussion of a number of topics that are covered in later chapters. The discussion might begin by asking what purpose the map is intended to fulfill. A map designed to aid a sequencing project might not be the same as one designed to enable individual genes to be cloned. If it is concluded that for sequencing purposes a physical map is more useful, and in fact a genetic map has little or no direct value (which is a reasonable inference to make from a reading of Chapter 4), then the discussion could turn to how easy or otherwise it would be to locate the genes in a genome sequence, and to assign functions to those genes, without any prior knowledge of where the genes are. These issues are discussed in Chapter 5.

Figure Tests
3.1. When the oligonucleotide is not hybridized to the target sequence, the fluorescent label and quenching molecule are next to each other and the fluorescence is quenched. When the oligonucleotide binds to a target sequence, the fluorescent label is located away from the quenching molecule. By controlling the hybridization conditions, the oligonucleotide will only bind to the target sequence if all the nucleotides are complementary.

3.3. This is orthogonal field alternation gel electrophoresis, in which the electric field alternates between the pairs of electrodes. The DNA molecules move down through the gel, but each change in the field forces the molecules to realign. Shorter molecules realign more quickly than longer ones and so migrate more rapidly through the gel. The overall result is that molecules much longer than those separated by conventional gel electrophoresis can be resolved.

Chapter 4: Sequencing Genomes

Multiple Choice Questions
4.1-b; 4.3-a; 4.5-c; 4.7-b; 4.9-c; 4.11-d; 4.13-c; 4.15-c

Short Answer Questions
4.1. The dideoxynucleotides lack a 3'–hydroxyl group, and when dideoxynucleotides are incorporated into the DNA, strand synthesis stops.

4.3. Yes, this is possible. The PCR product is purified and thermal cycle sequencing performed with one of the PCR primers used as the primer for the sequencing reactions.

4.5. Automated sequencers with multiple capillaries working in parallel can read up to 96 different sequences in a 2-hour period, which means that with an average of 750 bp per individual experiment, 864 kb of information can be generated per machine per day. This enables data needed to sequence an entire genome to be generated in a period of weeks.

4.7. The redundancy is required as the clones for the sequencing project are randomly generated and sequenced; thus, to ensure complete coverage of the genome it is necessary to sequence a large number of nucleotides.

4.9. Clone fingerprinting can be based on restriction enzyme patterns, repetitive DNA fingerprints, repetitive DNA PCR, and STS content mapping.

4.11. Shotgun sequencing of complex eukaryotic genomes can result in segments of DNA, possibly including genes or parts of genes, being omitted from the draft sequence. There is also a greater chance that sequence errors will not be recognized.

In-depth Problems

4.1. The text at the start of Section 4.1 states that "chain termination sequencing has gained preeminence for several reasons, not least being the relative ease with which the technique can be automated," and this is the major part of the answer to this question. A consideration of why the chemical degradation technique has proven intractable to automation is a good way of raising awareness of the problems inherent in the development of automated technology. A second drawback of the chemical degradation method is the toxicity of the chemicals that are used, toxicity being an unavoidable property of any compound that is able to bind and modify DNA molecules. Although not an academic point, this question can therefore be used to introduce risk and safety aspects of laboratory molecular biology.

4.3. Section 4.2.2 covers the clone contig approach, but a critical evaluation is more usefully based on a comparison between the clone contig and whole-genome shotgun approaches, in particular as these have been applied to the human genome. From this comparison it becomes obvious that a rigorous clone contig project is relatively time-consuming, but that it is currently the only way to ensure an error rate of less than one in 104 nucleotides, the figure set as the acceptable maximum for a "finished" sequence.

4.5. The central issue is the tension between the company's right to protect its investment, which is accepted without question in most other aspects of commercial activity, and the less-well-defined rights of the individual(s) whose genes were used by the company during the research that led to development of the drug. A variety of views are possible, and a professional and reasonable justification of the view expressed is the most important part of the answer to this question.

Figure Tests

4.1. For most sequencing experiments a universal primer is used, this being one that is complementary to the part of the vector DNA immediately adjacent to the point into which new DNA is ligated. The same universal primer can therefore give the sequence of any piece of DNA that has been ligated into the vector. It is possible to extend the sequence in one direction by synthesizing an internal primer, designed to anneal at a position within the insert DNA. An experiment with this primer will provide a second short sequence that overlaps the previous one.

4.3. This is chemical degradation sequencing and it is useful when there are problems with standard, chain termination sequencing reactions (due to blockage of DNA polymerase or altered migration of sequencing products during electrophoresis).

Chapter 5: Understanding a Genome Sequence

Multiple Choice Questions
5.1-b; 5.3-b; 5.5-a; 5.7-b; 5.9-b; 5.11-d; 5.13-b

Short Answer Questions

5.1. Computers can readily scan all six reading frames of a DNA sequence for ORFs. In addition, as a random DNA sequence would possess a stop codon at least every 100–200 bp and most genes contain more than this number of codons, it is fairly straightforward to identify coding sequences in bacterial genomes that lack introns and other significant noncoding sequences.

5.3. Computer programs can be modified to screen for codon bias, exon–intron boundaries, and upstream regulatory sequences of genes.

5.5. Some genes contain optional exons and may encode different-sized mRNA molecules. Also, it is possible that not all of the genes present in a DNA fragment will be expressed in the cells from which the RNA is isolated.

5.7. Orthologous genes are homologous genes present in different organisms and paralogous genes are homologous genes present in the same organism.

5.9. If the biochemical activity of the gene product is known in another species, it may provide clues to the function of the gene in humans. Other organisms can also be used in experimental analyses to study gene function.

5.11. Comparative genomics, identifying expressed sequences, and transposon tagging can be used to determine if short ORFs are genuine genes.

In-depth Problems

5.1. This is a difficult question but a consideration of the basic principles enables progress to be made. Section 5.1.1 makes it clear that identifying exon–intron boundaries is the major hindrance to eukaryotic gene location by sequence inspection. It is probably reasonable to conclude that the consensus sequences currently assigned to these sites are as accurate as they can be, as they are based on comparisons between many exon–intron boundaries in many organisms. Hence it could be argued that sequence inspection on its own will never be a means of identifying a gene sequence with certainty. The discussion must therefore center on the potential of homology searching, both for gene location and assignment of function. The key issues are probably the extent to which homology searching will become more powerful and more accurate as a greater number of sequences are added to the databases, and whether the potential of comparative genomics, as used with *Saccharomyces cerevisiae* and related yeasts, can be realized with other organisms. To address the second point, consider how closely related two genomes need to be for comparative genomics to be useful, and whether it is likely that pairs or groups of genomes with the required degree of relatedness will become available among, for example, mammals, in the foreseeable future.

5.3. The sequence is from the human myoglobin protein. The other sequences identified by the BLAST search are orthologs of the human protein.

Figure Tests

5.1. The computer program would search for the exon–intron boundary and identify the intron sequence.

5.3. Most of the regulatory signals that control gene expression are contained in the region of DNA upstream of the ORF, so the GFP gene will now display the same expression pattern as the test gene. The expression pattern of this gene can therefore be determined by examining the organism for the presence of GFP.

Chapter 6: Understanding How a Genome Functions

Multiple Choice Questions
6.1-b; 6.3-c; 6.5-a; 6.7-b; 6.9-a; 6.11-c; 6.13-b

Short Answer Questions

6.1. To understand how the genome as a whole operates within the cell, specifying and coordinating the various biochemical activities that take place. These global studies of genome activity must address not only the genome itself but also the transcriptome and proteome.

6.3. If two different mRNAs have similar sequences then they may cross-hybridize to each other's specific probe on the array. This often happens when two or more paralogous genes are active in the same tissue. The transcriptome then contains a group of related mRNAs, each of which is able to hybridize to some extent with different members of the gene family. Distinguishing the relative amounts of each mRNA, or even being certain which particular mRNAs are present, can then be difficult. To solve this problem it is necessary to design a DNA chip that carries oligonucleotides that are specific for the particular sequences that are unique to each member of a family of paralogs.

6.5. Genes that display similar expression profiles are likely to be ones with related functions. These can be identified by hierarchical clustering, which involves comparing the expression levels of every pair of genes in every transcriptome that has been analysed, and assigning a value that indicates the degree of relatedness between those genes. These data can be expressed as a dendrogram, in which genes with related expression profiles are clustered together. The dendrogram gives a clear visual indication of the functional relationships between genes.

6.7. Examination of the transcriptome gives an accurate indication of which genes are active in a particular cell, but gives a less accurate indication of the proteins that are present. This is because the factors that influence protein content include not only the amount of mRNA that is available, but also the rate at which the mRNAs are translated into protein and the rate at which the proteins are degraded.

6.9. The cloning vector used in phage display is designed so that a new gene that is cloned into it is expressed in such a way that its protein product becomes fused with one of the phage coat proteins. The phage protein therefore carries the foreign protein into the phage coat, where it is "displayed" in a form that enables it to interact with other proteins that the phage encounters.

6.11. It is hoped that when metabolomics reaches maturity it will be possible to use the information to design drugs that treat diseases by reversing or mitigating the particular flux abnormalities that occur in the disease state. Metabolic profiling could also indicate any unwanted side effects of drug treatment, enabling modifications to be made to the chemical structure of the drug, or to its mode of use, so that these side effects are minimized.

In-depth Problems

6.1. Section 6.1.2 describes how microarrays are used to compare transcriptomes in two or more tissues or in the same tissue under different conditions, and an exploration of the Further Reading will provide more detail and specific examples. A discussion of the difficulties in applying a cDNA sequencing approach such as SAGE to comparisons between transcriptomes ensures that the key differences between the sequencing and microarray approaches are appreciated, and emphasizes the utility of microarray technology.

6.3. The first part of the question is relatively straightforward, the text giving one example from *E. coli* (the lactose permease and β-galactosidase). The second part of the question—proteins that have physical but not functional interactions—is less easy to answer but there are examples, such as molecular chaperones (Section 13.3.1), which form physical interactions with proteins in order to assist those proteins to fold, and the proteins of the proteasome (Section 13.4), which similarly form physical interactions with proteins being degraded. Understanding the specificity of these types of interactions (e.g. which specific proteins are folded by which individual chaperones) is a major goal of protein research and hence these types of interactions are equally as interesting as those with a functional basis.

6.5. This is an open-ended question suitable for a class or small group discussion, possibly using a key paper, such as Kirschner, M.W. (2005) The meaning of systems biology. *Cell* **121:** 503–504, as the starting point.

Figure Tests

6.1. A cDNA preparation is labelled with a fluorescent marker and hybridized to the microarray. The label is detected by confocal laser scanning and the intensity converted into a pseudocolor spectrum.

6.3. In the first dimension, the proteins are separated by isoelectric focusing. The gel is then soaked in sodium dodecyl sulfate, rotated by 90° and a second electrophoresis, separating the proteins according to their sizes, carried out at right angles to the first.

Chapter 7: Eukaryotic Nuclear Genomes

Multiple Choice Questions

7.1-b; 7.3-c; 7.5-d; 7.7-c; 7.9-b; 7.11-b; 7.13-b

Short Answer Questions

7.1. Complete digestion of human chromatin with nucleases reveals that DNA sequences of 146 bp are protected from digestion. Partial nuclease digestion gives rise to DNA fragments of 200 bp and multiples thereof.

7.3. Minichromosomes are shorter than macrochromosomes, but have a much higher gene density.

7.5. Telomeres mark the ends of chromosomes and enable the cell to distinguish a real end from one caused by chromosome breakage.

7.7. A typical region of a human chromosome will have few genes (most of which will contain introns), several repeated sequences, and a large amount of nonrepetitive, nongenic DNA. Yeast chromosomes have higher gene densities, with very few genes containing introns, and have few repeated sequences and much less nongenic DNA.

7.9. Gene catalogs can be based on the known functions of genes, but such catalogs are incomplete because in most genomes many genes have unknown functions. Gene catalogs that are based on the identities of protein domains coded by genes are more comprehensive as these include many genes whose specific functions are unknown.

7.11. A conventional pseudogene has become inactivated due to mutation, while a processed pseudogene arose by reinsertion of a cDNA copy of an mRNA.

In-depth Problems

7.1. This question looks forward to much of the material covered in Chapter 10—Accessing the Genome. From first principles, it should be clear that genes that are present in regions of highly packaged chromatin are likely to be inaccessible to the proteins responsible for activating and transcribing a gene, and that the precise positioning of nucleosomes might also be important in determining the degree of access that there is to a gene. The text at the end of Section 7.1.1 indicates that chemical modification of histones is important in determining chromatin structure, and this topic can be usefully previewed at this stage prior to detailed treatment in Chapter 10.

7.3. Figure 7.13 should be the starting point and the first objective should be the definition of "intergenic DNA" and an appreciation that this excludes all the sequences contained within genes (coding regions and introns), related to genes (e.g., pseudogenes, gene fragments), or required for gene activity (e.g., the regions immediately upstream of genes). Traditionally, however, intergenic DNA includes some functional sequences such as origins of replication (Section 15.2.1) and sequences that attach chromosomes to the nuclear substructure (Section 10.1.2). Once this fact has been recognized, the next question is whether any of the substantial repetitive component has a function. From Section 7.1.2 it should be clear that at least some of the satellite and minisatellite DNA is functional, and the last paragraph of Chapter 7 indicates that much of the interspersed repetitive DNA has transpositional activity, but to what extent can this be considered a "function"?

Figure Tests

7.1. The figure shows part of the human karyogram. The chromosomes are distinguished by their size, the location of the centromere, and the banding patterns present after staining.

7.3. The figure shows a processed pseudogene, which is not functional because it is derived from an mRNA, and hence lacks the nucleotide sequences needed to switch on and regulate gene expression.

Chapter 8: Genomes of Prokaryotes and Eukaryotic Organelles

Multiple Choice Questions

8.1-d; 8.3-a; 8.5-c; 8.7-a; 8.9-c; 8.11-d; 8.13-c

Short Answer Questions

8.1. Eukaryotic chromosomes are linear and have histone proteins that are involved in packaging the DNA. Eukaryotic chromosomes have multiple origins of replication and contain centromeres and telomeres. The *E. coli* chromosome is a circular molecule that contains a single origin of replication, is packaged by supercoiling, and lacks centromeres and telomeres.

8.3. HU proteins are structurally different from histones, but like histones form tetramers around which the DNA is wound.

8.5. Prokaryotic genomes are very gene dense, contain very short regions of intergenic DNA, and lack introns and repetitive DNA sequences.

8.7. Because it is parasitic many of the nutritional needs of this bacterium are provided by its host and its genome therefore lacks many genes coding for proteins involved in biosynthetic pathways.

8.9. Because DNA can be exchanged between different species through lateral gene transfer.

In-depth Problems

8.1. The text in Section 8.1.1 strongly suggests that the traditional view of the prokaryotic genome as a single, circular DNA molecule should indeed be abandoned. A discussion of what definition should be adopted will probably not reach a firm conclusion (as no such conclusion has yet been reached by microbial geneticists), but the exercise is instructive in that it requires a clear distinction to be made between plasmids and genomes.

8.3. Section 8.2.3 suggests that the answer to this question is "no," and the discussion should cover the main points addressed in this Section: the general difficulty in applying species concepts devised for eukaryotes to prokaryotes; the substantial differences that genome sequencing has revealed between the gene contents of strains traditionally regarded as members of a single species; and the complications presented by lateral gene transfer.

Figure Tests

8.1. The *E. coli* chromosome is attached to a protein core from which supercoiled DNA loops extend. If a DNA break occurs in one loop, then only that loop loses its supercoiling.

8.3. The genes in this figure are present in an operon and hence will be transcribed into a single mRNA molecule.

Chapter 9: Virus Genomes and Mobile Genetic Elements

Multiple Choice Questions

9.1-c; 9.3-a; 9.5-b; 9.7-d; 9.9-c; 9.11-c; 9.13-c; 9.15-b

Short Answer Questions

9.1. Viruses are obligate parasites dependent on host cells for reproduction. Viruses lack many of the components essential for the viability of cellular organisms; all viruses use their host's ribosomes, and not all viruses have genes for DNA or RNA polymerases.

9.3. These are genes that share nucleotide sequences, but code for different proteins. The nucleotide sequences of overlapping genes are translated in different reading frames.

9.5. Only bacteriophages have capsids of the head-and-tail type. Eukaryotic viruses, especially those that infect animals, may be covered with a lipid membrane.

9.7. A transposon is a segment of DNA that can move from one position to another within a genome.

9.9. The long interspersed nuclear elements (LINEs) make up over 20% of the human genome and a full-length element contains two genes, one of which codes for reverse transcriptase. The short interspersed nuclear elements (SINEs) have the highest copy number of any sequence within the human genome and lack genes. To transpose they must make use of reverse transcriptases synthesized from LINEs.

9.11. Active DNA transposons are more common in plant genomes than in the human genome. Some plant transposons work together, as is seen with the Ac/Ds family discovered by Barbara McClintock. The Ac element codes for a transposase that recognizes both Ac and Ds sequences.

In-depth Problems

9.1. A traditional question and one for which little guidance can be provided. The key to a productive discussion is not to get bogged down in a consideration of viruses, but to identify a meaningful definition of "life," and then to decide if this definition encompasses noncellular systems.

9.3. Selfish DNA is looked on as DNA that confers no benefit on a genome but is tolerated because there is no selective pressure to get rid of it. If this view is correct then possession of transposons is neither an advantage nor a disadvantage and so these elements are simply propagated along with the functional parts of the genome. See Orgel, L.E. and Crick, F.H.C. (1980) Selfish DNA: the ultimate parasite. *Nature* **284:** 604–607. Note that an argument against the benign nature of transposons is provided by the attempts that some organisms make to limit their activity, notably by methylating and hence inactivating these sequences (Section 17.3.3).

9.5. This question is answered in Section 17.3.2 when we consider the transposition process for an LTR retroelement and discover that replication of the element involves two switches, each from one LTR to the other, these switches ensuring that the complete sequence of the retroelement is copied (see Figure 17.18).

Figure Tests

9.1. From left to right: icosahedral, filamentous, head-and-tail.
9.3. Retrovirus infection.
9.5. The Ac and Ds elements were first characterized by Barbara McClintock. The Ac elements possess a transposase gene which is lacking in the Ds elements. The transposase encoded by the Ac elements is responsible for the transposition of both the Ac and Ds elements.

Chapter 10: Accessing the Genome

Multiple Choice Questions

10.1-d; 10.3-b; 10.5-a; 10.7-c; 10.9-b; 10.11-d; 10.13-a; 10.15-c

Short Answer Questions

10.1. Electron microscopy of cells prepared by treatment with DNase, to degrade the DNA, and salt extraction, to remove the histone proteins, has revealed the nuclear matrix—a complex network of protein and RNA fibrils. Fluorescent labeling of specific proteins has shown that activities such as RNA splicing are localized into distinct regions of the nucleus.

10.3. This suggests that these pairs of chromosomes occupy adjacent territories within the nucleus.

10.5. The positional effect refers to the variability in gene expression that occurs when a gene has been cloned in an eukaryotic host. It is due to the random nature of the insertion, which could place the gene into a region of open or highly packaged chromatin.

10.7. Insulators and LCRs can both overcome the positional effect when linked to genes inserted into eukaryotic cells. LCRs also stimulate expression of genes that are present within their functional domains; insulators are unable to do this.

10.9. HDACs repress gene expression by removing acetyl groups from histone proteins.

10.11. DNase I cannot cleave DNA that is inaccessible, for example because it is contained in highly packaged chromatin. Sites that are susceptible to DNase I cleavage are usually adjacent to genes that are being expressed.

In-depth Problems

10.1. A consideration of the procedures used to prepare cells for electron microscopy usually leads to the conclusion that structures present in living nuclei are likely to be lost and that artifacts that do not exist in living nuclei might be created. The counter argument is the general concordance between the view of the inside of the nucleus as developed by electron microscopy and the interpretations arising from the more recent and less destructive methods based on confocal microscopy.

10.3. Starting points are Strahl, B.D. and Allis, C.D. (2000) The language of covalent histone modifications. *Nature* **403:** 41–45; and Jenuwein, T. and Allis, C.D. (2001) Translating the histone code. *Science* **293:** 1074–1080.

10.5. This fascinating question is best tackled by exploring the relevant research literature, such as Lee, J.T. (2005) Regulation of X-chromosome counting by *Tsix* and *Xite* sequences. *Science* **309:** 768–771.

Figure Tests

10.1. The level of gene expression will be highest if the gene is inserted into the region of open chromatin. There is little or no expression of genes in regions of condensed chromatin.

10.3. The methylated CpG island is bound by a methyl-CpG-binding protein that is part of a histone deacetylase complex that inactivates the gene.

Chapter 11: Assembly of the Transcription Initiation Complex

Multiple Choice Questions

11.1-d; 11.3-b; 11.5-b; 11.7-d; 11.9-c; 11.11-c; 11.13-d; 11.15-b;

Short Answer Questions

11.1. The homeodomain is an extended helix-turn-helix motif made up of 60 amino acids which form four α-helices, numbers 2 and 3 separated by a β-turn, with number 3 acting as the recognition helix and number 1 making contacts within the minor groove.

11.3. In one type of modification assay, the DNA is treated with a nuclease, which cleaves all phosphodiester bonds except those protected by the bound protein. In the second type of assay the DNA is treated with a methylating agent. Those nucleotides protected by the bound protein will not be methylated.

11.5. Within the major groove, hydrogen bonds form between the nucleotide bases and the R groups of amino acids in the recognition structure of the protein, whereas in the minor groove hydrophobic interactions are more important. On the surface of the helix, the major interactions are electrostatic, between the negative charges on the phosphate component of each nucleotide and the positive charges on the R groups of amino acids such as lysine and arginine, although some hydrogen bonding also occurs.

11.7. The core promoter is the site where the transcription initiation complex is assembled. The upstream promoter elements are the attachment sites for

DNA-binding proteins that regulate assembly of the initiation complex.

11.9. The lactose repressor binds to the operator sequence of the lactose operon to prevent transcription. When lactose is present, its isomer allolactose binds to the repressor. When allolactose is bound, the structure of the repressor changes so that it can no longer bind to the operator.

11.11. The presence of alternative or multiple promoters enables two or more transcripts to be specified by a single gene. This results in a similar but not identical protein being synthesized, perhaps in different tissues or at different developmental stages, or possibly concurrently in the same cell.

In-depth Problems

11.1. There are various possibilities based on immobilization of cloned DNA fragments representing entire chromosome sequences, followed by application of purified binding protein or nuclear extracts, with binding detected by treatment of the microarray with a labeled antibody specific for the binding protein.

11.3. The answer to the first part of the question can be obtained from Sections 11.3.1 and 11.3.2 but the justification requires some additional thought. At the end of Section 11.3.1 there is a series of bullet points listing the principles of bacterial gene regulation and the text states that these principles apply also to eukaryotes. This is true and undoubtedly recognition of these principles has aided the development of our understanding of gene regulation in eukaryotes. However, consider also the possibility that transferring these principles from prokaryotes to eukaryotes might be unhelpful in that it could result in the importance of certain aspects of the regulation of transcription initiation in eukaryotes being unappreciated because no equivalent processes occur in bacteria.

11.5. Consider, as advantages of the module concept: the clear picture that emerges of the regulatory scenario to which a gene is subject; the distinction between modules of different types, again providing clarity; and the fact that these modules clearly exist. Consider, as disadvantages: the emphasis placed on DNA sequences when it is the binding proteins that are the actual regulators; the possibility that cooperativity between binding proteins is obscured; and the emphasis placed on the region immediately upstream of a gene, whereas important regulatory signals might be located at distant sites. Of these issues, the most important is perhaps the emphasis that the modular system places on the DNA. Not only is this misleading—the binding proteins being the active players in gene regulation—but it obscures the role of chromatin modification in gene regulation.

Figure Tests

11.1. The 434 repressor contains a helix-turn-helix motif. The second helix of the motif fits into the major groove of DNA and the amino acid side chains make specific contacts with the bases.

11.3. In both the major and minor grooves the chemical features are asymmetric and the orientation of the A–T pair can be identified by a binding protein.

11.5. The *E. coli* RNA polymerase recognizes the −35 box as its binding sequence. After attachment to the DNA, the transition from closed to open complex is initiated by breakage of base pairs in the AT-rich −10 box.

Chapter 12: Synthesis and Processing of RNA

Multiple Choice Questions
12.1-a; 12.3-a; 12.5-c; 12.7-b; 12.9-d; 12.11-c; 12.13-c; 12.15-c

Short Answer Questions

12.1. Rho attaches to the transcript and moves along the RNA towards the polymerase. If the polymerase continues to synthesize RNA then it keeps ahead of the pursuing Rho, but at the termination signal the polymerase stalls and Rho is able to catch up. Rho is a helicase, which means that it actively breaks base pairs, in this case between the template and transcript, resulting in termination of transcription.

12.3. Attenuation works by coupling transcription and translation and these processes are not coupled in eukaryotic organisms as RNA is transcribed in the nucleus and translation occurs in the cytoplasm.

12.5. The tRNA sequence within the precursor molecule adopts its base-paired cloverleaf structure and two additional hairpin structures form, one on either side of the tRNA. Processing begins with a cut by ribonuclease E or F forming a new 3′ end just upstream of one of the hairpins. Ribonuclease D, which is an exonuclease, trims seven nucleotides from this new 3′ end and then pauses while ribonuclease P makes a cut at the start of the cloverleaf, forming the 5′ end of the mature mRNA. Ribonuclease D then removes two more nucleotides, creating the 3′ end of the mature molecule. All mature tRNAs must end with the trinucleotide 5′–CCA–3′. With some pre-tRNAs this sequence is absent, or is removed by the processing ribonucleases. This occurs with most of those pre-tRNAs whose 3′ ends are created by an endonuclease called ribonuclease Z, which makes a cut adjacent to the first base pair in the tRNA cloverleaf and hence removes the region that would contain the terminal CCA. When the CCA is absent, it is added one or more template-independent RNA polymerases such as tRNA nucleotidyltransferase.

12.7. Bacterial mRNA degradation begins with removal of the 3′ terminal region, including the hairpin, by an endonuclease, either RNase E or RNase III, exposing a new end from which the exonucleases RNase II and PNPase can degrade the remainder of the molecule.

12.9. The first step in capping is addition of an extra guanosine to the extreme 5′ end of the RNA. The γ-phosphate of the terminal nucleotide is removed, as are the β and γ phosphates of the GTP, resulting in a 5′–5′ bond. The reaction is carried out by the enzyme guanylyl transferase. The second step of the capping reaction converts the new terminal guanosine into 7-methylguanosine by attachment of a methyl group to nitrogen number 7 of the purine ring, this modification catalyzed by guanine methyltransferase.

12.11. The snoRNAs base-pair to the pre-rRNA to identify the residues to be methylated or pseudouridinylated. For residues to be methylated, the base pairing occurs upstream of a D box.

In-depth Problems

12.1. This view is most eloquently described by von Hippel, P.H. (1998) An integrated model of the transcription complex in elongation, termination, and editing. *Science* **281:** 660–665, which should be used as the starting point of an evaluation of this interpretation of transcription.

12.3. This question looks forward to the discussion of the origins of introns in Section 18.3.2. The "introns late" hypothesis postulates that introns evolved relatively recently and are gradually accumulating in eukaryotic genomes.

According to this model, introns are absent in bacteria because the basic bacterial genome plan became established before the first introns appeared. In contrast, the "introns early" hypothesis states that introns are very ancient and are gradually being lost from eukaryotic genomes. The absence of introns in bacterial genomes is therefore a stumbling block for proponents of the "introns early" hypothesis but, as described in Section 18.3.2, there are ways around this problem.

12.5. The RNA world within which ribozymes were the only type of biological catalyst is described in Section 18.1.1. Why some ribozymes have persisted is unknown, but note that most of the functions of ribozymes described in Table 12.4 involve sequence-specific cleavage of RNA molecules. Recall that Technical Note 5.1 states that "the only major deficiency in the RNA toolkit is the absence of enzymes with the degree of sequence specificity displayed by the restriction endonucleases that are so important in DNA manipulations."

Figure Tests

12.1. An RNA hairpin is formed when the inverted palindrome of the terminator sequence is transcribed. The hairpin formation is favored over the DNA–RNA base pairing and results in fewer DNA–RNA contacts. When the A-rich sequence of the template is transcribed, there are several A–U base pairs that have only two hydrogen bonds each. These two factors weaken the interaction between the template and transcript and cause termination.

12.3. Cleavage of the 5′ splice site occurs by a transesterification reaction promoted by the hydroxyl group attached to the 2′ carbon of the adenosine nucleotide located within the intron sequence. The result of the hydroxyl attack is cleavage of the phosphodiester bond at the 5′ splice site, accompanied by formation of a new 5′–2′ phosphodiester bond linking the first nucleotide of the intron (the G of the 5′–GU–3′ motif) with the internal adenosine. This means that the intron has now been looped back on itself to create a lariat structure. Cleavage of the 3′ splice site and joining of the exons result from a second transesterification reaction, this one promoted by the 3′–OH group attached to the end of the upstream exon. This group attacks the phosphodiester bond at the 3′ splice site, cleaving it and so releasing the intron as the lariat structure, which is subsequently converted back to a linear RNA and degraded. At the same time, the 3′ end of the upstream exon joins to the newly formed 5′ end of the downstream exon, completing the splicing process.

12.5. The figure shows the RNA interference pathway. Double-stranded RNA molecules are degraded to short interfering RNAs (siRNAs) by the enzyme Dicer. The siRNAs attach to the mRNA, which is then cleaved by the RNA induced silencing complex (RISC).

Chapter 13: Synthesis and Processing of the Proteome

Multiple Choice Questions

13.1-b; 13.3-c; 13.5-c; 13.7-c; 13.9-d; 13.11-c; 13.13-c; 13.15-d

Short Answer Questions

13.1. Transfer RNAs form the link between the mRNA and the polypeptide that is being synthesized. This is both a physical link, tRNAs binding to both the mRNA and the growing polypeptide, and an informational link, tRNAs ensuring that the polypeptide being synthesized has the amino acid sequence that is denoted, via the genetic code, by the sequence of nucleotides in the mRNA.

13.3. Most errors are corrected by the aminoacyl-tRNA synthetase itself, by an editing process that is distinct from aminoacylation, involving different contacts with the tRNA.

13.5. The pre-initiation complex comprises the 40S subunit of the ribosome, a "ternary complex" made up of the initiation factor eIF-2 bound to the initiator tRNA^{Met} and a molecule of GTP, and three additional initiation factors, eIF-1, eIF-1A, and eIF-3.

13.7. The phosphorylation of the initiation factor eIF-2 results in a repression of translation initiation as it prevents the factor from binding the molecule of GTP that is required for bringing the initiator tRNA to the small ribosomal subunit.

13.9. The secondary structural motifs along the polypeptide chain form within a few milliseconds. This step is accompanied by the protein collapsing into a compact, but not folded, organization, with its hydrophobic groups on the inside, shielded from water. During the next few seconds or minutes, the secondary structural motifs interact with one another and the tertiary structure gradually takes shape, often via a series of intermediate conformations.

13.11. Inteins are capable of self-splicing so they can remove themselves from a protein.

In-depth Problems

13.1. A good starting point for tackling this difficult problem is Ribas de Pouplana, L. and Schimmel, P. (2001) Aminoacyl-tRNA synthetases: potential markers of genetic code development. *Trends Biochem. Sci.* **26:** 591–596.

13.3. The evolution of the genetic code has provoked argument ever since DNA was established as the genetic material back in the 1950s. Many geneticists support the "frozen accident" theory, which suggests that codons were randomly allocated to amino acids during the earliest stages of evolution, the code subsequently becoming "frozen" because any changes would result in widespread disruption of the amino acid sequences of proteins, but various lines of evidence suggest that the code might have evolved in a less random manner First, controversial experimental results suggest that at least some amino acids bind directly to RNAs containing the appropriate codons, this occurring in the absence of the tRNA that mediates the interaction in present-day cells. If correct, the implication is that there is some kind of chemical relationship between an amino acid and its codon(s). Second, the deviations from the standard code (see Table 1.3) indicate that the same codon reallocations have occurred more than once. If the relationship between codon and amino acid is entirely random, as suggested by the "frozen accident" theory, then we would not expect to see the same codon reallocations recurring on different occasions. See also Knight, R.D., Freeland, S.J. and Landweber, L.F. (1999) Selection, history and chemistry: the three faces of the genetic code. *Trends Biochem. Sci.* **24:** 241–247; Szathmáry, E. (1999) The origin of the genetic code: amino acids as cofactors in an RNA world. *Trends Genet.* **15:** 223–229; and Yarus, M., Caporaso, J.G. and Knight, R. (2005) Origins of the genetic code: the escaped triplet theory. *Annu. Rev. Biochem.* **74:** 179–198.

13.5. Relevant points include the following: the early realization that ribosomes comprise large and small subunits was critical in developing the initial models of the mechanics of protein synthesis; identification of the P-, A-, and E-sites was key to a more detailed understanding of translation; and structural studies underlie the work currently being done on the peptidyl transferase activity.

Figure Tests

13.1. An inosine residue could be present at nucleotide number 34. An inosine at this position can base-pair with A, C, or U in the mRNA, allowing a single tRNA molecule to recognize three different codons for an amino acid.

13.3. The CCdA-phosphate-puromycin is an analog of the transition state that occurs during the formation of peptide bonds and is bound by the ribosome at the peptidyl transferase active site. As no proteins are close to the CCdA-phosphate-puromycin molecule, this indicates that the formation of peptide bonds is not catalyzed by proteins.

Chapter 14: Regulation of Genome Activity

Multiple Choice Questions

14.1-d; 14.3-a; 14.5-a; 14.7-c; 14.9-d; 14.11-d; 14.13-c; 14.15-d

Short Answer Questions

14.1. Differentiation refers to the adoption of a specialized physiological role by a cell. This results from permanent changes in genome expression that alter the biochemical composition of the cell. Development is a series of coordinated changes that occur during the life history of a cell or organism. These changes may be temporary or permanent and must continue over a long period of time.

14.3. When glucose is transported into *E. coli*, the sugar transport protein IIA^{Glc} becomes dephosphorylated. The dephosphorylated form of IIA^{Glc} inhibits the enzyme adenylate cyclase, which produces cAMP. So in the presence of glucose, cAMP levels are low; when glucose is absent, cAMP levels are high.

14.5. MAP kinase is activated when it is phosphorylated by the Mek protein. The phosphorylated MAP kinase moves to the nucleus where it phosphorylates transcription activators producing a response that stimulates cell division.

14.7. Class switching results in a complete change in the type of immunoglobulin that the lymphocyte synthesizes. This requires a recombination event that deletes the Cμ and Cδ sequence along with the part of the chromosome between this region and the C_H segment specifying the class of immunoglobulin that the lymphocyte will now synthesize. For example, for the lymphocyte to switch to synthesis of IgG, the most prevalent type of immunoglobulin made by mature lymphocytes, then the deletion will place one of the Cγ segments, which specify the IgG heavy chain, at the 5′ end of the cluster. Class switching is distinct from V-D-J joining and the recombination event does not involve the RAG proteins.

14.9. σ^F is activated by release from a complex with SpoIIAB. This is controlled by SpoIIAA, which, when unphosphorylated, can also attach to SpoIIAB and prevent the latter from binding to σ^F. If SpoIIAA is unphosphorylated then σ^F is released and is active; when SpoIIAA is phosphorylated σ^F remains bound to SpoIIAB and so is inactive. In the mother cell, SpoIIAB phosphorylates SpoIIAA and so keeps σ^F in its bound inactive state. But in the prespore, SpoIIAB's attempts to phosphorylate SpoIIAA are antagonized by yet another protein, SpoIIE, and so σ^F is released and becomes active.

SpoIIE's ability to antagonize SpoIIAB in the prespore but not the mother cell derives from the fact that SpoIIE molecules are bound to the membrane on the surface of the septum. Because the prespore is much smaller than the mother cell, but the septum surface area is similar in both, the concentration of SpoIIE is greater in the prespore, and this enables it to antagonize SpoIIAB.

14.11. The *bicoid* gene is transcribed in maternal nurse cells and the mRNA is injected into the anterior end of unfertilized eggs. The *bicoid* mRNA remains in the anterior end of the egg cell, attached to the cytoskeleton by its 3′ untranslated end. Translation of the mRNA occurs after fertilization of the egg, and the Bicoid protein diffuses throughout the syncytium setting up a concentration gradient from the anterior end (high) to the posterior end (low).

In-depth Problems

14.1. This long essay topic requires additional reading, a good starting point being Berg, J.M., Tymoczko, J.L. and Stryer, L. (2006) *Biochemistry*, 6th Ed. W.H. Freeman, New York.

14.3. This question is most effectively approached by considering the extent to which our current understanding of development in higher eukaryotes is based on information which would not be available if experiments had never been carried out with *C. elegans* or *D. melanogaster*. One opinion might be that study of *C. elegans* has contributed substantially to our understanding of the molecular basis to RNA interference (Section 12.2.6), but that this understanding would perhaps not have been substantially delayed if *C. elegans* was not available. On the other hand, it would have been difficult to uncover the role of homeotic selector genes in higher eukaryotes (Section 14.3.4) without the prior knowledge of these genes in *D. melanogaster*. Other views are possible.

14.5. Unlike the equivalent question concerning studies of heredity (Chapter 3, In-depth Problem 3.3) there is no easy description of an ideal model organism for higher eukaryotic development. An argument could be constructed along the lines that the model should be the least complex eukaryote that displays the particular developmental features under study. Hence the ideal model is different for different aspects of development.

Figure Tests

14.1. The phenomenon is called diauxie and it results from catabolite repression. Glucose represses expression of the lactose operon through an indirect influence on the catabolite activator protein. This protein binds to a recognition sequence at various sites in the bacterial genome and activates transcription initiation at downstream promoters. Productive initiation of transcription at these promoters is dependent on the presence of bound catabolite activator protein: if the protein is absent then the genes controlled by the promoter are not transcribed. Glucose does not itself interact with the catabolite activator protein. Instead, glucose controls the level in the cell of cAMP. It does this by inhibiting the activity of adenylate cyclase, the enzyme that synthesizes cAMP from ATP. Inhibition is mediated by IIA^{Glc}, a component of a multiprotein complex that transports sugars into the bacterium. When glucose is being transported into the cell, IIA^{Glc} becomes dephosphorylated. The dephosphorylated version of IIA^{Glc} inhibits adenylate cyclase activity. This means that if glucose levels are high, the cAMP content of the cell is low. The catabolite activator protein can bind to its target sites only in the presence of cAMP, so when glucose is present the protein remains detached and the operons it controls are switched off.

14.3. During the early stage of B lymphocyte development, the immuno-globulin loci in its genome undergo rearrangements. Within the heavy-chain locus, these rearrangements link one of the V_H gene segments with one of D_H gene segments, and then link this V–D combination with a J_H gene segment. The end result is an exon that contains the complete open reading frame specifying the V, D and J segments of the immunoglobulin protein. This exon becomes linked to a C segment exon by splicing during the transcription process, creating a complete heavy-chain mRNA that can be translated into an immunoglobulin protein that is specific for just that one lymphocyte. A similar series of DNA rearrangements results in the lymphocyte's light-chain V-J exon being constructed, and once again splicing attaches a C segment exon when the mRNA is synthesized.

Chapter 15: Genome Replication

Multiple Choice Questions
15.1-c; 15.3-a; 15.5-b; 15.7-c; 15.9-b; 15.11-a; 15.13-a; 15.15-c

Short Answer Questions
15.1. The dispersive model of replication predicts that each daughter molecule will be made up partly of parental DNA and partly of newly synthesized DNA. In semiconservative replication each daughter molecule is made up of one parental strand and one newly synthesized strand. If replication is conservative, then one of the daughter double helices is made entirely of newly synthesized DNA and the other comprises the two parental strands.

15.3. Rolling circle replication initiates at a nick which is made in one of the parent polynucleotides. The free 3′ end that results is extended, displacing the 5′ end of the polynucleotide. Continued DNA synthesis "rolls off" a complete copy of the genome, and further synthesis eventually results in a series of genomes linked head to tail. These single-stranded, linear genomes are converted to double-stranded circular molecules by complementary strand synthesis, followed by cleavage at the junction points between genomes and circularization of the resulting segments.

15.5. At each end of the open, melted DNA region at the origin of replication, a prepriming complex of DnaB and DnaC proteins assembles. DnaB is a helicase that extends the single-stranded region at the origin, allowing other replication proteins to attach.

15.7. Primase synthesizes a primer comprising 8–12 ribonucleotides. The strand is then extended by DNA polymerase α, which adds the next 20 or so nucleotides (which may include some ribonucleotides). The remainder of the leading strand copy is synthesized by DNA polymerase δ.

15.9. A chromosome will become shortened if the extreme 3′ end of the lagging strand is not copied because the final Okazaki fragment cannot be primed, the natural position for the priming site being beyond the end of the template. The absence of this Okazaki fragment means that the lagging-strand copy is shorter than it should be. If the copy remains this length then when it acts as a parent polynucleotide in the next round of replication the resulting chromosome will be shorter than its grandparent. Shortening, although to a lesser extent, will also occur if the primer for the last Okazaki fragment is placed at the extreme 3′ end of the lagging strand, because this terminal RNA primer cannot be converted into DNA by the standard processes for primer removal.

15.11. When the gene encoding Cdc6p is repressed, the prereplication complexes (pre-RCs) are absent. When the gene is overexpressed, there are multiple genome replications with no mitosis.

In-depth Problems

15.1. The reason for this acceptance is summed up in Watson and Crick's famous conclusion to their paper in *Nature* announcing the discovery of the double helix structure of DNA: "It has not escaped our notice that the specific pairing we have postulated immediately suggests a possible copying mechanism for the genetic material" (see the start of Section 15.1). Once this point has been understood, consider the reasons why semiconservative replication was, in fact, *not* immediately accepted (Section 15.1.1).

15.3. The information provided in *Genomes 3* can be supplemented by the greater amount of detail in a research review such as Patel, S.S. and Picha, K.M. (2000) Structure and function of hexameric helicases. *Annu. Rev. Biochem.* **69**: 651–697.

15.5. Key references are Shay, J.W. and Wright, W.E. (2005) Senescence and immortalization: role of telomeres and telomerase. *Carcinogenesis* **26**: 867–874; and Shay, J.W. (2005) Meeting report: the role of telomeres and telomerase in cancer. *Cancer Res.* **65**: 3513–3517.

Figure Tests

15.1. The figure shows displacement replication. The double helix is disrupted in the D-loop by the presence of an RNA molecule base paired to one of the DNA strands. This RNA molecule acts as the starting point for synthesis of one the daughter polynucleotides. This polynucleotide is synthesized by continuous copying of one strand of the helix, the second strand being displaced and subsequently copied after synthesis of the first daughter genome has been completed.

15.3. DNA polymerase III lacks a $5' \rightarrow 3'$ exonuclease activity and so dissociates from the lagging strand when it reaches the next Okazaki fragment. Its place is taken by DNA polymerase I, which does have a $5' \rightarrow 3'$ exonuclease and so removes the primer, and usually the start of the DNA component of the Okazaki fragment, extending the 3' end of the adjacent fragment into the region of the template that is exposed. The two Okazaki fragments now abut, and the missing phosphodiester bond is put in place by a DNA ligase.

Chapter 16: Mutations and DNA Repair

Multiple Choice Questions
16.1-b; 16.3-b; 16.5-c; 16.7-b; 16.9-a; 16.11-b; 16.13-c; 16.15-b

Short Answer Questions

16.1. Mutations arise via errors that occur during genome replication and from the effects of mutagens, which are chemical or physical agents that react with DNA and change the structure of individual nucleotides.

16.3. A DNA polymerase can discriminate against an incorrect nucleotide when the nucleotide is first bound to the DNA polymerase, when it is moved to the active site of the enzyme, and when it is attached to the 3' end of the polynucleotide that is being synthesized.

16.5. Heat stimulates the hydrolysis of the β-*N*-glycosidic bond that attaches the base to the sugar component of the nucleotide. This occurs more frequently with purines than with pyrimidines and results in an AP (apurinic/apyrimidinic) or baseless site. The sugar–phosphate that is left is unstable and rapidly degrades, leaving a gap if the DNA molecule is double stranded. Approximately 10,000 AP sites are generated in each human cell per day, but these rarely lead to mutations because cells have effective systems for repairing gaps.

16.7. The trait is dominant because a heterozygous individual is only able to synthesize approximately 50% of the active apolipoprotein B-100 present in the cells of an unaffected person. This reduction results in the disease state. This is an example of haploinsufficiency.

16.9. Hypermutation is thought to result from conversion of some cytosine bases into uracils, by a cytosine deaminase, followed by excision of the uracils from the polynucleotide, by a uracil-DNA glycosylase, resulting in AP sites at these positions. In base excision repair each AP site would then be filled in by DNA polymerase β, to restore the original sequence, but in hypermutation the AP sites are not repaired. This means that during the next round of replication either of the four nucleotides could be placed in the daughter strand opposite each AP site. A further round of replication then stabilizes the mutation.

16.11. In NHEJ, a multicomponent protein complex directs a DNA ligase to the break. The complex includes Ku, which binds the DNA ends either side of the break. Individual Ku proteins have an affinity for one another, which means that the two broken ends of the DNA molecule are brought into proximity. Ku binds to the DNA in association with the DNA-PK$_{CS}$ protein kinase, which activates a third protein, XRCC4, which interacts with the mammalian DNA ligase IV, directing this repair protein to the double-strand break.

In-depth Problems

16.1. A transition (a purine-to-purine or pyrimidine-to-pyrimidine change) does not change the purine–pyrimidine orientation in the double helix. A transversion (a purine-to-pyrimidine or pyrimidine-to-purine change) reverses the purine–pyrimidine orientation.

16.3. As well as the papers cited in the relevant part of Further Reading, a key review is Cummings, C.J. and Zoghbi, H.Y. (2000) Trinucleotide repeats: mechanisms and pathophysiology. *Annu. Rev. Genomics Hum. Genet.* **1:** 281–328.

16.5. The source paper for addressing this question is White, O., Eisen, J.A., Heidelberg, J.F., *et al.* (1999) Genome sequence of the radioresistant bacterium *Deinococcus radiodurans* R1. *Science* **286:** 1571–1577.

Figure Tests

16.1. The figure depicts the insertion of two nucleotides into a microsatellite sequence in the DNA molecule. This is an example of replication slippage.

16.3. The enzyme DNA glycosylase removes the damaged base from the DNA. This creates a site without a base and the sugar–phosphate group at this site is excised by an AP endonuclease. The gap is then filled in by a DNA polymerase, and the final phosphodiester bond is put in place by DNA ligase. This is the base excision repair pathway.

Chapter 17: Recombination

Multiple Choice Questions

17.1-c; 17.3-d; 17.5-b; 17.7-d; 17.9-c; 17.11-d

Short Answer Questions

17.1. Recombination allows for major changes and extensive restructuring of genomes. In the absence of recombination, genomes would undergo little change and be fairly static structures.

17.3. In the double-strand break model, homologous recombination initiates with a double-strand break in one of the molecules. One strand in each half of the broken chromosome is shortened, leaving 3′ overhangs. One of the overhangs invades the other intact DNA molecule to set up a Holliday junction. The shortened DNA strands are extended by DNA polymerase, with the DNA synthesis in the region being converted using the DNA molecule that did not undergo the original double-strand break as a template.

17.5. Bacteria can acquire new genes via transformation, transduction, and conjugation. If the DNA that enters the cell is similar in sequence to part of the *E. coli* genome, then homologous recombination can take place, possibly inserting the foreign DNA into the *E. coli* chromosome.

17.7. The *att* sites each contain an identical 15 base pair core sequence. The core sequences are flanked by variable sequences: B and B′ (each just 4 bp in length) in the bacterial genome, and P and P′ in the phage DNA. P and P′ are both over 100 bp in length. Mutations in the core sequence inactivate the *att* site so that it can also no longer participate in recombination.

17.9. The first step in replication of a retroelement is synthesis of an RNA copy. This RNA molecule is then converted into double-stranded DNA. The first stage in this conversion is synthesis, by reverse transcription, of a single-stranded DNA copy of the RNA molecule. This strand synthesis reaction is primed by a tRNA that anneals to a site within the 5′ long terminal repeat of the RNA copy of the retroelement.

In-depth Problems

17.1. The aim of this question is to raise awareness that although the name "RecA" indicates the role of this protein in recombination, it is more correctly looked on as a single-strand binding protein, with additional ability to stimulate protease activity, that plays a variety of roles in bacterial molecular biology. Critical reading includes Kowalczykowski, S.C. and Eggleston, A.K. (1994) Homologous pairing and DNA strand-exchange proteins. *Annu. Rev. Biochem.* **63:** 991–1043; Michel, B. (2005) After 30 years of study, the bacterial SOS response still surprises us. *PLoS Biol.* **3:** e255; and Lusetti, S.L. and Cox, M.M. (2002) The bacterial RecA protein and the recombinational DNA repair of stalled replication forks. *Annu. Rev. Biochem.* **71:** 71–100.

17.3. The relevant information is provided by Pyle, A.M. (2004) DNA repair: big engine finds small breaks. *Nature* **432:** 157–158.

17.5. Section 17.3.3 indicates that DNA methylation is the main process recognized for minimization of transposon activity. The extensive literature on this subject can be difficult to unravel, but good starting points are Yoder, J.A., Walsh, C.P. and Bestor, T.H. (1997) Cytosine methylation and the ecology of intragenomic parasites. *Trends Genet.* **13:** 335–340; and Rabinowicz, P.D., Palmer, L.E., May, B.P., Hemann, M.T., Lowe, S.W., McCombie, W.R. and Martienssen, R.A. (2003) Genes and transposons are differentially methylated in plants, but not in mammals. *Genome Res.* **13:** 2658–2664.

Figure Tests

17.1. Site-specific recombination.
17.3. RecA.

Chapter 18: How Genomes Evolve

Multiple Choice Questions
18.1-d; 18.3-b; 18.5-a; 18.7-d; 18.9-b; 18.11-c; 18.13-d

Short Answer Questions

18.1. The Earth's early atmosphere contained very low levels of oxygen and had high amounts of ammonia and methane, differing significantly from the atmosphere of present-day Earth.

18.3. It is quite possible that more than one biological system evolved on the ancient Earth, even though all present-day organisms appear to derive from a single origin. The most likely scenario is that the predominant system was the first to develop the means to synthesize protein enzymes and therefore probably also the first to adopt a DNA genome. The greater catalytic potential and more accurate replication conferred by protein enzymes and DNA genomes would have given these cells a significant advantage compared with those still containing RNA protogenomes. The DNA–RNA–protein cells would have multiplied more rapidly, enabling them to out-compete the RNA cells for nutrients.

18.5. One of these transitions accompanied the arrival of the first eukaryotes about 1.4 billion years ago, these cells probably containing at least 10,000 genes (the minimum in modern eukaryotes) compared with the 5000 or fewer typical of prokaryotes. The second transition was associated with the arrival of the first vertebrates soon after the end of the Cambrian, these, like modern vertebrates, probably having at least 30,000 genes.

18.7. A retrogene results from conversion of an mRNA molecule to cDNA followed by reinsertion of this cDNA back into the genome. Normally the inserted cDNA would give rise to a pseudogene as it lacks its own promoter (and has no introns). But if the cDNA is inserted adjacent to the promoter of an existing gene then it could become active by subverting this promoter for own use.

18.9. Segment duplications can place exons from one gene adjacent to exons from another gene. If the exons are then transcribed the resulting mRNA could give rise to a novel gene product.

18.11. The "exon theory of genes" holds that introns were formed when the first DNA genomes were constructed, soon after the end of the RNA world. These genomes would have contained many short genes, each specifying a very small polypeptide, perhaps just a single structural domain. To aid the synthesis of a multidomain protein it is envisaged that groups of these short genes were positioned next to each other. The short genes became exons and the sequences between them became introns.

In-depth Problems

18.1. The relevant paper, as cited in Further Reading, is Vision, T.J., Brown, D.G. and Tanksley, S.D. (2000) The origins of genomic duplications in *Arabidopsis*. *Science* **290:** 2114–2117.

18.3. The first papers to reject this proposal were Salzberg, S.L., White, O., Peterson, J. and Eisen, J.A. (2001). Microbial genes in the human genome: lateral transfer or gene loss? *Science* **292:** 1903–1906; and Stanhope, M.J., Lupas, A., Italia, M.J., Koretke, K.K., Volker, C. and Brown, J.R. (2001) Phylogenetic analyses do not support horizontal gene transfers from bacteria to vertebrates. *Nature* **411:** 940–944. The question looks forward to some of the aspects of molecular phylogenetics covered in Chapter 19.

18.5. The question refers to *comparisons* between genome sequences. If interpreted literally then the information provided in Section 18.4 and in the papers cited in Further Reading leads to the conclusion that the answer is "no," at least for comparisons between the human and chimpanzee genomes. Might the answer be different when and if the gorilla and orangutan genomes are sequenced? Or if a complete sequence of Neandertal DNA could be obtained (Section 19.3.2)? If the question is interpreted less strictly and post-genomic analyses are allowed, then Section 18.4 indicates that it might be possible to determine at least some of the factors that make us human.

Figure Tests

18.1. When there is selective pressure on just one copy of a duplicated gene pair, the second copy can accumulate mutations that could lead to new activities or functions. Pseudogenes result from the pathway in which deleterious mutations occur in the second copy of the gene so that this copy degrades over time.

18.3. The top panel shows the generation of a new gene by domain duplication, as domain B is duplicated from the original gene. The bottom panel shows domain shuffling where domains from two distinct genes combine to form a new gene.

Chapter 19: Molecular Phylogenetics

Multiple Choice Questions

19.1-b; 19.3-a; 19.5-c; 19.7-b; 19.9-b; 19.11-c; 19.13-b

Short Answer Questions

19.1. Phenetics is a method of phylogenetic analysis that utilizes as many variables as is possible; the prevailing view prior to the use of phenetics was that phylogenies should be based on a limited number of characteristics that were thought to be important.

19.3. Ancestral character states were present in the remote common ancestor of a group of organisms whereas derived character states have evolved from the ancestral state in a more recent common ancestor.

19.5. An internal node in a gene tree represents the divergence of the ancestral gene into two alleles via mutation. An internal node in a species tree indicates a speciation event that occurred when an ancestral group split into two groups that could not interbreed. These mutational and speciation events are unlikely to have occurred at the same time.

19.7. Allele frequency is affected by natural selection and random genetic drift. Natural selection changes the frequency of alleles that impact on the fitness of an individual, while random genetic drift changes the frequency of alleles because of the random nature of birth, death, and reproduction.

19.9. DNA was extracted from 400 mg of Neandertal bone and PCRs were directed at what was expected to be the most variable part of the Neandertal mitochondrial genome. It was anticipated that the DNA would be degraded, so the sequence was built up in sections by carrying out nine overlapping PCRs, none amplifying more than 170 bp of DNA but together giving a total length of 377 bp. A phylogenetic tree was constructed to compare the sequence obtained from the Neandertal bone with the sequences of six mitochondrial DNA haplogroups from modern humans. The Neandertal sequence was positioned on a branch of its own, not linked directly to any of the modern human sequences. A multiple alignment was made in order to compare the Neandertal sequence with 994 sequences from modern humans. The Neandertal sequence differed from the modern sequences at an average of 27.2 ± 2.2 nucleotide positions whereas the modern sequences differed from each other at only 8.0 ± 3.1 positions. The degree of difference between Neandertal and modern European DNA is incompatible with the notion that modern Europeans are descended from Neandertals and strongly supports the Out of Africa hypothesis.

19.11. The first mitochondrial DNA studies indicated that Native Americans are descended from Asian ancestors and identified four distinct mitochondrial haplogroups among the population as a whole. The migration into North America was dated to between 15,000 and 8000 years ago. A more recent, comprehensive analysis of mitochondrial DNA pushed this migration back to between 25,000 and 20,000 years ago. The first studies of Y chromosomes gave a date of approximately 22,500 years ago to the "Native American Adam", the carrier of the Y chromosome that is ancestral to most, if not all, of the Y chromosomes in modern Native Americans. The implications of these various results are still being debated.

In-depth Problems

19.1. Although Figure 19.9 illustrates an example of when a gene tree is not the same as a species tree, coincidence between the two is clearly possible. The follow-up question is "how can you tell if a gene tree is giving an accurate representation of a species tree?" Now the answer demands consideration of, for example, the information summarized in Figure 19.19, which shows that different genes or collections of genes must be studied in order to obtain a consensus view.

19.3. Possible starting points for research into this problem are Ladoukakis, E.D. and Zouros, E. (2001) Recombination in animal mitochondrial DNA: evidence from published sequences. *Mol. Biol. Evol.* **18:** 2127–2131; and Meunier, J. and Eyre-Walker, A. (2001) The correlation between linkage disequilibrium and distance: implications for recombination in hominid mitochondria. *Mol. Biol. Evol.* **18:** 2132–2135. These are among the first papers to raise the possibility of mitochondrial recombination, and the debate should be followed forward by searching for later publications that cite one or both of these papers.

19.5. A detailed introduction to this topic is given by Richards, M. (2003) The Neolithic invasion of Europe. *Annu. Rev. Anthropol.* **32:** 135–162. To answer this question fully, the paper which provides the information in Figure 19.30 must be examined—Richards, M., Macaulay, V., Hickey, E., *et al.* (2000) Tracing European founder lineages in the Near Eastern mtDNA pool. *Am. J. Hum. Genet.* **67:** 1251–1276. This is a complex paper and if you are able to make progress with it then your study of *Genomes* has not been in vain.

Figure Tests

19.1. Figure (B) shows that the single toe of horses is a derived character state.

19.3. If the first speciation event is quickly followed by a second speciation event in one of the two resulting populations, then the branching order of the gene tree might be different from that of the species tree. This can occur if the genes in the modern species are derived from alleles that had already appeared before the first of the two speciation events.

Glossary

2-aminopurine A base analog that can cause mutations by replacing adenine in a DNA molecule.

2 μm circle A plasmid found in the yeast *Saccharomyces cerevisiae* and used as the basis for a series of cloning vectors.

–25 box A component of the bacterial promoter.

3′-OH terminus The end of a polynucleotide that terminates with a hydroxyl group attached to the 3′-carbon of the sugar.

3′ transduction Transfer of a segment of genomic DNA from one place to another caused by movement of a LINE element.

3′-untranslated region The untranslated region of an mRNA downstream of the termination codon.

30 nm chromatin fiber A relatively unpacked form of chromatin consisting of a possibly helical array of nucleosomes in a fiber approximately 30 nm in diameter.

5-bromouracil A base analog that can cause mutations by replacing thymine in a DNA molecule.

5′-P terminus The end of a polynucleotide that terminates with a mono-, di- or triphosphate attached to the 5′-carbon of the sugar.

5′-untranslated region The untranslated region of an mRNA upstream of the initiation codon.

(6–4) lesion A dimer between two adjacent pyrimidine bases in a polynucleotide, formed by ultraviolet irradiation.

(6–4) photoproduct photolyase An enzyme involved in photoreactivation repair.

α-helix One of the commonest secondary structural conformations taken up by segments of polypeptides.

β-*N*-glycosidic bond The linkage between the base and sugar of a nucleotide.

β-sheet One of the commonest secondary structural conformations taken up by segments of polypeptides.

β-turn A sequence of four amino acids, the second usually glycine, which causes a polypeptide to change direction.

γ-complex A component of DNA polymerase III comprising subunit γ in association with δ, δ′, χ and ψ.

κ-homology domain A type of RNA-binding domain.

π–π interactions The hydrophobic interactions that occur between adjacent base pairs in a double-stranded DNA molecule.

Acceptor arm Part of the structure of a tRNA molecule.

Acceptor site The splice site at the 3′ end of an intron.

Acidic domain A type of activation domain.

Acridine dye A chemical compound that causes a frameshift mutation by intercalating between adjacent base pairs of the double helix.

Activation domain The part of an activator that makes contact with the initiation complex.

Activator A DNA-binding protein that stabilizes construction of the RNA polymerase II transcription initiation complex.

Acylation The attachment of a lipid side chain to a polypeptide.

Ada enzyme An *Escherichia coli* enzyme that is involved in the direct repair of alkylation mutations.

Adaptor A synthetic, double-stranded oligonucleotide used to attach sticky ends to a blunt-ended molecule.

Adenine A purine base found in DNA and RNA.

Adenosine deaminase acting on RNA (ADAR) An enzyme that edits various eukaryotic mRNAs by deaminating adenosine to inosine.

Adenylate cyclase The enzyme that converts ATP to cyclic AMP.

A-DNA A structural configuration of the double helix, present but not common in cellular DNA.

Affinity chromatography A column chromatography method that makes use of a ligand that binds to the molecule being purified.

Agarose gel electrophoresis Electrophoresis carried out in an agarose gel and used to separate DNA molecules between 100 bp and 50 kb in length.

Alarmone One of the stringent response activators, ppGppp and pppGpp.

Alkaline phosphatase An enzyme that removes phosphate groups from the 5′ ends of DNA molecules.

Alkylating agent A mutagen that acts by adding alkyl groups to nucleotide bases.

Allele One of two or more alternative forms of a gene.

Allele frequency The frequency of an allele in a population.

Allele-specific oligonucleotide (ASO) hybridization The use of an oligonucleotide probe to determine which of two alternative nucleotide sequences is contained in a DNA molecule.

Allopolyploid   A polyploid nucleus derived from fusion between gametes from different species.

Alphoid DNA   The tandemly repeated nucleotide sequences located in the centromeric regions of human chromosomes.

Alternative polyadenylation   The use of two or more different sites for polyadenylation of an mRNA.

Alternative promoter   One of two or more different promoters acting on the same gene.

Alternative splicing   The production of two or more mRNAs from a single pre-mRNA by joining together different combinations of exons.

Alu   A type of SINE found in the genomes of humans and related mammals.

Alu-PCR   A clone fingerprinting technique that uses PCR to detect the relative positions of Alu sequences in cloned DNA fragments.

Amino acid   One of the monomeric units of a protein molecule.

Aminoacyl or A site   The site in the ribosome occupied by the aminoacyl-tRNA during translation.

Aminoacylation   Attachment of an amino acid to the acceptor arm of a tRNA.

Aminoacyl-tRNA synthetase   An enzyme that catalyzes the aminoacylation of one or more tRNAs.

Amino terminus   The end of a polypeptide that has a free amino group.

Amplification refraction mutation system (ARMS test)   A technique for SNP typing in which PCR is directed by a pair of primers, one covering the position of the SNP.

Ancestral character state   A character state possessed by a remote common ancestor of a group of organisms.

Ancient DNA   DNA preserved in ancient biological material.

Annealing   Attachment of an oligonucleotide primer to a DNA or RNA template.

Anticodon   The triplet of nucleotides, at positions 34–36 in a tRNA molecule, that base-pairs with a codon in an mRNA molecule.

Anticodon arm   Part of the structure of a tRNA molecule.

Antigen   A substance that elicits an immune response.

Antitermination   A bacterial mechanism for regulating the termination of transcription.

Antiterminator protein   A protein that attaches to bacterial DNA and mediates antitermination.

AP endonuclease   An enzyme involved in base excision repair.

Apomorphic character state   A character state that evolved in a recent ancestor of a subset of organisms in a group being studied.

AP (apurinic/apyrimidinic) site   A position in a DNA molecule where the base component of the nucleotide is missing.

Apoptosis   Programmed cell death.

Archaea   One of the two main groups of prokaryotes, mostly found in extreme environments.

Artificial gene synthesis   Construction of an artificial gene from a series of overlapping oligonucleotides.

Ascospore   One of the haploid products of meiosis in an ascomycete, such as the yeast *Saccharomyces cerevisiae.*

Ascus   The structure which contains the four ascospores produced by a single meiosis in the yeast *Saccharomyces cerevisiae.*

Attenuation   A process used by some bacteria to regulate expression of an amino acid biosynthetic operon in accordance with the levels of the amino acid in the cell.

AU–AC intron   A type of intron found in eukaryotic nuclear genes: the first two nucleotides in the intron are 5′–AU–3′ and the last two are 5′–AC–3′.

Autonomously replicating sequence (ARS)   A DNA sequence, particularly from yeast, that confers replicative ability on a nonreplicative plasmid.

Autopolyploid   A polyploid nucleus derived from fusion of two gametes from the same species, neither of which is haploid.

Autoradiography   The detection of radioactively labeled molecules by exposure of an X-ray-sensitive photographic film.

Autosome   A chromosome that is not a sex chromosome.

Auxotroph   A mutant microorganism that can grow only when supplied with a nutrient that is not needed by the wild type.

Backtracking   The reversal of an RNA polymerase a short distance along its DNA template strand.

Bacteria   One of the two main groups of prokaryotes.

Bacterial artificial chromosome (BAC)   A high-capacity cloning vector based on the F plasmid of *Escherichia coli.*

Bacteriophage   A virus that infects a bacterium.

Bacteriophage P1 vector   A high-capacity cloning vector based on bacteriophage P1.

Barcode deletion strategy   A method that has been developed for the large-scale screening of deletion mutations in *Saccharomyces cerevisiae.*

Barr body   The highly condensed chromatin structure taken up by an inactivated X chromosome.

Basal promoter The position within a eukaryotic promoter where the initiation complex is assembled.

Basal promoter element Sequence motifs that are present in many eukaryotic promoters and set the basal level of transcription initiation.

Basal rate of transcription The number of productive initiations of transcription occurring per unit time at a particular promoter.

Base analog A compound whose structural similarity to one of the bases in DNA enables it to act as a mutagen.

Base excision repair A DNA repair process that involves excision and replacement of an abnormal base.

Baseless site A position in a DNA molecule where the base component of the nucleotide is missing.

Base pair The hydrogen-bonded structure formed by two complementary nucleotides. When abbreviated to "bp", the shortest unit of length for a double-stranded DNA molecule.

Base pairing The attachment of one polynucleotide to another, or one part of a polynucleotide to another part of the same polynucleotide, by base pairs.

Base ratio The ratio of A to T, or G to C, in a double-stranded DNA molecule. Chargaff showed that the base ratios are always close to 1.0.

Base stacking The hydrophobic interactions that occur between adjacent base pairs in a double-stranded DNA molecule.

Basic domain A type of DNA-binding domain.

B chromosome A chromosome possessed by some individuals in a population, but not all.

B-DNA The commonest structural conformation of the DNA double helix in living cells.

Beads-on-a-string An unpacked form of chromatin consisting of nucleosome beads on a string of DNA.

Biochemical profiling The study of metabolomes.

Bioinformatics The use of computer methods in studies of genomes.

Biolistics A means of introducing DNA into cells that involves bombardment with high-velocity microprojectiles coated with DNA.

Biological information The information contained in the genome of an organism and which directs the development and maintenance of that organism.

Biotechnology The use of living organisms, often, but not always microbes, in industrial processes.

Biotinylation Attachment of a biotin label to a DNA or RNA molecule.

Bivalent The structure formed when a pair of homologous chromosomes lines up during meiosis.

BLAST An algorithm frequently used in homology searching.

Blunt end An end of a double-stranded DNA molecule where both strands terminate at the same nucleotide position with no single-stranded extension.

Bootstrap analysis A method for inferring the degree of confidence that can be assigned to a branch point in a phylogenetic tree.

Bootstrap value The statistical value obtained by bootstrap analysis.

Bottleneck A temporary reduction in the size of a population.

Branch A component of a phylogenetic tree.

Branch migration A step in the Holliday model for homologous recombination, involving exchange of polynucleotides between a pair of recombining double-stranded DNA molecules.

Buoyant density The density possessed by a molecule or particle when suspended in an aqueous salt or sugar solution.

CAAT box A basal promoter element.

Cap The chemical modification at the 5′ end of most eukaryotic mRNA molecules.

Cap binding complex The complex that makes the initial attachment to the cap structure at the beginning of the scanning phase of eukaryotic translation.

Capillary electrophoresis Polyacrylamide gel electrophoresis carried out in a thin capillary tube, providing high resolution.

Capping Attachment of a cap to the 5′ end of a eukaryotic mRNA.

Capsid The protein coat that surrounds the DNA or RNA genome of a virus.

CAP site A DNA-binding site for the catabolite activator protein.

Carboxyl terminus The end of a polypeptide that has a free carboxyl group.

CASPs (CTD-associated SR-like proteins) Proteins thought to play regulatory roles during splicing of GU–AG introns.

Catabolite activator protein A regulatory protein that binds to various sites in a bacterial genome and activates transcription initiation at downstream promoters.

Catabolite repression The means by which extracellular glucose levels dictate whether genes for sugar utilization are switched on or off in bacteria.

cDNA A double-stranded DNA copy of an mRNA molecule.

cDNA capture or cDNA selection Repeated hybridization probing of a pool of cDNAs with the objective of obtaining a subpool enriched in certain sequences.

Cell cycle   The series of events occurring in a cell between one division and the next.

Cell cycle checkpoint   A period before entry into S or M phase of the cell cycle, a key point at which regulation is exerted.

Cell-free protein synthesizing system   A cell extract containing all the components needed for protein synthesis and able to translate added mRNA molecules.

Cell senescence   The period in a cell lineage when the cells are alive but no longer able to divide.

Cell-specific module   Sequence motifs present in the promoters of eukaryotic genes that are expressed in just one type of tissue.

Cell transformation   The alteration in morphological and biochemical properties that occurs when an animal cell is infected by an oncogenic virus.

Centromere   The constricted region of a chromosome that is the position at which the pair of chromatids are held together.

Chain termination method   A DNA sequencing method that involves enzymatic synthesis of polynucleotide chains that terminate at specific nucleotide positions.

Chaperonin   A multisubunit protein that forms a structure that aids the folding of other proteins.

Character state   One of at least two alternative forms of a character used in phylogenetic analysis.

Chemical degradation sequencing   A DNA sequencing method that involves the use of chemicals that cut DNA molecules at specific nucleotide positions.

Chemical shift   The change in the rotation of a chemical nucleus, used as the basis of NMR.

Chi form   An intermediate structure seen during recombination between DNA molecules.

Chimera   An organism composed of two or more genetically different cell types.

Chi site   A repeated nucleotide sequence in the *Escherichia coli* genome that is involved in the initiation of homologous recombination.

Chloroplast   One of the photosynthetic organelles of a eukaryotic cell.

Chloroplast genome   The genome present in the chloroplasts of a photosynthetic eukaryotic cell.

Chromatid   The arm of a chromosome.

Chromatin   The complex of DNA and histone proteins found in chromosomes.

Chromatosome   A subcomponent of chromatin made up of a nucleosome core octamer with associated DNA and a linker histone.

Chromosome   One of the DNA–protein structures that contains part of the nuclear genome of a eukaryote.

Less accurately, the DNA molecule(s) that contain(s) a prokaryotic genome.

Chromosome painting   A version of fluorescent *in situ* hybridization in which the hybridization probe is a mixture of DNA molecules, each specific for different regions of a single chromosome.

Chromosome scaffold   A component of the nuclear matrix which changes its structure during cell division, resulting in condensation of the chromosomes into their metaphase forms.

Chromosome territory   The region of a nucleus occupied by a single chromosome.

Chromosome theory   The theory, first propounded by Sutton in 1903, that genes lie on chromosomes.

Chromosome walking   A technique that can be used to construct a clone contig by identifying overlapping fragments of cloned DNA.

***Cis*-displacement**   Movement of a nucleosome to a new position on a DNA molecule.

Clade   A group of monophyletic organisms or DNA sequences that include all of those in the analysis that are descended from a particular common ancestor.

Cladistics   A phylogenetic approach that stresses the importance of understanding the evolutionary relevance of the characters that are studied.

Class switching   A process that results in a complete change in the type of immunoglobulin synthesized by a B lymphocyte.

Cleavage and polyadenylation specificity factor (CPSF)   A protein that plays an ancillary role during polyadenylation of eukaryotic mRNAs.

Cleavage stimulation factor (CstF)   A protein that plays an ancillary role during polyadenylation of eukaryotic mRNAs.

Clone   A group of cells that contain the same recombinant DNA molecule.

Clone contig   A collection of clones whose DNA fragments overlap.

Clone contig approach   A genome sequencing strategy in which the molecules to be sequenced are broken into manageable segments, each a few hundred kb or few Mb in length, which are sequenced individually.

Clone fingerprinting   Any one of several techniques that compare cloned DNA fragments in order to identify ones that overlap.

Clone library   A collection of clones, possibly representing an entire genome, from which individual clones of interest are obtained.

Cloning vector   A DNA molecule that is able to replicate inside a host cell and therefore can be used to clone other fragments of DNA.

Closed promoter complex The structure formed during the initial step in assembly of the transcription initiation complex. The closed promoter complex consists of the RNA polymerase and/or accessory proteins attached to the promoter, before the DNA has been opened up by breakage of base pairs.

Cloverleaf A two-dimensional representation of the structure of a tRNA molecule.

Coactivator A protein that stimulates transcription initiation by binding nonspecifically to DNA or via protein–protein interactions.

Coding RNA An RNA molecule that codes for a protein; an mRNA.

Codominance The relationship between a pair of alleles which both contribute to the phenotype of a heterozygote.

Codon A triplet of nucleotides coding for a single amino acid.

Codon-anticodon recognition The interaction between a codon on an mRNA molecule and the corresponding anticodon on a tRNA.

Codon bias Refers to the fact that not all codons are used equally frequently in the genes of a particular organism.

Cohesin The protein that holds sister chromatids together during the period between genome replication and nuclear division.

Cohesive end An end of a double-stranded DNA molecule where there is a single-stranded extension.

Coimmunoprecipitation Isolation of all the members of a protein complex with an antibody specific for just one of those proteins.

Cointegrate An intermediate in the pathway resulting in replicative transposition.

Commitment complex The initial structure formed during splicing of a GU–AG intron.

Comparative genomics A research strategy that uses information obtained from the study of one genome to make inferences about the map positions and functions of genes in a second genome.

Competent Refers to a culture of bacteria that have been treated, for example, by soaking in calcium chloride, so that their ability to take up DNA molecules is enhanced.

Complementary Refers to two nucleotides or nucleotide sequences that are able to base-pair with one another.

Complementary DNA (cDNA) A double-stranded DNA copy of an mRNA molecule.

Composite transposon A DNA transposon comprising a pair of insertion sequences flanking a segment of DNA usually containing one or more genes.

Concatamer A DNA molecule made up of linear genomes or other DNA units linked head to tail.

Concerted evolution The evolutionary process that results in the members of a multigene family retaining the same or similar sequences.

Conditional-lethal mutation A mutation that results in a cell or organism able to survive only under permissive conditions.

Conjugation Transfer of DNA between two bacteria that come into physical contact with one another.

Conjugation mapping A technique for mapping bacterial genes by determining the time it takes for each gene to be transferred during conjugation.

Consensus sequence A nucleotide sequence that represents an "average" of a number of related but nonidentical sequences.

Conservative replication A hypothetical mode of DNA replication in which one daughter double helix is made up of the two parental polynucleotides and the other is made up of two newly synthesized polynucleotides.

Conservative transposition Transposition that does not result in copying of the transposable element.

Constitutive control Control over bacterial gene expression that depends on the sequence of the promoter.

Constitutive heterochromatin Chromatin that is permanently in a compact organization.

Constitutive mutation A mutation that results in continuous expression of a gene or set of genes that is normally subject to regulatory control.

Context-dependent codon reassignment Refers to the situation whereby the DNA sequence surrounding a codon changes the meaning of that codon.

Contig A contiguous set of overlapping DNA sequences.

Contour clamped homogeneous electric fields (CHEF) An electrophoresis method used to separate large DNA molecules.

Conventional pseudogene A gene that has become inactive because of the accumulation of mutations.

Convergent evolution The situation that occurs when the same character state evolves independently in two lineages.

Core enzyme The version of *Escherichia coli* RNA polymerase, subunit composition $\alpha_2\beta\beta'$, that carries out RNA synthesis but is unable to locate promoters efficiently.

Core octamer The central component of a nucleosome, made up of two subunits each of histones H2A, H2B, H3 and H4, around which DNA is wound.

Co-repressor A molecule that represses expression of a gene or operon by binding to a repressor protein and enabling the repressor to attach to the operator.

Core promoter   The position within a eukaryotic promoter where the initiation complex is assembled.

Cosmid   A high-capacity cloning vector consisting of the λ *cos* site inserted into a plasmid.

***cos* site**   One of the cohesive, single-stranded extensions present at the ends of the DNA molecules of certain strains of λ phage.

Cotransduction   Transfer of two or more genes from one bacterium to another via a transducing phage.

Cotransformation   Uptake of two or more genes on a single DNA molecule during transformation of a bacterium.

CpG island   A GC-rich DNA region located upstream of approximately 56% of the genes in the human genome.

CREB   An important transcription factor.

Crossing-over   The exchange of DNA between chromosomes during meiosis.

Cryptic splice site   A site whose sequence resembles an authentic splice site and which might be selected instead of the authentic site during aberrant splicing.

Cryptogene   One of several genes in the trypanosome mitochondrial genome which specify abbreviated RNAs that must undergo pan-editing in order to become functional.

CTD-associated SR-like protein (CASP)   A type of protein thought to play a regulatory role during splicing of GU–AG introns.

C-terminal domain (CTD)   A component of the largest subunit of RNA polymerase II, important in activation of the polymerase.

C terminus   The end of a polypeptide that has a free carboxyl group.

C-value paradox   The nonequivalence between genome size and gene number that is seen when comparisons are made between some eukaryotes.

Cyanelle   A photosynthetic organelle that resembles an ingested cyanobacterium.

Cyclic AMP (cAMP)   A modified version of AMP in which an intramolecular phosphodiester bond links the 5′ and 3′ carbons.

Cyclin   A regulatory protein whose abundance varies during the cell cycle and which regulates biochemical events in a cell-cycle-specific manner.

Cyclobutyl dimer   A dimer between two adjacent pyrimidine bases in a polynucleotide, formed by ultraviolet irradiation.

Cys_2His_2 finger   A type of zinc-finger DNA-binding domain.

Cytochemistry   The use of compound-specific stains, combined with microscopy, to determine the biochemical content of cellular structures.

Cytosine   One of the pyrimidine bases found in DNA and RNA.

Dark repair   A type of nucleotide excision repair process that corrects cyclobutyl dimers.

D arm   Part of the structure of a tRNA molecule.

Deadenylation-dependent decapping   A process for degradation of eukaryotic mRNAs that is initiated by removal of the poly(A) tail.

Deaminating agent   A mutagen that acts by removing amino groups from nucleotide bases.

Degeneracy   Refers to the fact that the genetic code has more than one codon for most amino acids.

Degradosome   A multienzyme complex responsible for degradation of bacterial mRNAs.

Delayed-onset mutation   A mutation whose effect is not apparent until a relatively late stage in the life of the mutant organism.

Deletion mutation   A mutation resulting from deletion of one or more nucleotides from a DNA sequence.

Denaturation   Breakdown by chemical or physical means of the noncovalent interactions, such as hydrogen bonding, that maintain the secondary and higher levels of structure of proteins and nucleic acids.

Dendrogram   A tree that is drawn to indicate the relationships between, for example, a group of transcriptomes.

***De novo* methylation**   Addition of methyl groups to new positions on a DNA molecule.

Density gradient centrifugation   A technique in which a cell fraction is centrifuged through a dense solution, in the form of a gradient, so that individual components are separated.

Deoxyribonuclease   An enzyme that cleaves phosphodiester bonds in a DNA molecule.

Derived character state   A character state that evolved in a recent ancestor of a subset of organisms in a group being studied.

Development   A coordinated series of transient and permanent changes that occurs during the life history of a cell or organism.

Diauxie   The phenomenon whereby a bacterium, when provided with a mixture of sugars, uses up one sugar before beginning to metabolize the second sugar.

Dicer   The ribonuclease that plays a central role in RNA interference.

Dideoxynucleotide   A modified nucleotide that lacks the 3′ hydroxyl group and so terminates strand synthesis when incorporated into a polynucleotide.

Differential centrifugation   A technique that separates cell components by centrifuging an extract at different speeds.

Differential splicing The production of two or more mRNAs from a single pre-mRNA by joining together different combinations of exons.

Differentiation The adoption by a cell of a specialized biochemical and/or physiological role.

Dihybrid cross A sexual cross in which the inheritance of two pairs of alleles is followed.

Dimer A protein or other structure that comprises two subunits.

Diploid A nucleus that has two copies of each chromosome.

Directed evolution A set of experimental techniques that is used to obtain novel genes with improved products.

Direct readout The recognition of a DNA sequence by a binding protein that makes contacts with the outside of a double helix.

Direct repair A DNA repair system that acts directly on a damaged nucleotide.

Direct repeat A nucleotide sequence that is repeated twice or more frequently in a DNA molecule.

Discontinuous gene A gene that is split into exons and introns.

Dispersive replication A hypothetical mode of DNA replication in which both polynucleotides of each daughter double helix are made up partly of parental DNA and partly of newly synthesized DNA.

Displacement replication A mode of replication which involves continuous copying of one strand of the helix, the second strand being displaced and subsequently copied after synthesis of the first daughter strand has been completed.

Distance matrix A table showing the evolutionary distances between all pairs of nucleotide sequences in a dataset.

Distance method A rigorous mathematical approach to alignment of nucleotide sequences.

Disulfide bridge A covalent bond linking cysteine amino acids on different polypeptides or at different positions on the same polypeptide.

D-loop An intermediate structure formed during the Meselson-Radding model for homologous recombination. Also, a region of approximately 500 bp where the double helix is disrupted by the presence of an RNA molecule base-paired to one of the DNA strands, and which acts as the origin for the displacement mode of replication.

DNA Deoxyribonucleic acid, one of the two forms of nucleic acid in living cells; the genetic material for all cellular life forms and many viruses.

DNA adenine methylase (Dam) An enzyme involved in methylation of *Escherichia coli* DNA.

DNA bending A type of conformational change introduced into a DNA molecule by a binding protein.

DNA-binding motif The part of a DNA-binding protein that makes contact with the double helix.

DNA-binding protein A protein that attaches to a DNA molecule.

DNA chip A high-density array of DNA molecules used for parallel hybridization analyses.

DNA cloning Insertion of a fragment of DNA into a cloning vector, and subsequent propagation of the recombinant DNA molecule in a host organism.

DNA cytosine methylase (Dcm) An enzyme involved in methylation of *Escherichia coli* DNA.

DNA-dependent DNA polymerase An enzyme that makes a DNA copy of a DNA template.

DNA-dependent RNA polymerase An enzyme that makes an RNA copy of a DNA template.

DNA glycosylase An enzyme that cleaves the β-*N*-glyco-sidic bond between a base and the sugar component of a nucleotide as part of the base excision and mismatch repair processes. The name is a misnomer and should be *DNA glycolyase*, but the incorrect usage is now embedded in the literature.

DNA gyrase A Type II topoisomerase of *Escherichia coli*.

DNA ligase An enzyme that synthesizes phosphodiester bonds as part of DNA replication, repair and recombination processes.

DNA marker A DNA sequence that exists as two or more readily distinguished versions and which can therefore be used to mark a map position on a genetic, physical, or integrated genome map.

DNA methylation Refers to the chemical modification of DNA by attachment of methyl groups.

DNA methyltransferase An enzyme that attaches methyl groups to a DNA molecule.

DNA photolyase A bacterial enzyme involved in photore-activation repair.

DNA polymerase An enzyme that synthesizes DNA.

DNA polymerase I The bacterial enzyme that completes synthesis of Okazaki fragments during genome replication.

DNA polymerase II A bacterial DNA polymerase involved in DNA repair.

DNA polymerase III The main DNA replicating enzyme of bacteria.

DNA polymerase α The enzyme that primes DNA replication in eukaryotes.

DNA polymerase γ The enzyme responsible for replication of the mitochondrial genome.

DNA polymerase δ The main eukaryotic DNA replicating enzyme.

DNA repair The biochemical processes that correct

mutations arising from replication errors and the effects of mutagenic agents.

DNA replication   Synthesis of a new copy of the genome.

DNase I hypersensitive site   A short region of eukaryotic DNA that is relatively easily cleaved with deoxyribonuclease I, possibly coinciding with positions where nucleosomes are absent.

DNA sequencing   The technique for determining the order of nucleotides in a DNA molecule.

DNA shuffling   A PCR-based procedure that results in directed evolution of a DNA sequence.

DNA topoisomerase   An enzyme that introduces or removes turns from the double helix by breakage and reunion of one or both polynucleotides.

DNA transposon   A transposon whose transposition mechanism does not involve an RNA intermediate.

DNA tumor virus   A virus with a DNA genome, able to cause cancer after infection of an animal cell.

Domain   A segment of a polypeptide that folds independently of other segments; also the segment of a gene coding for such a domain.

Domain duplication   Duplication of a gene segment coding for a structural domain in the protein product.

Domain shuffling   Rearrangement of segments of one or more genes, each segment coding for a structural domain in the gene product, to create a new gene.

Dominant   The allele that is expressed in a heterozygote.

Donor site   The splice site at the 5′ end of an intron.

Dot matrix   A method for aligning nucleotide sequences.

Double helix   The base-paired double-stranded structure that is the natural form of DNA in the cell.

Double heterozygote   A nucleus that is heterozygous for two genes.

Double homozygote   A nucleus that is homozygous for two genes.

Double restriction   Digestion of DNA with two restriction endonucleases at the same time.

Double-strand break repair   A DNA repair process that mends double-stranded breaks.

Double stranded   Comprising two polynucleotides attached to one another by base pairing.

Double-stranded RNA-binding domain (dsRBD)   A common type of RNA-binding domain.

Downstream   Towards the 3′ end of a polynucleotide.

Dynamic allele-specific hybridization (DASH)   A solution hybridization technique used to type SNPs.

Electrophoresis   Separation of molecules on the basis of their net electrical charge.

Electrostatic interactions   Ionic bonds that form between charged chemical groups.

Elongation factor   A protein that plays an ancillary role in the elongation step of transcription or translation.

Elongator   A yeast protein, possibly with histone acetyltransferase activity, involved in the elongation phase of transcription.

Elution   The unbinding of a molecule from a chromatography column.

Embryonic stem (ES) cell   A totipotent cell from the embryo of a mouse or other organism.

End-labeling   The attachment of a radioactive or other label to one end of a DNA or RNA molecule.

End-modification   The chemical alteration of the end of a DNA or RNA molecule.

End-modification enzyme   An enzyme used in recombinant DNA technology that alters the chemical structure at the end of a DNA molecule.

Endogenous retrovirus (ERV)   An active or inactive retroviral genome integrated into a host chromosome.

Endonuclease   An enzyme that breaks phosphodiester bonds within a nucleic acid molecule.

Endosymbiont theory   A theory that states that the mitochondria and chloroplasts of eukaryotic cells are derived from symbiotic prokaryotes.

Enhanceosome   A structure formed by DNA bending that comprises a collection of proteins involved in activation of the RNA polymerase II transcription initiation complex.

Enhancer   A regulatory sequence that increases the rate of transcription of a gene or genes located some distance away in either direction.

Episome   A plasmid that is able to integrate into the host cell's chromosome.

Episome transfer   Transfer between cells of some or all of a bacterial chromosome by integration into a plasmid.

E site   A position within a bacterial ribosome to which a tRNA moves immediately after deacylation.

Ethidium bromide   A type of intercalating agent that causes mutations by inserting between adjacent base pairs in a double-stranded DNA molecule.

Ethylmethane sulfonate (EMS)   A mutagen that acts by adding alkyl groups to nucleotide bases.

Euchromatin   Regions of a eukaryotic chromosome that are relatively uncondensed, thought to contain active genes.

Eukaryote   An organism whose cells contain membrane-bound nuclei.

Excision repair   A DNA repair process that corrects various types of DNA damage by excising and resynthesizing a region of polynucleotide.

Exit site   A position within a bacterial ribosome to which a tRNA moves immediately after deacylation.

Exon   A coding region within a discontinuous gene.

Exonic splicing enhancer (ESE)   A nucleotide sequence that plays a positive regulatory role during splicing of GU–AG introns.

Exonic splicing silencer (ESS)   A nucleotide sequence that plays a negative regulatory role during splicing of GU–AG introns.

Exon-intron boundary   The nucleotide sequence at the junction between an exon and an intron.

Exon skipping   Aberrant splicing in which one or more or exons are omitted from the spliced RNA.

Exon theory of genes   An "introns early" hypothesis that holds that introns were formed when the first DNA genomes were constructed.

Exon trapping   A method, based on cloning, for identifying the positions of exons in a DNA sequence.

Exonuclease   An enzyme that removes nucleotides from the ends of a nucleic acid molecule.

Exosome   A multiprotein complex involved in degradation of mRNA in eukaryotes.

Exportin   A protein involved in transport of molecules out of the nucleus.

Expressed sequence tag (EST)   A cDNA that is sequenced in order to gain rapid access to the genes in a genome.

Expression proteomics   The methodology used to identify the proteins in a proteome.

Extein   The functional component of a discontinuous protein.

External node   The end of a branch in a phylogenetic tree, representing one of the organisms or DNA sequences being studied.

Extrachromosomal gene   A gene in a mitochondrial or chloroplast genome.

Facultative heterochromatin   Chromatin that has a compact organization in some, but not all, cells, thought to contain genes that are inactive in some cells or at some periods of the cell cycle.

FEN1   The "flap endonuclease" involved in replication of the lagging strand in eukaryotes.

Fiber-FISH   A specialized form of FISH that enables high marker resolution.

Field inversion gel electrophoresis (FIGE)   An electrophoresis method used to separate large DNA molecules.

Finished sequence   A chromosome sequence that is almost complete, defined for human chromosomes as one that covers at least 95% of the euchromatin with an error rate of less than one in 10^4 nucleotides.

Fitness   The ability of an organism or allele to survive and reproduce.

Fixation   Refers to the situation that occurs when a single allele reaches a frequency of 100% in a population.

Flow cytometry   A method for the separation of chromosomes.

FLpter value   The unit used in FISH to describe the position of a hybridization signal relative to the end of the short arm of the chromosome.

Fluorescence recovery after photobleaching (FRAP)   A technique used to study the mobility of nuclear proteins.

Fluorescent *in situ* hybridization (FISH)   A technique for locating markers on chromosomes by observing the hybridization positions of fluorescent labels.

Flush end   An end of a double-stranded DNA molecule where both strands terminate at the same nucleotide position with no single-stranded extension.

fMet   *N*-formylmethionine, the modified amino acid carried by the tRNA that is used during the initiation of translation in bacteria.

Folding domain   A segment of a polypeptide that folds independently of other segments.

Folding pathway   The series of events, involving partially folded intermediates, that results in an unfolded protein attaining its correct three-dimensional structure.

Footprinting   A range of techniques used for locating bound proteins on DNA molecules.

Fosmid   A high-capacity vector carrying the F plasmid origin of replication and a λ *cos* site.

F plasmid   A fertility plasmid that directs conjugal transfer of DNA between bacteria.

Fragile site   A position in a chromosome that is prone to breakage because it contains an expanded trinucleotide repeat sequence.

Frameshifting   The movement of a ribosome from one reading frame to another at an internal position within a gene.

Frameshift mutation   A mutation resulting from insertion or deletion of a group of nucleotides that is not a multiple of three and which therefore changes the frame in which translation occurs.

Functional analysis   The area of genome research devoted to identifying the functions of unknown genes.

Functional domain   A region of eukaryotic DNA around a gene or group of genes that can be delineated by treatment with deoxyribonuclease I.

Functional RNA   RNA that has a functional role in the cell; i.e. RNAs other than mRNA.

Fusion protein   A protein that consists of a fusion of two polypeptides, or parts of polypeptides, normally coded by separate genes.

G1 phase   The first gap period of the cell cycle.

G2 phase   The second gap period of the cell cycle.

Gain-of-function mutation   A mutation that results in an organism acquiring a new function.

Gamete   A reproductive cell, usually haploid, that fuses with a second gamete to produce a new cell during sexual reproduction.

Gap genes   Developmental genes that play a role in establishing positional information within the *Drosophila* embryo.

Gap period   One of two intermediate periods within the cell cycle.

GAPs (GTPase activating proteins)   A set of proteins that are intermediates in the Ras signal transduction pathway.

GC box   A type of basal promoter element.

GC content   The percentage of nucleotides in a genome that are G or C.

Gel electrophoresis   Electrophoresis performed in a gel so that molecules of similar electrical charge can be separated on the basis of size.

Gel retardation analysis   A technique that identifies protein-binding sites on DNA molecules by virtue of the effect that a bound protein has on the mobility of the DNA fragments during gel electrophoresis.

Gel stretching   A technique for preparing restricted DNA molecules for optical mapping.

Gene   A DNA segment containing biological information and hence coding for an RNA and/or polypeptide molecule.

Gene cloning   Insertion of a fragment of DNA, containing a gene, into a cloning vector, and subsequent propagation of the recombinant DNA molecule in a host organism.

Gene conversion   A process that results in the four haploid products of meiosis displaying an unusual segregation pattern.

Gene expression   The series of events by which the biological information carried by a gene is released and made available to the cell.

Gene flow   The transfer of a gene from one organism to another.

Gene fragment   A gene relic consisting of a short isolated region from within a gene.

General recombination   Recombination between two homologous double-stranded DNA molecules.

General transcription factor (GTF)   A protein or protein complex that is a transient or permanent component of the initiation complex formed during eukaryotic transcription.

Gene substitution   The replacement of an allele that at one time was fixed in the population by a second allele, this second allele arising by mutation and increasing in frequency until itself reaching fixation.

Gene superfamily   A group of two or more evolutionarily related multigene families.

Genes-within-genes   Refers to a gene whose intron contains a second gene.

Gene therapy   A clinical procedure in which a gene or other DNA sequence is used to treat a disease.

Genetic code   The rules that determine which triplet of nucleotides codes for which amino acid during protein synthesis.

Genetic footprinting   A technique for the rapid functional analysis of many genes at once.

Genetic linkage   The physical association between two genes that are on the same chromosome.

Genetic mapping   The use of genetic techniques to construct a genome map.

Genetic marker   A gene that exists as two or more readily distinguished alleles and whose inheritance can therefore be followed during a genetic cross, enabling the map position of the gene to be determined.

Genetic profile   The banding pattern revealed after electrophoresis of the products of PCRs directed at a range of microsatellite loci.

Genetic redundancy   The situation that occurs when two genes in the same genome perform the same function.

Genetics   The branch of biology devoted to the study of genes.

Gene tree   A phylogenetic tree that shows the evolutionary relationships between a group of genes or other DNA sequences.

Genome   The entire genetic complement of a living organism.

Genome expression   The series of events by which the biological information carried by a genome is released and made available to the cell.

Genome-wide repeat   A sequence that recurs at many dispersed positions within a genome.

Genomic imprinting   Inactivation by methylation of a gene on one of a pair of homologous chromosomes.

Genotype   A description of the genetic composition of an organism.

Gigabase pair   1,000,000 kb; 1,000,000,000 bp.

Global regulation   A general down-regulation in protein synthesis that occurs in response to various signals.

Glutamine-rich domain   A type of activation domain.

Glycosylation   The attachment of sugar units to a polypeptide.

GNRPs (guanine nucleotide-releasing proteins) A set of proteins that are intermediates in the Ras signal transduction pathway.

Green fluorescent protein A protein that is used to label other proteins and whose gene is used as a reporter gene.

Group I intron A type of intron found mainly in organelle genes.

Group II intron A type of intron found in organelle genes.

Group III intron A type of intron found in organelle genes.

GTPase activating proteins (GAPs) A set of proteins that are intermediates in the Ras signal transduction pathway.

GU–AG intron The commonest type of intron in eukaryotic nuclear genes. The first two nucleotides of the intron are 5′–GU–3′ and the last two are 5′–AG–3′.

Guanine One of the purine nucleotides found in DNA and RNA.

Guanine methyltransferase The enzyme that attaches a methyl group to the 5′ end of a eukaryotic mRNA during the capping reaction.

Guanine nucleotide releasing proteins (GNRPs) A set of proteins that are intermediates in the Ras signal transduction pathway.

Guanylyl transferase The enzyme that attaches a GTP to the 5′ end of a eukaryotic mRNA at the start of the capping reaction.

Guide RNA A short RNA that specifies the positions at which one or more nucleotides are inserted into an abbreviated RNA by pan-editing.

Hairpin A stem-loop structure made up of a base-paired stem and non-base-paired loop, which can form in a single-stranded polynucleotide that contains an inverted repeat.

Hammerhead An RNA structure with ribozyme activity that is found in some viruses.

Haplogroup One of the major sequence classes of mitochondrial DNA present in the human population.

Haploid A nucleus that has a single copy of each chromosome.

Haploinsufficiency The situation where inactivation of a gene on one of a pair of homologous chromosomes results in a change in the phenotype of the mutant organism.

Haplotype An individual mitochondrial DNA sequence.

Helicase An enzyme that breaks base pairs in a double-stranded DNA molecule.

Helix-loop-helix motif A dimerization domain commonly found in DNA-binding proteins.

Helix-turn-helix motif A common structural motif for attachment of a protein to a DNA molecule.

Helper phage A phage that is introduced into a host cell in conjunction with a related cloning vector, in order to provide enzymes and other proteins required for replication of the cloning vector.

Heterochromatin Chromatin that is relatively condensed and is thought to contain DNA that is not being transcribed.

Heteroduplex A DNA–DNA or DNA–RNA hybrid.

Heteroduplex analysis Transcript mapping by analysis of DNA–RNA hybrids with a single-strand-specific nuclease such as S1.

Heterogenous nuclear RNA (hnRNA) The nuclear RNA fraction that comprises unprocessed transcripts synthesized by RNA polymerase II.

Heteropolymer An artificial RNA comprising a mixture of different nucleotides.

Heterozygosity The probability that a person chosen at random from the population will be heterozygous for a particular marker.

Heterozygous A diploid nucleus that contains two different alleles for a particular gene.

Hierarchical clustering A method for analyzing transcriptomes based on comparisons between the expression levels of pairs of genes.

High mobility group N (HMGN) protein A group of nuclear proteins that influence chromatin structure.

High-performance liquid chromatography (HPLC) A column chromatography method with many applications in biochemistry.

Histone One of the basic proteins found in nucleosomes.

Histone acetylation Modification of chromatin structure by attachment of acetyl groups to core histones.

Histone acetyltransferase (HAT) An enzyme that attaches acetyl groups to core histones.

Histone code The hypothesis that the pattern of chemical modification on histone proteins influences various cellular activities.

Histone deacetylase (HDAC) An enzyme that removes acetyl groups from core histones.

Holliday structure An intermediate structure formed during recombination between two DNA molecules.

Holocentric chromosome A chromosome that does not have a single centromere but instead has multiple kinetochores spread along its length.

Holoenzyme The version of the *Escherichia coli* RNA polymerase, subunit composition $\alpha_2\beta\beta'\sigma$, that is able to recognize promoter sequences.

Homeodomain A DNA-binding motif found in many proteins involved in developmental regulation of gene expression.

Homeotic mutation   A mutation that results in the transformation of one body part into another.

Homeotic selector gene   A gene that establishes the identity of a body part such as a segment of the *Drosophila* embryo.

Homologous chromosomes   Two or more identical chromosomes present in a single nucleus.

Homologous genes   Genes that share a common evolutionary ancestor.

Homologous recombination   Recombination between two homologous double-stranded DNA molecules, i.e. ones which share extensive nucleotide sequence similarity.

Homology searching   A technique in which genes with sequences similar to that of an unknown gene are sought, the objective being to gain an insight into the function of the unknown gene.

Homoplasy   The situation that occurs when the same character state evolves independently in two lineages.

Homopolymer   An artificial RNA comprising just one nucleotide.

Homopolymer tailing   The attachment of a sequence of identical nucleotides (e.g. AAAAA) to the end of a nucleic acid molecule, usually referring to the synthesis of single-stranded homopolymer extensions on the ends of a double-stranded DNA molecule.

Homozygous   A diploid nucleus that contains two identical alleles for a particular gene.

Horizontal gene transfer   Transfer of a gene from one species to another.

Hormone response element   A nucleotide sequence upstream of a gene that mediates the regulatory effect of a steroid hormone.

Housekeeping protein   A protein that is continually expressed in all or at least most cells of a multicellular organism.

Hsp70 chaperone   A family of proteins that bind to hydrophobic regions in other proteins in order to aid their folding.

Hub   A protein that has many interactions within a protein interaction map.

Human Genome Project   The publicly funded project responsible for one of the draft human genome sequences and which continues to study the functions of human genes.

Hybrid dysgenesis   The event that occurs when females from laboratory strains of *Drosophila melanogaster* are crossed with males from wild populations, the offspring resulting from such crosses being sterile and having chromosomal abnormalities and other genetic malfunctions.

Hybridization   The attachment to one another, by base pairing, of two complementary polynucleotides.

Hybridization probing   A technique that uses a labeled nucleic acid molecule as a probe to identify complementary or homologous molecules to which it base-pairs.

Hydrogen bond   A weak electrostatic attraction between an electronegative atom such as oxygen or nitrogen and a hydrogen atom attached to a second electronegative atom.

Hydrophobic effects   Chemical interactions that result in hydrophobic groups becoming buried inside a protein.

Hypermutation   An increase in the mutation rate of a genome.

Illegitimate recombination   Recombination between two double-stranded DNA molecules which have little or no nucleotide sequence similarity.

Immunocytochemistry   A technique that uses antibody probing to locate the position of a protein in a tissue.

Immunoelectron microscopy   An electron microscopy technique that uses antibody labeling to identify the positions of specific proteins on the surface of a structure such as a ribosome.

Immunoscreening   The use of an antibody probe to detect a polypeptide synthesized by a cloned gene.

Importin   A protein involved in transport of molecules into the nucleus.

Imprint control element   A DNA sequence found within a few kb of clusters of imprinted genes, which mediates the methylation of the imprinted regions.

Incomplete dominance   Refers to a pair of alleles, neither of which displays dominance, the phenotype of a heterozygote being intermediate between the phenotypes of the two homozygotes.

Indel   A position in an alignment between two DNA sequences where an insertion or deletion has occurred.

Inducer   A molecule that induces expression of a gene or operon by binding to a repressor protein and preventing the repressor from attaching to the operator.

Inferred tree   A tree obtained by phylogenetic analysis.

Informational problem   The problem tackled by the early molecular biologists concerning the nature of the genetic code.

Inhibition domain   The part of a eukaryotic repressor that makes contact with the initiation complex.

Inhibitor-resistant mutant   A mutant that is able to resist the toxic effects of an antibiotic or other type of inhibitor.

Initiation codon   The codon, usually but not exclusively 5′–AUG–3′, found at the start of the coding region of a gene.

Initiation complex   The complex of proteins that initiates transcription. Also the complex that initiates translation.

Initiation factor A protein that plays an ancillary role during initiation of translation.

Initiation of transcription The assembly upstream of a gene of the complex of proteins that will subsequently copy the gene into RNA.

Initiation region A region of eukaryotic chromosomal DNA within which replication initiates at positions that are not clearly defined.

Initiator (Inr) sequence A component of the RNA polymerase II core promoter.

Initiator tRNA The tRNA, aminoacylated with methionine in eukaryotes or *N*-formylmethionine in bacteria, that recognizes the initiation codon during protein synthesis.

Inosine A modified version of adenosine, sometimes found at the wobble position of an anticodon.

Insertional editing A less extensive form of pan-editing that occurs during processing of some viral RNAs.

Insertional inactivation A cloning strategy whereby insertion of a new piece of DNA into a vector inactivates a gene carried by the vector.

Insertion mutation A mutation that arises by insertion of one or more nucleotides into a DNA sequence.

Insertion sequence A short transposable element found in bacteria.

Insertion vector A λ vector constructed by deleting a segment of nonessential DNA.

Instability element A sequence present in yeast mRNAs that affects degradation.

Insulator A segment of DNA that acts as the boundary point between two functional domains.

Integrase A Type I topoisomerase that catalyzes insertion of the λ genome into *Escherichia coli* DNA.

Integron A set of genes and other DNA sequences that enable plasmids to capture genes from bacteriophages and other plasmids.

Intein An internal segment of a polypeptide that is removed by a splicing process after translation.

Intein homing The conversion of a gene coding for a protein that lacks an intein into one coding for an intein-plus protein, catalyzed by the spliced component of the intein.

Interactome network A map showing the interactions between all or some of the proteins in a proteome.

Intercalating agent A compound that can enter the space between adjacent base pairs of a double-stranded DNA molecule, often causing mutations.

Intergenic region The regions of a genome that do not contain genes.

Internal node A branch point within a phylogenetic tree, representing an organism or DNA sequence that is ancestral to those being studied.

Internal ribosome entry site (IRES) A nucleotide sequence that enables the ribosome to assemble at an internal position in some eukaryotic mRNAs.

Interphase The period between cell divisions.

Interphase chromosome A chromosome, present in a cell during the period between cell divisions, which adopts a relatively uncondensed chromatin structure.

Interspersed repeat A sequence that recurs at many dispersed positions within a genome.

Interspersed repeat element PCR (IRE-PCR) A clone fingerprinting technique that uses PCR to detect the relative positions of genome-wide repeats in cloned DNA fragments.

Intramolecular base pairing Base pairing that occurs between two parts of the same DNA or RNA polynucleotide.

Intrinsic terminator A position in bacterial DNA where termination of transcription occurs without the involvement of Rho.

Intron A noncoding region within a discontinuous gene.

Intron homing The conversion of a gene lacking an intron into one that contains an intron, catalyzed by a protein coded by that intron.

Introns early The hypothesis that introns evolved relatively early and are gradually being lost from eukaryotic genomes.

Introns late The hypothesis that introns evolved relatively late and are gradually accumulating in eukaryotic genomes.

Inverted repeat Two identical nucleotide sequences repeated in opposite orientations in a DNA molecule.

***In vitro* mutagenesis** Techniques used to produce a specified mutation at a predetermined position in a DNA molecule.

***In vitro* packaging** Synthesis of infective λ phages from a preparation of λ proteins and a concatamer of λ DNA molecules.

Ion exchange chromatography A method for separating molecules according to how tightly they bind to electrically charged particles present in a chromatographic matrix.

Iron-response element A type of response module.

Isoaccepting tRNAs Two or more tRNAs that are charged with the same amino acid.

Isochore A segment of genomic DNA that has a uniform base composition which differs from that of the adjacent segments.

Isoelectric focussing Separation of proteins in a gel that contains chemicals which establish a pH gradient when the electrical charge is applied.

Isoelectric point The position in a pH gradient where the net charge of a protein is zero.

Isotope   One of two or more atoms that have the same atomic number but different atomic weights.

Isotope coded affinity tag (ICAT)   Markers, containing normal hydrogen and deuterium atoms, used to label individual proteomes.

Janus kinase (JAK)   A type of kinase that plays an intermediary role in some types of signal transduction involving STATs.

Junk DNA   One interpretation of the intergenic DNA content of a genome.

Karyogram   The entire chromosome complement of a cell, with each chromosome described in terms of its appearance at metaphase.

Karyopherin   A protein involved in transport of RNA out of or into the nucleus.

Kilobase pair (kb)   1000 base pairs.

Kinetochore   The part of the centromere to which spindle microtubules attach.

Klenow polymerase   A DNA polymerase enzyme, obtained by chemical modification of *Escherichia coli* DNA polymerase I, used primarily in chain termination DNA sequencing.

Knockout mouse   A mouse that has been engineered so that it carries an inactivated gene.

Kornberg polymerase   The DNA polymerase I enzyme of *Escherichia coli.*

Kozak consensus   The nucleotide sequence surrounding the initiation codon of a eukaryotic mRNA.

Lactose operon   The cluster of three genes that code for enzymes involved in utilization of lactose by *Escherichia coli.*

Lactose repressor   The regulatory protein that controls transcription of the lactose operon in response to the presence or absence of lactose in the environment.

Lagging strand   The strand of the double helix which is copied in a discontinuous fashion during genome replication.

Lariat   Refers to the lariat-shaped intron RNA that results from splicing a GU–AG intron.

Latent period   The period between injection of a phage genome into a bacterial cell and the time when cell lysis occurs.

Lateral gene transfer   Transfer of a gene from one species to another.

Leader segment   The untranslated region of an mRNA upstream of the initiation codon.

Leading strand   The strand of the double helix which is copied in a continuous fashion during genome replication.

Leaky mutation   A mutation that results in partial loss of a characteristic.

Lethal mutation   A mutation that results in death of the cell or organism.

Leucine zipper   A dimerization domain commonly found in DNA-binding proteins.

Ligase   An enzyme that synthesizes phosphodiester bonds as part of DNA replication, repair, and recombination processes.

LINE (long interspersed nuclear element)   A type of genome-wide repeat, often with transposable activity.

LINE-1   One type of human LINE.

Linkage   The physical association between two genes that are on the same chromosome.

Linkage analysis   The procedure used to assign map positions to genes by genetic crosses.

Linker DNA   The DNA that links nucleosomes: the "string" in the "beads-on-a-string" model for chromatin structure.

Linker histone   A histone, such as H1, that is located outside of the nucleosome core octamer.

Locus   The chromosomal location of a genetic or DNA marker.

Locus control region (LCR)   A DNA sequence that maintains a functional domain in an open active configuration.

Lod score   A statistical measure of linkage as revealed by pedigree analysis.

Long patch repair   A nucleotide excision repair process of *Escherichia coli* that results in excision and resynthesis of up to 2 kb of DNA.

Loss-of-function mutation   A mutation that reduces or abolishes a protein's activity.

LTR element   A type of genome-wide repeat typified by the presence of long terminal repeats (LTRs).

Lysis   The disruption of a bacterial cell by lysozyme, such as occurs at the end of the infection cycle of a lytic bacteriophage.

Lysogenic infection cycle   The type of bacteriophage infection that involves integration of the phage genome into the host DNA molecule.

Lysozyme   A protein used to destabilize the bacterial cell wall prior to DNA purification.

Lytic infection cycle   The type of bacteriophage infection that involves lysis of the host cell immediately after the initial infection, with no integration of the phage DNA molecule into the host genome.

Macrochromosome   One of the larger gene-deficient chromosomes seen in the nuclei of chickens and various other species.

MADS box   A DNA-binding domain found in several transcription factors involved in plant development.

Maintenance methylation Addition of methyl groups to positions on newly synthesized DNA strands that correspond with the positions of methylation on the parent strand.

Major groove The larger of the two grooves that spiral around the surface of the B-form of DNA.

Major histocompatibility complex (MHC) A mammalian multigene family coding for cell surface proteins and including several multiallelic genes.

Map A chart showing the positions of genetic and/or physical markers in a genome.

MAP kinase A signal transduction pathway.

Mapping reagent A collection of DNA fragments spanning a chromosome or the entire genome and used in STS mapping.

Marker A distinctive feature on a genome map. Also, a gene, carried by a cloning vector, that codes for a distinctive protein product and/or phenotype and so can be used to determine if a cell contains a copy of the cloning vector.

Mass spectrometry An analytical technique in which ions are separated according to their charge-to-mass ratios.

Maternal-effect gene A *Drosophila* gene that is expressed in the parent and whose mRNA is subsequently injected into the egg, after which it influences development of the embryo.

Mating type The equivalent of male and female for a eukaryotic microorganism.

Mating-type switching The ability of yeast cells to change from a to α mating type, or vice versa, by gene conversion.

Matrix-assisted laser desorption ionization time-of-flight (MALDI-TOF) A type of mass spectrometry used in proteomics.

Matrix-associated region (MAR) An AT-rich segment of a eukaryotic genome that acts as an attachment point to the nuclear matrix.

Maturase A protein, coded by a gene in an intron, thought to be involved in splicing.

Maximum parsimony method A method for reconstruction of phylogenetic trees.

Mediator A protein complex that forms a contact between various activators and the C-terminal domain of the largest subunit of RNA polymerase II.

Megabase pair (Mb) 1000 kb; 1,000,000 bp.

Meiosis The series of events, involving two nuclear divisions, by which diploid nuclei are converted to haploid gametes.

Melting Denaturation of a double-stranded DNA molecule.

Melting temperature (T_m) The temperature at which the two strands of a double-stranded nucleic acid molecule or base-paired hybrid detach as a result of complete breakage of hydrogen bonding.

Meselson-Stahl experiment The experiment which showed that cellular DNA replication occurs by the semiconservative process.

Messenger RNA (mRNA) The transcript of a protein-coding gene.

Metabolic engineering The process by which changes are made to the genome by mutation or recombinant DNA techniques in order to influence the cellular biochemistry in a pre-determined way.

Metabolic flux The rate of flow of metabolites through the network of pathways that make up the cellular biochemistry.

Metabolomics The study of metabolomes.

Metagenomics Studies of the mixture of genomes present in a particular habitat.

Metaphase chromosome A chromosome at the metaphase stage of cell division, when the chromatin takes on its most condensed structure and features such as the banding pattern can be visualized.

Methyl-CpG-binding protein (MeCP) A protein that binds to methylated CpG islands and may influence acetylation of nearby histones.

MGMT (O^6-methylguanine-DNA methyltransferase) An enzyme involved in the direct repair of alkylation mutations.

Microarray A low-density array of DNA molecules used for parallel hybridization analysis.

MicroRNA A class of short RNAs involved in regulation of gene expression in eukaryotes, and which act by a pathway similar to RNA interference.

Microsatellite A type of simple sequence length polymorphism comprising tandem copies of, usually, di-, tri- or tetranucleotide repeat units. Also called a simple tandem repeat (STR).

Minichromosome One of the smaller, gene-rich chromosomes seen in the nuclei of chickens and various other species.

Minigene The name given to the pair of exons carried by a cloning vector used in the exon-trapping procedure.

Minimal medium A medium that provides only the minimum nutritional requirements for growth of a microorganism.

Minisatellite A type of simple sequence length polymorphism comprising tandem copies of repeats that are a few tens of nucleotides in length. Also called a variable number of tandem repeats (VNTR).

Minor groove The smaller of the two grooves that spiral around the surface of the B-form of DNA.

Mismatch A position in a double-stranded DNA molecule where base pairing does not occur because the nucleotides are not complementary; in particular, a non-base-paired position resulting from an error in replication.

Mismatch repair   A DNA repair process that corrects mismatched nucleotide pairs by replacing the incorrect nucleotide in the daughter polynucleotide.

Missense mutation   An alteration in a nucleotide sequence that converts a codon for one amino acid into a codon for a second amino acid.

Mitochondrial genome   The genome present in the mitochondria of a eukaryotic cell.

Mitochondrion   One of the energy-generating organelles of eukaryotic cells.

Mitosis   The series of events that results in nuclear division.

Model organism   An organism which is relatively easy to study and hence can be used to obtain information that is relevant to the biology of a second organism that is more difficult to study.

Modification assay   A range of techniques used for locating bound proteins on DNA molecules.

Modification interference   A technique used to identify nucleotides involved in interactions with a DNA-binding protein.

Modification protection   A technique used to identify nucleotides involved in interactions with a DNA-binding protein.

Molecular biologist   A person who studies the molecular life sciences.

Molecular chaperone   A protein that helps other proteins to fold.

Molecular clock   A device based on the inferred mutation rate that enables times to be assigned to the branch points in a gene tree.

Molecular combing   A technique for preparing restricted DNA molecules for optical mapping.

Molecular evolution   The gradual changes that occur in genomes over time as a result of the accumulation of mutations and structural rearrangements resulting from recombination and transposition.

Molecular life sciences   The area of research comprising molecular biology, biochemistry, and cell biology, as well as some aspects of genetics and physiology.

Molecular phylogenetics   A set of techniques that enable the evolutionary relationships between DNA sequences to be inferred by making comparisons between those sequences.

Monohybrid cross   A sexual cross in which the inheritance of one pair of alleles is followed.

Monophyletic   Refers to two or more organisms or DNA sequences that are derived from a single ancestral organism or DNA sequence.

M phase   The stage of the cell cycle when mitosis or meiosis occurs.

mRNA processing   The chemical and physical modification events that occur after synthesis of an mRNA.

mRNA surveillance   An RNA degradation process in eukaryotes.

Multicopy   A gene, cloning vector, or other genetic element that is present in multiple copies in a single cell.

Multicysteine zinc finger   A type of zinc-finger DNA-binding domain.

Multidimensional protein identification technique (MudPIT)   A technique that combines various chromatography methods in order to isolate intact protein complexes.

Multigene family   A group of genes, clustered or dispersed, with related nucleotide sequences.

Multiple alignment   An alignment of three or more nucleotide sequences.

Multiple alleles   The different alternative forms of a gene that has more than two alleles.

Multiple hit or multiple substitution   The situation that occurs when a single nucleotide in a DNA sequence undergoes two mutational changes, giving rise to two new alleles, both of which differ from each other and from the parent at that nucleotide position.

Multipoint cross   A genetic cross in which the inheritance of three or more markers is followed.

Multiregional evolution   A hypothesis that holds that modern humans in the Old World are descended from *Homo erectus* populations that left Africa over 1 million years ago.

Mutagen   A chemical or physical agent that can cause a mutation in a DNA molecule.

Mutagenesis   Treatment of a group of cells or organisms with a mutagen as a means of inducing mutations.

Mutant   A cell or organism that possesses a mutation.

Mutasome   A protein complex that is constructed during the SOS response of *Escherichia coli*.

Mutation   An alteration in the nucleotide sequence of a DNA molecule.

Mutation scanning   A set of techniques for detection of mutations in DNA molecules.

Mutation screening   A set of techniques for determining if a DNA molecule contains a specific mutation.

Natural selection   The preservation of favorable alleles and the rejection of injurious ones.

N-degron   An N-terminal amino acid sequence that influences the degradation of a protein in which it is found.

Neighbor-joining method   A method for reconstruction of phylogenetic trees.

Nick A position in a double-stranded DNA molecule where one of the polynucleotides is broken as a result of the absence of a phosphodiester bond.

Nitrogenous base One of the purines or pyrimidines that form part of the molecular structure of a nucleotide.

N-linked glycosylation The attachment of sugar units to an asparagine in a polypeptide.

Nonchromatin region The space separating the chromosome territories within a nucleus.

Noncoding RNA An RNA molecule that does not code for a protein.

Nonhomologous end-joining (NHEJ) Another name for the double-strand break repair process.

Nonpenetrance The situation whereby the effect of a mutation is never observed during the lifetime of a mutant organism.

Nonpolar A hydrophobic (water-hating) chemical group.

Nonsense-mediated RNA decay (NMD) A process for degradation of eukaryotic mRNAs that is initiated by the presence of an internal termination codon.

Nonsense mutation An alteration in a nucleotide sequence that changes a triplet coding for an amino acid into a termination codon.

Nonsynonymous mutation A mutation that converts a codon for one amino acid into a codon for a second amino acid.

Northern blotting The transfer of RNA from an electrophoresis gel to a membrane prior to northern hybridization.

Northern hybridization A technique used for detection of a specific RNA molecule against a background of many other RNA molecules.

N terminus The end of a polypeptide that has a free amino group.

Nuclear genome The DNA molecules present in the nucleus of a eukaryotic cell.

Nuclear magnetic resonance (NMR) spectroscopy A technique for determining the three-dimensional structure of large molecules.

Nuclear matrix A proteinaceous scaffold-like network that permeates the cell.

Nuclear pore complex The complex of proteins present at a nuclear pore.

Nuclear receptor superfamily A family of receptor proteins that bind hormones as an intermediate step in modulation of genome activity by these hormones.

Nuclease An enzyme that degrades a nucleic acid molecule.

Nuclease protection experiment A technique that uses nuclease digestion to determine the positions of proteins on DNA or RNA molecules.

Nucleic acid The term first used to describe the acidic chemical compound isolated from the nuclei of eukaryotic cells. Now used specifically to describe a polymeric molecule comprising nucleotide monomers, such as DNA and RNA.

Nucleic acid hybridization Formation of a double-stranded hybrid by base pairing between complementary polynucleotides.

Nucleoid The DNA-containing region of a prokaryotic cell.

Nucleolus The region of the eukaryotic nucleus in which rRNA transcription occurs.

Nucleoside A purine or pyrimidine base attached to a five-carbon sugar.

Nucleosome The complex of histones and DNA that is the basic structural unit in chromatin.

Nucleosome remodeling A change in the conformation of a nucleosome, associated with a change in access to the DNA to which the nucleosome is attached.

Nucleotide A purine or pyrimidine base attached to a five-carbon sugar, to which a mono-, di-, or triphosphate is also attached. The monomeric unit of DNA and RNA.

Nucleotide excision repair A repair process that corrects various types of DNA damage by excising and resynthesizing a region of a polynucleotide.

Nucleus The membrane-bound structure of a eukaryotic cell in which the chromosomes are contained.

Octamer module A basal promoter element.

Okazaki fragment One of the short segments of RNA-primed DNA synthesized during replication of the lagging strand of the double helix.

Oligonucleotide A short synthetic single-stranded DNA molecule.

Oligonucleotide-directed mutagenesis An *in vitro* mutagenesis technique in which a synthetic oligonucleotide is used to introduce a predetermined nucleotide alteration into the gene to be mutated.

Oligonucleotide hybridization analysis The use of an oligonucleotide as a hybridization probe.

Oligonucleotide-ligation assay (OLA) A technique for SNP typing, which depends on ligation of two oligonucleotides which anneal adjacent to one another, one covering the position of the SNP.

O-linked glycosylation The attachment of sugar units to a serine or threonine in a polypeptide.

Open promoter complex A structure formed during assembly of the transcription initiation complex consisting of the RNA polymerase and/or accessory proteins attached to the promoter, after the DNA has been opened up by breakage of base pairs.

Open reading frame (ORF) A series of codons starting with an initiation codon and ending with a termination

codon. The part of a protein-coding gene that is translated into protein.

Operational taxonomic unit (OTU)   One of the organisms being compared in a phylogenetic analysis.

Operator   The nucleotide sequence to which a repressor protein binds to prevent transcription of a gene or operon.

Operon   A set of adjacent genes in a bacterial genome, transcribed from a single promoter and subject to the same regulatory regime.

Optical mapping   A technique for the direct visual examination of restricted DNA molecules.

ORF scanning   Examination of a DNA sequence for open reading frames in order to locate the genes.

Origin of replication   A site on a DNA molecule where replication initiates.

Origin recognition complex (ORC)   A set of proteins that binds to the origin recognition sequence.

Origin recognition sequence   A component of a eukaryotic origin of replication.

Orphan family   A group of homologous genes whose functions are unknown.

Orthogonal field alternation gel electrophoresis (OFAGE)   An electrophoresis system in which the field alternates between pairs of electrodes set at an angle of 45°, used to separate large DNA molecules.

Orthologous   Refers to homologous genes located in the genomes of different organisms.

Outgroup   An organism or DNA sequence that is used to root a phylogenetic tree.

Out of Africa   A hypothesis which holds that modern humans evolved in Africa, moving to the rest of the Old World between 100,000 and 50,000 years ago, displacing the descendants of *Homo erectus* that they encountered.

Overlapping genes   Two genes whose coding regions overlap.

P1-derived artificial chromosome (PAC)   A high-capacity vector that combines features of bacteriophage P1 vectors and bacterial artificial chromosomes.

Paired-end reads   Minisequences from the two ends of a single cloned fragment.

Pair-rule genes   Developmental genes that establish the basic segmentation pattern of the *Drosophila* embryo.

Pan-editing   The extensive insertion of nucleotides into an abbreviated RNA, resulting in a functional molecule.

Paralogous   Refers to two or more homologous genes located in the same genome.

Paranemic   Refers to a helix whose strands can be separated without unwinding.

Paraphyletic   In a phylogenetic tree, a group of sequences or taxa that excludes some members of a clade.

Pararetrovirus   A viral retroelement whose encapsidated genome is made of DNA.

Parental genotype   The genotype possessed by one or both of the parents in a genetic cross.

Parsimony   An approach that decides between different phylogenetic tree topologies by identifying the one that involves the shortest evolutionary pathway.

Partial linkage   The type of linkage usually displayed by a pair of genetic and/or physical markers on the same chromosome, the markers not always being inherited together because of the possibility of recombination between them.

Partial restriction   Digestion of DNA with a restriction endonuclease under limiting conditions so that not all restriction sites are cut.

Pedigree   A chart showing the genetic relationships between the members of a human family.

Pedigree analysis   The use of pedigree charts to analyze the inheritance of a genetic or DNA marker in a human family.

P element   A DNA transposon of *Drosophila*.

Pentose   A sugar comprising five carbon atoms.

Peptide bond   The chemical link between adjacent amino acids in a polypeptide.

Peptide nucleic acid (PNA)   A polynucleotide analog in which the sugar–phosphate backbone is replaced by amide bonds.

Peptidyl or P site   The site in the ribosome occupied by the tRNA attached to the growing polypeptide during translation.

Peptidyl transferase   The enzyme activity that synthesizes peptide bonds during translation.

Permissive conditions   Conditions under which a conditional-lethal mutant is able to survive.

PEST sequences   Amino acid sequences that influence the degradation of proteins in which they are found.

Phage   A virus that infects a bacterium.

Phage display   A technique for identifying proteins that interact with one another.

Phage display library   A collection of clones carrying different DNA fragments, used in phage display.

Phagemid   A cloning vector comprising a mixture of plasmid and phage DNA.

Phenetics   A classificatory approach based on the numerical typing of as many characters as possible.

Phenotype   The observable characteristics displayed by a cell or organism.

Philadelphia chromosome   An abnormal chromosome

resulting from a translocation between human chromosomes 9 and 22, a common cause of chronic myeloid leukemia.

Phosphodiesterase A type of enzyme that can break phosphodiester bonds.

Phosphodiester bond The chemical link between adjacent nucleotides in a polynucleotide.

Phosphorimaging An electronic method for determining the positions of radioactive markers in a microarray or on a hybridization membrane.

Photobleaching A component of the FRAP technique for studying protein mobility in the nucleus.

Photolithography A technique that uses pulses of light to construct an oligonucleotide from light-activated nucleotide substrates.

Photolyase An *Escherichia coli* enzyme involved in photoreactivation repair.

Photoproduct A modified nucleotide resulting from treatment of DNA with ultraviolet radiation.

Photoreactivation A DNA repair process in which cyclobutyl dimers and (6–4) photoproducts are corrected by a light-activated enzyme.

Phylogeny A classification scheme that indicates the evolutionary relationships between organisms.

Physical mapping The use of molecular biology techniques to construct a genome map.

Pilus A structure involved in bringing a pair of bacteria together during conjugation; possibly the tube through which DNA is transferred.

Plaque A zone of clearing on a lawn of bacteria caused by lysis of the cells by infecting bacteriophages.

Plasmid A usually circular piece of DNA often found in bacteria and some other types of cell.

Plectonemic Refers to a helix whose strands can only be separated by unwinding.

Plesiomorphic character state A character state possessed by a remote common ancestor of a group of organisms.

Point mutation A mutation that results from a single nucleotide change in a DNA molecule.

Polar A hydrophilic (water-loving) chemical group.

Polyacrylamide gel electrophoresis Electrophoresis carried out in a polyacrylamide gel and used to separate DNA molecules between 10 and 1500 bp in length.

Polyadenylate-binding protein A protein that aids poly(A) polymerase during polyadenylation of eukaryotic mRNAs, and which plays a role in maintenance of the tail after synthesis.

Polyadenylation The addition of a series of As to the 3′ end of a eukaryotic mRNA.

Polyadenylation editing A form of editing that occurs with many animal mitochondrial RNAs, resulting in a termination codon being formed by adding a poly(A) tail to an mRNA that ends with the nucleotides U or UA.

Poly(A) polymerase The enzyme that attaches a poly(A) tail to the 3′ end of a eukaryotic mRNA.

Poly(A) tail A series of A nucleotides attached to the 3′ end of a eukaryotic mRNA.

Polymer A compound made up of a long chain of identical or similar units.

Polymerase chain reaction (PCR) A technique that results in exponential amplification of a selected region of a DNA molecule.

Polymorphic Refers to a locus that is represented by a number of different alleles or haplotypes in the population as a whole.

Polynucleotide A single-stranded DNA or RNA molecule.

Polynucleotide kinase An enzyme that adds phosphate groups to the 5′ ends of DNA molecules.

Polypeptide A polymer of amino acids.

Polyphyletic A group of DNA sequences that are derived from two or more distinct ancestral sequences.

Polyprotein A translation product consisting of a series of linked proteins which are processed by proteolytic cleavage to release the mature proteins.

Polypyrimidine tract A pyrimidine-rich region near the 3′ end of a GU–AG intron.

Polysome An mRNA molecule that is being translated by more than one ribosome at the same time.

Positional cloning A procedure that uses information on the map position of a gene to obtain a clone of that gene.

Positional effect Refers to the different levels of expression that result after insertion of a gene at different positions in a eukaryotic genome.

Postreplication complex (post-RC) A complex of proteins, derived from a pre-RC, that forms at a eukaryotic origin of replication during the replication process and ensures that the origin is used just once per cell cycle.

Postreplicative repair A repair process that deals with breaks in daughter DNA molecules that arise as a result of aberrations in the replication process.

POU domain A DNA-binding motif found in a variety of proteins.

Preinitiation complex The structure comprising the small subunit of the ribosome, the initiator tRNA plus ancillary factors that forms the initial association with the mRNA during protein synthesis. Also, the structure that forms at the core promoter of a gene transcribed by RNA polymerase II.

Pre-mRNA The primary transcript of a protein-coding gene.

Prepriming complex   A complex of proteins formed during initiation of replication in bacteria.

Prereplication complex (pre-RC)   A protein complex that is constructed at a eukaryotic origin of replication and enables initiation of replication to occur.

Pre-RNA   The initial product of transcription of a gene or group of genes, subsequently processed to give the mature transcript(s).

Pre-rRNA   The primary transcript of a gene or group of genes specifying rRNA molecules.

Prespliceosome complex   An intermediate in the splicing pathway for a GU–AG intron.

Pre-tRNA   The primary transcript of a gene or group of genes specifying tRNA molecules.

Pribnow box   A component of the bacterial promoter.

Primary structure   The sequence of amino acids in a polypeptide.

Primary transcript   The initial product of transcription of a gene or group of genes, subsequently processed to give the mature transcript(s).

Primase   The RNA polymerase enzyme that synthesizes RNA primers during bacterial DNA replication.

Primer   A short oligonucleotide that is attached to a single-stranded DNA molecule in order to provide a start point for strand synthesis.

Primosome   A protein complex involved in genome replication.

Principal component analysis   A procedure that attempts to identify patterns in a large dataset of variable character states.

Prion   An unusual infectious agent that consists purely of protein.

Processed pseudogene   A pseudogene that results from integration into the genome of a reverse-transcribed copy of an mRNA.

Processivity   Refers to the amount of DNA synthesis that is carried out by a DNA polymerase before dissociation from the template.

Programmed frameshifting   The controlled movement of a ribosome from one reading frame to another at an internal position within a gene.

Programmed mutation   The possibility that under some circumstances an organism can increase the rate at which mutations occur in a specific gene.

Prokaryote   An organism whose cells lack a distinct nucleus.

Proliferating cell nuclear antigen (PCNA)   An accessory protein involved in genome replication in eukaryotes.

Proline-rich domain   A type of activation domain.

Promiscuous DNA   DNA that has been transferred from one organelle genome to another.

Promoter   The nucleotide sequence, upstream of a gene, to which RNA polymerase binds in order to initiate transcription.

Promoter clearance   The completion of successful initiation of transcription that occurs when the RNA polymerase moves away from the promoter sequence.

Promoter escape   The stage in transcription during which the polymerase moves away from the promoter region and becomes committed to making a transcript.

Proofreading   The 3′→5′ exonuclease activity possessed by some DNA polymerases which enables the enzyme to replace a misincorporated nucleotide.

Prophage   The integrated form of the genome of a lysogenic bacteriophage.

Protease   An enzyme that degrades protein.

Proteasome   A multisubunit protein structure that is involved in the degradation of other proteins.

Protein   The polymeric compound made of amino acid monomers.

Protein electrophoresis   Separation of proteins in an electrophoresis gel.

Protein engineering   Various techniques for making directed alterations in protein molecules, often to improve the properties of enzymes used in industrial processes.

Protein folding   The adoption of a folded structure by a polypeptide.

Protein interaction map   A map showing the interactions between all or some of the proteins in a proteome.

Protein profiling   The methodology used to identify the proteins in a proteome.

Protein–protein crosslinking   A technique that links together adjacent proteins in order to identify proteins that are positioned close to one another in a structure such as a ribosome.

Proteome   The collection of functioning proteins synthesized by a living cell.

Proteomics   A variety of techniques used to study proteomes.

Protogenome   An RNA genome that existed during the RNA world.

Protoplast   A cell from which the cell wall has been completely removed.

Prototroph   An organism that has no nutritional requirements beyond those of the wild type and which can grow on minimal medium.

Pseudogene   An inactivated and hence nonfunctional copy of a gene.

PSI-BLAST   A modified and more powerful version of the BLAST algorithm.

Punctuation codon A codon that specifies either the start or the end of a gene.

Punnett square A tabular analysis for predicting the genotypes of the progeny resulting from a genetic cross.

Purine One of the two types of nitrogenous base found in nucleotides.

Pyrimidine One of the two types of nitrogenous base found in nucleotides.

Pyrosequencing A novel DNA sequencing method in which addition of a nucleotide to the end of a growing polynucleotide is detected directly by conversion of the released pyrophosphate into a flash of chemiluminescence.

Quantitative PCR A PCR method that enables the number of DNA molecules in a sample to be estimated.

Quaternary structure The structure resulting from the association of two or more polypeptides.

RACE (rapid amplification of cDNA ends) A PCR-based technique for mapping the end of an RNA molecule.

Radiation hybrid A collection of rodent cell lines that contain different fragments of a second genome, constructed by a technique involving irradiation and used as a mapping reagent, for example in studies of the human genome.

Radioactive marker A radioactive atom incorporated into a molecule and whose radioactive emissions are subsequently used to detect and follow that molecule during a biochemical reaction.

Radiolabeling The technique for attaching a radioactive atom to a molecule.

Random genetic drift The process that leads to alleles gradually changing their frequency in a population.

Ras A protein involved in signal transduction.

Reading frame A series of triplet codons in a DNA sequence.

Readthrough mutation A mutation that changes a termination codon into a codon specifying an amino acid, and hence results in readthrough of the termination codon.

RecA An *Escherichia coli* protein involved in homologous recombination.

RecBCD enzyme An enzyme complex involved in homologous recombination in *Escherichia coli*.

Recessive The allele that is not expressed in a heterozygote.

Reciprocal strand exchange The exchange of DNA between two double-stranded molecules, occurring as a result of recombination, such that the end of one molecule is exchanged for the end of the other molecule.

Recognition helix An α-helix in a DNA-binding protein, one that is responsible for recognition of the target nucleotide sequence.

Recombinant A progeny member that possesses neither of the combinations of alleles displayed by the parents.

Recombinant DNA molecule A DNA molecule created in the test tube by ligating pieces of DNA that are not normally joined together.

Recombinant DNA technology The techniques involved in the construction, study and use of recombinant DNA molecules.

Recombinant protein A protein synthesized in a recombinant cell as the result of expression of a cloned gene.

Recombinase A diverse family of enzymes that catalyze site-specific recombination events.

Recombination A large-scale rearrangement of a DNA molecule.

Recombination frequency The proportion of recombinant progeny arising from a genetic cross.

Recombination hotspot A region of a chromosome where crossovers occur at a higher frequency than the average for the chromosome as a whole.

Recombination repair A DNA repair process that mends double-stranded breaks.

Regulatory control Control over bacterial gene expression that depends on the influence of regulatory proteins.

Regulatory mutant A mutant that has a defect in a promoter or other regulatory sequence.

Release factor A protein that plays an ancillary role during termination of translation.

Renaturation The return of a denatured molecule to its natural state.

Repetitive DNA A DNA sequence that is repeated two or more times in a DNA molecule or genome.

Repetitive DNA fingerprinting A clone fingerprinting technique that involves determining the positions of genome-wide repeats in cloned DNA fragments.

Repetitive DNA PCR A clone fingerprinting technique that uses PCR to detect the relative positions of genome-wide repeats in cloned DNA fragments.

Replacement vector A λ vector designed so that insertion of new DNA is by replacement of part of the nonessential region of the λ DNA molecule.

Replica plating A technique for transfer of colonies from one Petri dish to another, such that their relative positions on the surface of the agar medium are retained.

Replication factor C (RFC) A multisubunit accessory protein involved in eukaryotic genome replication.

Replication factory A large structure attached to the nuclear matrix; the site of genome replication.

Replication fork The region of a double-stranded DNA molecule that is being opened up to enable DNA replication to occur.

Replication licensing factors (RLFs) A set of proteins that regulate genome replication, in particular by ensuring that only one round of genome replication occurs per cell cycle.

Replication mediator protein (RMP) A protein responsible for detachment of single-strand binding proteins during genome replication.

Replication origin A site on a DNA molecule where replication initiates.

Replication protein A (RPA) The main single-strand binding protein involved in replication of eukaryotic DNA.

Replication slippage An error in replication that leads to an increase or decrease in the number of repeat units in a tandem repeat such as a microsatellite.

Replicative form The double-stranded form of the M13 DNA molecule found within infected *Escherichia coli* cells.

Replicative transposition Transposition that results in copying of the transposable element.

Replisome A complex of proteins involved in genome replication.

Reporter gene A gene whose phenotype can be assayed and which can therefore be used to determine the function of a regulatory DNA sequence.

Resin A chromatography matrix.

Resolution Separation of a pair of recombining double-stranded DNA molecules.

Response module A sequence motif found upstream of various genes that enables transcription initiation by RNA polymerase II to respond to general signals from outside of the cell.

Restriction endonuclease An enzyme that cuts DNA molecules at a limited number of specific nucleotide sequences.

Restriction fragment length polymorphism (RFLP) A restriction fragment whose length is variable because of the presence of a polymorphic restriction site at one or both ends.

Restriction mapping Determination of the positions of restriction sites in a DNA molecule by analyzing the sizes of restriction fragments.

Restrictive conditions Conditions under which a conditional-lethal mutant is unable to survive.

Retroelement A genetic element that transposes via an RNA intermediate.

Retrogene A gene duplicate that arises by insertion of a pseudogene adjacent to the promoter of an existing gene.

Retrohoming A process during which an excised intron, comprising single-stranded RNA, inserts directly into an organelle genome prior to being copied into double-stranded DNA.

Retroposon A retroelement that does not have LTRs.

Retrotransposition Transposition via an RNA intermediate.

Retrotransposon A genome-wide repeat with a sequence similar to an integrated retroviral genome and possibly with retrotransposition activity.

Retroviral-like element (RTVL) A truncated retroviral genome integrated into a host chromosome.

Retrovirus A virus with an RNA genome that integrates into the genome of its host cell.

Reverse transcriptase A polymerase that synthesizes DNA on an RNA template.

Reverse transcriptase PCR (RT-PCR) PCR in which the first step is carried out by reverse transcriptase, so RNA can be used as the starting material.

Rho A protein involved in termination of transcription of some bacterial genes.

Rho-dependent terminator A position in bacterial DNA where termination of transcription occurs with the involvement of Rho.

Ribbon-helix-helix motif A type of DNA-binding domain.

Ribonuclease An enzyme that degrades RNA.

Ribonuclease D An enzyme involved in processing pre-tRNA in bacteria.

Ribonuclease MRP An enzyme involved in processing eukaryotic pre-rRNA.

Ribonuclease P An enzyme involved in processing pre-tRNA in bacteria.

Ribonucleoprotein (RNP) domain A common type of RNA-binding domain.

Ribose The sugar component of a ribonucleotide.

Ribosomal protein One of the protein components of a ribosome.

Ribosomal RNA (rRNA) The RNA molecules that are components of ribosomes.

Ribosome One of the protein–RNA assemblies on which translation occurs.

Ribosome binding site The nucleotide sequence that acts as the attachment site for the small subunit of the ribosome during initiation of translation in bacteria.

Ribosome recycling factor (RRF) A protein responsible for disassembly of the ribosome at the end of protein synthesis in bacteria.

Ribozyme An RNA molecule that has catalytic activity.

RNA Ribonucleic acid, one of the two forms of nucleic acid in living cells; the genetic material for some viruses.

RNA-dependent DNA polymerase   An enzyme that makes a DNA copy of an RNA template; a reverse transcriptase.

RNA-dependent RNA polymerase   An enzyme that makes an RNA copy of an RNA template.

RNA editing   A process by which nucleotides not coded by a gene are introduced at specific positions in an RNA molecule after transcription.

RNA induced silencing complex (RISC)   A complex of proteins which cleaves and hence silences an mRNA as part of the RNA interference pathway.

RNA interference (RNAi)   An RNA degradation process in eukaryotes.

RNA polymerase   An enzyme that synthesizes RNA on a DNA or RNA template.

RNA polymerase I   The eukaryotic RNA polymerase that transcribes ribosomal RNA genes.

RNA polymerase II   The eukaryotic RNA polymerase that transcribes protein-coding and snRNA genes.

RNA polymerase III   The eukaryotic RNA polymerase that transcribes tRNA and other short genes.

RNA silencing   An RNA degradation process in eukaryotes.

RNA transcript   An RNA copy of a gene.

RNA world   The early period of evolution when all biological reactions were centered on RNA.

Rolling circle replication   A replication process that involves continual synthesis of a polynucleotide which is "rolled off" of a circular template molecule.

Rooted   Refers to a phylogenetic tree that provides information on the past evolutionary events that have led to the organisms or DNA sequences being studied.

S1 nuclease   An enzyme that degrades single-stranded DNA or RNA molecules, including single-stranded regions in predominantly double-stranded molecules.

SAP (stress activated protein) kinase   A stress-activated signal transduction pathway.

Satellite DNA   Repetitive DNA that forms a satellite band in a density gradient.

Satellite RNA   An RNA molecule some 320–400 nucleotides in length which does not encode its own capsid proteins, instead moving from cell to cell within the capsid of a helper virus.

Scaffold   A series of sequence contigs separated by sequence gaps.

Scaffold attachment region (SAR)   An AT-rich segment of a eukaryotic genome that acts as an attachment point to the nuclear matrix.

Scanning   A system used during initiation of eukaryotic translation, in which the preinitiation complex attaches to the 5′-terminal cap structure of the mRNA and then scans along the molecule until it reaches an initiation codon.

Secondary channel   A channel that leads from the surface of the bacterial RNA polymerase to the active site within the protein complex.

Secondary structure   The conformations, such as α-helix and β-sheet, taken up by a polypeptide.

Second messenger   An intermediate in a certain type of signal transduction pathway.

Sedimentation analysis   The centrifugal technique used to measure the sedimentation coefficient of a molecule or structure.

Sedimentation coefficient   The value used to express the velocity at which a molecule or structure sediments when centrifuged in a dense solution.

Segmented genome   A virus genome that is split into two or more DNA or RNA molecules.

Segment polarity genes   Developmental genes that provide greater definition to the segmentation pattern of the *Drosophila* embryo established by the action of the pair-rule genes.

Segregation   The separation of homologous chromosomes, or members of allele pairs, into different gametes during meiosis.

Selectable marker   A gene carried by a vector and conferring a recognizable characteristic on a cell containing the vector or a recombinant DNA molecule derived from the vector.

Selective medium   A medium that supports the growth of only those cells that carry a particular genetic marker.

Selfish DNA   DNA that appears to have no function and apparently contributes nothing to the cell in which it is found.

Semiconservative replication   The mode of DNA replication in which each daughter double helix is made up of one polynucleotide from the parent and one newly synthesized polynucleotide.

Sequenase   An enzyme used in chain termination DNA sequencing.

Sequence contig   A contiguous DNA sequence obtained as an intermediate in a genome sequencing project.

Sequence skimming   A method for rapid sequence acquisition in which a few random sequences are obtained from a cloned fragment, the rationale being that if the fragment contains any genes then there is a good chance that at least some of them will be revealed by these random sequences.

Sequence tagged site (STS)   A DNA sequence that is unique in the genome.

Serial analysis of gene expression (SAGE)   A method for studying the composition of a transcriptome.

Sex cell   A reproductive cell; a cell that divides by meiosis.

Sex chromosome   A chromosome that is involved in sex determination.

Shine-Dalgarno sequence   The ribosome binding site upstream of an *Escherichia coli* gene.

Short interfering RNA (siRNA)   An intermediate in the RNA interference pathway.

Short patch repair   A nucleotide excision repair process of *Escherichia coli* that results in excision and resynthesis of about 12 nucleotides of DNA.

Short tandem repeat (STR)   A type of simple sequence length polymorphism comprising tandem copies of, usually, di-, tri- or tetranucleotide repeat units. Also called a microsatellite.

Shotgun approach   A genome sequencing strategy in which the molecules to be sequenced are randomly broken into fragments which are then sequenced individually.

Shuttle vector   A vector that can replicate in the cells of more than one organism (e.g. in *Escherichia coli* and in yeast).

Signal peptide   A short sequence at the N terminus of some proteins that directs the protein across a membrane.

Signal transduction   Control of cellular activity, including genome expression, via a cell-surface receptor that responds to an external signal.

Silencer   A regulatory sequence that reduces the rate of transcription of a gene or genes located some distance away in either direction.

Silent mutation   A change in a DNA sequence that has no effect on the expression or functioning of any gene or gene product.

Similarity approach   A rigorous mathematical approach to alignment of nucleotide sequences.

Simple sequence length polymorphism (SSLP)   An array of repeat sequences that display length variations.

SINE (short interspersed nuclear element)   A type of genome-wide repeat, typified by the Alu sequences found in the human genome.

Single-copy DNA   A DNA sequence that is not repeated elsewhere in the genome.

Single nucleotide polymorphism (SNP)   A point mutation that is carried by some individuals of a population.

Single orphan   A single gene, with no homolog, whose function is unknown.

Single-strand binding protein (SSB)   One of the proteins that attach to single-stranded DNA in the region of the replication fork, preventing base pairs forming between the two parent strands before they have been copied.

Single stranded   A DNA or RNA molecule that comprises just a single polynucleotide.

Site-directed hydroxyl radical probing   A technique for locating the position of a protein in a protein–RNA complex, such as a ribosome, by making use of the ability of Fe(II) ions to generate hydroxyl radicals which cleave nearby RNA phosphodiester bonds.

Site-directed mutagenesis   Techniques used to produce a specified mutation at a predetermined position in a DNA molecule.

Site-specific recombination   Recombination between two double-stranded DNA molecules that have only short regions of nucleotide sequence similarity.

Slippage   The translocation of a ribosome along a short noncoding nucleotide sequence between the termination codon of one gene and the initiation codon of a second gene.

SMAD family   A group of proteins involved in signal transduction.

Small cytoplasmic RNA (scRNA)   A type of short eukaryotic RNA molecule with various roles in the cell.

Small nuclear ribonucleoprotein (snRNP)   Structures involved in splicing GU–AG and AU–AC introns and in other RNA processing events, comprising one or two snRNA molecules complexed with proteins.

Small nuclear RNA (snRNA)   A type of short eukaryotic RNA molecule involved in splicing GU–AG and AU–AC introns and in other RNA processing events.

Small nucleolar RNA (snoRNA)   A type of short eukaryotic RNA molecule involved in chemical modification of rRNA.

Somatic cell   A nonreproductive cell; a cell that divides by mitosis.

Sonication   A procedure that uses ultrasound to cause random breaks in DNA molecules.

SOS response   A series of biochemical changes that occur in *Escherichia coli* in response to damage to the genome and other stimuli.

Southern hybridization   A technique used for detection of a specific restriction fragment against a background of many other restriction fragments.

Species tree   A phylogenetic tree that shows the evolutionary relationships between a group of species.

S phase   The stage of the cell cycle when DNA synthesis occurs.

Spliced leader RNA (SL RNA)   A transcript which donates a leader segment to several RNAs by trans-splicing.

Spliceosome   The protein–RNA complex involved in splicing GU–AG or AU–AC introns.

Splicing   The removal of introns from the primary transcript of a discontinuous gene.

Splicing pathway   The series of events that converts a discontinuous pre-mRNA into a functional mRNA.

Spontaneous mutation   A mutation that arises from an error in replication.

SR-like CTD-associated factor (SCAF)   Proteins thought to play regulatory roles during splicing of GU–AG introns.

SR protein   A protein that plays a role in splice-site selection during splicing of GU–AG introns.

STAT (signal transducer and activator of transcription)   A type of protein that responds to binding of an extracellular signaling compound to a cell surface receptor by activating a transcription factor.

Stem cell   A progenitor cell that divides continually throughout the lifetime of an organism.

Stem-loop structure   A structure made up of a base-paired stem and non-base-paired loop, which can form in a single-stranded polynucleotide that contains an inverted repeat.

Steroid hormone   A type of extracellular signaling compound.

Steroid receptor   A protein that binds a steroid hormone after the latter has entered the cell, as an intermediate step in modulation of genome activity.

Sticky end   An end of a double-stranded DNA molecule where there is a single-stranded extension.

Stringent response   A biochemical and genetic response initiated in *Escherichia coli* when the bacterium encounters poor growth conditions such a low levels of essential amino acids.

Strong promoter   A promoter that directs a relatively large number of productive initiations per unit time.

Structural domain   A segment of a polypeptide that folds independently of other segments. Also, a loop of eukaryotic DNA, predominantly in the form of the 30 nm chromatin fiber, attached to the nuclear matrix.

STS mapping   A physical mapping procedure that locates the positions of sequence tagged sites (STSs) in a genome.

Stuffer fragment   A DNA fragment contained within a λ vector that is replaced by the DNA to be cloned.

Substitution mutation   Commonly used as a synonym for a point mutation.

SUMO   A protein related to ubiquitin.

Supercoiling   A conformational state in which a double helix is overwound or underwound so that superhelical coiling occurs.

Superwobble   The extreme form of wobble that occurs in vertebrate mitochondria.

Suppressor mutation   A mutation in one gene that reverses the effect of a mutation in a second gene.

S value   The unit of measurement for a sedimentation coefficient.

Syncytium   A cell-like structure comprising a mass of cytoplasm and many nuclei.

Synonymous mutation   A mutation that changes a codon into a second codon that specifies the same amino acid.

Synteny   Refers to a pair of genomes in which at least some of the genes are located at similar map positions.

Systems biology   An approach to biology that attempts to link metabolic pathways and subcellular processes with genome expression.

T4 polynucleotide kinase   An enzyme that adds phosphate groups to the 5′ ends of DNA molecules.

TψC arm   Part of the structure of a tRNA molecule.

TAF and initiator-dependent cofactor (TIC)   A type of protein involved in initiation of transcription by RNA polymerase II.

Tandem-affinity purification (TAP)   A method for isolating protein complexes that makes use of a test protein with a C-terminal extension that binds to calmodulin.

Tandemly repeated DNA   DNA sequence motifs that are repeated head to tail.

Tandem repeat   Direct repeats that are adjacent to each other.

TATA-binding protein (TBP)   A component of the general transcription factor TFIID, the part that recognizes the TATA box of the RNA polymerase II promoter.

TATA box   A component of the RNA polymerase II core promoter.

Tautomeric shift   The spontaneous change of a molecule from one structural isomer to another.

Tautomers   Structural isomers that are in dynamic equilibrium.

TBP-associated factor (TAF)   One of several components of the general transcription factor TFIID, playing ancillary roles in recognition of the TATA box.

TBP domain   A type of DNA-binding domain.

T-DNA   The portion of the Ti plasmid that is transferred to the plant DNA.

Telomerase   The enzyme that maintains the ends of eukaryotic chromosomes by synthesizing telomeric repeat sequences.

Telomere   The end of a eukaryotic chromosome.

Telomere binding protein (TBP)   A protein that binds to and regulates the length of a telomere.

Temperate bacteriophage   A bacteriophage that is able to follow a lysogenic mode of infection.

Temperature-sensitive mutation   A type of conditional-lethal mutation, one that is expressed only above a threshold temperature.

Template   The polynucleotide that is copied during a strand synthesis reaction catalyzed by a DNA or RNA polymerase.

Template-dependent DNA polymerase   An enzyme that synthesizes DNA in accordance with the sequence of a template.

Template-dependent DNA synthesis   Synthesis of a DNA molecule on a DNA or RNA template.

Template-dependent RNA polymerase   An enzyme that synthesizes RNA in accordance with the sequence of a template.

Template-dependent RNA synthesis   Synthesis of an RNA molecule on a DNA or RNA template.

Template-independent DNA polymerase   An enzyme that synthesizes DNA without the use of a template.

Template-independent RNA polymerase   An enzyme that synthesizes RNA without the use of a template.

Template strand   The polynucleotide that acts as the template for RNA synthesis during transcription of a gene.

Terminal deoxynucleotidyl transferase   An enzyme that adds one or more nucleotides to the 3′ end of a DNA molecule.

Termination codon   One of the three codons that mark the position where translation of an mRNA should stop.

Termination factor   A protein that plays an ancillary role in termination of transcription.

Terminator sequence   One of several sequences on a bacterial genome involved in termination of genome replication.

Territory   The region of a nucleus occupied by a single chromosome.

Tertiary structure   The structure resulting from folding the secondary structural units of a polypeptide.

Test cross   A genetic cross between a double heterozygote and a double homozygote.

Thermal cycle sequencing   A DNA sequencing method that uses PCR to generate chain-terminated polynucleotides.

Thermostable   Able to withstand high temperatures.

Thymine   One of the pyrimidine bases found in DNA.

Tiling array   A collection of oligonucleotide probes, each targeting a different position along a chromosome or a part of a chromosome.

Ti plasmid   The large plasmid found in those *Agrobacterium tumefaciens* cells able to direct crown gall formation on certain species of plants.

T_m   Melting temperature.

Tn3-type transposon   A type of DNA transposon that does not have flanking insertion sequences.

Topology   The branching pattern of a phylogenetic tree.

Totipotent   Refers to a cell that is not committed to a single developmental pathway and can hence give rise to all types of differentiated cell.

Trailer segment   The untranslated region of an mRNA downstream of the termination codon.

Transcript   An RNA copy of a gene.

Transcription   The synthesis of an RNA copy of a gene.

Transcription bubble   The non-base-paired region of the double helix, maintained by RNA polymerase, within which transcription occurs.

Transcription-coupled repair   A nucleotide excision repair process that results in repair of the template strands of genes.

Transcription factory   A large structure attached to the nuclear matrix; the site of RNA synthesis.

Transcription initiation   The assembly, upstream of a gene, of the complex of proteins that will subsequently copy the gene into RNA.

Transcriptome   The entire mRNA content of a cell.

Transcript-specific regulation   Regulatory mechanisms that control protein synthesis by acting on a single transcript or a small group of transcripts coding for related proteins.

***Trans*-displacement**   Transfer of a nucleosome from one DNA molecule to another.

Transduction   Transfer of bacterial genes from one cell to another by packaging in a phage particle.

Transduction mapping   The use of transduction to map the relative positions of genes in a bacterial genome.

Transfection   The introduction of purified phage DNA molecules into a bacterial cell.

Transfer RNA (tRNA)   A small RNA molecule that acts as an adaptor during translation and is responsible for decoding the genetic code.

Transfer-messenger RNA (tmRNA)   A bacterial RNA involved in protein degradation.

Transformant   A cell that has become transformed by the uptake of naked DNA.

Transformation   The acquisition by a cell of new genes by the uptake of naked DNA.

Transformation mapping   The use of transformation to map the relative positions of genes in a bacterial genome.

Transforming principle   The compound, now known to be DNA, responsible for transformation of an avirulent *Streptococcus pneumoniae* bacterium into a virulent form.

Transgenic mouse   A mouse that carries a cloned gene.

Transition A point mutation that replaces a purine with another purine, or a pyrimidine with another pyrimidine.

Translation The synthesis of a polypeptide, the amino acid sequence of which is determined by the nucleotide sequence of an mRNA in accordance with the rules of the genetic code.

Translational bypassing A form of slippage in which a large part of an mRNA is skipped during translation, elongation of the original protein continuing after the bypassing event.

Translocation The attachment of a segment of one chromosome to another chromosome. Also, the movement of a ribosome along an mRNA molecule during translation.

Transposable element A genetic element that can move from one position to another in a DNA molecule.

Transposable phage A bacteriophage that transposes as part of its infection cycle.

Transposase An enzyme that catalyzes transposition of a transposable genetic element.

Transposition The movement of a genetic element from one site to another in a DNA molecule.

Transposon A genetic element that can move from one position to another in a DNA molecule.

Transposon tagging A gene isolation technique that involves inactivation of a gene by movement of a transposon into its coding sequence, followed by the use of a transposon-specific hybridization probe to isolate a copy of the tagged gene from a clone library.

Trans-splicing Splicing between exons that are contained within different RNA molecules.

Transversion A point mutation that involves a purine being replaced by a pyrimidine, or vice versa.

Trinucleotide repeat expansion disease A disease that results from the expansion of an array of trinucleotide repeats in or near to a gene.

Triplet binding assay A technique for determining the coding specificity of a triplet of nucleotides.

Triplex A DNA structure comprising three polynucleotides.

Trisomy The presence of three copies of a homologous chromosome in a nucleus that is otherwise diploid.

tRNA nucleotidyltransferase The enzyme responsible for the posttranscriptional attachment of the triplet 5′–CCA–3′ to the 3′ end of a tRNA molecule.

***trp* RNA-binding attenuation protein (TRAP)** A protein involved in attenuation regulation of some operons in bacteria such as *Bacillus subtilis*.

True tree A phylogenetic tree that depicts the actual series of evolutionary events that led to the group of organisms or DNA sequences being studied.

Truncated gene A gene relic that lacks a segment from one end of the original, complete gene.

Tus The protein that binds to a bacterial terminator sequence and mediates termination of genome replication.

Twintron A composite structure made up of two or more Group II and/or Group III introns embedded in each other.

Two-dimensional gel electrophoresis A method for separation of proteins used especially in studies of the proteome.

Type 0 cap The basic cap structure, consisting of 7-methylguanosine attached to the 5′ end of an mRNA.

Type 1 cap A cap structure comprising the basic 5′-terminal cap plus an additional methylation of the ribose of the second nucleotide.

Type 2 cap A cap structure comprising the basic 5′-terminal cap plus methylation of the riboses of the second and third nucleotides.

Ubiquitin A 76-amino-acid protein which, when attached to a second protein, acts as a tag directing that protein for degradation.

Unequal crossing-over A recombination event that results in duplication of a segment of DNA.

Unequal sister chromatid exchange A recombination event that results in duplication of a segment of DNA.

Unit factor Mendel's term for a gene.

Unrooted Refers to a phylogenetic tree that merely illustrates relationships between the organisms or DNA sequences being studied, without providing information about the past evolutionary events that have occurred.

Upstream Towards the 5′ end of a polynucleotide.

Upstream control element A component of an RNA polymerase I promoter.

Upstream promoter element Components of a eukaryotic promoter that lie upstream of the position where the initiation complex is assembled.

Uracil One of the pyrimidine bases found in RNA.

U-RNA A uracil-rich nuclear RNA molecule including the snRNAs and snoRNAs.

UvrABC endonuclease A multienzyme complex involved in the short patch repair process of *Escherichia coli*.

van der Waals forces A particular type of attractive or repulsive noncovalent bond.

Variable number of tandem repeats (VNTR) A type of simple sequence length polymorphism comprising tandem copies of repeats that are a few tens of nucleotides in length. Also called a minisatellite.

Vegetative cell A nonreproductive cell; a cell that divides by mitosis.

Viral retroelement   A virus whose genome replication process involves reverse transcription.

Viroid   An RNA molecule 240–375 nucleotides in length which contains no genes and never becomes encapsidated, spreading from cell to cell as naked DNA.

Virulent bacteriophage   A bacteriophage that follows the lytic mode of infection.

Virus   An infective particle, composed of protein and nucleic acid, that must parasitize a host cell in order to replicate.

Virusoid   An RNA molecule some 320–400 nucleotides in length which does not encode its own capsid proteins, instead moving from cell to cell within the capsid of a helper virus.

V loop   Part of the structure of a tRNA molecule.

Wave of advance   A hypothesis which holds that the spread of agriculture into Europe was accompanied by a large-scale movement of human populations.

Weak promoter   A promoter that directs relatively few productive initiations per unit time.

Whole-genome shotgun approach   A genome sequencing strategy which combines random shotgun sequencing with a genome map, the latter used to aid assembly of the master sequence.

Wild type   A gene, cell or organism that displays the typical phenotype and/or genotype for the species and is therefore adopted as a standard.

Winged helix-turn-helix   A type of DNA-binding domain.

Wobble hypothesis   The process by which a single tRNA can decode more than one codon.

X inactivation   Inactivation by methylation of most of the genes on one copy of the X chromosome in a female nucleus.

X-ray crystallography   A technique for determining the three-dimensional structure of a large molecule.

X-ray diffraction   The diffraction of X-rays that occurs during passage through a crystal.

X-ray diffraction pattern   The pattern obtained after diffraction of X-rays through a crystal.

Yeast artificial chromosome (YAC)   A high-capacity cloning vector constructed from the components of a yeast chromosome.

Yeast two-hybrid system   A technique for identifying proteins that interact with one other.

Z-DNA   A conformation of DNA in which the two polynucleotides are wound into a left-handed helix.

Zinc finger   A common structural motif for attachment of a protein to a DNA molecule.

Zoo blotting   A technique that attempts to determine if a DNA fragment contains a gene by hybridizing that fragment to DNA preparations from related species, on the basis that genes have similar sequences in related species and so give positive hybridization signals, whereas the regions between genes have less similar sequences and so do not hybridize.

Zygote   The cell resulting from fusion of gametes during meiosis.

Acknowledgments

Genomes 3 is written from the published literature and draws on the work of many research groups. Where a figure has been reproduced directly from another source, the author has acknowledged this in the legend. In addition, the author wishes to acknowledge that various figures have been redrawn from, or are based upon, information given in the following books, papers and internet resources.

Figure 3.25: Oliver, S.G., van der Aart, Q.J.M., Agostini-Carbone, M.L., *et al.* (1992) The complete DNA sequence of yeast chromosome III. *Nature* **357**: 38–46.

Figure 5.19: Ponting, C.P. (1997) Tudor domains in proteins that interact with RNA. *Trends Biochem. Sci.* **22**: 51–52.

Figure 5.28: Dujon, B. (1996) The yeast genome project: what did we learn? *Trends Genet.* **12**: 263–270.

Figure 6.7: Leung, Y.F. and Cavalieri, D. (2003) Fundamentals of cDNA microarray data analysis. *Trends Genet.* **19**: 649–659.

Figure 6.23: Kalir, S. and Alon, U. (2004) Using a quantitative blueprint to reprogram the dynamics of the flagella gene network. *Cell* **117**: 713–720.

Figure 7.5: Strachan, T. and Read, A.P. (2004) *Human Molecular Genetics*, 3rd Ed. Garland, London.

Figure 7.12: GenBank entry HUMHBB.

Figure 7.15B: Oliver, S.G., van der Aart, Q.J.M., Agostini-Carbone, M.L., *et al.* (1992) The complete DNA sequence of yeast chromosome III. *Nature* **357**: 38–46.

Figure 7.15C: Adams, M.A., Celniker, S.E., Holt, R.A., *et al.* (2000) The genome sequence of *Drosophila melanogaster*. *Science* **287**: 2185–2195.

Figure 7.15D: SanMiguel, P., Tikhonov, A., Jin, Y.-K., *et al.* (1996) Nested retrotransposons in the intergenic regions of the maize genome. *Science* **274:** 765–768.

Figure 7.16: Venter, J.C., Adams, M.D., Myers, E.W., *et al.* (2001) The sequence of the human genome. *Science* **291**: 1304–1351.

Figure 7.17: IHGSC (International Human Genome Sequencing Consortium) (2001) Initial sequencing and analysis of the human genome. *Nature* **409**: 860–921.

Figure 7.18: IHGSC (International Human Genome Sequencing Consortium) (2001) Initial sequencing and analysis of the human genome. *Nature* **409**: 860–921.

Figure 8.4: Sinden, R.R. and Pettijohn, D.E. (1981) Chromosomes in living *Escherichia coli* cells are segregated into domains of supercoiling. *Proc. Natl Acad. Sci. USA* **78**: 224–228.

Figure 8.7: Blattner, F.R., Plunkett, G., Bloch, C.A., *et al.* (1997) The complete genome sequence of *Escherichia coli* K-12. *Science* **277**: 1453–1462.

Figure 8.10: Ochman, H., Lawrence, J.G. and Groisman, E.A. (2000) Lateral gene transfer and the nature of bacterial innovation. *Nature* **405**: 299–304.

Figure 10.4: Williams, R.R.E. (2003) Transcription and the territory: the ins and outs of gene positioning. *Trends Genet.* **19**: 298–302.

Figure 10.15: Bernstein, B.E., Kamal, M., Lindblad-Toh, K., *et al.* (2005) Genomic maps and comparative analysis of histone modifications in human and mouse. *Cell* **120**: 169–181.

Figure 11.2: Travers, A. (1993) *DNA–Protein Interactions*. Chapman & Hall, London.

Technical Note 11.1, Figure T11.1, Parts B & C: Campbell, N. (1993) *Biology*, 2nd Ed. Benjamin Cummings, San Francisco, CA.

Technical Note 11.1, Figure T11.1, Part D: Zubay, G. (1997) *Biochemistry*, 4th Ed. McGraw-Hill, New York, NY.

Figure 11.3: Travers, A. (1993) *DNA–Protein Interactions*. Chapman & Hall, London.

Figure 11.4: Travers, A. (1993) *DNA–Protein Interactions*. Chapman & Hall, London.

Figure 11.5: Travers, A. (1993) *DNA–Protein Interactions*. Chapman & Hall, London.

Figure 11.6: Travers, A. (1993) *DNA–Protein Interactions*. Chapman & Hall, London.

Figure 11.12: Kielkopf, C.L., White, S., Szewczyk, J.W., Turner, J.M., Baird, E.E., Dervan, P.B. and Rees, D.C. (1998) A structural basis for recognition of AT and TA base pairs in the minor groove of B-DNA. *Science* **282**: 111–115.

Figure 11.13: Travers, A. (1993) *DNA–Protein Interactions*. Chapman & Hall, London.

Figure 11.20: Xie, X., Kokubo, T., Cohen, S.L., *et al.* (1996) Structural similarity between TAFs and the heterotetrameric core of the histone octamer. *Nature* **380**: 316–322.

Figure 12.3: Korzheva, N., Mustaev, A., Kozlov, M., Malhotra, A., Nikiforov, V., Goldfarb, A. and Darst, S.A. (2000) A structural model of transcription elongation. *Science* **289**: 619–625.

Figure 12.14: Nickels, B.E. and Hochschild, A. (2004) Regulation of RNA polymerase through the secondary channel. *Cell* **118**: 281–284.

Figure 12.29: Stark, H., Dube, P., Lührmann, R. and Kastner, B. (2001) Arrangement of RNA and proteins in the spliceosomal U1 small nuclear ribonucleoprotein particle. *Nature* **409**: 539–542.

Figure 12.34: Graveley, B.R. (2001) Alternative splicing: increasing diversity in the proteomic world. *Trends Genet.* **17**: 100–107.

Figure 12.39: Burke, J.M., Belfort, M., Cech, T.R., *et al.* (1987) Structural conventions for Group I introns. *Nucleic. Acids Res.* **15**: 7217–7221.

Figure 13.3: Freifelder, D. (1986) *Molecular Biology*, 2nd Ed. Jones and Bartlett, Sudbury, MA.

Figure 13.12: Heilek, G.M. and Noller, H.F. (1996) Site-directed hydroxyl radical probing of the rRNA neighborhood of ribosomal protein S5. *Science* **272**: 1659–1662.

Figure 13.20: Nissen, P., Hansen, J., Ban, N., Moore, P.B. and Steitz, T.A. (2000) The structural basis of ribosome activity in peptide bond synthesis. *Science* **289**: 920–930.

Figure 15.33: Raghuraman, M.K., Winzeler, E.A., Collingwood, D., *et al.* (2001) Replication dynamics of the yeast genome. *Science* **294**: 115–121.

Figure 17.8: Rafferty, J.B., Sedelnikova, S.E., Hargreaves, D., *et al.* (1996) Crystal structure of DNA recombination protein RuvA and a model for its binding to the Holliday junction. *Science* **274:** 415–421.

Figure 18.15: Wolfe, K.H. and Shields, D.C. (1997) Molecular evidence for an ancient duplication of the entire yeast genome. *Nature* **387**: 708–713.

Figure 18.16: Eichler, E.E. (2001) Recent duplication, domain accretion and dynamic mutation of the human genome. *Trends Genet.* **17**: 661–669.

Figure 18.21: Jiang, N., Bao, Z., Zhang, X., Eddy, S.R. and Wessler, S.R. (2004) Pack-MULE transposable elements mediate gene evolution in plants. *Nature* **431**: 569–573.

Figure 19.6: Strachan, T. and Read, A.P. (2004) *Human Molecular Genetics*, 3rd Ed. Garland, London.

Figure 19.8: Li, W.-H. (1997) *Molecular Evolution*. Sinauer, Sunderland, MA.

Figure 19.9: Li, W.-H. (1997) *Molecular Evolution*. Sinauer, Sunderland, MA.

Figure 19.20: Wain-Hobson, S. (1998) 1959 and all that. *Nature* **391:** 531–532.

Figure 19.30: Richards, M., Macauley, V., Hickey, E., *et al.* (2000) Tracing European founder lineages in the Near Eastern mtDNA pool. *Am. J. Hum. Genet.* **67**: 1251–1276.

Figure 19.31: Jobling, M.A., Hurles, M.E. and Tyler-Smith, C. (2004) *Human Evolutionary Genetics: Origins, Peoples and Disease*. Garland, London.

Index

Note: Entries which are simply page numbers refer to the main text. Other entries have the following abbreviations immediately afer the page number: F, Figure; G, Glossary; T, Table; TN, Technical Note